Treatment of Contaminated Soil

Springer

Berlin
Heidelberg
New York
Barcelona
Hong Kong
London
Milan
Paris
Singapore
Tokyo

Rainer Stegmann • Gerd Brunner •
Wolfgang Calmano • Gerhard Matz (Eds.)

Treatment of Contaminated Soil

Fundamentals, Analysis, Applications

With 278 Figures and 83 Tables

EDITORS:

Prof. Dr.-Ing. Rainer Stegmann
TU Hamburg-Harburg
Arbeitsbereich Abfallwirtschaft
Harburger Schloßstraße 36
21071 Hamburg
Germany

Prof. Dr.-Ing. Gerd Brunner
TU Hamburg-Harburg
Arbeitsbereich Verfahrenstechnik II
Eißendorferstraße 38
21073 Hamburg
Germany

Prof. Dr.-Ing. Wolfgang Calmano
TU Hamburg-Harburg
Arbeitsbereich Umweltschutztechnik
Eißendorferstraße 40
21073 Hamburg
Germany

Prof. Dr.-Ing. Gerhard Matz
TU Hamburg-Harburg
Arbeitsbereich Elektrotechnik I
Harburger Schloßstraße 20
21071 Hamburg
Germany

ISBN 3-540-41736-2 Springer-Verlag Berlin Heidelberg New York

Library of Congress Cataloging-in-Publication Data applied for

Die Deutsche Bibliothek - CIP Einheitsaufnahme
Treatment of contaminated soil: fundamentals, analysis, applications / ed.: Rainer Stegmann ...- Berlin; Heidelberg; New York; Barcelona; Hong Kong; London; Milan; Paris; Singapore; Tokyo: Springer 2001
ISBN 3-540-41736-2

Springer-Verlag Berlin Heidelberg New York
a member of BertelsmannSpringer Science+Business Media GmbH
http://www.springer.de

Printed in Germany

Cover Design: Erich Kirchner, Heidelberg
Typesetting: Camera-ready by the editors

SPIN: 10892093 30/3111 – 5 4 3 2 1 – Printed on acid free paper

Contents

Introduction

Fundamental Aspects

Chemical Analysis of Contaminated Soils

List of Contributors

Alexander, M., Prof. Em.
Department of Crop and Soil Sciences and Institute of Comparative and Environmental Toxicology, Cornell University
Ithaca, NY, 14853, USA

Ahlf, W., Dr. habil. *
Department of Environmental Science and Technology,
Technical University of Hamburg-Harburg
Eissendorferstraße 40, 21073 Hamburg, Germany

Altrogge, M., Dr.
Institute of Organic Chemistry, University of Hamburg
Martin-Luther-King-Platz 6, 20146 Hamburg, Germany

Annweiler, E., Dr. *
Institute of Biogeochemistry and Marine Chemistry, University of Hamburg
Bundesstraße 55, 20146 Hamburg Germany

Antranikian, G., Prof. Dr. *
Institute of Technical Microbiology, Technical University of Hamburg-Harburg
Denickestraße 15, 21073 Hamburg, Germany

Biernoth, G., Dr.
Institute of Biochemistry and Food Chemistry, University of Hamburg
Grindelallee 117, 20146 Hamburg, Germany

Bjerg, P.L., Associate Professor, Ph.D.
Department of Environmental Science and Engineering, Groundwater Research Centre, Building 115, Technical University of Denmark
DK-2800 Lyngby, Denmark

Brunner, G., Prof. Dr.-Ing. *
Department of Thermal Process Engineering,
Technical University of Hamburg-Harburg
Eissendorferstraße 38, 21073 Hamburg, Germany

Bundt, J., Dr. *
Chemical Laboratory Lübeck GmbH
Hochofenstraße 23–25, 23561 Lübeck, Germany

Calmano, W., Prof. Dr.-Ing. *
Department of Environmental Science and Technology,
Technical University of Hamburg-Harburg
Eissendorferstraße 40, 21073 Hamburg, Germany

Christensen, T.H., Prof. Dr.
Department of Environmental Science and Engineering, Groundwater Research Centre, Building 115, Technical University of Denmark
DK-2800 Lyngby, Denmark

Cohrs, I., Dipl.-Ing. *
Department of Waste Management, Technical University of Hamburg-Harburg
Harburger Schloβstraβe 37, 21079 Hamburg

Dankwarth, F., Dipl.-Min. *
Department of Environmental Science and Technology,
Technical University of Hamburg-Harburg
Eissendorferstraβe 40, 21073 Hamburg, Germany

Deppe, U., Dipl. Biol. *
Institute of Technical Microbiology Technical University of Hamburg-Harburg
Denickestraße 15, 21073 Hamburg, Germany

Eschenbach, A., Dr. *
Institute for Physical and Chemical Analysis,
GKSS National Research Centre,
Max-Planck-Strasse, 21502 Geesthacht, Germany

Feitkenhauer, H., Dr.-Ing. *
Institute of Bioprocess and Biochemical Engineering, Technical University of Hamburg-Harburg
Denickestraße 15, 21073 Hamburg, Germany

Firus, A., Dr.-Ing. *
Bayer AG, ZT-TE
51368 Leverkusen, Germany

Förstner, U., Prof. Dr. *
Department of Environmental Science and Technology,
Technical University of Hamburg-Harburg
Eissendorferstraβe 40, 21073 Hamburg, Germany

Francke, W., Prof. Dr. Dr. hc *
Institute of Organic Chemistry, University of Hamburg
Martin-Luther-King-Platz 6, 20146 Hamburg, Germany

Frühling, W., Dipl.-Biol. *
Department of Environmental Science and Technology,
Technical University of Hamburg-Harburg
Eissendorferstraße 40, 21073 Hamburg, Germany

Garms, C., Dr. *
Institute of Organic Chemistry, University of Hamburg
Martin-Luther-King-Platz 6, 20146 Hamburg, Germany

Gerth, J., Dr. sc.agr. *
Department of Environmental Science and Technology,
Technical University of Hamburg-Harburg
Eissendorferstraße 40, 21073 Hamburg, Germany

Golinske, D., Dipl.-Chem.
Institute of Organic Chemistry, University of Hamburg
Martin-Luther-King-Platz 6, 20146 Hamburg, Germany

Gröngröft, A., Dr.
Institute for Soil Science, University of Hamburg
Allende-Platz 2, 20146 Hamburg, Germany

Harms H., Prof. Dr.
Swiss Federal Institute of Technology Lausanne,
Department of Rural Engineering
1015 Lausanne, Switzerland

Haupt, E.T.K., Dr. *
Institute for Inorganic und Applied Chemistry, University of Hamburg
Martin-Luther-King-Platz 6, 20146 Hamburg, Germany

Hebenbrock, S., Dr.-Ing. *
Institute of Technical Microbiology Technical University of Hamburg-Harburg
Denickestraße 15, 21073 Hamburg, Germany

Heerenklage, J., Dipl-Ing. *
Department of Waste Management, Technical University of Hamburg-Harburg
Harburger Schloβstraße 37, 21079 Hamburg, Germany

Höhne, J., Dipl.-Ing. *
Institute for Physical and Chemical Analysis,
GKSS National Research Centre,
Max-Planck-Strasse, 21502 Geesthacht, Germany

Hund-Rinke, K., Dr.
Fraunhofer-Institute for Environmental Chemistry and Ecotoxicology
Auf dem Aberg 1, 57392 Schmallenberg, Germany

Hupe, K., Dr.-Ing. *
Consultants for Waste Management, Prof. R. Stegmann and Partner
Nartenstraße 4a, 21079 Hamburg, Germany

Kasche, V., Prof. Dr. *
Department of Biotechnology II, Technical University of Hamburg-Harburg
Denickestraße 15, 21071 Hamburg, Germany

Käcker, T., Dr. *
Lohmann Animal Health GmbH
Heinz-Lohmann-Straße 4, 27472 Cuxhaven, Germany

Kästner, M., PD Dr. habil. *
Department of Remediation Research,
UFZ Centre for Environmental Research Leipzig-Halle
Permoserstraße 15, 04318 Leipzig, Germany

Kiene, A., Dipl. Biol.
Institute for Soil Science, University of Hamburg
Allende-Platz 2, 20146 Hamburg, Germany

Kjeldsen, P., Associate Professor, Ph. D.
Department of Environmental Science and Engineering, Groundwater Research Centre, Building 115 Technical University of Denmark
DK-2800 Lyngby, Denmark

Koning, M., Dipl.-Ing. *
Department of Waste Management, Technical University of Hamburg-Harburg
Harburger Schloβstraße 37, 21079 Hamburg, Germany

Kördel, W., Dr.
Fraunhofer-Institute for Environmental Chemistry and Ecotoxicology
Auf dem Aberg 1, 57392 Schmallenberg, Germany

Kranz, O., Dr.
Bran + Luebbe GmbH
Werkstraße 42, 2844 Norderstedt, Germany

Lotter, S., Dr.-Ing. *
Bundesverband der Deutschen Zementindustrie e.V.
Immenhof, 22087 Hamburg, Germany

Lüth, J. C., Dipl.-Ing. *
Department of Waste Management, Technical University of Hamburg-Harburg
Harburger Schloβstraße 37, 21079 Hamburg, Germany

Mahro, B., Prof. Dr. *
Institute for Environmental Technology, Hochschule Bremen
Neustadtwall 30, 28199 Bremen, Germany

Malerius, O., Dipl.-Ing. *
Chemical Engineering I, Technical University of Hamburg-Harburg
Denickestraße 15, 21073 Hamburg, Germany

Matz, G., Prof. Dr.-Ing. *
Environmental Measurement Technology,
Technical University of Hamburg-Harburg
Harburger Schloβstraße 20, 21079 Hamburg, Germany

Mangold, S.
Department of Environmental Science and Technology,
Technical University of Hamburg-Harburg
Eissendorferstraße 40, 21073 Hamburg, Germany

Märkl, H., Prof. Dr.-Ing. *
Institute of Technical Microbiology and Bioprocess and Bioengineering,
Technical University of Hamburg-Harburg
Denickestraße 15, 21073 Hamburg, Germany

Mescher, H.
Institute for Environmental Technology, Hochschule Bremen
Neustadtwall 30, 28199 Bremen, Germany

Meyer, S., Dr. *
Institute of Biochemistry and Food Chemistry, University of Hamburg
Grindelallee 117, 20146 Hamburg, Germany

Michaelis, W., Prof. Dr. *
Institute of Biogeochemistry and Marine Chemistry, University of Hamburg
Bundesstraße 55, 20146 Hamburg, Germany

Miehlich, G., Prof. Dr. *
Institute for Soil Science, University of Hamburg
Allende-Platz 2, 20146 Hamburg, Germany

Misch, B., Dipl.-Ing. *
Merck Darmstadt
Frankfurter Straße 250, 64293 Darmstadt, Germany

Müller, R., Prof. Dr. *
Department of Biotechnology II, Technical University of Hamburg-Harburg
Denickestraße 15, 21071 Hamburg, Germany

Neeβe, T., Prof. Dr.
Department of Environmental Process Engineering and Recycling,
University of Erlangen-Nürnberg
Paul-Gordan-Straße 3, 91052 Erlangen, Germany

Neumann-Hensel, H., Dr.
Dr. Fintelmann und Dr. Meyer
Handels- und Umweltschutzlaboratorien GmbH
Mendelssohnstr. 15 D, 22761 Hamburg, Germany

Niemeyer, B., Prof. Dr.-Ing. *
Institute for Physical and Chemical Analysis
GKSS National Research Centre,
Max-Planck-Strasse, 21502 Geesthacht, Germany

Nowak, K., Dr.-Ing. *
Bremer Sonderabfall Beratungsgesellschaft mbH
28195 Bremen, Germany

Nünnecke, D., Dr. *
Institute of Organic Chemistry, University of Hamburg
Martin-Luther-King-Platz 6, 20146 Hamburg, Germany

Petersen, D., Dipl.-Chem.
Institute of Organic Chemistry, University of Hamburg
Martin-Luther-King-Platz 6, 20146 Hamburg, Germany

Reimers, C., Dr. *
Department of Waste Management, Technical University of Hamburg-Harburg
Harburger Schloβstraße 37, 21079 Hamburg, Germany

Richnow, H.H., Dr. habil. *
Department of Remediation Research,
UFZ Centre for Environmental Research Leipzig-Halle
Permoserstraße 15, 04318 Leipzig, Germany

Rulkens, W.H., Prof. Dr.
Subdepartment of Environmental Technology, Wageningen University
P.O. Box 8129, 6700 EV Wageningen, The Netherlands

Schmidt, J., Dipl.-Ing. *
Chemical Engineering I, Technical University of Hamburg-Harburg
Denickestraße 15, 21073 Hamburg, Germany

Schröder, W., Dr.
Environmental Measurement Technology, Technical University of Hamburg-Harburg
Harburger Schloβstraße 20, 21079 Hamburg, Germany

Springael, D., Dr.
Flemisch Institute of Technological Research, Environmental Technology
2400 Mol, Belgium

Stegmann, R., Prof. Dr.-Ing. *
Department of Waste Management, Technical University of Hamburg-Harburg
Harburger Schloβstraße 37, 21079 Hamburg, Germany

Steinhart, H., Prof. Dr. Dr. *
Institute of Biochemistry and Food Chemistry, University of Hamburg
Grindelallee 117, 20146 Hamburg, Germany

Stichnothe, H., Dr.
Department of Environmental Science and Technology,
Technical University of Hamburg-Harburg
Eissendorferstraße 40, 21073 Hamburg, Germany

Stieber, M., Dr.
Water Technology Center
Karlsruher Straβe 84, 76139 Karlsruhe, Germany

Tiehm, A., Dr.
Water Technology Center
Karlsruher Straβe 84, 76139 Karlsruhe, Germany

Thoenes, H.W., Prof. Dr.
Advisory Scientific Council for Soil Protection at the GBMU
Aufm Kampe 39, 42279 Wuppertal, Germany

Thöming, J., Dr. *
Universidade Federal do Rio Grande do Sul
Department of Chemical Engineering
Rua Marechal Floriano, 501 apt 81
90020-061 Porto Alegre RS, Brasil

Voss, J., Prof. Dr. *
Institute of Organic Chemistry, University of Hamburg
Martin-Luther-King-Platz 6, 20146 Hamburg, Germany

Waller, E., Dr.
Im Hollergrund 79 F
28357 Bremen, Germany

Werther, J., Prof. Dr.-Ing. *
Department of Chemical Engineering I, Technical University of Hamburg-Harburg
Denickestraße 15, 21073 Hamburg, Germany

Wick, L.Y., Dr.
Swiss Federal Institute of Technology Lausanne,
Department of Rural Engineering
1015 Lausanne, Switzerland

Wienberg, R., Dr.
Umwelttechnisches Labor Dr. R. Wienberg
Gotenstraβe 4, 20097 Hamburg, Germany

Wilichowski, M., Prof. Dr.-Ing. *
Department of Mechanical Engineering/Process and Environmental Engineering,
Hochschule Wismar, University of Technology, Business and Design
Philipp-Müller-Straße, Postfach 1210, 23952 Wismar, Germany

Winterberg, R., Dr.
Plambeck ContraCon GmbH
Peter Henlein Straβe 2–4, 27472 Cuxhaven, Germany

* Participant of the *Sonderforschungsbereich 188* (research centre) founded by the German Research Foundation (Deutsche Forschungsgemeinschaft DFG)

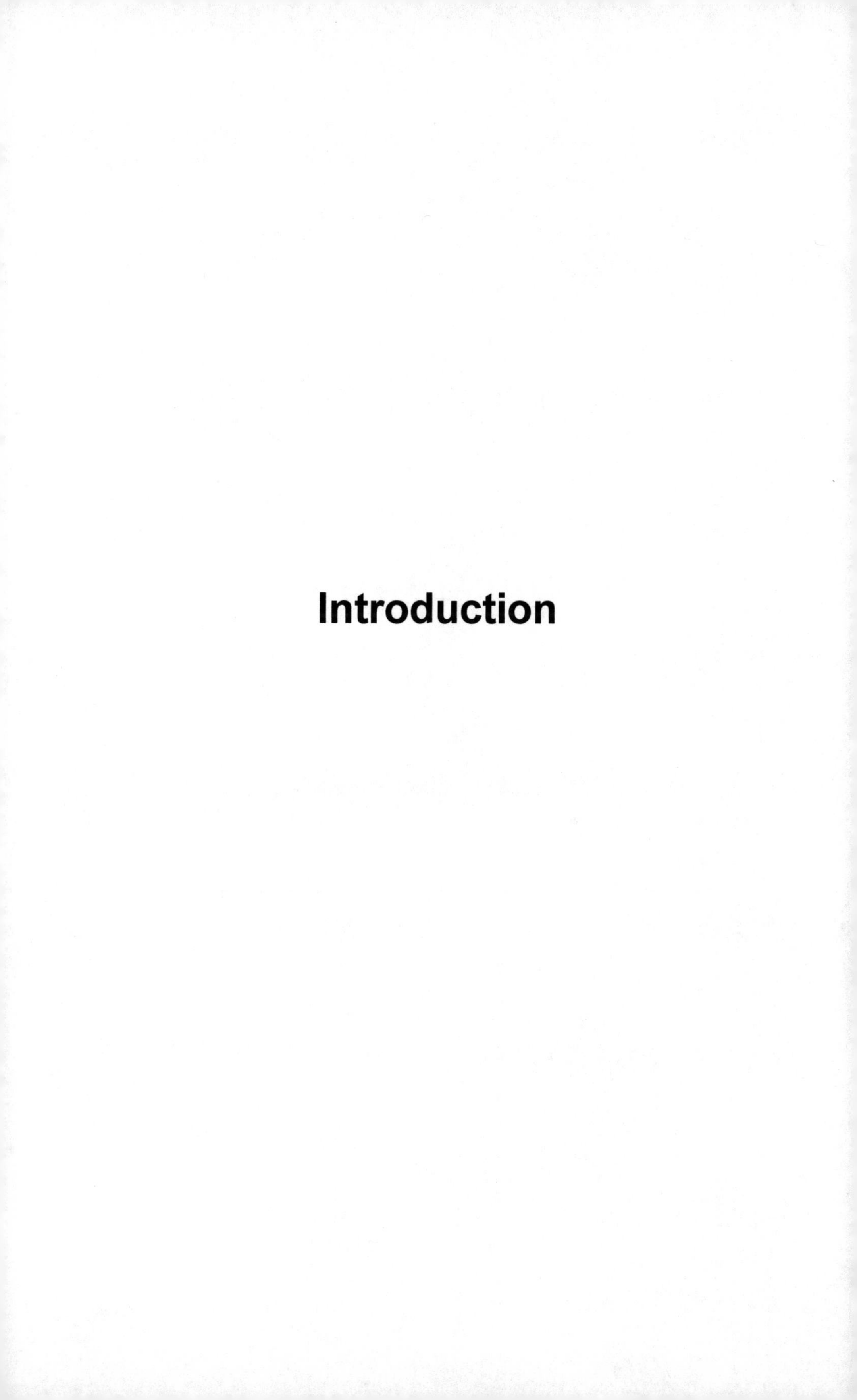

Introduction

1 Introduction

R. Stegmann
Technical University of Hamburg-Harburg, Department of Waste Management, Harburger Schloßstr. 37, 21079 Hamburg, Germany

1.1 Research Centre "Treatment of Contaminated Soil"

Anthropogenic activities have resulted in contaminated soils covering significant areas of land. In the Eighties people recognised the size and the consequences of this problem. Initially, treatment and remediation processes were developed mainly by the industry. Very often those approaches were very pragmatic and there was a lack of a scientific basis for the processes, and a need for optimisation and further development. This gap has been closed in the meantime, but there is of course a need for further investigations.

There are four main alternatives for the treatment of contaminated soils:

- Leave the contamination as it is, but restrict the utilisation of the land.
- Complete or partial encapsulation of the contamination.
- Excavation of the contaminated soil and landfilling.
- Treatment of the contaminated soil in-situ (e.g. without excavation) or ex-situ, either at an onsite or central plant.

In the long term, the only alternative that makes sense is the decontamination of the polluted soil. Only by this means the problem can be solved without transferring it to the future; the soil needs to be used again without any restrictions. Of course economic aspects are of vital importance and therefore the above described optimum solution cannot always be achieved, and compromises have to be made.

At present the "intrinsic remediation" is discussed intensively in the scientific world, where natural attenuation processes are "used" in order to minimise the adverse effects of the contamination. Intensive research is needed in this field in order to determine the possibilities and limits of this approach (Chapter 35, 36, 37).

In the actual remediation, mechanical, thermal and biological processes are usually practiced. The state of the art is characterised by a multitude of procedures. This situation has been achieved by intensive worldwide research where – as already mentioned – processes have been optimised and further developed. It was essential to adapt and further develop the chemical analytical methods and the monitoring processes for contaminated soil. In addition, the treatment goals have been elaborated and defined, as toxicological and ecotoxicological target values are now available on a scientific basis.

"Technical soil protection" has been developed into a newly acknowledged scientific discipline, where an integrated cooperation among scientists from different disciplines of engineering, chemistry, microbiology, soils, geology and environmental planning is essential.

The content of this book is mainly based on the results elaborated in the Research Centre of the German Research Foundation (Sonderforschungsbereich (SFB) der Deutschen Forschungsgemeinschaft (DFG) "Treatment of Contaminated Soil") which has been funded for a period of 12 years. It is well supplemented by contributions from Institutes and Organisations which are in close contact with the SFB. In the following sections some approaches, results and overall conclusions are presented, which – among other issues – are described in detail in the chapters of the book.

1.2 Approach

In order to be able to describe the basic processes the integrated research was performed in the first phase using soils that were artificially contaminated with mineral oil and/or polycyclic aromatic hydrocarbons (PAHs). In the next phases "real" contaminated soil was also investigated. Different kinds of technologies, consisting of thermal, mechanical, chemical and biological processes separately or in combination, were applied. An important part was the new and further development of chemical analytical procedures, as well as ecotoxicity tests, for contaminated soils.

Although a great deal of basic research has been performed, the practical application was always kept in mind, and in some cases realised.

1.3 Analytical Aspects

The analytical part in the research centre was very strong. Only if the processes can be described sufficiently, can adequate conclusions be drawn. In addition, analytical procedures in the soil matrix are very complex, with a great variety of matrix effects. The analytical problems of the analysis of bound residues (especially in the humic matrix) has already been highlighted. Also the analytical description of metabolites and minor components has been stressed. As can be concluded from the contributions in this book, very successful work has been done and many of the analytical questions have been answered. But, not only very highly sophisticated analytical procedures are necessary in the field of treatment of contaminated soils; in addition a simple, highly reproducible process describing analytical methods is necessary. Also, it is necessary to reduce the amount of toxic chemicals to be used for analytical purposes. In this context, thin-layer chromatography has been further developed, and the gas chromatography method for the cumulative quantification of mineral oil contaminants, which has become a DEV/ISO method, has been tested to a significant degree (see also Appendix).

A strong component in the further development of analytical methods in the DFG-Research-Centre was the Fast Mobile Analysis (Chapter 10). In minutes gas chromatographic and mass-spectrometric analysis can be made in the laboratory or in a mobile lab placed in a van, with a very high reproducibility. By this means, fast and accurate analysis can be performed in the field. These methods have been further developed for routine practical application.

Important to note, is the quality assurance of the chemical analysis which is, based on our experience, of highest priority. Regular control tests within the whole group, as well as exchange of experience, are necessary and are also a basis for collaboration.

Beside chemical analysis, biotests are used for the general description of the effect of contaminated soil and its treatment. Biotests are always problematic, since often only one test organism is used. This organism may be especially sensitive to one compound but less sensitive to another. For this and other reasons in the DFG-Research-Centre, an ecotoxicological test and evaluation system has been developed consisting of different kinds of organism, with different reactions to the great variety of pollutants (see Chapter 11).

1.4 Biological Processes

As can be seen from several contributions in the book, bioavailibility, which is the key to biodegradation, was intensively investigated. In order to distinguish between degradation, volatilisation, biomass production and physical/chemical fixation, carbon balances have to be made by using test procedures as developed in the DFG-Research-Centre (Chapter 22, 26). Very good experience has also been gained by setting up balances by means of ^{13}C-tracer marked organic substances. This technology has the advantage of using a tracer with easily managed radiation. In addition, the qualitative and quantitative description of bound residues becomes possible (Chapter 16, 17).

During the biological degradation of contaminants in soil, metabolites which appear within short or longer time frames are produced; these may be more toxic than the original substances. These metabolites often react with the macromolecular organic soil matrix and produce integral parts of the humic fraction. Adequate analytical methods have been developed to also describe the kind of binding in the humic matrix, which gives indications to the long term behaviour of these substance (Chapter 16, 17). The main question in this regard, is whether these bindings are stable or not. In the latter case, these bound residues will be released at some time into the environment, in their original form.

Strong efforts have been made in the DFG-Research-Centre in order to increase the bioavailability of substances, e.g. by means of ozone treatment, transfer from solid into slurry phase, and mechanical pre-treatment. The results show varying success. In general high energy input will be necessary in order to achieve success. In this context, if they are not bioavailable, the question was discussed as to how far it makes sense to extract the pollutants out of the bound residues. A new approach has also been developed for enhancing biological degradation processes, and increasing bioavailability, by means of selecting thermophilic and hyperthermophilic microorganisms from nature. Their efficiency in biodegradation of organic pollutants needs to be described. Some of these isolated organisms are able to degrade mono-, di- and polyaromatic compounds. Often the growth rates are still low, but there seems to be great potential for application in this field. As an example, a phenol degrading organism has been isolated, it is much more effective than any currently known organisms in mixed cultures (Chapter 24).

For the biological degradation of organic pollutants, the milieu conditions have been optimised in the SFB. Now, optimum water content, oxygen content, temperature, adequate additives etc. can be selected (Chapter 22, 26).

Based on the extensive research in the DFG-Research-Centre, mainly at laboratory scale, a full scale process has been developed using high windrows (3 – 5 m height) with forced aeration. It is regulated on the basis of the actual oxygen consumption of the bacteria. The windrows are prepared at optimum water content, which is also monitored and regulated, and by the addition of adequate additives which ensure sufficient nutrients, organisms and structure. Due to a self-heating process as a result of the biological degradation, the temperature rises and enhances the process. The windrows should be operated under a roof in order to avoid the production of polluted water (Chapter 26).

1.5 Soil Washing

Soil washing has been successfully practised for many years. Since the pollutants are attached mainly to the fine particles, and as much soil as possible should be reused, it is important to separate as far as possible the very fine particles. Instead of having, as usual, a separation of particles < 70 µm, the separation of particles < 20 µm is now possible. By these means the concentration factor of the pollutants increases and less material has to be disposed of (Chapter 28).

1.6 Heavy Metal Removal

A new process has been developed for the treatment of soil which is polluted by heavy metals. They are extracted out of the soil by means of organic acids as a "heavy metal foam". Electro-dialysis is used, for the treatment of the produced wastewater, so that it can be reused in a second washing step. This system is at present on a pilot scale and needs application to full scale (Chapter 30). In this context also, electrolysis processes have been investigated (Chapter 34).

1.7 Thermal Treatment

Residues from soil washing e.g. can be further treated by means of extraction with "critical water" at temperatures of 374 °C and pressure of 22.1 MPa. This system has been developed on a laboratory scale and has proven to be of general applicability (Chapter 31).

Another method which has been further developed to technical scale in the DFG-Research-Centre, can be used, for instance, also for organic residues concentrated in the fine soil particles as a result of soil washing. This so called "steam stripping" process consists mainly of a pipe reactor in which the polluted fine particles are fed as a suspension. Due to the energy input the water is spontaneously evaporated, and the organic pollutants are transferred into the steam phase. Subsequent to the decontamination steps, the treated soil and the separated organic pollutant phase are obtained. Good experience has so far been gained with soils polluted with mineral oil and low condensed PAHs. The advantages of this process are the relatively low operation temperatures of around 300°C, and low off-gas volumes. In addition, the fine fraction from soil washing can be reused without the need for a further dewatering step (Chapter 32, 33).

1.8 Natural Attenuation

In the final stage of the DFG-Research-Centre, two projects investigating natural attenuation at an arsenic contaminated soil site have been started. On the basis of a large number of monitoring points, and intensive sample characterisation in the laboratory, a descriptive model shall be developed. The hypothesis is, that part of the arsenic is moving into the ground water, and part of it is taken up by the plants existing on the site (Chapter 37).

1.9 Final Remarks

The activities highlighted in this chapter result mainly from the DFG-Research-Centre. After having read this chapter, the reader should be curious to read the following chapters where the above mentioned, and additional subjects, are presented in detail. Further information can be obtained from the great number of publications provided in the References Sections. In order to make this book more complete, the results from the DFG-Research Centre are complimented by external researchers from other scientific institutions with which the DFG-Research-Centre has close contacts.

Fundamental Aspects

2 Hazardous Waste Contaminated Sites and Soil

H.W. Thoenes
Advisory Scientific Council for Soil Protection at the GBMU, Aufm Kampe 39, 42279 Wuppertal, Germany

2.1 Problems Relating to Soil Awareness

The relationship of our society towards soil and its functions can be seen highlighted on the photograph showing the festive ceremony of the turning of the first sod in Leipzig, Germany. There is the pleasure to have finally achieved the use of a site for the construction of an industrial production plant. The participants have equipped themselves with high yellow rubber boots. They obviously associated soil with dirt and mud. Shoes and trousers had to be protected against the environmental medium soil in its natural form. So far concerning the introduction of a highlight for soil consciousness in politics and society.

The opinions and attitudes of general public with regard to environmental questions are brought together under the heading of environmental awareness. In our society there are different views concerning the place and functions of environmental awareness (SRU 1978), since the basic attitudes of individuals and the various social groups towards the need for environmental awareness are not homogeneous.

The importance placed on forms of environmental pollution represents a subjective estimate of the risks and advantages involved. However, risks are not judgement-free attributes of a particular problem area; the way they are perceived varies between individuals and groups (SRU 1987).

The attitude of the public towards environmental questions has been followed in Germany for years. The social significance of environmental protection is certainly still a noteworthy factor in problems of social policy.

The question of the need to protect not only air and bodies of water, but also soil did not play a part in earlier surveys. Only in the surveys carried out in 1996 and 1998 was the state of the soil included in the section dealing with perceived improvements in our environment (UBA 1998). According to this, 17 % of those questioned in Germany as a whole and 23 % in the New Federal States see considerable progress being made in improving the state of the soil. The subject of soil awareness itself is, however, not dealt with in surveys carried out within the population.

The "Sachverständigenrat für Umweltfragen" (SRU) (Council of Experts for Environmental Issues) had already proposed a definition of the term "environmental awareness" in 1978. Environmental awareness is defined as the insight in the threat against the natural foundations of life of mankind resulting from own activities in combination with the willingness to find remedies. This is all about the awareness as a conscious feeling for the different outer, that is anthropogenic, influences that cause an action. Recently terms as "environmental attitudes" and "environmental behaviour" are commonly used.

The term soil awareness is relatively recent. It has been clearly established by the "Wissenschaftlicher Beirat der Bundesregierung Globale Umweltveränderungen (WBGU)" (Scientific Advisory Council of the Federal Government on Global Changes to the Environment) and its content was examined in the 1994 annual report in terms of an interdisciplinary discussion on economic, ecological and social aspects of soil use with the related consequences. During the course of this, the question of the value placed on soil quality in our society was also analysed.

The analysis of the WBGU showed that the cultural function of soil in combination with its social function is determined by the spatial behaviour of the human beings, that is inevitably soil bound itself. As a consequence results a different perception and esteem of soil, that is determined by the particular societal backgrounds. In this context these are mainly the societal structures and the handling of soil, which bring about differing esteems. As an antipode there are on one side the farmers, miners as well as allotment-holders and on the other side the inhabitants of cities without a garden on their own. For most citizens and inhabitants of cities soil has become a matter of course; they do not feel they need the soil and its natural functions for survival – in contrary to the vital need of air and water. The perception is mostly visual and in the course of their hikes and walks they observe the beauty of the landscape and the kind of cultivation, but also take care not to dirty their trousers.

The result, which is still of relevance today, is that the soil on which we stand and live, which we use to cultivate our food, which is a major partner for the ground water and hence for our drinking water and on which we build houses, factories and roads is hardly or only partially perceived and taken into consideration by the wider public. In view of this, the phrase "the forgotten soil" has been coined. DOSSO (1999) established that the word "soil" is used in so many different ways that we must assume that the concept of soil does not stern from a common cultural background and that it does not carry a commonly understood and accepted meaning. This statement also makes clear what difficulties exist in developing a well-founded and sustainable basis for sustainable soil use as the heart of sustained development within our society.

With this situation having been established the representatives of environmental, soil and geological sciences have to ask themselves, why they took on questions of biodegradation with its consequences in the public discussion only relatively late and in this way supported directly or indirectly the "soil oblivion".

With regard to the question of the influences that affect soil awareness, knowledge, perception, experience, as well as opinions, feelings and in particular value systems, play a major part. They determine the predisposition to act and behave in certain ways as well as the modes of action and behaviour adopted when dealing with the soil in terms of prevention and restoration. Within this multifunctional network of relationships, the values held in relation to the environment play a particularly important role in determining environmental awareness. In the body of guiding notions relating to preventive soil protection, the "Wissenschaftlicher

Beirat Bodenschutz" (Scientific Advisory Council for Soil Protection) has incorporated the notion "soil is valuable" (WBB 2000). Everything possible must be done to ensure that the ideal of soil protection is rendered not only politically, but also socially acceptable.

Despite all the efforts of the soil sciences and of many associations and institutions, the complexity of natural soil functions, and their interrelationship with the various uses to which the soil is put, is still barely accessible to the wider public. It is also largely unknown how scientific discoveries and descriptions of these are assimilated into the body of everyday knowledge. The fact that the wider public and the media pay little attention to the soil is connected to a large extent to the normally low level of public perception in relation to soil.

2.2 The Role of Hazardous Waste Contaminated Sites

Attention is paid to soil and areas of land when reasonably-priced sites are being sought for building projects, and in particular when damaging changes to the soil involve a subjectively threatening situation for people, leading to the need for hazard control. In particular, residential areas on former landfills and sites with historical contamination arouse wide interest among the public and in the media. In these cases the perception of soil is reinforced by public concern.

With the more than 300,000 areas of suspected contamination sites in Germany, there is a large and varied perception potential on a local and regional level. This potential must be exploited to a far greater extent than up to today to make clear to the public in the discussion on areas of suspected contamination sites the significance of soil functions for our life. The relevant sciences also have a duty to play a part in this. There must be repeated, systematic and scientific presentation of the problem in a form comprehensible to the public. This would form the basis for enhancing sensitivity to soil and its functions, essential as they are for man's survival. Without this, the future of a restorative soil policy will be uncertain. Because of a lack of soil awareness, there is no public acceptance as regards the major task of restoring as far as possible natural soil functions which have been impaired below the hazard threshold or of supporting natural regeneration processes. This also means that there is not enough "political pressure" and therefore little readiness to provide public funding for the development of appropriate measures.

More effective, i.e. more systematic, use should be made of the chance to strengthen and confirm soil awareness. This should be done using the problems connected with contaminated sites and the revitalisation of building land, sites and even regions. The regulatory framework is laid down in the German Soil Protection Act. This framework is filled out by the executive. But in addition there is an individual responsibility for the protection of the soil. In the context of each local or regional discussion regarding contaminated sites, it should be highlighted how valuable and important non-contaminated soil is for present and future generations. The information conveyed in the discussion must be presented in the form of concrete knowledge. It will thus be possible, by means of discussions, to develop a certain sensitivity towards the protection and conservation of the soil and its functions.

For such discussions, that are mainly to be led on-site by experts of soil treatment, there has to be established a methodology that convinces the public and the media in the frame of an interdisciplinary co-operation. For this natural scientists and engineers out of the field of examination of contaminated sites and techniques for remediation together with social scientists and environmental psychologists have to develop ways and possibilities for the variety of cases in practice, that result in an improvement of soil awareness.

2.3 Conclusion

When it comes to sharpening soil awareness in our society – an important prerequisite for soil protection and sustainable development – the values attached to the soil and its, in human survival terms, vital functions play a major role. For this reason, the existence of more than 300,000 of suspected contamination sites in Germany, each contaminated site identified as involving health hazards and each instance of ground water and plant pollution due to soil contamination should be exploited, along with all the remediation measures taken, in order to highlight the value of the soil and its intact functions.

Contaminated sites cause concern in our society, and concern enhances perception with respect to soil protection and hence soil awareness.

References

DOSSO M (1999) Bodenbewußtsein in Öffentlichkeit und Erziehung, Dokumentation der internationalen Tagung der Evangelischen Akademie, Tutzing, November 1999. Munich, Süddeutsche Zeitung, Germany, p 81

Preisendörfer P (1999) Umwelteinstellungen und Umweltverhalten in Deutschland. Opladen, Leske & Budrich

SRU (Rat von Sachverständigen für Umweltfragen) (1978) Umweltgutachten 1978, Chapter 2.1.2. Verlag W Kohlhammer, Stuttgart, Germany

SRU (Rat von Sachverständigen für Umweltfragen) (1987) Umweltgutachten 1987, Chapter 1.2. Verlag W Kohlhammer, Stuttgart, Germany

UBA (1998) Bevölkerungsumfrage "Umweltbewußtsein in Deutschland 1991–1998." Umweltbundesamt, Berlin, Germany

WBB (2000) Wege zum vorsorgenden Bodenschutz. E Schmidt-Verlag, Berlin, Germany

WBGU (1994) Wissenschaftlicher Beirat der Bundesregierung Globale Umweltveränderungen. Jahresgutachten 1994, Chapter 1.3.1.7, Economica-Verlag, Bonn, Germany

3 Do Contaminated Soils Have to Be Decontaminated?

G. Miehlich
Institute of Soil Science, University of Hamburg, Allende-Platz 2, 20146 Hamburg, Germany

3.1 Introduction

Twelve years ago, when the investigations within the research centre "treatment of contaminated soils" began, it was the common consensus that decontamination of contaminated soils is preferred to protection measures like cover systems. Meanwhile the situation has changed thoroughly, because in Germany the federal Soil Protection Act (BBodSchG) and the federal Soil Protection and Contaminated Sites Ordinance (BBodSchV) came into force. The opinion of practitioners is that there is no preference of decontamination to protection measures. With few exceptions, cheaper methods of remediation are available in accordance with the law, so that the title of the paper is somewhat rhetorical.

This paper describes the actual significance of soil contamination in Germany. It comprises the description of the legal situation, the extent of soil decontamination, the effects of soil decontamination on soil functions and the possibilities of restoration of soil functions following decontamination.

3.2 Legal Situation in Germany

In the following text the sentences are extracted from the official translation of the federal soil protection act (BBodSchG) and of the federal soil protection and contaminated sites ordinance (BBodSchV) as published by the Federal Ministry for the Environment, Nature Conservation and Nuclear Safety.

In the BBodSchG it is laid down that

> the party who caused a harmful soil change or a contaminated site, and his universal successor, as well as the relevant property owner and the occupant of the relevant real property, shall be obligated to remediate the soil and contaminated sites (§4 (3) BBodSchG). Harmful soil changes within the meaning of this act are harmful impacts on soil functions that are able to bring about hazards, considerable disadvantages or considerable nuisances for individuals or the general public (§2 (3) BBodSchG).

For the investigation and assessment of suspect sites two types of values are defined in the BBodSchG (§ 8 (1)):

1. Values which, if exceeded, shall mean that the investigation with respect to the individual case in question is required, taking the relevant soil use into account, to determine whether a harmful soil change or site contamination exists (trigger values),
2. Values for impacts or pollution which, if exceeded, shall normally signal the presence of a harmful soil change or site contamination, taking the relevant soil use into account, and to mean that measures are required (action values).

In the BBodSchV action and trigger values are given for three pathways: soil – human being (direct contact), soil – food plant, soil – groundwater. Values are fixed for selected contaminants and soil use within the pathways (e.g. direct contact: playgrounds, residential areas, parks and recreational facilities, industrial and commercial real properties). In appendix 1 of the BBodSchV investigation methods, sampling, analytical procedures and quality assurance are prescribed.

The soil or the contaminated site shall be remediated

> in such a manner that no hazards, considerable disadvantages or considerable nuisances for individuals or the general public occur in the long term (§4 (3) BBodSchG). As part of fulfilment of obligations relative to the soil and to contaminated sites ... the permissible use of the piece of land under planning law, and the resulting protection requirements, shall be taken into account, as far as this is compatible with the protection of the soil functions....

This means, that remediation values vary with planned use of the site and the role of soil functions (see below) is "somewhat unclarified" (Sanden and Schoeneck 1998).

Within the BBodSchG decontamination and "securing measures that permanently prevent spread of pollutants" are equal forms of remediation (§4 (3) BBodSchG, Holzwarth et al. 2000). Also

> suitable covering of soils that underwent a harmful change or of contaminated sites by means of a soil layer or sealing can be taken into consideration as stabilising measure (§5 (4) BBodSchV).

Where remediation measures

> are not possible or cannot be reasonably required, other protection and restriction measures shall be carried out (§4 (3) BBodSchG). On areas used for agricultural and forestry purposes, especially protective measures through adjustment of use and management of soils as well as changes in the quality of soil come into consideration (§5 (5) BBodSchV).

As shown, the BBodSchG and the BBodSchV provide several possibilities for the remediation of contaminated soils and sites. Mainly, costs decide the procedure of remediation.

3.3 Extent of Soil Decontamination in Germany

The share of soils cleaned up is very small compared to the portion of contaminated soils that are remediated by either above ground or below ground deposition. With an exploited capacity of approximately 65 % of the existing German decontamination plants, in 1997 only 1.2 MT of soil were remediated biologically, 0.85 MT chemically/physically and 0.1 MT were decontaminated thermically (Schmitz and Andel 1997). Quantitatively, only oil-contaminated soils play a larger role. The very small portion of soil clean-up compared to securing and deposition can be attributed to economical and not to technological difficulties (Lotter and Stegmann 1999). Although costs for decontamination treatment fell significantly during recent years in Germany, costs for dumping in landfills and in exploited mines are so low, that decontamination treatment is still more expensive.

There is an increasing tendency to reuse soil material according to LAGA-directions (LAGA 1997). In the city of Hamburg, for example, 50 % of the excavated soil materials are reused without decontamination, 20 % are reused after decontamination and 20 % are deposited in landfills (Kilger 1999). The scientific advisory board "soil protection" of the federal Ministry for the Environment, Nature Conservation and Nuclear Safety recommends alignment of the LAGA recommendations with the values given in the BBodSchV (Bachmann and Thoenes 2000).

3.4 Effects of Soil Decontamination on Soil Functions

Purpose of the BBodSchG is

> to protect or restore the functions of the soil on an permanent sustainable basis. ... Where impacts are made on the soil, disruptions of its natural functions and of its function as an archive of natural and cultural history should be avoided as far as possible (§1 BBodSchG).

The soil performs natural functions:

1. As a basis for life and a habitat for people, animals, plants and soil organisms,
2. As part of natural systems, especially by means of its water and nutrient cycles,
3. As a medium for decomposition, balance and restoration as a result of its filtering, buffering and substance-converting properties, and especially groundwater protection (§2 BBodSchG).

Every decontamination method influences the natural soil functions and its function as an archive of natural and cultural history. According to remediation "also the consequences of the interference, in particular for soils and waters, must be considered" (§5 (1) BBodSchV, details Fehlau et al. 2000).

With regard to the degree of disturbance of soil functions there are considerable differences among the decontamination methods. Methods without soil excavation (e.g. soil air stripping, hydraulic measures) may influence natural soil functions adversely by change of soil structure, swelling and shrinking or change of the water balance. However, methods based on soil excavation (e.g. thermal treatment, soil washing and biological treatments) influence soil functions to a greater extent. On the remediated site the excavation of topsoil may cause a reduction of soil organic matter, nutrients and CEC with the consequence that the filtering, buffering and substance-converting properties are reduced. The soil completely loses its function as an archive of natural and cultural history. The decontaminated soil materials change their properties to different degrees dependent on treatment. Thermal treatment for instance, leads to an almost complete loss of all properties important for the natural soil functions. During soil washing the fine particles of the soil materials are removed which adversely influences the natural soil function. Biological treatments cause only small changes to soil properties (e.g. enrichment of nutrients). Various proposals exist for the consideration of soil protection aspects during the choice of decontamination procedures (DVWK 1996; Litz and Smettan 1997).

Furthermore, it has to be considered that decontamination procedures themselves may affect soils or other compartments of the ecosystem beyond the actual remediation site (e.g. effects of emissions on soils, waters, the atmosphere and the biosphere by the decontamination plant).

3.5 Restoration of Soil Functions after Decontamination

Following larger decontamination measures based on soil excavation, it is possible to almost fully restore natural soil functions by redeposition of soil materials. Soils bearing very distinct characteristics can be established through a suitable selection of properties of the redeposited materials. Some of the natural soil functions can be reestablished very quickly after recultivation (site characteristics) whereas others take tens to hundred of years for regeneration (e.g. habitat function, soil structure, soil type).

The work on recultivation of opencut mining sites and waste sites has yielded extensive experience regarding the restoration of soil properties (Dumbeck 1997; Melchior 1998; Tresselt 2000). The results show that restoration of chemical soil properties (e.g. pH, organic matter, nutrients) is much easier than the restoration of physical soil properties (e.g. pore volume, pore size distribution, soil structure).

The selection of the appropriate soil texture is of great importance in order to create a suitable habitat for flora and fauna (water supply, root penetrability, biomass for decomposers) and also to establish maximum capacity for filtering, buffering and substance transformation. Soil installation has to be carried out using suitable techniques and machines in order to avoid compaction leading to air shortage and conditions of low redox potentials (Tresselt 2000).

The remediated materials vary in their suitability for recultivation depending on the decontamination procedure applied (Holz 1997; Goetz et al. 1997):

- Thermal treatment leads to such a change in properties that these materials may only be used as filling material in deeper soil layers.
- Due to their coarse texture, washed soil materials are suitable for dry and nutrient-poor soils only.
- Biologically decontaminated materials are suitable for a wide range of applications.

References

Bachmann G, Thoenes HW (2000) Wege zum vorsorgenden Bodenschutz. Bodenschutz und Altlasten 8. Erich Schmidt Verlag, Berlin, Germany

DVWK (1996) Sanierung kontaminierter Böden, DVWK Schriften 116. Wirtschafts- und Vertragsgesellschaft Gas und Wasser mbH, Bonn, Germany

Dumbeck G (1997) Zur Entwicklung des gegenwärtigen Kenntnisstandes der Rekultivierung landwirtschaftlicher Nutzflächen im rheinischen Braunkohlenrevier. In: Felix-Henningsen P, Wegener HR (eds) Festschrift für Prof Dr Tamás Harrach zum 60 Geburtstag. Universität Gießen, Germany, pp 9–20

Fehlau KP, Hilger B, König W (2000) Vollzugshilfe Bodenschutz und Altlastensanierung – Erläuterungen zur Bundes-Bodenschutz- und Altlastenverordnung. Bodenschutz und Altlastensanierung 7. Erich Schmidt Verlag, Berlin, Germany

Goetz D, Bauske B, Claussen A, Gläseker W, Holz C (1997) Bodenkundliche Untersuchungen zu thermischen, chemischen und biologischen Bodenreinigungsverfahren. Umweltbundesamt Projektträger Abfallwirtschaft und Altlastensanierung, Berlin, Germany

Holz C (1997) Hochtemperaturbehandeltes Bodenmaterial – Eigenschaften, Alterungsverhalten und Verwendungsmöglichkeiten als Rekultivierungshilfsstoff. Hamburger Bodenkundliche Arbeiten 35

Holzwarth F, Radtke H, Hilger B, Bachmann G (2000) Bundes-Bodenschutzgesetz/Bundes-Bodenschutzverordnung – Handkommentar. Bodenschutz und Altlasten 5, Erich Schmidt Verlag, Berlin, Germany

Kilger R (1999) Sanierung der städtischen Flächen – eine Zwischenbilanz. Altlastensanierung in Hamburg – dekontaminieren oder sichern? Umweltbehörde Hamburg, Germany

LAGA (1997) Länderarbeitsgemeinschaft Abfall, Anforderungen an die stoffliche Verwertung von mineralischen Abfällen – Technische Regeln. In: Mitteilungen der Länderarbeitsgemeinschaft Abfall Nr 20, Erich Schmidt Verlag, Berlin, Germany

Litz N, Smettan U (1997) Berücksichtigung der Schutzwürdigkeit kontaminierter Böden bei der Auswahl angepaßter Sanierungsverfahren. Mitteilgn Dtsch Bodenkundl Ges 85/II: 745–748

Lotter S, Stegmann R (1999) Stand der Technik und Wissenschaft bei der Dekontamination von Altlasten. Altlastensanierung in Hamburg – dekontaminieren oder sichern? Umweltbehörde Hamburg, Germany

Melchior S (1998) Ansätze zur Gestaltung von Rekultivierungsschichten in Abdecksystemen für Altdeponien und Altlasten. In: Stief K, Engelmann B (eds) Abfallwirtschaft in Forschung und Praxis 107, Erich Schmidt Verlag, Berlin, Germany, pp 161–180

Sanden J, Schoeneck S (1998) Bundes-Bodenschutzgesetz – Kurzkommentar. C F Müller Verlag, Heidelberg, Germany

Schmitz HJ, Andel P (1997) Bodenbehandlungsanlagen: Die Jagd auf den Boden wird härter. TerraTech 5: 17–31

Tresselt K (2000) Feldversuche zur Wirksamkeit von Oberflächenabdichtungssystemen mit Dichtungen aus Hafenschlick. Hamburger Bodenkundliche Arbeiten 46 (in press)

References

4 An Overview of Soil and Sediment Treatment Research in the Netherlands

W.H. Rulkens
Subdepartment of Environmental Technology, Wageningen University, P.O. Box 8129, 6700 EV Wageningen, The Netherlands

4.1 Introduction

Soil and sediment pollution is a serious environmental problem in The Netherlands. Over the past 10 to 15 years, awareness of the problem, and the policy and strategy to tackle the problem have radically changed. Initially the approach to tackle the problem of polluted soils was primarily focused on the clean-up of soil after excavation. This approach was consistent with government policy of that time, whose aim was the maintenance or restoration of soil multifunctionality. The result was the development and application of intensive and relatively expensive methods such as thermal treatment and soil washing/wet classification, that were capable of achieving the target values set for clean soil. At that time biological treatment methods and in-situ treatment were not considered feasible. However, inventories revealed that the polluted sites in The Netherlands numbered approximately 100,000 and contained more than 200 million tons of contaminated soil. Moreover, the problems associated with cleaning up these sites were found to be highly diverse. It became clear that for both technical and financial reasons, it would be impossible to clean up all these sites to the target values. This realisation resulted in a shift in policy. Rather than try to clean-up all sites completely, the approach shifted to reducing the risk to humans and the ecosystem. Due to this shift in policy, other techniques which were not able to clean-up the soil to the target values were becoming important for practical application. The result was, among other things, an increasing application of in-situ treatment techniques, biological treatment techniques and isolation methods. However, this approach did not fully solve the problem, mainly due to the high costs. This has recently led to a second shift in soil remediation policy. This second shift has resulted in a remediation approach that now also includes a direct link with the intended use of the remediated site. Also, the financial basis for the remediation activity has been broadened by integration of soil remediation into other social and industrial activities such as building, residing, and rural development. Altering the system of financing will increase the market dynamics of remediation activities.

Over the past two decades substantial research efforts have been made in The Netherlands to develop new, and to improve existing soil remediation techniques. The shifts in governmental policy to tackle the problem of soil pollution have strongly influenced the research issues regarding soil remediation. Initially, research was mainly focused on the development of clean-up techniques for excavated soil, whereas at the present time research into the development of remediation techniques is mainly focused on intensive and extensive in-situ biological clean-up and in-situ natural degradation processes.

Awareness of a polluted sediments problem is more recent than that of polluted soil. The problem of polluted sediments is for a large part comparable with that of polluted soils. Remediation techniques developed for polluted soils are often applicable to polluted sediments, although process conditions, costs, and the like may vary a great deal. Nevertheless research into the development and application of polluted sediment treatment technologies found a sound scientific and technological basis in polluted soil research.

Over the past two decades there have been eight research programmes focused on the development and application of soil and sediment remediation techniques (Rogaar 1996). A substantial part of these research programmes was financed by the government. Aims, research issues, type of research and intended participants in the various programmes varied. Most of these programmes have concluded, or will conclude in 2000. In addition, most of these programmes did not deal only with the development and application of soil remediation techniques. Some encompassed issues also related to general soil science or integrated technological and non-technological aspects of soil or sediment pollution. The majority of the research results are public. However, for some programmes, especially those where industry was involved in a competitive context, the availability of the results is restricted.

The most important programmes, of which also the results are for a large part publicly available, are:

- NOVEM (Stimulation Regulation Environmental Technology, Netherlands Organisation for Energy and Environment),
- NOBIS (Dutch Research Programme In-Situ Bioremediation),
- PGBO (Integrated Programme on Soil Research),
- POSW (Development Programme for Treatment Processes for Polluted Sediments).

From these research programmes some general information about the aim and research area as well as some specific results will be mentioned in the following paragraphs. It has to be emphasised that the presented results cover only a minor part of all results obtained with these research programmes. In addition, this chapter will conclude with a short discussion about future research issues of interest.

4.2
NOVEM (Stimulation Regulation Environmental Technology, Netherlands Organisation for Energy and Environment)

4.2.1
Introduction

The general aim of NOVEM in the context of the Stimulation Regulation Environmental Technology is to develop, demonstrate and implement innovative technologies to tackle pollution problems effectively. This stimulation programme came into effect on 1 January 1989. Within the framework of this programme, proposals for the development of soil remediation technologies could be submitted and remunerated up to 1 January 1998. During the entire period 60 research projects dealing with polluted soil (a few also with sediments) were executed (NOVEM 2000).

4.2.2
Overview of the Research Programme

The main research issues within this programme were:

- In-situ clean-up of polluted soil:
  - Physical/chemical techniques (application of radiofrequencies, venting, extraction, stripping, use of surfactants).
  - Microbiological techniques (biorestoration, biosparging, bioventing, bioscreens).
  - Combination of physical/chemical and microbiological techniques (biodielectrical systems).
- Treatment of excavated soil:
  - Particle separation (application of hydrocyclones, foam separation).
  - Thermal treatment (halogenated hydrocarbons).
  - Physical/chemical treatment (solvent extraction, supercritical extraction, electro-chemical degradation, leaching).
  - Microbiological treatment (landfarming, slurry reactors, composting systems).
  - Combination of physical/chemical and microbiological treatment (gamma radiation in combination with biological degradation).
- Isolation and immobilisation of polluted soil (immobilisation heavy metals, application of polymergel).
- Measuring and monitoring (early warning systems).

The research results have been summarised in a special report which refers to detailed reports dealing with the separate research projects (NOVEM 2000).

4.2.3
Examples of Research Projects

Removal of Hexachlorocyclohexane (HCH) from Soil (NOVEM 2000)

Regarding the removal of HCH, several research projects have been executed. One project focused on the simultaneous aerobic biodegradation of β-HCH and polycyclic aromatic hydrocarbons (PAHs) from soils and sludges in a bioreactor. Photochemical oxidation was used as a pre-treatment step. It was found that β-HCH was not biodegradable in this system while for the PAHs a limited biodegradation was found. Photochemical pre-treatment had a small effect on the biodegradation of PAHs but had no effect on the degradation of β-HCH. Another project, in which the application of gamma radiation was investigated, showed that by using gamma radiation, β-HCH can be degraded to tetrachloro-cyclohexene, a compound that is biodegradable. This observation will likely lead to a new technology that can be applied in practice. A third project dealt with in-situ removal of HCH from soil by extraction with lye. The investigation showed that for loamy soil with a high organic matter content, the method is not technologically feasible due to the low permeability of the soil. However the method seems technologically and financially feasible for sandy soils polluted with HCH or other type of pesticides. Leaching with a hot aqueous solution of sodium chloride (70 °C) results in an acceleration of the removal efficiency of HCH from permeable soils by a factor 10. However, compared with a conventional leaching method, the remediation process is twice as expensive.

In-Situ Biorestoration of Soil Polluted with Volatile Chlorinated Aliphatic Hydrocarbons under Influence of Different Redox Conditions (NOVEM 2000)

Based on the results of a lab-scale investigation, a remediation concept was developed for in-situ biorestoration of a real site polluted with chloro-ethylenes. This concept integrates an anaerobic dechlorination of perchloro-ethylene and an aerobic co-metabolic mineralisation of the lower chloro-ethylenes such as trichloro-ethylene, *cis*-dichloro-ethylene and vinylchloride. Anaerobic degradation was stimulated by the injection of methanol. The aerobic degradation step was stimulated by the injection of pressurised air and the injection of the co-substrate phenol in the groundwater flow just outside the anaerobic zone. It was found that within six months about 90 % of the perchloro-ethylene, originally present, was anaerobically degraded to *cis*-dichloro-ethylene, and even to vinylchloride and ethylene. It was observed that after passing the aerobic zone, the concentrations of the anaerobic degradation products in the groundwater were relatively low. Also in the air, removed from the soil, only small amounts of chlorinated compounds could be measured. It can be concluded that the investigated remediation concept is technically, financially and environmentally feasible for in-situ remediation of sites polluted with chlorinated solvents. It is an alternative to the conventional pump and treat methods.

4.3 NOBIS (Dutch Research Programme In-Situ Bioremediation)

4.3.1 Introduction

The aim of the Dutch Research Programme In-situ Bioremediation (NOBIS) is to develop, evaluate and demonstrate effective innovative strategies, methods and techniques for in-situ bioremediation of polluted soil (CUR/NOBIS 1995). Biological in-situ soil remediation techniques are expected to contribute to a significant reduction in costs, especially in those situations where they are put into practice on a large scale and in an extensive manner. It is a research programme focused on the market demand. The content of the programme is based on project proposals, sent in by consortia in which problem owners co-operate with consultancy firms, scientific institutes, such as universities, and contractors. The programme started in 1994 and runs until 2000. The programme comprises three types of projects:

- Implementation projects. These are practical projects on location with a characteristic combination of soil, contamination, and use. The projects are aimed to demonstrate successful in-situ bioremediation projects in practise and to investigate specific bottlenecks in the field.
- Feasibility projects which may lead to implementation and/or research projects. The aim of these projects is to generate new developments and to test their feasibility.
- Research projects which are application oriented and which are connected to the implementation projects. The aim of these projects is to investigate general gaps in know-how for implementation projects, and to develop additional know-how for innovative concepts and techniques.

4.3.2 Overview of Research Issues

Based on the above mentioned types of projects, the research programme of NOBIS can be subdivided into the following six research issues (CUR/NOBIS 1997, 1998, 2000):

- Risk and risk assessments.
- Extensive remediation concepts (natural attenuation, enhanced intrinsic biodegradation, enhanced biodegradation with extensive technologies, bioscreens, landfarming).
- Intensive remediation concepts.
- Combi-remediation.
- Measurements and monitoring (field characterisation, bio-assays, monitoring remediation progress, developments of measurement instruments).
- Decision support systems.

As already mentioned the NOBIS programme will conclude in 2000. The final result of NOBIS is that biological in-situ remediation has been placed unconditionally on the soil map (CUR/NOBIS 2000; Vermeulen and Verheul 1999).

4.3.3
Examples of Research Projects

Bioscreens, Technical Possibilities and Economic Potential, Aimed at Implementation (CUR/NOBIS 1999a)

This study was aimed to evaluate the technical and economic feasibility of bioscreens for the Dutch situation. A bioscreen is a permeable screen positioned in the soil perpendicular to the flow direction of a polluted groundwater flow. The aim of the bioscreen is to stimulate biodegradation of organic pollutants in groundwater that passes the bioscreen. In the project special attention was paid to the implementation of bioscreens as a controlling measure on sites contaminated with a mobile pollution. From the study it can be concluded that bioscreens can be technically realised in the Netherlands, however the market is limited. From the several modifications of bioscreens, the bioreactive trench for shallow contamination has the highest potential. For deep contamination the biologically activated zone has the largest potential. Application of bioscreens must be focused on large industrial sites where the source of contamination cannot readily be removed and where control of the contamination is an absolute necessity. The study makes clear that there is a need for solving some practical bottlenecks which might hamper a successful implementation. The most important ones are clogging of infiltration equipment, and the reliability and durability of the degradation process.

Benzene Degradation in a Strongly Reducing Soil (CUR/NOBIS 1999b)

This project involved a site characterised by benzene pollution in a sulphate-reducing to methanogenic environment. The aim of the project was a further characterisation of the polluted site and the soil, and to investigate whether degradation of benzene had already taken place, and whether this degradation could be stimulated by the application of minimal quantities of oxygen and/or the electron acceptors nitrate and sulphate. In laboratory tests it was observed that no intrinsic degradation of benzene occurred. The application of sulphate as an electron acceptor did not lead to benzene degradation. However application of nitrate as an electron acceptor resulted in a complete degradation of benzene. Furthermore, degradation also took place in samples to which a small quantity of oxygen had been added. These results can be considered as a breakthrough in the field of anaerobic biological degradation of benzene.

Combi-Remediation: Combined Biodegradation of Chlorinated Solvents and BTEX by Engineered Mixing of Ground Water Plumes (CUR/NOBIS 1999c)

This project dealt with a contaminated site in which the ground water contained a contaminated plume with mono-aromatic compounds such as benzene, toluene, ethylbenzene and xylene (BTEX) and a plume with trichloro-ethylene (TRI) and perchloro-ethylene (PER). The aim of this project was to investigate the possibilities of a so-called combi-remediation, which is based on the concept that BTEX might serve as fuel for microorganisms in the degradation process of PER and TRI. Extensive field characterisation of soil and ground water showed that indeed, compared with the separate biodegradation, an improved biodegradation of both type of pollutants could be obtained when the plumes of BTEX and PER/TRI were mixed. However, a complicating factor in proving the concept of improved biodegradation was the presence of mineral hydrocarbons, a substantial part of the total amount of dissolved organic carbons, in the BTEX plume. The degradation of PER and TRI may be affected by the presence of those hydrocarbons or their transformation products. In the observed process of combi-remediation no distinction could be made between dechlorination induced by BTEX or dechlorination induced by the total amount of dissolved organic carbon. Notwithstanding these complications, the results are of practical importance. Simultaneous occurrence of BTEX (and mineral hydrocarbons) and PER and TRI is a very common phenomenon. More research is needed to unravel this process.

4.4 PGBO (Integrated Programme on Soil Research)

4.4.1 Introduction

The Integrated Programme on Soil Research (PGBO) started in 1995 and concluded in 1999 (Rogaar et al. 1999). It was the continuation of the former Programme on Soil Research (SPBO). This programme was initiated in 1986 and concluded in 1994. It was aimed at the development of fundamental soil knowledge, with emphasis on soil pollution aspects, establishing co-operation and network building. It comprised more than 113 research projects. The results of more than 80 of these projects are given in 7 summary reports, the titles of which are given below:

- The fate of organic pollutants in soil and sediments and the development of soil remediation techniques.
- Adaptation and selection mechanisms of natural and genetically modified soil microorganisms.
- Biological availability and transformations of organic compounds in soil and sediment systems.
- Spatial variability of soil contamination and the consequences for environmental risk assessment.

- Remediation and isolation techniques for soils and sediments.
- Soil structure and transport processes – implications for water, gases, nutrients, pesticides, and contaminants in soils.
- Speciation and bioavailability of heavy metals, sulphur, phosphorus and cyanide in soils and sediments.

The aim of PGBO was to embed the acquired knowledge of SPBO into practical application. Due to budgetary constraints the programme of PGBO was limited to the execution of so-called definition studies and feasibility studies. The aim of these definition studies was to study, for a lot of relevant issues dealing with soil and soil pollution, the state of the art and possible needs for further research.

4.4.2
Overview of Research Issues

The main research issues of PGBO are (Rogaar 1996; Rogaar et al. 1999; PGBO 1999):

- Applicability of immobilising agents for the remediation of heavy metal polluted soils in the Netherlands.
- Feasibility of in-situ sediment treatment.
- Phytoremediation: Ready for use in The Netherlands?
- Volatile chlorinated aliphatic hydrocarbons: remediation options and research needs.
- Stimulation programmes for soil and sediment research in The Netherlands.
- Monitoring and control of soil and groundwater.
- Characterisation methods for assessment of biological remediation of soil and sediments.
- Classification of soil pollution.
- Assessing risks from soil pollution: inventory of bottlenecks and possible solution.
- Location specific ecological risks: a basic approach for function oriented assessment of soil pollution.
- Use of bio-assays for assessment of soil pollution.
- Ecotoxicological risk assessment of polluted sediments and soils.
- Decision support for soil-related working processes for regional and local authorities.
- Management options for phosphate leaking agricultural soils.
- Soil indicators for critical source areas of phosphorous leaching.
- Applicability of the concept of soil vulnerability as a tool for decision makers in the area of environmental planning and soil remediation.
- Views on developing Nature on contaminated soils.
- Communication on soil remediation.
- The identification of the supply of, and the demand for, knowledge regarding the behaviour and the consequences of contamination in the deeper subsoil.

4.4.3 Examples of Research Projects

Applicability of Immobilising Agents for the Remediation of Heavy Metal Polluted Soils in the Netherlands (Koopmans et al. 1998)

The aim of this feasibility study was to evaluate the applicability of immobilisation agents for the remediation of soils polluted with heavy metals. From the literature several immobilising agents were identified which might be promising for application to the Dutch situation. The most important agents were beringite, aluminium- and aluminium 13-montmorillonite and manganese-oxide. Beringite is a waste product of a coalmine. It has a high affinity for Zn, Cd, Pb and Cu and should therefore, in principle, be suitable for immobilisation of these metals. However, addition of beringite causes a pH increase in the soil which results in the dissolution of organic material. Because Cu has a stronger affinity to adsorb to soluble organic matter than to beringite, this agent is not suitable for the immobilisation of Cu. Aluminium- and aluminium montmorillonite and manganese-oxide are, in principle, also suitable to immobilise heavy metals such as Ni, Zn, Cd and Pb. The immobilisation effect can be enhanced by adding lime to increase the pH. Under reducing conditions, often the case in wet areas, manganese-oxide will dissolve, and hence heavy metals sorbed to manganese-oxide will be remobilised. This property makes manganese-oxide less suitable. From the study it became clear that the final choice of an immobilising agent has to be based on the effectiveness of the agent and stability of the effects in the field. However experience under field conditions is rather scarce.

In-Situ Aquatic Sediment Remediation: Imaginable and Feasible? (Van der Gun and Joziasse 1999)

The aim of this feasibility study was to investigate how far the vision, developed for in-situ remediation of polluted terrestrial soils, could be applied to in-situ treatment of (aquatic) sediments. The possibilities for in-situ remediation of polluted sediments strongly depend on the local situation. That is, the options are dependent on, among others, the type and speciation of the contaminants, the extent of the pollution, functions, sizes and current conditions of the waterway. The study indicates a number of methods and techniques that might be considered in different situations. For the most relevant options, a provisional feasibility judgement is given. These options include stimulating microbiological degradation of organic contaminants, concentrating metals into vegetation, immobilisation of metals or organic contaminants by adsorption to various materials, binding in cement, vitrification and reducing the advective dispersion of contaminants towards groundwater or surface water by capping the soil or by hydrological isolation of the waterway. In specific cases refraining from any action can come forward as the best option in connection with risk reduction.

Generally, it can be concluded from the study that the feasibility of the options identified is hard to estimate. Especially the effects of the application of the various options on the aquatic ecosystem will have to be taken into account. In specific cases, where a conventional approach encounters serious difficulties, an investigation dedicated to the prevailing conditions will have to give a decisive judgement on the feasibility of an alternative (in-situ) approach. If certain options appear promising for application in practice, a pilot-scale experiment will be necessary as a subsequent step.

4.5 POSW (Development Programme for Treatment Processes for Contaminated Sediments)

4.5.1 Introduction

The development Programme for Treatment Processes for Contaminated Sediments (POSW), starting up in 1989 and running until 1996, was aimed at the development of ecologically sound dredging and processing techniques, to be used in the remediation and reuse of polluted sediments. The POSW-programme consisted of two stages. The first stage of POSW was an exploratory phase focused on the inventory and preliminary research into methods, which might be useful for sediment remediation. The results of this stage are mentioned in the report Development Programme Treatment Processes for Polluted Aquatic Sediments in the Netherlands, Phase I (1989–1990) (POSW I 1992). In this first stage a number of promising processes were identified. In the second stage of the programme, POSW Stage II, running from 1990–1996, the research was aimed at the further development of processes and methods, identified as promising, and on the technical applicability of these processes and methods. This technical applicability had to be demonstrated in practise, as part of an integrated remediation chain. Attention was also paid to the economic and environmental consequences of the several types of techniques as part of entire clean-up chains.

4.5.2 Overview of the Research Programme

The main research issues of the POSW Stage II programme were (POWS II, Final Report 1997):

- Pre-dredging survey and dredging (establishing pre-dredging surveys, fast screening methods, optimisation of the dredging, control and monitoring systems, practical experience in pilot remediation).
- Separation of sludge into subflows (hydrocyclone separation, upstream separation, settling, flotation, dewatering of fine fractions, practical experience in pilot remediation).
- Thermal and chemical treatment methods (thermal desorption, incineration, wet oxidation, solvent extraction).
- Biological treatment (landfarming, greenhouse farming, slurry treatment in bioreactors).

- Immobilisation of pollutants in products (melting, sintering, practical experience in pilot remediation).
- Assessment of the environmental effects of processing chains (based on life cycle analysis, LCA).
- Selection of methods for large scale processing.
- Scenarios for large scale processing, varying from “natural” processes in treatment plants (sedimentation, dewatering, landfarming and ripening) to maximum deployment of classifying and polishing methods.
- Scenarios and characteristics (costs, environmental effects, spatial claims).
- Conditions and control (financial means, administrative agreements and promoting measures).

Results of POSW are presented in a summary report (POSW II, Final Report 1997) and in several additional reports dealing in detail with each of the main research issues of the research programme. From this summary report a few short abstracts, which can be considered as characteristic for the research programme, have been taken and mentioned in the next paragraph.

4.5.3 Examples of Research Projects

Methods of Separation According to Grain Size (POSW II, Final Report 1997)

This research was aimed at optimising the existing methods for the processing of sludge. The technology presently available and the practical experience gained by contractors in the decontamination of sediments render it possible to extract building materials (class 1 of the Building Materials Decree) from sandy sludge. Apart from the positive ecological effects of processing – less dumping space needed, saving on the extraction of primary materials – the processing itself has negative side-effects. The separation of sand is energy-consuming and requires water to dilute the input. The water is recycled during the process, but any surplus will have to be treated, either locally or in a purification plant elsewhere.

The Possibilities of Sludge Flotation (POSW II, Final Report 1997)

Within the POSW programme research was executed into the possibilities to purify the fine fraction – i.e. the residual flow after the separation of sand – by means of flotation. This was tried initially in the pilot remediation of Elburg, later in the framework of the project “Flotation North and South-Holland provinces”, in which POSW participated along with water boards, provincial authorities and Rijkswaterstaat. Unlike the flotation of sand, the heavy metal contents of silt remain practically unchanged after flotation, because such substances are bonded very tightly to clay minerals. That is why flotation, when deployed to purify the fine fraction, only affects the content of organic pollutants. The fine fraction, however, is generally contaminated with mineral oil and PAHs to such an extent that, after treatment, the concentrations still exceed the provisional warning value.

Pilot Remediation Elburg Harbour (POSW II, Part 5, Main Report 1995)

The objective of this pilot remediation was to remove a minimum of 90 % of the pollutants present (PAHs and mineral oil) in the harbour, and to process the sludge in such a way that at least 50 % of the material could be recycled. The pilot remediation succeeded in doing so. It was the first of three decontaminations carried out in the framework of POSW-II and the requirements of precision dredging were new at that time. The contractor developed a special grab and improved controlling equipment to meet the requirements. The sand in this project was separated by hydrocyclone. No upgrading of an existing method was involved here, but Rijkswaterstaat gained experience in drawing up specifications which demand a certain quality level for the operating method. The contractor succeeded in meeting these high standards: 53 % of the dredged spoil could be separated as sand clean enough to be used as construction material in a housing programme of the town of Den Helder. A further 32 % of extracted sand could be marked as filtering sand and embankment fill in a deposit site for polluted soil.

Biological Treatment in Slurry Reactors (POSW II, Final Report 1997)

Treatment in bioreactors involves introducing the sludge as slurry in tanks, where it is constantly kept moving and aerated. The slurry can be treated batch-wise or in a continuous process. Within the framework of POSW, two types of bioreactors were tested on a scale of some cubic metres, and were deemed suitable: the Slurry Decontamination Process (SDP, continuous system) and the Aeration Basin (batch system). The latter is suitable for the treatment of fine fractions from a hydrocyclone, the first for the treatment of both separated and non-separated sludge. The treatment in bioreactors is on the whole four to ten times as expensive as landfarming, but the required retention times are decreased commensurately. The results achieved in this shorter period remain similar. Costs are expected to decrease further with up-scaling of the treatment plants and commercialisation of the market. Emissions into the air are not an ecologically problematic feature of bioreactors. The treatment involves the recycling of water, the surplus of which can be discharged – after purification if need be. Another environmental aspect is the energy consumption of this method.

Pilot Remediation Petroleum Harbour (POSW II, Final Report 1997)

The Amsterdam Petroleum Harbour was selected for a pilot remediation emphasising the biological treatment of sludge. To treat this sediment, contaminated by mineral oil (to 20,000 mg kg^{-1}) and PAHs (to 1000 mg kg^{-1}), a continuous type of bioreactor was utilised. Pre-conditions were quality requirements for the final product and a decontamination period of 100 days. A total of 5000 m^3 of sludge was dredged and processed. Pre-treatment with a hydrocyclone produced a coarse and a fine fraction. The former was subjected to flotation, whereas the fine fraction was dewatered to the desired density. It was next biologically treated in a continuous process in a series of bioreactors and finally dewatered. Tests with a few cubic metres of material showed that PAHs and oil could be removed from the sludge by 92 % and 76 %, respectively.

4.6 Future Research

All research programmes mentioned in the previous paragraphs have concluded. A new organisation, SKB (Foundation for Knowledge Development and Knowledge Transfer) has been set up for the continuation of soil research in The Netherlands (Vermeulen 1999). Similar to NOBIS the approach of SKB is to set up and finance research and demonstration projects executed by consortia of problem owners, research institutes, consultancy firms and problem solvers. Also support will be given to long-term fundamentally and strategically oriented research. SKB will not only focus on the development of innovative technology development but will also promote the application of existing knowledge by means of education and training programmes.

As already mentioned in the introduction, soil remediation activities are becoming more and more part of other social and industrial activities such as redevelopment of rural, municipal and industrial areas, construction of railways, building of tunnels, etc. The aim of this approach is to eliminate stagnation in soil remediation operations. In agreement with this approach it is expected that innovative technologies will be developed that pay special attention to measuring and monitoring systems for in-situ application, to biotic and abiotic degradation screens for in-situ removal of pollutants from ground water and to natural attenuation. It is also expected that attention will be given to the problem of NAPLS and DNAPLS, immobilisation as a treatment technique, and the possibilities of phytoremediation. Long-term fundamental research will primarily be focused on processes regarding pollutants and the effects of these pollutants in the subsoil. Regarding the pollutants, it is intended that the primarily attention will be given to mobile pollutants such as BTEX, mineral oil, and TRI and PER.

In additional to SKB, which is primarily focused on polluted soils, a new advising organisation AKWA (Advising and Knowledge Centre Aquatic Sediments) has been established (AKWA 1998; Groen 1999). AKWA is primarily a government-advising organisation. Its main field of operation is the design and execution of projects with respect to remediation of polluted sediments. This organisation also has the task to underpin government policy regarding polluted sediments. Depending on knowledge gaps and bottlenecks, research into sediment treatment will be stimulated.

References

AKWA (1998) Nieuwsbrief, Advies- en Kenniscentrum Waterbodems (Newsletter, Advise- and Knowledge Centre Sediments) (in Dutch). nr 1, May 1998, RIZA, PO Box 17, 8200 AA Lelystad, The Netherlands

CUR/NOBIS (1995) Dutch Research Programme In-Situ Bioremediation. NOBIS, PO Box 420, 2800 AK Gouda, The Netherlands

CUR/NOBIS (1997) Survey of current projects as of October 1997. NOBIS, PO Box 420, 2800 AK Gouda, The Netherlands

CUR/NOBIS (1998) Jaarverslag (Annual Report) (in Dutch). NOBIS, PO Box 420, 2800 AK Gouda, The Netherlands

CUR/NOBIS (1999a). Bioschermen, technische mogelijkheden en markt potentie, gericht op implementatie. (Bioscreens, technical possibilities and economic potential, aimed at implementation). NOBIS Report nr 96-1-01 (in Dutch). NOBIS, PO Box 420, 2800 AK Gouda, The Netherlands

CUR/NOBIS (1999b) Benzeenafbraak in een sterk reducerende bodem. Fase In: Eindrapport (Benzene degradation in a strongly reducing soil. Phase I: End report). NOBIS Report nr 96-3-05 (in Dutch). NOBIS, PO Box 420, 2800 AK Gouda, The Netherlands

CUR/NOBIS (1999c) Combi-remediatie; gecombineerde saneringsaanpak van CKW en BTEX. (Combi-remediation; Combined biodegradation of chlorinated solvents and BTEX by engineered mixing of ground water plumes.) Phase I. Report nr 97-1-15. (in Dutch). NOBIS, P.O. Box 420, 2800 AK Gouda, The Netherlands

CUR/NOBIS (2000) CD ROM (in Dutch). NOBIS, PO Box 420, 2800 AK Gouda, The Netherlands

Groen KP (1999) Advies- en Kenniscentrum Waterbodems – AKWA (Advise and Knowledge Centre Sediments – AKWA) (in Dutch). Bodem 2: 52–53

Japenga J (1999) Fytoremediering: Klaar voor gebruik in Nederland (Phytoremediation: Ready for use in The Netherlands) (in Dutch). The Netherlands Integrated Soil Research Programme, PO Box 37, 6700 AA Wageningen, The Netherlands

Koopmans GF, Chardon WJ, Bril J, De Ruiter PC, Dolfing J (1998) Applicability of Immobilizing Agents for the Remediation of Heavy Metal Polluted Soils in The Netherlands. The Netherlands Integrated Soil Research Programme, PO Box 37, 6700 AA Wageningen, The Netherlands

NOVEM (2000) De gids voor bodemland. Tien jaar Stimuleringsregeling Milieutechnologie voor de verontreinigde bodem (Guide for "soilland". Ten year Stimulation Regulation Environmental Technology) (in Dutch). Visser-Westerweele E P C (ed) The Netherlands

PGBO (1999) Reports. CD-ROM, ISBN 90-73270-41-3 (in Dutch). The Netherlands Integrated Soil Research Programme, PO Box 37, 6700 AA Wageningen, The Netherlands

POSW I (1992) Development Programme Treatment Processes for Polluted Aquatic Sediments in the Netherlands. Phase I. (1989–1990). PO Box 17, 8200 AA Lelystad, The Netherlands

POSW II (1995) Development Programme for Treatment Processes for Contaminated Sediments. Deel 5. Evaluatie van de waterbodemsanering van de haven van Elburg. Hoofdrapport (Part 5. Evaluation report of sediment remediation of the harbour of Elburg) (in Dutch) (1995). RIZA Report nr 95.051, ISBN 90 369 01456, PO Box 17, 8200 AA Lelystad, The Netherlands

POSW II (1997) Development Programme for Treatment Processes for Contaminated Sediments. Final Report. RIZA Report nr 97.051, ISBN 90 369 50 97 X, PO Box 17, 8200 AA Lelystad, The Netherlands

Rogaar H (1996) Stimuleringsprogramma's voor bodem – en waterbodemonderzoek in Nederland anno 1996. (Stimulation programmes for soil and sediment research in The Netherlands in the year 1996) (in Dutch). The Netherlands Integrated Soil Research Programme, PO Box 37, 6700 AA Wageningen, The Netherlands

Rogaar H, Ouboter PSH, Van Veen HJ (1999) PGBO en de kunst van het netwerkonderhoud. (PGBO and the ability of network maintenance) (in Dutch) (1999). Bodem 2: 54–56

Van der Gun JHJ, Joziasse J (1999) In-situ waterbodem sanering: voorstelbaar en haalbaar? (In-situ aquatic sediment remediation: imaginable and feasible) (in Dutch). The Netherlands Integrated Soil Research Programme, PO Box 37, 6700 AA Wageningen, The Netherlands

Vermeulen H (1999) SKB maakt vliegende start. (SKB makes flying start) (in Dutch). Bodem 2: 48–49

Vermeulen H, Verheul J(1999) NOBIS en het resultaat van vier jaar werk. Biologische in-situ sanering definitief op de (bodem)kaart. (NOBIS and the result of four year effort. Biological in-situ remediation unconditional on the (soil map) (in Dutch). Bodem 2: 57–59

Chemical Analysis of Contaminated Soils

5 How much Analytical Work Do We Need?

H. Steinhart, T. Käcker, S. Meyer, and G. Biernoth
Institute of Biochemistry and Food Chemistry, University Hamburg, Grindelallee 117, 20146 Hamburg, Germany

5.1 Introduction

To determine the contaminating substances it is necessary to validate the analysis of contaminated soil, and it is also necessary to monitor the change in composition of such soil during decontamination procedures by using suitable analytical methods.

The results of the analysis of contaminated soil make it possible to estimate the hazard potential of the situation, and they also influence the choice of measures to be taken. Ideally, complete and reliable analytical values are indispensable to an accurate assessment of the contamination. Moreover, the effects of soil cleaning should be followed up analytically to characterise remaining dangers to microorganisms, plants, animals and humans.

5.2 What Is to Be Measured?

Mineral oil and tar oil are two of the main sources of soil contamination. Mineral oils consist of different types of hydrocarbons, mainly of the aliphatic type, but even up to about 25 % are aromatic compounds, the latter comprising also polar, heterocyclic and high-molecular weight compounds.

The composition of the different fractions of mineral oil is, furthermore, highly complex. Within the group of aliphatic compounds *n*-alkanes and paraffins are present, as are *iso*- and cycloalkanes and cyclic isoprenoids such as steranes and hopanes. These cyclic compounds comprise between 30 and 50 % of the weight and the *iso*-alkanes between 15 and 35 %. The less volatile aromatic fraction comprises alkylised, mainly a methylised fraction, and other core containing aromatic compounds such as benzenes, biphenyls, naphthalenes, fluorenes, phenanthrenes, pyrenes, chrysenes, benzo-fluorenes and benzo-pyrenes; some of these have mutagenic and carcinogenic potential. Naphtho-aromatics like alkylised indane and tetraline are also part of the aromatic fraction, which comprises between 5 and 20 % of the weight. The polar fraction of mineral oils, between 2 and 15 %, consists of heterocyclic compounds such as dibenzofuranes, dibenzothio-

Table 5.1. Composition of mineral oils

Structure type	Compound
aliphatic hydrocarbons	*iso*-alkanes, including isoprenoids (e.g. phytane and pristane) => 15–35 %
	cycloalkanes/naphthenes (e.g. alkylised cyclopentanes and hexanes), cyclic isoprenoids (e.g. steranes and hopanes) => 30–50 %
aromatic hydrocarbons	alkylised, especially methylised aromatic compounds of one up to 5 cores, (e.g. benzene, biphenyls, naphthalines, fluorenes, phenanthrenes, pyrenes, chrysenes, benzofluorenes, benzopyrenes, naphthenaromatic compounds) => 5–20 %
polar compounds	heterocyclic compounds (e.g. dibenzofuranes, dibenzo-thiophenes, carbazols)
	polar hydrocarbons (e.g. alcohols, phenols and carboxylic acids)
	high-molecular asphaltene-like compounds => 2–15 %

phenes and carbazols; of polar hydrocarbons such as alcohols, phenols and carboxylic acids; and of high-molecular asphaltene-like compounds.

This complex composition of mineral oils is summarised in table 5.1. Furthermore, hydrocarbons in soil undergo biotic and abiotic degradation processes and form more or less stable degradation products.

5.3 How Is it to Be Measured?

These compounds have to be separated by extraction from the contaminated soil and to be analysed either in total, or individually, to give insight into the degree of the contamination, or into the effect of the cleaning procedures. The complete analysis comprises the analysis of the major as well of the minor components. Especially the analysis of the minor components requires high analytical effort, with application of analytical methods like:

- Solid phase extraction (SPE),
- Gas chromatography with flame ionisation detection (GC-FID),
- Gas chromatography with mass spectrometric detection (GC-MSD),
- High performance liquid chromatography with diode-array detection (HPLC-DAD),
- ^{13}C-cross polarisation magic angle-spinning-nuclear magnetic resonance spectroscopy (^{13}C-CPMAS-NMR).

The contaminants are separated into different fractions by different SPE procedures with different adsorbents and different extraction solvents (Bundt et al. 1991). The individual contaminants are identified and quantified within their fraction by the appropriate method. In the case of degradation products or their parent compounds, which may form bound residues with soil organic matter, ^{13}C-CPMAS-NMR is a useful technique to characterise the kind of binding. This is in contrast to a determination of the different classes of compounds in total, which may give an overview of the situation but does not create a basis for the estimation of the hazardous situation.

5.4 What Are the Statements to Be Made by Analytical Work?

Both the extent and the nature of the soil contamination in a given situation have to be clarified by analysis. To this end the contamination is to be characterised more or less exactly, steps of decontamination carried out are to be followed up, and ultimately a judgement on the success of the total decontamination is to be given. In order to accomplish this, the major and minor compounds have to be identified and quantified over a wide range of concentration, toxicologically dangerous compounds have to be quantified even down to the $\mu g\ kg^{-1}$-range, and degradation products, i.e. compounds produced from the contaminants by biotic or abiotic processes, have to be monitored.

5.5 Analysis of Component Groups and/or of Individual Compounds?

Can summary group analysis produce such statements? Is it sufficient? The advantage of this type of analysis is that it renders a rapid and simple estimate of the type of contamination. Examples of such analytical procedures are, for example, the determination of the weight loss by soil glowing, extraction of the organic material, UV- and IR-spectrometric analysis, some kinds of biotests, and fingerprinting by GC. With these methods it is possible to get an overview of the type and extent of the contamination. In the next step it is necessary to characterise the qualitative and quantitative composition of the contamination, to observe the chemical transformations of the contaminants within the soil, and to judge the danger potential on the basis of knowledge of the soil's toxic components.

Therefore, it is usually desirable that an analysis of the individual contaminant compounds is carried out. First of all, a representative soil has to be collected and the contaminants have to be extracted by Soxhlet or ultra-sonically. Fig. 5.1 shows a GC-MSD-SCAN-chromatogram of a raw extract of a soil contaminated by a mineral oil.

The extracts are subsequently cleaned up and separated by SPE. Finally, the compounds are identified and quantified by GC-MSD, HPLC-DAD, or GC-FID (see fig. 5.2).

For instance, a diesel fuel as contaminant was separated by SPE (Bundt et al. 1991) into the fractions aliphates (71.7 %), monoaromates (16.4 %), diaromates (8.0 %) and polyaromates (3.9 %) (fig. 5.3). The fractions were analysed by GC-FID thereby identifying and quantifying the individual compounds. The dangerous compounds were mainly concentrated within the fraction of polyaromates.

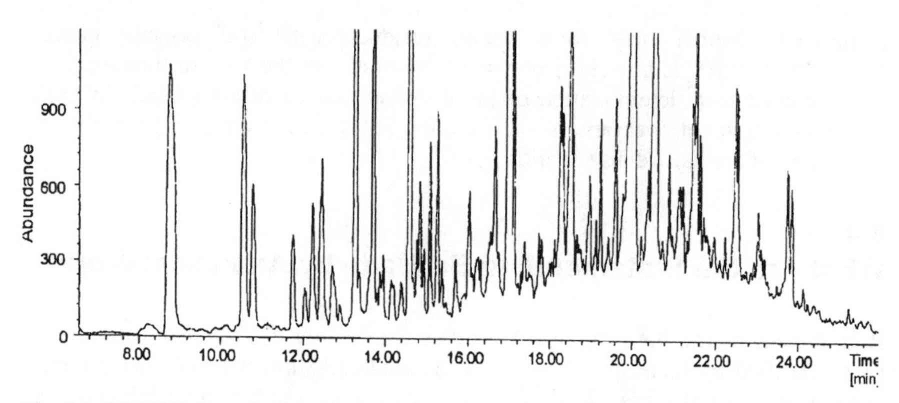

Fig. 5.1. GC-MSD-Scan-chromatogram of a raw extract from soil contaminated by a mineral oil

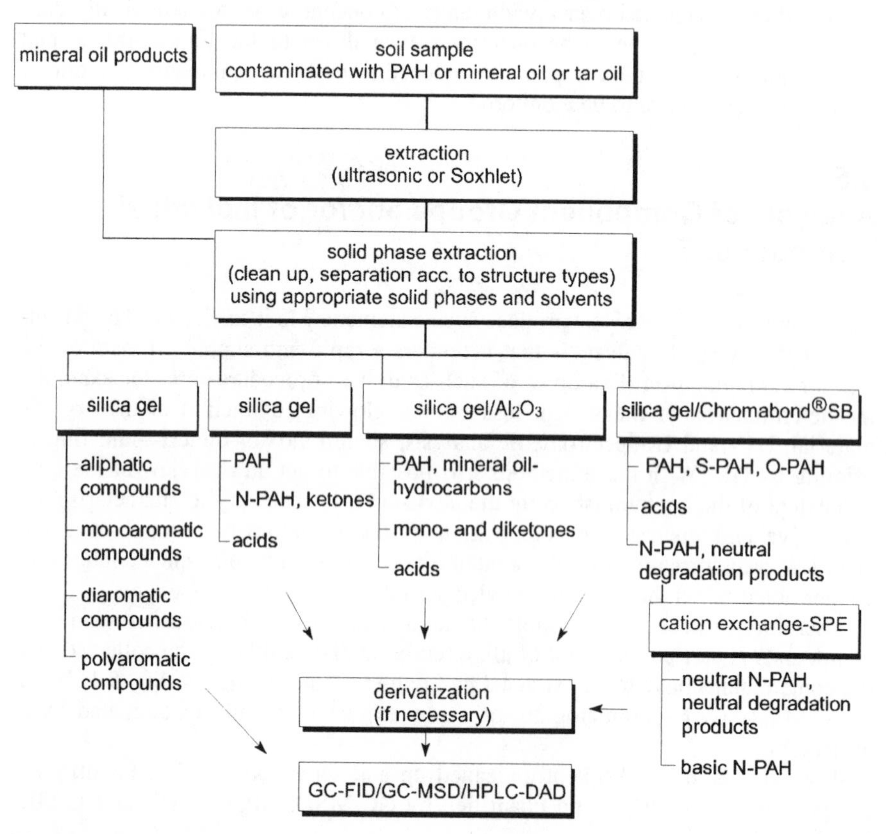

Fig. 5.2. Analysis scheme for characterisation of mineral oils or soil contaminated by mineral oils

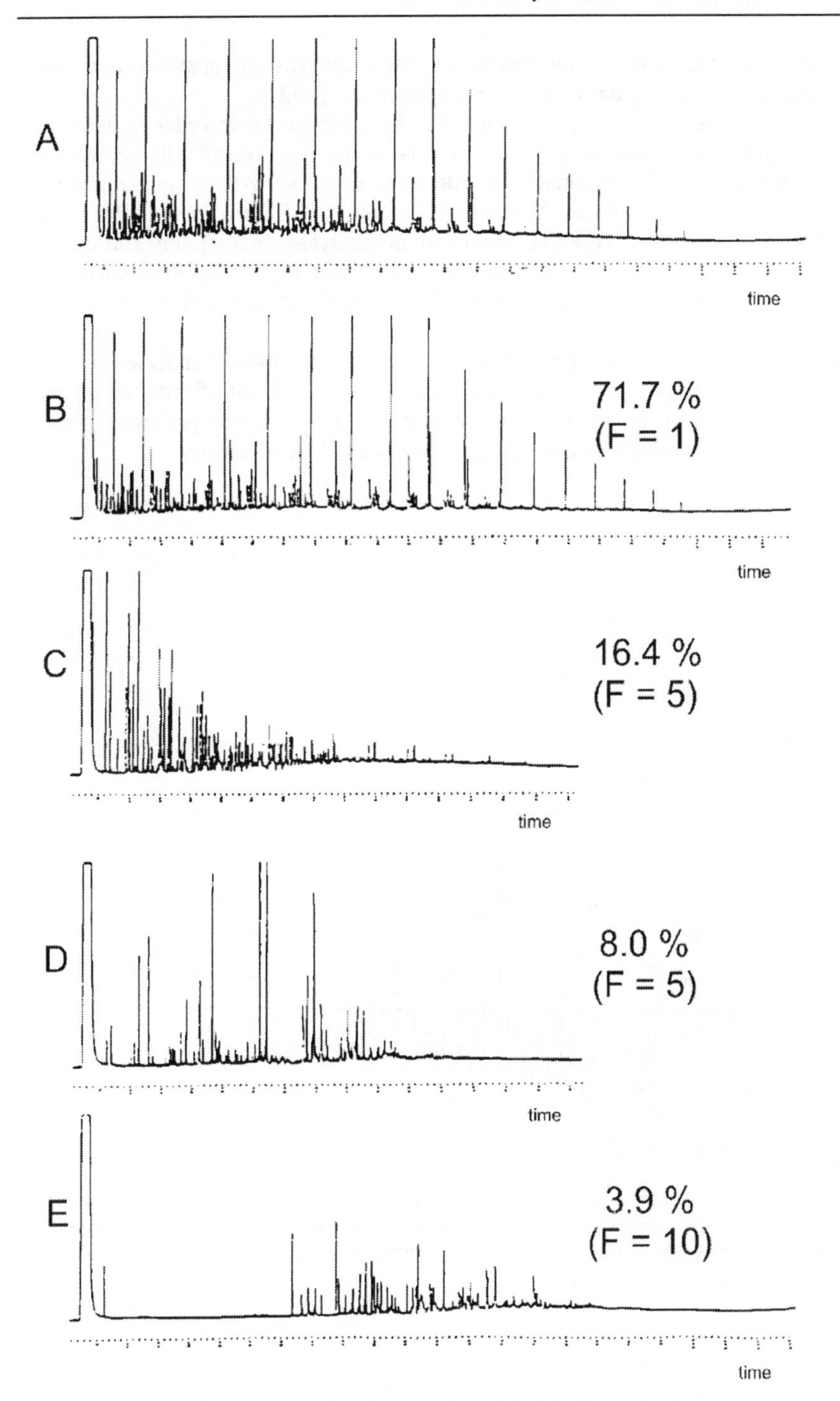

Fig. 5.3. Gas chromatogram of a diesel fuel before (A) and after separation into fractions (B: 71.7 % aliphatic compounds, C: 16.4 % monoaromatic compounds, D: 8.0 % diaromatic compounds, E: 3.9 % polyaromatic compounds) F = enrichment factor

Within an unused motor lubricating oil, the material to be analysed contained 84 polycyclic aromatic hydrocarbons (Paschke et al. 1992).

Because some minor compounds are the most dangerous, in order to judge the real potential of danger, it is necessary within the fraction of polyaromates to analyse the individual compounds. In order to analyse the PAHs in a contaminated soil it is customary to apply Method 610 of the US Environmental Protection Agency (EPA). Hereby, 16 PAHs from soil are analysed qualitatively and quantitatively. Yet it seems doubtful whether in a given case these results are sufficient. Clearly, it seems to make sense to extend the number of analysed PAHs to more than 16.

It is to be recommended that group and individual compound analyses be combined in order firstly to estimate quickly the type and amount of order of the soil contamination by group analysis; and secondly to judge the real potential of danger on the basis of results of an analysis of the individual compounds.

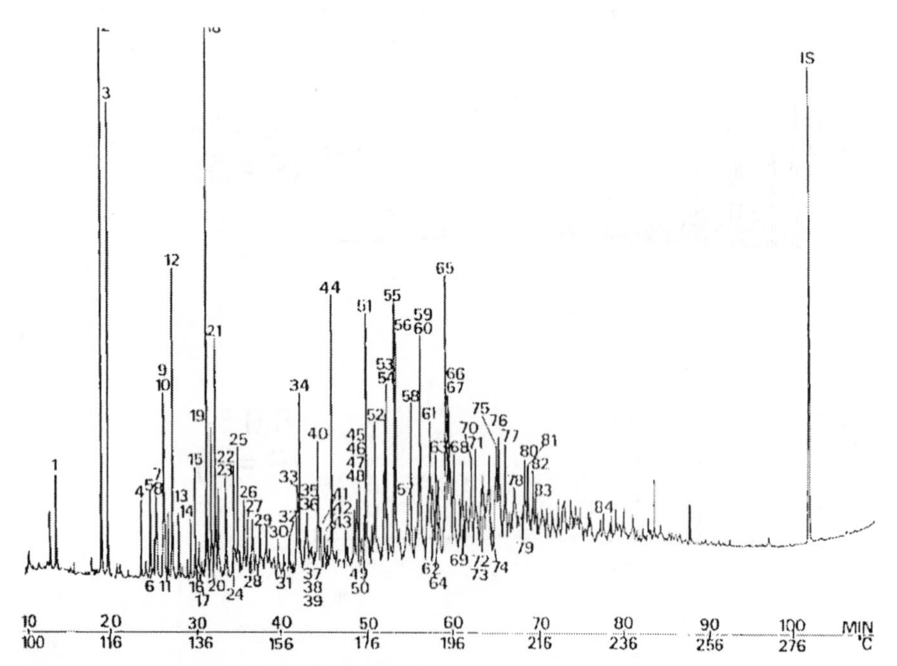

Fig. 5.4. GC-chromatogram of the PAH-fraction of an unused motor lubricating oil (84 identified compounds); aromatic compound fraction: < 0.04 %; composition: naphthalenes, phenanthrenes, biphenyls, fluorenes, dibenzothiophenes, alkyl-derivatives; IS: internal standard

5.6
What about Degradation Products?

In the contaminated soil the pollutants undergo biotic and abiotic degradation processes, the degree depending on type of contaminants, on type of soil, on soil conditions such as water and air content, pH and temperature, bioavailability of contaminants, and on intensity of microbiological activities. Therefore, in addition to the contaminants, degradation products (Meyer et al. 1999) are present in the contaminated soil. Some selected PAH and their degradation products are summarised in fig. 5.6. (Wischmann et al. 1996; Wischmann and Steinhart 1997).

The different types of hydrocarbons such as alkanes, aromatic compounds and condensed cycloalkanes that are degraded at different rates (Langbehn and Steinhart 1995) as shown schematically in fig. 5.5.

These degradation products can accumulate in the soil over time. Metabolites of PAH may be more toxic than their initial compounds. Therefore, they have to be analysed.

Because the analysis of individual metabolite compounds is time- and cost-consuming, so-called indicator metabolites have been defined according to their

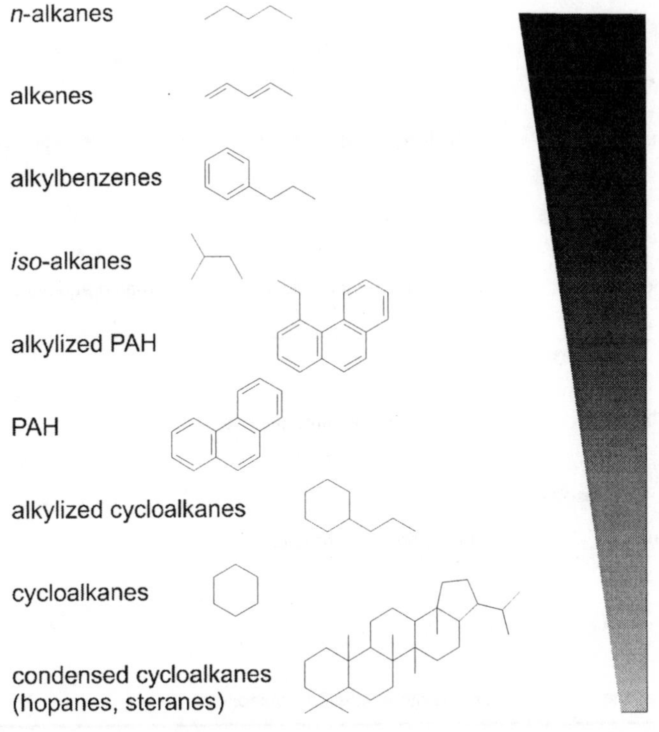

Fig. 5.5. Schematic display of the degradation rates of different types of hydrocarbons.

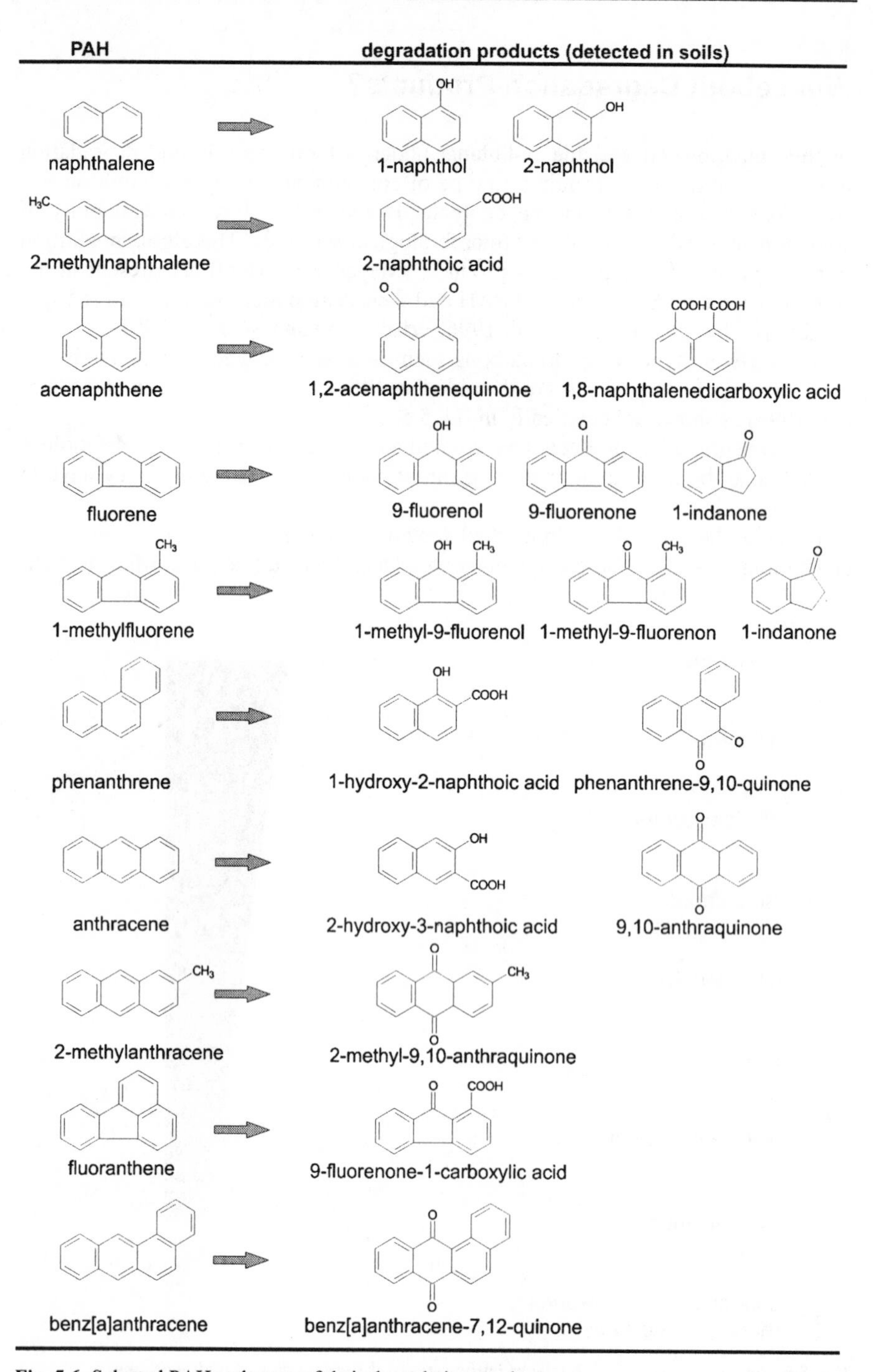

Fig. 5.6. Selected PAH and some of their degradation products

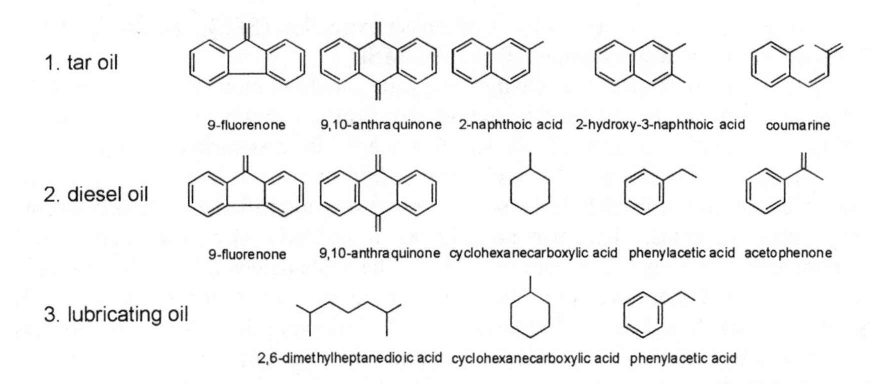

Fig. 5.7. Screening method for indicator metabolites in soil of tar oil, diesel oil, and lubricating oil

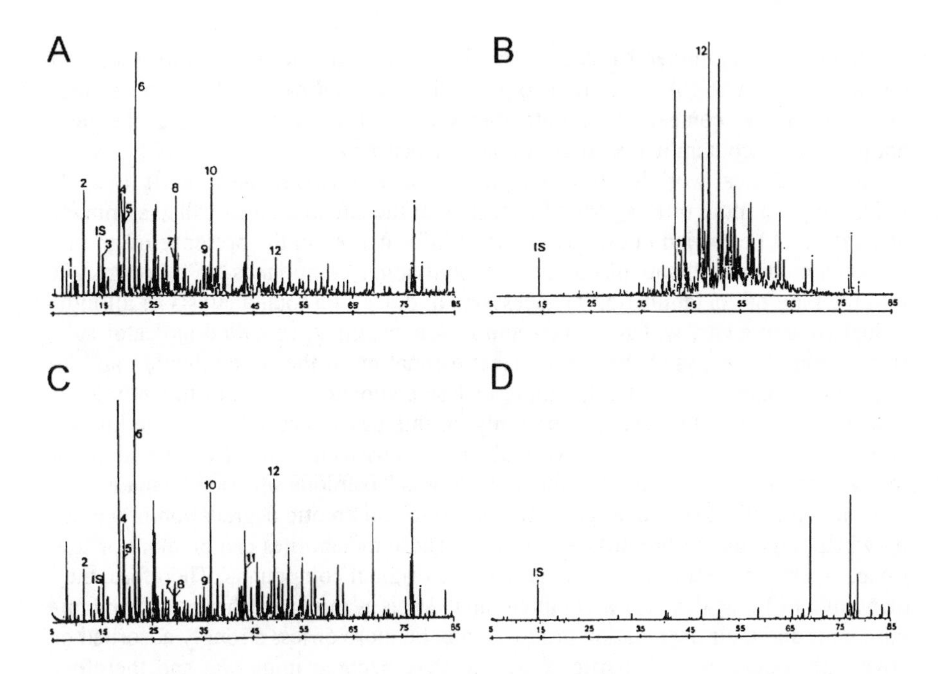

Fig. 5.8. GC-MSD-chromatograms of extracted acid fractions from soil contaminated with 5 % lubricating oil, after 1 (A), 3 (B) and 13 weeks (C) of microbiological degradation and sterile soil 13 weeks after contamination (D)

frequency in soil contaminants and have been analysed. Fig. 5.7 shows examples of possible indicator metabolites of tar oil, diesel oil, and lubricating oil.

The method of analysis of indicator metabolites is comparable to that of the original contaminants, namely extraction of the contaminated soil, separation of different groups of metabolites by solid phase extraction (SPE), and finally determination of the indicator compounds by GC-MSD.

Fig. 5.8 demonstrates the change in soil contamination representing GC-chromatograms of an acid fraction extract from soil contaminated with 5 % lubricating oil, after 1, 3 and 13 weeks of storage. In comparison, a GC-MSD-chromatogram of sterile soil 13 weeks after addition of oil is shown in fig. 5.8 D. The acids isolated from the sterile soil samples were identified as naturally occurring carboxylic acids. There was no evidence of oxidative decomposition by soil components (e.g. iron or manganese oxides). The acids detected in the biologically active soil must therefore result from microbiological oxidation (Langbehn and Steinhart 1994; Langbehn and Steinhart 1995). Following the degradation process the chromatograms A–C show decomposition of non-cyclic carboxylic acids (1–8), and an enrichment of monoaromatic acids (12).

5.7 Summary

Contamination by mineral oils and tar oils are two of the main sources of soil contamination with PAH. In order to judge the type and extent of the contamination, and in the context of decontamination measures to be taken, a chemical analysis of the contaminants proves to be indispensable.

Here, analytical work has to be adapted to the problems in question. If it is sufficient to get a more or less general overview of the situation, a so-called summary analysis may be carried out to give general information on the presence of groups of contaminants without information on individual compounds. If more insight into the type of contamination is necessary, in place of a total analysis of all individual contaminants, which is time- and cost-consuming, so-called indicator substances may be analysed. Yet, if an exact estimation of the contamination and its potential of danger is essential, a more or less complete analysis of the individual contaminants should be carried out. Only on this basis, that means on the knowledge of the occurrence of the individual compounds with their different individual toxicity, can a safe estimation of the more or less hazardous situation be made.

Contaminants in soil undergo different biotic and abiotic degradation processes by which degradation products are formed. These metabolites can be more or less toxic, in some cases even more toxic than the original compounds. Therefore, they also have to be analysed if a complete analysis is wanted. Furthermore, the contaminants and their degradation products may be more or less strongly adsorbed or covalently bound by soil particles so that they become immobile and therefore biologically unavailable. These so-called bound residues reduce the toxicity of the contaminated soil. Therefore it has to be considered whether an applied analytical method includes these bound compounds or not.

So far, for practical and economic reasons, contaminated soil is analysed by summary analysis such as determination of the weight loss after glowing of a soil sample, extraction of the organic material and UV- and IR-spectrometric analysis, some kinds of biotests, and fingerprinting by GC. For example, DIN method 38409-H18, developed for the determination of hydrocarbons in water, is applied for the determination of hydrocarbons in soil.

In contrast to these quick summary analysis methods, the analysis of more or less all individual contaminating compounds, including the minor components in concentrations down to the $\mu g\ kg^{-1}$-range, requires much greater effort. These methods comprise separation and identification techniques like SPE, GC-FID, GC-MSD, HPLC-DAD, and in the case of bound residues ^{13}C-CPMAS-NMR. This enables identification and quantification of the whole pattern of the contaminating individuals present in the soil and thereby judgement of the danger of the contamination, or the effect of decontamination.

It may be concluded that the quick summary analysis on the one hand, as well as the methods of determination of individual compounds on the other hand, have to be considered for determination of contaminated soils, depending on the questions to be answered. It is not sufficient, to judge and treat contaminated soils only on the basis of results of a quick summary analysis. However it is not always necessary to do all analytical work in the context of contaminated soils by complete individual analysis.

References

Bundt J, Herbel W, Steinhart H, Franke S, Francke W (1991) Structure-type Separation of diesel fuels by solid phase extraction and identification of the two- and three-ring aromatics by capillary GC-mass spectrometry. J High Res Chromatogr 14: 91–98

Bundt J, Herbel W, Steinhart H (1992) Determination of polycyclic aromatic sulphur heterocyclics (PASH) in diesel fuel by high performance liquid chromatography and photodiode-array detection. J High Res Chromatogr 15: 682–685

Langbehn A, Steinhart H (1994) Determination of organic acids and ketones in contaminated soils. J High Res Chromatogr 17: 293–298

Langbehn A, Steinhart H (1995) Biodegradation studies of hydrocarbons in soils by analysing metabolites formed. Chemosphere 30: 855–867

Meyer S, Cartellieri S, Steinhart H (1999) Simultaneous determination of PAHs, Hetero-PAHs (N,S,O) and their degradation products in creosote-contaminated soils. Method development, validation, and application to hazardous waste sites. Anal Chem 71: 4023–4029

Meyer S, Steinhart H (2000) Effects of heterocyclic PAHs (N,S,O) on the biodegradation of typical tar oil PAHs in a soil/compost mixture. Chemosphere 40: 359–367

Meyer S, Steinhart H (2000) Fate of PAHs and hetero-PAHs during biodegradation in a model soil/compost system – formation of extractable metabolites. Water Air and Soil Pollution (submitted)

Paschke A, Herbel W, Steinhart H, Franke S, Francke W (1992) Determination of mono- to tetracyclic aromatic hydrocarbons in lubricating oil. J High Res Chrom 15: 827–833

Wischmann H, Steinhart H, Hupe K, Montresori G, Stegmann R (1996) Degradation of selected PAHs in soil/compost and identification of intermediates. Intern J Environ Anal Chem 64: 247–255

Wischmann H, Steinhart H (1997) The formation of PAH oxidation products in soils and soil/compost mixtures. Chemosphere 35: 1681–1698

6 Methods for the Determination of Mineral Oil Hydrocarbons in Soil Materials for the Evaluation of Remediation Measures

J. Bundt
Chemical Laboratory Lübeck GmbH, Hochofenstraße 23–25, 23561 Lübeck, Germany

6.1 Introduction

In the majority of cases the currently contaminated and problem sites result from a pollution with mineral oils and their derived products. For the investigation and the registration of such mineral oil contaminated materials, as well as for the selection of suitable measures of treatment and remediation, the physico-chemical analysis is a major requirement.

The examination of samples of soil material with regard to total petroleum hydrocarbons has been advanced in recent years to a level based on Standards. The method commonly used to the mid Nineties, in the absence of specific draft guidance, was the DIN 38 409 Part 18 (DEV H-18), which was originally designed for water. This happened without further establishment of particular processing steps or boundary conditions. The soil material is extracted with 1,1,2-trichlorotrifluoroethane (freon R113), with the extract being purified on an activated aluminium-oxide column and is subsequently measured using infrared spectroscopy. However, this cumulative parameter does not give any further information about the molecular composition of a mineral oil contamination, nor does it describe the substance-specific changes during a remediation measure. For a deeper assessment, it is thus imperative to call in more precise analytical procedures. For only with the corresponding detailed substance-based knowledge can a reliable risk assessment for the local environment, and an extensive statement about the quality of the employed soil treatment measure, be made.

In June 1994 the gaschromatographic method ISO/TR 11046 had its appearance as an alternative, but this routine still uses the extremely ozone layer damaging freon for extraction. Its production has at present nearly stopped. Further developments of this method are the ISO/TC190/SC3/WG6 (as a working draft), which suggests an extraction with an acetone/heptane mixture, and the NEN 5733 as well as the CEN/TC292/WG5.

6.2 Substantial Composition of Mineral Oils

To highlight the analytical problem, the composition of mineral oils will be initially outlined in this chapter.

Petroleum and its derivations (mineral oils) are natural multi-substance mixtures composed of mainly aliphatic and aromatic hydrocarbons, of which the boiling ranges can extend over several hundreds of degrees (Tissot and Welte 1984). Only the boiling range of petrol (up to 200 °C) with its about 500 compounds can presently be approximately structurally determined (Matisova et al. 1985). This variety of compounds becomes even more complex throughout far-reaching petrochemical processes (cracking, reformation, hydrofining, mixing with additives etc.), and the biotic and abiotic processes through which it passes in the environment. To characterise a mineral oil accurately remains a great challenge for a modern chemical analysis.

In general the variety of different hydrocarbons can be divided into the following three substance classes of similar molecular structure, the so-called structure types (the quantitative identifications are valid for crude oils and can vary according to provenience and genese (Gunkel 1988)):

1. Aliphatic hydrocarbons (45–85 %):
   - *n*-alcanes (*n*-paraffines),
   - *iso*-alcanes (partly of isopronoid structure as for phytane and pristane),
   - cycloalcanes (naphthene, alkylised cyclopentanes and -hexanes, cyclic isoprenoides, such as sterane and hopane).
2. Aromatic hydrocarbons (5–20 %):
   - alkalised (for the most part methylised) 1 up to 5 aromatic rings,
   - naphthene aromates (such as alkalised indanes and tetralines).
3. Polar residue (2–15 %):
   - *S*-, *N*- and *O*-containing heterocyclic aromatic compounds,
   - polar hydrocarbons (such as alcohols, phenols, carbon acids),
   - high molecular weight asphaltenic structures.

For the assessment of a microbiological treatment measure attention should be directed to the determination of persistent, polycyclic aromatic hydrocarbons (PAHs). Regardless of the problem, if this compound class can be biodegraded at all, the question of the metabolic pathway is of a special interest, as it is the reactive metabolic products that are of a carcinogenic and/or mutagenic risk.

The PAH profiles of mineral oils differ strongly from those of tar oils and those resulting from the complete thermal destruction of organic materials. While during thermal degradation some hundreds of different PAH structures can develop via acetylene intermediates, mineral oils consist of a very complex mixture of isomers of alkalised PAHs of just some principal bodies. In petroleum most PAHs originate during the so-called coalification over a period of millions of years out of existing biogen structures (such as terpenes and carotonoides). The greater the depth, in which processes take place, is accompanied by greater pressure and temperature. Grimmer et al. (1983) detected in crude oils of different origin; 226 PAHs, furanes, thiophenes and carbazoles. The majority of these compounds are methylderivates of the parent substances naphthalene, biphenyl, fluorine, dibenzofurane, dibenzothiophene, phenanthrene, fluoranthene, pyrene, benzo[a]fluorine, benzo[b]naphtha[2,1-d]thiophene, benzo[a]anthracene, chrysene, benzo[b]fluoranthene and benzo[e]pyrene.

6.3 Characteristics of Biodegraded Mineral Oils

The biodegradation of hydrocarbons under aerobic environment conditions follows a preferred pathway, that is characterised by some structural features (Atlas 1988; Rehm 1988).

6.3.1 Molecular Weights

With the molecular weight increasing, apolarity is augmented; this leads to a decrease in water solubility and therefore in bioavailibility. The mass transport, mass transfer and the substance absorption are the limiting factors for biodegradation. *n*-Alkanes with chain lengths of C10 up to about C30 are principally degraded relatively readily via β-oxidation, which is an integral part of the heterotrophic metabolism. Higher molecular alkanes, that can exist in solid form as waxes, are however hardly degraded.

Alkanes containing 1 up to 9 C-atoms (petrols) have a bactericide effect (membrane toxic) and can thus only be decomposed by a few specialised microorganisms. In these cases stripping effects are the essential paths of degradation.

6.3.2 Molecular Structure

Isoprenoide and other branched alkanes are not readily degradable, because a normal degradation via β-oxidation is not possible at the branching point. Also the alicyclic hydrocarbons undergo, due to their structure, more complex degradation processes. Among the *n*-alkylcycloalkanes, those with a pair C-atom in the side chain are particularly resistant.

6.3.3 Grade of Saturation

Alkenes are not as readily metabolised as the respective *n*- and *iso*-alkanes, because of the epoxides, that are formed during the degradation process, having a toxic effect on the micro-organisms.

6.3.4
Grade of Condensation

The degradability of aromatic hydrocarbons decreases with an increasing grade of condensation and with a decrease in water solubility respectively. PAHs up to 4 rings can be metabolised. For those of higher condensation grades a cometabolic transformation is more likely to occur.

6.3.5
Grade of Substitution and Position of the Substituents

The readiness of biodegradation decreases with increasing grade of alkylation. The position of the substituents has a crucial influence on the biodegradability. β-substituted aromatics are for example, much more sensitive to microbial aggression than the respective α-isomers. Double substituents in ortho-position also make a microbial aggression more difficult (Rowland et al. 1986).

6.4
Analysis of Mineral Oil Contamination and Biological Remediation Measures

6.4.1
Scientific Aims

The knowledge regarding single processes during the alteration of mineral oils in the soil is still very incomplete. For in this complex ecosystem the whole spectrum of possible biotic and abiotic reaction processes, in dependent on the prevailing environment conditions, can run simultaneously. The scientific aim for assessment and control of a biological remediation measure is the construction of a complete balance of the degradation pathways, in order to be able to give a valuable statement about the elimination and immobilisation paths that have taken place. The following analytical questions have to be settled as a prerequisite for this:

- Grade of complete mineralisation into water and carbon dioxide.
- Grade of stripping effects.
- Type and toxicological assessment of the remaining substance concentration.
- Type and concentration of possible metabolic intermediate and end products (metabolites).
- Grade of incorporation into biomass (natural attenuation).
- Grade of incorporation into the humic substance structures and of sorption onto organic and inorganic soil matrix.

6.4.2
Steps of Examination

6.4.2.1
Sampling

The first and decisive step for the accuracy of an analysis is the type of sampling carried out. The heterogeneity of the medium soil and the inhomogeneous distribution of the contaminant mineral oil in this medium due to the slow transport processes in soil, require a high degree of care when working out a sampling strategy. Only in this way can it be guaranteed that an approximately representative image of the situation has been developed. The actual sampling should be preceded by an intensive "historical" investigation. Further exploration is carried out on the basis of detailed scanning examinations, with support from techniques such as soil vapour measurements, prospecting, pile core sampling, drilling, accompanied by geoelectrical and seismic procedures. In addition the soil characteristics have to be recorded, and an examination of the geological and hydrogeological structure of the soil should be carried out.

The sample size is determined on the one hand by the individual damage situation, but on the other hand also by the grain size distribution in the soil (see DIN 18123). Because the mineral oil hydrocarbons are preferably enriched in the fine grain fraction, great care needs to be taken to get a representative proportion of fine fraction to coarse fraction from the original sample.

6.4.2.2
Preparation of Samples

The preparation of samples is a further critical point in soil analysis and has consequently a crucial influence on the accuracy of the examination result.

After the sampling process the soil materials should be filled into glass flasks, that can be tightly closed, with the sample material reaching up to the top of the flask, in order to keep the loss of highly volatile compounds as small as possible. The transport and storage of the samples should be carried out under cool conditions and under exclusion of light.

Further steps of sample preparation are conditioning processes such as division, mixing, crushing, grinding or sieving. A commonly used technique is the sieving through < 2 mm, where the coarse skeleton partition is rejected. This procedure is only authorised for materials with a small coarse fraction, because the respective results finally refer only to the fine fraction. For a higher coarse fraction it seems thus to be useful to crush the whole sample in a crushing machine to < 2 mm and subsequently analyse it as a whole. For all of these procedures it has to be kept in mind, that a high loss of volatile compounds (BTXE-aromatics) is likely to occur.

6.4.2.3
Extraction of the Soil Sample

The complex composition of mineral oils with their wide spectrum of polarity, solubility, sorption and molecular mass as well as the complex and varying soil matrix, make it difficult to develop a universally applicable scheme for extraction. The completeness of extraction is on the one hand dependent on the type of mineral oil, and on the other hand on the prevailing soil conditions. The selection of the extraction procedure and the extracting agent consequently directly affects the results of the analysis.

The actual soil extraction is in the most cases preceded by a drying of the material. The drying is done to prevent the formation of emulsions and should allow an easier access to the contaminants, that are shielded by interstitial water or by the hydrate shell of the soil particles.

The most commonly used drying method is the grinding with fused sodium sulfate mainly in combination with the Soxhlet-extraction. In this way a sample material is produced that is as pourable as possible to prevent preferential routing and channelling effects by the extracting agent during the subsequent extraction.

Other drying methods such as air, drying closet (105 °C), vacuum or freeze-drying are much more time-consuming and lead to the loss of highly volatile mineral oil components. Our own experiments however, have demonstrated that for the extraction from soils of natural humidity, higher yields can be achieved.

The most commonly applied extraction measures are the cold extraction via shaking or ultrasound as well as the hot extraction on the backflow, in a Soxtec-device or in a Soxhlet-apparatus.

In our Institute the ultrasonic extraction is favoured because of its decisive advantages, namely; saving of time, the little extracting agent consumption, the reduced loss of highly volatile and thermo-sensitive compounds, as well as the effective extraction yield. The high-frequency excess pressure leads to cavitation phenomenona, that cause a rupture of the soil agglomerates and an enhanced substance exchange. When employing these methods important boundary conditions have to be taken into account, which are of importance for the consistency of the method. These are mainly the position and strength of the ultrasonic field, the water level of the ultrasonic bath, the depth of immersion of the sample flasks, the temperature stability as well, as the type and the quantity of the extracting agent. For the fine-grained soil material containing expandable clay minerals a preceding dispersion with a 0.4 M tetra-sodiumpyrophosphate solution has proved to be worthwhile. Soil samples, that are rich in humic substances, or samples containing large quantities of organic additives (such as compost) should be subjected to an alkaline hydrolysis with methanolic potassium hydroxide-solution after extraction. This treatment dissolves humic substances and at the same time releases diffused as well as sorbed hydrocarbons. In addition, existing ester compounds and so-called "charge-transfer"-complexes, which are formed mainly by PAHs containing chinoid terminal groups of humic substances, are being cracked.

As extracting agents in most cases nonpolar (*n*-hexane, cyclohexane, petroleum) up to solvents of medium polarity (toluol, dichloromethane, 1,1,2-trichlorotrifluoroethane, butylmethylether (TBME), THF) are used. In addition, mixtures of solvents such as heptane/acetone or hexane/acetone/toluol are applied. When choosing an appropriate extracting agent the endeavour is always to find the optimum between effective extraction yield and minimisation of inclusion of accompanying substances. In this context halogenated solvents have proven to be worthwhile.

In recent years promising extraction methods using supercritical fluids (SFE = Supercritical Fluid Extraction) as well as extraction using excess pressure (ASE = Accelerated Solvent Extraction) have been developed.

The advantage of supercritical fluids (such as CO_2 and N_2O) is to be seen principally in the fact that they show densities and solubility characteristics of fluids, but also have the higher diffusion coefficients and low viscosities of gases. This facilitates a considerably faster mass transport (Hawthorne 1990). Extraction times of SFE are consequently clearly shorter than those of the classic fluid extraction. The dissolving capacity, i.e. the polarity can be directly controlled via pressure (and density respectively) or by use of modifiers. Supercritical fluids are gaseous at ambient temperature, so that the concentration of the soil extracts is much easier. The production of liquid wastes and the exposure of the laboratory staff with hazardous solvents are both minimised.

Extraction results from a soil sample having been spiked with polar hydrocarbons (possible metabolites) showed that this compound class cannot be extracted sufficiently using conventional extraction methods (Soxhlet or ultrasonic). These results demonstrate that aromatic hydroxy and dihydroxy compounds are partly bound to the soil matrix to a point that a quantitative determination cannot be guaranteed. Only the applied keto and chinon compounds could be detected with satisfactory yields. In the discussion about metabolites the analysis should guarantee that these unhealthy metabolites do not stay in the treated soil (Langbehn and Steinhart 1995; Wischmann and Steinhart 1997).

The question has to be asked, should it be the aim of an extraction to collect the whole contaminant mass as well as their metabolites in a soil sample, or would it not be more sensible to find a description of the mobility of these compounds only via an aqueous extract. On the other hand the question of the long-term behaviour of this substance class has to be settled; i.e. to be able to take into account the remobilisation events in the future of at present tightly bound residues. For example, this can be caused by a change in the pH value in the soil (Richnow et al. 1999).

6.4.2.4 Examination Methods

Principally, two types of physico-chemical analysis can be distinguished: The determination of cumulative parameters and single substance analysis. For a first assessment of the damage situation the following examination methods for cumulative parameters are recommended:

- ignition loss,
- organic extract,
- TOC (total organic carbon),

- column chromatographic structural type with subsequent gravimetric determination of the fractions,
- biotests.

The disadvantages of these methods lie in the fact that neither a statement can be given about the molecular composition nor about the substantial changes in the course of a remediation measure. A better founded statement can be made by using the following methods:

- gas chromatography (CGC-FID or CGC-MS),
- high pressure liquid chromatography (HPLC),
- thin layer chromatography (TLC),
- SFC (super critical fluid chromatography).

Cumulative Parameters. Recent mineral oil damage can be detected via the odour, while older damage cases can only be detected using the appropriate measuring methods.

The ignition loss of the soil sample shows the total content of organic substance (humic and organic contaminants) in the soil. In connection with a gravimetric determination of the organic extract, a first hint can be found about the quantity of the contamination with total petroleum hydrocarbons. The same is true for the application of the parameter TOC, which has the disadvantage, though, of the small sample size. Further hints about the character of these hydrocarbons can be won via detection of IR and UV spectra.

The quantitative determination of hydrocarbons via infrared spectrometry is carried out in general accordance to the rules in the DIN 38409-H18. It has to be stated though that the application range of this method is restricted to the examination of "hardly contaminated up to contaminated water". For the transfer of this method onto soil samples, it is recommended to assume that the metrology-related details about sample preparation are fixed in the norm in order to obtain at least comparable results. Especially the column chromatographic treatment of the extract should be carried out exactly as described, and should neither be left out nor be changed in a way that filtering is only carried out using aluminium-oxide or even only by shaking with aluminium-oxide. This altered proceedure has a direct effect on the measurement results. In addition the following important aspects should be taken into account when applying the infrared-method:

- The soil is in comparison to water a much more complex and strongly varying matrix, so that the results of the measurements are strongly dependent on the type of soil and on the extraction method having been applied.
- For small contaminant concentrations, i.e. at the end of a remediation measure, a differentiation between soil component hydrocarbons and mineral oil hydrocarbons is not possible any more. This can, for a given remediation guide value, have a strong influence on the duration and success of a remediation measure.
- The extinction of the aromatic CH-stretching vibration at 3030 cm^{-1} has in comparison to the CH_2- (2924 cm^{-1}) and CH_3-bands (2958 cm^{-1}) a perceptibly smaller intensity. For high contents of aromatic hydrocarbons there are systematically determined values that are too low. It is thus advisable to carry out the

calibration, when possible, with the prevailing mineral oil contamination and not as suggested in the norm, with the saturated hydrocarbon squalane.

- The chromatographic purification via aluminium-oxide for the separation of lipophilic substances, that do not belong to the group of hydrocarbons, has some risks. With this method, polar metabolites, high molecular weight components, and a part of the aromatic hydrocarbons are adsorbed and in this way are withdrawn from the determination. This leads to some especially relevant substances, from a toxicological point of view, not being considered. The existence of phenoltype metabolites could in addition be examined via determination of the phenolindices in the extract or in the leachate.
- The proportion of hydrocarbons, measured using the IR-method, to the organic extract can be called in as a first hint for the age estimation of a contamination. Relatively recent damage cases show a proportion of about 1:2, while older damages show for the most part a clearly higher proportion (Sellner and Stachel 1992). This is caused by the fact that with increasing degradation, the partition of polar hydrocarbons increases, which is not considered when using the H-18 method.

The separation into groups of the hydrocarbons gives further information about the change of a mineral oil contamination over the time, i.e. during a biological treatment measure. For this the soil extracts are subjected to liquid chromatography on a preparative level, and the fractions aliphates, aromatics and polar residues are determined gravimetrically (Püttmann 1988). During biodegradation there is a shift from aliphatic to polar compounds in this component distribution.

Biological testing methods for the assessment of contaminants seems to be a sensible supplement to the analytical spectrum for risk assessment and for the evaluation of biological treatment measures. Testing methods are partially available (such as luminescent bacteria test, plant cell test, germ and grow tests with turnip, root elongation test with cress, bioassays), but still have to be adapted to the different soil conditions and types of contamination (Rönnpagel 1995).

Analysis in Detail. For a well founded evaluation of soil contamination with mineral oils, the application of more complex analytical procedures is imperative. The capillary gaschromatographic analysis with flame ionization detector (FID) is suitable to give an overview of the existence and concentration of single compounds. Via the boiling range (procedure of the simulated distillation) statements can be made about the type of mineral oil. Such a "finger-print"-analysis gives in addition a first hint on the grade of degradation of a mineral oil contamination. This can be read out of a reduced *n*-alkane partition and a reduction in the amount of highly volatile compounds. This leads to a relative accumulation of hardly degradable, branched, cyclic and aromatic hydrocarbons, which shows in a clear rise in the baseline in the gaschromatogram. The peaks that are protruding in a chromatogram of a biodegraded middle destillate are for the most part isoprenoid compounds such as farnesane, norpristane, pristane and phytane.

The concentration proportions of *n*-C17 to pristane and *n*-C18 to phytane decrease continuously with progressing microbial (aerobic) degradation and are thus used as an aid for the age estimation of mineral oil sites. At the same time the gaschromatogram gives important hints about the mobility of the prevailing hydrocarbons, and consequently hints at a groundwater risk. Mixtures with a predominant partition of low-boiling point hydrocarbons (up to about *n*-tentacosane (*n*-C25)) are regarded as more mobile and thus more of a groundwater risk than those having a predominant partition of higher-boiling point hydrocarbons.

The analytical characterisation of, from a toxicological point of view, relevant hydrocarbons requires an additional preparation step. In the literature, there can be found different methods for their isolation, which are in most cases of a column chromatographic nature, which are at times preceded by steps of precipitation or liquid-liquid-extraction. Methods of the adsorption and distribution chromatography, as well as techniques of the halfpreparative HPLC and MPLC, the flash-chromatography as well as solid-phase extraction are used.

With the solid-phase extraction (SPE) a fast, simple and cheap method is available to isolate aromatic hydrocarbons out of the complex mineral oil matrix (Bundt et al. 1991). This type of fractionation is suitable for mineral oils with a relatively high partition of aromatics such as crude oils and their derivates (petrol, petroleum, gasoils). Heavy oils (motor and lubricating oils) with little aromatics require an additional accumulation step. Here the aromatic hydrocarbons are at first separated from the high partition of aliphates via liquid-liquid-extraction and subsequently treated via solid-phase extraction (Paschke et al. 1992).

An aromate profile, that has been won in this way, makes it possible to characterise accurately the risk potential and the grade of degradation of a mineral oil via GC-FID and MS. Fig. 6.1 shows the GC-FID-chromatogram of a di- and polyaromatic fraction of a diesel fuel after a pre-separation via solid-phase extraction.

While gaschromatography is commonly used in the mineral oil analysis, the HPLC technique is at present only barely applied. The advantages of the HPLC technique is to be seen in the shorter duration of the analysis, the registration of low vaporable highly molecular and polar compounds as well as in the selective and sensitive detection of aromatic compounds (Bundt et al. 1992). Without further preparation steps, the degradation behaviour of aromatic hydrocarbons can be determined and the appearance of polar metabolites can be recognised. The latter elute in the reversed-phase-chromatography in the front chromatogram area and are thus separated off the other compounds. The application of aphotodiodearray-detector allows first statements about the type of metabolites via registration of UV spectra. The HPLC chromatogram of a diesel fuel is shown in fig. 6.2.

The SFC is an interesting development in the chromatographic measurement technique. It connects the advantages of gaschromatography with those of HPLC. Via combination of detectors such as UV with FID or FTIR important information can be won about the molecular composition of a mineral oil contaminated site.

The advantages of the thin-layer chromatography are its speed, the simultaneous separation of different samples without further preparation, the small apparatus requirements and the simple operation (Reimers 1999).

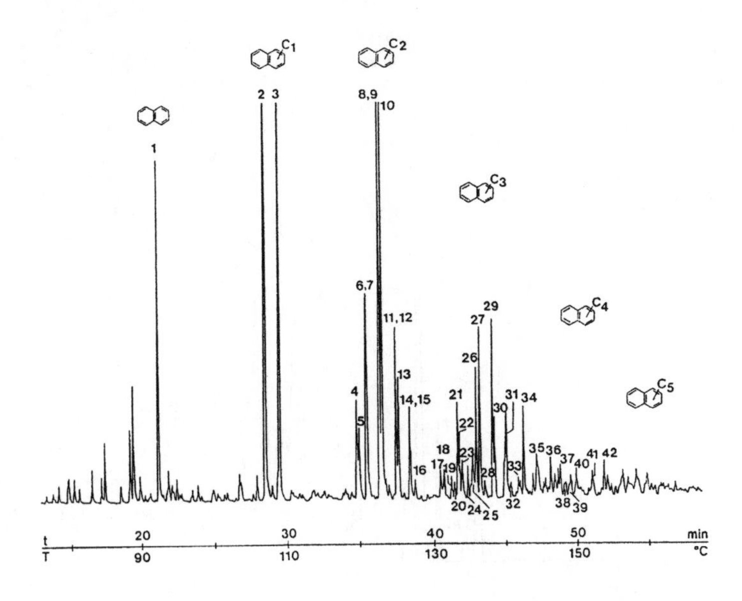

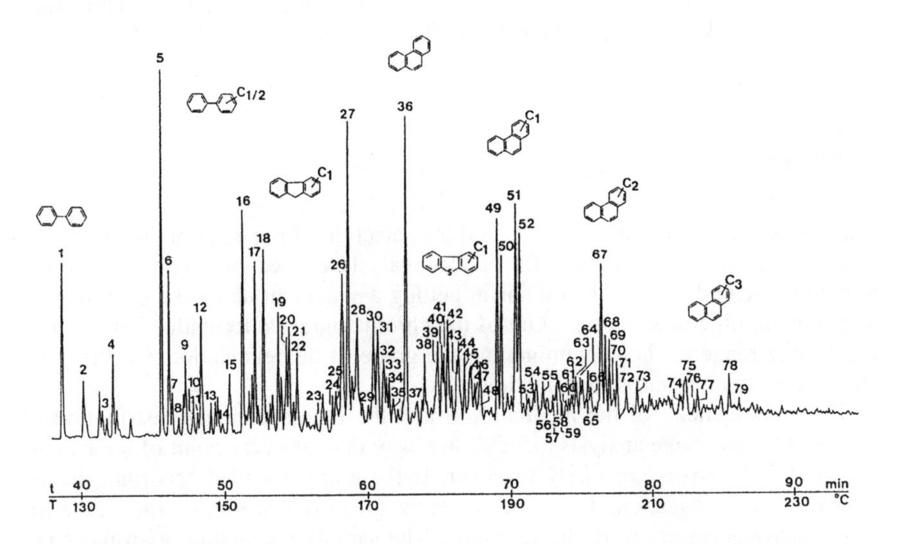

Fig. 6.1. Gaschromatogram of the di- and polyaromatic fraction of a diesel fuel after a structural separation by solid phase extraction (the identified peaks are shown in Bundt et al. (1991))

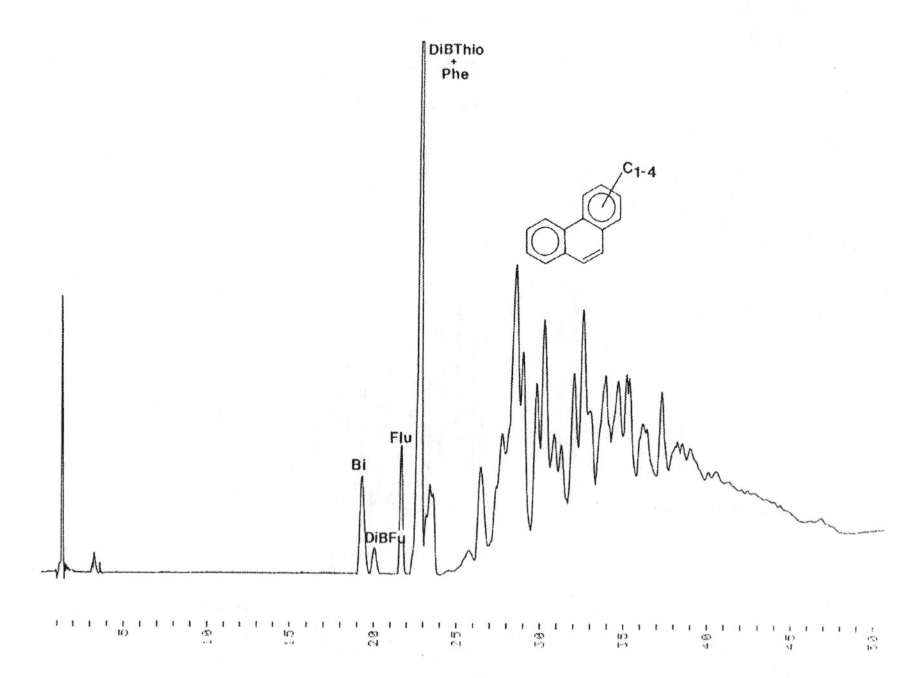

Fig. 6.2. HPLC-chromatogram (UV=252 nm) of the polyaromatic fraction of a diesel fuel (Bi = biphenyl; DiBFu = dibenzofurane; Flu = fluorene; Phe = phenanthrene; DiBThio = dibenzothiophene. The other peaks are alkylated isomeres of these aromatic mother substances.)

6.4.2.5
Summary

The determination of risk potentials and the selection of remediation measures are strongly dependent on the results of the analysis carried out. The gaschromatographic method is clearly superior in getting a statement about the composition of a mineral oil contamination. Out of the chromatogram information can be read the boiling range of the contamination, the grade of biodegradation and the existence of further organic contamination.

The heterogeneity of the medium soil and the inhomogeneous distribution of the contaminant make analysis difficult in a way that, at every point of the examination system, errors are likely to occur. If these are regarded according to the rules of error propagation, it is obvious that the greatest influence on the total error of the analysis results from the sampling. The sample preparation as well as the pulping and extraction methods are to be seen as important process steps, that are afflicted with a great error probability. This leads to the fact that with the different "institute methods", different results are obtained for the same question. Values, that are not comprehensible, reproducible or even not plausible, cannot be used for a statement (Pudill et al. 1991).

In the first place, standardised methods need to be developed and measures of quality assurance to guarantee the comparability of measurement results need to be agreed. Such an international standardisation is strived for in the frame of the ISO/TC 190 – Soil Quality.

With this standardisation proceeding however, it is not guaranteed that accurate results are necessarily supplied, but at least they are comparable. If the remediation aims are determined according to the same analysis methods, this standardised proceedure can be seen as a sensible compromise. Besides, for hydrocarbons there should be more detailed analysis, to obtain a well founded base for the assessment of risk potentials and remediation success.

References

Atlas, RM (1988) Biodegradation of Hydrocarbons in the Environment. In: Omenn GS (ed) Environmental Biotechnology, New York pp 81–95, 211–222

Bundt J, Herbel W, Steinhart H, Franke S, Francke W (1991) Structure-Type Separation of Diesel Fuels by Solid Phase Extraction and Identification of the Two- and Three-Ring Aromatics by Capillary GC-mass Spectrometry. J High Resolut Chromatogr 14: 91–98

Bundt J, Herbel W, Steinhart H (1992) Determination of Polycyclic Aromatic Sulphur Heterocycles (PASH) in Diesel Fuel by High-Performance Liquid Chromatography and Photodiode-Array-Detection. J High Resolut Chromatogr 15: 682–685

Grimmer G, Jacob J, Naujack KW (1983) Profile of the Polycyclic Hydrocarbons from Crude Oils. Part 3, Fresenius Z Anal Chem 314: 29–36

Gunkel W (1988) Ölverunreinigungen der Meere und Abbau der Kohlenwasserstoffe durch Mikroorganismen. In: Schweisfurth R (ed) Angewandte Mikrobiologie der Kohlenwasserstoffe in Industrie und Umwelt. Ehningen, pp 18–36

Hawthorne SB (1990) Analytical-Scale Supercritical Fluid Extraction. Anal Chem 62: 633A–642A

Langbehn A, Steinhart H (1995) Biodegradation studies of hydrocarbons in soils by analyzing metabolites formed. Chemosphere 30: 855–868

Matisova E, Krupcik J, Cellar P, Kocan A (1985) Quantitative Analysis of Hydrocarbons in Gasolines by Capillary Gas-Liquid Chromatography. J Chromatogr 346: 177–190

Paschke A, Herbel W, Steinhart H, Franke S, Francke W (1992) Determination of Mono- to Tetracyclic Aromatic Hydrocarbons in Lubrication Oil. J High Resolut Chromatogr 15: 827–833

Püttmann W (1988) Analytik des mikrobiellen Abbaus von Mineralöl in kontaminierten Böden. In: Wolf K (ed) Altlastensanierung '88, Bd 1, Dordrecht, pp 189–199

Pudill R, Müller HW, Zöllner U (1991) Altlasten – eine Herausforderung für den Analytiker. Umwelt-Technologie 2: 19–37

Rehm HJ (1988) Mikrobiologie und Biochemie der Kohlenwasserstoffe. In: Schweisfurth R (ed) Angewandte Mikrobiologie der Kohlenwasserstoffe in Industrie und Umwelt. Ehningen, pp 1–17

Reimers C (1999) Screening mineral oil contaminated soils by thin-layer chromatography. GIT, Laboratory Journal 1/99: 42–43

Richnow HH, Eschenbach A, Mahro B, Kästner M, Annweiler E, Seifert R, Michaelis W (1999) Formation of Nonextractable Soil Residues: A Stable Isotope Approach Environ Sci Technol 33: 3761–3767

Rönnpagel K, Liß W, Ahlf W (1995) Microbial bioassays to assess the toxicity of solid associated contaminants. Ecotoxicology and Environmental safety 31: 99–103

Rowland SJ, Alexander R, Kagi RI, Jones DM, Douglas AG (1986) Microbial Degradation of Aromatic Components of Crude Oils. Org Geochem 9: 153–161

Sellner M, Stachel B (1992) Untersuchungsmethoden bei der Erkundung und der Sanierung mineralölkontaminierter Böden. Beitrag zum 10. DECHEMA-Fachgespräch Umweltschutz, Leipzig

Tissot BP, Welte DH (1984) Petroleum Formation and Occurrence, 2. Auflage. Berlin/Heidelberg

Wischmann H, Steinhart H (1997) The formation of PAH oxidation products in soils and soil compost mixture. Chemosphere 8: 1681–1698

7 Determination of the TPH- and PAH-Content of Mineral Oil Contaminated Soils by Thin-Layer Chromatography

C. Reimers
Technical University of Hamburg-Harburg, Department of Waste Management, Harburger Schloβstraße 37, 21079 Hamburg, Germany

7.1 Introduction

Chemical analysis takes a major part in the treatment of contaminated soils. Results from analyses are called in for the determination and subsequent assessment of contaminated soils. Furthermore they give information about the success of a soil treatment measure and ultimately indicate the aim of the remediation measure.

The aim of analyses of environmentally relevant substances is not only the ability to determine, in an accurate manner, lower and lower concentration ranges. Analyses also strive for the determination of an ever widening range of chemical substances which may be potentially very harmful. Frequently this results in more expensive and elaborate investigations. If the economic aspects of such analytical methods are examined, however, it can be clearly seen that only the sensible use of devices, time and materials etc. result in the desired economy of use (Hein and Kunze 1995).[2]

In order to better manage the increased number of analyses, the samples can be divided during a first screening step into "critical" and "uncritical" samples[2]. Furthermore, the complexity of the soil matrix usually demands a difficult and time-consuming preparation of the samples including clean-up and accumulation steps. Depending on the analytical question, a more simple preparation of samples via less elaborate processing and fractionation steps is possible. This is because in most cases it is sufficient to determine the order of magnitude of the contamination, rather than absolute quantity.

A semi-quantitative assay is often used to establish whether the level is markedly below, just reaches or exceeds a particular threshold concentration (Gottwald and Heinrich 1998). Group parameters enable rapid conclusions to be drawn on the overall degree of contamination (Hellmann 1975) and indicate where further high sensitivity investigations for individual substances should be conducted. Considerable savings on costs and working time can be made this way.[2]

The examination and preparation of samples from contaminated sites is habitually problematic for there are some parameters for which no standardised analysis procedures do exist. This leads to a variety of different methods being applied.

The values resulting from samples having gone through different procedures of preparation, though, are not necessarily comparable. Standardisation of measuring principles can be achieved by adopting standardised procedures from other fields, such as the analysis of water, but without any further adaptation of sample preparation.

Many researchers have studied the development and application of methods that can be applied quickly and are environmentally friendly. The main focus of this article are screening methods for the determination of total petroleum hydrocarbons and polycyclic aromatic hydrocarbons.

In addition to the quantification of the total petroleum hydrocarbons (TPH), the determination of the polycyclic aromatic hydrocarbons (PAH) is an important assay in petroleum analysis. The concentration of hydrocarbons derived from petroleum is determined up to now as a summation parameter by means of IR-spectroscopy using fluorochlorinated hydrocarbons according to DEV H18 (water) (DIN 38409 part 18 1981) and the LAGA-guidance KW/85 (1993).

Generally, chromatographic methods characterised by high selectivity and sensitivity of detection are used to quantify PAH e.g. gas-chromatography or high performance liquid chromatography.[2] One universal method of analysis, that is becoming more and more important in the field of environmental research, is thin-layer chromatography.[3] This is thanks to the dramatic progress made with automation and its suitability for a wide range of different uses. Thin-layer chromatography is available as a very flexible and sensitive method, not only for routine analysis: it is also suitable for preliminary semi-quantitative investigations. Information on the presence of petroleum contamination and PAH in soil can rapidly be obtained. It can therefore be used as a screening method for the determination the type of contamination.[3]

7.2 Total Petroleum Hydrocarbons

The determination of total petroleum hydrocarbons was, until recently, carried out using the IR-spectroscopy method according to DEV H18 (DIN 38409 part 18 1981) and the LAGA guideline KW/85 (1993). It can be seen as a great disadvantage, though, that perhalogenated solvents must be used and that a differentiation between different oils is hardly possible and worse between anthropogenic or biogenic hydrocarbons. Information about the boiling range of a contamination is not possible either. Such information is only possible after at least partial separation of the complex hydrocarbon mixture. Because of these disadvantages, for some time there has already been great effort to develop more meaningful and eco-friendlier methods. The IR-spectroscopic method according to DEV H18 is replaced by a gas-chromatographic method according to DEV H53 (1998). Another possibility for the determination of the total petroleum hydrocarbons can be seen in the thin-layer chromatography (TLC).

Using thin-layer chromatography total petroleum hydrocarbons have already been determined qualitatively (Weisheit and Enl 1973) as well as semi-quantitatively (Goebgen and Brockmann 1977). A reference to this method can be found in the DEV H18.

7.2.1 Method

A thin-layer chromatographic method for the determination of total petroleum hydrocarbons (aliphatic compounds and naphtenes) of medium and low volatility in soil samples is presented which is based on the separation and quantitative determination of the hydrocarbons. The extraction of the soil samples is effected by means of an *n*-hexane-acetone mixture supported by ultrasonics. The evaluation can be made visually or densitometrically by means of a thin-layer scanner. The applicability and restrictions of the method are shown. The advantage of the method lies in its low requirements pertaining to apparatus and chemicals, as well as in its speed. This makes it suitable for quick in-situ analyses, such as the assessment of the mineral oil content of a contaminated site.[1]

7.2.1.1 Determination of Total Petroleum Hydrocarbons

No universally applicable method exists for the extraction of mineral oil contamination from soil samples. The extraction is carried out with a mixture of *n*-hexane and acetone (1:1) with ultrasonic support, a relatively environmentally friendly extracting agent and time efficient. A time-consuming Soxhlet-extraction as well as the use of chlorinated solvents is not undertaken (Schmidt 1995; Bundt et al. 1990).[1]

The samples are subsequently centrifuged. Spreading of the extracts is done manually with fixed-volume capillaries or automatically by spraying with spreading devices. High performance thin-layer chromatography (HPTLC) silica 60 phases are used. After chamber saturation the plates are developed in *n*-hexane, dried and immersed into detection agents. The evaluation can be made visually or by means of a densitometre (see fig. 7.1) (Reimers et al. 1998).

The determination principle is based on the separation of the aliphatic hydrocarbons from other substance groups like aromatic hydrocarbons, lipids and other compounds (see fig. 7.2).

As early as 1977 Goebgen and Brockmann (1977) have examined total petroleum hydrocarbons by means of TLC. In contrast to these investigations the analysis described here uses high-performance thin-layer chromatography plates with a concentration zone (HPTLC). These make it possible to obtain a higher separating capacity with smaller spot diameters because of smaller grain sizes and more narrow grain size distribution. In addition lower detection limits are possible because of a better concentration compared to other commonly used thin-layer plates. The detection limit for hydrocarbons is indicated to be an absolute of 2 µg squalane by Goebgen and Brockmann (1977), while by means of HPTLC 0.1 µg squalane can be detected.[1]

For manually spread extracts in particular the concentration zones of the HPTLC-plates have proven to be advantageous by focusing of the spread substances. Moreover, matrix components which cause disturbance (noticeable by a brown to yellow colour of appearing spots) can be held back in the concentration zone. Uncharged soil samples show no disturbances.[1]

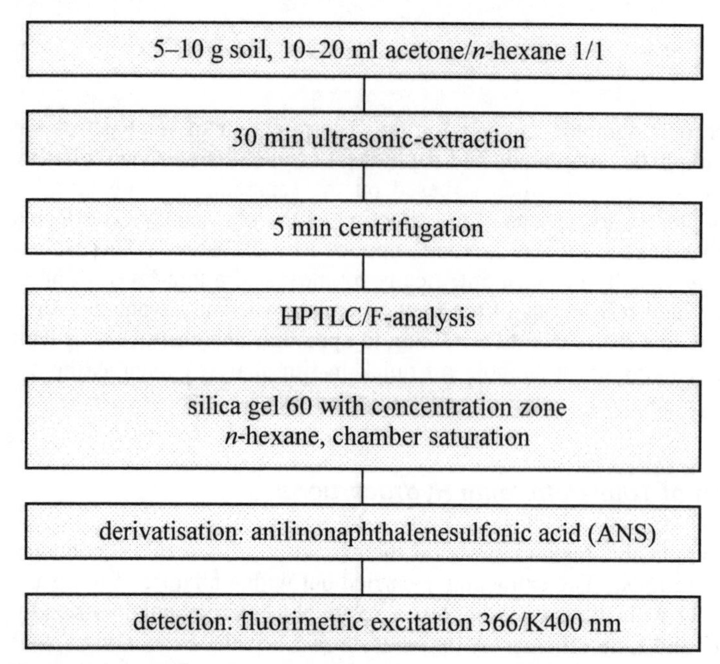

Fig. 7.1. Determination of the TPH-content in soils using HPTLC/F

Weakly polar and non polar components of biogenic origin can be determined as total petroleum hydrocarbons. Tests on uncontaminated soil materials do not show disturbances. As we have demonstrated on butter, margarine and vegetable oil, typical lipids are not chromatographed, i.e. they stay at the starting point. The total petroleum hydrocarbons show as a spot after dyeing (hRf-value of about 80). Polycyclic aromatics (PAH) can be viewed in UV-light for quantitative estimations before the immersion. They cannot be assigned to a single "group-Rf-value" though (hRf-value 25–65). According to Murr (1990) a differentiation between different oils that are rich in aromatics and others that are poor in aromatics can be made. These can be characterised according to their Rf-values on one hand and visually according to the differently coloured zones on the other hand.[1]

7.2.1.2 Detection Agents

A prerequisite for the in-situ detection of aliphates is that they are made visible by suitable colouring agents. Their number is small due to the high inertness of saturated hydrocarbons. Suitable detection agents are anilinonaphthalenesulfonic acid (Gitler 1972), bromothymol blue (Goebgen and Brockmann 1977), sodium fluoresceinate (Mamlock 1981), as well as berberine chloride (Jork et al. 1990).[3] The detection limit for squalane and lubricating oil is 0.1 µg and for diesel fuel 0.4 µg.

The hydrocarbons dyed with bromothymol blue can be distinguished and assessed directly after immersion in daylight as yellow spots on a deep-blue background or after complete drying of the plate as dark spots on olive-coloured to

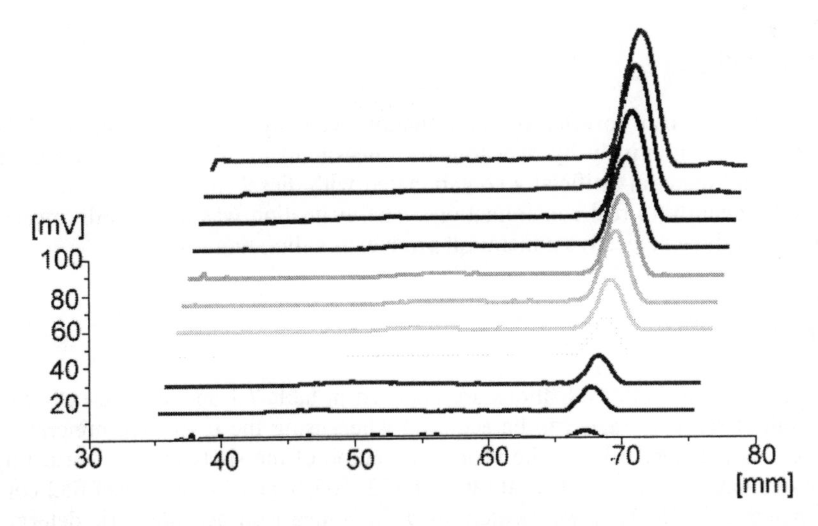

Fig. 7.2. Separation of aliphatic hydrocarbons from other substance groups and determination as summation parameters. Detection with anilinonaphthalenesulfonic acid and fluorescence excitation (366/400 nm) (from G.I.T. Laboratory Journal 1/99, reproduced with permission, copyright G.I.T. Verlag GmbH, 1999).

blue-grey background. The estimation on the dry plate is facilitated by means of a more steady background colour and its application is preferred. On the freshly immersed layers the colours change quickly which makes it more difficult to compare the spots.

Substances dyed with anilinonaphthalenesulfonic acid are clearly distinguishable in a humid as well as in a dry state in UV-light of 366 nm as light spots on dark ground or at 254 nm as dark spots on green fluorescent background. For a detection by means of anilinonaphthalenesulfonic acid the plate has to be heated up to 105 °C before immersion, otherwise any colouring cannot be obtained. This can be disadvantageous for the examination of volatile components. Any deficiencies in concentration for the materials could not be observed having been examined in comparison to the other colouring reagents where no heating is necessary. The colouring stays for a few days and then the plate can be immersed again.

Using sodium fluoresceinate the hydrocarbons show as orange spots on yellow background (daylight). In UV-light (254 and 366 nm) they can be distinguished as dark spots on a light green fluorescent background. After drying of the reagent the spots fade, but can be dyed again.

A berberine chloride solution causes the hydrocarbons to show as light yellow spots on a green fluorescent background at 366 nm. Berberine chloride is the only reagent used and consequently the plates can be treated either after processing or directly before the spreading of the samples, as they are not chromatographed during the processing with *n*-hexane. The impregnation of the plates by means of immersion in the detection reagent before processing causes a better contrast formation.[1]

7.2.1.3
Evaluation

The evaluation can be effected semi-quantitatively by visual estimation of the content or quantitatively by densitometric measurement. The methods were applied to soil material artificially contaminated with diesel fuel and lubricating oil and subsequently to real contaminations. The evaluation was made both towards squalane and the respective mineral oil product as calibrating substances.[1]

7.2.2
Results

The results from both evaluations are opposed in table 7.1 and show clearly that the highest retrieval rates can be acquired when using the respective mineral oil product as the standard. For the visual evaluation of the spots for the lubricating oil contaminated sample retrieval rates of 102–105 % and for the diesel fuel contaminated soils 78–93 % were determined. In comparison densitometric determination retrieval rates are 78–88 % for the lubricating oil contaminated samples and of 78–91 % for the diesel fuel contaminated soils. Because the artificially contaminated samples have already been stored for a longer duration interactions can occur with the soil matrix caused by sorption effects or by desorption losses (this could have already partly occurred during the desorption process) and this can lead to lower retrieval rates.

The results of the semi-quantitative visual evaluation and those of the densitometric determination show a great degree of compliance. According to Hezel (1973) one has to be prepared for variations of up to 20 % when using the visual evaluation.

As an analogue to Goebgen and Brockmann (1977) squalane was used as the reference substance. If, instead of the respective mineral oil product another calibration substance is used, there can occur considerable variations at times. Whereas squalane is a suitable calibration substance for lubricating oil contaminations, for diesel fuel contamination retrieval rates of only 8–10 % are attained.[1]

Goebgen and Brockmann (1977) stress in opposition to this the validity of their method for contamination with middle distillates, such as diesel fuel. They calibrated exclusively towards squalane. They did not carry out comparative examinations with diesel fuel solutions, so that deficiencies in concentration could not be recognised.

The low retrieval rate is caused by the composition of the diesel fuel. Diesel consists of up to about 30 % aromatic and other compounds (Bundt et al. 1990), that are separated on the plate from the aliphates. The aliphatic components are relatively volatile, so that a proportion cannot be detected on the plate. Thus the selection of the standards should be done very carefully. The standard should be identical, or at least as similar as possible to the contamination with respect to its composition. This is especially true for substances with a high content of aromatics, that are also less suitable for IR-spectroscopy (DIN 38409 Part 18, 1981). For unknown contamination it is advisable to use different standards or a mixture of diesel fuel and lubricating oil (1:1) respectively. Also the application of residual oil as a standard has proven to be worthwhile. It is sensible to report the standard

Table 7.1. Content in aliphates of the soils (dry matter) having been contaminated with lubricating oil and diesel fuel, semiquantitative estimation and densitometric evaluation, fluorescence detection 366/400 nm, calibration towards the respective mineral oil product and squalane, detection anilinonaphthalenesulfonic acid (254 nm), bromothymol blue (daylight) (from Umweltwissenschaften und Schadstoff-Forschung 4/98, reproduced with permission, copyright ecomed verlagsgesellschaft AG & Co.KG, 1998).

Visual evaluation						Densitometric evaluation			
	L $[g\,kg^{-1}]$	L $[g\,kg^{-1}]$	RR [%]	S $[g\,kg^{-1}]$	RR [%]	L $[g\,kg^{-1}]$	RR [%]	S $[g\,kg^{-1}]$	RR [%]
1[a]:	9.0	9.4 ± 1.1	105 ± 12	7.4 ± 0.8	82 ± 9	7.9 ± 0.3	88 ± 4	6.2 ± 0.2	69 ± 3
1[b]:	9.0	9.3 ± 0.9	102 ± 10	7.5 ± 1.0	83 ± 11	7.2 ± 0.3	80 ± 4	6.4 ± 0.2	71 ± 3
2[a]:	7.7	18.2 ± 1.6	102 ± 9	15.6 ± 1.3	88 ± 7	14.4 ± 0.3	81 ± 2	13.1 ± 0.2	74 ± 1
2[b]:	16.3	17.0 ± 2.2	104 ± 14	15.6 ± 1.5	96 ± 9	12.8 ± 0.1	78 ± 1	11.9 ± 0.2	73 ± 1

Visual evaluation						Densitometric evaluation			
	D $[g\,kg^{-1}]$	D $[g\,kg^{-1}]$	RR [%]	S $[g\,kg^{-1}]$	RR [%]	D $[g\,kg^{-1}]$	RR [%]	S $[g\,kg^{-1}]$	RR [%]
1[a]:	94.1	80.9 ± 4.7	86 ± 5	7.4 ± 0.8	10 ± 1	79.9 ± 1.3	85 ± 1	12.8 ±0.4	14 ± 1
1[b]:	84.9	78.9 ± 7.6	93 ± 9	7.5 ± 1.0	10 ± 2	76.9 ± 2.3	91 ± 3	8.6 ±0.2	10 ± 1
2[a]:	72.8	64.7 ± 3.6	89 ± 5	15.6 ± 1.3	9 ± 1	60.4 ± 1.0	83 ± 1	6.8 ±0.3	9 ± 1
2[b]:	47.8	37.3 ± 5.3	78 ± 11	15.6 ± 1.5	8 ± 1	37.3 ± 1.7	78 ± 4	5.4 ±0.2	11 ± 1

1.: pseudo-vergleyte Parabraunerde (according to German definitions)
[a] schwach lehmiger Sand, 1.1 weight % organic carbon.
[b] stark lehmiger Sand, 0.14 weight % organic carbon.
2. Gley:
[a] schwach schluffiger Ton, 2.3 weight % organic carbon.
[b] schluffig toniger Lehm, 11.0 weight % organic carbon.
RR retrieval rate, *L* lubricating oil, *S* squalane, *D* diesel fuel.

together with the results. The visual comparison of the thin-layer chromatograms of samples of unknown contamination and standards allows conclusions to be drawn on the selection of suitable standards. Contamination that is rich in aromatics show, for example, significant differences in comparison to contamination that is poor in aromatics and are detectable in UV-light before immersion. Higher concentrations are also detectable after immersion by means of colouring reagents (UV, daylight).[1]

This method proved to be disadvantageous due to the considerable loss of hydrocarbons with a low boiling point. In most cases sufficiently accurate results are obtainable, so that the method is also suitable for rapid on-site analysis, such as the assessment of the mineral oil content of a contaminated site.[1]

7.2.3 Comparison of Thin-Layer Chromatography and IR-Spectroscopy

With consideration of a suitable choice of standard the method was applied to real soil contamination. The samples investigated were taken from sites contaminated with waste oil or unknown substances. The results are shown in table 7.2.

Furthermore, investigations carried out by Koning (1999) were accompanied by analysis of the TPH-content. During these analyses the degradation of total petroleum hydrocarbons in windrows under different experiment conditions was determined and the influence of additives and oxygen was examined. The TPH-content of this series of experiments was determined in parallel by IR-spectroscopy, TLC and GC (Reimers et al. 2000).

As can be seen in fig. 7.3 the results of both methods correlate well. The calibration of the TPH content was effected towards the isolated contamination as a standard substance (IR-spectroscopy as well as TLC).

The comparison of results of the above methods can show differential values, with the contents having been measured according to the DEV H18-method usually being well below those of other methods. This can be due to the different sorbents for the clean-up of the extracts and the extracting agents.

A comparison of methods has also been carried out in the frame of a cooperative test by Giese (1996). The TLC-method took part in this cooperative test concerning the "Determination of organic summation parameters in soils" as a "special method". The results fitted well into the overall context.

7.2.4 Conclusions

The HPTLC proved to be a simple and quick method for the examination of middle and low volatile mineral oil contamination in soils. Because mineral oil contamination consists, to the great extent, of hydrocarbons it is possible to get information about the total contamination via the content in aliphates taking into con sideration the occurrence of polycyclic aromatics. A decisive factor in analysis is the selection

Table 7.2. Densitometric determination of the content of total petroleum hydrocarbons in real contamination, calibration towards lubricating oil, comparison with IR-measurement, detection anilinonaphthalenesulfonic acid, fluorescence measurement (366/400 nm)

Samples	TLC [g kg^{-1}]	IR [g kg^{-1}]
1	22.60	20.9
2	15.60	18.9
3	0.080	n.d.
4	33.0	28.8
5	0.110	0.11
6	1.200	1.30
7	0.400	0.52
8	0.30	0.24

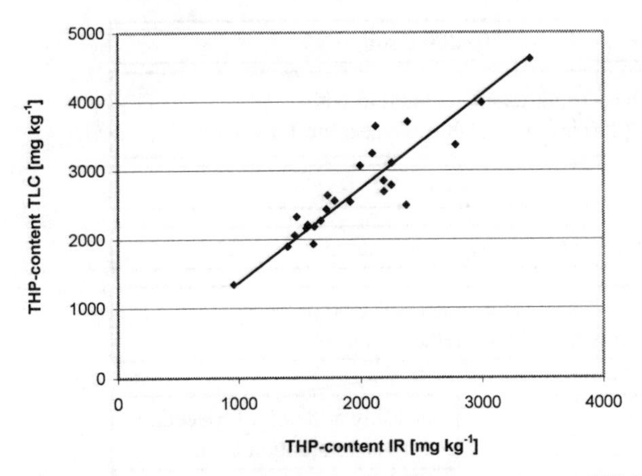

Fig. 7.3. Comparison of the results obtained by IR-spectroscopy and TLC

of a suitable calibration substance, when using this method as well as the IR-method according to DEV H18 (1981). This can be interpreted from the results of the analyses of artificially contaminated soil materials.[1]

Thin-layer chromatography is very flexible concerning the choice of solvents and thus independent of the use of halogenated compounds. This method is characterised by a low consumption of solvents and fast analyses of many samples on one plate, both remarkable features. Furthermore the clean-up of the extracts via column chromatography is becoming superfluous.

For medium distillates and contamination with a high content in aromatic substances in particularly, the quantities retrieved were, in part, much too small when the chosen standards were not suitable. The standard should be identical or similar in composition with the contamination as far as possible. For this reason it is sensible to state not only the results but also the reference substance. A disadvantage is the loss of hydrocarbons with a low boiling point. In most cases the results that can be obtained are sufficiently exact so that the method is also suitable for rapid on-site analyses, e.g. to evaluate the mineral oil content of an abandoned contaminated site. The results correlate well when compared to the IR-spectroscopy.

7.3
Polycyclic Aromatic Hydrocarbons (PAH)

The determination of PAH is normally carried out by means of high performance and detection sensitive methods of analysis such as GC or HPLC. A standardised method for the determination does not yet exist. In addition to a variety of so-called "institute regulations" there are some more regulations that differ with respect to the type of extraction, the solvents, clean-ups and principles of analysis. A selection of commonly applied methods can be seen in table 7.3.

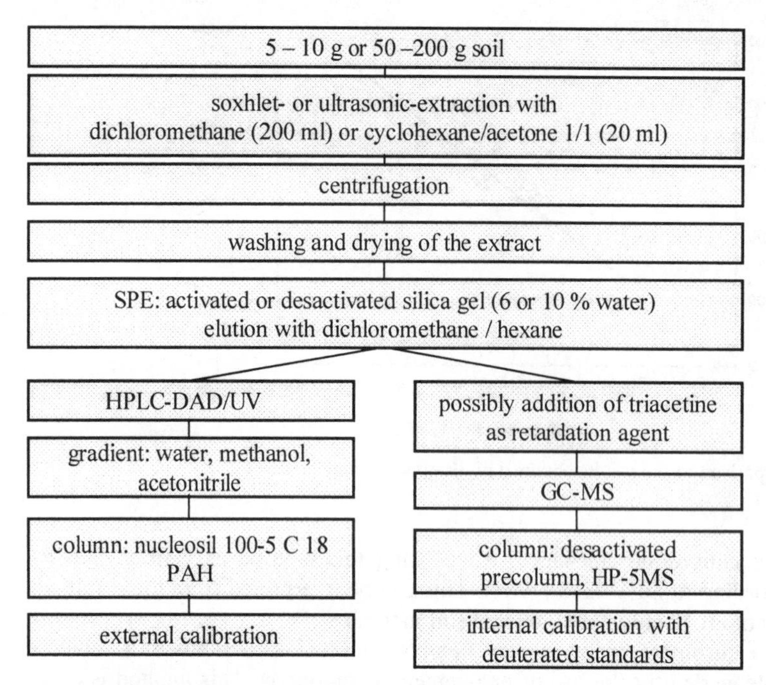

Fig. 7.4. Determination of PAH by HPLC and GC

In the context of our investigations, methods were developed and applied that are based on HPLC (see Bundt et al. 1990, 1991, 1992; Bundt and Stegmann 1993; Bundt and Steinhart 1993), GC (Paschke 1993; Paschke et al.1992) and TLC (Reimers et al. 2000). An example of the determination of PAHs by means of the HPLC and GC is demonstrated in the flowchart in fig. 7.4.

For the determination of PAH it is not always necessary to use an elaborate analysis by means of GC or HPLC. Thin-layer chromatography is not only suitable for routine analysis, but also for pre-investigation on a semi-quantitative scale. For characterisation of contamination the thin-layer chromatography can be applied as a screening method for PAH (Reimers et al. 2000).

7.3.1 Method

The soil is extracted with *n*-hexane and acetone (1/1; v/v) by ultrasonic treatment. The extracts are washed twice with demineralised water and are dried with sodium sulfate. Subsequently the extracts are purified using silica gel. The analysis is performed on HPTLC RP 18-plates (see fig. 7.5).

The best separation of PAH could be yielded by developing RP18-phases stepwise in the three solvents 1. *n*-hexane, 2. *n*-hexane/tert. butyl methyl ether (8/2; v/v) and 3. acetonitrile/dichloromethane/water (9/1/1; v/v/v) at -20 °C in saturated chambers. The above described separation system allows a separation of the PAH into eight groups (fig. 7.6).

Table 7.3. Commonly applied methods for the determination of PAH

Information about regulation	Method
Extraction by means of dichloromethane, clean-up step not necessarily important, HPLC/UV or F-detection or GC/FID	EPA-method 610 (1982)
Soxhlet-extraction by means of toluol, chromatographic clean-up on florisil, GC/MS Ultrasonic extraction by means of acetonitrile, eventually clean-up of extracts on modified silica gel, HPLC-UV/DAD/F	Landesumweltamt NRW, Merkblatt 1, 1994
Extraction via shaking by means of acetone without drying of the soil, addition of petrolether, elimination of the acetone, column clean-up on Al_2O_3, reception in acetonitrile, HPLC-UV/DAD/F (minimum of 0.01 mg kg^{-1} for each component) Soxhlet-extraction by means of toluene and air-drying of the soil, a toluenealiquote is analysed, HPLC-UV/DAD/F (minimum of 0.1 mg kg^{-1} for each component)	DIN ISO 13877 (2000)
Extraction with a mixture of water/acetone/petrolether in the presence of NaCl, clean-up on polystyrol, additionally on silica gel and on polydextrane, sephadex, GC/FID/MS or HPLC-UV/DAD/F	Handbuch Altlasten, Bd. 7, Hessische Landesanstalt für Umwelt, VDLUFA-Methodenbuch, Band VII (1996)
Extraction by means of cyclohexane, as an alternative filtration via silica gel, HPLC/F (6 PAH soil leachate)	DIN 38407-8 (10/95)
Soxhlet-extraction by means of cyclohexane or hexane, clean-up silica gel, HPLC/F (6 PAH)	DIN 38414-21 DEV S21 (sludge, sediment)
Extraction by means of hexane, eventually clean-up of extract on Al_2O_3 or silica gel, HPTLC/F (6 PAH according to German drinking water regulation)	DIN 38407, DEV F7 (water)

After developing the plates in saturated double-walled chambers, were are air-dried and dipped into a mixture consisting of *n*-hexane/paraffin. The PAH could either be observed by UV light or be evaluated by densitometric scanning and selective fluorescence detection using different excitation wavelengths and edge filters. The results are interpreted as semi-quantitative because of their group or sum character respectively.

The analyses contain the EPA-PAH without naphthalene and acenaphthylene. In order to get a sufficient separation, different stationary and mobile phases were examined. Several chromatographic systems have been described (Baranowska et al. 1994; Hellmann 1979b; DIN 38 407 1997). Experiments with various sorbent materials and selected different PAH were carried out. Although the authors analysed soil, ground-water, mineral oil, sediments, as well as exhaust, they could not offer an universally applicable solution for the determination of PAH. The investigation resulted in many different methods, but each only offers a solution for a very specific problem.

For the stabilisation and fluorescence enhancement of PAH Hellmann (1983)

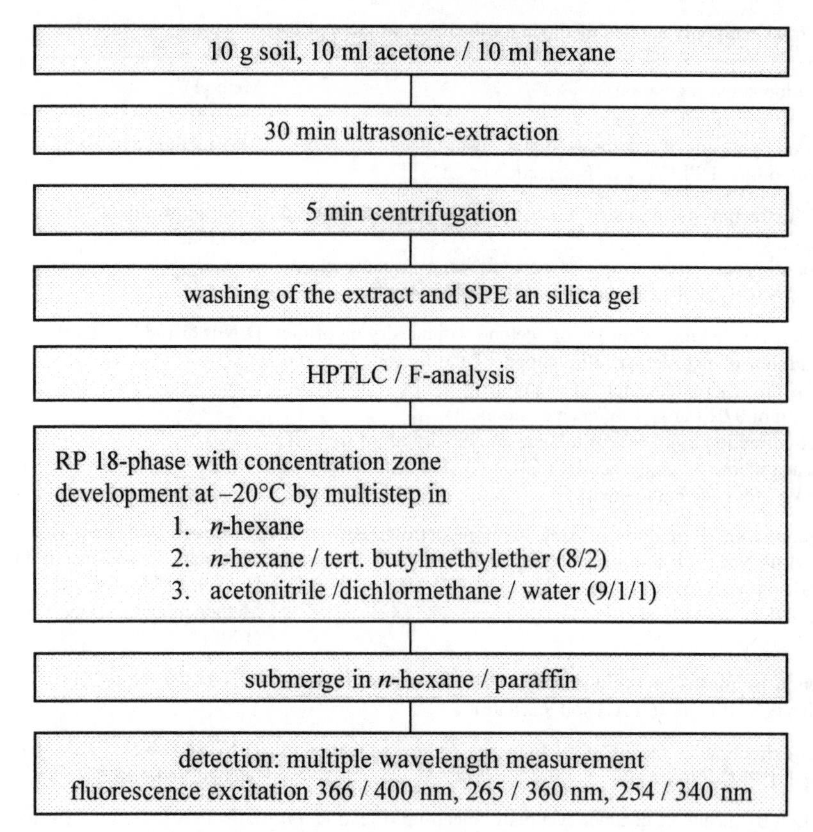

Fig. 7.5. Screening of PAH-content in soils by HPTLC/F

applies a coating of paraffin over the stationary phases. The fluorescence intensity increases depending on the concentration of paraffin. In our experiment we plunged the plates into *n*-hexane/paraffin (4:1) and achieved an enhancement by a factor of 4 to 5.

7.3.1.1
Selective Initiation

Hellmann (1975) described a thin-layer chromatographic method for the simple routine determination of PAH by means of fluorescence spectroscopy. From evaluation of the initiating wavelength he ascertained the dependence of fluorescence intensity of PAH (Hellmann 1979a). Therefore he uses the differentiated fluorescence excitation for the determination of not only selected aromatics, but also of entire PAH-groups. Six PAHs according to the German drinking water regulation (DIN 38407) are determined in a similar manner. RP-18 phases are suited for the determination of PAHs in combination with differentiated fluorescence excitation. The initiating wavelengths and suitable filters are shown in table 7.4. Different wavelengths can also be employed.

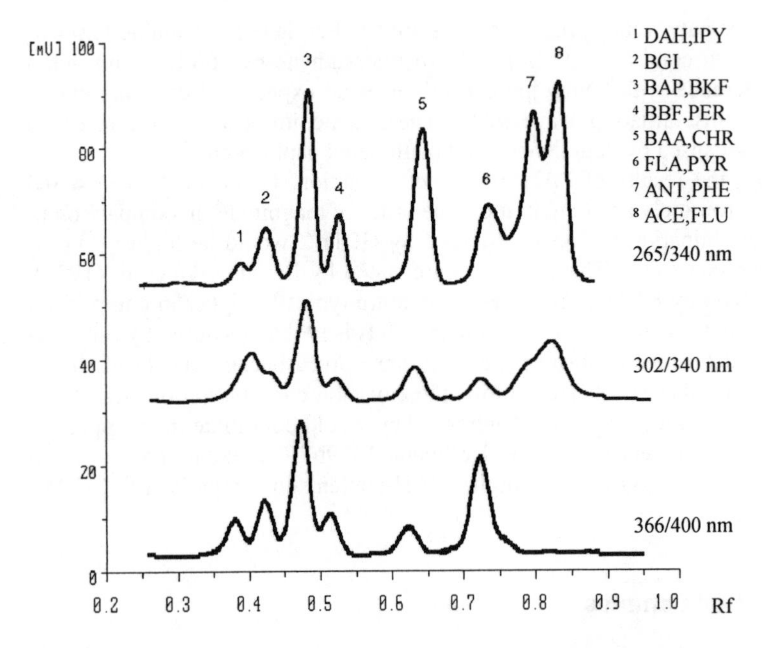

Fig. 7.6. Separation of PAH into 8 groups on RP-18 phases by HPTLC. Fluorescence excitation with various wavelengths (abbreviations see table 7.5) (from G.I.T. Laboratory Journal 2/00, reproduced with permission, copyright G.I.T. Verlag GmbH, 2000)

Table 7.4. Selected wavelengths for the quantitative determination of PAH on RP-18 phases in soil samples

PAH	Wavelengths [nm]	Filter [nm]
• benzo[a]pyrene/benzo[k]fluoranthene, benzo[b]fluoranthene, benzo[ghi]perylene,indeno[1,2,3-cd]pyrene, fluoranthene	366	K 400
• benzo[a]anthracene/chrysene	254	K 340
• pyrene/fluoranthene, anthracene/phenanthrene	265	M 360
• dibenz[ah]anthracene/indeno[1,2,3-cd]pyrene	302	K 400
• acenaphthene/fluorene	265	K 340

K cut-off filter, *M* narrow band-pass filter

7.3.1.2 Quantitative Determination

With the described method the PAH compounds can be determined in the ng-range. Acenaphthene and fluorene are not quantified because they could only be detected with varying accuracy. These variations may be due to the interference caused by other polycyclics present in the samples. Perylene for example is sensitively detected in combination with benzo[b]fluoranthene by 366/K400 nm whereas an interference with benz[e]pyrene is not to be expected. Benz[e]pyrene is insensitive to this initiating wavelength. There is no influence of fluorescence intensity being caused by naphthalene and methylated naphthalene.

Concerning the results of PAH group determination, Hellmann (1983) stated that they were routinely accurate in the same order of magnitude. In comparison to a separate determination the results yielded by HPTLC would be higher – up to 30–50 % in certain cases. These findings are stated by Baranowska et al. (1994). Analyses of PAH by HPTLC gives results that are typically higher in comparison to the GC/MS data. The deviation of both methods is probably caused by the presence of other determined substances and an unresolved background level. If the concentration of other substances is low, the emission contribution of these compounds could be neglected. To the contrary, they would contribute in an appreciable manner to fluorescence intensity (Hellmann 1979b). Our examinations of soil samples did not show significantly higher PAH-contents in comparison to GC/MS (see table 7.5).

7.3.2 Analysis of Soil Samples

The method was used to verify the procedure. First a certified PAH standard (fig. 7.7), then various real-word soil-samples were analysed. Furthermore the PAH concentrations were determined by GC/MS (see table 7.5).

The results of the certified standard determined by HPTLC correspond very well to the given concentrations with exception of the results for anthracene/phenanthrene. These deviate appreciably. This deviation is due to different PAH contained in the standard (e.g. methylated compounds) which interfere with the anthracene/phenanthrene. Their emission contribution could not be neglected as stated by Hellmann (1979b). The concentrations of these compounds are similar in order of magnitude to those of the PAH compounds of interest. In comparison to the analysed soil samples we could not state such interferences.

7.3.3 Comparison to other Analytical Methods

Thin-layer chromatography is an excellent method for fast and reliable analyses of various compounds. It allows a rapid analysis of a greater number of samples in comparison to HPLC and GC analyses. Furthermore it is possible to track the separation by UV-light. In the case of PAH analyses it is facilitated by their different fluorescence colours. Comparison of the results yielded by HPTLC and GC/MS shows good agreement.

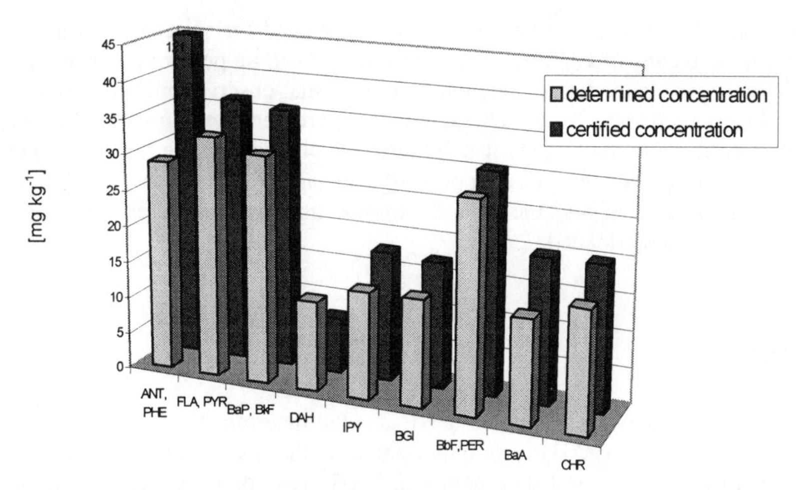

Fig. 7.7. PAH concentration of a certified standard in comparison to the concentration determined by HPTLC (abbreviations see table 7.5)

Table 7.5. PAH concentrations of soils: comparison of HPTLC and GC/MS analyses. Soil 1 consists of EPA PAH and TPH, coarse clay 5 %, soil 2 consists of EPA PAH, TPH and cyanides, coarse clay 5 %

Substance	Soil 1 HPTLC [mg kg^{-1}]	Soil 1 GC/MS [mg kg^{-1}]	Soil 2 HPTLC [mg kg^{-1}]	Soil 2 GC/MS [mg kg^{-1}]
Indeno[1,2,3-cd]pyrene (IPY)	1.7	1.7	1.8	1.2
Benzo[ghi]perylene (BGI)	0.9	1.0	1.0	1.3
Benzo[a]pyrene (BaP), Benzo[k]fluoranthene (BkF)	6	5.6	5.2	5.6
Benzo[b]fluoranthene (BbF)	4.6	3.3	5.9	3.6
Perylene (PER)		PER n.d.		PER n.d.
Benzo[a]anthracene (BaA)	>2.7	8.1	<11	6.9
Chrysen (CHR)	>3.6	6.7	7.1	6.3
Pyrene (PYR), Fluoranthene (FLA)	119	99	>27	105
Phenanthrene (PHE), Anthracene (ANT)	146	136	>54	163

7.3.4 Conclusions

The described method is demonstrated to be easy and rapid in handling, requires small amounts of solvents and equipment and is qualitative as well as quantitative about composition and concentration of contaminants. The method is well suited for the separation and semi-quantitative determination of PAH. In addition, conclusions are possible concerning the distribution and concentration of the contamination. Using this method clues can be obtained for selective and advanced analyses by means of GC or HPLC. The analysis can be performed quantitatively by fluorescence excitation or qualitatively by visual observation. The visual identification is facilitated by the characteristic fluorescence colours of PAH. When evaluating the chromatograms it must be remembered that the method is not always selective, such that overlapping of substances cannot be completely excluded. This is especially the case for multi-component mixtures and when the thin layer is heavily loaded.

7.4 Summary

In the field of analysis of contaminated sites the determination of polycyclic aromatic hydrocarbons (PAH) is as important as is the quantification of total petroleum hydrocarbons. The determination of PAH does not always call for elaborate and extensive analysis. By means of screening methods, a quickly obtainable overview over a contamination can be gained. Methods have been developed and applied that can be used in a fast and efficient manner in practice.

Two different methods were applied for the determination of the total petroleum hydrocarbons for comparison with the IR-spectroscopic method according to DEV H18. These are thin-layer chromatography and gas-chromatography.

Being easy in handling and requiring relatively low expenditure in time and money, thin-layer chromatography is an excellent method, not only for routine analyses but it is also suitable for pre-investigations on a semi-quantitative scale. The preparation of the samples is simple and the operation costs are low and it allows a rapid analysis of a great number of samples. Another advantage is the almost unlimited field of application of mobile phases and the manifold of detection possibilities. With their help, mineral oil contamination can be determined via the content of aliphates because they consist, for the most part, of hydrocarbons. A decisive factor when applying this method for analysis (as for the IR-method according to DEV H18) is the selection of a suitable calibration substance.

On the other hand, the total petroleum hydrocarbons are determined by means of the gas-chromatographic method according to DEV H53, or according to ISO/TC/190/SC3/6 respectively. A hydrocarbon index is indicated that corresponds to substances lying between C9 or C10 and C40. This method is advantageous in its use of eco-friendlier solvents. Statements can be made about the type of contamination, as well as about the substantial changes in the course of a remediation measure.

For the determination of PAH a thin-layer chromatographic screening method with selective fluorescence excitation was applied. This method is suitable for a qualitative as well as for a semi-quantitative analysis of PAH in soil materials. It allows a separation of the compounds into 8 PAH-groups that contain each two PAH at a maximum. The results are judged as being semi-quantitative due to of their grouping characteristics.

References

Baranowska I, Wieslaw S, Wasilewski P (1994) J planar chromatogr 7: 137–141

Bestimmung (1994) von polycyklischen aromatischen Kohlenwasserstoffen (PAK) in Bodenproben. Landesumweltamt NRW (Merkblatt 1)

Bestimmung (1998) von Mineralöl-Kohlenwasserstoffen in Feststoffen aus dem Altlastenbereich Teil 2 (vorläufige Methode). Hessische Landesanstalt für Umwelt, Wiesbaden (Analyseverfahren im Handbuch Altlasten Band 7)

Bundt J, Herbel W, Steinhart H (1990) Mineralölanalytik von Bodenkontaminationen unter besonderer Berücksichtigung von aromatischen Kohlenwasserstoffen. In: Stegmann R, Franzius V (eds) Reinigung kontaminierter Böden, Hamburger Berichte Band 3, Economia Verlag GmbH, Bonn, Germany, pp 49–64

Bundt J, Herbel W, Steinhart H (1992) Determination of polycyclic aromatic sulphur heterocycles (PASH) in diesel fuel by highperformance liquid chromatography and photo-diode-array-detection, J High Resolution Chromatogr (HRC) 15: 682–685

Bundt J, Herbel W, Steinhart H, Franke S, Francke W (1991) Structuretype separation of Diesel fuels by solid phase extraction and identification of the two- and threering aromtics by capillary GC-mass spectrometry. J High Resolution Chromatogr (HRC) 14: 91–98

Bundt J, Stegmann R (1993) Bestimmung von PAKs in Teerölen und deren Altlasten mittels HPLC. GIT Fachz Lab 2: 64–69

Bundt J, Steinhart H (1993) Analytische Verfahren zur Bestimmung der Gehalte an Kohlenwasserstoffen und PAK in kontaminierten Böden. In: Stegmann R (ed) Bodenreinigung Hamburger Berichte, 6, Economica Verlag, Bonn, Germany, pp 39–50

CEN/TC929/WG5 (2000) Bestimmung des Gehalts an Kohlenwasserstoffen von C10 bis C40 mittels Gaschromatographie

DEV H53 (1998) Bestimmung des Kohlenwassestoffindex Teil 4: Verfahren nach Lösungsmittelextraktion und Gaschromatographie

DIN 38 407 Teil 7 (Entwurf) (1997) Bestimmung von 6 polycyclischen aromatischen Kohlenwasserstoffen (PAK) in Trink- und Mineralwasser mittels Hochleistungs-Dünnschicht-Chromatographie (HPTLC) (F7)

DIN 38 407 Teil 8 (1995) Bestimmung von 6 polycyclischen aromatischen Kohlenwasserstoffen (PAK) in Wasser mittels Hochleistungs-Flüssigkeitschromatographie (HPLC) mit Fluoreszenzdetektion (F8)

DIN 38 409 H18 (1981) Bestimmung von Kohlenwasserstoffen. Normenausschuß Wasserwesen im DIN Deutsches Institut für Normung e V

DIN 38 414 Teil 21 (Jahreszahl) Bestimmung von 6 polycyclischen aromatischen Kohlenwasserstoffen (PAK) mittels Hochleistungs-Flüssigkeitschromatographie (HPLC) und Fluoreszenzdetektion (S 21)

DIN ISO 13877 (2000) Bestimmung von polycyklischen aromatischen Kohlenwasserstoffen

EPA-mehod 610 (1982) Polynuclear Aromatic Hydrocarbons. United States Environmental Protection Agency, Environmental Monitoring and Support Laboratory Cincinnati

Giese G (1996) Ringversuch zur Bestimmung von Summenparametern in Böden. Umweltbehörde Hamburg

Giese G (1999) Mineralölkohlenwasserstoffe in Böden und Abfällen: Ersatz der H18-Methode durch eine FCKW-freie, gaschromatographische Alternative. Umweltbehörde Hamburg, unpublished

Gitler C (1972) Use of ANS to Detect Phospholipids and Apolar Molecules in Chromatograms. Anal Biochem 50: 324–325

Goebgen HG, Brockmann J (1977) Die Anwendung der Dünnschichtchromatographie in der Abwasseranalytik am Beispiel der Bestimmung schwerflüchtiger Mineralöl-Kohlenwasserstoffe. Vom Wasser 48: 167–178

Gottwald W, Heinrich KH (1998) UV/VIS-Spektroskopie für Anwender. Wiley-VCH Verlag, Weinheim, Germany

Handbuch Altlasten (1996) Hessische Landesanstalt für Umwelt, VDLUFA-Methodenbuch, Band VII

Hein H, Kunze W (1995) Umweltanalytik mit Spektrometrie und Chromatographie. 2. Auflage, VCH-Verlagsgesellschaft, Weinheim, Germany

Hellmann H (1975) Vereinfachte routinemäßige Bestimmung von polycyclischen Aromaten. Fresenius Z Anal Chem 275: 109–113

Hellmann H (1979a) Fresenius Z Anal Chem 295: 24–29

Hellmann H (1979b) Fresenius Z Anal Chem 295: 388–392

Hellmann H (1983) Fresenius Z Anal Chem 314: 125–128

Hezel U (1973) Direkte quantitative Photometrie an Dünnschichtchromatogrammen. Angew Chem 85: 334–342

ISO/TC/190/3/6 (1998) First working draft Soil quality Determination of mineral oil content by gas-chromatography.

Jork H, Funk, Fischer, Wimmer (1990) Thin-Layer Chromatography. Reagents and Detection Methods. Volume 1a, VCH-Verlag, Weinheim, Germany, pp 213–215

Koning M, Braukmeier J, Cohrs I, Lüth J, Stegmann R (1999) Untersuchungen zur Optimierung der biologischen Bodenhaltung im Mietenverfahren im Hinblick auf eine gezielte Wiederverwertung des zu behandelnden Bodenmaterials. DBU Projekt 09815

KW/85 (1993) Bestimmung des Gehaltes an Kohlenwasserstoffen in Abfällen. Länderarbeitsgemeinschaft Abfall (LAGA) (in Richtlinien für das Vorgehen bei physikalischen und chemischen Untersuchungen im Zusammenhang mit der Beseitigung von Abfällen)

Lepom P (1996) Bestimmung von Mineralölkontaminationen in Böden und Abfall – Stand der Normung bei ISO und CEN. Umweltbundesamt Berlin

Mamlock M (1981) Berberine Hydrochloride for Detection in Thin-Layer Chromatography. J Chromatogr Sci 19: 53

Murr J (1990) Erarbeitung eines raschen und sicheren Analysenverfahrens zur zweifelsfreien Identifikation von Mineralölkontaminationen in Umweltproben. Gewässerschutz, Wasser, Abwasser 121, Aachen, Germany

Nagy S, Nordby HE (1972) Long Chain Hydrocarbon Profiles of Ducan Grapefruit, Dancy Mandarin und Their Hybrids. Lipids 7: 722–727

NEN 5733 (1995) Soil. Determination of mineral oil content in soil and sediment by gas chromatography. (draft)

Paschke A (1993) Beitrag zur Untersuchung von polycyclischen aromatischen Kohlenwasserstoffen in Schmierölen und Mineralölaltlasten. Dissertation, Universität Hamburg, unpublished

Paschke A, Herbel W, Steinhart H, Franke S, Francke W (1992) Determination of Mono- to Tetracyclic Aromatic Hydrocarbons in Lubricating Oil. J High Resolution Chromatogr (HRC) 15: 827–832

Reimers C, Schmidt B, Stegmann R, Franke W (1998) Bestimmung von Mineralölkohlenwasserstoffen in Bodenproben mit Hilfe der Dünnschichtchromatographie. Umweltwissenschaften und Schadstoff-Forschung 4/98: 200–204

Reimers C, Zielonka B, Stegmann R, Steinhart H (2000) Screening of PAH-contaminated soil samples using thin-layer chromatography. GIT 4: 93–94

Reimers C, Koning M, Cohrs I, Scholz A (2000) Vergleichende Untersuchungen von Mineralölkohlenwasserstoffen in Bodenproben mit Hilfe der IR-Spektroskopie, Gaschroma-tographie und Dünnschichtchromatographie, Technische Universität Harburg, Arbeitsbereich Abfallwirtschaft, unveröffentlicht

Schmidt B (1995) Gaschromatographische und Dünnschichtchromatographische Untersuchungen zur Bestimmung von Mineralölkontaminationen in Böden. Diplomarbeit, Universität Hamburg

Weisheit W, Enl H (1973) Nachweis und Bestimmung von Kohlenwasserstoffen in Leinölen und Leinölkitten. Seifen, Fette, Wachse 99: 711–714

8 Solid State NMR Spectroscopy: a Tool to Study the Reaction of Humic Acids with ^{13}C-Labelled Polyaromatic Hydrocarbons (PAHs)

E.T.K. Haupt[1], T. Käcker[2] and H. Steinhart[2]
[1]Institute for Inorganic and Applied Chemistry, University of Hamburg, Martin-Luther-King-Pl. 6, 20146 Hamburg, Germany
[2]Institute for Biochemistry and Food Chemistry, University of Hamburg, Martin-Luther-King-Pl. 6, 20146 Hamburg, Germany

In memoriam Dr. Hans-Otto Kalinowski

8.1 Introduction

Nuclear Magnetic Resonance Spectroscopy (NMR) is the most prominent method to study structural features of many types of chemical substances, especially organic compounds. Consequently this method has been a standard tool to gain insight to the structures of fulvic and humic acids and other humic fractions for many years. From the viewpoint of NMR, two different targets will be discussed: the investigation of soluble compounds with high resolution NMR and the study of insoluble material with Solid State NMR. Good compilations of both types of investigation are given elsewhere (Nanny et al. 1997; Wershaw and Mikita 1987). For technical and practical reasons, this choice has important consequences for studies of fulvic and humic acids. This contribution concentrates on Solid State NMR studies.

Solid State NMR spectra, in particular, are not as well structured as spectra in solutions because humic materials are complex mixtures of numerous compounds. A direct consequence of this is that many of the successful results of high-resolution studies are non transferable. As a result the idea was born to use isotopic enrichment to study selected compounds in the soil environment, either to get more insight into the structures of some of the constituents of humic material, or the fate of contaminants. Although many of the studies were also performed with solution spectra, the use in solid state studies has grown permanently.

8.2 Investigations on Real and Artificial Material

Solid State NMR spectrum of humic substances concentrate predominantly on ^{13}C-CPMAS (*C*ross *P*olarisation, *M*agic *A*ngle *S*pinning; for details see Fyfe 1983)

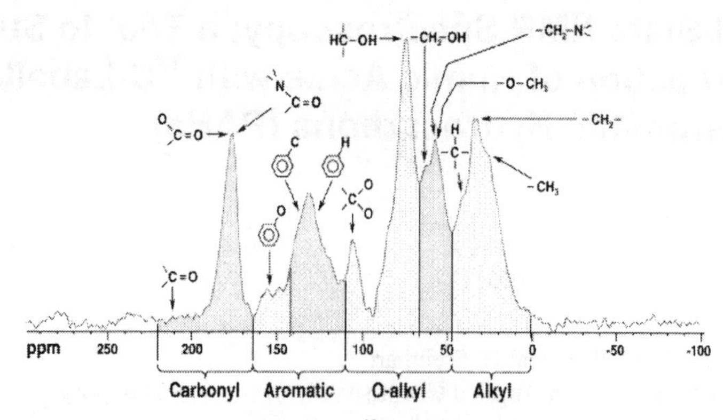

Fig. 8.1. Characteristic chemical shift areas in ^{13}C-CPMAS-spectra of humic acids

studies, although other nuclei like ^{15}N also deliver valuable results (Knicker et al. 1997). One of the reasons is that the carbon (or nitrogen) nucleus exhibits characteristic areas of chemical shift (fig. 8.1) with respect to different classes of chemical functional groups.

Therefore, it appears to be a simple task to detect changes in the organic part of the material or to determine signals from additives or contaminants. Because the spectra are not as well resolved as high-resolution spectra, in reality this task is very difficult, especially when the concentration differences of the components of interest are minimal. To circumvent this problem, the application of ^{13}C enriched compounds is useful. The aim is to neglect, more or less, changes in the background signal from the organic soil material and to concentrate only on the observation of the signal(s) from the enriched carbon positions riding on top of the spectrum. Successful studies require highly enriched compounds (fig. 8.2) to reduce the measurement time. In our studies, we had access to synthesised compounds which were specifically labelled for the purposes of this study (Francke 2000; Richnow et al. 2000). The labels were introduced in phenanthrene and fluoranthene (scheme 8.1) at positions which are expected to be involved in the process of bound residue formation.

It is obvious from fig. 8.2 that the labelling strategy has enormous advantages compared to the application of natural occurring material. However, the basic

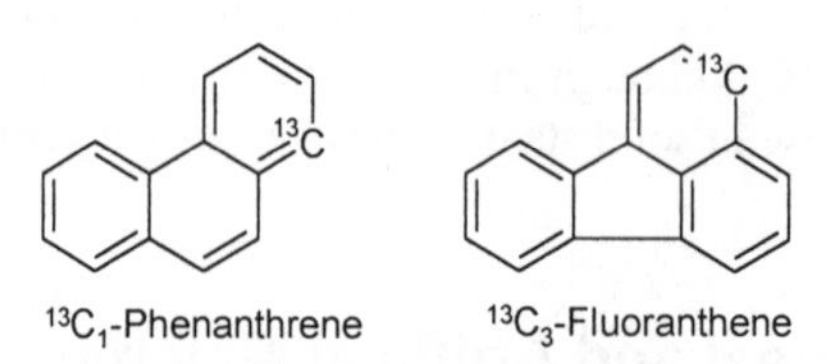

Scheme 8.1. Labelled phenanthrene and fluoranthene

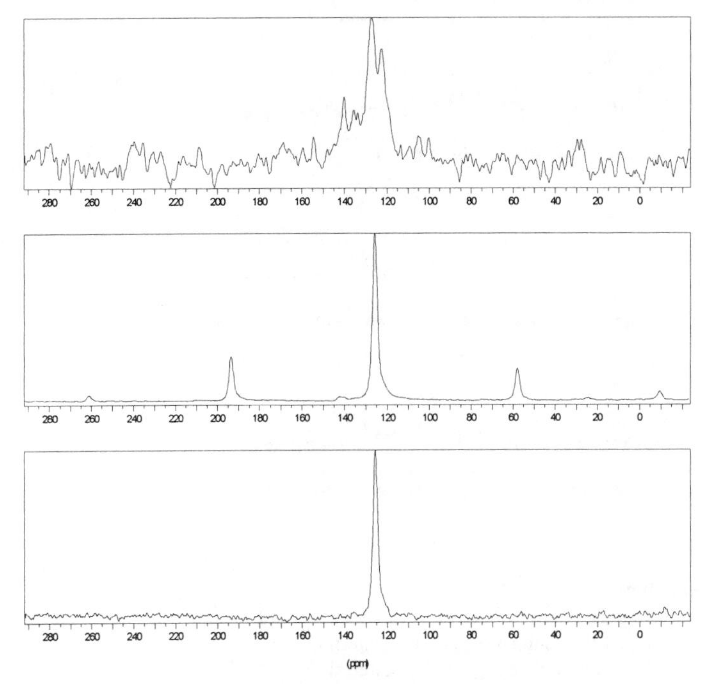

Fig. 8.2. The influence of rotation frequency and isotopic enrichment on the quality (S/N) and measurement time of ^{13}C-CPMAS spectra:
upper trace: fluoranthene, natural abundance, rotation = 5,200 Hz, 24 h;
middle trace: fluoranthene, isotopically enriched at C_3, rotation = 5,200 Hz, 3 h;
lower trace: same as middle, but rotation = 12,200 Hz, 10 min.

concept of transferring this result to natural soil samples fails. Fig. 8.3 b, c demonstrate that even with relatively high concentrations of the enriched marial only broad lines from the contaminants are achieved. Thus, too much of the enriched material is necessary to achieve reliable results.

Additionally, fig. 8.3 demonstrates that with our soil material (Ah soil) we could not detect a carbon spectrum. This situation is undesirable because failures in the experimental settings or the instrument performance may be overseen. Thus, a background signal has the advantage to act like an indicator to assure correct experimental conditions.

Several reasons come into account to explain the results in fig. 8.3:

- the content of organic material is too low to yield a background spectrum,
- the contaminant signal is broadened by:
  differences in the surroundings,
  paramagnetic impurities.

Thus, we concentrated our investigations on extracted humic acid material achieved under special conditions to avoid or reduce the above mentioned

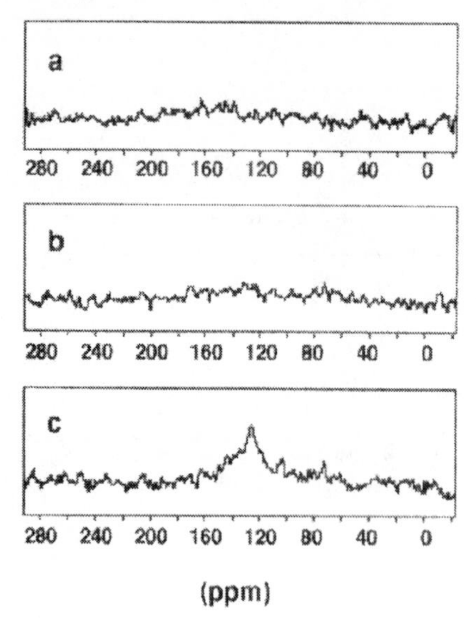

Fig. 8.3. ^{13}C enrichment vs. natural soil sample:
upper trace: Ah soil, pure
middle trace: 1 mg $^{13}C_3$-fluoranthene (g Ah dry material)$^{-1}$
lower trace: 5 mg $^{13}C_3$-fluoranthene (g Ah dry material)$^{-1}$

disadvantages. The details of the extraction procedure and the preparation of the samples are extensively described by Käcker et al. (2000a, 2000b). Following these procedures, more or less slight differences in the spectra can be observed.

8.3 Difference Spectroscopy

The visualisation of the differences in the single spectra causes a problem (fig. 8.4) and can be enhanced by the difference spectroscopy method. Two techniques are known: direct formation of the difference by subtracting original data during the measurement (de Groot et al. 1988) or the calculation of differences of the separated spectra after the measurement. A detailed discussion of the advantages and disadvantages of each approach is given by Sanders and Marsh (1982).

While the first difference method is often preferred in high resolution NMR, in the present case we preferred the second method. This was due to very long measurement times, even to detect the separate spectra and to be certain that the experimental set-up was stable during the measurement. Hence we avoided the measurement of an additional difference spectrum.

The processing of the difference from separate spectra is done visually on the computer and has the disadvantage that some personal influence from the operator may be introduced into the result. Therefore, a very careful inspection has to be performed to get reliable data. On the other hand, this inspection allows the

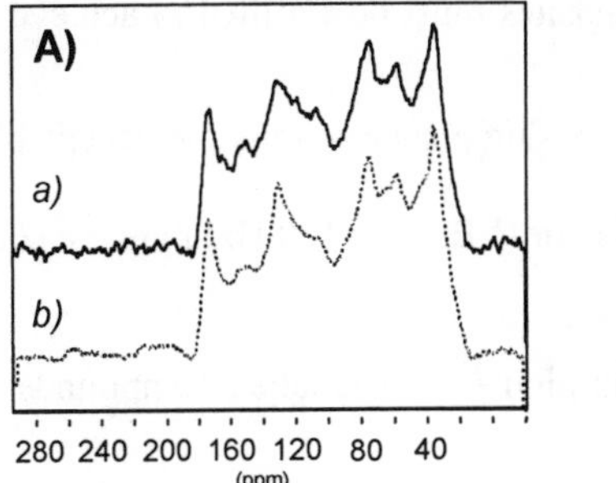

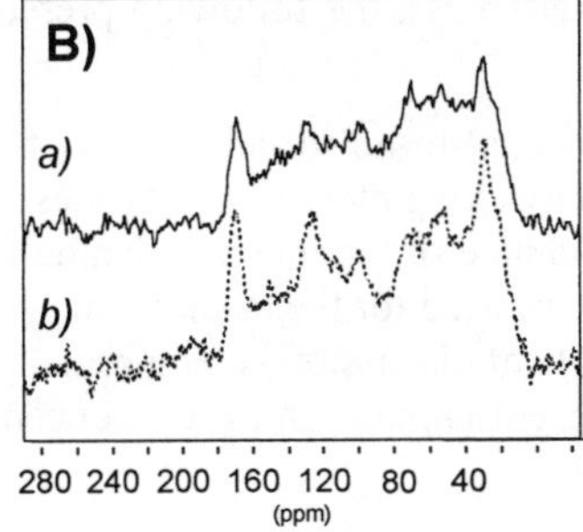

Fig. 8.4. ^{13}C CPMAS spectra of extracted humic acids (HS): A) a) HS from Ah/compost material b) same as a), spiked with $^{13}C_1$-phenanthrene, both after 87 days; B) a) HS from Ah/compost material b) same as a), spiked with $^{13}C_3$-fluoranthene, both after 300 days

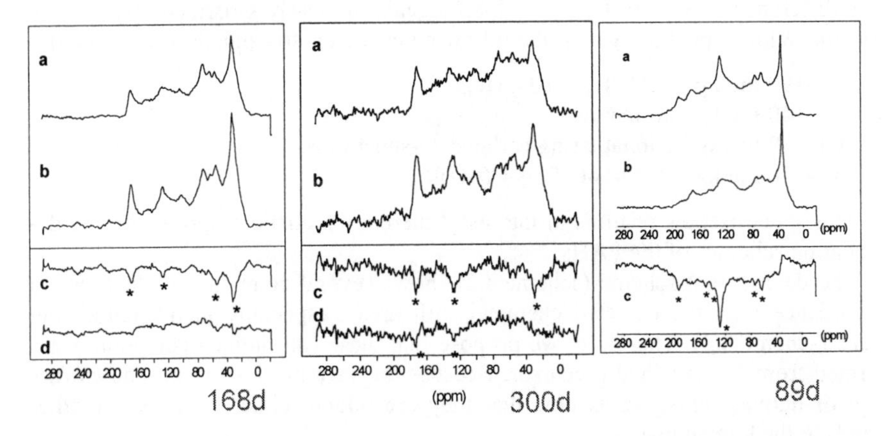

Fig. 8.5. ^{13}C CPMAS spectra of different types of extracted humic acids; left: Ah soil (after 168 days), middle: Ah/compost material (after 300 days), right: artificial soil with humic acid (ACROS, after 89 days). Traces a: original material; traces b: spiked with labelled compounds (left + right: phenanthrene; middle: fluoranthene) traces c: differences with respect to the background; traces d: differences with respect to peaks.

careful operator to neglect doubtful data, because the result is no longer a more or less "black box" spectrum that normally results from direct difference measurements.

The traces c and d in fig. 8.5 demonstrate the problem: if the difference is built by fitting the background signals (traces c), the result differs from those where the most prominent signals are used as a reference (traces d).

Although the process of calculation of the differences bears some problems, in the present study we had the advantage of having three independent materials available. Nevertheless, we observe in all three cases similar signals in the difference spectra more or less in the same spectral area. We conclude that this is due to the same chemical processes and therefore trust the reliability of the data.

Nevertheless, the following prerequisites must be fulfilled to achieve qualified data:

- the signal-to-noise ratio (S/N) of the single spectra must be acceptable, which requires long measurement times,
- the choice of the line broadening factor (LB ~ 50–100 Hz) must be equal in the spectra used for the calculation,
- a critical discussion is required,
- high enrichment and high concentration of the labelled compounds is necessary.

8.4 Results

The difference spectra in fig. 8.5 exhibit signals in clearly separated areas of the spectra. With respect to fig. 8.1, the following characteristic signals are observed:

- 170–190 mg kg^{-1} -COOH, -CHO, etc.,
- 120–140 mg kg^{-1} aromatics,
- 60– 80 mg kg^{-1} aliphatic ring carbons, O-substituted,
- 30– 40 mg kg^{-1} R(-OR)-C-CH_2-C(=O)R.

This spectroscopic result is of interest if there are realistic proposals for the degradation schemes of the PAHs.

The degradation scheme (scheme 8.2) shows several compounds which are in accordance with the detected chemical shift areas, especially in the initialising steps (scheme 8.3). Therefore, we propose that these (or similar) compounds are formed from the enriched precursor. Because they are not lost during the extraction of human acids, we assume that they are bound, either by ester- or other bonds to the human matrix.
To verify the shifts, model compounds can be used for the search in spectroscopic databases. We choose model compounds from the degradation scheme and used R to replace the humic matrix, where R is set to methyl in the database search. Slight differences may be introduced in this way, but in all cases the interesting carbons are several bonds separated from the R-bond. It is well-known from ^{13}C-NMR that in these cases the replacement of R by other groups is not very relevant to the chemical shifts and we expect, that in our case the matrix influences are much bigger then the influences by changing the R's. It is shown, that the resulting data correspond well with our detected signals (fig. 8.6) (Specinfo 1999).

Scheme 8.2 phenanthrene degradation pathways

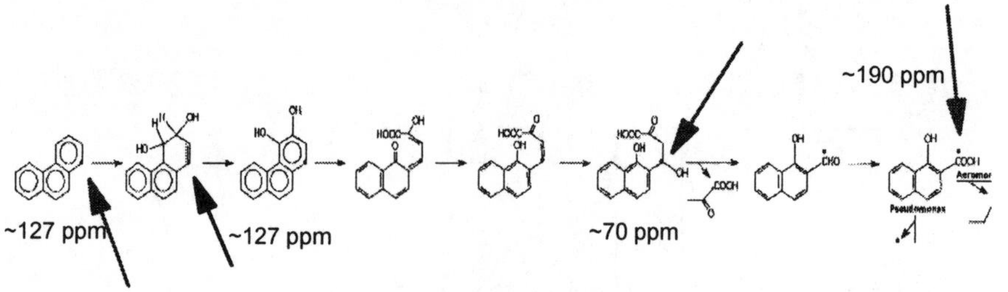

Scheme 8.3. Initial part of phenanthrene degradation pathways (Käcker 2000a)

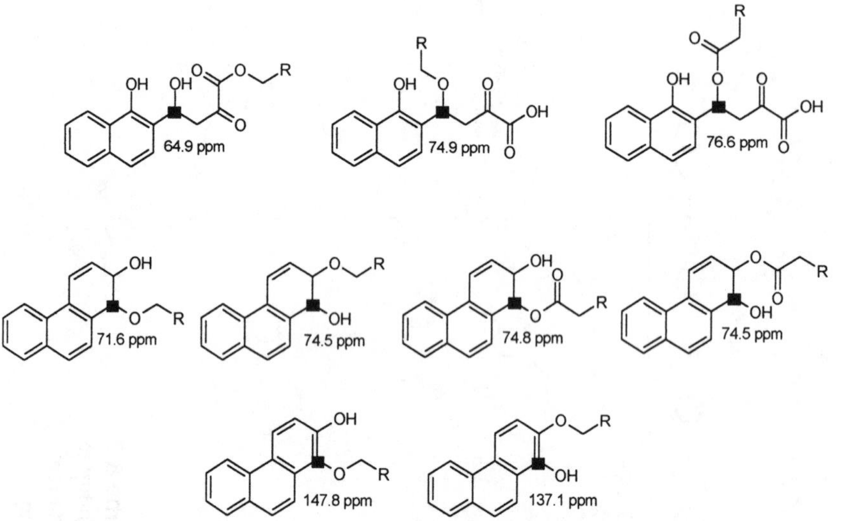

Fig. 8.6. Model compounds with ^{13}C NMR chemical shifts expected from database search (compare to fig. 8.5, right)

8.5 Verification by an Independent Method

We have demonstrated for the first time, that ^{13}C-CPMAS-NMR spectroscopy is a valuable tool to follow the fate of ^{13}C labelled PAHs in the soil matrix. Nevertheless, the process of difference spectroscopy is critical. Until now, the data available for PAHs is given here. Therefore it is especially desirable to prove the formation of bound residues by an independent method, e.g. to avoid results which stem from simple inclusion or absorption processes. We choose the method of alkaline hydrolysis with $Na^{18}OH$, which has been introduced by Richnow et al. (1994). In combination with the use of ^{13}C enriched PAHs this should deliver an independent proof of the detection of formerly bounded material. For this purpose, the isolated humic acid material is hydrolysed under basic conditions with $Na^{18}OH$, resulting in a cleavage of ester-bonds. Thus, the formerly bounded residues are split off from the humic material and should bear the ^{18}O labelling however, in our case the ^{13}C labels should also be present. The doubly labelled residues are then detected via GC-MSD and strengthen the NMR results. The details of the procedure are given elsewhere (Käcker 2000b).

8.6 Quantitative Aspects

In principle NMR is a quantitative method (well-known from basic courses) because the area under the signals corresponds to the relative number of atoms detected. In reality, this feature is only easily achieved in ^{1}H-High-Resolution-NMR. In heteronuclear NMR, several additional parameters, such as the efficiency of decoupling, increased relaxation times, Nuclear Overhauser Enhancements and others have to be observed which make most of these exact measurements critical and time-consuming. Additional problems arise in Solid-State-NMR, for example, with different contact times for nuclei in different parts of the "molecules" and a homogenous excitation of all nuclei in different frequency channels (^{1}H, hetero). Thus, in principle quantitative measurements are possible, but have to be performed very carefully by well trained personnel. In the special case of humic substances, the situation is further complicated because of the broad and overlapping lines and long measuring times even for standard spectra. In the case of difference spectroscopy, additional complicating parameters come into account. Thus, in general one has to mention that the obtained spectra give no absolute quantitative results despite their use in very special cases.

In the present study, some other quantitative factors are of importance to the results. Firstly, the studies are strongly dependent on the amount of organic carbon in the starting material.

We used three types of materials: a) Ah: 1.1 % C_{org}, b) Ah/compost: 2.6 % C_{org} and c) model soil (containing Acros humic acid): 0.43 % C_{org}. The spectra are obtained from the extracted humic acids with an estimated C content of ~50 %. Nevertheless, the result in the difference spectra is more or less the same in all three types of spectra (fig. 8.5), but the quality is increased for the artificial material. This material has been introduced by the following reasons: a) to avoid undesirable components which increase in the linewidth of the signals and therefore decrease the signal/noise-ratio (see fig. 8.2) in the NMR spectrum and b) to achieve a situation, where possibly reduced amounts of enriched material are necessary to gain a ratio of $^{13}C_{\text{activity PAH}}\,(^{13}C_{\text{activity soil}})^{-1} > 1$. Obviously, these ideas were successfully implemented and the introduction of the artificial material has no consequences for the reaction mechanisms under investigation.

The second point is the amount of enrichment in the PAH's. In our study, we used material with > 99 % (phenanthrene and fluoranthene) enrichment. Even in this case, the concentrations of the contaminants must be relatively high, in any case much higher than real contamination.

The more interesting question is to estimate how much of the enriched material has to be used to gain reliable results with Solid State NMR spectroscopy. Results are more or less available only for ^{13}C and ^{15}N NMR (Knicker et al. 1997). In the present investigation we have concluded that a ratio of 1/~4 $^{13}C_{activity\text{-}PAH}$ $(^{13}C_{activity\text{-}soil})^{-1}$ is necessary to obtain spectra in moderate times, but a ratio of 1.5/1 $^{13}C_{activity\text{-}PAH}$ $(^{13}C_{activity\text{-}soil})^{-1}$ is desirable to obtain a quality which is necessary for difference spectra.

8.7 Technical Prerequisites

In our experience a 300 MHz (7 Tesla) (or better) NMR-spectrometer with CPMAS-equipment is necessary. Depending on the availability of probeheads, 7 mm or 4 mm o.d. rotors can be used. We have found, that the 4 mm rotors are sufficient for these studies. Because of the increased homogeneity, the loss in S/N with respect to the 7 mm rotors is negligible. In contrast a significant advantage lies in the reduced amount of substances required (150–200 mg humic acid). This makes the extraction procedure much easier and reduces the enormous costs for the enriched materials. Nevertheless, measuring times of 12 h and more are standard, especially when the difference spectroscopy method is to be applied. As we have found, an increase in rotation frequencies has no positive effect in the case of humic material, in contrast to the results for a pure enriched compound (fig. 8.2). Thus, spinning rates of 5 KHz are sufficient. The operators should have qualified experience with Solid State NMR to be able to achieve reproducible spectra and to recognise technical problems and instabilities that arise during the measurement. With the broad spectra of humic acids this is sometimes more difficult than with standard CPMAS spectra of good to excellent resolution. The reproducibility of the spectrometer settings should be checked from time to time to avoid misinterpretations of the data. Additional experience with humic acid spectra from other sources shows that the spectra depend strongly on the extraction and handling procedure. Thus, for studies of the type described here, it is advantageous when the samples are always prepared by the same person and are also measured by the same operator. Under these circumstances, a high level of reproducibility can be achieved.

8.8 ViewSpec

Normally the ^{13}C-CPMAS-NMR spectra of humic materials yield broad, barely structured spectra which have to be analysed preferably by a visual inspection of the total shape of the spectrum. On the other hand, many scientists who use NMR as an analytical tool are not specialists of this method. It follows that they are not trained to use the numerous and dedicated features of modern NMR processing software. Therefore, in the course of the SFB 188, a simple program has been developed which allows even the inexperienced user the ability to viewi and compare up to three spectra and to easily make comments and remarks during the process of inspection and interpretation. The program is very similar to well-

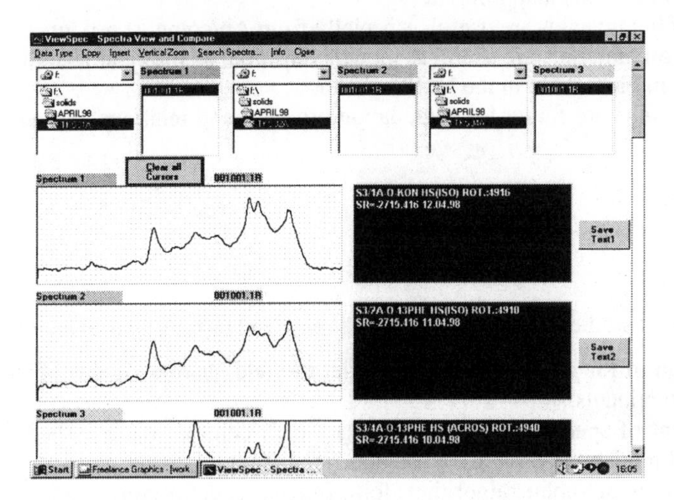

Fig. 8.7. Screenshot of ViewSpec

known simple picture viewers and text-editors and it should cause no problems to a standard PC user (fig. 8.7).

At present, processed spectra can be loaded from different sources of processing software. For these types of studies it is advantageous to always measure the standard spectra with the same spectral conditions, especially with the same chemical shift range. Then, up to three spectra can be loaded and compared visually together with a short text. This text can be either the spectrum title, or additions which have been added during the interpretation of the spectra. Some handling of the spectra such as increase in vertical or horizontal scale is possible, but is performed in a combined manner so that the user always sees the same spectral region for all three spectra. Text can be added as in a simple notepad. After closing the program, all the actual entries are saved and loaded again during the next start of ViewSpec, so that the scientist starts again at exact the situation as he left the program. Some additional features like a standard search function to find the files on the disk or a copy/paste function which transfers the spectra/text data directly into a standard word processing program are available.

At present, the program (Haupt and Haupt 2000) is freeware and can be downloaded at http://haupt1.chemie.uni-hamburg.de/~nmr/viewspec/.

8.9 Conclusions

- The investigation of the fate of ^{13}C-labelled PAH´s and the study of the resulting bound residues with ^{13}C-CPMAS-NMR spectroscopy is successful.
- In spite of labelling, concentrations above realistic conditions are still necessary: $^{13}C_{activity,\ PAH}\ (^{13}C_{activity,\ soil})^{-1} \sim 1.5/1$.

- Real Ah-soil is (still) inappropriate.
- Extracts of humic acids are useful, especially from Ah/compost mixtures.
- Artificial soil materials are useful as models for initial studies and benefits from the quality improvement in the NMR spectra.
- Interesting hints are found for degradation products and relations to theoretical degradation.

8.10 Outlook

From the present studies, the following points may lead to future investigations:

- optimisation of the use of real soil material to yield sharper signals of the enriched contaminants,
- development of special NMR-experiments which enhance the enriched molecules and diminish the background signals (if visible),
- development of automatic algorithms for difference spectroscopy,
- establishing of a centralised laboratory for studies of humic acids with NMR spectroscopy to allow a qualified use of expensive equipment.

References

de Groot HJM, Copie V, Smith SO, Allen PJ, Winkel C, Lugtenburg J, Herzfeld J, Griffin RG (1988) Magic-angle-sample-spinning NMR difference spectroscopy. J Magn Res 77: 251–257

Francke W (2000) chapter 2.4, this book

Fyfe CA (1983) Solid State NMR for chemists. CRC Press, Guelph, Ontario, Canada

Haupt A, Haupt ETK (2000) ViewSpec – easy to use program to view and compare (NMR)-spectra, details at http://haupt1.chemie.uni-hamburg.de/

Käcker T, Haupt ETK, Steinhart H (2000) Structural characterization of humic acid-bound PAH residues in soil by ^{13}C-CPMAS-NMR-spectroscopy: evidence of covalent bonds. Chemosphere: submitted

Käcker T (2000 a) personal communication, extracted from literature data

Käcker T (2000 b) Bildung und Verbleib von PAK-Abbauprodukten in einfach und mischkontaminierten Bodenmaterialien – Einfluss von Schwermetallen auf den PAK-Abbau und ^{13}C-CPMAS-NMR-Studien an PAK/Huminstoff-Komplexen. Dissertation, Universität Hamburg, Germany (www.sub.uni-hamburg.de/disse/202/DISSTK.pdf)

Knicker H, Fründ R, Lüdemann H-D (1997) Characterization of nitrogen in plant composts and native humic material by natural-abundance ^{15}N CPMAS and solution NMR spectra. p 272 in Nanny et al. (1997)

Nanny MA, Minear RA, Leenheer JA (1997) Nuclear Magnetic Resonance Spectroscopy in Environmental Chemistry. Oxford University Press, Oxford, UK

Richnow HH, Seifert R, Hefter J, Kästner M, Mahro B, Michaelis W (1994) Metabolites of xenobiotica and mineral oil constituents linked to macromolecular organic matter in polluted environments. Org Geochem 22: 671–681

Richnow HH, Annweiler E, Koning M, Lüth J-C, Stegmann R, Garms C, Francke W, Michaelis M (2000) Tracing the transformation of labelled [1-^{13}C]phenanthrene in a soil bioreactor. Environ Poll 108: 91–101

Sanders JKM, Marsh JD (1982) NMR double resonance – the use of difference spectroscopy. Progr NMR Spectr 15: 353–400

9 Syntheses of ^{13}C-Labelled Polycyclic Aromatic Compounds

C. Garms, W. Francke
Institute for Organic Chemistry, University of Hamburg, Martin-Luther-King-Platz 6, 20146 Hamburg, Germany

9.1 Introduction

The use of stable isotopes, especially ^{13}C, is well established for investigation of metabolic pathways or biosysntheses, as a tool in diagnostic medicine and to generally trace the fate of organic compounds in the environment. Excellent techniques to monitor the isotope in labelled compounds include nuclear magnetic resonance (^{13}C-NMR-spectroscopy) (Marshall 1983; Breitmaier and Voelter 1987) and (combustion) isotope-ratio-monitoring mass spectrometry (IRM-MS) (Merrit et al. 1994; Wong et al. 1995; Newman 1996; Brenna 1999).

In general labelled model compounds are required to trace biotic and abiotic mechanisms in the degradation of polycyclic aromatic hydrocarbons (PAH), to screen the involvment of PAH in the formation of bound residues in the soil as well as to investigate the chemical interactions between PAH and the humus matrix.

In this paper we report on the design and completion of efficient and economic syntheses of ^{13}C-labelled PAH. The following 13 compounds were synthesised (for chemical structures see fig. 9.1): [1-^{13}C]-naphthalene (**1**), [1-^{13}C]-naphthol (**2**), [α-^{13}C]-1,2-dimethylnaphthalene (**3**), [5-^{13}C]-acenaphthylene (**4**), [5-^{13}C]-acenaphthene (**5**), [9-^{13}C]-anthracene (**6**), [1-^{13}C]-phenanthrene (**7**), [1-^{13}C]-phenanthren-1-ol (**8**), [4-^{13}C]-phenanthrene (**9**), [9-^{13}C]-phenanthrene (**10**), [3-^{13}C]-fluoranthene (**11**), [3a-^{13}C]-pyrene (**12**) and [7-^{13}C]-benzo[a]pyrene (**13**).

9.2 Results and Discussion

In our syntheses we attempted to follow a generally applicable route which involves anellation of a suitable carboxylic acid (or its acid chloride) by intramolecular Friedel-Crafts acylation followed by aromatisation. The ^{13}C-label was introduced into the corresponding carboxylic acid during a late step via reaction of a Grignard compound (prepared from the corresponding

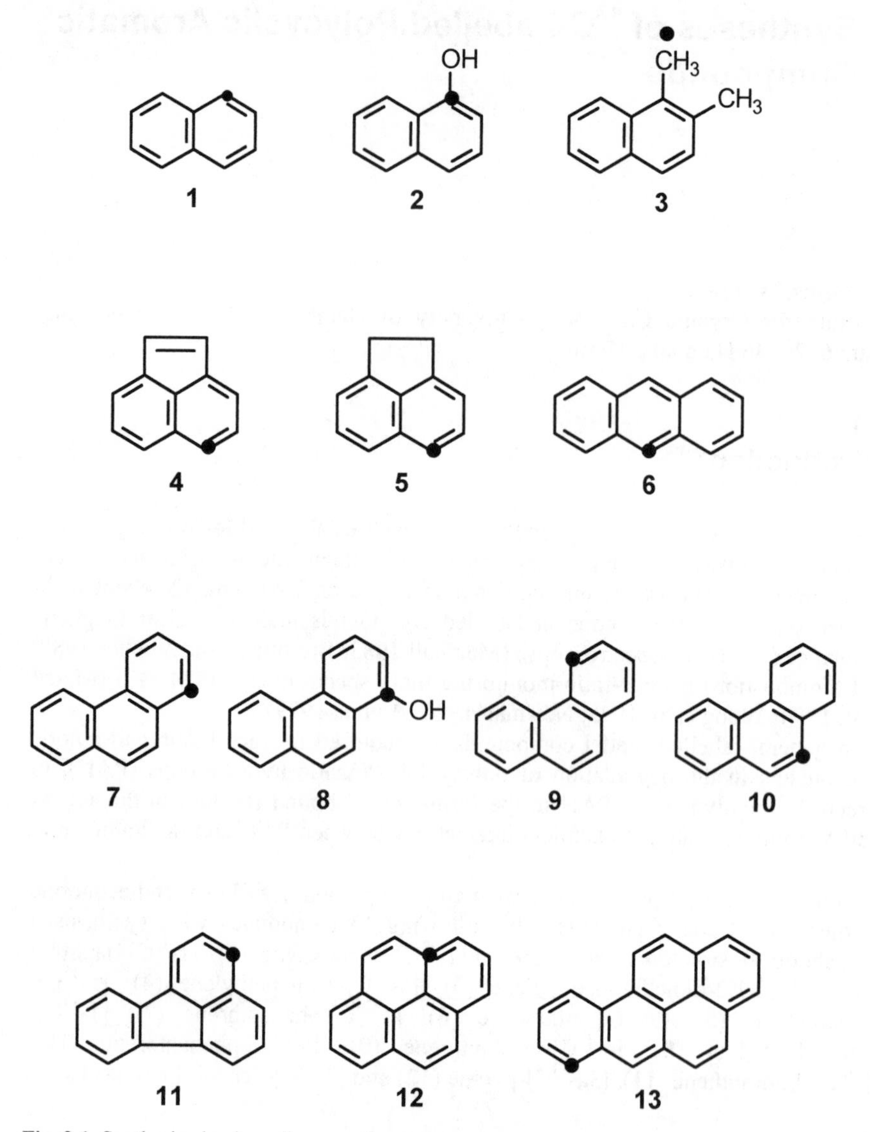

Fig. 9.1. Synthesized polycyclic aromatic compounds

bromoalkylaromates) and ^{13}C-carbon dioxide (prepared from commercially available sodium ^{13}C-carbonate). In this step we preferred ^{13}C-carbon dioxide rather than ^{13}C-cyanide as an alternative C 1-unit, as it is cheaper, easy to handle, and immediately yields the desired acid upon work up. The Friedel-Crafts acylation was carried out either with acid using polyphosphoric acid or with the acid chloride using tin tetrachloride. After removal of the carbonyl group

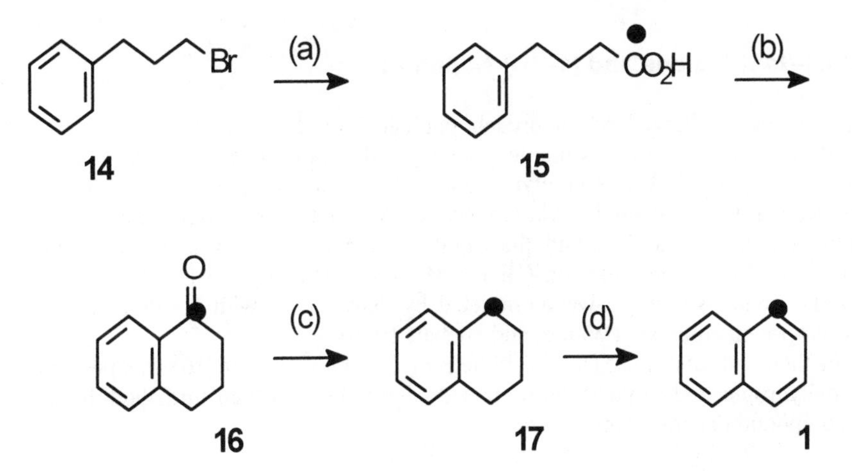

Fig. 9.2. (a) 1. Mg [Et_2O], 2. $^{13}CO_2$, -78 °C; (b) PPA, Δ; (c) NH_2NH_2/KOH [triethylene glycol]; *(d)* DDQ [THF]

generated upon Friedel-Crafts reaction, (usually Wolff-Kishner reaction or a sequence of the reduction with lithium aluminiumtetrahydride followed by deoxygenation of the benzylic alcohol upon catalytic hydrogenation) subsequent aromatisation was achieved with 2,3-dichloro-5,6-dicyano-*p*-benzoquinone (DDQ).

In the following, details of the syntheses are described separately for each target compound.

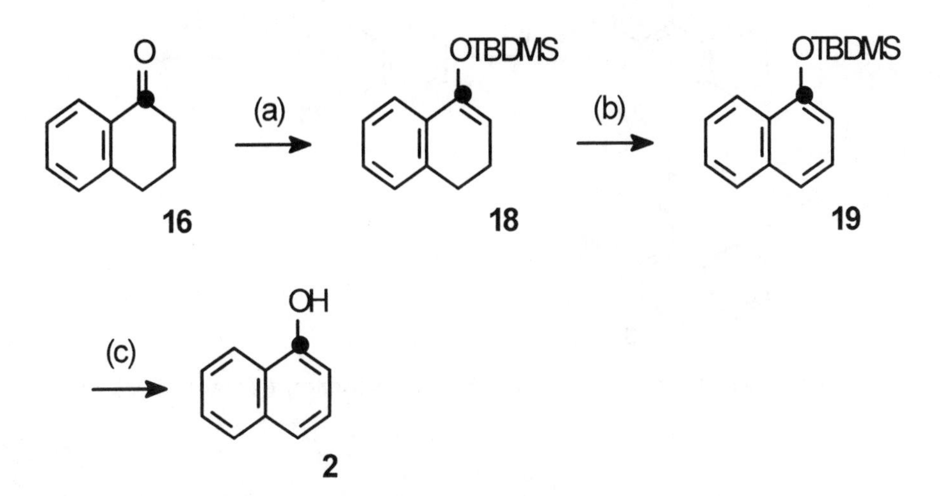

Fig. 9.3. (a) 1. NaH [DMF]; 2. TBDMSCl, (b) DDQ [THF], (c) HF, TBAF [CH_3CN]

9.2.1 [1-^{13}C]-Naphthalene and [1-^{13}C]-Naphthol

Our synthesis followed the approach published by Staab and Haenel (1970) (fig. 9.2). Commercially available 3-phenyl-1-bromopropane (14) was transformed to [1-13C]-4-phenylbutyric acid (15) using Grignard reaction. Intramolecular Friedel-Crafts acylation was carried out with polyphosphoric acid (Uhlig and Snyder 1960), and the obtained tetralone (16) was deoxygenated through Wolff-Kishner reaction (Huang-Minlon 1946) to yield [1-^{13}C]-tetralene (17). The synthesis of (1) was completed by dehydration with 2,3-dichloro-5,6-dicyano-*p*-benzoquinone (Walker and Hiebert 1966).

For the synthesis of (**2**) (fig. 9.3), the protected enolether of (**16**) was prepared and dehydrogenated to yield the protected naphthol (**19**) which furnished the target compound (**2**) after deprotection.

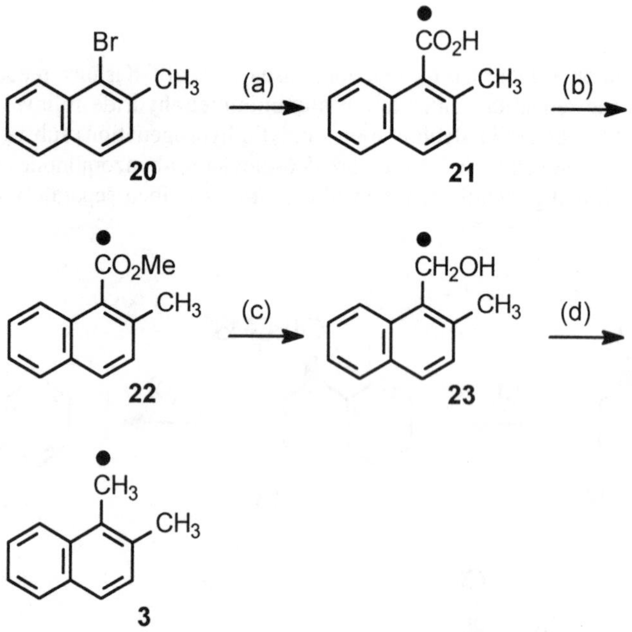

Br CH3 (a) → CO2H CH3 (b) →
20 21
CO2Me CH3 (c) → CH2OH CH3 (d) →
22 23
CH3 CH3
3

Fig. 9.4. (a) 1. Mg [Et_2O]; 2. $^{13}CO_2$, (b) $(CH_3O)_2SO_2$, K_2CO_3 [Aceton], (c) $LiAlH_4$ [Et_2O], (d) H_2, Pd/C [CH_3CO_2H/EtOH]

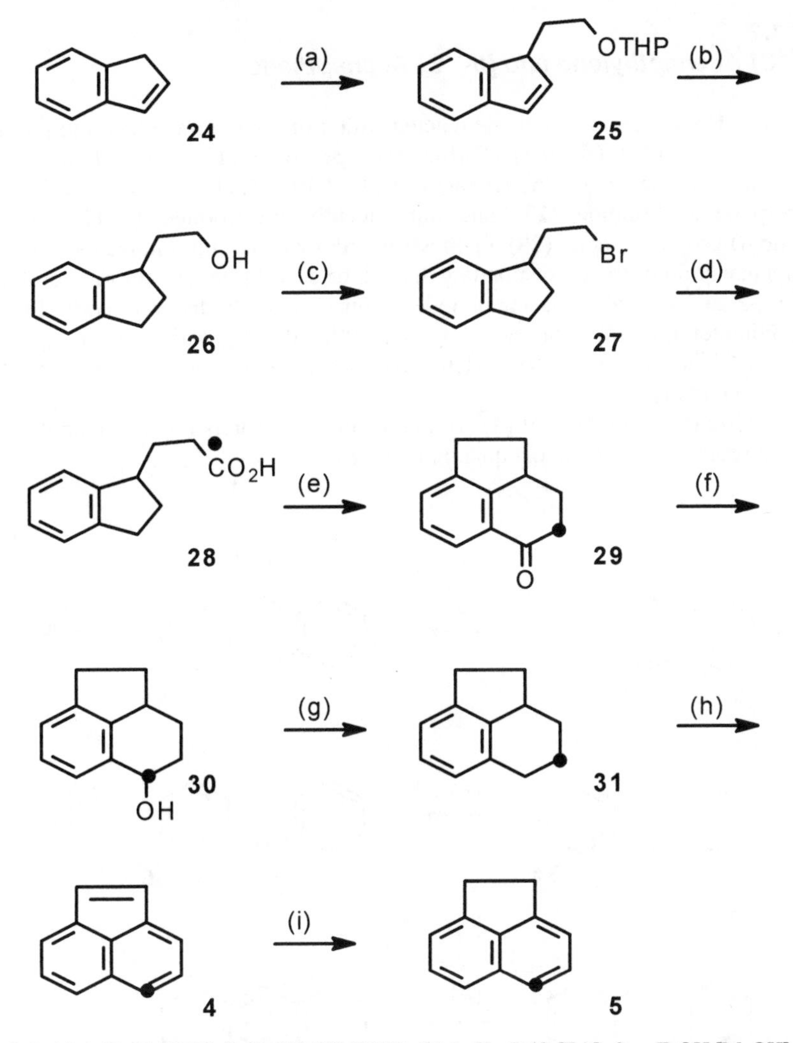

Fig. 9.5. (a) 1. BuLi [THF]; 2. $BrCH_2CH_2OTHP$, (b) 1. H_2, Pt/C [EA]; 2. *p*-TsOH [MeOH], (c) Ph_3PBr_2 [CH_2Cl_2], (d) 1. Mg [Et_2O]; 2. $^{13}CO_2$, (e) 1. $(COCl)_2$ [CH_2Cl_2]; 2. $SnCl_4$ [benzene], (f) $LiAlH_4$ [Et_2O], (g) H_2, Pd/C [CH_3COOH/EtOH], (h) DDQ [THF], (i) H_2, Pd/C [EtOH]

9.2.1.1
[α-^{13}C]-1,2-Dimethylnaphthalene

For the synthesis of (3) (fig. 9.4) commercially available 1-bromo-2-methylnaphthalene (**20**) was converted to ^{13}C-2-methyl-1-naphthoic acid (**21**) via Grignard reaction. Preparation of the methylester (**22**) followed by reduction to the corresponding benzylic alcohol (**23**) with lithium aluminiumtetrahydride and deoxygenation upon catalytic hydrogenation furnished (**3**) in good yield.

9.2.1.2
[5-^{13}C]-Acenapthylene and [5-^{13}C]-Acenapthene

Indene (**24**) was deprotonated and reacted with protected bromoethanol to produce the derivative (**25**) (fig. 9.5). Partial hydrogenation and deprotection yielded 2-(1-indanyl)-ethan-1-ol (**26**) (Bisagni et al. 1970) which was converted to the corresponding bromide (**27**) and subsequently transformed to [1-^{13}C]-3-(1-indanoyl)-propionic acid (**28**) upon Grignard reaction. Intramolecular Friedel-Crafts acylation of the corresponding acid chloride with tin tetrachloride provided [5-^{13}C]-2,2a,3,4-tetrahydroacenaphthylen-5-one (**29**). Reduction with lithium aluminiumtetrahydride followed by catalytic deoxygenation gave [5-^{13}C]-1,2,2a,3,4,5-hexahydroacenaphthylene (**31**) which was aromatised in situ with DDQ to yield (**4**).

Catalytic hydrogenation of (**4**) with palladium on charcoal using ethanol as the solvent produced (**5**) in almost quantitative yield.

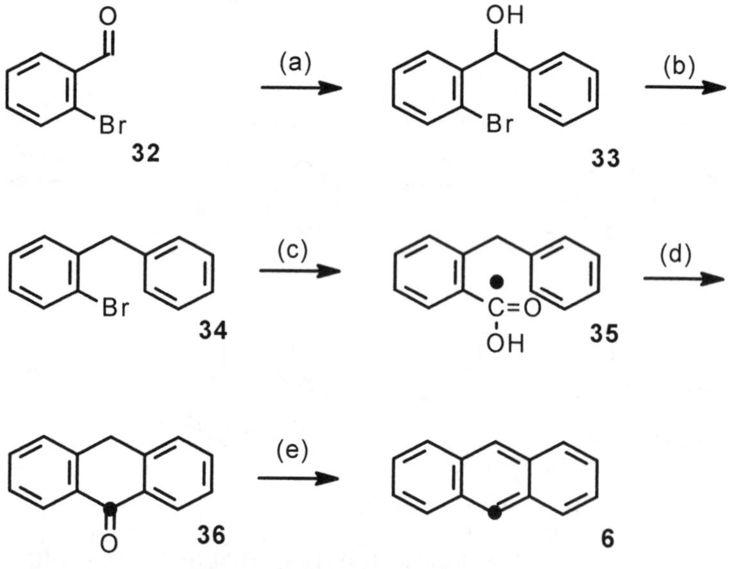

Fig. 9.6. (a) PhLi [Et_2O/cyclohexane], (b) I_2, P_{red} [CH_3COOH] (c) 1. Mg [Et_2O]; 2. $^{13}CO_2$, (d) 1. $(COCl)_2$ [CH_2Cl_2]; 2. $SnCl_4$ [benzene], (e) 1. $NaBH_4$, 2. CH_3CO_2H

9.2.1.3
[9-^{13}C]-Anthracene

Our synthesis followed the procedure described by Isaev et al. (1972) (fig. 9.6). Commercially available 2-bromobenzaldehyde (**32**) was reacted with phenyllithium to yield the diphenylcarbinol (**33**) which was deoxygenated to the bromide (**34**). Grignard reaction to ^{13}C-2-benzylbenzoic acid (**35**) introduced the label.

Intramolecular Friedel-Crafts acylation of the acid chloride with tin tetrachloride followed by deoxygenation of the resulting [9-^{13}C]-anthrone (**36**) completed the synthesis of (**6**).

9.2.1.4
[1-^{13}C]-Phenanthrene and [1-^{13}C]-Phenanthren-1-ol

Our synthesis is a modification of published procedures (Agranat and Shish 1976; Soderquist et al. 1992) (fig. 9.7).

Commercially available 1-methylnaphthalene (**37**) was deprotonated and reacted with protected bromoethanol. After deprotection of the resulting derivative (**38**), the corresponding alcohol (**39**) was transformed to its bromide (**40**) with triphenylphosphonium dibromide. Again, the label was introduced via Grignard reaction and the resulting carboxylic acid (**41**) was cyclised via the

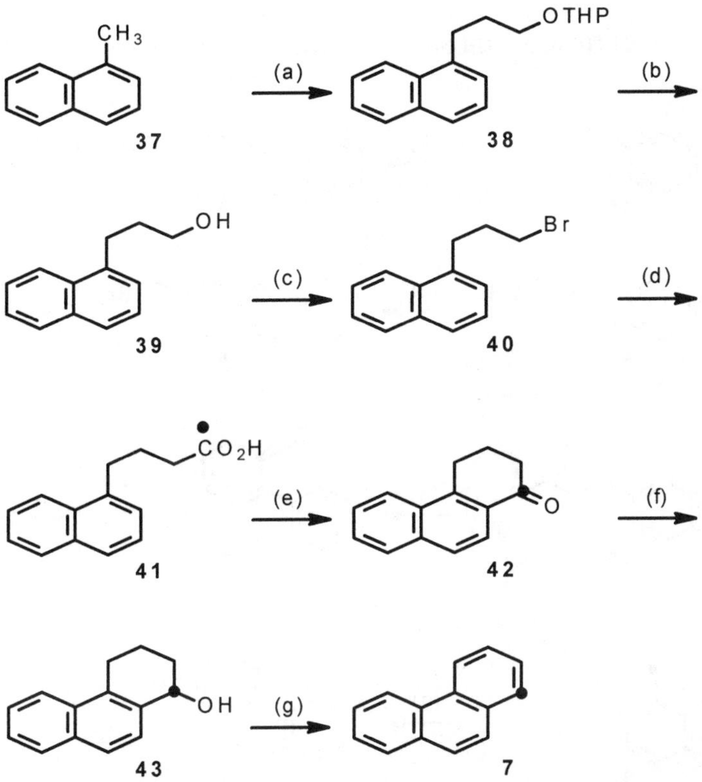

Fig. 9.7. (a) 1. BuLi [TMEDA/Et_2O]; 2. $BrCH_2CH_2OTHP$, (b) *p*-TsOH [MeOH], (c) Ph_3PBr_2 [CH_2Cl_2], (d) 1. Mg [Et_2O]; 2. $^{13}CO_2$, (e) 1. $(COCl)_2$ [CH_2Cl_2]; 2. $SnCl_4$ [C_6H_6], (f) $LiAlH_4$ [Et_2O], (g) 1. H_2, Pd/C [EtOH, CH_3COOH]; 2. DDQ [THF]

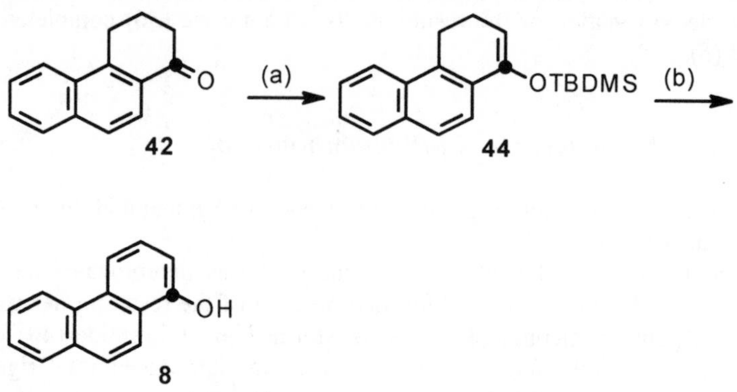

Fig. 9.8. (a) 1. NaH [DMF]; 2. TBDMSCl, (b) 1. DDQ; 2. HF/TBAF [THF]

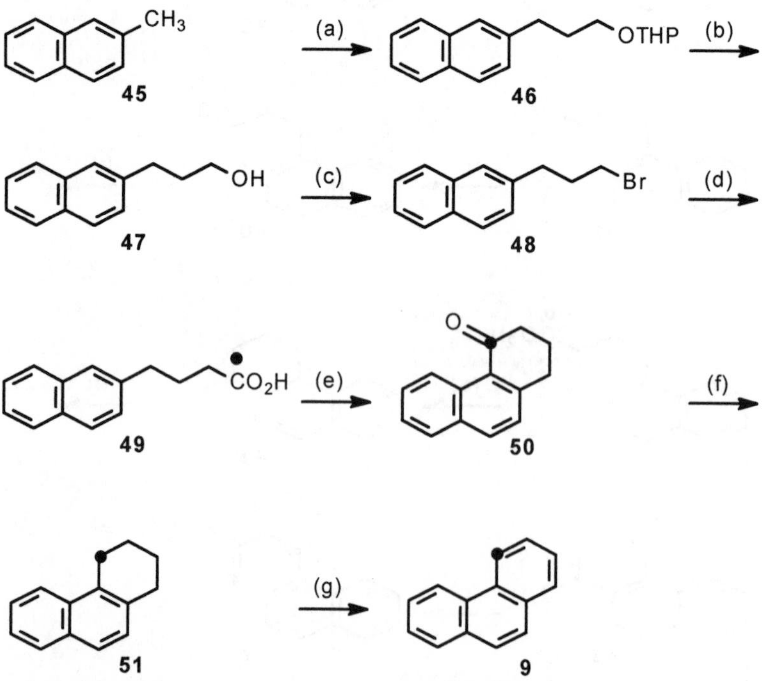

Fig. 9.9. (a) 1. BuLi [TMEDA/Et_2O]; 2. $BrCH_2CH_2OTHP$, (b) *p*-TsOH [MeOH], (c) Ph_3PBr_2 [CH_2Cl_2], (d) 1. Mg [Et_2O]; 2. $^{13}CO_2$, (e) 1. $(COCl)_2$ [CH_2Cl_2]; 2. $SnCl_4$ [benzene], (f) 1. $LiAlH_4$ [Et_2O]; 2. H_2 (Pd/C) [CH_3COOH/EtOH], (g) DDQ [THF]

acid chloride to yield the ketone (**42**). Transformation to the target compound (**7**) via the benzylic alcohol (**43**) followed essentially the deoxygenation/aromatisation steps as described for acenaphthylene (**4**).

The transformation of the ketone (**42**) to [1-^{13}C]-phenanthren-1-ol (**8**) via the protected enolether (**44**) followed essentially the same sequence as described for [1-^{13}C]-naphthol (**2**) (fig. 9.8).

9.2.1.5
[4-^{13}C]-Phenanthrene

Starting from 2-methylnaphthalene (**45**) the synthesis of (**9**) followed the same procedure as for (**7**) (fig. 9.9). Deprotonation of (**45**) and elongation of the side chain yielded the bromide (**48**) via (**46**) and 3-(2-naphthyl)-propan-1-ol (**47**). Grignard reaction of (**48**) introduced the label to yield [1-^{13}C]-4-(2-naphthyl)-butyric acid (**49**). Transformation of (**49**) to (**8**) via (**50**) and (**51**) followed the procedure outlined for the synthesis of (**7**).

9.2.1.6
[9-^{13}C]-Phenanthrene

For the synthesis of (**10**) we employed the rearrangement of 9-fluorenyl carbinol to phenanthrene (Brown and Bluestein 1949; Goehler and Schütte 1970) (fig. 9.10).

Fluorene (**52**) was deprotonated and directly carboxylated to yield ^{13}C-9H-9-fluorene carboxylic acid (**53**). Esterification followed by reduction furnished the labelled carbinol (**54**) which could be almost quantitatively rearranged to (**9**) upon the action of phosphorous oxide in xylene at 150 °C.

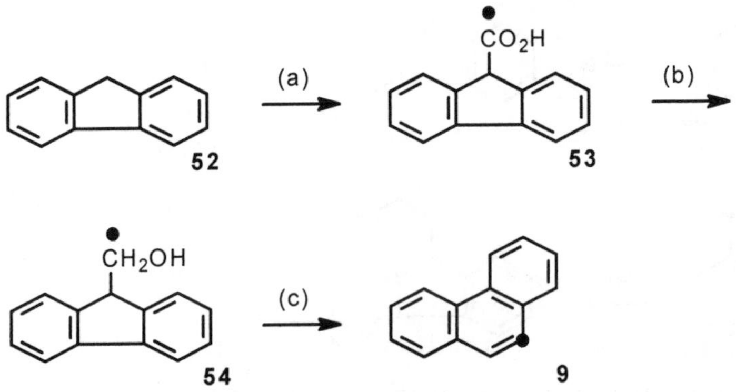

Fig. 9.10. (a) 1. BuLi [THF]; 2. $^{13}CO_2$, (b) 1. $(CH_3O)_2SO_2$, K_2CO_3 [acetone]; 2. $LiAlH_4$ [Et_2O], (c) P_2O_5 [xylene]

9.2.1.7
[3-^{13}C]-Fluoranthene

Our synthesis of (**11**) used the procedure described by McDowell et al. (1962) (fig. 9.11). Fluorene (**52**) was deprotonated and alkylated with protected bromoethanol to yield (**55**). After deprotection 2-(9-fluorenyl)-ethanol (**56**) was obtained which was converted to the corresponding bromide (**57**). Carboxylation via Grignard reaction introduced the label. Subsequent steps from [1-^{13}C]-3-(9-fluorenyl)-propionic acid (**58**) to the target compound (**11**) via the ketone (**59**) and [3-^{13}C]-1,2,3,10b-tetrahydrofluoranthene (**60**) followed essentially our standard procedure outlined above.

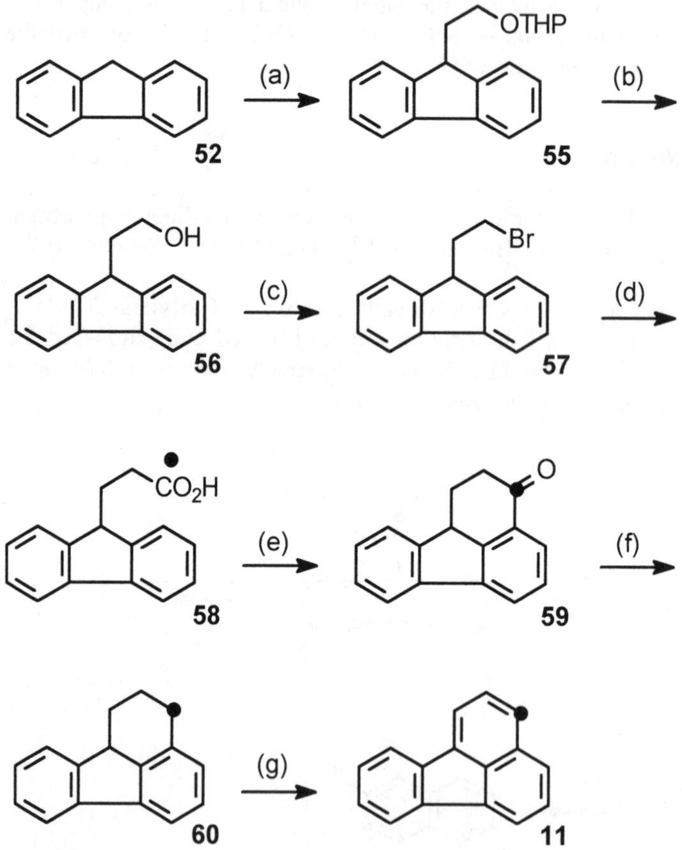

Fig. 9.11. (a) 1. BuLi [THF]; 2. $BrCH_2CH_2OTHP$, (b) *p*-TsOH [MeOH], (c) Ph_3PBr_2, (d) 1. Mg [Et_2O]; 2. $^{13}CO_2$, (e) $[H_3PO_4]_x$, (f) NH_2NH_2 / KOH [triethylene glycole], (g) DDQ [THF]

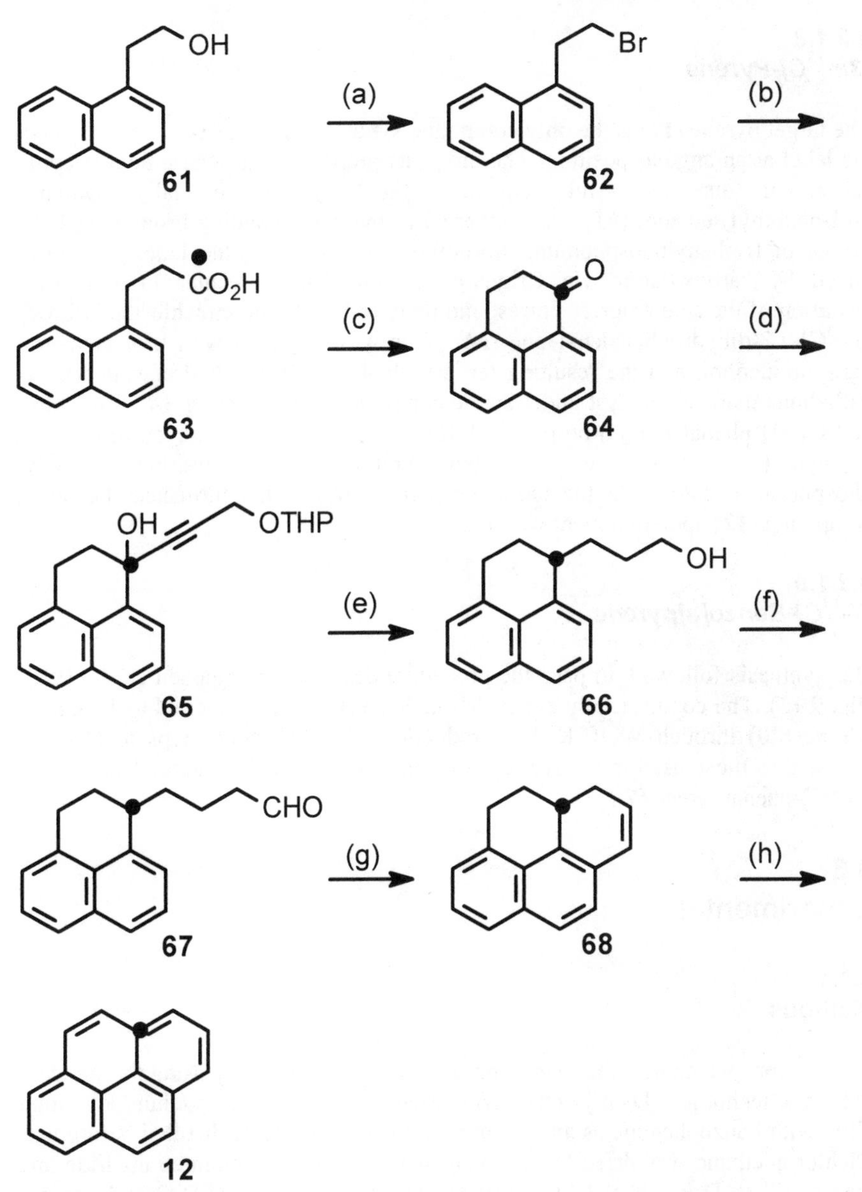

Fig. 9.12. (a) Ph_3PBr_2, (b) 1. Mg [Et_2O]; 2. $^{13}CO_2$, (c) 1. $(COCl)_2$ [CH_2Cl_2]; 2. $SnCl_4$, (d) $LiCCHCH_2OTHP$ [THF/TMEDA], (e) H_2, Pd/C [MeOH/CH_3COOH], (f) PDC, (g) 1. PPA, (h) DDQ [THF]

9.2.1.8
[3a-^{13}C]-Pyrene

The target pyrene (**12**) is the only compound synthesised in this study which keeps the label at an angular position. Therefore, its preparation needed a double application of our successful sequence (fig. 9.12). Commercially available 2-(1-naphthyl)-ethanol (**61**) was converted to the corresponding bromide (**62**) by means of triphenylphosphonium dibromide. Subsequently, the label was introduced by carboxylation via Grignard reaction. Intramolecular Friedel-Crafts acylation of the acid chloride corresponding to (**63**) with tin tetrachloride afforded [1-^{13}C]-2,3-dihydrophenalen-1-one (**64**). This was alkylated with protected propargylic alcohol, and the resulting tertiary alcohol (**65**) was hydrogenated using palladium/charcoal catalyst under acidic conditions. The resulting 3-(2,3-dihydro-[1-^{13}C]-1H-phenalen-1-yl)-propan-1-ol (**66**) was oxidised to the corresponding aldehyde (**67**). The latter was subjected to intramolecular cyclisation with polyphosphoric acid to yield the tetrahydropyrene (**68**) which furnished the target compound (**12**) upon treatment with DDQ.

9.2.1.9
[7-^{13}C]-Benzo[a]pyrene

Our synthesis followed, in part, the procedure described by Klassen et al. (1983) (fig. 9.13). The commercially available aldehyde (**69**) was converted to 1-methylpyrene (**70**) through Wolff-Kishner reduction. The following steps to (**13**) are identical to those used in the sequence leading from 1-methylnaphthalene (**37**) to [1-^{13}C]-phenanthrene (**7**).

9.3
Experimental

9.3.1
Methods

All reactions which demand inert conditions were performed by using the standard Schlenck technique. Diethyl ether was dried by stirring with sodium/potassium alloy with benzophenone as an indicator at 25 °C and freshly distilled before use. Dichloromethane was dried by refluxing with lithiumaluminiumtetrahydride and then distilled. Dry N,N,N',N'-tetramethylethylenediamine (TMEDA) and tetrahydrofuran (THF) were purchased from Fluka. Flash column chromatography was performed on silica gel (230–400 mesh, Merck).

NMR spectra were run on an AMX-400 (Bruker) using tetramethylsilane as the internal standard.

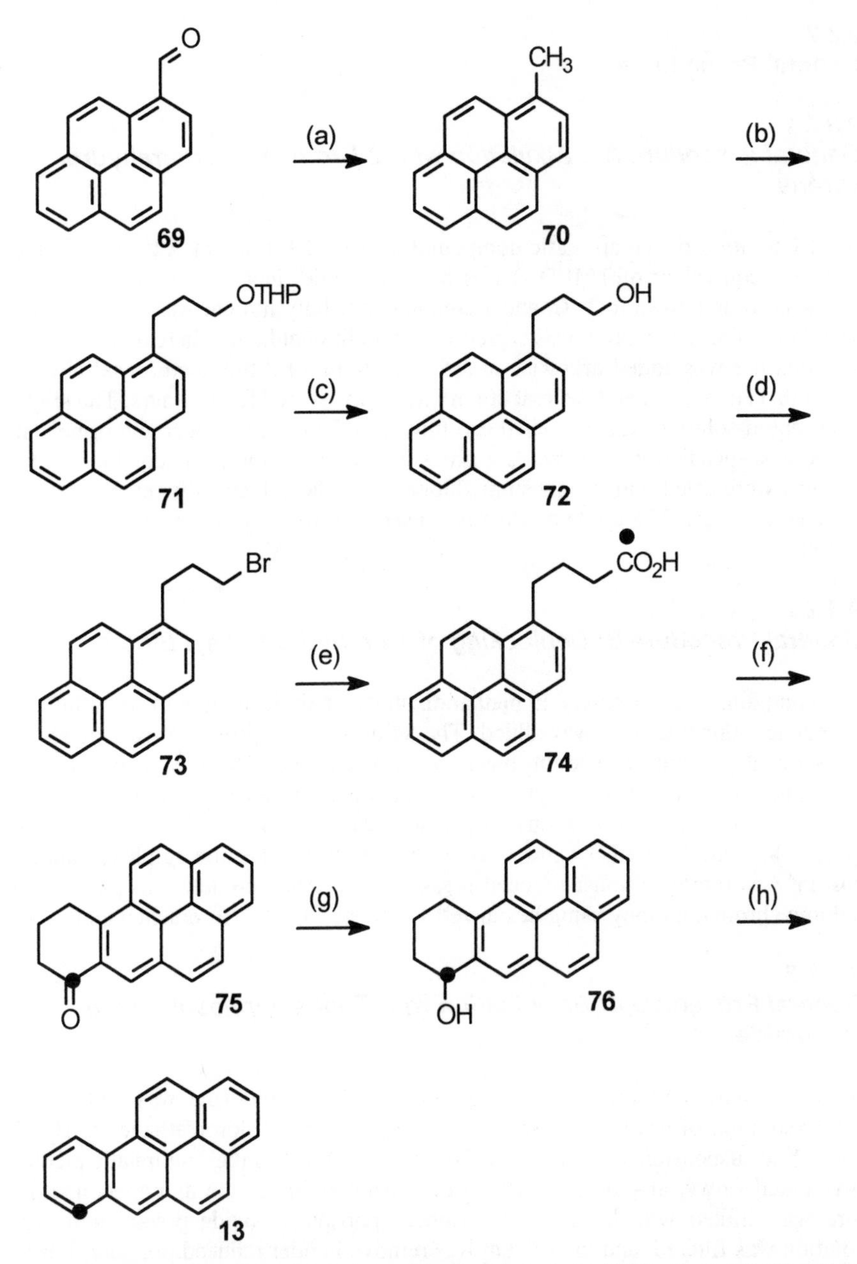

Fig. 9.13. (a) NH_2NH_2, KOH [triethylene glycol], (b) 1. BuLi [THF]; 2. $BrCH_2CH_2OTHP$, (c) p-TsOH [MeOH], (d) 2. Ph_3PBr_2, (e) 1. Mg [THF]; 2. $^{13}CO_2$, (f) 1. $(COCl)_2$ [CH_2Cl_2]; 2. $SnCl_4$ [Benzol], (g) $LiAlH_4$ [Et_2O/THF], (h) 1. Δ; 2. DDQ [THF]

9.3.2 General Procedures

9.3.2.1 General Procedure A1: Alkylation with 2-Bromoethoxytetrahydro-pyrane

A 2 M solution of the aromatic compound in dry THF (variant **A2**: diethyl ether with one equivalent of TMEDA) was prepared under inert conditions. The solution was cooled down to 5 °C, and 1 equivalent of butyllithium was added slowly. The dark coloured solution was stirred for an additional hour. Thereafter, the reaction mixture was added dropwise to a 5 M solution of 2-bromoethoxytetrahydropyran in THF at 5 °C and the reaction mixture was stirred for 10 hours. The slightly coloured solution was quenched carefully by adding some water and brine. The aqueous suspension was extracted 4 times with diethyl ether. The combined ether extracts were dried with magnesium sulphate and the solvent was removed under reduced pressure. The crude product was used in further syntheses without purification.

9.3.2.2 General Procedure B: Deblocking of Tetrahydropyranyl Ethers

The compound was dissolved in methanol, and a catalytic amount of recrystallised 4-toluene sulphonic acid was added. The solution was stirred for 10 h at 25 °C. Subsequently 1 gram of sodium bicarbonate was added. The solvent was evaporated under reduced pressure, and the solution was concentrated to one third. After diluting with brine, the aqueous suspension was extracted 4 times with diethyl ether. The combined ether extracts were dried with magnesium sulphate, and the solvent was removed under reduced pressure. The crude product was purified by column chromatography using hexane/ethyl acetate (5:1 v:v) as eluenteluent.

9.3.2.3 General Procedure C: Bromination with Triphenylphosphonium Dibromide

A 10 M solution of bromine (1.1 eq.) in dry dichloromethane was added to a 1.5 M solution of triphenylphosphane (1.1 eq.) in dry dichloromethane at -10 °C. The white suspension was stirred for 30 minutes, and then the ar-aliphatic alcohol was added slowly at -10 °C. After 8 h of stirring at 25 °C, the homogenous mixture was diluted with hexane until triphenylphosphane oxide precipitated. The solution was filtered, and the solvent was removed under reduced pressure. Purification of the crude product was done by column chromatography using hexane as eluent.

9.3.2.4
General Procedure D: Grignard Reaction with Carbon-13 Dioxide

A 4 M solution of the ar-aliphatic bromide (2 eq.) in dry diethyl ether was added dropwise to activated magnesium (2.2 eq.) under a nitrogen atmosphere. After addition, the mixture was refluxed for 30 minutes and cooled down to -78 °C. The Grignard reagent was treated with carbon-13 dioxide which was prepared *in situ* by reaction of conc. sulphuric acid and ^{13}C-sodium carbonate (1 eq.). After 2 h of stirring at -78 °C, the reaction mixture was slowly warmed up to 25 °C. A rubber balloon was used to compensate for pressure. The reaction mixture was stirred 10 h at a constant 25 °C. Excess Grignard reagent was destroyed by adding water. Concentrated hydrochloric acid was added to dissolve remnants of magnesium. Subsequently, the aqueous solution was extracted 4 times with diethyl ether, and the organic extract was washed with 1 N aqueous sodium hydroxide solution. The alkaline solution was extracted three times with diethyl ether, and the aqueous solution was acidified using hydrochloric acid. The pure ^{13}C-labelled carboxylic acid was obtained after extraction with diethyl ether and removal of the solvent under reduced pressure.

9.3.2.5
General Procedure E1: Intramolecular Friedel-Crafts-Acylization with Polyphosphoric Acid (PPA)

A suspension of the labelled compound in polyphosphoric acid was heated for 1 h to 180 °C. The reaction mixture was dissolved in water, and the aqueous suspension was washed four times with diethyl ether. The combined ether extracts were dried with magnesium sulphate, and the solvent was removed under reduced pressure. The crude product was purified by column chromatography using hexane/ethyl acetate (10:1 v:v) as eluent.

9.3.2.6
General Procedure E2: Intramolecular Friedel-Crafts-Acylization with $SnCl_4$

The labelled ar-aliphatic carboxylic acid was converted into the corresponding acid chloride by reaction with 5 eq. of oxalyl chloride in dry dichlormethane. The labelled acid chloride was dissolved in benzene after removal of the solvent and the excess of oxalylchloride under reduced pressure. The solution was cooled down to 5 °C, and 1 eq. of tin tetrachloride was added in one portion. After stirring the mixture for 20–30 minutes, the mixture was poured into cold 6 M hydrochloric acid. The aqueous solution was extracted four times with diethyl ether, and the combined organic extracts were dried with magnesium sulphate. The solvent was removed under reduced pressure, and the crude product was purified by column chromatography using hexane/ethyl acetate (10:1 v:v) as eluent.

9.3.2.7
General Procedure F: Wolff-Kischner Reduction

The carbonyl compound was reduced using the standard Huang-Minlon variant of the reduction with 2 eq. of hydrazine hydrate and 5 eq. of potassium hydroxide in triethylene glycol. After refluxing of the reaction mixture for 1 hour, the mixture was cooled down to 25 °C and poured into water. The aqueous phase was saturated with sodium chloride and extracted four times with diethyl ether. The combined extracts were dried over magnesium sulphate, and the solvent was removed under reduced pressure. The crude product was purified by column chromatography using hexane as eluent.

9.3.2.8
General Procedure G: Reduction with $LiAlH_4$

A solution of the carbonyl compound in dry diethyl ether was prepared. The solution was cooled down to 5 °C and slowly 1.5 eq of $LiAlH_4$ was added. The grey suspension was stirred for approx. 10 h. Quenching of the reaction mixture was done by adding water at 5 °C under vigorous stirring. The colourless aqueous suspension was acidified with phosphoric acid until the white precipitate was completely dissolved. After extraction of the aqueous phase with diethyl ether, the combined organic phases were dried with magnesium sulphate. The solvent was removed under reduced pressure, and the crude product was used in further syntheses without purification.

9.3.2.9
General Procedure H: Hydrogenolysis of Benzylic Alcohols

A catalytic amount of palladium on charcoal (10 %) was added to a solution of the benzylic alcohol in ethanol/acetic acid (10:1 v:v), and the suspension was hydrogenated. The solution was stirred 12 h to 24 h at 25 °C. After filtration and removal of the solvent under reduced pressure, the crude product was used in further syntheses without purification.

9.3.2.10
General Procedure I: Aromatisation with DDQ

Under a nitrogen atmosphere, a solution of the compound in dry THF was prepared and 2,3-Dichloro-5,6-dicyano-*p*-benzoquinone (1.2 eq. per double bond) was added in one portion. The colour of the solution changed instantly to black. The solution was refluxed for 1.5 h and then poured into a 4 N aqueous solution of sodium hydroxide. The mixture was stirred for 15 minutes. Subsequently, the aqueous suspension was saturated with sodium chloride and extracted 4 times with diethyl ether. The combined organic extracts were dried with magnesium sulphate, and the solvent was removed under reduced pressure. Finally, the crude product was purified by column chromatography using hexane as eluent.

9.3.2.11 General Procedure J: Conversion of Ketones to Enol-tert.-butyl-dimethylsiloxanes

A solution of tert-butyldimethylsilylchloride (TBDMSCl, 1.1 eq.) in DMF was added dropwise at 5 °C to a suspension of the ketone and sodium hydride (1.1 eq.) in dry N,N-dimethylformamide (DMF) under a nitrogen atmosphere. The reaction mixture was stirred for 2 hours, and then the mixture was poured into water. Subsequently, the aqueous suspension was saturated with sodium chloride. After extraction with diethyl ether, the combined organic phases were dried with magnesium sulphate and then the solvent was removed under reduced pressure. Purification of the crude product was performed by column chromatography using hexane as eluent.

9.3.2.12 General Procedure K: Deblocking of tert.-Butyldimethylsilylether

A solution of the compound, TBAF (1.0 eq.) and aqueous HF (1.0 eq.) in THF was stirred for 8 hours at 25 °C. The mixture was diluted with brine, and the aqueous suspension was extracted four times with diethyl ether. The combined organic phases were dried over magnesium sulphate, and then the solvent was removed under reduced pressure. Finally, the crude product was purified by column chromatography using hexane/ethyl acetate (5:1) as eluent.

9.3.3 Syntheses

9.3.3.1 [1-^{13}C]-4-Phenyl Butyric Acid (15)

According to general procedure D, 23.9 g (0.12 mol) of 3-phenyl-1-bromopropane (**14**) was converted to the corresponding Grignard reagent and treated with ^{13}C-carbon dioxide, generated from 6.42 g ^{13}C-sodium carbonate (60 mmol) to give 4.13 g (42 % yield, calculation based on ^{13}C-sodium carbonate) of the colourless, labelled carboxylic acid.

^{1}H-NMR: δ [ppm] = 1.93–2.03 (m, 2H, H-3), 2.35 (dt, 2H, H-2), 2.67 (t, 2H, H-4), 7.14–7.31 (m, 5H, Ph-H); ^{3}J(H-2, H-3) = 7.4 Hz, ^{3}J(H-3, H-4) = 7.6 Hz, ^{2}J(H-2, C-1) = 7.4 Hz

^{13}C-NMR: δ [ppm] = 26.11, 33.04 (d), 34.95, 126.06, 128.43, 128.48, 141.18, **180.11;** ^{1}J(C-1.C-2) = 55.2 Hz

9.3.3.2 [1-^{13}C]-3,4-Dihydro-2H-naphthalen-1-one (16)

4.12 g of [1-^{13}C]-4-phenyl butyric acid (**15**) (25 mmol) was treated with PPA according to general procedure E1 to give 2.16 g (65 % yield) of an oily residue.

^{1}H-NMR: δ [ppm] = 2.06–2.14 (m, 2H, H-3), 2.61 (dt, 2H, H-2), 2.92 (t, 2H, H-4), 7.21 (d, 1H, H-5), 7.27 (dd, 1H, H-7), 7.43 (ddd, 1H, H-6), 8.01 (ddd, 1H,

1H, H-8); ^3J(H-2, H-3) = 6.6 Hz, ^2J(H-2, C-1) = 6.1 Hz, ^3J(H-3, H-4) = 6.6 Hz, ^3J(H-5, H-6) = 7.6 Hz, ^3J(H-6, H-7) = 7.6 Hz, ^4J(H-6, H-8) = 1.5 Hz, ^3J(H-7, H-8) = 7.1 Hz, ^3J(H-8, C-1) = 4.1 Hz

^{13}C-NMR: δ [ppm] = 23.29, 29.36 (d), 38.84 (d), 126.67 (d), 126.76, 132.26, 133.06, 136.25, 144.19 (d), **199.51**; ^1J(C-1, C-2) = 40.7 Hz, ^3J(C-1, C-4) = 3.1 Hz, ^2J(C-1, C-4a) = 3.1 Hz, ^4J(C-1, C-6) = 2.0 Hz, ^2J(C-1, C-8) = 4.1 Hz, ^1J(C-1, C-8a) = 50.9 Hz

9.3.3.3
[1-^{13}C]-1,2,3,4-Tetrahydronaphthalene (17)

A mixture of 2.16 g [1-^{13}C]-3,4-dihydro-2H-naphthalen-1-one (**16**) (16 mmol), potassium hydroxide (5.61 g; 100 mmol) and hydrazine hydrate (2.40 g; 75 mmol) was treated according to general procedure F to give 1.07 g (50 % yield) of the labelled tetraline.

^{1}H-NMR: δ [ppm] = 2.20 (t, 4H, H-2/H-3), 2.76, (dt, 4H, H-1/H-4), 7.00–7.10 (m, 4H, H-5/H-6/H-7/H-8); ^3J(H-1, H-2) = 7.6 Hz, ^1J(H-1, C-1) = 125.7 Hz;

^{13}C-NMR: δ [ppm] = 23.10 (d), **29.28**, 125.29, 128.68 (d), 137.02 (d); ^1J(C-1, C-2) = 40.3 Hz; ^1J(C-1, C-8a) = 50.6 Hz; ^3J(C-1, C-7) = 3.0 Hz

9.3.3.4
[1-^{13}C]-Naphthalene (1)

A solution of 1.02 g of [1-^{13}C]-1,2,3,4-tetrahydronaphthalene (**17**) (8 mmol) and 13.62 g DDQ (60 mmol) in dry THF was refluxed according to general procedure I to give 509 mg of pure colourless [1-^{13}C]-naphthalene.

^{1}H-NMR: δ [ppm] = 7.45–7.50 (m, 4H, H-2/H-3/H-5/H-6), 7.82–7.86 (m, 4H, H-1/H-4/H-5 /H-8); ^1J(H-1, C-1) = 157.6 Hz

^{13}C-NMR: δ [ppm] = 126.08 (d), **128.01**, 133.45 (d), ^1J(C-1, C-2) = 68.0 Hz, ^1J(C-1, C-8a) = 55.6 Hz, ^2J(C-1, C-3) = 3.6 Hz

9.3.3.5
[1-^{13}C]-1-(tert.-Butyl-dimethylsiloxy)-3,4-dihydronaphthalene (18)

A solution of 2.20 g [1-^{13}C]-3,4-dihydro-2H-naphthalen-1-one (**16**) (14.9 mmol) in dry 20 ml DMF was treated with sodium hydride (870 mg; 30 mmol) and 5.00 g (33 mmol) TBDMSCl according to general procedure J to give 3.90 g (yield: 99 %) of the labelled silylenolether.

^{13}C-NMR: δ [ppm] = 4.68, 18.28, 22.16, 25.86, 28.16 (d), 104.77 (d), 121.80, 126.09 (d), 126.87 (d), 127.18, **148.20**; ^1J(C-1, C-2) = 80.4 Hz, ^2J(C-1, C-3) = 5.1 Hz, ^2J(C-1, C-4a) = 3.1 Hz, ^1J(C-1, C-8a) = 61 Hz, ^3J(C-1, C-4) = 4.1 Hz, ^3J(C-1, C-6) = 3.9 Hz

9.3.3.6
[1-^{13}C]-1-tert.-Butyldimethylsilyloxynaphthalene (19)

Under a nitrogen atmosphere [1-^{13}C]-1-(tert.-butyl-dimethylsiloxy)-3,4-dihydronaphthalene (**18**) (3.90 g; 14.9 mmol) was dissolved in dry THF and 5.10 g DDQ

(22.5 mmol) added in one portion. The black reaction mixture was treated as described in general procedure I to give 3.56 g (yield: 92 %) of [1-^{13}C]-1-tert.-butyldimethylsilyloxynaphthalene.

Calculated for $C_{16}H_{22}OSi$: C: 74.46 %, H: 8.55 %; found: C: 73.08 %, H: 8.43 %

^{1}H-NMR: δ [ppm] = 0.31 (s, 6H, Si-C**H**), 1.12 (s, 9H, C-C**H**), 6.86 (dd, 1H; H-2), 7.31 (dd, 1H, H-3), 7.42–7.48 (m, 3H, H-4/H-6/H-7), 7.77–7.81 (m, 1H, H-5), 8.16–8.21 (m, 1H, H-8); ^{3}J(H-2, H-3) = 7.6 Hz, ^{2}J(C-1, H-2) = 2.5 Hz, ^{3}J(H-3, H-4) = 8.1 Hz

^{13}C-NMR: δ [ppm] = -4.25, 18.45, 25.90, 112.55 (d), 120.87 (d), 122.58, 125.06, 125.86 (d), 126.08, 127.57 (d), 127.91 (d), 134.98 (d), **152.02**; ^{1}J(C-1,C-2) = 70.2 Hz, ^{1}J(C-1,C-8a) = 65.1 Hz, ^{2}J(C-1,C-4a) = 2.0 Hz, ^{3}J(C-1,C-5) = 4.1 Hz, ^{3}J(C-1, C-4) = 7.1 Hz

9.3.3.7
^{13}C-2-Methyl-1-naphthalenecarboxylic acid (21)

According to general procedure D, a Grignard solution, prepared from 25.0 g 1-bromo-2-methylnaphthalene (**20**) (114 mmol, purity: 90 %) and 2.0 g magnesium (82 mmol) in dry THF (75 ml), was treated with carbon-13 dioxide generated from reaction of 5.35 g ^{13}C-sodium carbonate (50 mmol) and conc. sulphuric acid. After extraction, 7.61 g (yield: 81 %; calculation based on sodium ^{13}C-carbonate) of the labelled carboxylic acid was obtained.

^{1}H-NMR (DMSO-d_6): δ [ppm] = 2.48 (s, 3H, -C**H**), 7.41 (d, 1H, H-3), 7.50 (dd, 1H, H-6), 7.56 (dd, 1H, H-7), 7.82 (d, 1H, H-5), 7.90 (d, 1H, H-8), 7.93 (d, 1H, H-4), 13.55 (s, 1H, -COO**H**); ^{3}J(H-3, H-4) = 8.1 Hz, ^{3}J(H-5, H-6) = 8.1 Hz, ^{3}J(H-6, H-7) = 7.1 Hz, ^{3}J(H-7, H-8) = 8.1 Hz

^{13}C-NMR: δ [ppm] = 19.94, 124.78, 125.83, 127.29, 128.36, 128.74, 128.79, 129.05, 129.33, 131.51, 131.90, **170.85**

9.3.3.8
^{13}C-(2-Methyl-naphthalen-1-yl)-methanol (23)

A suspension of 6.08 g potassium carbonate and 7.61 g ^{13}C-2-methyl-1-naphthalenecarboxylic acid (**21**) (44 mmol) in 100 ml acetone was heated under reflux for 10 min. and then treated with 4.2 ml dimethylsulphate (44 mmol). After refluxing for 1 hour the mixture was poured into 100 ml of brine and extracted with diethyl ether. The combined organic extracts were dried over magnesium sulphate, and the solvent was removed under reduced pressure. The crude product was reduced with lithium aluminiumtetrahydride as described in general procedure G to give 5.00 g of the benzylic alcohol (yield: 73 %).

^{1}H-NMR: δ [ppm] = 1.58 (s, 1H, OH), 2.60 (s, 3H, -C**H**), 5.15 (d, 2H, C**H**OH), 7.31 (d, 1H, H-3), 7.44 (m, 1H, H-6), 7.53 (m, 1H, H-7), 7.72 (d, 1H, H-4), 7.81 (d, 1H, H-5), 8.18 (d, 1H, H-8); ^{1}J(C**H**OH, **C**HOH) = 142.4 Hz, ^{3}J(H-3, H-4) = 8.7 Hz, ^{3}J(H-5, H-6) = 8.1 Hz, ^{3}J(H-7, H-8) = 8.7 Hz

^{13}C-NMR: δ [ppm] = 20.16 (d), **58.87**, 123.89 (d), 125.37, 127.05, 128.95, 128.97, 129.64 (d), 132.76 (d), 133.03, 133.13 (d), 135.08 (d); ^{3}J(**C**H, **C**HOH) =

3.1 Hz, ^{3}J(**C**HOH, C-2) = 2.0 Hz, ^{1}J(C-1, **C**HOH) = 48.8 Hz, ^{3}J(**C**HOH, C-8) = 3.1 Hz, ^{3}J(**C**HOH, C-4a) = 3.1 Hz, ^{3}J(**C**HOH, C-3) = 3.1 Hz

9.3.3.9
[α-^{13}C]-1,2-Dimethylnaphthalene (3)

In accordance to general procedure H, palladium on charcoal (20 mg) was added in one portion to a solution of 2.30 g ^{13}C-(2-methylnaphthalene-1-yl)-methanol (**23**) in 20 ml ethanol and 2 ml acetic acid. The solution was then hydrogenated at 4 Mpa for 72h at 25 °C. After purification 2.02 g (yield: 97 %) of the product was obtained.

Calculated for $C_{12}H_{12}$: C: 92.30 %, H: 7.69 %; found: C: 91.74 %, H: 7.98 %

^{1}H-NMR: δ [ppm] = 2.46 (s, 3H, β-C**H**), 2.65 (s, 3H, α-C**H**), 7.31 (d, 1H, H-3), 7.44 (m, 1H, H-6), 7.53 (m, 1H, H-7), 7.72 (d, 1H, H-4), 7.81 (d, 1H, H-5), 8.18 (d, 1H, H-8); ^{1}J(α-C**H**, α-**C**H) = 142.0 Hz, ^{3}J(H-3, H-4) = 8.7 Hz, ^{3}J(H-5, H-6) = 8.1 Hz, ^{3}J(H-7, H-8) = 8.7 Hz

^{13}C-NMR: δ [ppm] = **14.76**, 16.82 (d), 123.89 (d), 125.37, 127.05, 128.95, 128.97, 129.64 (d), 132.76 (d), 133.03, 133.13 (d), 135.08 (d); ^{3}J(α-**C**H, β-**C**H) = 3.1 Hz, ^{3}J(α-**C**H, C-2) = 2.0 Hz, ^{1}J(C-1, **C**H) = 48.5 Hz, ^{3}J(**C**H, C-8) = 3.1 Hz, ^{3}J(**C**H, C-4a) = 3.1 Hz, ^{3}J(**C**H, C-3) = 3.1 Hz

9.3.3.10
[1-^{13}C]-Naphthol (2)

According to general procedure K, a solution of [1-^{13}C]-1-tert.-butyldimethyl-silyloxy-naphthalene (**19**) (1.94 g; 7.5 mmol) in THF and TBAF (5.5 mol) and 1 ml of HF was stirred at 25 °C to obtain 596 mg (yield: 55 %) of [1-^{13}C]-naphthol (colourless crystals).

^{1}H-NMR: δ [ppm] = 5.29 (d, 1H, -O**H**), 6.80 (dd, 1H, H-2), 7.30 (dd, 1H, H-3), 7.44 (d, 1H, H-4), 7.47-7.51 (m, 2H, H-6/H-7), 7.78–7.84 (m, 1H, H-5), 8.14–8.21 (m, 1H, H-8); ^{2}J(O**H**, C-1) = 3.6 Hz, ^{3}J(H-2, H-3) = 7.6 Hz, ^{2}J(H-2, C-1) = 2.5 Hz, ^{3}J(H-3, H-4) = 8.6 Hz

^{13}C-NMR: δ [ppm] = 108.16 (d), 120.28 (d), 121.09, 123.88 (d), 124.84 (d), 125.40, 126.03, 127.25 (d), 134.33 (d), **151.50**; ^{1}J(C-1, C-2) = 70.2 Hz, ^{1}J(C-1, C-8a) = 65.1 Hz, ^{2}J(C-1, C-4a) = 2.0 Hz, J = 4.1 Hz, J = 7.1 Hz, J = 4.1 Hz

9.3.3.11
2-[2-(1-Indenyl)-ethoxy]-tetrahydropyran (25)

Alkylation of freshly distilled indene (**24**) (27.8 g; 239 mmol) was successful using general procedure A1. After extraction and purification, 52.5 g (yield: 215 mmol) of the tetrahydropyranylether was obtained.

^{1}H-NMR: δ [ppm] = 1.50–1.92 (m, 6H, THP-H-2/THP-H-3/THP-H-4), 2.88–2.94 (m, 2H, C**H**CHO-), 3.34–3.38 (m, 2H, CHC**H**O-), 3.49–4.13 (m, 4H, THP-H-1/Ind-H-5), 4.68 (dd, 1H, THP-H-1), 6.27–6.30 (m, 1H, Ind-H-2), 7.19-(ddd, 1H, Ind-H-5), 7.22 (dd, 1H, Ind-H-6), 7.24–7.26 (m, 1H, Ind-H-3), 7.40 (d, 1H, Ind-H-7), 7.45 (d, 1H, Ind-H-4)

^{13}C-NMR: δ [ppm] = 19.93, 25.88, 28.64, 31.00, 38.28, 62.64, 63.75, 99.55, 119.39, 124.23, 124.94, 126.88, 129.37, 130.06, 141.63, 145.78

9.3.3.12
2-(1-Indanyl)-ethan-1-ol (26)

Treatment of 2-[2-(1-indenyl)-ethoxy]-tetrahydropyran (**25**) (52.5 g; 215 mmol) according to the general procedures H and B gave 31.5 g of the alcohol (yield: 90 %).

^{1}H-NMR (500 M Hz): δ [ppm] = 1.55 (s, 1H, -O**H**), 1.65–1.75 (m, 2H, H-β), 2.14 (dtt, 1H, H-2a), 2.32 (ddtt, 1H, H-2e), 2.89 (ddt, 2H, H-3), 3.25 (ddt, 1H, H-1), 3.80 (dt, 2H, H-α), 7.10–7.23 (m, 4H, H-4/H-5/H-6/H-7); ^{3}J(H-1, H-2a) = 7.5 Hz, ^{3}J(H-1, H-2e) = 1.6 Hz, ^{3}J(H-1, H-β) = 5.2 Hz, ^{3}J(H-2e, H-1) = 1.6 Hz, ^{3}J(H-2e, H-3) = 4.6 Hz, ^{2}J(H-2a, H-2e) = -14.9 Hz, ^{4}J(H-2e, H-β) = 3.1 Hz, ^{5}J(H-2a, H-α) = 0.5 Hz, ^{4}J(H-2a, H-β) = 1.6 Hz, ^{3}J(H-2a, H-3) = 8.3 Hz, ^{4}J(H-2a, H-α) = 2.6 Hz, ^{5}J(H-3, H-α) = 0.5 Hz, ^{4}J(H-3, H-β) = 1.0 Hz, ^{3}J(H-α, H-β) = 7.1 Hz

^{13}C-NMR: δ [ppm] = 30.12, 32.65, 38.31, 41.88, 61.88, 123.93, 124.95, 126.54, 126.96, 144.31, 147.43

9.3.3.13
1-(2-Bromoethyl)-indane (27)

Bromination of 2-(1-indanyl)-ethan-1-ol (**26**) (31.5 g; 194 mmol) according to the general procedure C gave 23.4 g of 1-(2-bromoethyl)-indane (yield: 53 %).

^{1}H-NMR: δ [ppm] = 1.69 (dtt, 2H, H-2ax), 1.99 (ddtt, 2H, H-2eq), 2.41–2.29 (m, 2H, H-β), 2.91 (dddt, 2H, H-3), 3.29–3.37 (m, 1H, H-1), 3.52 (ddt, 2H, H-α), 7.14–7.25 (m, 4H, H-4/H-5/H-6/H-7); J(H-1, H-α) = 0.5 Hz, J(H-1, H-3) = 0.5 Hz, J(H-1, H-β) = 7.5 Hz, J(H-1, H-2eq) = 5.6 Hz, J(H-1, H-2ax) = 4.1 Hz, J(H-2eq, H-α) = 1.5 Hz, J(H-2eq, H-3) = 6.1 Hz, J(H-2eq, H-β) = 4.1 Hz, J(H-2ax, H-2eq) = -14.2 Hz, J(H-2ax, H-3) = 7.0 Hz, J(H-2ax, H-β) = 1.0 Hz, J(H-α, H-β) = 7.6 Hz

^{13}C-NMR: δ [ppm] = 31.26, 31.60, 31.92, 38.02, 43.13, 123.56, 124.61, 126.21, 126.66, 143.79, 145.90

9.3.3.14
[1-^{13}C]-3-(1-Indanyl)-propionic acid (28)

According to general procedure D, a Grignard solution, prepared by reaction of 1-(2-bromoethyl)-indane (**27**) (23.4 g; 103 mmol) and magnesium (2.91 g; 120 mmol), was treated with carbon-13 dioxide produced from a mixture of 3.35 g sodium ^{13}C-carbonate (9.3 mmol) and conc. sulphuric acid. After extraction, 1.79 g of the labelled carboxlic acid (yield: 30 %, calculation based on ^{13}C-sodium carbonate) was obtained.

^{1}H-NMR: δ [ppm] = 1.63–1.84 (m, 2H, H-3), 2.15–2.35 (m, 2H, Indan-H-2), 2.39–2.55 (m, 2H, H-2), 2.79–2.99 (m, 2H, Indan-H-3), 3.11–3.20 (m, 1H, Indan-H-1), 7.11–7.24 (m, 4H, Indan-H-4/H-5/H-6/H-7)

^{13}C-NMR: δ [ppm] = 30.05, 31.37, 31.95 (d), 32.84, 44.43, 124.08, 124.99, 126.60, 127.03, 144.38, 146.71, **180.63**; ^{1}J(C-1, C-2) = 40.7 Hz

9.3.3.15
[5-^{13}C]-2,2a,3,4-Tetrahydroacenaphtylen-5-one (29)

Cyclization of 1.78 g [1-^{13}C]-3-(1-indanyl)-propionic acid (**28**) (9.3 mmol) was undertaken by using general procedure E2 to obtain 1.01 g of the ketone (yield: 63 %).

^{1}H-NMR: δ [ppm] = 1.66–1.85 (m, 2H, H-3), 2.31–2.49 (m, 2H, H-4), 2.55–2.68 (m, 1H, H-2ax), 2.71–2.80 (m, 1H, H-2eq), 2.89–3.10 (m, 2H, H-1), 3.21–3.32 (m, 1H, H-2a), 7.27 (dd, 1H, H-7), 7.43 (d, 1H, H-8), 7.70 (dd, 1H, H-6); J(H-6, H-7) = 7.6 Hz, J(H-7, H-8) = 7.1 Hz, J(H-6, C-5) = 3.0 Hz

^{13}C-NMR: δ [ppm] = 31.20, 33.03, 34.75, 39.70 (d), 41.97, 123.22, 127.74 (d), 129.70, 130.09 (d), 144.01, 153.10, **198.94**; J(C-4,C-5) = 41.4 Hz, J(C-5a, C-5) = 51.9 Hz, J(C-5a, H-7) = 3.3 Hz

9.3.3.16
[5-^{13}C]-Acenaphthylene (4)

Conversion of [5-^{13}C]-2,2a,3,4-tetrahydroacenaphtylen-5-one (**29**) (264 mg; 1.5 mmol) to [5-^{13}C]-acenaphthylene was done by using the general procedures G and H. After purification, 103 mg of the desired product (yield: 45 %) was obtained.

^{1}H-NMR: δ [ppm] = 7.08 (s, 2H, H-1/H-2), 7.54 (ddd, 1H, H-4 und dd, 1H, H-7), 7.68 (dd, 1H, H-3 und d, 1H, H-8), 7.80 (dd, 1H, H-5 und dd, 1H, H-6); ^{3}J(H-3, H-4) = 6.6 Hz, ^{3}J(H-3, C-5) = 6.6 Hz, ^{3}J(H-4, H-5) = 8.1 Hz, ^{2}J(H-4, C-5) = 1.5 Hz, ^{1}J(H-5, C-5) = 159.7 Hz, ^{3}J(H-6, H-7) = 8.1 Hz, ^{3}J(H-7, H-8) = 6.6 Hz

9.3.3.17
[5-^{13}C]-Acenaphthene (5)

[5-^{13}C]-acenaphthylene (**4**) (50.1 mg; 0.30 mmol) was dissolved in ethanol, and 5 mg of palladium on charcoal (10 %) was added. The solution was hydrogenated (0.5 MPa) for 2 hours. After filtration and removal of the solvent under reduced pressure, 49.6 mg of colourless crystals were obtained (yield: 98 %).

^{1}H-NMR: δ [ppm] = 3.39 (s, 4H, H-1/H-2), 7.24–7.29 (m, 2H, H-3/H-8), 7.43 (dd, 2H, H-4/H-7), 7.57 (dd, 1H, H-5), 7.58 (dd, 1H, H-6); ^{3}J(H-3, H-4) = 8.1 Hz, ^{3}J(H-4, H-5) = 8.1 Hz, ^{1}J(H-5, C-5) = 160.2 Hz, ^{2}J(H-4, C-5) = 3.6 Hz, ^{2}J(H-6, C-5) = 4.1 Hz

9.3.3.18
(2-Bromophenyl)-phenyl-methanol (33)

Under a nitrogen atmosphere, a solution of 29.6 g 2-bromobenzaldehyde (**32**) (160 mmol) in 200 ml dry diethyl ether was prepared. At 5 °C, 100 ml of a 1.6 M solution of phenyllithium in diethyl ether/cylcohexane (70:30 v:v) was added dropwise. After 5 hours of stirring, the mixture was quenched carefully with water and then diluted with water. The aqueous suspension was neutralised using sodium dihydrogenphosphate and extracted 4 times with diethyl ether. The combined organic extracts were dried over magnesium sulphate, and the solvent was removed under reduced pressure. The crude product (31.20 g; yield: 74 %) was used for the next step without purification.

^{1}H-NMR: δ [ppm] = 5.82 (s, 2H, -C**H**-), 7.08 (ddd, 1H, H-5), 7.10–7.14 (m, 1H, H-4'), 7.17–7.23 (m, 4H, H-2/H-3/H-4/H-3'), 7.26–7.31 (m, 2H, H-2'), 7.56 (dd, 1H, H-6); ^{4}J(H-4, H-6) = 1.6 Hz, ^{3}J(H-5, H-6) = 7.6 Hz, ^{3}J(H-4, H-5) = 7.9 Hz, ^{4}J(H-3, H-5) = 1.6 Hz

9.3.3.19
2-Benzyl-1-bromobenzene (34)

To a suspension of 5.0 g iodine (19.7 mmol) and 5.0 g phosphor (38 mmol) in 100 ml acetic acid, 31.20 g of (2-bromophenyl)-phenylmethanol (**33**) (119 mmol) was added in one portion. The mixture was heated under reflux for 32 h. Subsequently, the mixture was filtered, and the filtrate was neutralised carefully with a solution of sodium hydroxide (70 g; 1.7 mol) in 250 ml water. Sodium bisulphite was added to the aqueous suspension until the dark colour disappeared. The aqueous suspension was extracted 4 times with diethyl ether. The combined organic phases were dried over magnesium sulphate, and the solvent was removed under reduced pressure. After purification by column chromatography (eluent: hexane), 23.44 g (yield: 80 %) of the bromide was obtained.

^{1}H-NMR: δ [ppm] = 4.11 (s, 2H, -C**H**-), 7.08 (ddd, 1H, H-5), 7.10-7.14 (m, 1H, H-4'), 7.17–7.23 (m, 4H, H-2/H-3/H-4/H-3'), 7.26–7.31 (m, 2H, H-2'), 7.56 (dd, 1H, H-6); ^{4}J(H-4, H-6) = 1.6 Hz, ^{3}J(H-5, H-6) = 7.6 Hz, ^{3}J(H-4, H-5) = 7.9 Hz, ^{4}J(H-3, H-5) = 1.6 Hz

^{13}C-NMR: δ [ppm] = 41.72, 124.90. 126.25, 127.45, 127.88, 128.47, 128.99, 131.08, 132.85, 139.48, 140.37

9.3.3.20
^{13}C-2-Benzyl-benzoic acid (35)

According to the general procedure D, a Grignard solution, prepared from 23.44 g 2-benzyl-1-bromobenzene (**34**) (94.8 mmol) and 4.00 g magnesium (165 mmol) in 50 ml dry diethyl ether, was treated with carbon-13 dioxide, generated from 5.00 g ^{13}C-sodium carbonate (46 mmol) and conc. sulphuric acid. After extraction, 3.50 g (yield: 36 %; calculation based on ^{13}C-sodium carbonate) of the labelled carboxylic acid was obtained.

^{1}H-NMR: δ [ppm] = 4.45 (s, 2H, -C**H**-), 7.13–7.38 (m, 7H, H-3/H-4/H-5/H-2'/H-3'), 7.43–7.48 (m, 1H, H-4'), 8.02–8.06 (m, 1H, H-6)

^{13}C-NMR: δ [ppm] = 39.86, 126.02, 126.36 (d, J = 4.1 Hz), 127.98, 128.36, 128.70, 129.07, 131.69 (d, J = 2.0 Hz), 131.78 (d, J = 6.1 Hz), 133.00, 140.71, 143.45, **172.44**

9.3.3.21
[9-^{13}C]-Anthrone (36)

According to the general procedure E2, 3.50 g of ^{13}C-2-benzylbenzoic (**35**) (16.4 mmol) acid was converted to the corresponding acid chloride and then treated with tin tetrachloride to give 1.85 g (yield: 58 %) of the labelled ketone after work up.

^{1}H-NMR: δ [ppm] = 4.38 (s, 2H, H-10), 7.47 (dd, 4H, H-2/H-4/H-5/H-7), 7.60 (ddd, 2H, H-3/H-6), 8.35–8.38 (m, 2H, H-1/H-8); ^{3}J(H-2, H-3) = 7.6 Hz, ^{3}J(H-3, H-4) = 7.9 Hz, ^{2}J(H-3, C-9) = 1.3 Hz

^{13}C-NMR: δ [ppm] = 32.38 (d), 127.02, 127.62, 128.44, 132.04, 132.73, 141.86, **184.29**; ^{3}J(C-2, C-9) = 2.9 Hz, ^{3}J(C-4, C-9) = 2.9 Hz, ^{1}J(C-9, C-9a) = 54.9 Hz, ^{3}J(C-9, H-10) = 3.2 Hz, ^{3}J(C-9, C-10a) = 2.9 Hz

9.3.3.22
[9-^{13}C]-Anthracene (6)

Under a nitrogen atmosphere, 1.85 g [9-^{13}C]-anthrone (**36**) was dissolved in 20 ml ethylene-glycoldimethylether and 1.00 g sodium borohydride (26 mmol) was added in one portion. After stirring for 30 min at 25 °C, the suspension was cooled to 5 °C. Subsequently, 10 ml of methanol was added dropwise and then an additional amount of 0.5 g sodium borohydride was added. The mixture was stirred for 12 hours. The mixture was acidified to pH 4 by adding 30 ml acetic acid and stirred for 1 hour. The mixture was diluted with 200 ml water and filtered. The aqueous suspension was extracted 4 times with dichloromethane. After drying of the combined organic extracts over magnesium sulphate, the solvent was removed under reduced pressure. Recrystallisation in toluene gave 1.41 g (yield: 83 %) of pure [9-^{13}C]-anthracene.

^{1}H-NMR: δ [ppm] = 7.42–7.48 (m, 4H, H-2/H-3/H-6/H-7), 7.96–8.03 (m, 4H, H-1/H-4/H-5/ H-8), 8.42 (d, 1H, H-9), 8.42 (s, 1H, H-10); ^{1}J(C-9, H-9) = 158.7 Hz^{13}C-NMR: δ [ppm] = 125.3 (d + d), 126.21, 128.16, 131.68 (d + s); ^{2}J(C-1, C-9) = 2.0 Hz, ^{3}J(C-2, C-9) = 2.0 Hz, ^{4}J(C-3, C-9) = 6.1 Hz, ^{1}J(C-9, C-9a) = 59.9 Hz

9.3.3.23
2-(3-Naphth-1-yl)-propoxy-tetrahydropyran (38)

According to general procedure A2, 21.3 g of 1-methylnaphthalene (**37**) (150 mmol) was deprotonated with butyllithium in TMEDA/Et_2O and then the dark red solution was added dropwise to a mixture of 2-(2-bromoethoxy)-tetrahydropyran (29.2 g; 139 mmol) in THF. Without further purification, 30.4 g of 2-(3-naphth-1-yl-propoxy)-tetrahydropyran was obtained.

MS: m/z (%) = 270 (1), 186 (27), 168 (28), 167 (18), 155 (10), 154 (10), 153 (43), 152 (17), 142 (67), 141 (100), 139 (11), 128 (16), 127 (10), 115 (50), 85 (50), 84 (42), 67 (10), 63 (10), 57 (15), 56 (18), 55 (64), 54 (16), 43 (15), 41 (32), 39 (26)

9.3.3.24
3-(1-Naphthyl)-propan-1-ol (39)

Deblocking of 30.4 g 2-(3-naphth-1-yl-propoxy)-tetrahydropyran (**38**) was done in methanol with catalytic amounts of p-toluenesulphonic acid according to general procedure B to give 20.0 g (yield: 95 %) of the pure alcohol.

^{1}H-NMR: δ [ppm] = 1.84 (s, 1H, -O**H**), 1.96–2.04 (m, 2H, H-2), 3.15 (t, 2H, H-3), 3.71 (t, 2H, H-1), 7.32 (d, 1H, Nap-H-2), 7.38 (dd, 1H, Nap-H-3), 7.43–7.52 (m, 2H, Nap-H-6/H-7), 7.70 (d, 1H, Nap-H-4), 7.84 (dd, 1H, Nap-H-5), 8.05 (dd, 1H, Nap-H-8); ^{3}J(H-1, H-2) = 6.3 Hz, ^{3}J(H-2, H-3) = 7.6 Hz, ^{3}J(Nap-H-2, Nap-H-3) = 7.1 Hz, ^{3}J(Nap-H-3, Nap-H-4) = 8.1 Hz, ^{3}J(Nap-H-5, Nap-H-6) = 7.1 Hz, ^{4}J(Nap-H-5, Nap-H-7) = 1.5 Hz, ^{4}J(Nap-H-6, Nap-H-8) = 1.5 Hz, ^{4}J(Nap-H-7, Nap-H-8) = 1.5 Hz

^{13}C-NMR: δ [ppm] = 29.16, 33.50, 62.55, 123.77, 125.47, 125.66, 125.80, 125.99, 126.68, 128.78, 131.84, 133.91, 137.95

9.3.3.25
1-(3-Bromo-propyl)-naphthalene (40)

Bromination of 3-(1-naphthyl)-propan-1-ol (**39**) (20.0 g; 107 mmol) was performed according to general procedure C to obtain 17.0 g of the bromide (yield: 64 %).

^{1}H-NMR: δ [ppm] = 2.30 (m, 2H, H-2), 3.25 (t, 2H, H-3), 3.46 (t, 2H, H-1), 7.36 (dd, 1H, Nap-H-2), 7.40 (dd, 1H, Nap-H-3), 7.45–7.55 (m, 2H, Nap-H-6/H-7), 7.73 (d, 1H, Nap-H-4), 7.86 (d, 1H, Nap-H-5), 8.04 (d, 1H, Nap-H-8); ^{3}J(H-1, H-2) = 6.3 Hz, ^{3}J(H-2, H-3) = 7.3 Hz, ^{3}J(Nap-H-2,Nap-H-3) = 7.1 Hz, ^{4}J(Nap-H-2, Nap-H-4) = 1.5 Hz, ^{3}J(Nap-H-3, Nap-H-4 = 7.6 Hz, ^{3}J(Nap-H-5, Nap-H-6) = 8.1 Hz, ^{3}J(Nap-H-7, Nap-H-8) = 8.1 Hz

^{13}C-NMR: δ [ppm] = 31.17, 33.39, 33.56, 123.60, 125.51, 125.57, 125.98, 126.40, 127.03, 128.85, 131.71, 133.95, 136.60

9.3.3.26
[1-^{13}C]-4-(1-Naphthyl)-butyric acid (41)

According to general procedure D, a solution of 1-(3-bromo-propyl)-naphthalene (**40**) (35.1 g; 141 mmol) was treated with magnesium (3.65 g; 150 mmol) under a nitrogen atmosphere and treated with carbon-13 dioxide, prepared from ^{13}C-sodium carbonate (7.49 g; 70 mmol) at -78 °C. After extraction, 6.36 g of the pure labelled carboxylic acid (yield: 49 %, calculation based on ^{13}C-sodium carbonate) was obtained.

^{1}H-NMR: δ [ppm] = 2.06–2.16 (m, 2H, H-2), 2.47 (dt, 2H, H-1), 3.15 (t, 2H, H-3), 7.31–7.42 (m, 2H, Nap-H-2/H-3), 7.45–7.55 (m, 2H, Nap-H-6/H-7), 7.73 (d, 1H, Nap-H-4), 7.85 (d, 1H, Nap-H-5), 8.05 (d, 1H, Nap-H-8); ^{1}J(H-1, C-1) = 7.1 Hz, ^{3}J(H-1, H-2) = 7.1 Hz, ^{3}J(H-2, H-3) = 7.6 Hz, ^{3}J(Nap-H-3, Nap-H-4) = 8.1 Hz, J(Nap-H-5, Nap-H-6 = 7.6, J(Nap-H-7, Nap-H-8) = 8.6 Hz

^{13}C-NMR: δ [ppm] = 25.37, 33.11, 33.65, 123.60, 125.41, 125.44, 125.83, 126.14, 126.84, 128.72, 131.71, 133.84, 137.21, **178.92**

9.3.3.27
[1-^{13}C]-3,4-Dihydro-2H-phenanthren-1-one (42)

According to general procedure E2, 1.00 g of [1-^{13}C]-4-(1-naphthyl)-butyric acid (**41**) was converted to the corresponding acid chloride by reaction with oxalyl chloride (2.72 g; 21.6 mmol) in dry dichloromethane. The acid chloride was cyclized by treatment with tin tetrachloride (1.20 g; 4.6 mmol) in dry benzene. After purification, 680 mg (yield: 75 %) of the pure cyclic ketone was obtained.

^{1}H-NMR: δ [ppm] = 2.29 (tt, 2H, H-3), 2.72 (dt, 2H, H-2), 3.37 (t, 2H, H-4), 7.54–7.62 (m, 2H, H-6/H-7), 7.73 (d, 1H, H-9), 7.84 (m, 1H, H-10), 8.07–8.15 (m, 2H, H-5/H-8); ^{1}J(H-2,C-1) = 6.6 Hz, ^{3}J(H-2,H-3) = 6.6 Hz, ^{3}J(H-3, H-4) = 6.1 Hz, ^{3}J(H-9, H-10) = 8.7 Hz

^{13}C-NMR: δ [ppm] = 22.75, 25.65 (d), 38.37 (d), 122.76, 124.81, 126.67, 126.91 (d), 128.28, 128.39, 131.37, 131.41, 135.28 (d), 135.71, **198.16** (C-1); ^{1}J(C-1, C-2) = 40.7 Hz; ^{1}J(C-1, C-10a) = 50.9 Hz, ^{3}J(C-1, C-9) = 3.1 Hz, ^{3}J(C-1, C-4) = 3.1 Hz

9.3.3.28
[1-^{13}C]-1,2,3,4-Tetrahydro-phenanthren-1-ol (43)

A suspension of $LiAlH_4$ (360 mg; 9.5 mmol) and [1-^{13}C]-3,4-dihydro-2H-phenanthren-1-one (**42**) (680 mg; 3.5 mmol) in diethyl ether was stirred according to general procedure G to obtain 628 mg of the benzylic alcohol (yield: 90 %).

^{1}H-NMR: δ [ppm] = 1.82–2.01 (m, 4H, H-2/H-3), 2.91 (t, 2H, H-4), 4.62 (dt, 1H, H-1), 7.11 (d, 1H, H-10), 7.23–7.36 (m, 2H, H-6/H-7), 7.47 (d, 1H, H-9), 7.64–7.75 (m, 2H, H-5/H-8); ^{1}J(C-1, H-1) = 126.1 Hz, ^{3}J(H-3, H-4) = 5.6 Hz, ^{3}J(H-1, H-2) = 6.1 Hz, ^{3}J(H-9, H-10) = 8.7 Hz

9.3.3.29
[1-^{13}C]-Phenanthrene (7)

[1-^{13}C]-1,2,3,4-tetrahydrophenanthren-1-ol (**43**) was treated with hydrogen according to general procedure H, and then the crude product was aromatised with DDQ (2.61 g; 11.5 mmol) according to general procedure I. After purification, 401 mg of [1-^{13}C]-phenanthrene (yield: 70 %) was obtained.

^{1}H-NMR: δ [ppm] = 7.58–7.72 (m, 4H, H-2/H-3/H-6/H-7), 7.74 (d, 2H, H-9/H-10), 7.89 (dd, 1H, H-8 und ddd, 1H, H-1), 8.70 (d, 2H, H-4/H-5); ^{1}J(H-1, C-1) = 156.7 Hz, ^{3}J(H-1, H-2) = ^{3}J(H-7, H-8) = 7.6 Hz, ^{3}J(H-1, H-3) = 1.5 Hz, ^{3}J(H-10, C-1) = 2.5 Hz

^{13}C-NMR: δ [ppm] = 122.66 (s und d), 126.54 (s und d), 126.77, **128.52**, 130.28, 132.03 (s und d); ^{1}J(C-1, C-2) = 58.0 Hz, ^{3}J(C-1, C-4) = 8.1 Hz, ^{1}J(C-1, C-10a) =57.0 Hz

9.3.3.30
[1-^{13}C]-1-(tert.-Butyldimethylsiloxy)-3,4-dihydrophenanthrene (44)

According to general procedure J, a suspension of 100 mg [1-^{13}C]-3,4-dihydro-2H-phenanthren-1-one (**42**) (0.5 mmol) and 20 mg sodium hydride (0.8 mmol) in

5 ml dry DMF was treated with 75 mg tert.-butyldimethylsilylchloride (0.5 mmol) in 2 ml dry DMF at -20 °C to obtain 85 mg (yield: 55 %) of the labelled silyl enolether.

^{1}H-NMR: δ [ppm] = 0.27 (s, 6H, Si-C**H**), 1.01 (s, 9H, C-C**H**), 2.54 (dt, 2H, H-3), 3.31 (t, 2H, H-4), 5.81 (dt, 1H, H-2), 7.29 (d, 1H, H-10), 7.37–7.52 (m, 3H, H-9/H-6/H-7), 7.73–7.97 (m, 2H, H-5/H-8); ^{3}J(H-2, H-3) = 6.6 Hz, ^{3}J(H-3, H-4) = 6.1 Hz, ^{2}J(C-1, H-2) = 2.5 Hz, ^{3}J(H-3, H-4) = 8.1 Hz, ^{3}J(H-9, H-10) = 8.7 Hz

9.3.3.31
[1-^{13}C]-Phenanthren-1-ol (8)

According to general procedure I, a mixture of 85 mg [1-^{13}C]-1-(tert.-butyldimethylsiloxy)-3,4-dihydrophenanthrene (**44**) (0.27 mmol) and DDQ (120 mg; 0.54 mmol) in 5 ml dry THF was heated under reflux for thirty minutes. The crude product was deblocked using general procedure K to give 27 mg (yield: 51 %) of the labelled phenol.

^{1}H-NMR: δ [ppm] = 5.03 (d, 1H, -O**H**), 7.43 (dd, 1H, H-2), 7.67–7.84 (m, 5H, H-3/H-6/H-7/H-9/H-10), 8.25 (d, 1H, H-8), 8.62 (dd, 1H, H-4), 8.99 (d, 1H, H-5); ^{2}J(O**H**, C-1) = 3.5 Hz, ^{3}J(H-2, H-3) = 7.6 Hz, ^{2}J(C-1, H-2) = 2.5 Hz, ^{3}J(H-3, H-4) = 8.6 Hz, ^{3}J(C-1, H-4) = 6.1 Hz, ^{3}J(H-7, H-8) = 7.6 Hz, ^{3}J(H-5, H-6) = 8.1 Hz

^{13}C-NMR: δ [ppm] = 111.90 (d), 115.02 (d), 117.26 (d), 121.41, 123.31, 123.82, 126.55, 127.64, 127.73, 129.08, 130.12, 131.93, 133.27, **154.71**; ^{1}J(C-1, C-2) = 67.0 Hz, ^{1}J(C-1, C-10a) = 66.3 Hz, ^{1}J(C-1, C-4) = 8.4 Hz

9.3.3.32
3-(2-Naphthyl)-propan-1-ol (47)

2-Methylnaphthalene (**45**) (14.3 g; 101 mmol) was deprotonated with butyllithium and alkylated with 2-(2-bromoethoxy)-tetrahydropyran according to general procedure A2. After extraction, 11.0 g of the alcohol (yield: 59 %) was obtained.

^{1}H-NMR: δ [ppm] = 1.94–2.03 (m, 2H, H-2), 2.28 (s, 1H, -OH), 2.88 (t, 2H, H-3), 3.69 (t, 2H, H-1), 7.36 (dd, 1H, Nap-H-3), 7.42–7.51 (m, 2H, Nap-H-6/H-7), 7.66 (d, 1H, Nap-H-1), 7.77–7.86 (m, 3H, Nap-H-4/H-5/H-8); ^{3}J(H-1, H-2) = 6.6 Hz, ^{3}J(H-2, H-3) = 7.6 Hz, ^{4}J(Nap-H-1,Nap-H-3) = 1.5 Hz, ^{3}J(Nap-H-2, Nap-H-3) = 6.6 Hz

^{13}C-NMR: δ [ppm] = 32.73, 34.57, 62.58, 125.71, 126.48, 126.95, 127.84, 127.96, 128.15, 128.48, 132.56, 134.19, 139.94

9.3.3.33
2-(3-Bromo-propyl)-naphthalene (48)

Bromination of 3-(2-naphthyl)-propan-1-ol (**47**) (11.0 g; 59 mmol) was undertaken by reaction with triphenylphosphonium dibromide according to general procedure C to give 5.67 g of the pure ar-aliphatic bromide (yield: 39 %).

^{1}H-NMR: δ [ppm] = 2.21 (m, 2H, H-2), 2.94 (t, 2H, H-1), 3.42 (t, 2H, H-3), 7.33 (dd, 1H, Nap-H-3), 7.40–7.47 (m, 2H, Nap-H-6/H-7), 7.64 (d, 1H, Nap-H-1), 7.75–7.83 (m, 3H, Nap-H-4/H-5/H-8); ^{3}J(H-1, H-2) = 7.1 Hz, ^{3}J(H-2, H-3) = 6.6 Hz, ^{4}J(Nap-H-1, Nap-H-3 = 2.0 Hz, ^{3}J(Nap-H-3, Nap-H-4) = 6.5 Hz

^{13}C-NMR: δ [ppm] = 33.48, 34.45, 34.53, 125.77, 126.46, 127.19, 127.57, 127.86, 128.04, 128.55, 132.56, 134.02, 138.42

9.3.3.33.1
[1-^{13}C]-4-(2-Naphthyl)-butyric acid (49)

A Grignard reagent, prepared from treatment of 2-(3-bromo-propyl)-naphthalene (**47**) (5.80 g; 22 mmol) and magnesium (560 mg; 22 mmol) in diethyl ether. Carboxylation of the reagent was undertaken by using carbon-13 dioxide generated from ^{13}C-sodium carbonate (1.17 g; 11 mmol) gave 1.20 g of the labelled carboxlic acid (yield: 51 %, calculation based on ^{13}C-sodium carbonate).

^{1}H-NMR: δ [ppm] = 1.99–2.12 (m, 2H, H-2), 2.38 (dt, 2H, H-1), 2.84 (t, 2H, H-3), 7.33 (dd, 1H, Nap-H-3), 7.40–7.47 (m, 2H, Nap-H-6/H-7), 7.64 (d, 1H, Nap-H-1), 7.75–7.83 (m, 3H, Nap-H-4/H-5/H-8); ^{3}J(H-1, C-1) = 7.1 Hz, ^{3}J(H-1, H-2) = 7.1 Hz, ^{3}J(H-2, H-3) = 7.6 Hz, ^{4}J(Nap-H-1, Nap-H-3) = 2.0 Hz, ^{3}J(Nap-H-3, Nap-H-4) = 6.5 Hz

^{13}C-NMR: δ [ppm] = 26.08, 33.18 (d), 35.15, 125.27, 125.98, 126.64, 127.18, 127.46, 127.62, 128.05, 132.12, 133.59, 138.67, **178.79**; ^{1}J(C-1, C-2) = 56 Hz

9.3.3.34
[4-^{13}C]-2,3-Dihydro-1H-phenanthren-4-one (50)

According to general procedure E1, [1-^{13}C]-4-(2-naphthyl)-butyric (**49**) acid was treated with PPA to give 500 mg of the cyclic ketone (yield: 90 %).

^{1}H-NMR: δ [ppm] = 2.29 (tt, 2H, H-2), 2.72 (dt, 2H, H-3), 3.06 (t, 2H, H-1), 7.15 (d, 1H, H-10), 7.54–7.62 (m, 2H, H-6/H-7), 7.68–7.73 (m, 2H, H-8/H-9), 8.63 (d, 1H, H-5); ^{2}J(H-3, C-4) = 6.6 Hz, ^{3}J(H-2, H-3) = 6.6 Hz, ^{3}J(H-3, H-4) = 6.1 Hz, ^{3}J(H-5, H-6) = 7.9 Hz, ^{3}J(H-9, H-10) = 8.5 Hz

^{13}C-NMR: δ [ppm] = 22.75, 28.15 (d), 37.87 (d), 122.73, 125.43, 126.27 (d), 127.75, 128.28, 128.39, 130.61, 131.82, 132.98 (d), 139.11, **198.12** (C-4); ^{1}J(C-3, C-4) = 40.7 Hz; ^{1}J(C-4, C-4a) = 50.9 Hz, ^{3}J(C-4, C-10) = 3.9 Hz, ^{3}J(C-1, C-4) = 3.1 Hz

9.3.3.35
[4-^{13}C]-1,2,3,4-Tetrahydrophenanthrene (51)

Reduction and hydrogenolysis of [4-^{13}C]-2,3-dihydro-1H-phenanthren-4-one (**50**) (500 mg; 2.5 mmol) according to the general procedures G and H gave 346 mg (yield: 76 %) of the hydrocarbon.

^{1}H-NMR: δ [ppm] = 1.82–1.92 (m, 4H, H-2/H-3), 2.96 (t, 2H, H-1), 2.97 (dt, 2H, H-4), 7.31–7.73 (m, 6H, H-5/H-6/H-7/H-8/H-9/H-10), ^{3}J(H-1, H-2) = 6.1 Hz, ^{3}J(H-3, H-4) = 5.6 Hz, ^{1}J(H-4, C-4) = 126.1 Hz

^{13}C-NMR: δ [ppm] = 23.30 (d), 25.68, **29.78**, 30.46, 122.68, 124.88, 126.65, 126.97, 128.79, 128.98, 130.80, 131.95, 132.13 (d), 134.23; ^{1}J(C-3, C-4) = 41.2 Hz, ^{4}J(C-1, C-4) = 17.9 Hz, ^{1}J(C-4, C-4a) = 57.6 Hz

9.3.3.36
[4-^{13}C]-Phenanthrene (9)

According to general procedure I, a solution of 346 mg [4-^{13}C]-1,2,3,4-tetrahydrophenanthrene (**50**) was treated with 1.08 g DDQ (4.8 mmol). After purification, 280 mg of the labelled phenanthrene was obtained.

^{1}H-NMR: δ [ppm] = 7.58–7.72 (m, 4H, H-2/H-3/H-6/H-7), 7.74 (d, 2H, H-9/H-10), 7.89 (dd, 1H, H-8/H-1), 8.70 (dd, 1H, H-4 und d, 1H, H-5); ^{1}J(H-4, C-4) = 156.2 Hz, ^{3}J(H-1, H-2) = ^{3}J(H-7, H-8) = 7.6 Hz, ^{3}J(H-1, H-3) = 1.5 Hz

^{13}C-NMR: δ [ppm] = **122.66**, 126.54 (s und d), 126.77, 128.52 (s and d), 130.28, 132.03 (s and d); ^{1}J(C-3, C-4) = 58.0 Hz, ^{3}J(C-1, C-4) = 8.1 Hz, ^{1}J(C-4, C-4a) = 57.0 Hz

9.3.3.37
^{13}C-9H-9-Fluorenecarboxylic acid (53)

Under a nitrogen atmosphere, a solution of 6.64 g fluorene (**52**) (40 mmol) in 100 ml dry THF was prepared, and 25 ml of methyllithium (1.6 M solution in hexane; 40 mmol) was added dropwise at 25 °C. After 1 hour of stirring, the dark coloured solution was treated with carbon-13 dioxide, produced from 2.14 g ^{13}C-sodium carbonate and conc. sulphuric acid, at -78 °C. The solution was stirred for additional 6 hours and then quenched carefully with water. The mixture was diluted with brine and extracted four times with diethyl ether. The combined organic extracts were dried over magnesium sulphate, and the solvent was removed under reduced pressure to obtain 3.05 g (yield: 72 %) of the labelled carboxylic acid.

^{1}H-NMR: δ [ppm] = 4.89 (d, 1H, H-9), 7.34 (dd, 2H, H-2/H-7), 7.43 (dd, 2H, H-3/H-6), 7.68 (d, 2H, H-1/H-8), 7.76 (d, 2H, H-4/H-5); ^{3}J(H-1, H-2) = 7.6 Hz, ^{3}J (H-2, H-3) = 7.1 Hz, ^{3}J(H-3, H-4) = 7.6 Hz, ^{2}J(H-9, $\underline{C}$OOH) = 10.2 Hz

^{13}C-NMR: δ [ppm] = 53.39 (d), 120.53, 126.13, 127.90, 128.81, 140. ; ^{1}J(C-9, $\underline{C}$OOH) = 56.8 Hz, ^{2}J(C-9a, $\underline{C}$OOH) = 2.0 Hz

9.3.3.38
^{13}C-9H-9-Fluorenylmethanol (54)

A suspension of 2.94 g potassium carbonate (21 mmol) and 3.00 g ^{13}C-9H-9-fluorenecarboxylic acid (**53**) (14.1 mmol) in 50 ml acetone was heated under reflux for 10 min. and then treated with 2.35 g dimethylsulphate (21 mmol). After refluxing for 1 hour, the mixture was poured into 100 ml of brine and extracted with diethyl ether. The combined organic extracts were dried over magnesium sulphate, and the solvent was removed under reduced pressure. The crude product was reduced with lithium aluminiumtetrahydride according to the general procedure G to yield 2.20 g of benzylic alcohol (yield: 71 %).

^{1}H-NMR: δ [ppm] = 1.60 (s, 1H, OH), 3.95 (t, 1H, Flu-H-9), 4.17(dd, 2H, C$\underline{\mathbf{H}}$OH), 7.32–7.44 (m, 4H, Flu-H-2/H-3/H-6/H-7), 7.57 (d, 2H, Flu-H-1/H-8), 7.80 (d, 2H, Flu-H-4/H-5); ^{1}J($\underline{\mathbf{C}}$HOH, C$\underline{\mathbf{H}}$OH) = 143.4 Hz, ^{3}J(C$\underline{\mathbf{H}}$OH, Flu-H-9) = 6.1 Hz, ^{3}J(Flu-H-3,Flu-H-4) = J(Flu-H-5, Flu-H-6) = 7.6 Hz, ^{3}J(Flu-H-1, Flu-H-2) = J(Flu-H-7, Flu-H-8) = 7.3 Hz

^{13}C-NMR: δ [ppm] = 41.59 (d), **70.00,** 119.98, 124.44, 127.00, 127.16, 140.94, 146.81; ^{1}J(Flu-C-9, CHOH) = 38.5 Hz

9.3.3.39
[9-^{13}C]-Phenanthrene (10)

To solution of 200 mg ^{13}C-9H-9-fluorenylmethanol (**54**) in 15 ml xylene, 500 mg P_4O_{10} was added in one portion. The mixture was then heated for 45 min. at 150 °C. After filtration, the solvent was removed under reduced pressure, and the crude product was purified using column chromatography (eluent: hexane). 175 mg (yield: 97 %) of the labelled phenanthrene was obtained.

^{1}H-NMR: δ [ppm] = 7.58–7.72 (m, 4H, H-2/H-3/H-6/H-7), 7.74 (dd, 1H, H-9 und d, 1H, H-10), 7.89 (dd, 1H, H-8/H-1), 8.70 (dd, 1H, H-4 und d, 1H, H-5); ^{1}J(H-9, C-9) = 156.4 Hz, ^{3}J(H-1, H-2) = ^{3}J(H-7, H-8) = 7.6 Hz, ^{3}J(H-1, H-3) = 1.5 Hz

^{13}C-NMR: δ [ppm] = 122.66, 126.54, **126.73**, 128.52, 130.72 (d), 132.03; ^{1}J(C-9,C-9a) = 59.0 Hz

9.3.3.40
2-[2-(9H-fluoren-9-yl)-ethoxy]-tetrahydropyran (55)

According to general procedure A1, fluorine (**51**) (56.5; 0.34 mol; dissolved in 500 ml THF) was deprotonated with butyllithium and added dropwise to a solution of 71.1 g 2-(2-bromoethoxy)-tetrahydropyran (0.34 mol) in 50 ml THF to obtain 80.1 g of 2-[2-(9H-Fluoren-9-yl)-ethoxy]-tetrahydropyran (yield: 80 %).

H-NMR: δ [ppm] = 1.45–1.88 (m, 6H, THP-H-2/H-3/H-4), 2.32 (dt, 2H, H-2), 3.48 (t, 2H, H-1), 3.64–3.71 (m, 1H, THP-H-5eq), 3.88–3.95 (m, 1H, THP-H-5ax), 4.16 (t, Flu-H-9), 4.93 (m, THP-H-1), 7.57 (d, 2H, Flu-H-1/H-8), 7.80 (d, 2H, Flu-H-4/H-5); ^{3}J(H-1, H-2) = 6.6 Hz, ^{3}J(H-2, Flu-H-9) = 6.0 Hz, ^{3}J(Flu-H-3,Flu-H-4) = J(Flu-H-5, Flu-H-6) = 7.6 Hz, ^{3}J(Flu-H-1, Flu-H-2) = J(Flu-H-7, Flu-H-8) = 7.3 Hz

9.3.3.41
2-(9H-Fluorenyl)-ethanol (56)

Deblocking of 2-[2-(9H-fluoren-9-yl)-ethoxy]-tetrahydropyran (**55**) (80.1 g; 0.27 mol) was performed according to general procedure B to give 53.2 g (yield: 93 %) of the alcohol.

MS (m/z): 210 (29), 192 (89), 178 (46), 165 (100), 152 (11)

^{1}H-NMR: δ [ppm] = 1.19 (s, 1H, OH), 2.34 (dt, 2H, H-2), 3.64 (t, 2H, H-1), 4.17 (t, 1H, Flu-H-9), 7.32–7.44 (m, 4H, Flu-H-2/H-3/H-6/H-7), 7.57 (d, 2H, Flu-H-1/H-8), 7.80 (d, 2H, Flu-H-4/H-5); ^{3}J(H-1, H-2) = 6.6 Hz, ^{3}J(H-2, Flu-H-9) = 6.0 Hz, ^{3}J(Flu-H-3,Flu-H-4 = J(Flu-H-5, Flu-H-6) = 7.6 Hz, ^{3}J(Flu-H-1, Flu-H-2) = J(Flu-H-7, Flu-H-8) = 7.3 Hz

^{13}C-NMR: δ [ppm] = 35.74, 44.59, 60.23, 119.98, 124.44, 127.00, 127.16, 140.94, 146.81

9.3.3.42
9-(2-Bromoethyl)-fluorene (57)

Bromination of 50.6 g 2-(9H-fluorenyl)-ethanol (**56**) was carried out according to general procedure C. After purification, 48.6 g of the bromide (yield: 67 %) was obtained.

MS (m/z): 274, 272 (34), 191 (12), 178 (14), 165 (100), 96 (10), 83 (11)

^{1}H-NMR: δ [ppm] = 2.47–2.55 (m, 2H, H-2), 3.30 (t, 2H, H-1), 4.17 (t, 1H, Flu-H-9), 7.31–7.42 (m, 4H, Flu-H-2/H-3/H-6/H-7), 7.53 (d, 2H, Flu-H-1/H-8), 7.76 (d, 2H, Flu-H-4/H-5); ^{3}J(H-2,Flu-H-9) = 6.1 Hz, ^{3}J(H-1, H-2) = 7.6 Hz, ^{3}J(Flu-H-3, Flu-H-4) = ^{3}J(Flu-H-5, Flu-H-6) = 7.1 Hz, , ^{3}J(Flu-H-1, Flu-H-2) = ^{3}J(Flu-H-7, Flu-H-8) = 8.1 Hz

^{13}C-NMR: δ [ppm] = 30.78, 37.02, 46.74, 120.48, 124.74, 125.40, 127.56, 127.83, 141.48, 146.16

9.3.3.43
[1-^{13}C]-(9H-Fluorenyl)-propionic acid (58)

The Grignard reagent of 9-(2-bromoethyl)-fluorene (**57**) (48.6 g; 178 mmol) was prepared according to general procedure D and treated with ^{13}C-carbon dioxide generated from 8.56 g ^{13}C-sodium carbonate (80 mmol). After extraction, 9.64 g of the labelled carboxylic acid (yield: 50 %, calculation based on ^{13}C-sodium carbonate) was obtained.

^{1}H-NMR: δ [ppm] = 1.93–2.00 (m, 2H, H-3), 2.39–2.46 (m, 2H, H-2), 4.09 (t, 1H, Flu-H-9), 7.30 (dd, 2H, Flu-H-2/H-7), 7.37 (dd, 2H, Flu-H-3/H-6), 7.50 (d, 2H, Flu-H-1/H-8), 7.75 (d, 2H, Flu-H-4/H-5); ^{3}J(Flu-H-1, Flu-H-2) = ^{3}J(Flu-7, Flu-8) = 7.6 Hz, ^{3}J(Flu-H-2, Flu-H-3) = ^{3}J(Flu-H-6, Flu-H-7) = 7.6 Hz, ^{3}J(Flu-H-3, Flu-H-4) = ^{3}J(Flu-H-5, Flu-H-6) = 7.6 Hz, ^{3}J(Flu-H-9, H-2 = 5.3 Hz

^{13}C-NMR: δ [ppm] = 27.26, 28.96 (d), 46.09, 119.981, 124.25, 127.10, 127.34, 141.32, 145.80, **177.84**; ^{1}J(C-1,C-2) = 56.0 Hz

9.3.3.44
[3-^{13}C]-1,10b-Dihydro-2H-fluoranthen-3-one (59)

According to general procedure E1, 4.63 g [1-^{13}C]-(9H-fluorenyl)-propionic acid (**58**) was heated in 25 ml PPA to obtain 3.00 g (yield: 70 %) of [3-^{13}C]-1,10b-dihydro-2H-fluoranthen-3-one.

^{1}H-NMR: δ [ppm] = 1.95–2.07 (m, 2H, H-1), 2.51–2.63 (m, 2H, H-2), 4.09 (t, 1H, H-10b), 7.26–7.38 (m, 3H, H-8/H-9/H-5), 7.55 (d, 1H, H-10), 7.68 (d, 1H, H-4), 7.84 (d, 1H, H-7), 7.96 (d, 1H, H-6); ^{3}J(H-2, H-10b) = 6.0 Hz, ^{3}J(H-4, H-5) = 7.9 Hz, ^{3}J(H-5, H-6) = 7.6 Hz, ^{3}J(H-7, H-8) = 7.6 Hz, ^{3}J(H-9, H-10) = 7.3 Hz

^{13}C-NMR: δ [ppm] = 30.09, 33.05, 36.24 (d), 126.06, 126.73 (d), 127.93, 127.95, 128.96, 132.36, 136.41, 136.53, 136.85 (d), 141.90, 142.02, **198.50**; ^{1}J(C-2, C-3) = 40.1 Hz, ^{1}J(C-3, C-3a) = 50.9 Hz, ^{3}J(C-3, C-5) = 3.1 Hz

9.3.3.45
[3-^{13}C]-1,2,3,10b-Tetrahydrofluoranthene (60)

3.00 g [3-^{13}C]-1,10b-Dihydro-2H-fluoranthen-3-one (**60**) was reduced according to general procedure F to give 2.24 g (yield: 80 %) of the labelled product.

^{1}H-NMR: δ [ppm] = 2.08–2.23 (m, 4H, H-1/H-2), 2.85 (dt, 2H, H-3), 4.08 (t, 1H, H-10b), 7.14 (d, 1H, H-4), 7.27–7.39 (m, 3H, H-5/H-8/H-9), 7.54–7.67 (m, 2H H-6/H-10), 7.84 (d, 1H, H-7); ^{1}J(C-3, H-3) = 126.1 Hz, ^{3}J(H-2, H-3) = 7.6 Hz, ^{3}J(H-2, H-10b) = 6.0 Hz, ^{3}J(H-4, H-5) = 7.9 Hz, ^{3}J(H-7, H-8) = 7.6 Hz

^{13}C-NMR: δ [ppm] = 31.09 (d), **32.18**, 34.47, 41.83, 125.10, 126.50, 126.96, 127.70, 127.93, 128.25, 129.15, 136.41, 136.51, 139.81 (d), 141.77, 141.92; ^{1}J(C-2, C-3) = 40.3 Hz; ^{1}J(C-3, C-3a) = 50.3 Hz

9.3.3.46
[3-^{13}C]-Fluoranthene (11)

Under a nitrogen atmosphere, 13.6 g of DDQ (60 mmol) was added in one portion to a solution of [3-^{13}C]-1,2,3,10b-tetrahydrofluoranthene (**60**) (2.20 g; 10.7 mmol) in dry THF (50 ml) according to general procedure I to obtain 550 mg (yield: 21 %) of [3-^{13}C]-fluoranthene.

^{1}H-NMR: δ [ppm] = 7.34–7.38 (m, 2H, H-8/H-9), 7.61 (dd, 2H, H-2/H-5), 7.82 (dd, 1H, H-3 und dd, 1H, H-4), 7.88–7.93 (m, 4H, H-1/H-6/H-7/H-10); ^{3}J(H-1, H-2) = 6.9 Hz, ^{3}J(H-2, H-3) = 8.2 Hz, ^{1}J(H-3, C-3) = 159.5 Hz, ^{3}J(H-4, H-5) = 8.2 Hz, ^{3}J(H-4, C-3) = 3.8 Hz

^{13}C-NMR: δ [ppm] = 120.00 (d), 121.47, **126.90**, 127.90 (d), 129.97 (d), 132.37, 136.92 (d), 139.42; ^{3}J(C-1, C-3) = 2.4 Hz, ^{1}J(C-2, C-3) = 66.6 Hz, ^{1}J(C-3, C-3a) = 57.5 Hz, ^{3}J(C-3, C-5) = 4.8 Hz, ^{3}J(C-3, C-6a) = 2.4 Hz, ^{3}J(C-3, C-10b) = 8.5 Hz

9.3.3.47
1-(2-Bromoethyl)-naphthalene (62)

According to general procedure C, 2-(1-naphthalene)-1-ethanol (**61**) (50 g; 0.29 mol) was converted to the corresponding bromide (61.5 g; yield: 90 %).

^{1}H-NMR: δ [ppm] = 3.54–3.65 (m, 4H, -C**H**C**H**Br), 7.30 (d, 1H, H-2), 7.36 (dd, 1H, H-3), 7.42–7.51 (m, 2H, H-6/H-7), 7.82 (d, 1H, H-5), 7.94 (d, 1H, H-8); ^{3}J(H-2, H-3) = 6.6 Hz, ^{3}J(H-3, H-4) = 8.1 Hz, ^{3}J(H-5, H-6) = 7.6 Hz, ^{3}J(H-7, H-8) = 8.1 Hz

^{13}C-NMR: δ [ppm] = 31.87, 36.76, 123.08, 126.31, 125.71, 126.79, 126.93, 127.77, 128.95, 131.54, 133.95, 134.86

9.3.3.48
[1-^{13}C]-3-(1-Naphthyl)-propionic acid (63)

A solution of the Grignard reagent, prepared from 18.8 g 1-(2-bromoethyl)-naphthalin (**62**) (80 mmol) and 1.42 g magnesium in dry diethyl ether, was treated with carbon-13 dioxide, produced from 4.28 g ^{13}C-sodium carbonate (40 mmol), along with general procedure D to obtain 4.25 g (yield: 53 %; calculation based on ^{13}C-sodium carbonate) of the labeled carboxylic acid.

^{1}H-NMR: δ [ppm] = 2.72 (dt, 2H, H-1), 3.38 (dt, 2H, H-2), 7.33–7.40 (m, 2H, Nap-H-2/H-3), 7.43–7.53 (m, 2H, Nap-H-6/H-7), 7.70 (d, 1H, Nap-H-4), 7.83 (d, 1H, Nap-H-5), 8.02 (d, 1H, H-8), 11.85 (s, 1H, COO**H**); ^{3}J(H-1, H-2) = 7.6 Hz, ^{2}J(H-1, **C**OOH) = 8.1 Hz, ^{3}J(H-2, **C**OOH) = 3.6 Hz, ^{3}J(Nap-H-3, Nap-H-4 = 7.6 Hz, ^{3}J(Nap-H-5, Nap-H-6 = 8.7 Hz, ^{3}J(Nap-H-7, Nap-H-8 = 8.7 Hz

^{13}C-NMR: δ [ppm] = 27.93, 34.95 (d), 123.33, 125.49, 125.73, 125.95, 126.83, 128.69, 131.47, 133.66, 136.68, 136.72, **178.80**; ^{1}J(C-1, C-2) = 56.0 Hz

9.3.3.49
[1-^{13}C]-2,3-Dihydrophenalen-1-one (64)

According to the general procedure E2, 4.25 g of [1-^{13}C]-3-(1-naphthyl)-propionic acid (**63**) (21 mmol) was first converted to the corresponding acid chloride and then treated with tin tetrachloride in benzene to obtain 2.25 g (yield: 59 %) of the tricyclic ketone.

^{1}H-NMR: δ [ppm] = 3.17–3.23 (m, 2H, H-2), 3.44 (t, 2H, H-3), 7.44–7.52 (m, 2H, H-4/H-5), 7.60 (dd, 1H, H-8), 7.80 (d, 1H, H-6), 8.09 (dd, 1H, H-9), 8.20 (dd, 1H, H-7); ^{3}J(H-1, H-2) = 7.1 Hz, ^{3}J(H-8, H-9) = 8.1 Hz, ^{3}J(H-7, H-8) = 7.1 Hz, ^{3}J(H-5, H-6) = 7.6 Hz, ^{4}J(H-7, H-9) = 1.5 Hz

^{13}C-NMR: δ [ppm] = 29.83, 43.38 (d), 123.55, 126.54, 126.94, 128.35, 128.43, 129.02, 130.80 (d), 132.05 (d), 133.26, 133.65, **198.28**; ^{1}J(C-1, C-2) = 40.5 Hz, ^{1}J(C-1, C-9a) = 54.0 Hz, ^{3}J(C-1, C-6a) = 3.1 Hz, ^{3}J(C-1, C-8) = 3.1 Hz

9.3.3.50
1-[3-(Tetrahydroypran-2-yloxy)-prop-1-inyl]-[1-^{13}C]-2,3-dihydro-1H-phenalen-1-ol (65)

Under a nitrogen atmosphere, a solution of 701 mg 2-prop-2-inyloxy-tetrahydropyran (5 mmol) in 10 ml dry THF/TMEDA (1:1 v:v) was prepared and 5 mmol of butyllithium (1.6 M solution in hexane; 3.1 ml) was added slowly. Subsequently, the solution was stirred 10 min. at 25 °C, and then a solution of 916 mg of [1-^{13}C]-2,3-dihydrophenalen-1-one (**64**) (5 mmol) in dry THF was added. The mixture was stirred for 12 h and then poured into water. After extraction of the aqueous suspension, the combined organic phases were dried with magnesium sulphate, and the solvent was removed under reduced pressure. 851 mg of the product (yield: 52 %) was obtained after purification.

^{1}H-NMR: δ [ppm] = 1.43–1.92 (m, 7H, THP-H-2/H-3/H-4/-O**H**), 2.33–2.39 (m, 2H, Phel-H-2), 3.09 (t, 2H, Phel-H-3), 3.63–3.99 (m, 2H, THP-H-5), 4.23 (s, 2H, H-3), 4.67–4.77 (m, 1H, THP-H-1), 7.17–7.69 (m, 6H, Phel-H-4/H-5/H-6/H-7/H-8/H-9); ^{3}J(Phel-H-2, Phel-H-3) = 7.1 Hz

9.3.3.51
3-(2,3-Dihydro-[1-^{13}C]-1H-phenalen-1-yl)-propan-1-ol (66)

According to the combination of the general procedures B and H, 400 mg of 1-[3-tetrahydropran-2-yloxy)-prop-1-inyl]-[1-^{13}C]-2,3-dihydro-1H-phenalen-1-ol

(**65**) (1.2 mmol) was deblocked and deoxygenated to obtain 250 mg (yield: 90 %) of the corresponding alcohol.

^{1}H-NMR: δ [ppm] = 1.56 (s, 1H, -O**H**), 1.69–1.80 (m, 2H, H-2), 2.03–2.15 (m, 4H, H-3/ Phel-H-2), 3.08 (mc, 1H, Phel-H-1), 3.13 (t, 2H, Phel-H-3), 4.12 (m, 2H, H-1), 7.20.–7.30 (m, 2H, Phel-H-4/Phel-H-9), 7.35–7.44 (m, 2H, Phel-H-5/Phel-H-8), 7.65–7.72 (m, 2H, Phel-H-6/Phel-H-7); ^{1}J(Phel-C-1, Phel-H-1) = 126.7 Hz, ^{3}J(Phel-H-2, Phel-H-3) = 6.1 Hz

^{13}C-NMR: δ [ppm] = 30.42, 33.19 (d), 34.19, 37.28 (d), **39.16**, 63.46, 125.72, 125.90, 126.30, 129.84 (d), 131.26; ^{1}J(Phel-C-1, Phel-C-2) = 40.7 Hz, ^{1}J(Phel-C-1, C-3) = 40.7 Hz, ^{1}J(Phel-C-1, Phel-C-9a) = 48.8 Hz

9.3.3.52 *[10a-^{13}C]-1,9,10,10a-Tetrahydropyrene (68)*

To a suspension of 2.5 g pyridinium dichromate in 20 ml dry dichloromethane was added 250 mg of 3-(2,3-Dihydro-[1-^{13}C]-1H-phenalen-1-yl)-propan-1-ol (**66**), and the reaction mixture was stirred 12 h at 25 °C. After filtration, the solvent was removed under reduced pressure. 25 ml of PPA was added to the crude product, and the suspension was heated to 150 °C for 15 min. Subsequently, the mixture was poured into ice, and the aqueous suspension was extracted 4 times with diethyl ether. The organic phase was dried over magnesium sulphate, and the solvent was removed under reduced pressure. After purification by column chromatography (eluent: hexane), 62 mg (yield: 27 %) of the product was obtained.

^{1}H-NMR: δ [ppm] = 2.01–2.05 (m, 2H, H-10), 2.26–2.31 (m, 2H, H-1), 2.85 (mc, 2H, H-10a), 3.01 (t, 2H, H-9), 5.76–5.82 (m, 1H, H-2), 6.71 (d, 1H, H-3), 7.07–7.22 (m, 3H, H-4/H-7/H-8), 7.42–7.51 (m, 2H, H-5/H-6); ^{1}J(C-1, H-1) = 127.1 Hz, ^{3}J(H-2, H-3) = 8.7 Hz, ^{3}J(H-9, H-10) = 6.1 Hz

^{13}C-NMR: δ [ppm] = 34.42, **34.92**, 36.41 (d), 37.34 (d), 122.93, 125.63, 126.09, 126.15, 126.65, 126.92 (d), 128.16 (d), 129.56, 129.87, 131.19, 132.58, 134.81; ^{1}J(C-1, C-10a) = 40.7 Hz, ^{1}J(C-10, C-10a) = 40.7 Hz, ^{1}J(C-10a,10b) = 48.8 Hz, ^{3}J(C-3, C-10a) = 3.1 Hz

9.3.3.53 *[3a-^{13}C]-Pyrene (12)*

According to general procedure I, a solution of 62 mg [10a-^{13}C]-1,9,10,10a-tetrahydropyren (**68**) and 200 mg DDQ (0.88 mmol) in 5 ml dry THF was refluxed for thirty minutes to give 20 mg (yield: 33 %) of the labelled PAH.

^{1}H-NMR: δ [ppm] = 7.69–7.74 (m, 4H, H-4/H-5/H-9/H-10), 7.80–7.84 (m, 2H, H-2/H-7), 8.02–8.07 (m, 4H, H-1/H-3/H-6/H-8)

^{13}C-NMR: δ [ppm] = **126.24**, 126.33, 126.58, 127.26 (s + d), 131.93 (s + d); ^{1}J(C-3a, C-10a) = 57.0 Hz, ^{1}J(C-3,C-3a) = 58.0 Hz

9.3.3.54
1-Methylpyrene (70)

The Wolff-Kischner-Reduction of 1-pyrenecarbaldehyde (**69**) (25.0 g; 109 mmol) was performed according to general procedure F to obtain 21.4 g of 1-methylpyrene (yield: 91 %).

^{1}H-NMR: δ [ppm] = 2.97 (3H, s, -C**H**), 7.85–8.23 (m, 9H, H-2/H-3/H-4/H-5/H-6/H-7/H-8/ H-9/H-10);

^{13}C-NMR: δ [ppm] = 34.56, 123.30, 124.74, 124.85, 124.90, 125.10, 125.11, 125.84, 126.66, 127.27, 127.32, 127.50, 128.69, 129.89, 130.91, 131.44, 136.14

9.3.3.55
3-(1-Pyrenyl)-propan-1-ol (72)

After alkylation of 1-methylpyrene (**70**) (30.9 g; 143 mmol) with 2-(2-bromoethoxy)-tetrahydropyran (29.9 g; 143 mmol) in accordance with general procedure A1, the crude product was deblocked following the general procedure B to give 27.8 g (yield: 75 %) of the pure alcohol.

^{1}H-NMR: δ [ppm] = 2.10–2.18 (m, 2H, H-2), 3.47 (t, 2H, H-1), 3.80 (t, 2H, H-3), 7.88–8.34 (m, 9H, Pyr-H-2/H-3/H-4/H-5/H-6/H-7/H-8/H-9/H-10); ^{3}J(H-1, H-2) = 7.6 Hz; ^{3}J(H-2, H-3) = 6.1 Hz

^{13}C-NMR: δ [ppm] = 29.65, 34.56, 62.41, 123.33, 124.74, 124.85, 124.90, 125.01, 125.11, 125.84, 126.66, 127.27, 127.32, 127.50, 128.69, 129.89, 130.91, 131.44, 136.14

9.3.3.56
1-(3-Bromopropyl)-pyrene (73)

Bromination of 27.8 g 3-pyrenyl-propan-1-ol (**72**) (107 mmol) according to general procedure C gave 25.0 g of colourless crystals.

^{1}H-NMR: δ [ppm] = 2.31–2.39 (m, 2H, H-2), 3.42–3.49 (m, 4H, H-1/H-3), 7.82–8.24 (m, 9H, Pyr-H-2/H-3/H-4/H-5/H-6/H-7/H-8/H-9/H-10)

^{13}C-NMR: δ [ppm] = 31.59, 33.48, 34.37, 123.09, 124.83, 124.93, 125.00, 125.10, 125.88, 126.80, 127.39, 127.44, 127.49, 128.67, 130.07, 130.83, 131.38, 134.71

9.3.3.57
[1-^{13}C]-4-(1-Pyrenyl)-butyric acid (74)

A Grignard solution, prepared from reaction of 1-(3-bromopropyl)-pyrene (**73**) (10.0 g; 33 mmol) and magnesium (0.80 g; 33 mmol) in dry THF, was diluted with diethyl ether and treated with carbon-13 dioxide, generated from ^{13}C-sodium carbonate (1.80 g; 17 mmol) to obtain 2.60 g (yield: 65 %, calculation based on ^{13}C-sodium carbonate) of the labelled carboxylic acid.

^{1}H-NMR (DMSO-d_6): δ [ppm] = 1.96–2.10 (m, 2H, H-2), 2.38–2.46 (m, 2H, H-1), 3.28–3.42 (m, 2H, H-3), 7.91–8.49 (m, 9H, Pyr-H-2/H-3/H-4/H-5/H-6/H-7/H-8/H-9/H-10)

^{13}C-NMR: δ [ppm] = 26.74, 31.97, 33.25 (d), 123.35, 124.10, 124.19, 124.72, 124.87, 126.04, 126.45, 127.18, 127.37, 127.41, 127.71, 127.95, 128.10, 129.28, 130.36, 130.83, **174.26;** ^{1}J(C-1, C-2) = 59 Hz

9.3.3.58
[7-^{13}C]-9,10-Dihydro-7(8H)-benzo[a]pyrenone (75)

According to the general procedures E2, the cyclic ketone was prepared from [1-^{13}C]-4-(1-pyrenyl)-butyric acid (**74**) (11 mmol) to give 1.12 g (yield: 34 %) as slightly coloured crystals.

^{1}H-NMR: δ [ppm] = 2.34–2.43 (m, 2H, H-9), 2.87 (t, 2H, H-8), 3.61 (t, 2H, H-10), 7.93–8.29 (m, 7H, H-1/H-2/H-3/H-4/H-5-/H-11/H-12), 8.82 (s, 1H, H-6); ^{3}J(H-1, H-2) = 6,1 Hz, ^{3}J(H-2, H-3) = 6.6 Hz

^{13}C-NMR: δ [ppm] = 23.12, 26.14, 38.89 (d), 123.19, 123.28, 124.45, 125.25, 125.41, 127.08, 127.22, 127.41, 127.97, 128.35,129.49, 129.59 131.59, 132.11, 137.67, 137.70, **199.16**; ^{1}J(C-7, C-8) = 41 Hz

9.3.3.59
[7-^{13}C]-Benzo[a]pyrene (13)

According to the combination of the general procedures G, H and I, [7-^{13}C]-9,10-dihydro-7(8H)-benzo[a]pyrenone (**75**) (660 mg; 2.4 mmol) was deoxygenated and aromatisized to yield 108 mg of [7-^{13}C]-benzo[a]pyrene (18 %).

^{1}H-NMR (DMSO-d_6): δ [ppm] = 7.82–8.62 (m, 10H, H-2/H-3/H-4/H-5/H-7/H-8/H-9/H-10/H-11/H-12), 8.71 (d, 1H, H-1), 9.23 (s, 1H, H-6), ^{3}J(H-1, H-2) = 5.1 Hz

^{13}C-NMR: δ [ppm] = 122.43, 124.73, 125.56, 126.36, 127.59, 127.70, **128.73**, 138.85

References

Agranat I, Shish YS (1976) The scope of the Haworth-Synthesis. J Chem Educ 53: 488–493

Bisagni E, Pépin JJ, André-Louisfert J (1970) Dérivés partiellement hydrogénés du diméthyl-7,12-benzo[a]anthracène. – Préparation des matières premières. Bull Soc Chim Fr 37: 3023–3031

Breitmaier E, Voelter W (1987) ^{13}C-NMR-Spectroscopy, Methods and Applications. Weinheim, VCH. Germany

Brenna JF (1999) High-precision gas isotope-ratio-mass spectrometry: Recent Advances in instrumentation and biomedical applications. Acc Chem Res 27: 340–346

Brown WG, Bluestein B (1949) Dehydration of 9-Fluorenylcarbinol: A new Synthesis of Phenanthrene. J Amer Chem Soc 62: 3256–3257

Goehler KD, Schütte HR (1970) Synthesen radioaktiv-markierter Verbindungen. Darstellung von 9-Fluorenyl-[9,10-^{14}C]-n-butylester. Z Chem 10: 190–191

Huang-Minlon (1946) A simple modification of the Wolff-Kischner Reduction. J Am Chem Soc 68: 2487–2488

Isaev IS, Buraev VI, Regvhukin AI, Kaptyag VA (1972) Complexes of aromatic hydrocarbons with metal halides and hydrogen halides. X. Valence state of a certain atom to which a proton had been added. J Org Chem USSR 8: 1479–1484

Klassen SE, Daub GH, VanderJagt DL (1983) Carbon-13 Labelled Benzo[a]pyrene and Derivatives. 4. Labelling the 7-10 Positions. J Org Chem 23: 4361–4366
Marshall JL (1983) Carbon-Carbon and Carbon-Proton NMR Couplings. Deerfield Beach, Florida, VCH
McDowell BL, Smolinsky G, Rapaport H (1962) 1,2-Dihydrocyclopenta[jk]fluorene. J Amer Chem Soc 84: 3531–3538
Merrit DA, Brand WA, Hayes JM (1994) Isotope-ratio-monitoring gas chromatography mass spectrometry – methods for isotopic calibration. Org Geochem 21: 53–58
Newman A (1996) The precise world of isotope ratio MS. Analytical Chem 68: A373–A377
Soderquist A, Hughes CD, Horton WJ, Facelli JC, Grant DM (1992) Carbon-13 Chemical Shifts Tensors in Aromatic Compounds. 3. Phenanthrene and Triphenylene. J Am Chem Soc 114: 2826–2832
Staab HA, Haenel M (1970) [1-^{13}C]-Naphthalin: Synthese, NMR-Spektren, ESR-Spektrum des Radikalanions und Autoisomerisationsversuche. Chem Ber 103: 1095–1100
Uhlig F, Snyder HR (1960) Polyphosphoric acid as a reagent in organic chemistry. Adv Org Chem 1: 35–81
Walker D, Hiebert JD (1966) 2,3-Dichloro-5,6-Dicyanobenzoquinone and its reaction. Chem Rev: 153–195
Wong WW, Hackey D, Zhang S, Clarke LL (1995) Accuracy and precision of gas chromatography combustion isotope ratio mass spectroscopy for stable carbon isotope ratio measurements. Rapid Comm Mass Spectrom 9: 1007–1011

10 Field Analytical Techniques for the Analysis of Contaminated Soil

G. Matz, W. Schröder
Environmental Measurement Technology, Technical University of Hamburg-Harburg, Harburger Schloβstr. 20, 21079 Hamburg, Germany

10.1 Introduction

The contamination of soil from former industrial sites varies across all kinds of inorganic and organic chemical products, and thus different analytical instruments (SITE 1996; Overton et al. 1996; Matz 1993b) and procedures are required to determine the toxic potential of a soil sample. The contaminating chemicals vary over an extremely wide range of volatility and solubility, they are typically distributed inhomogeneously and the soil material is heterogeneous. This variability prompts the following question, that should more often be asked by the decision-makers dealing with the treatment of large amounts of contaminated soil:

10.1.1 Is there a Representative Soil Sample?

Two facts are the reason for the answer, "no":

- Semivolatile substances like hydrocarbons and polycyclic hydrocarbons; these were of main interest in our reserch centre SFB and are spilled as liquids like mineral or tar oil. They flow through the soil dependent on the soil particle size and capillary forces, soil pressure and, are partly dissolved and dispersed by surface or ground water and may remain in the soil for decades under the influence of biological degradation. So this contamination is normally found as an extremely inhomogeneous distribution in distances of some centimetres with variations in concentration over some orders of magnitude (Hart et al. 1997) (see Chapter 10.4.1).
- Even if a single mid size sample or a mixed sample from different spots of for example 1 kg, is mixed thoroughly and thought to be homogenised, the analytical results show wide variations (Ramsey et al. 1997; Grant et al. 1997) (see Chapter 10.3.3), as been demonstrated in many interlaboratory comparisons.

  So the second question is compulsory:

10.1.2
How Many Samples and Analytical Results do We Need to Characterise a Site?

"As many as possible to receive a reliable assessment". The number can only be predefined by a regulation or convention, and this has to be related to the soil characteristics. Statistics based on the analytical results of contaminated soil like the mean of all results, the percentiles or standard deviation may then be used to characterise the whole site. But statistics need data. Decisions based on only some analytical results may be extremely misleading (see Chapter 10.4.2). The examples prove that analytical precision as required for many laboratory procedures is not the first priority, nevertheless the quality of analysis should be comparable, assured and controlled (see Chapter 10.3.3).

10.1.3
Why Field Analysis?

Taking samples in the field, and analysing them by analytical procedures in the laboratory, is time consuming. By laboratory analysis detailed information is often not received until some weeks after the work on the contaminated area has been completed. Thus it is not possible to directly react to the result of the analysis in the field. On the other hand, if samples are not analysed immediately, but filled in a sample vial, transported and analysed later, then sample loss and further microbiological degradation may lead to false results.

In order to produce information on field samples in real time, new strategies and analytical procedures based on mobile instruments have been developed. These so called field screening methods are based on simple instrumentation; for example chemical sensors, photo-ionisation detectors or gas chromatographs, but also on more complex mobile instruments like X-ray fluorescence spectrometers (XRFS) and hyphenated techniques like gas chromatography/mass spectrometry (GC/MS).

It takes a long time from the development of analytical methods to their practical use in the field under accepted circumstances and standardised methods. To find acceptance of these methods the US Environmental Protection Agency EPA has instituted a program, the Environmental Technology Verification Program (ETV, www.epa.gov/etv) to verify the performance of innovative technical solutions to problems that threaten human health or the environment. The ETV was created to substantially accelerate the entrance of new environmental technologies into the domestic and international marketplace.

10.2
Field Analytical Instruments

Through the ETV since 1994 a broad selection of mobile instruments has been evaluated in relation to their specifications, analytical procedures, comparison of field results, with laboratory results and advantages in field use. These technologies are grouped in:

- *PCB Field Analytical Technologies*: these are 6 different products like immunoassay or enzymatic kits and other simple sensors especially to do PCB semi-quantitative screening analysis. Similar devices exist for HC, PAH and PCP.
- *Well-Head Monitoring-VOCs*: these 5-gas analysers, like gas chromatographs, IR-photometer and mass spectrometer, are mainly designed for the analysis of volatile organic compounds.
- *Cone Penetrometer-Deployed-Sensor Technologies*: the well known analytical cone penetrometers are 2 laser induced fluorescence spectrometers (LIF). This technology combines the high sensitivity of the fluorescence effect of fluorophores like many PAHs, and measurement directly during the push of the penetrometer into the ground (Hart et al. 1997). Via glass fibre, UV-laser light is guided through a window positioned at the penetrometer´s tip where the fluorescence takes place and back to the spectrometer on the surface. With cone penetrometers the contamination may be qualitatively analysed quasi-continuously (10 datapoints per second). Typically the push is carried out with a speed of 1 m min^{-1} (Hart et al. 1997) down to 10 m. It is mainly used for ground analysis of former petrol stations. The statement of the US EPA report is that the LIF technique does not provide species-specific quantification but can be used as a field screening, qualitative method which can also produce semi-quantitative results at concentrations within two orders of magnitude of its detection limit for fluorescent fuel hydrocarbons. The estimated cost of using the LIF system varies between US$ 36 and US$ 60 per metres depending upon whether the operators provide a turnkey operation or the customer provides field deployment assistance. Under normal condition, 60 metres of pushes can be advanced per day. The advantage of a cone penetrometer is the direct and fast screening sum parameter analysis without sampling, having regard for a poorer analytical quality achieved in case of fluorescent cross contamination. In the prototype stage are some other techniques, which have been deployed in the cone penetrometer; for instance XRFS (Elam et al. 1998), electrochemical sensors (Adams et al. 1995), microscopes (Liebermann et al. 1998) and membrane separators as inlet to GC (Sacks et al. 1996) or GC/MS (Gorshtreyn et al. 1999).
- *Field Portable X-Ray Fluorescence Analyser Technologies*: 7 hand portable energy dispersive XRF techniques for soil samples containing heavy metals.
- *Portable Gas Chromatograph – Mass Spectrometer Technologies*: 2 systems.

10.3 Gas Chromatography – Mass Spectrometry (GC/MS)

GC/MS is the most versatile and sensitive technology for the identification and quantification of organic compounds. Results from field analysis are directly comparable to those achieved by the standard methods in the laboratory (Schuetz et al. 1995). Automated testing offices (Simmonds et al. 1995) or fully equipped mobile laboratories (Li et al. 1995; Hemberger et al. 1991) are often used to operate standard laboratory instruments in the field. Other laboratory instruments have been modified to make them more suitable for field use. Such devices as the completely redesigned field portable Spectra Trak GC/MS (Schuetz et al. 1995) (50 kg without power supply) can be operated without the need for a protective laboratory enclosure. With these devices, usually equipped with a direct MS-inlet system and GC capillary columns, standard laboratory procedures (30 min) can be carried out.

Newly invented methods based on more portable and self-supporting GC/MS systems that are adapted to mobile, on-the-go measurement operation (McClennen et al. 1994; Hemond 1991; Sinha et al. 1991) have also been reported.

On the other hand, very rugged mass spectrometers that operate under all environmental conditions have been developed for military use (Matz et al. 1982). The analytical methods discussed here are based on the Bruker Daltonik MEM (Robbat et al. 1992), the detector for chemical warfare agents used by the German army. Combined with a gas chromatograph, fast analytical methods have boosted the performance of this device, making pollutant analysis in the environment feasible. This article summarises the development of the mobile GC/MS analytical procedures and applications at the Technical University of Hamburg-Harburg over the last decade.

10.3.1 The Mobile GC/MS Instrument

The procedures described here have been established mainly on MEM instruments and are now adapted to the succeeding model EM 640 (Baykut 1995).

The MEM consists of a quadrupole MS with a membrane separator inlet, an electron impact ion source operating at 70 eV, and an ion getter pump. Further features are its modest power consumption (less than 600 watts), simple operation and low maintenance requirements. In the procedures described here mass spectra in full scan mode from m/z 1–400 are recorded. The data system processes and stores the measured data. The mass spectra are a good match to those of commercial databases. Thus identification of unexpected compounds can be done by comparison with the NIST mass spectra database (NIST), with help of the Bruker peak finding and mass spectra deconvolution algorithm. This allows the so-called Non-Target-Screening asides standard analysis.

Two GC modules are available. They can easily be mounted in quick exchange with plug-in connectors. Carrier gas is drawn into the MS over the membrane separator, the GC column and the injector. Mainly, ambient air is used as carrier gas. Temperature programs run from 40 °C to 240 °C. Decomposition of the stationary phase does not occur for several hundreds of analyses. The short capillaries can easily be exchanged. This is necessary if the internal standard signal indicates decomposition of the stationary phase.

10.3.1.1 GC Module for VOCs

Volatile organic compounds (VOCs) are analysed on a GC module with a thermal desorber and a short thick film column (DB1, i.d. 0.32 mm, phase 5 µm, length 5-10 m, J&W Scientific Inc., No. 123/035). Tenax adsorption tubes are used (SKC No.: 226-35-03, glass tubes with 150 mg Tenax TA). The tubes are heated for 30 s at 240 °C in the stop flow carrier-gas supply mode. Then column flow is started to perform the injection. The injection is terminated after 15 s with countercurrent gas flow, and the GC/MS run is started. Column heating is carried out in a small oven at a rate of 1.5 °C s^{-1}.

The thermal desorber can additionally be fitted with a needle adapter to suck in volatile samples, for example as in a headspace injector. Continuous injection is possible when countercurrent gas flow is switched off during the MS acquisition.

10.3.1.2
GC Module for SVOCs

The GC module for semivolatile organic compounds (SOCs) contains a short column (SE-54, i.d. 0.32 mm, phase 0.25 µm, length 3.5 m, NORDION Finland No. 083 054) that is resistively heated directly by means of a metal capillary shielding of the GC column. A special device, the foil stamp injector, has been constructed as an injector. For injection a 10–50 µl aliquot of extract is applied onto a Teflon foil and rapidly heated between two hot stamps pressed together pneumatically. Evaporated compounds are sucked onto the column through a fine metal gauze, and nonvolatiles remain on the disposable injection foil and do not contaminate the capillary column. If the foil stamp injector is contaminated by dirt from the normally uncleaned extracts, the gauze can be changed within seconds without interruption of the GC run.

Table 10.1. Overview of the analytical methods for the mobile GC/MS system

Sample type	Sample preparation	GC separation	Typical limit of detection	References
Air	adsorption on Tenax [1 L in 1 min]	temperature program to 240 °C [5 min]	0.1 ppm (benzene, chlorobenzene)	(Schröder et al. 1989)
Soil	supersonic extraction [3 min]	isothermal 220 °C [2 min]	5–10 ppm (PAH, semiquantitative)	(Schröder et al. 1995a, 1995b; Matz et al. 1993a)
	supersonic extraction [3 min]	temperature program to 240 °C [7 min]	5–10 ppm (SOC, PAH, e.g. benzpyrene)	
	headspace vial [5 min]		5–10 ppm (VOC)	
Water	spray and trap [5 min]	temperature program to 240 °C [7 min]	0.001 ppm (vinylchloride, tetrachloroethene)	(Matz et al. 1993b)
	liquid/liquid microextraction [5 min]	temperature program to 240 °C [7 min]	0.1 ppm (PAH, DDT)	(DIN 1981a)
Surface	direct thermal desorption or wipe samples [1 min]	isothermal or temperature program to 240 °C [2–5 min]	0.1–5 µg cm^{-2} (HCH, PCP)	(Matz et al. 1991)

10.3.2 Analytical Procedures

A variety of procedures with emphasis on short duration and easy use under field conditions have been developed for air, soil and water. An overview over the procedures is given in table 10.1. In detail the analysis shall be described for soil extract analysis of semivolatile compounds like PAH and mineral oil.

10.3.2.1 Analysis of Semivolatile Organic Compounds (SOCs) like PAH

Fast non target screening analysis.

New methods with mobile GC/MS systems (Hemberger et al. 1991; Virkki et al. 1995; Schuetz et al. 1995) on short columns and fast GC runs for high sample throughput, have lowered the costs per sample to prices near those of sum parameter screening analyses. Besides the powerful detection of single compounds, groups of pollutants and multi-compound mixture pollutants e.g. petroleum hydrocarbons, PCBs or chlorinated naphthalenes are recorded as more or less structured "humps". The molecular structure of the mixture components can be determined from the superimposed mass spectra, while the area of the unresolved signals leads to quantitative results as a sum parameter.

SOCs in soil are extracted supersonically with acetone and injected without clean-up on the foil stamp injector. Extraction with acetone has proven to be very efficient (Marvin et al. 1992) and is not influenced by the moisture content of the sample (Dunnivant et al. 1988). Furthermore, in the GC/MS run the signal of water (m/z 18) can be used to quantify the moisture content. For high-speed screening of extracts within approximately 200 s per sample, the GC column is kept isothermally at 240 °C. Under these conditions no significant separation takes place, as shown in fig. 10.1. The lower part displays the logarithmic chromatogram of 8 consecutive MS analyses over 24 minutes. The upper track represents the total ion current and the other tracks show the intensities of significant selected ion peaks (internal standards and PAH). Overlapping full-scan mass spectra (see upper part) are recorded in 1-s intervals.

By this method, samples with organic soil contaminants are quickly identified and selected for more detailed subsequent analysis. Unlike conventional procedures designed for the determination of aggregate parameters (e.g., total hydrocarbons) these fast GC/MS runs allow the identification of major individual contaminants in the sample by the mass spectrum. A first estimate of concentrations present, for example, in terms of trace, medium or high levels can be done.

To get more information about individual constituents, temperature-programmed GC/MS runs are required (see fig. 10.2). If 7–10 min.-runs are used, groups of isomers can be separated, and detailed information on the pattern of contamination is obtained. Substances of medium to low volatility are quantified in the range of 1–1000 mg kg^{-1} within approximately 3–5 min.

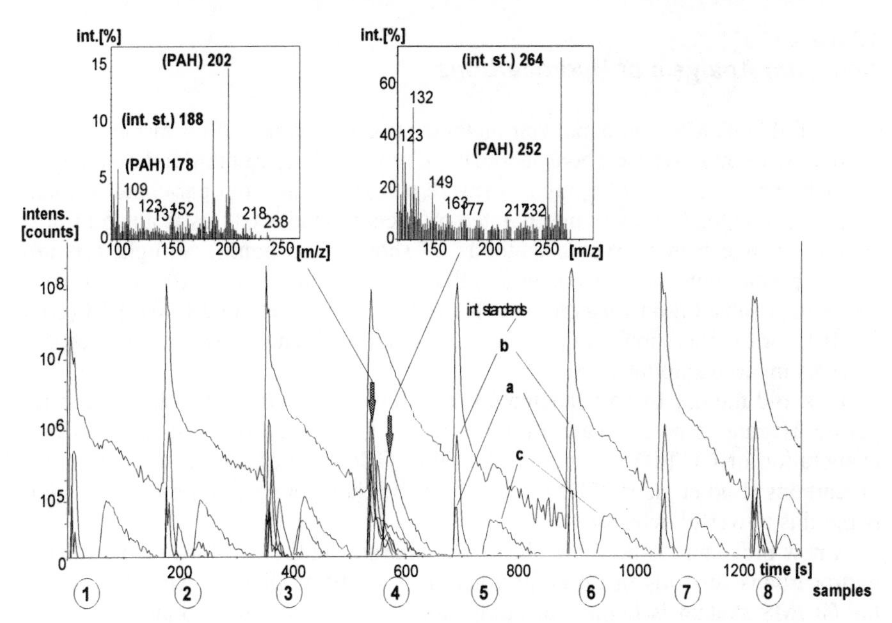

Fig. 10.1. Fast GC/MS-screening run for SOC (PAH), 8 analyses in 24 min

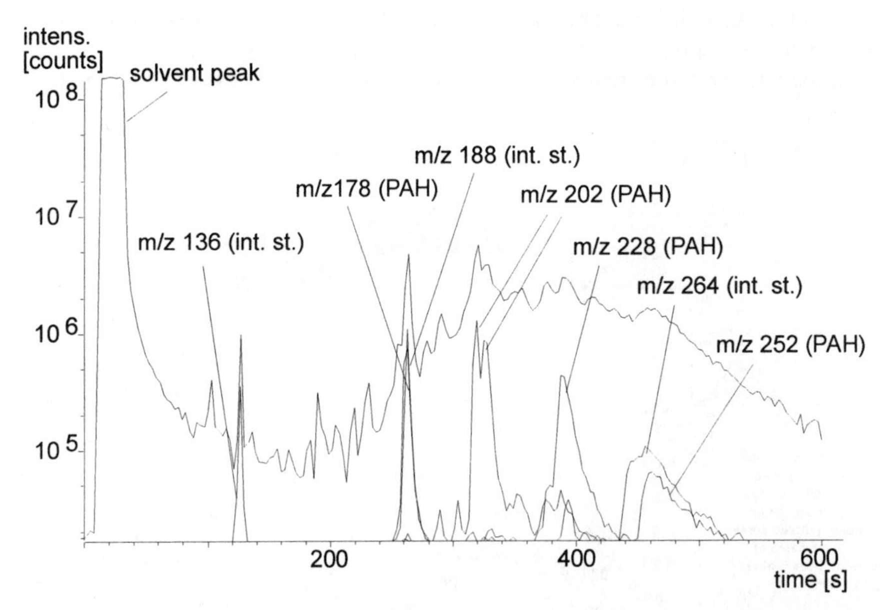

Fig. 10.2. GC/MS run of a soil extract containing PAH, total ion current (45–400 m/z) and selected ion traces

10.3.2.2
New Sum Analysis of Hydrocarbons

A new field GC/MS sum parameter method has been developed for fast evaluation of total petroleum hydrocarbons, and can be run parallelly to the standard analysis (see Chapter 10.3.2.1) solely by a software program. This sum parameter evaluation of chromatographically unseparated multicompound mixtures is an additional advantage of chromatographic methods on short GC columns. Straight forward sample preparation and analysis in less than 10 min. makes this method also interesting as a substitution for common IR methods (DIN method 38 409 H18 (DIN 1981b)) for hydrocarbon sum parameter estimation which use banned halogenated solvents in large amounts.

To avoid the use of halogenated solvents alternative procedures have been reported. Hexane is an alternative for extraction. Quantification is done by solvent evaporation or GC/FID (Draper et al. 1996). GC/MS is also used to analyse hydrocarbons (Luo et al. 1995) but high costs and time consuming procedures hampered this powerful detector.

A new procedure is described here. To calibrate the GC/MS for total petroleum hydrocarbons five oils of different volatility have been injected. The response of the GC/MS system is higher for compounds of lower volatility than for volatile compounds. To express the volatility of each oil calibrated, the so-called volatility range number (VRN) is introduced. It is the ratio of the signal integral of substances eluting before the internal standard d10-anthracene (area with striped pattern) and the complete integral (striped area and grey area). The VRN corresponds linearly to the response calculated from the calibration (see fig. 10.3).

In the GC runs from real samples the VRN calculation is used to find the best

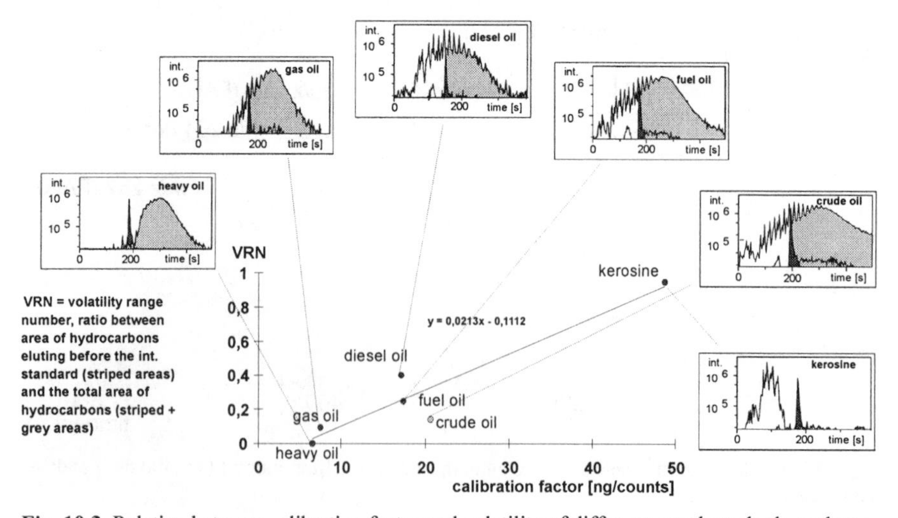

Fig. 10.3. Relation between calibration factor and volatility of different petroleum hydrocarbons

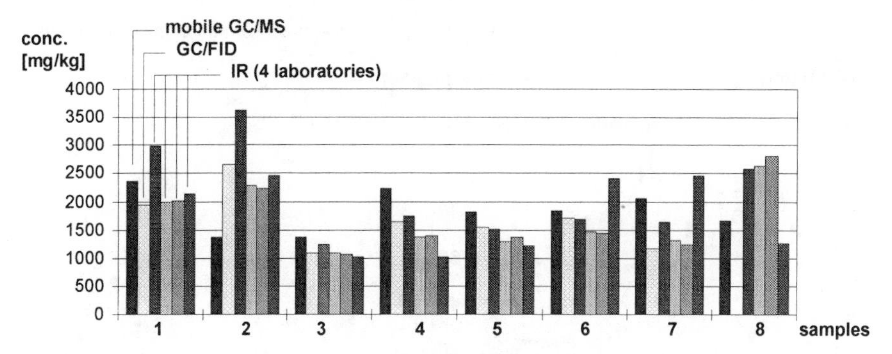

Fig. 10.4. Interlaboratory comparison based on the parallel analyses of 8 soil samples

matching calibration curve. In addition to the quantification of total hydrocarbons, an estimation of the original product is possible by the shape of the GC run.

The interlaboratory comparison has proven the reliability of this method and has shown good comparability with IR or GC/FID procedures (see fig. 10.4).

10.3.3 Quality Assurance and Control

To ascertain the reproducibility of analyses performed with these procedures attention has to be given to quality assurance by the following measures:

- *mass scale tuning*:
  automatically done daily with FC77 used as calibration gas.
- mass scale and sensitivity check:
  the mass scale tuning and sensitivity are controlled by software. Because air is used as carrier gas in the MEM, the argon signal (m/z 40) is used to do a detailed check of the MS function. For optimum comparability spectra need to be obtained with good stability. For the evaluation of the spectra a special peak finding algorithm is used (Sander 1995). Co-eluting or at least partial overlapping are separated into discrete spectra by software. In this manner no distortion of the MS from the peak shape occurs, and clean spectra are obtained comparable with spectra of a commercial spectra database (NIST).
- use of disposables:
  cross contamination is effectively eliminated, because all parts used for sampling, sample preparation and injection are disposable.
- internal standards:
  deuterated compounds like d_6-benzene, d_8-toluene, d_{10}-xylene and d_8-naphthalene dissolved in methanol, are added to the Tenax cartridges for VOC-analysis before sampling. For SOC analysis deuterated naphthalene, anthracene, pyrene and benzpyrene are dissolved in acetone and added to the extract. Because the internal standards undergo the complete sample preparation process, their MS signals can be used for quality checks of sampling efficiency, extraction and injection.

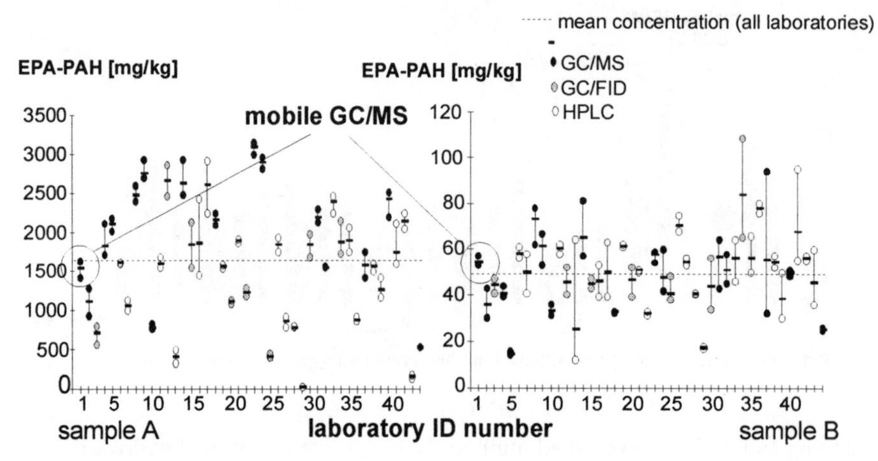

Fig. 10.5. Results of an interlaboratory comparison of two soil samples containing PAH EPA-PAH: sum of 16 PAH as defined in EPA method 610 (1984)

- reference samples and control chart:
  for quality control a reference soil sample containing 14 substances is analysed twice a day. The results are displayed in a control chart (Wagner 1992) as response factors normalised on the surrogate. Thus, errors in the analytical procedure can be detected and corrected immediately.
- interlaboratory comparison:
  external quality control takes place from time to time through interlaboratory comparison. All of these show good results obtained by the on-site GC/MS-analyses (Matz 1993a). Fig. 10.5 compares the results of 44 laboratories. Here two homogenised contaminated soils containing PAH have been analysed. The results from the MEM (Lab. no.1) demonstrate the good quality of these fast field analytical methods.

10.4 Experience from Field Analysis

10.4.1 Site Investigation

Soil contamination at industrial sites consists normally of very inhomogeneous layers of contaminated material. Excavation, construction of new buildings and land filling with clean material produce sites with hot spots of high contamination next to clean areas. Multisample analysis can help to obtain information on the specific appearance of regions of contaminated soil during excavation.

Fig. 10.6 shows the sampling map for the analysis of 55 soil samples. A dark layer has been declared – as expected – to be highly contaminated. On the other hand, soils of the same colour have proved to be clean, whereas highly contaminated regions without visual abnormality were found (compare fig. 10.6 and

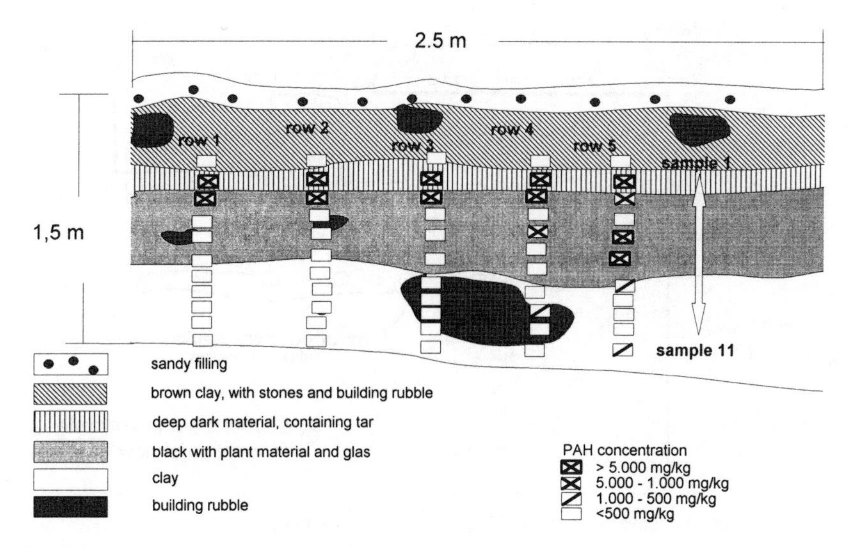

Fig. 10.6. Sampling map for classification of soil contamination by multiple analysis

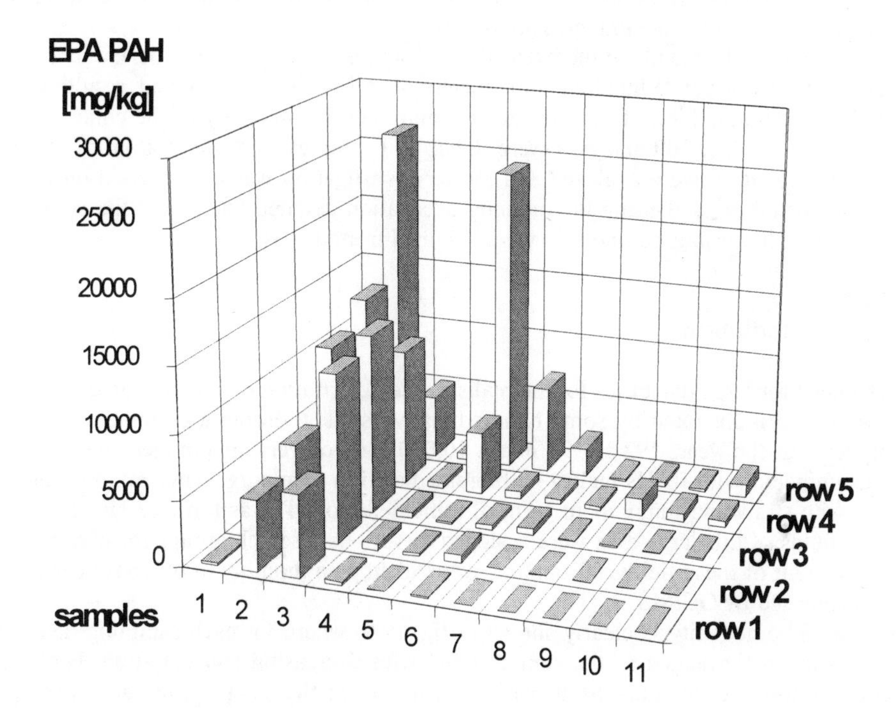

Fig. 10.7. Results of multisample analyses

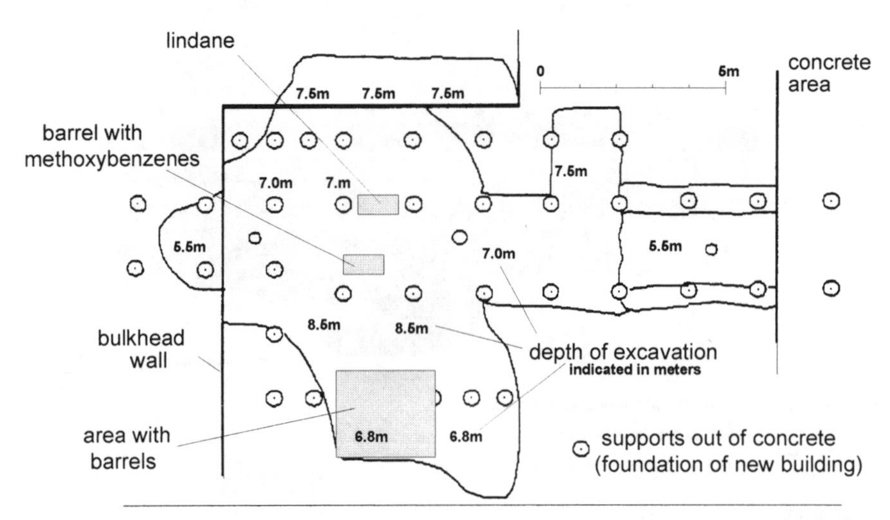

Fig. 10.8. Map of an excavated area, directed by the results of analyses

fig. 10.7). Guided by these analytical results in the field, subsequent excavation could be reduced to contaminated areas only.

In fig. 10.8 the results of the excavation of an unknown site over a period of 11 days are shown. Seven hundred cubic meters of soil have been removed, resulting in the decontamination of the site exactly following the boundary of the contamination. More than 300 analyses were done. Soil samples from the bottom of the actual excavation were analysed and further investigation was only carried on in contaminated areas. During the ongoing excavation detailed data about the composition of hazardous compounds were obtained from the GC/MS runs.

10.4.2 Soil Remediation

The need for large databases for the realistic characterisation of contaminated soils has been demonstrated by some hundred analyses taken during a bioremediation process. In the years '92 to '94 about 10,000 tons of very inhomogeneous soil contaminated with PAHs had to be bioremediated in Hamburg. After mixing the soil with straw and fungi about 1 km of "shielded biobed" was formed. Two specific fields of this large area were assigned to various research groups for investigation. Out of these fields, 60 samples were taken on a monthly to quarterly basis and analysed by GC/MS.

The inhomogenity is clearly shown in fig. 10.9, where for each sampling date all results are arranged as the sum of PAH with decreasing concentration. High concentrations originating from particulate matter in the sample can cause pronounced shifts in the average value. If this detailed information is not available the lack of data can cause a severe misinterpretation of the overall result. 26 months after the start of the cleaning process, there is obviously a small decrease in the average concentration. But by statistical analysis the real effect of the remediation

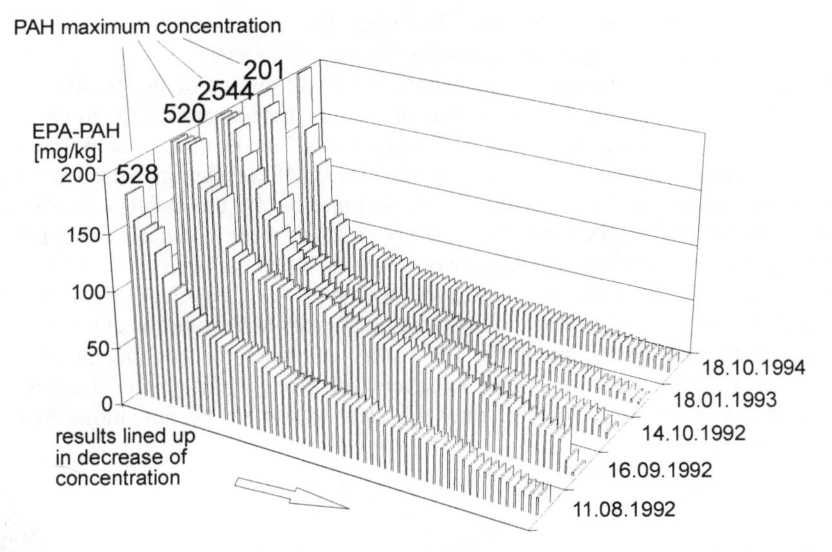

Fig. 10.9. Decrease of the sum of PAHs over a period of 26 months

process can be demonstrated. The most important results of the bioremediation process become visible when the ratios of the different PAH fractions over time are plotted. As an example, see fig. 10.10, where all concentrations of substances with molecular mass 252 are displayed against those with mass 178. Mass 252

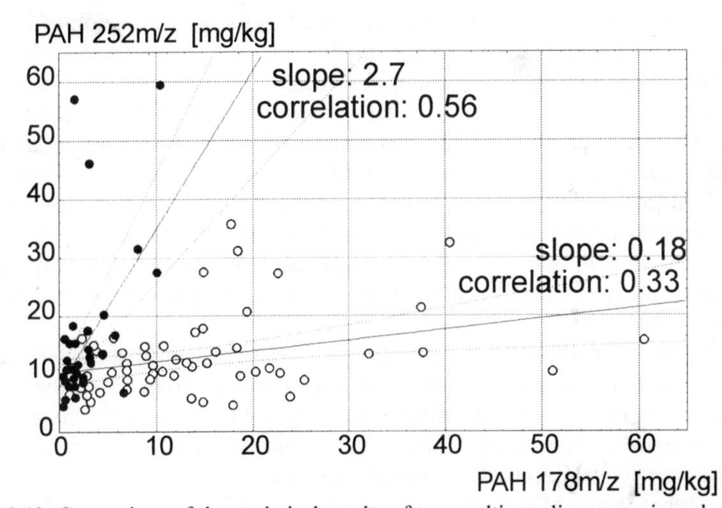

Fig. 10.10. Comparison of the analytical results of two multisampling campaigns during bioremediation of PAH contaminated soil in the beginning (light circles) and 26 months later (dark circles)

represents all isomers of benzpyrenes including the very toxic benz(a)pyrene. Mass 178 represents the less toxic phenanthrene and anthracene.

At the start of the remediation in August '92 (light circles in fig. 10.10), the slope of the regression line is 0.18, with a correlation coefficient of 0.33. In October '94 the distribution of the m/z 252/178 ratio of PAH changed drastically into a slope of 2.7 and a correlation factor of 0.56. The small correlation indicates the high inhomogeneity of the material. But, more importantly, the changes in PAH composition are indicated clearly. Either evaporation of more volatile compounds or biodegradation could have been the reason for this effect.

In any case, the benzpyrenes – substances with high toxic potential – remain in the soil and no significant biological degradation appears to have been achieved. If only aggregate parameters (as in fig. 10.9) were available for qualification this remediation would have been judged as successful. Conversely, this detailed investigation raises the question of whether this cleaned soil has become more toxic than before.

10.5 Conclusion

Fast field screening methods have proven to be useful tools in the management of contaminated sites and for the detection of hazardous compounds in air, water and soil. They can be used as target as well as non target screening methods, because the GC/MS run delivers mass spectra even from non expected substances. The procedures discussed here have been in practical use since 1990. Many different analytical projects have been carried out in Northern Germany on industrial and waste dump sites, wells and in bioremediation processes. The mobile GC/MS is used for site investigation and remediation as well as for the protection of workers against toxic vapours and gaseous compounds.

In routine operation by two persons, by these GC/MS procedures with GC-separation allow 30 samples, and by fast screening without GC-separation 80 samples, to be analysed per day. The comparison with results of samples analysed in commercial laboratories has shown the high reliability of the results obtained in the field. Fast field analyses may greatly reduce the analytical costs without loss of quality. A further advantage of getting the analytical results immediately on site is the cost saving regarding the complete work on contaminated sites; no time delay by laboratory analysis occurs and the ongoing work on site can be lead by the analytical results.

It has further been shown that petroleum hydrocarbons sum parameter evaluation can be generated as a byproduct of multi-compound screening analysis. Fast GC/MS analyses of total petroleum hydrocarbons in the field can be regarded as an alternative to IR methods, which have to be replaced for environmental protection due to their consumption of halogenated solvents.

In the future, further standardisation and regulations by the authorities are necessary to enable complete acceptance for these new methods.

References

Adams JW, Cespedes ER, Cooper SS, Davis WM (1995) Development and testing of cone penetrometer sensor probe for in situ detection of explosive contaminants. Field Screening Methods for Hazardous Wastes and Toxic Chemicals 1: 491–501

Baykut G (1995) Modular sampling and inlet systems for mobile environmental mass spectrometry. Trends in analytical chemistry 14: 10–23

DIN 38 409 Teil 18, DIN Deutsche Industrie Normen (1981b), Deutsche Einheitsverfahren zur Wasser-, Abwasser- und Schlammuntersuchung, Summarische Wirkungs- und Stoffkenngrößen, Bestimmung von Kohlenwasserstoffen (H18). Beuth Verlag, Berlin

DIN 38 409 Teil 13, DIN Deutsche Industrienormen (1981a), Deutsche Einheitsverfahren zur Wasser-, Abwasser- und Schlammuntersuchung, Summarische Wirkungs- und Stoffkenngrößen (Gruppe H), Bestimmung von polycyclischen aromatischen Kohlenwasserstoffen (PAK) in Trinkwasser (H13-1 bis 3). Beuth Verlag, Berlin

Draper WM, Dhaliwal JS, Perera SK (1996) Determination of Diesel Fuel and Motor Oil in Water and Wastes by a Modified Diesel-Range Organic Petroleum Hydrocarbon Method. Journal of AOAC International 79: 508–519

Dunnivant FM, Elzerman AW (1988) Determination of Polychlorinated Biphenyls in Sediments, Using Sonication Extraction and Capillary Column Gas Chromatography-Electron Capture Detection with Internal Standard Calibration. Journal of AOAC International 71: 551–556

Elam WT, Adams JW, Hudson KR, McDonald B, Gilfrich JV (1998) Subsurface measurement of soil heavy-metal concentrations with the SCAPS X-Ray fluorescence (XRF) metals sensor. Field Analytical Chemistry and Technology 2: 97–102

Gorshtreyn A, Smarason S, Robbat A (1999) Speciation of subsurface contaminants by cone penetrometry gas chromatography/mass spectrometry. Environ Sci Technol 33: 2474–2480

Grant CL, Jenkins TF, Mudambi AR (1997) Comparison of Environmental Chemical Results for Split Samples Analyzed in Different Laboratories. Journal of AOAC International 80: 1129–1138

Hart SJ, Chen YM, Kenny JE, Lien BK, Best TW (1997) Field Demonstration of a Multichannel Fiber-Optic Laser-Induced Fluorescence System in a Cone Penetrometer Vehicle. Field Analytical Chemistry and Technology 1: 343–355

Hemberger PH, Alarid JE, Cameron D, Leibmann CP, Cannon TM, Wolf MA, Kaiser RE (1991) A Transportable Turnkey Gas Chromatograph-Ion Trap Detector for Field Analysis of Environmental Samples. Int J Mass Spectrom Ion Processes 106: 299–313

Hemond HF (1991) A backpack-portable mass spectrometer for measurement of volatile compounds in the environment. Rev Sci Instrument 62: 1420–1425

Li K, Fingas MF, Belanger JMR, Paré JRJ (1995) The vehicle-portable analytical system of the Emergencies Science Division of Environment Canada. J of Hazardous Materials 43: 141–154

Lieberman SH, Knowles DS (1998) Cone penetrometer deployable in situ video microscope for characterizing sub-surface soil properties. Field Analytical Chemistry and Technology 2: 127–132

Luo Z, Hsia Y, Xie K (1995) Analysis of Diesel Components in Soil and Water Conrtaminated by Semivolatile Synthetic Organic Compounds. J Chromat Science 33: 263–267

Marvin CH, Allan L, McCarry BE, Bryant DW (1992) A Comparison of Ultrasonic Extraction and Soxhlet Extraction of PAH from Sediments and Air Particulate Material. Int J Env Anal Chem 49: 221–230

Matz G, Trinks H (1982) Mobile Mass Spectrometer System for the Analysis of Organic Trace Gases. International Journal of Mass Spectrometry and Ion Physics 43: 79–82

Matz G, Schröder W (1991) Fast On Site Analysis Using a Mobile Mass-Spectrometer, Final report on "Neue Verfahren und Methoden zur Sanierung von Altlasten am Beispiel der Deponie Georgswerder Hamburg". Environmental Protecting Agency Hamburg, p 98

Matz G, Schröder W, Flachowsky J (1993a) On-site investigation of contaminated soil by GC-MS and EXDRF-techniques. In Arend F, Annokkée GJ, Bosman R, van den Brink (eds) Contaminated Soil '93, pp 657–664

Matz G, Kessners P (1993b) Spray and Trap Method for Water Analysis by Thermal Desorption Gas Chromatography/Mass Spectrometry in Field Applications. Anal Chem 65: 2366–2371

Matz G, Schröder W (1993c) Fast GC/MS-Analysis of Contaminated Soil: Routine Field Screening in Hamburg. presented on the Interational Symposium "Field Screening Methods for Hazardous Wastes and Toxic Chemicals", Las Vegas 2, pp 963–969

Matz G, Kübler J, Schröder W (1995) Multisampling, fast field GC-MS-analysis and statistics for site investigation and remediation of inhomogeneous soil contamination. In: Van den Brink WJ, Bosman R, Arend F (eds) Contaminated Soil '95, Kluwer Academic Publisher, pp 233–234

McClennen WH, Arold NS, Meuzelaar HLC (1994) Field-portable hyphenated instrumentation: the birth of the tricorder? Trends in analytical chemistry 13: 286–293

NIST/EPA/NIH Mass Spectral Database, US Department of Commerce

Overton EB, Dharmasena HP, Ehrmann U, Carney KR (1996) Trends and Advances in Portable Analytical Instrumentation. Field Analytical Chemistry and Technology 1: 87–92

Ramsey MH (1997) Measurement Uncertainty Arising from Sampling: Implications for the Objectives of Geoanalysis. Analyst 122: 1255–1260

Robbat A, Liu T, Abraham BM (1992) Evaluation of a Thermal Desorption Gas Chromatograph/Mass Spectrometer: On-Site Detection of Polychlorinated Biphenyls at a Hazardous Waste Site. Anal Chem 64: 356–364

Sacks R, Klemp M, Akard M (1996) High-Speed Capillary Column GC for Rapid Screening of Gasoline to Diesel Range Organic Compounds. Field Analytical Chemistry and Technology 1: 97–102

Sander P (1995) Application of Fuzzy Logic-finding Algorithm for GC/MS-Analysis. Proc of the 43nd ASMS Conference on Mass Spectrometry and Allied Topics, p 724

Schröder W, Matz G (1989) Schnelle Analytik mit einem mobilen Massenspektrometer bei Transportunfällen im Hamburger Hafen. presented on the International Environment Congress, Umweltbehörde Hamburg, pp 355–357

Schröder W, Matz G (1995a) Boden-GC/MS-Screening im Gelände. TerraTech 4/95: 32–36

Schröder W, Matz G (1995b) Feld-Analytik mit GC/MS. WLB Wasser, Luft und Boden 95/1–2: 47–50

Schuetz SP, Solinski PJ, Mickunas DB, Humphrey AM, Turpin RD (1995) Comparison of data quality produced by an on-site field GC/MS and an off-site permanent laboratory GC/MS: support of a cleanup action at an inactive drum recycling facility. J of Hazardous Materials 43: 67–75

Simmonds PG, O'Doherty SO, Nickless G, Sturrock GA, Swaby R, Knight P, Ricketts J, Woffendin G, Smith R (1995) Automated Gas Chromatograph/Mass Spectrometry for Routine Atmospheric Field Measurements of the CFC Replacement Compounds, the Hydrofluorocarbons and Hydrochlorofluorocarbons. Anal Chem 67: 717–723

Sinha MP, Gutnikov G (1991) Development of a Miniaturized Gas Chromatograph-Mass Spectrometer with a Microbore Capillary Column and an Array Detector. Anal Chem 63: 2012–2016

SITE Superfund innovative technology evaluation Technology Profiles (1996) US EPA, EPA/540/R-97/502

Virkki VT, Ketola RA, Ojala M, Kotiaho T, Komppa V, Grove A, Facchetti S (1995) On-Site Envionmental Analysis by Membrane Inlet Mass Spectrometry. Anal Chem 67: 1421–1425

Wagner RJ (1992) Ringversuche im Rahmen der Analytischen Qualitätssicherung. Tm Technisches Messen 59: 167–172

Ecotoxicological Assessment of Soils

11 Advances in Biological Assessment of Chemical-Impacted Soils

W. Ahlf, W. Frühling, H. Neumann-Hensel
Department of Environmental Science and Technology, Technical University of Hamburg-Harburg, Eissendorfer Str. 40, 21073 Hamburg, Germany

11.1 Introduction

Ecotoxicology has a long history of engaging in the development of biological assessment tools applicable to ecological hazard and risk evaluation. The Proceedings of the 1st International Conference "Environmental Bioassay Techniques and their Application" contained 67 papers, but only two were dealing with soils (Munawar et al. 1989). While numerous compilations of aquatic and soil test methods are readily available for evaluating toxicity as an ecological effect, fewer test methods are readily accessible for evaluating soil contamination. Here, as a result of a 12 year-study, we present biological assessment tools applicable to soil contamination evaluation, with a central focus on methods that are currently available for evaluating soil toxicity. Soil toxicity is one potential indicator of adverse effects associated with soils, caused by chemicals released to the environment.

It is important to note, that the responses of test organisms vary in their sensitivity to single compounds. Thus, a battery of bioassays is typically used for the detection of potential adverse effects of complex mixtures of contaminants (Keddy et al. 1995). Toxicity is a dose-related concept, and the ecotoxicology of a particular contaminant depends on its fate and how it becomes distributed in ecosystems. In this regard, we have to consider the specific properties of the pollutants in question. Is it mobile, in a geochemical sense, because of aqueous solubility, volatility, or is it associated with solid phases? These different exposure routes require a selection of suitable test organisms. Based on further considerations related to bioassay application, it is obvious that rapid, inexpensive methods are needed.

Microorganisms are major components of the soil ecosystem and play a key role in organic matter degradation and nutrient cycling. This ability is connected with the decomposition of environmental organic pollutants. Bacteria can serve as indicators of soil pollution, where the indicators are base on measurements of microbial populations and their activity. However, it is sometimes difficult to distinguish between natural fluctuations in microbial populations and activity, and those due to toxicant input. We have developed assays which, with consideration of biological test methods using plants and animals, meet many of the above requirements (Rönnpagel et al. 1995; Traunspurger et al. 1997). This paper proposes a strategy for determining the potential environmental impact of contaminants in soil, which also provides information on the magnitude of the degradation potential.

11.2 Strategy

Ecotoxicological testing should be conducted on an appropriate, limited battery of species, end points, and exposure routes. If this is carefully done and the results are interpreted as an integrative assessment, there is a clear opportunity to prioritise areas of most concern. In a conceptual soil toxicity exposure model, organisms from different trophic levels can be exposed to toxicants in either of two ways: directly from solid-bound contaminants or via the soil interstitial water. Elutriates are often used as a surrogate for soil interstitial water to assess groundwater hazard. The Elutriate Test was developed as a leaching procedure primarily to determine the mobility of contaminants subject to release when solid waste was in contact with water. The use of elutriates as toxicant solutions has facilitated the testing of standard bioassay organisms, such as *Scenedesmus subtilis* and *Vibrio fisheri*. In some cases the extent and severity of environmental contamination were adequately determined with elutriate tests at hazardous waste sites (Miller et al. 1985). However, elutriates (water extracts) enable detection of a part of multiple contamination, due to the different solubility of each contaminant in water. Water elutriation could underestimate the types and concentrations of bioavailable organic contaminants present (Liss and Ahlf 1997).

An additional approach is to use organisms which have contact with the contaminated solids. It is recommended that within a set of screening tests, soil-living organisms should be preferred (Keddy et al. 1995). At present, toxicological testing is routinely conducted at individual organism level. Three organism-level measurements are required to provide information on the stability of populations and higher levels of organisation: survival, growth, and reproduction (Chapman 1991). If an organism can fulfil all of these integrative functions, then it is not being adversely affected. Tests for whole soils use organisms, such as nematoda or collembola need to be included. Many experiments have been conducted to evaluate the toxicity of chemicals to earthworms and to identify the pollutants that may adversely affect the growth and reproduction of this valuable species in the soil ecosystem. The 14-day earthworm survival test is commonly used as an endpoint. The earthworm reproduction test using *Eisenia fetida* proved to be an appropriate replacement for the earthworm survival test. The test duration must be extended to three to five weeks. These test procedures are time consuming and expensive, and therefore not well suited for screening contaminated sites. Consequently, short-term tests are needed to measure biolgical endpoints characteristic of chronic effects associated with soil contamination. Nematodes are the most abundant and species-rich organisms of the metazoa in soil. Traunspurger et al. (1997) presented a life cycle-test using soil medium, it needs to run for only 72 h. The nematode *Caenorhabditis elegans* has been used in tests for contaminants in the liquid phase, and in whole soil samples. The development of first-stage larval worms to the reproductive stage, the number of eggs per adult, and the number of offsprings per worm are useful parameters for the ecotoxicological interpretation of test data.

Ideally, soil toxicity to bacteria should be examined using a representative soil bacterium and conducted in the soil. A contact bioassay using *Bacillus subtilis* or *Arthrobacter globiformis* has been developed using the inhibition of dehydrogenase enzyme activity as an endpoint (Rönnpagel et al. 1995). An advantage is the exposure time of 2 h in comparison to the short generation times of the bacteria, which allows toxic effects to be expressed on growing cells for approximately two generations.

In-situ alterations occur in the resident microbial community, in relation to different kinds of stress. Methods of microbial ecology designed to detect changes in soil communities should be incorporated in a comprehensive survey. Newer methods such as fatty acid analysis, carbon substrate utilisation or genomic analysis, show considerable promise to improve the speed and precision of soil microbial toxicity tests. At present, a combination of bioassays, and a description of the microflora using enzyme activities, should be worthwhile for the interpretation of the measured effects. Studies with soil microcosms demonstrate that bioassays do not lead to an overestimation of hazard (Rönnpagel et al. 1998). Only soil materials in which soil processes, or at least parts of them, are disturbed have an impact on the test organisms. Therefore performed bioassays gave no false positive results. Conversely, it is not possible to exclude an impact on soil microflora if the employed test battery does not indicate any response.

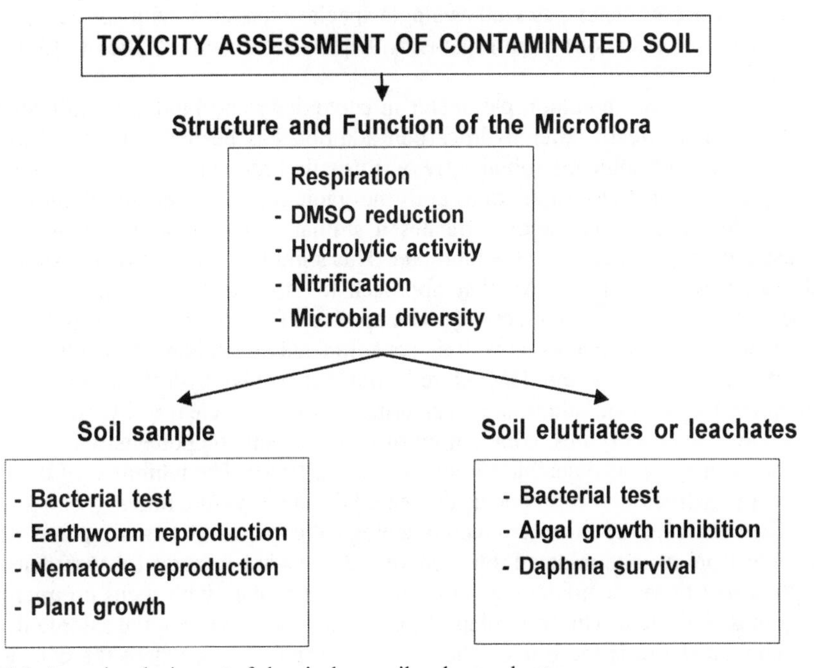

Fig. 11.1. Assessing the impact of chemicals on soil and groundwater

This means false negative results are possible, although a test battery is used with indigenous and highly sensitive organisms and exposure scenarios. Identification of risks for highly sensitive nitrification needs highly sensitive bioassays for effective detection. Contact assays are more ecologically relevant for adsorbing compounds, and therefore more sensitive to them. Effects of highly water-soluble substances, which hardly adsorb soil microflora, are well represented by the elutriate tests. In this case, sensitive elutriate tests give a better estimation than contact assays with less susceptible species. Alternatively, a comparison of dose-response curves for soil activities and bioassays may indicate if a quantitative evaluation of effects is possible. Prognosis of effect concentrations presumes that the types of reactions of contaminants in soil and representative bioassays match, or are at least connected in a way that allows deduction. Before a discriminating prognosis of contaminant effects for soil microflora is possible with bioassay results, as suggested by van Beelen and Doelman (1997), this problem must be solved. Fig. 11.1 outlines the proposed strategy.

11.3 Application of a Battery of Tests

The aim of the study was to quantify toxic effects of hydrocarbon-contaminated soils. A short term experiment over four weeks was designed to quantify toxic effects on organisms exposed to soil, contaminated with a new synthetic lubricant, in a concentration of 4 % (w/w). Consequently, the experimental approach includes some parameters, as well as single species bioassays, with special consideration on (or effect on) microorganisms. Effects are examined after 1, 15 and 29 days (fig. 11.2).

Respiration is a common parameter in ecotoxicity and biodegradation studies. A CO_2 burst directly after hydrocarbon addition has been frequently observed. Because heterotrophic microbial activity is usually limited by organic C, up to 5 % oil application causes stimulation, until inorganic nutrients become limiting. The respiration data in our study indicates a similar trend for a stimulation effect caused by the addition of a new lubricant. This stimulation is also demonstrated by the solid-phase bioassay. Another approach to estimate harmful effects on soil microflora is via enzymatic activity. The applied FDA method has been described for laboratory experiments as well as contaminated site studies. In contrast to the results of respiration and DMSO reduction, microbial hydrolytic activity was inhibited by lubricant addition. Considering that FDA is cleaved by esterases and other hydrolytic enzymes, competition of hydrocarbon components with FDA for the same enzymes is plausible (Song and Bartha 1990). The inhibition of the FDA assay by hydrocarbon components decreased during exposure time.

The data from the short-term ecotoxicological experiment was compared with a sample from a contaminated site, where 2.7 % (w/w) mineral oil was detected. Only a part of the ecotoxicological responses are shown, with focus on microbial respones (fig. 11.3). The solid-phase bioassay demonstrates that the sample from a contaminated site is more toxic than new lubricant in soil, where the concentrations are comparable. The result of the algal growth inhibition test suggested that by-products of the degradation, or other unknown chemicals, increase the toxicity of the water soluble fraction. Enzyme activity of the microflora was measurable, but on a low level. Thus, the toxicity data must be seen as supplementary to enzyme activity, indicating in this case a depressed biological degradation.

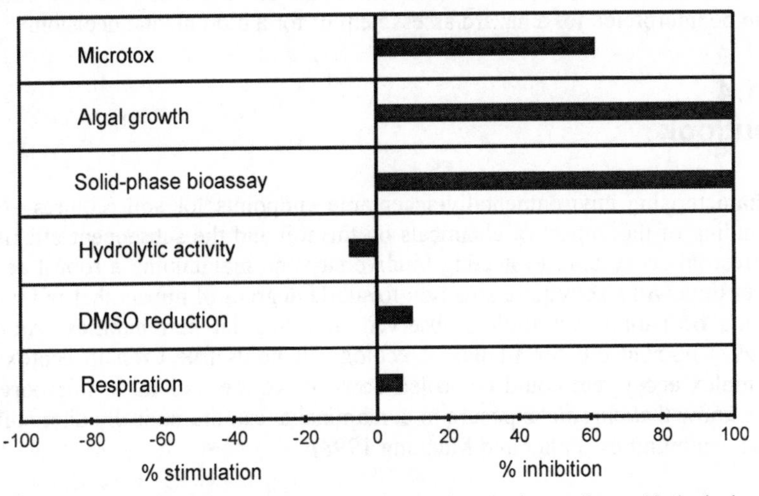

Fig. 11.2. Impact on biological tests of 4 % (w/w) lubricant in a A_h-layer of an orthic luvisol

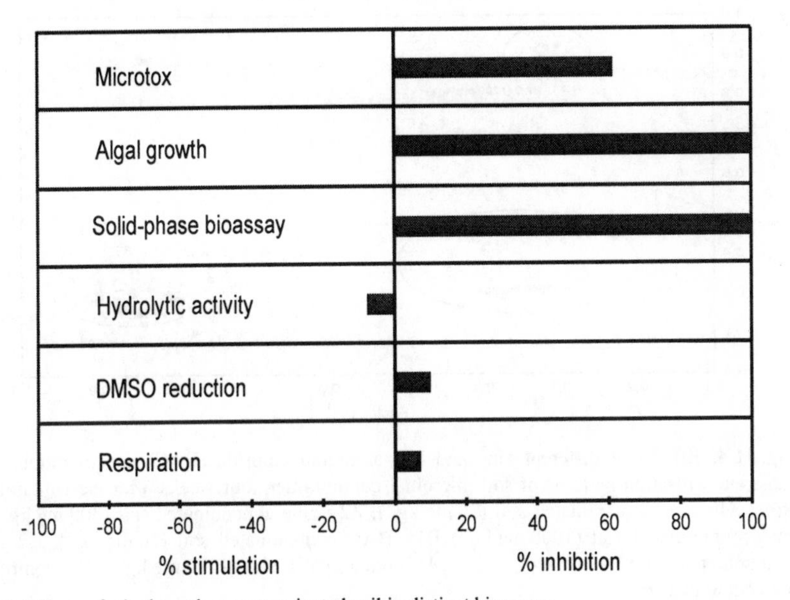

Fig. 11.3. Effects of a hydrocarbon-contaminated soil in distinct bioassays

A tiered examination with the described tests is recommended for evaluation of contaminated sites (Neuman-Hensel et al. 1999). The approach begins with a set of bioassays that helps to identify what is important to protect, and to identify the impact on the important groundwater pathway. If these tests indicate no meaningful interpretation, investigations of the second or third level are required. Several biological assessment tools are recommended to test soil toxicity on soil microflora, soil animals and plants. Furthermore, the method describes how information can be interpreted for a hazard assessment or for a bioremedial decision.

11.4 Outlook

Characterising environmentally-acceptable endpoints for soil requires an understanding of the impact of chemicals on the soil and the subsequent effects on the terrestrial ecosystem. Protecting biodiversity and maintaining a robust ecosystem over time will also require attention to subtle degrees of impact that fall within the limits of natural variability, observed in terrestrial communities. As Fenchel (1992) pointed out for studies in ecology, it holds true even in ecotoxicology, complex ecosystem could be scaled down if we are looking at microorganisms. We know that chronic exposure to contaminants can dramatically alter soil microbial communities (Fuller and Manning 1998).

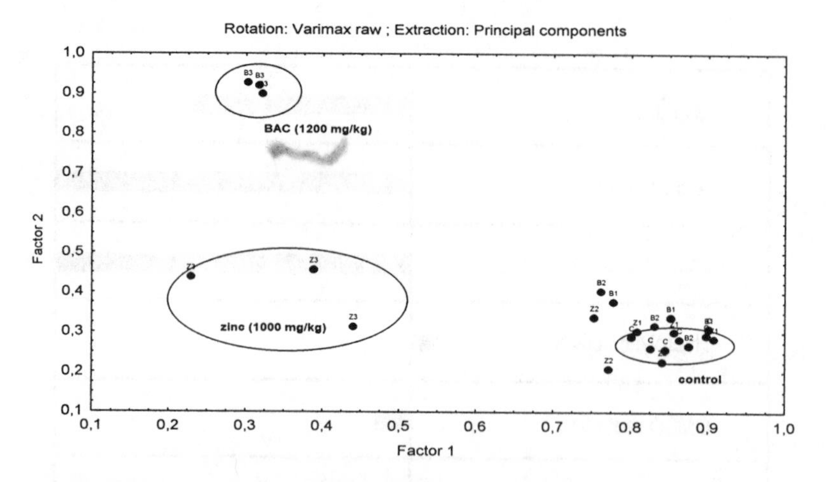

Fig. 11.4. Effects of different zinc and benzalkonium chloride (BAC) concentrations on the substrate utilisation patterns of soil microbial communities four weeks after the contamination. Note: Z1= zinc contaminated soil (30 mg kg^{-1}), Z2= zinc contaminated soil (300 mg kg^{-1}), Z3= zinc contaminated soil (1000 mg kg^{-1}) B1= BAC contaminated soil (50 mg kg^{-1}); B2 = BAC contaminated soil (500 mg kg^{-1}), B3= BAC contaminated soil (1200 mg kg^{-1}), C = control without contamination.

Microorganisms existing in a polluted soil ecosystem provide a robust measure of the impact of toxicants on the soil ecosystem. Using indigenous microorganisms to assess soil contamination is also advantageous with regard to environmental relevance. If we consider microbial communities as functional units, it should be possible to characterise them on the basis of the metabolic properties of the microbial taxa involved. Consequently, the substrate utilisation pattern of the total community can reflect a cause-and-effect relationship between contaminant levels and microbial response. Wünsche et al. (1995) and other authors used the Biolog system as a tool for detecting shifts in microbial communities under changing environmental conditions. In fig. 11.4, a principal component analysis plot of substrate utilisation patterns is shown comparing soils spiked with different concentrations of zinc or benzalkonium chloride (BAC). It is obvious that high contamination of soil samples with either 1000 mg zinc $\cdot$ kg^{-1} or 1200 mg BAC $\cdot$ kg^{-1} lead to considerable changes of metabolic patterns in the bacterial community. In comparison with the uncontaminated control, treatments with lower concentration of both chemicals only have a slight influence on the patterns. This example gives an illustration that microbial communities in soil are sensitive to contaminants. At present, these experiments give results which can be classified. Therefore, the Biolog system is a rapid and cheap screening technique to detect stressed microbial soil communities in cases where an appropriate control is available. However, the interpretation and fundamental ecological insight into the microbial communities of complex ecosytems are limited by unknown variables (Konopka et al. 1998).

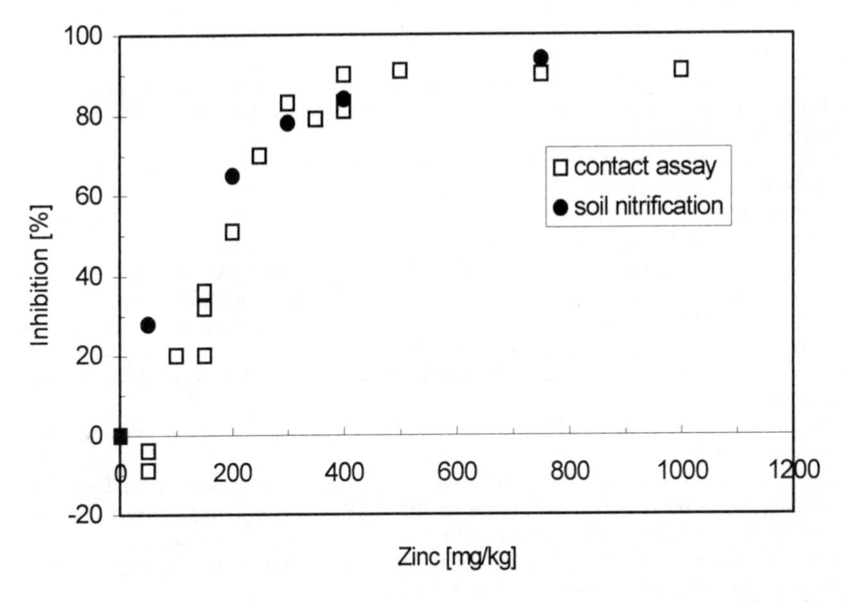

Fig. 11.5. Effect of zinc-spiked soil on ammoium oxidation in the contact assay with test bacteria *Nitrosomonas communis* Nm2 and in comparison on autochthone ammonium oxidation in soil

A way to a better understanding of the impact of contaminants on microbial communities is a comparison of single species tests, with naturally derived microbial communities in microcosms. Rönnpagel et al. (1995) demonstrated that chemicals, which cannot be extract by water can act as toxicants in suspension with soilds and bacteria. Therefore, microbial assays, testing the whole soil, seemed to be a more realistic procedure to reflect the situation of the indigenous organisms in contaminated sites.

In fig. 11.5 the dose-effect relationships of ammonium oxiding bacteria supports this proposal in an impressive way. The ammonium oxidation in a single species contact assay, with the *Nitrosomonas communis* strain as test organisms, was affected by zinc contamination on nearly the same level as the ammonium oxidation of the resident soil microorganisms.

These results indicate that soil microorganisms are very sensitive to zinc toxicity and that single species and community results are consistent and comparable, if the exposure conditions are considered. These new tools promise accurate measurements of the impact of contaminants on the soil microbial community. Comparisons with field data information will provide information on how natural stressors can confound the response of the ecological receptors.

References

Beelen van P, Doelman P (1997) Significance and Application of Microbial Toxicity Tests in Assessing Ecotoxicological Risks of Contaminants in Soil and Sediment. Chemosphere 34: 455–495

Chapman PM (1991) Environmental Quality Criteria. Environ Sci Technol 25:1353–1359

Fenchel T (1992) What can Ecologists Learn from Microbes: Life beneath a Square Centimetre of sediment Surface. Funct Ecol 6: 499–507

Fuller ME, Manning Jr JF (1998) Evidence for differnetial Effects of 2,4,6-Trinitrotoluene and Other Munitions Compounds on Specific Subpopulations of Soil Microbial Communities. Environ Toxicol Chem 17: 2185–2195

Keddy CJ, Greene JC, Bonnell MA (1995) Review of Whole-Organism Bioassays: Soil, Freshwater sediment, and Freshwater Assessment in Canada. Ecotoxicol Environ Saf 30: 221–251

Konopka A, Oliver L, Turco RF (1998) The Use of Carbon Substrate Utilisation Patterns in Environmental and Ecological Microbiology. Microbial Ecology 35: 103–115

Liss W, Ahlf W (1997) Evidence from Whole-sediment, Porewater, and Elutriate Testing in Toxicity Assessment of Contaminated Sediments. Ecotoxicol Environ Saf 36: 140–147

Miller WE et al. (1985) Comparative Toxicology of laboratory Organisms for Assessment of Hazardous Waste Sites. J Environ Qual 14: 569–574

Munawar M et al. (1989) Environmental Bioassay Techniques and Their Application. Kluwer Academic Publishers

Neumann-Hensel H et al. (1999) Natural Attenuation of Petroleum Products in Soils. Vom Wasser 93: 185–193

Rönnpagel K, Janssen EW, Ahlf W (1998) Asking for the Indicator Function of Bioassays Evaluating Soil Contamination: Are Bioassay Results Reasonable Surrogates of Effects on Soil Microflora? Chemosphere 36: 1291–1304

Rönnpagel K, Liß W, Ahlf W (1995) Microbial Bioassays to Assess the Toxicity of Solid-Associated Contaminants. Ecotoxicol Environ Saf 31: 99–103

Song HG, Bartha R (1990) Effect of Jet Fuel Spills on the Microbial Community of Soil. Appl Environ Microbiol 56: 646–651

Traunspurger W et al. (1997) Ecotoxicological Assessment of Aquatic Sediments with Caenorhabditis elegans (Nematoda). Environ Toxicol Chem 16: 245–250

Wünsche L, Brüggemann L, Babel W (1995) Determination of substrate utilisation patterns of soil microbial communities: An approach to assess population changes after hydrocarbon pollution. FEMS Microbiology Ecology 17: 295–306

12 Ecotoxicological Assessment of Soils – Bioavailability from an Ecotoxicological Point of View

W. Kördel and K. Hund-Rinke
Fraunhofer-Institute for Environmental Chemistry and Ecotoxicology, Auf dem Aberg 1, 57392 Schmallenberg, Germany

12.1 Introduction

It is the aim of the Federal Soil Protection Act (Bundes-Bodenschutzgesetz, BBodSchG 1998) to protect or to restore the functions of the soil in a sustainable way. According to the BBodSchG (1998) the natural (ecological) functions of the soil include:

- the basis for life and habitat, for human beings, animals, plants and soil organisms (in one word: the biospherical function),
- as part of natural systems, especially as part of water and nutrient cycles,
- as a medium for decomposition, balance and restoration as a result of its filtering, buffering and substance-converting properties, especially for groundwater protection.

According to the BbodSchG § 2, page 3, a harmful soil change is an impairment of the soil functions that can lead to increased risk, applicable disadvantage or a considerable nuisance for a single person or for the general public. The existence and extent of a contaminant, or the combination of different contaminants, in a soil leads to an impairment of soil functions. This depends on the type of substance or the combination of substances respectively (substance-specific properties) as well as on its concentration, soil properties and the respective use of soil.

The derivation of the precautionary, trigger and action values indicated in the Soil Protection Ordinance (Bundesbodenschutzverordnung, BBodSchV) is substance-based, as commonly performed in toxicology. Presently soil values have been derived only for some chosen environmentally relevant contaminants concerning the potential risk of target protection (e.g. human health, groundwater) taking into consideration defined transfer pathways. Principally analogous substances and values can be derived for the transfer pathway soil → soil organisms. Results for some contaminants have already been published (Dreher et al. 2000). During the assessment of a soil exceedence values concerning a specific soil use or reuse of soil can be checked analytically applying standardised methods.

In reality it might not be sufficient to focus on contaminants that have been already edited and listed. This is especially true for treated soils that were contaminated by different and often unclear mixtures of contaminants. This is why there is a need for further investigation methods besides physical and chemical soil characterisation via fixed analysis methods and a chosen set of contaminants. These investigation methods should be able to indicate a general substance effect in order to include the presently omitted protection targets: "habitat function for soil organisms" and "retention function". In this context the determination of the exposure of organisms, and consequently the bioavailability, takes a major role concerning the test results obtained and subsequent assessment of soils.

An overview table regarding selection criteria for biological test systems depending on the intended reuse/use of soils has been developed by an ad hoc working group. This is entitled "methods for the toxicological/ecotoxicological assessment of soils" and is part of the DECHEMA in 1995.

Exisiting guidelines for the recycling of soils in addition to the BBodSchV include the technical standard DIN 19731 "Anforderungen an die Verwertung von Bodenmaterial" and the Technical Regulations "Anforderungen an die stoffliche Verwertung von mineralischen Reststoffen/Abfällen" (LAGA 1997). The essential protection targets for these soils can be safeguarded sufficiently via solid matter and eluate analysis as included in ecotoxicological/soil-biological tests. This makes it necessary to have a scheme for testing and evaluation whereby chemical analyses and biological tests come together for a scientifically founded assessment of soils according to the state-of-the-art. The test scale and the soil values to be used should be differentiated between intended reuses, namely:

- unrestricted open reuse,
- restricted open reuse,
- restricted installation with defined technical protection measures.

Table 12.1. Selection criteria for biological test systems depending on the intended reuse

Use of soil	Soil functions ↙ ↘		
	Retention function ↓	Habitat function ↓	↓
	(water pathway)	(plant)	(soil organisms)
• below sealed areas	no	no	no
• commercially and industrially used unsealed areas	yes	no	no
• landfill covering	yes	(yes)	no
• green areas, park and recreation areas	yes	(yes)	(yes)
• areas used in horticulture or agriculture	yes	yes	yes

According to DECHEMA, 1995

12.2 Evaluation Strategy under Inclusion of Ecotoxicological Tests

The primary aim of a remediation measure will usually be for a restricted open reuse of the treated soil materials. Nevertheless, a testing scheme will be required to demonstrate differences in the test scale for an unrestricted open reuse. An open unrestricted reuse means reuse in sensitive agriculture and gardening areas as well as in residential areas where children play and also food is cultured. Taking into account a sustainable soil management approach the analysis values for defined contaminants should fall into the range of precaution and background values (see reasoning of the Z 0-values, LAGA 1997).

Tables 12.2 and 12.3 give an overview of testing and evaluation strategies for an open reuse. For the open reuse the protection of the groundwater is a fundamental protection target. Test values for the evaluation of the eluate are, for example, indicated in the BBodSchV for metals and mineral oil hydrocarbons. The table provided by BBodSchV elucidates that values remain missing for priority organic compounds. In addition the LAGA (1997) contains values obtained from the analysis of eluates from soils graduated according to their different uses. It becomes clear that values for Z 0 and Z 1.1 only differ by a small amount because both types of use (unrestricted open and restricted open reuse) concern groundwater protection. This means that there should not be any difference between the ecotoxicological tests with soil eluates. Ecotoxicological tests are used as an "analytical instrument" for the determination of the summation parameter toxicity potential and consequently complete the chemical analysis. However, this is – as outlined above – based on an incomplete substance lists.

One problem may be soils polluted with high geogenic contents such as heavy metals. In practice these soils can be reused at the same site after remediation of the anthropogenic pollutants. In aquatic tests with soil eluates the geogenic heavy metal content can lead to an exceedence of the tolerated values for these tests. In this case it is recommended that the soil material on the site of the intended reuse should be examined in comparison to the soil material to be reused. With regard to sustainable soil management practice, the results of the ecotoxicological tests with the eluates of the soil material to be reused should fall at least into the range of values that are obtained with the eluates of the soil on the site (soil with geogenic high background values).

The intended use of the soil material determines the size of the test set for a determination of the potential effect on the biospherical function. For both types of reuse it is imperative to examine the influence on the soil (microbial respiration and nitrification). The microbial activity determines the filter function of soils and thus is an essential element of both, groundwater protection and biospherical function.

The present experience with soil organisms demonstrates that in general acute tests are not sensitive enough for lower concentrations of pollutants. A protection of the habitat function can only be acquired if sensitive endpoints such as reproduction are considered. The standardised plant test should be supplemented by further tests if soil materials are to be brought out onto sites that serve for food

Table 12.2. Test strategy with regard to unrestricted open reuse of soils

Protection target: Human health	Groundwater protection	Biospherical function
tests: chemical analysis of the solid matter compliance with given values (e.g. BBodSchV LAGA Z 2.0)	chemical analysis of the eluate compliance with given values (e.g. BBodSchV LAGA Z 0) in addition to ecotoxicological tests luminescent bacteria: $G_L \leq 8$ algae: $G_A \leq 4$ daphnia: $G_D \leq 4$	test set consisting of: • micro-organisms respiration nitrification • earthworm mortality reproduction (combined test) • collembola, reproduction • plants: germination biomass if necessary, further tests with regard to: • transfer of contaminants in crops (accumulation) • usefulness of the reused soil

Table 12.3. Testing strategy with regard to restricted open reuse of soils

Protection target: Human health	Groundwater protection	Biospherical function
chemical analysis of the solid matter compliance with given values (e.g. BBodSchV LAGA Z 1.1)	chemical analysis of the eluate compliance with given values (e.g. BBodSchV LAGA Z 1.1) in addition to ecotoxicological tests luminescent bacteria: $G_L \leq 8$ algae: $G_A \leq 4$ daphnia: $G_D \leq 4$ For soils with a high background contamination (e.g. according to LAGA Z 1.2) Examination of the eluate of the soil in comparison to the soil intended for a re-use. The results of the ecotoxicity tests should fall into the range of a standard soil.	Test set consisting of: • micro-organisms respiration nitrification • plants[a]: germination biomass • earthworm[a], mortality reproduction (combined test) • BBodSchV[a], reproduction

[a] possibility to reduce the test size

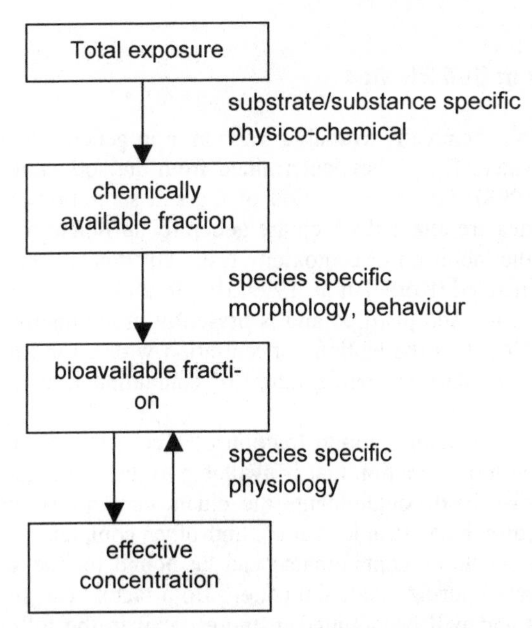

Fig. 12.1. Influence variables on the biologically effective concentration of a contaminant in the soil/test substrates (according to Herrchen et al. 1999)

production. However, these supplementary tests can be omitted for restricted reuse. If future soil data sets become available, consideration of a possible reduction in the test set should be made on the basis of these data sets. The aim of examination for the purposes of restricted reuse is to determine whether the biospherical function is properly considered.

12.3 Bioavailibility as a Key Term for the Composition of Ecotoxicological Test Sets and for the Evaluation of Results

12.3.1 Exposure and Bioavailibility

Exposure to pollutants is commonly divided into a chemically available fraction, the fraction available to the organisms (bioavailable fraction) and the fraction that is taken up by the organism (effective concentration).

The important terms and influence factors for the biologically effective concentration of contaminants in soils are shown in fig. 12.1.

12.3.2
Bioavailibility in Soil Eluates

In soil eluates the chemically available fraction is in general the concentration in the interstitial water. This is best determined from the soil saturation extract (Utermann et al. 1998). For the collection of the soil saturation extract only small volumes of eluates are attained which are generally sufficient for chemical analysis, but not for the inclusion of ecotoxicity tests. For this reason a soil/water ratio of 1:2 is recommended (Kördel et al. 1999; BMBF 2000). This approach has been experimentally tested and justified and is presently in the international standardisation process. Note that the elution concentration with a solid matter/water ratio of 1:2 is maximum starting concentration of contaminants when leaching commences.

The chemically available mobile fraction of a contaminant must not be automatically equated to the fraction that is available to the soil organisms (fig. 12.1). In addition to dissolved contaminants the eluate contains soluble soil contents such as DOM (dissolved organic matter) and other complexing agents. Furthermore, relatively insoluble contaminants can be bound to fine soil particles that have been separated during eluate collection. Both factors can significantly influence the results and will be outlined in more detail in the following section. A differentiation between bioavailable and biologically effective concentration is not routinely performed in ecotoxicity test systems.

12.3.3
Influence of DOM (Dissolved Organic Matter) on the Bioavailibility of Contaminants

Higher concentrations of DOM (dissolved organic matter) can occur in soils that have been supplemented with higher amounts of organic additives for the remediation purposes. Eluates of these soils can contain DOC-contents of more than 200 mg l^{-1}. DOM concentrations are known to be competent solutes for relatively insoluble organic compounds. This means that they increase their concentration in the eluate (Kördel et al. 1999; Rebhuhn 1996). Because of the binding of the contaminants to the DOM their bioavailability can be reduced to a significant extent.

An experiment was undertaken to demonstrate this effect whereby a PAH contaminated soil was eluated with water and two DOM-solutes containing different amounts of DOC (206 mg l^{-1} and 459 mg l^{-1}) in a soil/water-ratio of 1:2. The eluates were subsequently tested with daphnids.

As can be seen in table 12.4, the PAH content in the eluate increases from 3.1 mg l^{-1} up to 14.6 mg l^{-1}. The content in humic matter of the soil material leads to a higher DOC-value of 46 mg l^{-1} biogenic soluble humic mass. Tests with daphnids were carried out on these three eluates in two series. In the first series a concentration succession with dilution water was produced according to the testing guidelines. For each dilution stage not only the content in PAH decreased but also the content in DOC. This result demonstrates that the eluate 3a with the highest concentration of PAH also shows the greatest effect in the test with daphnids. This is clearly more toxic than the eluate 1a; i.e. less than 1 % of the base eluate

Table 12.4. Influence of the DOM on availability

a) Soil: SPAK 1a
1. Experimental model, dilution of samples with dilution water

Eluate	Extraction	PAH-content [mg l^{-1}]	DOC-content In eluate [mg l^{-1}]	Daphnia test [EC_{50}]	DOC at EC_{50} [mg l^{-1}]
1a	water	3.1	46	15.6	7
2a	DOM-solution DOC 206 mg l^{-1}	12.9	206	1.0	2
3a	DOM-solution DOC 459 mg l^{-1}	14.6	459	< 1.0	< 5

b) Soil: SPAK 1a
2. Experimental model, sample dilution with DOM-solution

Eluate	Extraction	PAH-content [mg l^{-1}l]	DOC-content in eluate [mg l^{-1}]	Daphnia test [EC_{50}]	DOC at EC_{50} [mg l^{-1}]
1b	water	3.1	46	22.7	46
2b	DOM-solution DOC 206 mg l^{-1}	12.9	206	12.7	206
3b	DOM-solution DOC 459 mg l^{-1}	14.6	459	17.6	459

3a is sufficient to immobilise half of the daphnid, whereas of the eluate 1a 15.6 % of the base solution are necessary to immobilise 50 % of the daphnids.

In the second series the respective DOC-concentrations were kept constant during the dilution stages; i.e. the dilution water contained additional DOM. The results do not show any clear differences in the EC_{50}-values; i.e. eluate 1b with a content in PAH of 3.1 mg l^{-1} shows a comparable toxicity as eluate 3b with a content in PAH of 14.6 mg l^{-1}. This experiment proves thus the increased complexity of DOM in addition to a significant decrease of the bioavailability of PAHs.

12.3.4 Influence of Particles on the Availability of Harmful Compounds

Relatively insoluble compounds are generally bound to the surface of particles. According to the specific intensity and effectiveness of clean-up steps, significantly different concentrations can be measured in soil eluates. Filtration of the eluates through filters with a pore size of less than 0.45 µm does not always result in a uniquely clear eluate especially for clay soils and for soils that are rich in humic substances (a clear eluate is commonly ascribed to have a turbidity ≤ 10 FNU). Therefore a turbidity measurement according to DIN EN 27027 as a quality control should be compulsory. To demonstrate the dependency of the

Table 12.5. Influence of turbidity on the contaminant concentration and bioavailability

a) Soil: SPAK 1b

Turbidity [FNU]	PAH-content in eluate [$\mu g\ l^{-1}$]	Luminescent bacteria test EC_{50} [%]
11	16.8	30.3
43	40.4	25.8
94	377.0	8.9

b) Soil: EPAK 3a

Turbidity [FNU]	PAH-content in eluate [$\mu g\ l^{-1}$]	Luminescent bacteria G_L-value
< 0.5	6.7	12
10.1	13.3	8
31.4	25.9	8

c) Soil: CTNT 1a

Turbidity [FNU]	TNT-content in eluate [$mg\ l^{-1}$]	Luminescent bacteria test EC_{50} [%]	Daphnids EC_{50} [%]
1.8	57.3	1.0	26.7
9.6	56.7	0.9	25.0
32.9	62.4	1.2	28.4

d) Soil: BMKW 1a

Turbidity [FNU]	mineral oil-content in eluate [$\mu g\ l^{-1}$]	Luminescent bacteria test EC_{50} [%]
1.7	103	35
9.5	418	40
32.2	1479	47

compound concentration in the eluate, eluates have been centrifuged to different degrees. The following table 12.5 clearly shows that the concentration of harmful compounds (e.g. PAH, mineral oils) with a low water solubility increases by a factor of 10 if micro-particles are present in the eluate. TNT and nitroaromatics in general are not effected due to their good solubility in water and are, therefore, not sorbed to particles.

Ecotoxicological tests were carried out on these eluates in order to determine, whether the fraction of contaminant being bound to particles is significantly less bioavailable than the dissolved fraction. As can be seen in table 12.5 the results of the tests with luminescent bacteria as well as the tests with daphnids prove that the turbidity and subsequently the fraction of contaminants being bound to particles have little influence on the test results. There was little variation between determined toxicities although the contaminant concentrations in the eluate were significantly different. This allows the conclusion to be drawn that in the aquatic tests only the dissolved contaminant fraction is bioavailable. These results have encouraged the authors to suggest that a higher turbidity should be tolerated for aquatic tests with soil eluates (perhaps ≤ 50 FNU) in order to facilitate the production of eluates and especially the turbidity measurements for biological laboratories.

Extensive experience of single institutes/laboratories and project cooperation respectively are presently available concerning practical testing of aquatic experiments. Many aquatic test systems such as luminescent bacteria inhibition test, luminescent bacteria growth test, algae test (growth) as well as daphnia test (immobilisation) have their origin in the testing of chemicals or the examination of wastewater. The suitability of these tests for the assessment of soil eluates could be proved in a ring test, which has been supported financially by the Deutsche Bundesstiftung Umwelt (DBU 2000).

12.3.5 Bioavailability in Soil Biological Tests

In soil biological tests it is possible for the chemically available fraction and other substances (e.g. heavy metals by humic substances) to be masked in a way that the uptake by soil organisms is impossible. Consequently differentiation must be made between the available and the bioavailable fraction (fig. 12.1). The necessity of such differentiation between total load, that which is "chemically" available and the bioavailable fraction is also emphasised by Eijsackers et al. (1997). In addition Peijnenburg et al. (1997) require that the bioavailibility of a substance in the soil has to be seen as a dynamic process, because the effective concentration is determined by species-specific physiologic processes such as uptake, accumulation and elimination. These processes are taken into account by the concept of the "lethal body burden" as a reference value for an effect. The dependency of the heavy metal uptake by different soil organisms on the soil characteristics was demonstrated by the working group at Peijnenburg (Peijnenburg et al. 1999 a, b; Vijver 2000). Furthermore, it is important that elaborate regulation mechanisms for essential metals are present in organisms that completely superimpose the correlation of metal uptake with soil characteristics.

For terrestrial test system it should be taken into account that the organisms exposure depends upon their way of life via different pathways. This also applies to organisms living in soil pores such as microorganisms and the meso-fauna mainly exposed via interstitial water. This is for the most part also valid for earthworms, but not for arthropods (Spurgeon 1997).

In general exposure via food uptake is of a greater importance for the epigaeic fauna. As soon as the interstitial water cannot be considered as the main exposure pathway, the estimation of the available contaminant fractions is practically not possible. This is because the respective quantities of the different exposure pathways (interstitial water, food, soil vapour) are unknown. Contaminants are not distributed evenly in soil. Fine particles usually contain higher concentrations, especially of organic compounds that are strongly bound to humic matter. For

Table 12.6. Exposition pathways for terrestrial organisms

Organism	Pore water	Soil air	Humus	Biomass
plants	++	(+)		
bacteria	++		(+)	
fungi	++		(+)	
Protozoa	++		(+)	
nematodes	++			+
enchytraeids	++		+	
earthworm	++	(+)	+	
camasina	+	+		++
isopods		+	++	
collembola	+	+	++	
carabids		+		++
staphylinids		+		++

++ = main pathways (modified according to Hammel 1998)
\+ = effects' pathways
(+) = subordinate pathway

Table 12.7. Cd and Pb uptake in Indian mustard (*B. juncea*) under addition of dissolving agents

Substance added		Soluble content [mg kg^{-1}]	Content in plant [mg kg^{-1}] DM
		Cadmium	
---		0.002	1.3
KCl	100 g m^{-2}	0.003	2.5
NTA	100 g m^{-2}	0.003	2.3
EDTA	160 g m^{-2}	1.684	4.2
		Lead	
---		1.9	112
NTA	240 g m^{-2}	17	156
EDTA	300 g m^{-2}	201	599

Felix (1999)

airborne contaminants there exists a significant concentration gradient between soil surface, top soil and deeper soil layers etc.

If the relevant contaminant concentrations are to be determined, extraction methods are to be selected in a way that they reflect the fraction of the contaminant that is available to the respective organism (group of organisms). This is only possible with limitations. If the interstitial water is the main exposition pathway the distribution of contaminants between soil solid phase, interstitial water (including fraction bound to DOM) and the soil air should be known.

In addition knowledge of the variance of the data in time and space is required. Methods for the estimation of the partition of inorganic and organic contaminants in the interstitial water are outlined in detail in the GDCh-monography "Anforderungen an physikalisch-chemische und biologische Testmethoden zur Einschätzung von Böden und Bodensubstraten" (GDCh 2000).

For strongly bound substances ($P_{ow} > 4$) the partition of the soil contaminants dissolved in the interstitial water decreases with increasing hydrophobia. This means that other exposure pathways such as direct contact of organisms with the soil matrix, uptake of soil and humic particles as well as of contaminated food increase in importance. The distribution between the specific uptake pathways is dependent to a large extent on the diet and the way of life of the organisms in question and is thus organism specific.

There is a great deal of knowledge about the extent to which complexing agents can increase the solubility of heavy metals. This has been successfully applied, for example, in phytoremediation. The results from Felix (1999) prove that the chemically available fraction (soluble content) is not to be equated to the bioavailable fraction (content in plant) (table 12.7). Complexing bound metals are clearly less bioavailable as shown by an observed increase in the chemical availability of a factor of 1000 and the bioavailability including uptake by plants by a factor of 4 for cadmium.

12.3.6 Consequences for the Experimental Set-Up

12.3.6.1 Selection of Test Organisms for a Test Set

According to table 12.4, a test set should cover important functions, and test organisms should be selected with regard to the important exposure pathways. Test set for the evaluation of soils/substrates with regard to their recycling and reuse should at least contain the groups listed in table 12.8.

When executing biological testing methods in the framework of soil quality assessments, the central problem is generally that there is no reference substrate available to be compared to the potentially contaminated soil in question. The relevance of this problem is emphasised by the fact that the variances between a

Table 12.8. Selection of test organisms

Organisms	Function	Important exposition pathway
micro-organisms	nutrients cycles degradation of organic contaminants	interstitial water
earthworms	soil structure and, soil productiveness	interstitial water (food)
collembola	contribution to formation of humic substances	food
plants	producers	interstitial water

specific measurement parameter at different sites can clearly be greater than between different contaminant concentrations. In ideal circumstances such a control and reference substrate (including reference soil) should correspond to the soil to be tested in all soil parameters and nutrient contents – with exception of the chemical contamination. But usually only one of the following possibilities is available (ISO 1999):

- a scientifically characterised "standardised" natural soil such as the standard soils of the LUFA Speyer (commonly used for pesticide registration);
- an artificial (in nearly all characteristics) predefined soil (OECD 1984, ISO 1996), for which it is hard to standardise the peat quality.

At present none of these possibilities can be recommended for all tests. Organism and experimental parameter specific solutions have to be selected. This was carried out successfully for the test organisms and systems (respiration, nitrification), earthworms (Eisenia fetida: mortality, reproduction), collembola (mortality, reproduction), Enchytrae (mortality, reproduction) and plants (germination, biomass). The suitability of the test systems including the evaluation parameters for the assessment of the results has been demonstrated in a project supported by the Deutschen Bundesstiftung Umwelt (DBU 2000).

12.4 Bioavailability of Bound Residues

During biodegradation of anthropogenic contaminants different compound classes can be bound to humic substances, especially during biological remediation measures, where additives consisting of readily biodegradable materials are used. The following table 12.9 gives an overview of substance groups and their binding mechanisms. A further outstanding example is the bioremediation of military contaminated sites where nitroaromatics are reduced and aminocompounds become covalently bound to humic substances.

For an assessment of the remediation success, the question is how stable these chemical bounds are, so that no effective and bioavailable contaminant concentrations will be released. It is also important to know whether during further reaction of the humic fraction (mainly where the contaminants are bound) contaminants can be released as the parent substance or as at least toxic metabolite.

In order to examine this there is a graded extraction scheme available (table 12.10).

Isotopically marked contaminants can be used for the determination of what portion of a compound has been bound during a remediation measure and whether bound contaminants could be remobilized during reuse. The following example shows the results of different extraction experiments by means of a case study of biological treatment of military contaminated sites. This model treatment was carried out with the addition of ^{14}C-TNT in 2 m^3 reactors under identical conditions (including turning and mixing processes) just as in a commercial-scale remediation. It can be seen that the treatment has lead to an outstanding success; i.e. more than 99 % of the initial concentration of TNT are not detectable analytically

Table 12.9. Characterisation of bound residues

Substance class	Potential bonds to humic substances
s-triazine	covalent bonds hydrogen bonds charge-transfer complexes ligand exchange
bipyridylium cations	ionic bonds charge-transfer complexes ligand exchange van der Waals forces
cationic surfactants	ionic bonds
amines	covalent bonds ionic bonds charge-transfer complexes
chlorinated aromatic hydrocarbons	covalent bonds hydrophobic interactions van der Waals forces
polycyclic aromatic hydrocarbons	covalent bonds van der Waals forces hydrophobic interactions

Wais (1998)

Table 12.10. Extraction methods for the release of bound residues

Method	Commentary
extraction with org. solvent or SFE	exhaustive extraction without dissolving the soil matrix
complexifying agents (e.g. EDTA) or sylilation (e.g. DMF/TMSCl)	dissolution of e.g. DOM complexes
NaOH-solution	extraction of humic substances and proteins decomposition of cells
acetone/HCl or DMF/HCl sulphuric acid tetrabutylammoniumhydroxide hydrolytic enzymes	hydrolytic splitting of bondings of organic substances

any more (not extractable with methanol) (table 12.11). Using water 2.2 % of the radioactivity can be washed out. As already stated, watery soil eluates contain dissolved soil components including DOM. Because the TNT was bound to the soil available organic substance during remediation, a smaller fraction (that fraction bound to DOM) is washed out with water. It cannot be concluded from the elution, therefore, that the radioactivity measured corresponds to TNT or its free metabolites. During further analytical preparation of the samples neither parent compounds nor free metabolites could be detected.

Table 12.11. Extraction experiments after TNT remediation

Method	Radioactivity in eluate [%]
watery eluate	2.2
watery eluate after freezing and thawing	4.1
0.1 m NaOH	36.1
5 % hot H_2SO_4	57.8
1.0 m HCl	13.3
0.2 m EDTA	25.3

This becomes even clearer when hydrolytic agents such as acids, bases or complexiing agents are used which lead to a dissolution of parts of the humic matrix. Hot 5 % H_2SO_4 hydrolyses humic components (e.g. hemi-cellulose, proteins, carbohydrates and fulvic acids) which distinguish themselves by the existence of aromatic structures. Consequently these are especially suitable for the covalent bonding of TNT-metabolites during remediation.

The measurement of the total radioactivity of such extracts cannot give any hint regarding the remobilisation of biologically active TNT-metabolites. However, it can give valuable information about bonding structures. Measurements of the total radioactivity have to be completed by the analytical identification of TNT and its metabolites. At times this can be very time consuming and laborious. Consequently biological tests are advantageous because effects can only be caused by bioavailable contaminant components. Sufficient expert knowledge is necessary, however, to guarantee a definitive assignment and to exclude side effects such as a high salt or DOM content.

The following example demonstrates these findings. After the biological treatment of a soil contaminated with explosives (see also table 12.11), ecotoxicological testswere carried out at the end of the remediation measure and at a later time after the treated soil material had spent about 5 months in an outside-lysimeter. As mentioned, above 99 % of the nitroaromatics had been eliminated during the remediation measure. However, there a significant toxicity remained as indicated by the earthworm and algae test. When taking a closer look at the analytical results it becomes clear that the toxicity measured can be caused by high salt and DOM loads. After having washed out these loads, partly in lysimeters by means of natural rainfall, the initial toxicity has vanished in the sampling undertaken in autumn. During examination specifically with regards to their content in nitroaromatics or bound residues, significant differences between these two samplings could not be observed .

It is known from research regarding bound residues that DOM especially can include intact initial molecules or biologically active metabolites and consequently can withdraw these from the extraction with organic solvents. These DOM-complexes can be broken up by means of sylilation without breaking covalent bonds.

Table 12.13 shows 2 examples where significant amounts of biologically active molecules are released by means of sylilation. It has to be taken into

Table 12.12. Examination of eluates

Examination	Sampling at the end of remediation (26.05.99)	Sampling during lysimeter experiments (18.10.99)
luminescent bacteria test EC_{50} [%]	62	--
G_L-value	8	6
daphnia test EC_{50} [%]	42	--
G_D-value	4	1
algae test EC_{50} [%]	28	--
G_A-value	16	2
nutrients in eluate:		
P [mg l^{-1}]	3	4
NH_4^+-N [mg l^{-1}]	72	4
NO_3^--N [mg l^{-1}]	50	35
DOC-content in eluate [mg l^{-1}]	541	271
ammonium oxidation [ng NO_2-N (g min)$^{-1}$]	89 ± 12	12 ± 1
basal respiration[µg (g h)$^{-1}$]	4 ± 0.1	4 ± 0.3
SIR [µg (g h)$^{-1}$]	12 ± 1.0	12 ± 2.3
earthworm mortality [%]	100	0

-- cannot be calculated

Table 12.13. Examples for the release of biologically active molecules out of bound residues (DOM)

Substance	Release with DMF/TMS Cl	Biological effect	Author
amitrol – DOM	30 %	cress – chlorosis	Spiteller (1998)
anilazin – DOM	27 %	?	
simazin – DOM (29 rot days)	30 %	-	Hartlieb (2000)
simazin – DOM (200 rot days)	20 %	-	

consideration, though, that only a smaller part of the existing radioactivity in the soil is bound to DOM (see table 12.11). After accumulation of the released substance an effect in the cress test could be observed in only one case (Spiteller 1998).

Finally, it should be stated that the measurement of the total radioactivity in different eluates of soils containing bound residues, do not allow any statements with regard to bioavailability, effect and thus risk. The spectrum of the biological test systems is suitable to clarification of whether effective metabolites are remobilised under different experimental conditions in amounts that can lead to an effect.

References

BMBF Michels J, Track T, Gehrke U, Sell D (2000) Biologische Verfahren zur Bodensanierung. BMBF-Veröffentlichung

DBU (2000) Ökotoxikologische und genotoxikologische Beurteilung von Böden und Substraten – Ringtestergebnisse. To be published by Erich Schmidt Verlag – Initiativen zum Umweltschutz

DECHEMA (1995) Biologische Testmethoden für Böden. Deutsche Gesellschaft für chemisches Apparatewesen, chemische Technik und Biotechnologie eV, Frankfurt a M, Germany

DIN 19371 (1998-05) Anforderungen an die Verwertung von Bodenmaterial

Dreher P, Hund K, Rüdel H (2000) Wirkungen ausgewählter Schadstoffe auf Bodenorganismen. UBA-Texte 8/00

Eijssackers HJP, Herrchen M, Strehlow R (1997) Major statements and conclusions. Bioavailability as a key property in terrestrial ecotoxicity assessment and evaluation. Fraunhofer IRB Verlag, Stuttgart, Germany

Felix H (1999) Sanierung von Schwermetall-belasteten Böden und Gewässern mit Hilfe von Pflanzen. In: Heiden S, Erb R, Warelmann J, Dierstein R (eds) Biotechnologie im Umweltschutz; Bioremediation: Entwicklungsstand – Anwendungen – Perspektiven. Erich Schmidt Verlag, Berlin, Germany, pp 154–165

GDCh (2000) Anforderungen an physikalisch-chemische und biologische Testmethoden zur Einschätzung von Böden und Bodensubstraten. GDCh-Monographie Bd. 20, Gesellschaft Deutscher Chemiker, Frankfurt am Main, Germany, ISBN 3-924763-89-5

Hammel W (1998) Exposure assessment under consideration of soil characteristics for ecotoxicological risk assessment. Intern Workshop "Ökotoxikologische Bewertungsmaßstäbe zur Beurteilung der stofflichen Belastung von Altlasten und Böden", Fraunhofer IUCT Schmallenberg

Hartlieb N (2000) Markierung organischer Schadstoffe während der Kompostierung und ihre Auswirkungen auf die stoffliche Bodenqualität. Thesis, Gesamthochschule Duisburg, Germany

Herrchen M, Kratz W, Marschner A, Necker U, Pieper S, Römbke J, Riepert F, Rück F, Terytze K, Throl C, Wilke B-M (2000) Eckpunkte zur Gefahrenbeurteilung des Wirkungspfades Bodenverunreinigungen – Bodenorganismen. Fachausschuss "Biologische Bewertung von Böden" der Fachgruppe 4 "Bodenfunktionen und -belastungen" des Bundesverband Boden (BVB), 83 S, publication under preparation

ISO/DIS 11268-2 (1996) Soil Quality – Effects of Pollutants on Earthworms (Eisenia fetida). Part 2: Determination of Effects on Reproduction. International Organization for Standardization – Draft International Standard, Genf, Switzerland

ISO/DIS 15799 (1999) Soil Quality – Guidance on the ecotoxicological characterization of soils and soil materials. International Organization for Standardization, Genf, Switzerland

Kördel W, Hund K (1999) Ziele und Bewertung biologischer Sanierungsverfahren – Gefahrenpotentiale und Bewertung umweltrelevanter Schadstoffe. In: Heiden S, Erb R, Warelmann J, Dierstein R (eds) Biotechnologie im Umweltschutz; Bioremediation: Entwicklungsstand – Anwendungen – Perspektiven. Erich Schmidt Verlag, Berlin, Germany, pp 22–32

Kördel W, Hund K, Klein W (1999) Erfassung und Bewertung von Mobilität und Bioverfügbarkeit bei der Beurteilung kontaminierter und gereinigter Böden. In: Heiden S (ed) Innovative Techniken der Bodensanierung. Spektrum-Verlag, Berlin, Germany

LAGA (1997) Anforderungen an die stoffliche Verwertung von mineralischen Reststoffen/Abfällen. In: Rosenkranz D, Bachmann G, Einsele G, Harreß H-M (eds) "Bodenschutz – Ergänzbares Handbuch", 29. Lfg.VII/99, Erich Schmidt Verlag, Berlin, Germany

OECD (Organisation for Economic Development) (1984) OECD-Guideline for Testing of Chemicals No. 207. Earthworm Acute Toxicity Test, Paris, France

Peijnenburg WJGM, Posthuma L, Eijsackers HJP, Allen HE (1997) Implementation of bioavailability for policy and environmental management purposes. Bioavailability as a key property in terrestrial ecotoxicity assessment and evaluation. Fraunhofer IRB Verlag, Stuttgart Germany

Peijnenburg WJGM, Posthuma L, Zweers PGPC, Baerselman R, De Groot AC, Van Veen RPM, Jager T (1999 a) Prediction of metal bioavailability in dutch field soils for the Oligochaeta Enchytraeus crypticus. Ecotoxicology and Environmental Safety 43: 170–186

Peijnenburg WJGM, Baerselman R, de Groot AC, Jager T, Posthuma L, van Veen RPM (1999 b) Relating environmental availability to bioavailability: Soil-Type-Dependent Metall Accumulation in the Oligochaete Eisenia andrei. Ecotoxicologicy and Environmental Safety 44: 294–310

Rebhuhn M, De Smedt F, Rwetäbula J (1996) Dissolved humic substances for remediation of sites contaminated by organic pollutants, Binding-desorption model prediction. Wat Res 26: 1645–1654

Spiteller M (1998) Fate and ecological significance of pesticide residues in soil and plant. Pesticide Bound Residues in Soil. Deutsche Forschungsgemeinschaft, Wiley-VCH, pp 161–165

Spurgeon DJ (1997) Can the uptake and toxicity of pollutants by soil invertebrates be described by simple one exposure route models based on pore water concentrations. Bioavailability as a key property in terrestrial ecotoxicity assessment and evaluation. Fraunhofer IRB Verlag, Stuttgart, Germany

Utermann J, Gäbler H-E, Hindel R, Kues J, Mederer J, Pluquet E (1998) Schwermetallgehalte im Bodenwasser – ein Vergleich von drei Extraktionsverfahren. Z angew Geol 44/4: 204–209

Vijver M (2000) The impact of soil characteristics on metal bioavailability. Thesis, Hogeschool Ijselland and RIVM

Wais A (1998) Non-extractable residues of organic xenobiotics in soils – a review. Pesticide Bound Residues in Soil. Deutsche Forschungsgemeinschaft, Wiley-VCH, pp 5–31

Bioremediation

13 Bioavailability – the Key Factor of Soil Bioremediation

B. Mahro[1], R. Müller[2] and V. Kasche[2]
[1]Institute for Environmental Technology, Hochschule Bremen, Neustadtwall 30, 28199 Bremen, Germany
[2]Section Biotechnology II, Technical University Hamburg-Harburg, Denickestr. 15, 21071 Hamburg, Germany

13.1 Hard to Avoid – Pollutant Residues in Soil

In the last two decades it has been shown that many pollutant compounds being found in soils or aqueous ecosystems may potentially be transformed by microorganisms. Transformation may be either completely into carbon dioxide and water, or at least into non-toxic metabolites (Klein 2000). These promising research results inspired many remediation companies to set up particular bioremediation approaches for the clean-up of such contaminated areas. The bio-enthusiasm of the early years, however, is now followed by a more realistic and sometimes even sceptical view of bioremediation. The major reason for this turn-around is that it has now become clear that results being obtained in the laboratory with artificially contaminated soils do not necessarily indicate what may happen actually in the field with soil from contaminated sites. With hydrophobic pollutants like PAH (Bossert et al. 1984; Erickson et al. 1993; Schaefer et al. 1995; Weissenfels et al. 1992) or some sorts of mineral oil (Angehrn et al. 1997; Bossert et al. 1984; Riis et al. 1998) in particular, it has been observed that even the degradation of compounds being completely mineralisable in the lab-culture may be incomplete in practical field bioremediation. Considerable residual concentrations of analytically detectable pollutants in the soil are subsequently left behind. An example for such a typical "hockey-stick-kinetic" (a term coined by M. Alexander, see Chapter 14) is shown in fig. 13.1.

In addition it was in the last decade it was discovered that the formation of bound or non-extractable residues (a process that had been known so far only for pesticides) may contribute significantly to the disappearance of the contaminants during bioremediation (Bollag 1992; Lenke et al. 1993; Mahro and Kästner 1993).

The observation that considerable parts of the pollutant pool may persist in soil is, of course, rather unsatisfactory at first sight. This is because it gives rise to concerns that such residual pollutants reservoirs may cause a continuous risk on the long run and that the incomplete clean-up might, therefore, possibly be a

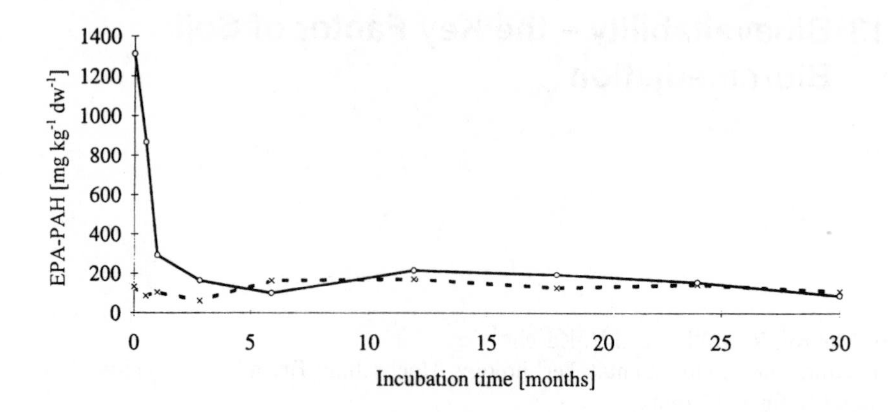

Fig. 13.1. Degradation of PAH in a bioreactor containing a tar-oil contaminated soil that was supplemented with the white-rot fungus *Pleurotus ostreatus* on straw (from Schaefer et al. 1995)

waste of money. This paper will introduce discussion on this important question and illustrate in more detail possible reasons for the occurrence of pollutant residues and possibilities to solve or to handle the problem. The following papers of this section will deepen further some aspects of the topic. This contribution will mainly focus on the extractable type of residues since the non-extractable type of residues will be addressed separately in another section of this volume.

13.2 Reasons for the Occurrence of Extractable Pollutant Residues

The occurrence of incomplete biodegradation of xenobiotics in soil may be due to a number of different reasons.

The first possible reason for incomplete pollutant degradation is that the pollution consists of a mixture of compounds that differ in their extent of biochemical degradability. In this case, the more degradable fraction of the pollutant mixture would be eliminated more quickly from soil than the more refractory part of the pollution he slow or non-degradable pollutant fraction may represent at the end, therefore, an even higher relative portion of the pollution than at the beginning of remediation. Fig. 13.2 shows an example of such a selective degradation consisting of the residual pattern of a complex PAH-contamination. The observations were made for a contaminated soil after 6 months of biological treatment with a mixture of white rot fungi and straw. The results show clearly that high-molecular weight PAHs like benzo[a]pyrene in particular have been omitted from biodegradation within the given time period. Other examples of such "hard nuts" are cycloalkanes or *iso*-paraffins but also more polar hydrocarbon fractions have been reported as possible fractions of residual hydrocarbon pollutant pools (Angehrn et al. 1997; Riis et al. 1996; 1998).

Several reasons are possible to explain the relative recalcitrance of particular compounds. One reason might be that the soil does not contain the appropriate microorganisms being able to attack the xenobiotics under the given environmental conditions or that there are even no such microorganisms in nature at all. However, examples for organic compounds that have been reported as completely non-degradable are rather rare. In most cases cooperative attack of different microorganisms may lead at least to a partial or slow degradation of such pollutants, as one may see for example with the high molecular weight PAH (see fig. 13.2 and Boonchan et al. 2000; Kottermann et al. 1998; Ye et al. 1996). A candidate for a truly non-degradable compound is probably 2,3,6,7-tetrachlordibenzodioxin (Müller 1992) which fortunately is not a common pollutant in bioremediation practice. Another theoretical explanation for to the formation of extractable pollution residues is that the degradation process may have come to an end due to the accumulation of toxic intermediates or due to another negative shift of the environmental conditions. The suicide inactivation of biodegradation process has been well known for a long time from liquid culture experiments and has found meanwhile also its way in biotechnology textbooks (Knackmuss 1997). However, it seems unlikely that this process plays an important role in soil. Both the cooperative character of microbial degradation processes in soil and the inactivation of the metabolites toxicity by sorption may reduce that problem in the soil environment.

The third and probably most important reason for the occurrence of pollutant residues becomes obvious if one considers the fact that incomplete biodegradation

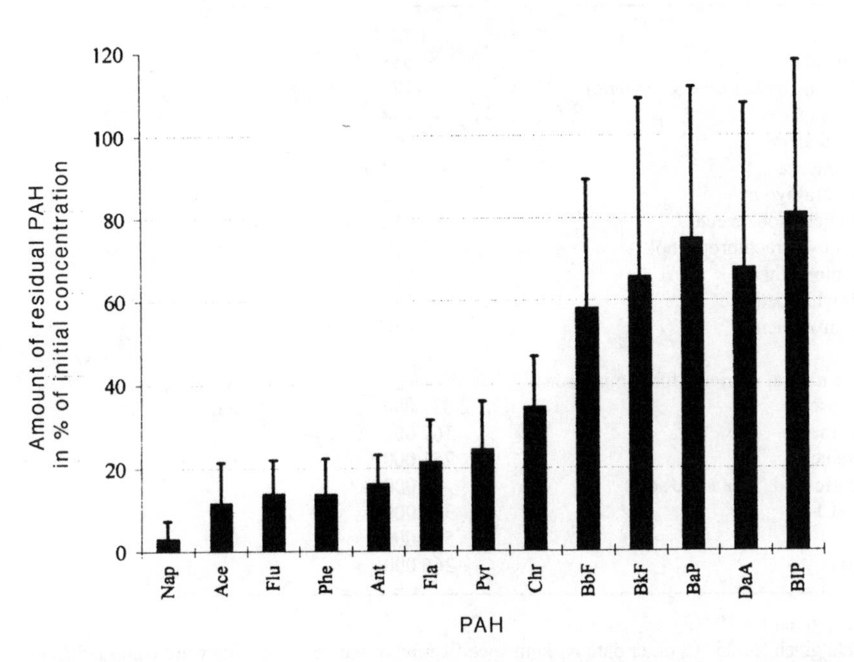

Fig. 13.2. Residual soil concentrations of different PAH in % of the initial concentration (Schäfer and Mahro, unpublished)

also comprises fractions of the pollutant mixtures that are known to be completely biodegradable under other conditions (in fig. 13.2 for example phenanthrene or pyrene). The halt of degradation is, in many similar cases, primarily due to a lack of bioavailability, which describes a failure or the decrease of rate by which a degrading microbial community may get a metabolic contact with the xenobiotic substrates (Mahro 2000). This lack of bioavailability may be related either to the chemistry of the compound (e.g. low aqueous solubility, high sorptivity at hydrophobic surfaces), to the soil (high amounts of sorptive surfaces, high fraction of micropores with reduced accessibility) or to both. The most important aspects of these parameters will be presented in more detail in the following but have also been reviewed more extensively by Harms (1998) and Mahro (2000).

The lack of bioavailability occurs primarily as a problem with hydrophobic soil pollutants. While the water solubility of many natural substances is in the range of "g l^{-1}", the water solubilities of most of the relevant soil pollutants range between "mg l^{-1}" and "µg l^{-1}" (see table 13.1). The low aqueous solubility of many hydrophobic pollutants for example, may either lead to a crystallisation of the compound once the maximum solubility is exceeded or to sorption onto surrounding hydrophobic surfaces. The rate by which a crystallised compound may then

Table 13.1. Aqueous solubilities of some important environmental pollutants. The aqueous solubility of some common natural nutrients is given for comparison (from Mahro 2000)

Substrates	Maximum solubility in water [mg l^{-1}]
Benzene	1 780[a]
Toluene	515[a]
1,2-Dimethylbenzene (o-Xylene)	175
Cyclohexane	58
Naphthalene	30
Anthracene	0.07
Benzo(a)pyrene	0.004
1,2 Dichlorobenzene	92–145
2,3,4,6-Tetrachlorophenol	100
Trichloroethane	950[a]
Tetrachloroethene	150[b]
Trinitrotoluene	140[b]
some natural nutrients for comparison	
Sucrose	2 039 000
Alanine	167 000
Glycine	251 000
Valeric acid (Pentanoic acid)	32 000
NH4Cl	372 000
Urea	519 000
FeSO4	266 000

[a] Data given for 20 °C.
[b] Data given for 25 °C; other data without specification of temperature; data were compiled from different Chemistry-textbooks or publications.

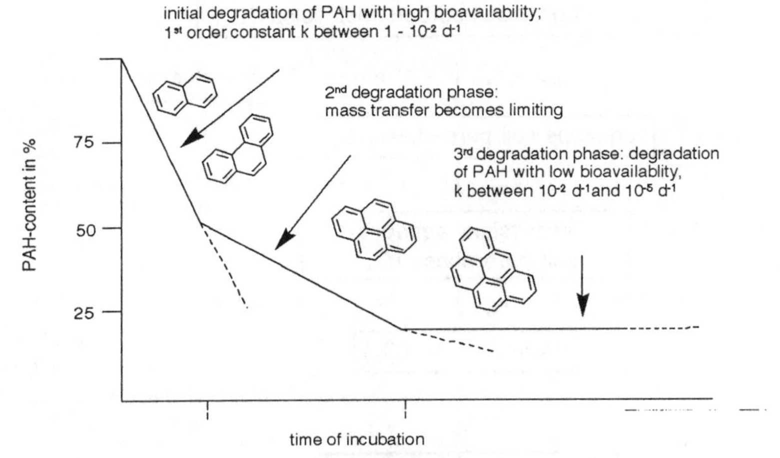

Fig. 13.3. Sequential order of degradation of different types of PAH in a complex pollutant mixture

redissolve in the aqueous phase is – due to the higher surface/volume ratio of small particles – indirectly correlated with the size of the crystals. The high sorption affinity of hydrophobic pollutants to the humic matrix may result in the low maximum water solubility of the pollutant even not being reached due to the physical and chemical constraints of the phase equilibrium. If – for example – the logarithmic soil adsorption coefficient (K_{oc}) for a compound like benzo[a]pyrene is about 5 it would require a total extractable concentration of 400 mg of benzo[a]pyrene kg^{-1} of soil to reach the maximum solubility of this compound in the soil water phase (4 $\mu g\ l^{-1}$)!

The negative impact of the low water solubility and high sorptivity on biodegradation rate can be seen clearly with PAH. The water solubility of PAH varies between about 30 mg l^{-1} for naphthalene and about 4 $\mu g\ l^{-1}$ for benzo[a]pyrene. Both the author's experiments and data being published elsewhere (e.g. Ericksson et al. 1993; Mueller et al. 1991; Tiehm et al. 1997) show that this difference in the water solubility is also closely correlated with the rate of degradation. This allows the sequence of the degradation of the individual PAHs to be sorted according to their water solubility (fig. 13.3).

The slow degradability of benzo[a]pyrene (fig. 13.2) is therefore probably less a biochemical problem than a reflection of the fact that compounds like benzo[a]pyrene are available to the bacteria in the accessible water phase only with very small amounts of substrate at a time.

In addition the availability of a pollutant substrate for microbial growth is very often not only a problem of the absolute amount of a substrate being present in the cells environment, but even more of rate by which a substrate may reach the cell along a given distance (mass transfer; Bosma et al. 1997; Harms 1998; Mahro and Schaefer 1998; Stucki and Alexander 1987). The distinct steps of mass transfer that one may distinguish along that path are summarised in fig. 13.4.

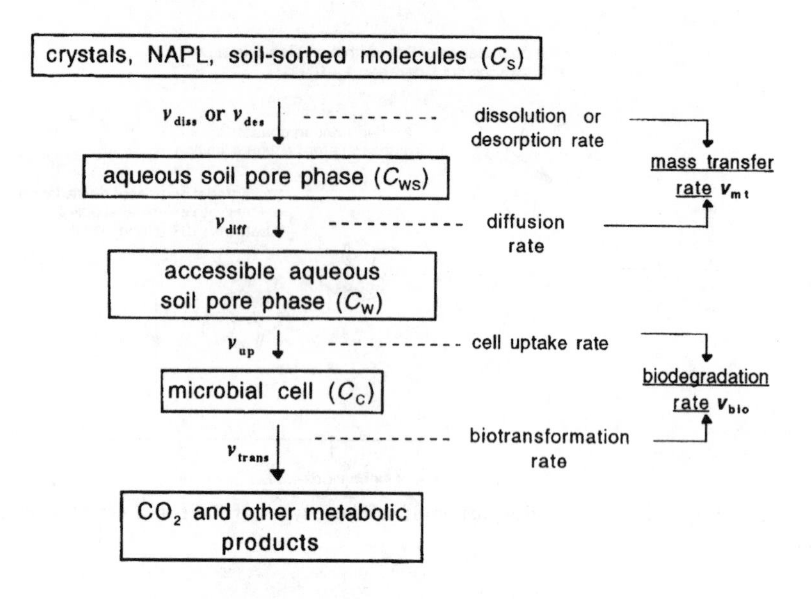

Fig. 13.4. Model of different kinetic steps in the overall process of pollutant biodegradation. The pollutant concentration C is different in the each pool, indicated by separate indices (C_S = source or soil pollutant concentration; C_{WS} = pollutant concentration in the aqueous phase close to the source; C_W = pollutant concentration in the accessible aqueous phase; C_C = pollutant concentration at or in the cell (from Mahro 2000)).

The first factor by which mass transfer may contribute to the lack of bioavailability is simply the fact that molecules are much smaller than microorganisms and therefore they may therefore diffuse into soil pores where bacteria have no access to due to their size (fig. 13.5).

The extent by which this factor may reduce biodegradation is primarily dependent on the length of the distance a molecule has to diffuse until it reaches the bacterial cell. The degradation rate of an inaccessible pollutant pool is therefore completely independent of its biochemical biodegradability and this might explain why also degradable pollutants like phenanthrene may contribute substantially to the detectable pollutant residue fraction.

An additional factor that may significantly influence the rate of mass transfer is the diffusion resistance the pollutant molecule is exposed to along its diffusion path. Soil sites that have been contaminated with so-called Dense Non-Aqueous Phase Liquids (DNAPL) like tar oil or pitch especially may be confronted with such problems (Efroymson and Alexander 1995; Ghoshal and Luthy 1995). The liquid character of DNAPLs like tar oil allows the pollutants to become spread in all accessible soil pores, thereby filling up parts of the available pore volume.

However, since the viscosity of the oily matrix, or even pitch, is much higher than that of water the diffusion coefficient (D) becomes much higher which – according to Fick's first law – reduces the diffusion rate (V_{Diff}):

$$V_{Diff} = dp/dt = -D * F * dc/ds \quad (13.1)$$

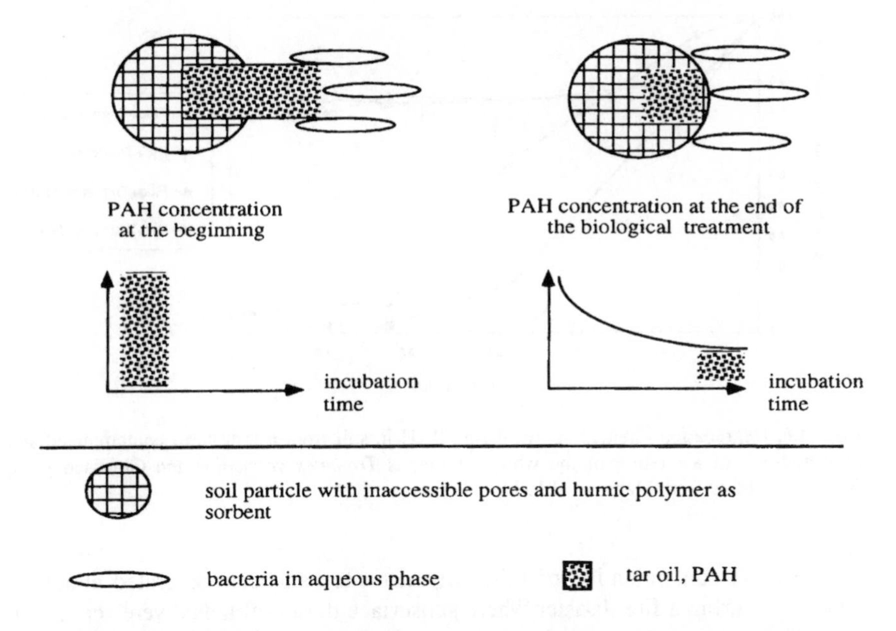

Fig. 13.5. Correlation between the occurrence of residual pollutant concentrations and a lack of accessibility of soil pores for microorganisms (from Mahro 2000)

where dp/dt represents the pollutant flow/time (mol time^{-1}) from C_{WS} to C_W, F is the diffusion surface (area), dc the pollutant concentration gradient (mol volume^{-1}), s the diffusion distance (length) and D is the diffusion coefficient (cm^2 sec^{-1}).

Summarising this brief discussion of possible reasons for the occurrence residual pollutant pools in soil bioremediation it can be concluded that bioavailability is actually the most relevant factor that determines, in most cases, the rate, the extent and thereby the success of bioremediation.

13.3 Approaches to Handle the Lack of Bioavailability

13.3.1 Avoidance of the Problem by Preliminary Tests

As shown, the lack of bioavailability may influence biodegradation in many cases but the lack of bioavailability does not necessarily occur in all cases. For example, it can be seen in fig. 13.6 that one may find even for a specific PAH-contaminated site (normally the typical candidate for the lack of bioavailability) that clean up can be undertaken completely without any significant amounts of residual pollutant concentrations.

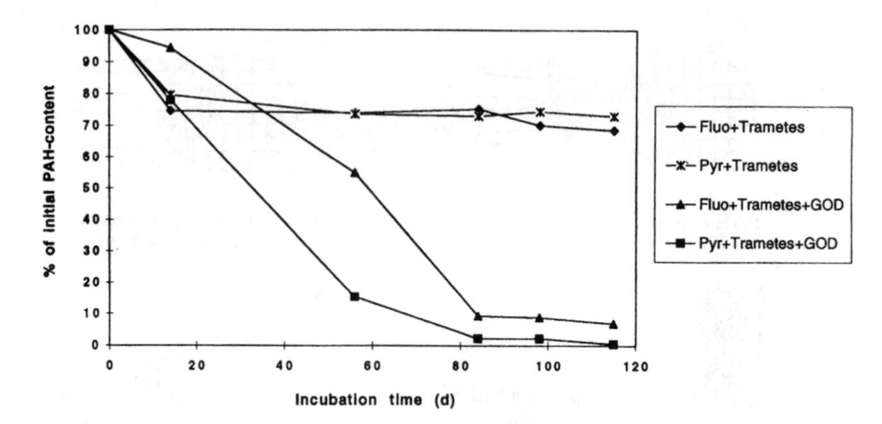

Fig. 13.6. Degradation kinetics of two 4-ring-PAH in soil from a long-term contaminated site after addition of a mixture of the white rot fungus *Trametes versicolor* and Glucoseoxidase (GOD) (modified from Mahro et al 1999)

The PAH contamination of this particular site had been generated about 50 years ago during a fire disaster where subsurface diesel oil tanks were set on fire by military actions. The fact that the pyrogenic PAH of this sandy site were therefore obviously not shielded within a (D)NAPL might explain the comparatively good degradation results with this PAH-contamination. This example also shows that it is necessary to test site by site the extent of the bioavailability problem in advance in order to avoid unpleasant surprises at the end of the bioremediation.

One possibility to carry out such preliminary tests is to design test procedures that simulate field conditions as close as possible. This approach must also include a very close look to the physical and spatial heterogeneity of the pollutants at the site itself since the pollutants may occur at different horizontal and vertical regions and under very different physical or chemical conditions.

As an additional approach, one could also try a "Best-Case"-test strategy. This employs test conditions where one tries to overcome all presumed bioavailability problems as well as possible (for example by the use of a slurry reactor). If one observes an incomplete clean-up even under such optimised lab-conditions it is highly unlikely that the field performance will be better.

Another interesting lab-scale approach to test bioavailability in advance is to use simple chemical oxidation tests; for example the oxidation of contaminated soil samples with persulfate. Such tests have been developed by Rulkens et al. (see Chapter 4) and the extent of measurable pollutant oxidation in these tests showed a good correlation with the degree of biological PAH-removal from soil later observed. Another significant advantage of such tests is, of course, that they can be run within hours rather than over days, weeks or even months like true biodegradation tests.

Another approach of interest to biotechnologists, especially for in situ treatments, is to predict the time necessary for pollutant transformation in a given matrix by modelling. One could theoretically evaluate kinetic constants for both, biodegradation and mass transfer in defined lab-experiments, (in more detail see

Mahro 2000). However, such modelling approaches are hampered primarily by the fact that one needs field data that describe precisely the spatial heterogeneity and distribution of the pollution, soil particle size, viscosity, microbial activity etc. for every site and every location. It seems rather unrealistic, therefore, to expect that one can model precise time scenarios for the bioremediation of the unsaturated soil zone of contaminated sites at the present time.

13.3.2 Options to Overcome the Lack of Bioavailability

Attempts to improve bioavailability by technical means should principally only be made if the pre-testing program has indicated at least the chance that biological means might be used successfully for the clean up of a contaminated soil site. Technical options that might help in this regard can be directed to both an increase of the low water solubility of the pollutants, and an increase of the slow mass transfer (principally diffusion rate in soils).

Strategies to optimise water solubility can focus either on a direct increase of the water solubility of the pollutants (e.g. by chemical pre-oxidation), or on the use of surfactants. Pre-oxidation of hydrophobic pollutants could be carried out by chemical means like the addition of ozone or hydrogen peroxide. The more widely adopted strategy is to use surfactants which may increase the absolute amount of pollutant molecules being present in the water volume by entrapment within surfactant micelles. However, the higher apparent solubility of the pollutants in the water phase may not necessarily increase the availability of the molecule for the microbial cell. To increase the overall rate of the pollutant uptake in the cell, the transfer of the pollutant along the path "surface sorbed pollutant molecule"→"micelle"→"cell membrane" must be faster than the transfer along the pathway "surface sorbed pollutant molecule"→"water phase"→"cell membrane". The discussion on the use of surfactants in soil remediation is still rather controversial. It has been shown by several investigators (Deschenes et al. 1996; Foght et al. 1989; Laha and Luthy 1991) that in some cases the occurrence of an additional micellar phase barrier may even slow down the mass transfer rate. Other authors, however, have shown that the application of surfactants may be more successful if parameters like surfactant toxicity, solubilising efficiency, loss due to soil-adsorption and the rate of surfactant degradation, are carefully considered in advance (Thibault et al. 1996; Tiehm 1994; Tiehm et al. 1997; Zhang and Miller 1995). Although the use of surfactants may speed up the degradation it cannot be guaranteed that complete elimination of the residual pollutant pool from the soil of a contaminated site will be reached. Tiehm et al. (1997) report, for example, that about 80 % of the high molecular weight PAH like benzo[a]pyrene remained in soil, even after surfactant application.

Another interesting aspect in the context of the discussion of water solubility is the question of whether, and under which circumstances, hydrophobic pollutants may be taken up directly via the membrane without any detour through the water bottleneck phase. It can be seen in fig. 13.7 – and related electron micrographs support that view (e.g. Marin et al. 1996) – that bacteria and other microorganisms like yeast can in fact colonise the surface of mineral oil and other hydrophobic pollutant surfaces directly, suggesting that the direct uptake of hydrophobic pollutants seems to be possible. This hypothesis is backed by reports in the literature that describe a direct microbial uptake mechanism for hydrophobic pollutants (Efroymson and Alexander 1991; Wodzinski and Larocca 1977). Direct uptake mechanisms may be supported by special cell surface features, cellular appendages like fimbriae (Käppeli et al. 1984; Rosenberg et al. 1982) or by liposome-mediated transport (Scott and Finnerty 1976).

However, it cannot be completely excluded that in these experiments the cells did not need at least a minimum layer of water between the cell and the NAPL. More convincing evidence is, therefore, a kinetic experiment like that one being carried out by Cuno and Wiesmann (Cuno 1996) which showed that the degradation rate of PAH in a bioreactor was greater than would be expected if the dissolution of the pollutant *in* and the uptake of PAH *from* water would have been involved in the process. These results are also backed by earlier findings that had shown that naphthalene and hexadecane can be used directly from an inert hydrophobic model NAPL like Heptamethylnonane, rather than from the aqueous phase only (Efroymson and Alexander 1991; Wodzinski and Larocca 1977).

The hypothesis that bacteria might circumvent the bottleneck water phase if they had a more direct access to the pollutants makes it even more urgent to try to overcome a lack of bioavailability in soil by reduction of the actual diffusion distance between the pollutants and the microorganisms. Since – according to Fick's 2nd law – the distance (s) that may be overcome by diffusion is not proportional to time (t), but proportional to the root of the time (s = diffusion constant * $\sqrt{t}$) every effort in this regard is warranted. The diffusion distance in soil may be best reduced if soil or pollutant particle clumps (e.g. tar pitch) are cracked, thereby allowing a more direct access for bacteria to the inner surfaces of these

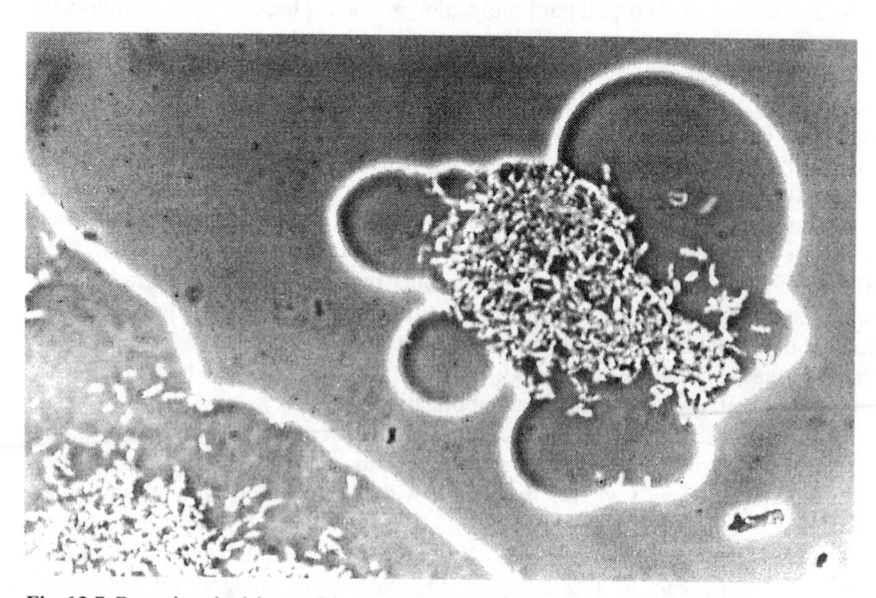

Fig. 13.7. Bacteria colonising on Diesel oil (from Mahro 2000)

particles. This may be best achieved if the soil is suspended and stirred in water (slurry reactor) or if the clumped material is cracked by ultrasonic treatment, as it was proposed by Koning et al. (see Chapter 26) or Tiehm et al. (see Chapter 20). The draw back of these techniques is, of course, that they require both expensive equipment, a high amount of additional energy and in addition, that they produce a considerable amount of waste water to be treated afterwards. Such techniques are therefore applicable in laboratory investigations (for example as "Best-Case"-tests as described above) but not in field applications, where high volumes of contaminated soil have to be cleaned up. Therefore, the preparation of coarse-grained soil material by mechanical means is probably the best one can do in the field to reduce the diffusion distances in soil/tar-particles.

13.4 Different Ways to Consider the Lack of Bioavailability

The preceding chapter has shown that it will not be easy to find the one and only sweeping tool to solve problems like the lack of bioavailability in soil bioremediation. The question that remains to be discussed is whether the lack of bioavailability must necessarily be considered as a problem or failure. From a toxicological point of view the lack of bioavailability of a toxic compound must be considered more as an advantage since it reduces the risk that the target organism is attacked by an overdose of the toxic compound. The different valuations of the term bioavailability can be recognised if one considers its different meanings in different fields of biosciences (table 13.2).

In the context of soil bioremediation, bioavailability can be considered, above all, as the chemical and physical status of an environmental pollution that determines the extent and rate by which a degrading microbial community may get in metabolic contact with the xenobiotic substrates (Mahro 2000). This means that the occurrence of a residual pollutant concentration in soil remediation is primar-

Table 13.2. Use of the term bioavailability in different fields of applied biological science (modified from Mahro 2000)

Field of Biosciences	Meaning of the Term Bioavailability
Agriculture:	Accessibility of the soil nutrients for plants in order to grow
Pharmacy:	Possibility of a drug to reach the receptor at the target cells in order to cause positive effects
Toxicology:	Possibility of a drug to reach the receptor at the target cells and to cause harmful effects
Nutrition physiology:	Suitability of the food to become processed by the human digestion tract
Biotechnology:	Status of accessibility of the fed substrates for microbial transformation into useful products
Environmental Biotechnology:	Status of accessibility of pollutants in the environmental matrix for microbial degradation

ily only a lack of time for mass transfer but does not, in most cases, describe a principal barrier for biodegradation (fig. 13.8). It is therefore mainly the engineer or biotechnologist for whom the lack of bioavailability is a problem since the extension of the time scale of the overall pollutant transformation decreases significantly the yield per time and volume of the bioremediation treatment.

However, if we cannot speed up mass transfer in a cost effective manner we have to assess the risk that occurs if we let the *residual* pollutants diffuse out as slowly as allowed by the physical or chemical conditions in soil. Provided that the pollutant pool has been stabilised by a preceding (bio-)remediation treatment (i.e. the pollution concentration has been reduced to an apparently stable residual plateau), the maximum absolute amount of pollutant that may become available by diffusion over time is primarily determined by the maximum water solubility of the pollutant (see table 13.1) and by the actual equilibrium concentration between the residual pollutant pool and the accessible pore space. The fraction of the residual pollutant pool that might be potentially mobilised at any given time, however, is so low that one can neglect its toxic impact on any environmental target of value. More importantly, one must also assume that even the low absolute pollutant amounts being released by diffusion from inaccessible soil pores may be degraded instantaneously as soon as they reach the accessible and microbially active soil phase. The goal to reduce the ecotoxic or toxic potential of a soil contamination seems, therefore, to be sufficiently achieved if the mobile fraction of the pollution (= bioavailable and therefore degradable pollutant fraction) has been eliminated from the soil.

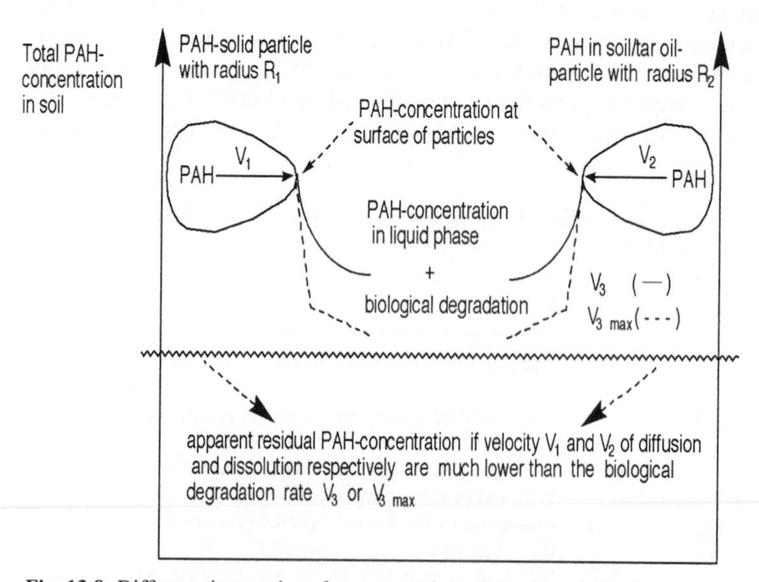

Fig. 13.8. Different time scales of mass transfer and biodegradation as a major reason for residual pollutant concentrations in soil

The practical conclusion of this discussion is that at this time it appears to be justifiable to set the targets of soil bioremediation more in accordance with an actual risk rather than on chemical detectability only. The pollutant residues that remain stable at the end of bioremediation, and which cannot be degraded any further in a reasonable time frame, do not represent such a severe risk since they are of very low mobility. To be on the safe side however, one should check eventual risks by a set of reliable and evaluated biotests, which measure the pollutants that actually reach an ecotest-organism, rather than to determine residual pollutant concentrations by organic solvent only.

The proposal to adopt the targets of bioremediation according to actual risk is certainly not quite satisfactory and surely only a second preference since a complete pollutant removal and the immediate restoration of a "clean soil" would certainly be more desireable. A cost/benefit analysis of the different available options suggests however that from an environmental perspective in particular, it is better that we recycle sites that contain low residual amounts of stabilised pollutants (brown fields) into use, rather than to use up new, ecologically more valuable green fields.

References

Angehrn D, Gälli R, Schluep M, Zeyer J (1997) Biologisch saniertes Bodenmaterial aus Mineralölschadensfällen: Abfall oder Produkt ? TerraTech 3/1997: 51–56

Bollag JM (1992) Decontaminating soil with enzymes. Environ Sci Technol 26: 1876–1881

Boonchan S, Britz ML, Stanley G (2000) Degradation and mineralization of high-molecular-weight polycyclic aromatic hydrocarbons by defined fungal-bacterial-cocultures. Appl Environ Microbiol 66: 1007–1019

Bosma TNP, Middeldorp PJM, Zehnder AJB (1997) Mass transfer limitation of biotransformation: Quantifying bioavailability. Environ Sci Technol 31: 248–252

Bossert I, Kachel MW, Bartha R (1984) Fate of hydrocarbons during oily sludge disposal in soil. Appl Environ Microbiol 47: 763–767

Cuno M (1996) Kinetische Untersuchungen zum biologischen Abbau von Mineralölen und polycyclischen aromatischen Kohlenwasserstoffen. Fortschr-Ber VDI Series 15 (Umwelttechnik), Nr 148, VDI-Verlag Düsseldorf

Deschenes L, Lafrance P, Villeneuve JP, Samson R (1996) Adding sodium dodecyl sulfate and *Pseudomonas aeruginosa* UG2 biosurfactants inhibits polycyclic aromatic hydrocarbon biodegradation in a weathered creosote-contaminated soil. Appl Microbiol Biotechnol 46: 638–646

Efroymson RA, Alexander M (1991) Biodegradation by an Arthrobacter species of hydrocarbons partitioned into an organic solvent. Appl Environ Microbiol 57: 1441–1447

Efroymson RA, Alexander M (1995) Reduced mineralization of low concentrations of phenanthrene because of sequestering in nonaqueous-phase-liquids. Environ Sci Technol 29: 515–521

Erickson DC, Loehr RC, Neuhauser EF (1993) PAH loss during bioremediation of manufactured gas plant site soils. Water Res 27: 911–919

Foght JM, Gutnick DL, Westlake DWS (1989) Effect of emulsan on biodegradation of crude oils by pure and mixed bacterial cultures. Appl Environ Microbiol 55: 36–42

Ghoshal S, Luthy RG (1996) Bioavailability of hydrophobic organic compounds from nonaqueous-phase-liquids: the biodegradation of naphthalene from coal tar. Environ Toxicol Chem 15: 1894–1900

Harms H (1998) Bioavailability of dioxin-like compounds for microbial degradation. In: Wittich RM (ed) Biodegradation of dioxins and furans. Springer Verlag, Heidelberg, Germany, pp 135–163

Käppeli O, Walther P, Mueller M, Fiechter A (1984) Structure of the cell surface of the yeast *Candida tropicalis* and its relation to hydrocarbon transport. Arch Microbiol 138: 279–292

Klein J (2000) Environmental processes II – Soil decontamination. In: Klein J (ed) (Rehm HJ, Reed G, Pühler A, Stadler P, (series eds)) "Biotechnology" a multi-volume comprehensive treatise, Vol 11 b, Wiley-VCH, Weinheim, Germany

Knackmuss HJ (1997) Abbau von Natur- und Fremdstoffen. In: Ottow JCG, Bidlingmaier W (eds) Umweltbiotechnologie. Gustav Fischer Verlag, Stuttgart, Germany, pp 39–80

Kottermann MJJ, Vis E, Field JA (1998) Successive mineralization and detoxification of benzo[a]pyrene by the white rot fungus *Bjerkandera* sp. strain BOS55 and indigenous microflora. Appl Environ Microbiol 64: 2853–2858

Laha S, Luthy RG (1991) Inhibition of phenanthrene mineralization by nonionic surfactants in soil-water systems. Environ Sci Technol 25: 1920–1930

Lenke H, Daun G, Bryniok D, Knackmuss HJ (1993) Biologische Sanierung von Rüstungsaltlasten. Spektrum der Wissenschaft 10/1993: 106–108

Mahro B (2000) Bioavailability of contaminants. In: Klein J (ed) (Rehm HJ, Reed G, Pühler A, Stadler P (series eds)) Environmental processes II – Soil decontamination. "Biotechnology" a multi-volume comprehensive treatise, Vol 11 b. Wiley-VCH, Weinheim, Germany, pp 61–88

Mahro B, Kästner M (1993) Der mikrobielle Abbau polyzyklischer aromatischer Kohlenwasserstoffe (PAK) in Böden und Sedimenten: Mineralisierung Metabolitenbildung und Entstehung gebundener Rückstände. Bioengineering 9: 50–58

Mahro B, Schaefer G (1998) Bioverfügbarkeit als limitierender Faktor des mikrobiellen Abbaus von PAK im Boden-Ursachen des Problems und Lösungsstrategien. Altlastenspektrum 7: 127–134

Mahro B, Schmidt L, Eschenbach A (1999) Möglichkeiten und Grenzen mikrobiologischer Verfahren bei der Sanierung kontaminierter Böden. In: Heiden S, Erb R, Warrelmann J, Dierstein R (eds) Biotechnologie im Umweltschutz Bioremediation: Entwicklungsstand – Anwendungen – Perspektiven. E Schmidt-Verlag, Berlin, Germany, pp 99–107

Marin M, Pedregosa A, Laborda F (1996) Emulsifier production and microscopical study of emulsions and biofilms formed by the hydrocarbon-utilizing bacteria *Acinetobacter calcoaceticus* MM5. Appl Microbiol Biotechnol 44: 660–667

Mueller JG, Lantz SE, Blattman BO, Chapman PJ (1991a) Bench-scale evaluation of alternative biological treatment processes for the remediation of pentachlorophenol- and creosote-contaminated materials: solid phase bioremediation. Environ Sci Technol 25: 1045–1055

Müller R (1992) Bacterial degradation of xenobiotics. In: Fry JC, Gadd GM, Herbert RA, Jones CW, Watson-Craik IA (eds) Microbial control of pollution, SGM Symposium Vol 48. Cambridge University Press, UK, pp 35–57

Riis V, Miethe D, Babel W (1998) Grenzen der Sanierbarkeit von Mineralölschäden. Altlastenspektrum 7: 214–218

Riis V, Miethe D, Möder M (1996) Analytical characterization of the persistent residues after the microbial degradation of mineral oils. Fresenius J Anal Chem 356: 378–384

Rosenberg M, Beyer EA, Delarea J, Rosenberg E (1982) Role of thin fimbriae in adherence and growth of *Acinetobacter calcoaceticus* RAG-1 on hexadecane. Appl Environ Microbiol 44: 929–937

Schaefer G, Hattwig S, Unterste-Wilms M, Hupe K, Heerenklage J, Lüth JC, Kästner M, Eschenbach A, Stegmann R, Mahro B (1995) PAH-degradation in soil: microbial activation or inoculation A comparative evaluation with different supplements and soil materials. In: van den Brink WJ, Bosman R, Arendt F (eds) Contaminated soil'95. Kluwer Academic Publ, The Netherlands, pp 415–416

Schneider J, Grosser R, Jayasimhulu K, Xue W, Warshaawsky D (1996) Degradation of pyrene benz(a)anthracene and benzo[a]pyrene by *Mycobacterium* sp strain RJGII-135, isolated from a former coal gasification site. Appl Environ Microbiol 62: 13–19

Scott CCL, Finnerty WR (1976) A comparative analysis of the ultrastructure of hydrocarbon-oxidizing microorganisms. J Gen Microbiol 94: 342–350

Stucki G, Alexander M (1987) Role of dissolution rate and solubility in biodegradation of aromatic compounds. Appl Environ Microbiol 53: 292–297

Thibault SL, Anderson M, Frankenberger WT (1996) Influence of surfactants on pyrene desorption and degradation in soils. Appl Environ Microbiol 62: 283–287

Tiehm A (1994) Degradation of polycyclic aromatic hydrocarbons in the presence of synthetic surfactants. Appl Environ Microbiol 60: 258–263

Tiehm A, Stieber M, Werner P, Frimmel FH (1997) Surfactant enhanced mobilization and biodegradation of polycyclic aromatic hydrocarbons in manufactured gas plant soil. Environ Sci Technol 31: 2570–2576

Weissenfels WD, Klewer HJ, Langhoff J (1992) Adsorption of polycyclic aromatic hydrocarbons (PAHs) by soil particles: influence on biodegradability and biotoxicity. Appl Microbiol Biotechnol 36: 689–696

Wodzinski RS, Larocca D (1977) Bacterial growth kinetics on diphenylmethane and naphthalene-heptamethylnonane mixtures. Appl Environ Microbiol 33: 660–665

Ye D, Siddiqi MA, Maccubin AE, Kumar S, Sikka H (1996) Degradation of polynuclear aromatic hydrocarbons by Sphingomonas paucimobilis. Environ Sci Technol 30: 136–142

Zhang Y, Miller RM (1995) Effect of rhamnolipid (biosurfactant) structure on solubilization and biodegradation of *n*-alkanes. Appl Environ Microbiol 61: 2247–2251

14 Bioavailability of Organic Compounds Sequestered in Soils

M. Alexander
Department of Crop and Soil Sciences and Institute of Comparative and Environmental Toxicology, Cornell University, Ithaca, NY, 14853, USA

The pollutants present in soil and sediments usually are not recent introductions, rather they were typically added one, two and frequently more than three decades ago. This fact was one of the concerns as we devised our initial studies. The studies were also prompted by a number of long- and medium-term field studies that showed that persistent pesticides (such as DDT, dieldrin, heptachlor and kepone) and a number of other compounds initially disappear from soil at reasonable rates, but the rate subsequently slows appreciably. Indeed, the subsequent disappearance is often so slow that the rate of decline in concentration sometimes cannot be estimated. Because the compounds were initially subject to biodegradation by microorganisms (as well as loss by volatilisation or other abiotic processes in some instances), clearly something was occurring in soil that was slowly but progressively making the compounds less and less bioavailable, at least to microorganisms (Alexander 1995).

To determine whether a loss of bioavailability was actually occurring, a study was conducted to determine whether this ageing resulted in a diminished accessibility to microorganisms. For this purpose, test compounds were added to soils that had been sterilised. Sterilisation was required to prevent degradation of the test substances during the ageing period. After various times of ageing, bacteria able to metabolise the chemicals were added, and either the conversion of the compound to CO_2 (in our earlier studies) or the quantity remaining was measured. Our first experiments showed that phenanthrene became progressively more resistant to microbial attack as it aged in two soils. Nevertheless, vigorous extractions showed that all of the compound was still present. In some manner, the molecules were being sequestered; i.e., hidden or inaccessible (Hatzinger and Alexander 1995). We have subsequently demonstrated that other compounds can be sequestered under comparable conditions; e.g., naphthalene (Kelsey and Alexander 1997), anthracene, fluoranthene, pyrene (Tang et al. 1998 a) and atrazine (Chung and Alexander 1998).

We then proceeded to assess whether ageing also reduces the bioavailability of persistent compounds to other groups of organisms. We observed that earthworms (*Eisenia foetida*) assimilated decreasing amounts of phenanthrene, naphthalene, anthracene, fluoranthene, pyrene and atrazine that had been previously aged for increasing periods of time in sterile soil. Nevertheless, vigorous methods of solvent extraction showed that most or all of the compounds were present in the soil throughout the ageing period and that they were not converted to covalently bound organic complexes. Therefore, the compounds were sequestered, although still present, and in this form were less accessible for earthworm assimilation (Kelsey and Alexander 1997; White et al. 1997; Tang et al. 1998 a).

The diminution in acute toxicity as chemicals age in soil was also evaluated. The test organisms were *Musca domestica* (house fly), *Drosophila melanogaster* (fruit fly) and *Blatella germanica* (German cockroach), and the toxicants were DDT and dieldrin. The initial concentrations added to soil caused 100 % mortality of all three species. However, mortality of the insects was markedly reduced after the two compounds had aged for 30 d, and longer periods of ageing further reduced the mortality. Indeed, dieldrin was no longer toxic to *D. melanogaster* and *B. germanica* after 120 d of ageing, and neither DDT nor dieldrin caused lethality to *M. domestica* after 270 d. Yet, 92.1 % of the dieldrin and 84.7 % of the DDT were still present as determined by vigorous extraction of the soils after 270 and 180 d, respectively (Robertson and Alexander 1998).

In each instance in the studies reported and in most of the studies summarised below, the type of vigorous extraction used for regulatory decisions overestimated the bioavailability. The overestimation became progressively larger as the compounds persisted for longer periods of time. Inasmuch as such organic pollutants have indeed been present in soils and sediments for long periods of time, the data suggest that regulatory decisions currently are based on values that do not reflect bioavailability.

Many genotoxicity assays make use of solvent extracts of soils (or sediments) or the aqueous phase in the presence of the particulate matter from environmental samples. Because the test organism in such assays does not have prolonged or known frequency of contact with the particulate matter or is exposed to a solvent extract that may not provide the same exposure as the particulate material containing the sorbed chemical (coupled with our observations that sorbed PAHs are bioavailable), we initiated a study designed to devise a solid-phase genotoxicity assay for organic compounds that are extensively sorbed in soil. The assay relies on measurement of the rate of mutation of a strain of *Pseudomonas putida* to rifampicin resistance (Alexander et al. 1999). The assay was then used to evaluate the possible ageing of two carcinogens, namely benzo[a]pyrene and 9,10-dimethyl-1,2-benzanthracene. Seven days of ageing resulted in a marked diminution in the mutation rates, and the number of mutants declined further in the next 8 d. A further decrease in genotoxicity was not observed with further ageing. Despite the large decline in genotoxicity, the concentration of the two PAHs did not decrease (as revealed by vigorous extraction) (Alexander and Alexander 1999). These findings extend to genotoxicity that which we previously observed, namely that ageing reduces bioavailability for biodegradation, assimilation by invertebrates and acute toxicity.

A test of the availability of aged chemicals in field samples was also conducted. The samples were from (a) experimental plots that were treated in 1949 with known concentrations of individual insecticides and (b) a sandy loam that had a major amount of contamination (about 30 yr earlier) from DDT and dieldrin. On the basis of either the concentrations or the percentages of the compound that were assimilated by *E. foetida*, only ca. 70, 88, 66, and 80 of DDT, DDE, DDT and total of those three compounds, respectively, were sequestered and unavailable to

E. foetida. DDT, DDE, and DDD were sequestered in soil samples from the waste-disposal site receiving the insecticides ca. 30 yr earlier (Morrison et al. 2000).

Engineered bioremediation or natural biodegradation (intrinsic bioremediation) often removes a large percentage of a pollutant in soil, but some frequently remains and appears to resist further microbial destruction. This is particularly true of PAHs but has also been observed with other classes of chemicals. We thus carried out an investigation to determine whether the reason for the lack of complete chemical destruction was the result of ageing and the consequent sequestration. PAHs were aged for 140–203 d in sterile soil and then bacteria capable of degrading the PAHs were added to the soil in the laboratory. As in other of our studies, ageing reduced the bioavailability of phenanthrene, anthracene, fluoranthene and pyrene for bioremediation and for assimilation by earthworms (*E. foetida*). The bench-scale bioremediation also markedly reduced the quantity of phenanthrene, fluoranthene and pyrene assimilated by the worms. Of particular interest was the finding that the smallest quantities of the three compounds were assimilated by the worms from soil in which the PAHs had been aged and then bioremediated. However, even after ageing and extensive bioremediation, small amounts of the PAHs were still available to the animals (Tang et al. 1998 a).

Extrapolation from published studies of DDT and lindane suggest that the percentages of these aged compounds that are bioavailable vary among soils, but only a few soils were evaluated and few or no properties of the soil were determined in these early investigations. We therefore decided to study sequestration and bioavailability in a variety of soils. In our first investigation, we found marked differences in the extent of sequestration, earthworm uptake and bacterial degradation of phenanthrene sequestered for 160 d in 7 soils differing appreciably in organic matter and clay content (White et al. 1997). Subsequently, we selected 16 soils with widely dissimilar properties, and each was analysed to determine physical and chemical properties that might be important for sequestration. At regular intervals during a 200 d period of ageing in sterile soil of phenanthrene and atrazine, the quantities sequestered were determined by assessing biodegradation by subsequently added bacteria and the amounts by a mild extraction procedure. The rate and extent of sequestration varied appreciably among the 16 soils. In some instances, sequestration was almost complete in 120 d, but it was still going on at the last sampling time (200 d) in other soils. In addition, the extent of sequestration of the two compounds in these 16 soils was not highly correlated (Chung and Alexander 1998).

As indicated above, published results of long-term monitoring studies show a decline in the rate of disappearance from soil of a number of pesticides applied in the field. There is ultimately a period with little or no chemical loss, the results suggesting that the compounds are no longer biodegraded because they have become sequestered. It is also evident from these published field data that the percentage of the originally applied pesticide that apparently became sequestered varied appreciably even with a single compound in different soils. Because sequestration is a time-dependent process, we hypothesised that the percentage of an organic compound that becomes sequestered is determined by the rate of its loss following its initial introduction into soil. The existence of a relationship between loss rate and sequestration is not surprising since the slower the disappearance of the compound, the more time for the sequestration to occur. To test this hypothesis, phenanthrene was added to samples of a single soil type, and various rates of biodegradation were obtained. An appreciable amount of the compound became unavailable to microorganisms when the initial rates of biodegradation were slow, and little remained if the rates were rapid, indicating that the percentage of a compound that will be sequestered is determined by the rate of its initial disappearance from soil. We also approximated the rate of disappearance and the amounts of DDT and aldrin-dieldrin remaining from the published field studies of 8 different soils. Our calculations indicated that the percentages of the initial concentrations of the insecticides remaining after several years were inversely related to the rates of chemical disappearance in the period shortly after the compounds were introduced into the soil, thus supporting our hypothesis (Nam and Alexander, unpublished data).

A study was performed to ascertain whether means could be found to increase the bioavailability of sequestered compounds. The purpose was to ultimately allow for their subsequent bioremediation, thereby resulting in a lower final concentration. Five methods were found. First, slurrying samples of soils or aquifer solids containing aged phenanthrene or di-(2-ethylhexyl) phthalate increased the rate and extent of their biodegradation by individual bacteria. Second, use of certain specialised microorganisms promoted further destruction of the PAH in slurries of soil. Third, the addition of a surfactant enhanced the degradation, although not all surfactants were beneficial. Fourth, supplementation with an unaged chemical (pyrene or anthracene) resulted in greater losses of aged phenanthrene even though that supplement was not itself utilised under the test conditions (White et al. 1999a). The effect of the additional compound appears to be a consequence of competitive displacement from a finite number of sorption sites, since pyrene added to soil with aged phenanthrene increased the physical availability of the latter PAH to a mild extractant. Fifth, certain methods involving a cyclic sequence of wetting and drying increased the subsequent biodegradation of aged phenanthrene (White et al. 1998; White et al. 1999b).

Two of the mechanisms initially proposed for the sequestration associated with ageing are the partitioning of the organic molecules into organic solids in soil and their entrapment within nanopores. Model solids and silica particles were used to assess the possible validity of these hypotheses. It was found that phenanthrene within solid alkanes containing 18 to 32 C atoms, three waxes or low-molecular-weight polycaprolactone, polyethylene and polypropylene was degraded by a strain of *Pseudomonas* more slowly and to a lesser extent than when the PAH was not within these nonporous solids. In contrast, although phenanthrene in 6.0- or 15-nm nanopores with nonsorptive surfaces was more slowly utilised by the bacterium than when not entrapped in these minute pores, the retardation was not as marked as when the compound was within the nonporous solids (Hatzinger and Alexander 1997).

On the other hand, a further series of experiments revealed that the bioavailability of phenanthrene was markedly reduced or entirely prevented if the nanopores, many of which are smaller than the smallest bacterium, have hydrophobic surfaces. Thus, little biodegradation of phenanthrene by a bacterium occurred if the substrate was sorbed to hydrophobic polystyrene beads containing 5- or 300- to 400-nm pores, coincident with the little desorption that was evident even after

240 h (Nam and Alexander 1998). Although the actual mechanism of the time-dependent reduction in bioavailability in an environment as complex as soil is difficult to establish, the findings show that the two proposed mechanisms are plausible.

Our data from investigations of ageing, as well as the results from many other laboratories (using either unaged chemicals or field samples), show clearly that assays for bioavailability of sequestered compounds should not be the chemical procedures (typically associated with an initial vigorous extraction) that are the basis for current assessments of exposure and risk. These procedures overestimate, often grossly so, the potential exposure. Although bioassays would appear to be the appropriate types of assays, such approaches are slow, often expensive and commonly do not have the precision needed for regulatory decisions. One approach that we have begun to evaluate is designed to find a mild extractant that removes from the soil (for analysis) the same amount of a compound that is available to one or more species. In our initial evaluations, it was found that the ageing-induced decline in bioavailability of phenanthrene and atrazine to earthworms (*E. foetida*) and a strain of *Pseudomonas* was paralleled by the progressive reduction in the quantities recovered by a series of mild organic extractants and that the percentage of the test chemicals available to earthworms or for microbial degradation was approximated by the percentage recovered by some mild extractants (Kelsey et al. 1997).

In an extension of this work, three additional PAHs were tested, and measurements were made not only of the amounts remaining after earthworm uptake or microbial degradation but also the quantities remaining. After showing that ageing of anthracene, fluoranthene and pyrene decreased the quantities removed by mild extraction with several organic solvents, it was demonstrated that the uptake of the PAHs by earthworms, wheat and barley was correlated with the quantities recovered from soil by the solvents (r values ranging from 0.89 to 0.99). Similarly, biodegradation was found to markedly diminish the percentage of aged anthracene removed from soil by a mild extraction with *n*-butanol or ethyl acetate (Tang and Alexander 1999). These findings, in conjunction with studies now in progress with *n*-butanol and tetrahydrofuran, indicate the potential for use of a mild extraction to predict PAH bioavailability.

Solid-phase extractants are also being evaluated as the basis for chemical assays to predict bioavailability. In one study, the solid-phase extractant was Tenax TA beads. Using both unaged DDT, DDE and DDD and the same three compounds aged in the field for ca. 30 or 49 yr or in the laboratory, we found that the quantities removed from the various soils by solid phase extraction with Tenax TA beads were highly correlated with the amounts assimilated by *E. foetida*, the correlation coefficients (r values) being 0.933, 0.980 and 0.995 for DDT, DDE, and DDE, respectively (Morrison et al. 2000). We have also used C18 membrane disks that are placed in suspensions of these soils, and subsequent extraction and analysis of the disks show r values of 0.921 or higher in correlating the disk assay with uptake by the earthworms (Tang and Alexander 1999b).

References

Alexander M (1995) How toxic are toxic chemicals in soil? Environ Sci Technol 29: 2713–2717

Alexander RR, Alexander M (1999) Genotoxicity of polycyclic aromatic hydrocarbons declines as they age in soil. Environ Toxicol Chem 18: 1140–1143

Alexander RR, Chung N, Alexander M (1999) Solid-phase genotoxicity assay for organic compounds in soil. Environ Toxicol Chem 18: 420–425

Chung N, Alexander M (1998) Differences in sequestration and bioavailability of organic compounds aged in dissimilar soils. Environ Sci Technol 32: 855–860

Hatzinger PB, Alexander M (1995) Effect of ageing of chemicals in soil on their biodegradability and extractability. Environ Sci Technol 29: 537–545

Hatzinger PB, Alexander M (1997) Biodegradation of organic compounds sequestered in organic solids or in nanopores within silica particles. Environ Toxicol Chem 15: 2215–2221

Kelsey JW, Alexander M (1997) Declining bioavailability and inappropriate estimation of risk of persistent compounds. Environ Toxicol Chem 15: 582–585

Kelsey JW, Kottler BD, Alexander M (1997) Selective chemical extractants to predict bioavailability of soil-aged organic chemicals. Environ Sci Technol 31: 214–217

Morrison DE, Robertson BK, Alexander M (2000) Bioavailability to earthworms of aged DDT, DDE, DDD, and dieldrin in soil. Environ Sci Technol 24: 709–713

Nam K, Alexander M (1998) Role of nanoporosity and hydrophobicity in sequestration and bioavailability: tests with model solids. Environ Sci Technol 32: 71–74

Robertson BK, Alexander M (1998) Sequestration of DDT and dieldrin in soil: Disappearance of acute toxicity but not the compounds. Environ Toxicol Chem 17: 1034–1038

Tang J, Alexander M (1999) Mild extractability and bioavailability of polycyclic aromatic hydrocarbons in soil. Environ Toxicol Chem 18: 2711–2714

Tang J, Carroquino MJ, Robertson BK, Alexander M (1998 a) Combined effect of sequestration and bioremediation in reducing the bioavailability of polycyclic aromatic hydrocarbons in soil. Environ Sci Technol 2: 3586–3590

Tang J, Robertson BK, Alexander M (1999 b) Chemical-extraction methods to estimate bioavailability of DDT, DDE, and DDD in soil. Environ Sci Technol 33: 4346–4351

White JC, Alexander M, Pignatello JJ (1999 a) Enhancing the bioavailability of organic compounds sequestered in soil and aquifer solids. Environ Toxicol Chem 18: 182–187

White JC, Hunter M, Pignatello JJ, Alexander M (1999 b) Increase in the bioavailability of aged phenanthrene in soils by competitive displacement with pyrene. Environ Toxicol Chem 18: 1728–1732

White JC, Kelsey JW, Hatzinger PB, Alexander M (1997) Factors affecting sequestration and bioavailability of phenanthrene in soils. Environ Toxicol Chem 16: 2040–2045

White JC, Quiñones–Rivera A, Alexander M (1998) Effect of wetting and drying on the bioavailability of organic compounds sequestered in soil. Environ Toxicol Chem 17: 2378–2382

15 Bacterial Strategies to Improve the Bioavailability of Hydrophobic Organic Pollutants

L.Y. Wick [1], D. Springael [2], H. Harms [1]
[1] Swiss Federal Institute of Technology Lausanne, Department of Rural Engineering, 1015 Lausanne, Switzerland
[2] Flemisch Institute for Technological Research, Environmental Technology, 2400 Mol, Belgium

Bioremediation of polluted soil mostly depends on indigenous micro-organisms. The principle of engineered bioremediation is to stimulate those micro-organisms, which are able to degrade xenobiotic substrates (Aelion et al. 1987). The most often reported genera in degradation of anthropogenic pollutants include *Ralstonia*, *Burkholderia*, *Comamonas*, *Arthrobacter*, *Mycobacterium*, *Nocardia*, fluorescent *Pseudomonas*, *Rhodococcus*, and *Sphingomonas* (Häggblom and Valo 1995; Neilson 1995; Commandeur et al. 1995). In response to the introduction of electron acceptors (like oxygen) or nutrients to soil, specific degrader organisms will multiply. However, in spite of a potentially metabolically active biomass, *in-situ* bioremediation of soils polluted with hydrophobic organic pollutants (HOC) frequently results in slow pollutant degradation rates, high residual concentrations and, as a consequence, in limited clean-up efficiencies (Zhang et al. 1998; Luthy et al. 1994). The unequal spatial distribution of micro-organisms and pollutants in combination with physically retarded substrate diffusion are nowadays generally accepted as key limiting factors for efficient biodegradation of hydrophobic contaminants in soil.

Extensive mechanical mixing of the soil, the addition of surfactants, and electrokinetic stimulation are considered as bioavailability-enhancing engineering solutions. Although many bacteria have been found to degrade and to live on HOC, still very little is known about the strategies of bacteria *themselves* to improve their access to hydrophobic pollutants in soils. In spite of the generally poor bioavailability of HOC some bacteria obviously subsist on these compounds; this has not been looked at in detail and might be of major interest for bioremediation of HOC-polluted matrices. Bacterial features may enhance the transfer of poorly bioavailable substrates;

- by reducing the mean distances between pollutants and bacteria, e.g. by adhesion to sorbents,
- by causing active transfer of the pollutant into the aqueous phase,

- by uncoupling the development and maintenance of active biomass from pollutant bioavailability,
- by using uptake systems with high specific affinity.

15.1 Bioavailability – a Question of Give and Take

15.1.1 Definition

Bioavailability for degradation is a dynamic process that is determined by the rate of mass transfer to microbial cells relative to their intrinsic catabolic activity (Bosma et al. 1997; Harms and Bosma 1997). According to Bosma et al. (1997), it is defined as the ratio of the capacity of an organism's or a population's environment, to provide a chemical to the capacity of the organism/population to transform that chemical. This definition points to the importance of mass fluxes for consumptive processes and distinguishes bioavailability for degradation from bioavailability for basically "non-consumptive" processes such as poisoning or inhibition. At high mass transfer rates, the overall degradation rate is controlled by the metabolic activity of the bacteria, i.e. the specific activity of the cells and the population density. Conversely, when the transport of the substrate decreases or the bacterial population grows, the mass transfer may become the factor that controls the degradation. In order to quantify bioavailability, a basic set of abiotic and biotic processes has to be considered;

- mass transfer from the pollutant source to the bacteria, e.g. the effective diffusivity of a chemical in a given environment,
- the spatial distribution of microbial populations and pollutants in soil,
- the rate of microbial substrate consumption,
- growth and decay of cells.

15.1.2 Mass Transfer-Limiting Factors in Soil

As a result of their hydrophobicity, HOC such as polycyclic aromatic hydrocarbons (PAH), polychlorinated biphenyls (PCB), dioxins, or some pesticides are only poorly water soluble. Higher molecular weight HOC normally exhibit higher octanol-water partitioning coefficients (K_{ow}) and lower aqueous solubilities. It is generally observed, that HOC strongly accumulate in the solid (Pignatello and Xing 1996) or non aqueous liquid phases of soils, which can be attributed to ad- and absorption by soil particles, dissolution in non-aqueous-phase-liquids (NAPL), and sequestration in soil micropores (Luthy et al. 1997; Guerin and Boyd 1992; Ogram et al. 1985; Rijnaarts et al. 1990; Beurskens et al. 1993; Efroymson and Alexander 1994; Gosh et al. 2000). In contrast, micro-organisms appear to degrade chemicals only when they are dissolved in water (Wodzinski and Johnson 1968; Wodzinski and Bertolini 1972; Wodzinski and Coyle 1974; Ogram et al. 1985; Cameotra et al. 1983; Thomas et al. 1986; Harms and Zehnder 1995;

Bouchez et al. 1995) and consequently, pollutant mass transfer becomes prerequisite for biodegradation. Many mass transfer-limited processes have been described, including sorption-related processes (Rijnaarts et al. 1990; Scow and Johnson 1997; Mihelcic and Luthy 1991; Ortega-Calvo et al. 1999), substrate dissolution (Volkering et al. 1992; Volkering et al. 1993; Mulder et al. 1998; Stucki and Alexander 1987; Gray et al. 1994), transport from the bulk liquid to the surfaces of porous media (Harms and Zehnder 1994; Tros et al. 1998) or combinations of various processes (Ramaswami et al. 1997; Ramaswami and Luthy 1997; Mulder et al. 1998). Mass transfer rates of pollutants are strongly affected by low solubility of the pollutant described by the following equation (eq. 15.1).

$$N = -D_P \cdot \delta C / \delta x \tag{15.1}$$

in which N is the mass flux of the pollutant (P) (kg m^{-2} s^{-1}), D_P the effective diffusion coefficient (m^2 s^{-1}) for P, δC the concentration difference between the maximal aqueous concentration C_{max} and the actual dissolved concentration C (kg m^{-3}), and x the space coordinate in direction of transport. Consequently, the maximum value of the gradient of the dissolved concentration over a certain distance in space is limited by the value of the maximum solubility of the pollutant, and by its molecular diffusion coefficient D_P. Diffusion coefficients of low molecular-weight-compounds in water are in the range of $5 \cdot 10^{-10} - 1 \cdot 10^{-9}$ m^2 s^{-1} (Weast 1984; Schwarzenbach et al. 1993), whereas effective diffusion coefficients in soil or sediments can be orders of magnitude lower ($9 \cdot 10^{-12}$–$2 \cdot 10^{-21}$ m^2 s^{-1} (Harms and Bosma 1997)). This results in typical average daily diffusion distances of about 1 cm d^{-1} in water and 1 µm d^{-1} or less in soil or sediments, respectively.

Although many bacteria are motile *in vitro*, soil bacteria are predominantly immobilised *in situ*. Penetration depths of 3–5 cm were observed when bacteria were applied to soil (Madsen and Alexander 1982; Edmonds 1976; Deware and Alexander 1995). Most soil bacteria are attached to the surfaces of soil particles (Costerton and Lappin-Scott 1989; Gray and Parkinson 1968), or live in pores with a mean diameter of 2 µm (Bosma et al. 1997; Kilbertus 1980; Postma and Veen van 1990), whereas metabolically active cells are likely to be in contact with the soil water (Tate III 1995). Because of spatial constraints, bacterial growth is likely to result in the formation of micro-colonies rather than in the release of single cells into the soil water leading to a largely heterogeneous distribution of soil bacteria. Assuming a typical population of 10^7 to 10^8 bacteria per gram of soil and micro-colonies of up to 100 cells, average distances of about 100 μm between micro-colonies in soil can be calculated (Bosma et al. 1997). Average distances between pollutant-degrading bacteria and a fresh contaminant point source are presumably even bigger. The combination of physically retarded substrate diffusion and unequal spatial distribution of micro-organisms is the principal factor limiting the bioavailability of HOC in the soil matrix.

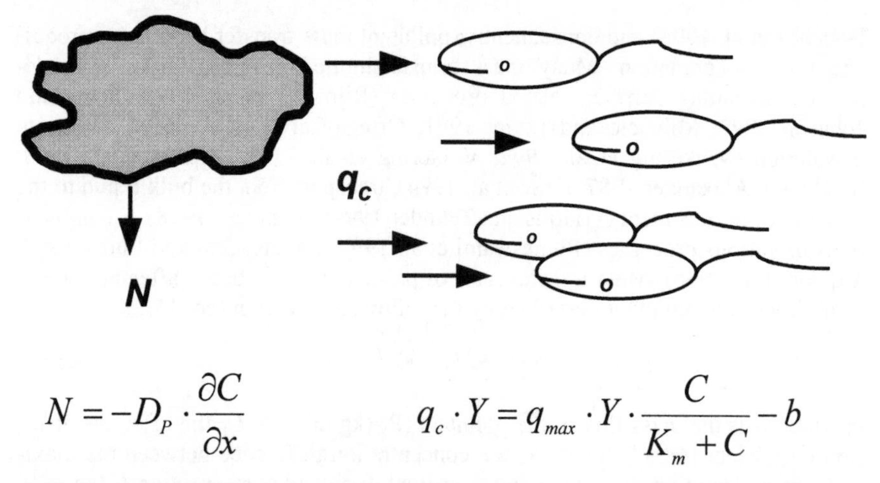

q_c
N
$N = -D_P \cdot \frac{\partial C}{\partial x}$
$q_c \cdot Y = q_{max} \cdot Y \cdot \frac{C}{K_m + C} - b$

Fig. 15.1. Schematic view of the processes jointly governing bioavailability: Due to bacterial uptake and degradation a pollutant concentration gradient between the bacterial surface and the bulk solution is created, which acts as a driving force for ongoing contaminant release from the soil matrix. Pollutant mass transfer N from the soil matrix to the degrader bacteria is controlled by Fickian diffusion whereas bacterial transformation (q_c) and growth ($q_c \cdot Y$) can be described by extended Monod kinetics (cf. text for further discussion of the equations).

15.1.3
Biodegradation as Mass Transfer-Controlling Factor

In the soil matrix two processes will jointly control the aqueous concentration of a pollutant. Matrix-sorbed or NAPL-dissolved HOC will dissolve to the soil water phase, thereby striving towards the equilibrium aqueous concentration C_{eq}, whereas bacterial uptake and transformation will counteract and simultaneously reduce the bulk concentration. Due to the dynamic character of these joint processes, the average aqueous pollutant concentration is not an appropriate measure for pollutant bioavailability (Bosma et al. 1997; Harms and Bosma 1997) (fig. 15.1). The concentration difference between the aqueous concentration at the cell surfaces C and C_{eq} acts as a driving force for ongoing dissolution. Quasi-steady state will be reached when the HOC mass transfer flux N equals the bacterial degradation flux q_c. Substrate uptake and degradation as a function of the bioavailable substrate concentration are often adequately described by whole-cell Michaelis-Menten kinetics, where the substrate flux through the cell membranes q_c is a function of the substrate concentration at the cell surface C. If the substrate serves as a carbon source, a fraction of the flux q_c, represented by the yield factor Y, will be used for the synthesis of biomass and growth of the bacterial culture. Inclusion of this factor converts the Michaelis-Menten equation into the Monod equation (fig. 15.1), where K_m is the half-maximum uptake rate-related concentration, q_{max} the maximal specific degradation flux, that has been further extended for the maintenance rate coefficient b, accounting for the substrate consumption for non-growth processes such as endogenous metabolism and compensation for cell decay (Uden 1967). At low mass transfer fluxes, for instance, the number of organisms that can be maintained may be too low, resulting in poor biodegradation and high residual pollutant concentrations (Schmidt et al. 1985). It has to be noted that the conversion of the mechanistic Michaelis-Menten equation into the empirical Monod equation can only be a rough approximation (Kovárová-Kovar and Egli 1998). However, based on this concept Bosma et al. (1997) could predict residual concentrations of heterogeneously distributed pollutants in the range of grams per litre at low effective diffusivities.

15.2 Bacteria as Bioavailability Enhancing Agents

Although limited bioavailability appears to be primarily a physical, i.e. mass transfer-controlled, process, different recent observations indicate that organism-specific bioavailability-enhancing strategies may exist and that generalisations about the bioavailability of sorbed, solid or dissolved substrates are inappropriate. Some bacterial strains degrade HOC, either sorbed to a solid matrix or dissolved in an organic phase, at higher rates than desorption or partitioning rates determined in the absence of bacteria would suggest (Bouchez et al. 1997; Calvillo and Alexander 1996; Ortega-Calvo and Alexander 1994; Lahlou and Ortega-Calvo 1999). Moreover, isolates able to degrade the same HOC compound, exhibited different rates of degradation of sorbed HOC (Crocker et al. 1995; Grosser et al. 2000; Guerin and Boyd 1992; Guerin and Boyd 1997; Tang et al. 1998). Other recent reports, using PAHs as HOC compounds, indicate that sorption-limited bioavailability plays an important role in the selection of HOC-degrading bacteria and that different HOC-degrading bacteria inhabiting the same soil may be adapted to different degrees of HOC bioavailability. In these studies, different bacterial strains or microbial assemblages were selected depending on how the compound was provided to the bacteria (Bastiaens et al. 2000; Friedrich et al. 2000; Tang et al. 1998). PAH sorbed to a solid phase enriched primarily for gram-positive bacteria belonging to the *Mycobacterium* phylogeny, whereas PAH provided in crystalline form selected predominantly for strains belonging to gram-negative genera like *Burkholderia* and *Sphingomonas* (Bastiaens et al. 2000; Friedrich et al. 2000). These results are in accordance with the observation that although PAHs form a group of structurally closely related chemicals, enrichments in which PAHs are provided as sole source of carbon and energy seem to lead to the recovery of bacteria whose phylogenetic position depend on the molecular weight and hence water solubility of the PAH compound used. Use of lower molecular weight and higher soluble PAHs like naphthalene seems to select especially for fluorescent *Pseudomonas*, use of medium soluble PAHs such as phenanthrene for *Sphingomonas* and *Mycobacterium*, whereas the use of higher molecular weight PAHs mainly for *Mycobacterium* (for a review, see Kanaly and Harayama 2000). Moreover, members of the phylogenetic groups of *Sphingomonas* and *Mycobacterium* or related actinomycetes have been identified not only as efficient degraders of polycyclic aromatic hydrocarbons but also other HOC exhibiting low aqueous solubilities (Balkwil et al. 1997; Kleespies et al. 1996; Linos et al. 2000). These results show that the physiology of a bacterium or its particular life style might play an important role in its adaptation to degrade HOCs by providing a strategy for enhancing the compound's bioavailability. Detailed studies on such bioavailability-promoting strategies are scarce. However, experimental evidence and theoretical considerations (*vide supra*) show that specific HOC-degrading bacteria may adapt to low substrate regimes by (a) increasing the diffusive substrate flux by increasing the aqueous phase substrate concentration, by shortening the distance between the substrate source and the bacteria or by developing high-affinity uptake systems (fig. 15.2), and (b) by decoupling physiological requirements from the availability of the contaminant.

15.2.1 Attachment to the Substrate Source

Reduction of the mean distance between pollutants and pollutant-degrading bacteria would considerably enhance mass transfer by steepening the diffusion gradients (eq. 15.1 and fig. 15.2A). Preferred bacterial adhesion to pure solid or liquid pollutants, NAPL-dissolved, and matrix-sorbed pollutants therefore is a powerful bacterial strategy to overcome mass transfer limitations. Under mass transfer limitation-creating laboratory conditions, biofilm formation and bacterial growth in close contact to poorly available solid PAH (Mulder et al. 1998; Tongpim and Pickard 1996), (fig. 15.3), liquid hydrocarbons and waxes (Bouchez et al. 1997; Goswami and Singh 1990; Rosenberg and Rosenberg 1981; Kirschner-Zilber et al. 1980; Thomas and Alexander 1987; McLee and Davies 1972) have been observed. Additionally, the importance of the close contact to the substrate was shown when the bacterial attachment to the substrate source was suppressed by use of non-toxic surfactants or adhesion-hindered mutants (Koch et al. 1991; Efroymson and Alexander 1991; Rosenberg and Rosenberg 1981; Stelmack et al. 1999). No activity of yeast was observed when adhesion to hydrocarbons was suppressed by addition of surfactants (Aiba et al. 1969).

Attachment to the hydrophobic substrate source might be a particular mechanism used by *Mycobacterium* to enhance bioavailability. As mentioned above *Mycobacterium* is well-known for its ability to degrade HOCs. Members of this genus and related strains exhibit particular hydrophobic cell surfaces due to the presence of compounds like mycolic acids in their outer cell wall. The structures are believed to stimulate attachment to hydrophobic surfaces like crystalline HOC, HOC sorbents or NAPLs containing dissolved HOC, and hence increase access to the HOC substrate. Recently, a rubber-degrading *Mycobacterium* strain and other related rubber degrading actinomycetes were found to be tightly attached to rubber and form biofilms and colony craters on the vulcanised rubber substrate (Linos et al. 2000). Cells were even directly embedded and merged into the rubber matrix. fig. 15.3 shows an electron scanning micrograph of a crystalline anthracene surface covered by *Mycobacterium* sp. LB501T growing on this PAH as sole carbon source. *M.* sp. LB501T was isolated using a new Teflon membrane-based extraction method that appears to select strongly hydrophobic bacteria (Bastiaens et al. 2000). The adhesion of bacteria is influenced by cell properties such as cell surface hydrophobicity, which can be expressed as water contact angle (Θ_w) (van der Mei et al. 1991). Bacteria exhibiting $\Theta_w < 30\,^{o}$ are hydrophilic, whereas Θ_w values of $> 70\,^{o}$ are considered hydrophobic. Preferences for hydrophobic surfaces may explain the observation that different PAH degrading bacteria were isolated when hydrophobic membranes were used to enrich such bacteria from soil instead of using a conventional water enrichment protocol (Bastiaens et al. 2000). A strong

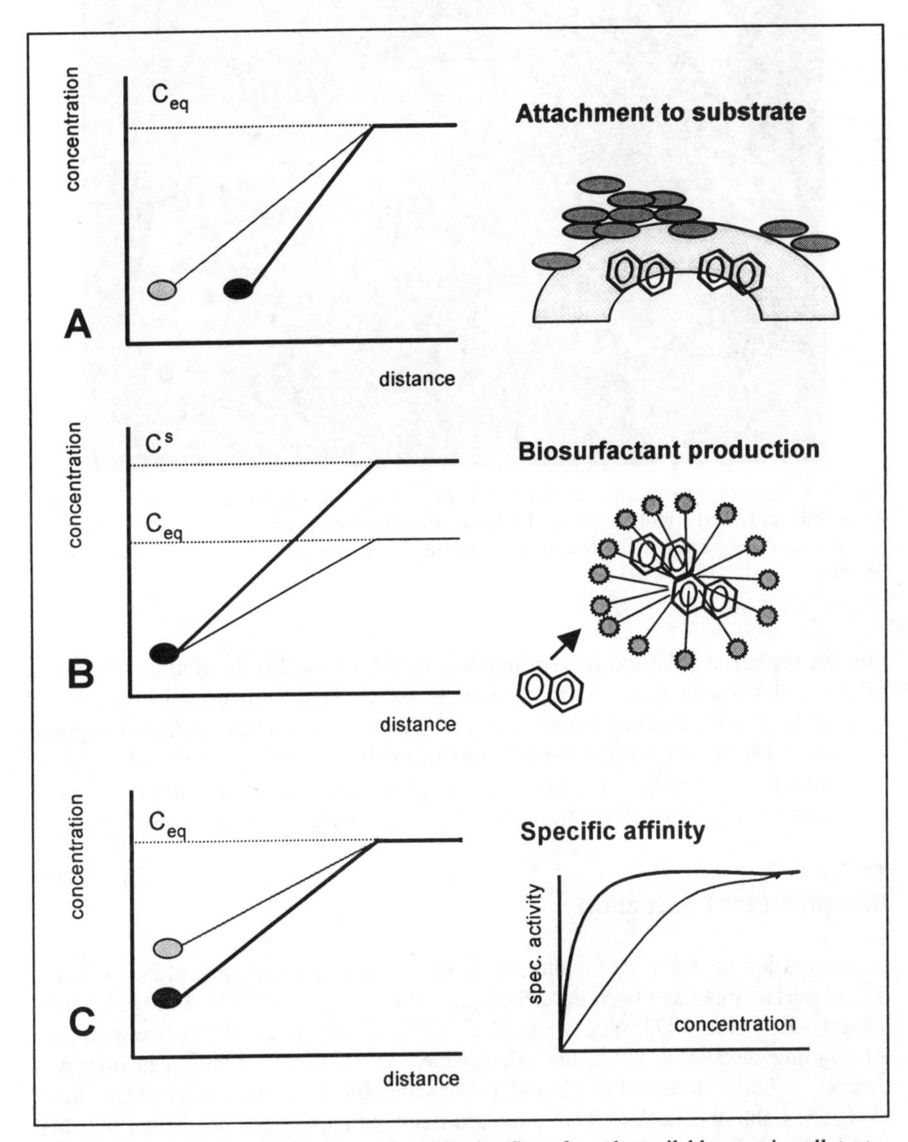

Fig. 15.2. Bacterial strategies to enhance the diffusive flux of poorly available organic pollutants. Optimisation of the diffusive flux, i.e. steepening of the concentration gradient can be obtained by shortening of the diffusion distance by attachment to the substrate (A), enhancing the dispersion and the (pseudo)solubility (C_s) of the contaminant by biosurfactant production (B), and reducing the cell surface concentration due to high substrate affinity (C).

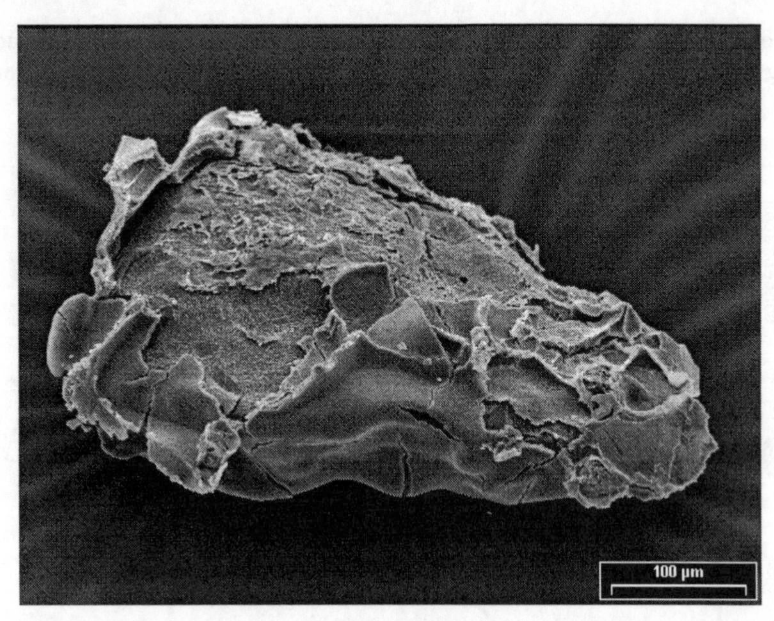

Fig. 15.3. Scanning electron micrograph of an anthracene crystal covered by a confluent *Mycobacterium* sp. LB501T biofilm. *M.* sp. LB501T grows in close contact to solid anthracene. The gap between the anthracene surface and the biofilm is an artefact due to the air drying of the sample

hint for preferred adhesion to hydrophobic surfaces in soil is the finding that 60 % of the soil bacteria were attached to particles covered with organic matter, although these particles contributed only 15 % to the total particle surface (Gray and Parkinson 1968). In transport experiments through soils and selected soil constituents such as sand, clay, and clay-humic acid complexes, *M.* sp. LB501T was in fact more retarded than five other PAH-degrading bacteria (Lahlou et al. 2000).

15.2.2 Biosurfactant Production

Many studies on the use, toxicity, and bioavailability-enhancing effect of (synthetic) surfactants have been described (e.g. Volkering et al. 1998; Liu et al. 1995; Tiehm 1994; Neu 1996; Noordman et al. 1998; Zhang et al. 1997). There is also an ongoing discussion about the effectiveness of surfactant addition in bioremediation, as both, increased (Tiehm and Fritzsche 1995) and inhibited (Volkering et al. 1995; Laha and Luthy 1991) biodegradation of HOC have been reported. Microbial biosurfactant and bioemulsifier production, however, is an often observed natural phenomenon. Glycolipids and phospholipids are two of the most common groups of biosurfactants, although the type, quantity and quality of microbial surfactant is influenced, among others, by the nature of the substrate (Willumsen and Karlson 1997). For instance, biosurfactants are usually produced by microorganisms growing on poorly soluble substrates (Neu 1996). The enhancement of biodegradation of HOC has been ascribed to the surfactant-mediated solubilisation or emulsification of sorbed, crystalline, or separate phase pollutants. It should be noted that surfactants do not increase the aqueous concentration of a compound, but form a so-called micellar pseudophase within the aqueous phase (Liu et al. 1995; Gray et al. 1994). This micellar phase accumulates the contaminant and may facilitate pollutant transport, by decreasing the distance to the active microorganisms. It has been shown, that the overall effect of surfactants on biodegradation is often the net influence of enhanced solubilisation in the micellar pseudophase and reduced substrate uptake from the surfactant micelles (Zhang et al. 1997). Bacteria may not have direct access to compounds inside the micelles (Willumsen et al. 1998), resulting in limited HOC-mass transfer from the micellar phase and limited HOC biodegradation, respectively.

Surfactants may also have an effect on the transport of bacteria and bacteriophages in porous media and sandy soil (Gross and Logan 1995; Li and Logan 1999; Ryan et al. 1999; Bai et al. 1997) likely by affecting the irreversible adsorption of cells on soil surfaces. For instance, addition of a rhamnolipid biosurfactant 5–20x above its critical micelle concentration (CMC) enhanced the transport of both hydrophobic and hydrophilic *Pseudomonas aeruginosa* cells through a sandy soil (Bai et al. 1997). At low biosurfactant concentrations below the CMC, however, this effect may be less important.

15.2.3 Specific Affinity

The rate, at which bacteria degrade a substrate that is available at low concentration depends on their specific affinity (a^o_A) towards this substrate. The specific affinity is defined as the ratio of the maximal rate of substrate uptake (q_{max}) and the half saturation constant K_m ($a^o_A = q_{max} K_m^{-1}$) (Button 1985) and equals the slope of the first order part of the activity-versus-concentration plot. Hence for a given substrate, a^o_A determines the efficiency of a bacterium to reduce the substrate concentration at its surface in relation to C_{eq}. High specific affinities therefore lead to efficient pollutant depletion at low concentrations at the cell surface, steeper concentration gradients, and higher substrate transfer rates (fig. 15.2C). High-affinity cells thus drive pollutant dissolution and desorption more efficiently than low-affinity cells (fig. 15.4). For instance, the efficient dibenzofuran and 3-chlorodibenzofuran degrader *Sphingomonas* sp. HH19k exhibited specific affinities of 62700 and 3200 L $g_{protein}^{-1} h^{-1}$, respectively. Both values are well above a^o_A-values for bacteria growing on well water soluble substrates listed in the literature (Button 1985).

15.2.4 Microbial Maintenance Requirements

Ample evidence shows that certain bacteria are well-suited for growth at very low substrate concentrations. Below some finite available substrate concentration, however, growth is thought to be impossible (Harms 1996). This concentration has been called the threshold concentration for growth (Schmidt et al. 1985).

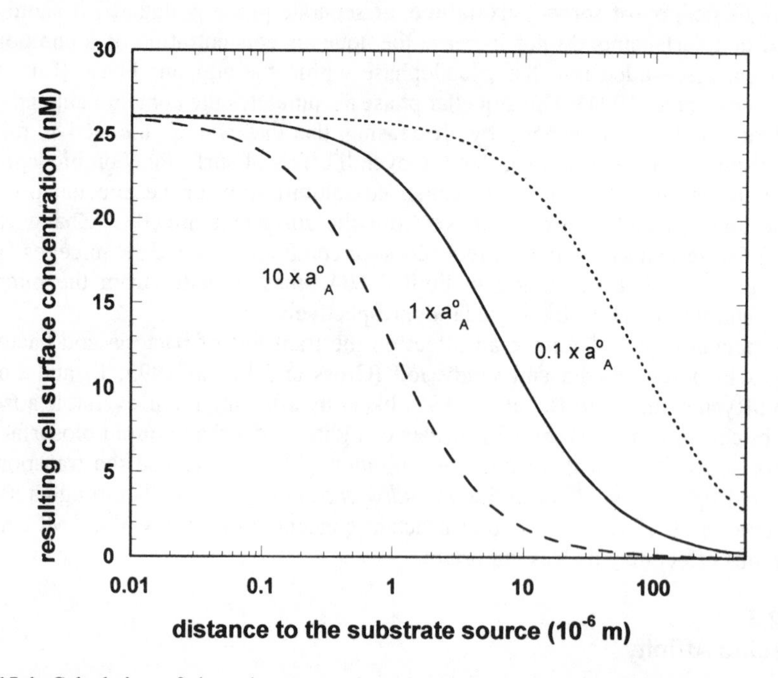

Fig. 15.4. Calculation of the substrate uptake driven near-cell-surface 3-chlorodibenzofuran-concentration of *Sphingomonas* sp. HH19k (solid line) and two other imaginary bacterial strains differing from HH19k by their 10-fold lower (short dashes) and 10-fold higher (long dashes) specific affinities, assuming linear pollutant diffusion to the cells (adapted from: Harms and Zehnder, 1994). High specific affinity leads to more efficient pollutant uptake, which in turn results in higher pollutant depletion, lower near cell surface concentrations, and higher substrate release rates.

Bosma et al. showed that the threshold concentration for growth in soil can be calculated from the effective diffusivity of the substrate and the bacterial maintenance coefficient *b* (Bosma et al. 1997). According to the definition of Pirt (Pirt 1965) the maintenance coefficient *b* reflects the "consumption of substrate which does not produce growth". A low maintenance rate characterises the ability of a micro-organism to survive and to grow on a very low amount of substrate consumed per unit time. The simultaneous utilisation of natural carbon sources and a pollutant can also be of extreme importance for the degradation of poorly bioavailable pollutants, since it partly uncouples the build-up and the maintenance of a pollutant-degrading population from the flux of the pollutant.

Although low "maintenance coefficients" do not directly affect the bioavailability, i.e. increase mass transfer to and into the cell respectively, they may form a specific ecological advantage of a pollutant degrading community to build up an active biomass. The less substrate an organism needs to survive, the better it is equipped to survive periods of famine and the better it is also capable of building up active biomass under a low-substrate regime, which may respond quickly to newly available pollutants and thus enhance the bioremediation efficiency. Interestingly, adhesion to surfaces accompanied by biofilm formation is seen as a general microbial strategy for survival as well as for utilisation of solid substrates especially in low-nutrient environments (Flemming 1998).

15.3 Conclusions for Bioremediation

Mass transfer limitations and microbial activity and diversity are important factors affecting the biodegradation of pollutants in soil. In the past engineered bioremediation focussed on the purely physical side of limited bioavailability, but neglected aspects of microbial ecology and physiology (Hinchee et al. 1995). Additionally, many studies have focussed on the intrinsic biochemical capacity, including genetic engineering of new catabolic pathways without considering physicochemical characteristics of the degrading bacteria. Most frequently however the indigenous microbial population in a contaminated area is capable of biodegrading HOC contaminants, but population density or activity is restricted by mass transfer or environmental factors and hence needs to be stimulated by addition of nutrients, air or homogenisation of the soil matrix. In order to enhance microbial activity, bioaugmentation, i.e. addition of efficient HOC-degrading specialists or targeted stimulation of endogenous HOC degraders may also be performed. Bioaugmentation however has to be considered not only as an addition/stimulation of metabolic functions, since HOC-degrading bacteria may exhibit particular bioavailability-enhancing strategies such as a specific affinity for the pollutant, modest maintenance requirements, abilities to co-utilise natural substrates, active or passive mobility, attachment to the substrate, or the ability to produce biosurfactants. In this context, it is thus crucial to know the physiological *and* the ecological *in-situ* needs of the HOC-bioavailability promoting bacterial community in order to find ideal biostimulation conditions in which these populations are active *in-situ* and *ex-situ*. Indeed, current engineering solutions to enhance bioavailability such as mechanical mixing or the addition of surfactants may not necessarily be appropriate to stimulate activity of specialised HOC degraders.

Acknowledgements

This study has been financially supported by EC Biotech programs (contracts BIO4-CT97-2015 and QLK3-CT-1999-00326) and the Swiss Federal Office for Education and Science (contracts 96.0404 and 99.0366). The support is greatly acknowledged.

References

Aelion CM, Swindoll CM, Pfaender FK (1987) Adaptation to and biodegradation of xenobiotic compounds by microbial communities from a pristine aquifer. Appl Environ Microbiol 53: 2212–2217

Aiba S, Moritz V, Someya J, Haung KL (1969) Cultivation of yeast cells by using n-alkanes as the sole carbon source. I Batch culture. J Ferm Technol 47: 203–210

Bai G, Brusseau ML, Miller RM (1997) Influence of a rhamnolipid biosurfactant on the transport of bacteria through a sandy soil. Appl Environ Microbiol 63: 1866–1873

Balkwil DL, Drake GR, Reeves RH, Frederickson JK, White DC, Ringelberg DB, Chandler DP, Romine MF, Kennedy DW, Spadoni CM (1997) Taxonomic study of aromatic-degrading bacteria from deep-terrestrial-subsurface sediments and description of *Sphingomonas aromaticivorans* sp nov, *Sphingomonas subterranea* sp nov, and *Sphingomonas stygia* sp nov Int J Syst Bacteriol 47: 191–201

Bastiaens L, Springael D, Wattiau P, Harms H, Wachter RD, Verachtert H, Diels L (2000) Isolation of new polycyclic aromatic hydrocarbon (PAH) degrading bacteria using PAH sorbing carriers. Appl Environ Microbiol 66: 1834–1843

Beurskens JEM, Dekker CGC, Johnkhoff J, Pompstra L (1993) Microbial dechlorination of hexachlorobenzene in a sedimentation area of Rhine river. Biogeochemistry 19: 61–81

Bouchez M, Blanchet D, Vandecasteele JP (1997) An interfacial uptake mechanism for the degradation of pyrene by a *Rhodococcus* strain. Microbiology 143: 1087–1093

Bosma TNP, Middeldorp PJM, Schraa G, Zehnder AJB (1997) Mass transfer limitation of biotransformation: Quantifying bioavailability. Environ Sci Technol 31: 248–252

Bouchez M, Blanchet D, Vandecasteele J-P (1995) Substrate availability in phenanthrene biodegradation: transfer mechanism and influence on metabolism. Appl Microbiol Biotechnol 43: 952–960

Button KD (1985) Kinetics of nutrient-limited transport and microbial growth. Microbiol Rev 49: 270–297

Calvillo YM, Alexander, M. (1996) Mechanisms of microbial utilization of biphenyl sorbed to polyacrylic beads. Appl. Microbiol. Biotechnol., 45: 383-390

Cameotra SS, Singh HD, Hazarika AK, Baruah JN (1983) Mode of uptake of insoluble solid substrates by microorganisms. II: Uptake of solid n-alkanes by yeast and bacterial species. Biotechnol Bioeng 25: 2945–2956

Crocker FH, Guerin WF, Boyd SA (1995) Bioavailability of naphthalene sorbed to cationic surfactant-modified smectite clay. Environ Sci Technol 29: 2593–2958

Commandeur LCM, Eyseren van HE, Opmeer MR, Govers HAJ, Parsons JR (1995) Biodegradation kinetics of highly chlorinated biphenyls by *Alcaligenes* sp. JB1 in an aerobic continuous culture system. Environ Sci Technol 29: 3038–3043

Costerton JW, Lappin-Scott HM (1989) Behavior of bacteria in biofilms. ASM news: 650–654

Deware M, Alexander M (1995) Bacterial transport and phenanthrene biodegradation in soil and aquifer sand. Soil Sci Soc Am J 59: 1316–1320

Edmonds RL (1976) Survival of coliform bacteria in sewage sludge applied to a forest clearcut and potential movement into groundwater. Appl Environ Microbiol 32: 537–546

Efroymson RA, Alexander M (1991) Biodegradation by an *Arthrobacter* species of hydrocarbon partitioned into an organic solvent. Appl Environ Microbiol 57: 1441–1447

Efroymson RA, Alexander M (1994) Role of partitioning in biodegradation of phenanthrene dissolved in nonaqueous-phase liquids. Environ Sci Technol 28: 1172–1179

Flemming HC (1998) Relevance of biofilms for biodeterioration of surfaces of polymeric materials. Polymer Degrad Stabil 59: 309–315

Friedrich M, Grosser RJ, Kern A, Inskeep WP, Ward DM (2000) Effect of model sorptive phases on phenanthrene biodegradation: Molecular analysis of enrichments and isolates suggests selection based on bioavailability. Appl Environ Microbiol 66: 2703–2710

Gosh U, Gillette JS, Luthy R, Zare RN (2000) Microscale location, characterization, and association of polycyclic aromatic hydrocarbons on harbor sediment particles. Environ Sci Technol 34: 1729–1736

Goswami P, Singh HD (1990) Different modes of hydrocarbon uptake by two *Pseudomonas* species. Biotechnol Bioeng 37: 1–11

Gray MR, Banerjee DK, Fedorak PM, Hashimoto A, Masliyah JH, Pickard MA (1994) Biological remediation of anthracene-contaminated soil in rotating bioreactors. Appl Microbiol Biotechnol 40: 933–940

Gray TRG, Parkinson D (1968) The Ecology of Soil Bacteria. Liverpool University Press, Liverpool, UK

Gross MJ, Logan BE (1995) Influence of different chemical treatments on transport of *Alcaligenes paradoxus* in porous media. Appl Environ Microbiol 61: 1750–1756
Grosser RJ, Friedrich M, Ward DM, Inskeep WP (2000) Effect of model sorptive phases on phenanthrene biodegradation: different enrichment conditions influence bioavailability and selection of phenanthrene-degrading isolates. Appl Environ Microbiol 66: 2695–2702
Guerin WF, Boyd SA (1992) Differential bioavailability of soil-sorbed naphthalene to two bacterial species. Appl Environ Microbiol 58: 1142–1152
Guerin WF, Boyd SA (1997) Bioavailability of naphthalene associated with natural and synthetic sorbents. Water Res 31: 1504–1512
Häggblom MM, Valo RJ (1995) Bioremediation of chlorophenol wastes. In: Microbial transformation and degradation of toxic organic chemicals (eds Young L, Cerniglia C), pp 389–434. Wiley-Liss, Inc, New York (NY)
Harms H (1996) Bacterial growth on distant naphthalene diffusing through water, air, water-saturated, and unsaturated porous media. Appl Environ Microbiol 62: 2286–2293
Harms H, Bosma TNP (1997) Mass transfer limitation of microbial growth and pollutant degradation. J Ind Microbiol Biotechnol 18: 97–105
Harms H, Zehnder AJB (1994) Influence of substrate diffusion on degradation of dibenzofuran and 3-chlorodibenzofuran by attached and suspended bacteria. Appl Environ Microbiol 60: 2736–2745
Harms H, Zehnder AJB (1995) Bioavailability of sorbed 3-chlorodibenzofuran. Appl Environ Microbiol 61: 27–33
Hinchee RE, Fredrickson J, Alleman BC (1995) Bioaugmentation for Site Remediation. Battelle Press, Columbus, Ohio
Kanaly RA, Harayama S (2000) Biodegradation of high-molecular-weight polycyclic aromatic hydrocarbons. J Bacteriol 182: 2059–2067
Kilbertus G (1980) Etude des microhabitats contenus dans les aggrégats du soil. Leur relation avec la biomass bactérienne et la taille des procaryotes présents. Rev Ecol Biol Soil 17: 543–557
Kirschner-Zilber I, Rosenberg E, Gutnick G (1980) Incorporation of 32P and growth of *Pseudomonas* UP-2 on n-teracosane. Appl Environ Microbiol 40: 1086–1093
Kleespies M, Kroppenstedt RM, Rainey FA, Webb LE, Stackebrandt E (1996) *Mycobacterium hodleri* sp. nov., a new member of the fast growing mycobacteria capable of degrading polycyclic aromatic hydrocarbons. Int J Syst Bacteriol 46: 683–687
Koch AK, Käppeli O, Fiechter A, Reiser J (1991) Hydrocarbon assimilation and biosurfactant production in *Pseudomonas aeruginosa* mutants. J Bacteriol 173: 4212–4219
Kovárová-Kovar K, and Egli T (1968) Growth kinetics of suspended microbial cells: from single-substrate-controlled growth to mixed-substrate kinetics. Microbiol Mol Biol Rev 62: 646–666
Laha S, Luthy RG (1991) Inhibition of phenanthrene mineralization by nonionic surfactants in soil-water systems. Environ Sci Technol 25: 1921–1930
Lahlou M, Harms H, Ortega J-J (2000) Influence of individual soil components on the transport of polycyclic aromatic hydrocarbon-degrading bacteria. Environ Sci Technol : in press
Lahlou M, Ortega-Calvo JJ (1999) Bioavailability of labile and desorption-resistant phenanthrene sorbed to montmorillonite clay containing humic acids. Environ Toxicol Chem 18: 2729–2735
Li Q, Logan BE (1999) Enhancing bacterial transport for bioaugmentation of aquifers using low ionic strength solutions and surfactants. Water Research 33: 1090–1100
Linos A, Berekaa MM, Reichelt R, Keller U, Schmitt J, Flemming HC, Kroppenstedt RM, Steinbuchel A (2000) Biodegradation of *cis*-1,4-polyisoprene rubbers by distinct Actinomycetes: Microbial strategies and detailled surface analysis. Environ Appl Microbiol 66 : 1639–1645
Liu Z, Jacobson AM, Luthy RG (1995) Biodegradation of naphthalene in nonionic surfactant systems. Appl Environ Microbiol 51: 145–151

Luthy RG, Aiken GR, Brusseau ML, Cunningham SD, Gschwend PM, Pignatello JJ, Reinhard M, Traina SJ, Weber WJ, Westall JC (1997) Sequestration of hydrophobic organic contaminants by geosorbents. Environ Sci Technol 31: 3341–3347

Luthy RG, Dzombak DA, Peters CA, Roy SB, Ramaswami A, Nakles DV, Nott BR (1994) Remediating tar-contaminated soils at manufactured gas plant sites. Environ Sci Technol 28: 266A–277A

Madsen EL, Alexander M (1982) Transport of Rhizobium and Pseudomonas through soil. Soil Sci Am J 46: 557–560

McLee AG, Davies SL (1972) Linear growth of a *Torulopsis* sp. on n-alkanes. Canad J Microbiol 18: 315–319

van der Mei HC, Rosenberg M, Busscher HJ (1991) Assessment of microbial cell surface hydrophobicity. In: Microbial Cell Surface Analysis (eds N Mozes, PS Handley, HJ Busscher, PG Rouxhet), pp 263–287. VCH Publishers Inc, New York, Weinheim, Cambridge

Mihelcic JR, Luthy RG (1991) Sorption and microbial degradation of naphthalene in soil water suspensions under denitrification conditions. Environ Sci Technol 25: 169–177

Mulder H, Breure AM, Andel JGV, Grotenhuis JT, Rulkens WH (1998) Influence of hydrodynamic conditions on naphthalene dissolution and subsequent biodegradation. Biotechnol Bioeng 57: 145–154

Mulder H, Breure AM, Honschooten Dv, Grotenhuis JT, Andel JGV, Rulkens WH (1998) Effect of biofilm formation by *Pseudomonas* 8909N on the bioavailability of solid naphthalene. Appl Microbiol Biotechnol 50: 277–283

Neilson AH (1995) An environmental perspective on the biodegradation of organochlorine xenobiotics. Int Biodeteriorat Biodegrad 37: 3–21

Neu TR (1996) Significance of bacterial surface-active compounds in interaction of bacteria with interfaces. Microbiol Reviews 60: 151–166

Noordman WH, Ji W, Brusseau ML, Janssen DB (1998) Effects of rhamnolipid biosurfactants on removal of phenanthrene from soil. Environ Sci Technol 32: 1806–1812

Ogram AV, Jessup RE, Ou LT, Rao PS (1985) Effects of sorption on biological degradation rates of (2,4-dichlorophenoxy)acetic acid in soils. Appl Environ Microbiol 49: 582–587

Ortega-Calvo JJ, Fesch C, Harms H (1999) Biodegradation of sorbed 2,4-dinitrotoluene in a clay-rich, aggregated porous medium. Environ Sci Technol 33: 3737–3742

Ortega-Calvo JJ, Alexander M (1994) Roles of bacterial attachment and spontaneous partitioning in the biodegradation of naphthalene initially present in nonaqueous-phase liquids. Appl Environ Microbiol 60: 2643–2646

Pignatello JJ, Xing B (1996) Mechanism of slow sorption of organic chemicals to natural particles. Environ Sci Technol 30: 1–11

Pirt SJ (1965) The maintenance energy of bacteria in growing cultures. Proc R Soc B 163: 224–231

Postma J, Veen van JA (1990) Habitable pore space and population dynamics of *Rhizobium leguminosarium* biovar *trifolii* introduced into soil. Microb Ecol 19: 149–161

Ramaswami A, Ghoshal S, Luthy RG (1997) Mass transfer and bioavailability of PAH compounds in coal tar systems. 2. Experimental Evaluations. Environ Sci Technol 31: 2268–2275

Ramaswami A, Luthy RG (1997) Mass transfer and bioavailability of PAH compounds in coal tar systems. 1. Model development. Environ Sci Technol 31: 2260–2267

Rijnaarts HHM, Bachmann A, Jumelet JC, Zehnder AJB (1990) Effect of desorption and intraparticle mass transfer on the aerobic biomineralization of a-hexachlorocyclohexane in a contaminated calcareous soil. Environ Sci Technol 24: 1349–1354

Rosenberg M, Rosenberg E (1981) Role of adherence in growth of *Acinetobacter calcoaceticus* RAG-1 on hexadecane. J Bacteriol 148: 51–57

Ryan JN, Elimelech M, Ard RA, Harvey RW, Johnson PR (1999) Bacteriophage PRD1 and silica colloid transport and recovery in iron oxide-coated sand aquifer. Environ Sci Technol 33: 63–73

Schmidt SK, Alexander M, Shuler ML (1985) Predicting threshold concentrations of organic substrates for bacterial growth. J Theor Biol 114: 1–8

Schwarzenbach RP, Gschwend PM, Imboden DM (1993) Environmental Organic Chemistry. John Wiley & Sons, Inc, New York, NY

Scow KM, Johnson CR (1997) Effect of sorption on biodegradation of soil pollutants. Adv Agron 58: 168–223

Stelmack PL, Gray MT, Pickard MA (1999) Bacterial adhesion to soil contaminants in the presence of surfactants. Appl Environ Microbiol 65: 163–168

Stucki G, Alexander M (1987) Role of dissolution and solubility in biodegradation of aromatic compounds. Appl Environ Microbiol 53: 292–297

Tang WC, White JC, Alexander M (1998) Utilization of sorbed compounds by microorganisms isolated for that purpose. Appl Microbiol Technol 49: 117–121

Tate III RL (1995) Soil Microbiology. John Wiley & Sons, Inc, New York, Chichester, Brisbane, Toronto, Singapore

Thomas JM, Alexander M (1987) Colonization and mineralization of palmitic acid by *Pseudomonas pseudoflava*. Microb Ecol 14: 75–80

Thomas JM, Yordy JR, Amador JA, Alexander M (1986) Rates of dissolution and biodegradation of water-insoluble organic compounds. Appl Environ Microbiol 52: 290–296

Tiehm A (1994) Degradation of polycyclic aromatic hydrocarbons in the presence of synthetic surfactants. Appl Environ Microbiol 60: 258–263

Tiehm A, Fritzsche C (1995) Utilization of solubilized and crystalline mixtures of polycycclic aromatic hydrocarbons by a *Mycobacterium* sp. Appl Microbiol Biotechnol 42: 964–968

Tongpim S, Pickard MA (1996) Growth of *Rhodococus* S1 on anthracene. Can J Microbiol 42: 289–294

Tros ME, Schraa G, Zehnder AJB, Bosma TNP (1998) Anomalies in the transformation of 3-chlorobenzoate in percolation columns with *Pseudomonas* sp. strain B13. Water Sci Tech 37: 89–96

Uden van N (1967) Transport-limited growth in the chemostat and its competitive inhibition; a theoretical treatment. Arch Microbiol 58: 145–154

Volkering F, Breure AM, Andel van JG (1993) Effect of micro-organisms on the bioavailability and biodegradation of crystalline naphthalene. Appl Microbiol Biotechnol 40: 535–540

Volkering F, Breure AM, Andel van JG, Rulkens WH (1995) Influence of nonionic surfactants on bioavailability and biodegradation of polycyclic aromatic hydrocarbons. Appl Environ Microbiol 61: 1699–1705

Volkering F, Breure AM, Rulkens WH (1998) Microbial aspects of surfactant use for biological soil remediation. Biodegradation 8: 401–417

Volkering F, Breure AM, Strekenburg A, Andel JGv (1992) Microbial degradation of polycyclic aromatic hydrocarbons: effect of substrate availability on bacterial growth kinetics. Appl Microbiol Biotechnol 36: 548–552

Weast RC (ed) (1984) Handbook of Chemistry and Physics, 64th edition. CRC Press, Inc, Boca Raton

Willumsen PA, Karlson U (1997) Screening of bacteria, isolated from PAH-contaminated soils, for production of biosurfactants and bioemulsifiers. Biodegradation 7: 415–423

Willumsen PA, Karlson U, Pritchard PH (1998) Response of fluoranthene-degrading bacteria to surfactants. Appl Microbiol Biotechnol 50: 475–483

Wodzinski RS, Bertolini D (1972) Physical state in which naphthalene and biphenyl are utilized by bacteria. Appl Microbiol 23: 1077–1081

Wodzinski RS, Coyle JE (1974) Physical state of phenanthrene for utilization by bacteria. Appl Microbiol 27: 1081–1084

Wodzinski RS, Johnson MJ (1968) Yields of bacterial cells from hydrocarbons. Appl Microbiol 16: 1886–1891

Zhang WX, Bouwer EJ, Ball WP (1998) Bioavailability of hydrophobic organic contaminants: effects and implications of sorption-related mass transfer on bioremediation. Ground Water Monit R 18: 126–138

Zhang Y, Maier WJ, Miller RM (1997) Effect of rhamnolipids on the dissolution, bioavailability, and biodegradation of phenanthrene. Environ Sci Technol 31: 2211–2217

16 Formation of Residues of Organic Pollutants Within the Soil Matrix – Mechanisms and Stability

M. Kästner, H. H. Richnow
Department of Remediation Research, Centre for Environmental Research Leipzig-Halle (UFZ)

16.1 Introduction

All anthropogenic organic chemicals form non-extractable residues to some extent when entering soils. This well-known phenomenon has been studied for many years especially in the field of soil agrochemistry. Similar processes have been observed during the bioremediation of oil-contaminated soils (see table 16.3) and the residue formation of toxic and carcinogenic polycyclic aromatic hydrocarbons is of particular concern. Apart from mineralisation, the formation of non-extractable residues is certainly the major sink of anthropogenic pollutants in soils. However, macromolecular non-extractable xenobiotic residues are difficult to examine by conventional analytical techniques. Therefore, formation, structure and significance of these macromolecular residues in the environment are still unknown today.

Soil "bound residues" of pesticides first received attention more than 30 years ago when Bailey and White (1964) demonstrated that a significant portion of pesticides is not readily biological degraded but bound to soil organic matter or clay minerals. Since then bound residues of pesticides have been shown to occur in plants, other biological materials, sediments and soils. During the scientific debate concerning pesticide residues, they have been the subject of several reviews addressing their significance, occurrence and formation in the environment (Kaufmann 1976; Klein and Scheunert 1982; Khan 1982; Roberts 1984; Führ 1987; Calderbank 1989; Northcott and Jones 1999). Regulatory aspects of bound residues in plants have been discussed as well (Kovacs 1986).

Agrochemicals accumulate in soil bound residues even when "good agriculture practices" are used. This fact is considered to be relevant in the context of health and environmental risk as well as in long-term quality management of soils. Today, bound residue formation is one of the criteria considered in the registration of agrochemicals and has some impact on regulatory aspects (Kovacs 1986; US Federal Register 1975; Domsch 1992).

First attempts to define bound residues were reported in the field of pesticide soil chemistry. According to the IUPAC definition, "non-extractable residues, (sometimes also referred to as *bound residues*, or *non-extracted residues*) in plants and soils are defined as chemical species originating from pesticides, used according to good agriculture practice, that are unextracted by methods which do not significantly change the chemical nature of these residues. These non-extractable residues are considered to exclude fragments recycled through metabolic pathways leading to natural products" (Roberts 1984). The European Union regulatory policy use a similar definition according to the extraction methods used for the preparation of bound residues (Craven 2000). The use of neutral or polar organic solvents is prescribed which do not alter the chemical structure of bound residues during the removal of extractable substances in the preparation procedure. Reactive solvents may cleave chemical bonds or cause chemical reactions and therefore should not be used. The remaining material is the bound residue fraction, sometimes referred to as non-extractable residues to make the distinction between bound residues and residual concentrations of persistent extractable chemicals in soils. Moreover, former pollutants may be completely transformed by metabolic processes to chemical structures identical to those of natural origin such as carbohydrates, amino acids and fatty acids. These compounds are not considered as bound residue, although parts of their carbon originate from the anthropogenic chemicals (Roberts 1984). In practice, bound residues are very difficult to distinguish from completely transformed compounds especially when radioactive labelled tracer substances are used to quantify the bound residue fraction without any structural assignments. Pesticide residues are studied far more intensively than those of other anthropogenic chemicals. More research is needed to trace the fate of other priority pollutants in the environment.

The formation of bound residues in soils is mainly attributed to physical and chemical interaction of xenobiotics including polycyclic aromatic hydrocarbons (PAH) with natural organic matter. The main factors influencing the bound residue formation include:

- the chemical reactivity of the anthropogenic chemical,
- the reactivity of the binding substrate (macromolecular matrix),
- microbial transformation processes, and
- the presence of catalysing agents like enzymes, clay minerals, reactive oxygen or iron and manganese species.

Non-extractable residues are considered to be a major sink for organic pollutants in soils (Bollag 1983). Their formation has been proposed to be a naturally occurring detoxification process (Bollag and Loll 1983; Berry and Boyd 1984) and the stimulation of binding of pollutants to humic substances has been proposed as an alternative strategy for the bioremediation of anthropogenic contamination (Berry and Boyd 1985; Bollag and Bollag 1990; Bollag 1992). Theoretically, the PAHs and their metabolites can interact with macromolecular organic matter in soils by non-covalent association (hydrophobic sorption, charge transfer complexes, hydrogen bonding) and by the formation of covalent bonds (ester, ether, carbon-carbon bonds) (Richnow et al. 1994). However, the overall process of residue formation is often considered to be a naturally occurring humification of anthropogenic pollutants, since the xenobiotic carbon is associated with natural macromolecular soil organic matter. Thus the xenobiotic carbon is sequestered within macromolecular organic matrices and is not easy to access by common chemical analytical techniques (Bollag and Loll 1983; Berry and Boyd 1984). With binding, the xenobiotic chemical loses its structural identity including their characteristic physical, chemical and biological properties.

Microbial activity stimulates the formation of bound residue in many cases and is therefore often associated with the biodegradation of the respective chemical. But once bound within residues, the bioavailability of the xenobiotic carbon in soils is strongly reduced. As a consequence, the actual toxicity of the chemical declines, but the persistence of the xenobiotic carbon is increased in the environment. The ecological consequences of bound residues have not been thoroughly evaluated. Moreover, in the context of ecological risk assessments, the long-term stability of bound residues and their remobilisation potential have to be assessed to evaluate bioremediation techniques. A detailed knowledge of the precise chemical structure and the type of linkage within bound xenobiotic moieties in soil may help to predict their behaviour and final fate in the environment. Therefore, the biogeochemistry and transformation of xenobiotics in soils hold keys to understanding the formation and fate of bound residues. This paper summarises present knowledge about bound residue formation in the first part and provides several results from the joint projects of the SFB 188 in the second. Further information can be gained from Chapter 26. Moreover, consequences for the bioavailability and the long-term fate of xenobiotics in the environment are discussed.

16.2 Formation of Residues in Soil

Large numbers of agrochemicals were investigated for their potential to build soil bound residues. All classes of studied pesticides form “bound residues” to some extent when entering soils (table 16.1). In many cases the parent chemical, its degradation products, or both are involved. The quantities of pesticide residues listed in some reviews range from a few percent up to more than 90 % of the applied substance (e.g. Klein and Scheunert 1982; Khan 1982; Khan and Dupont 1987).

Chlorinated hydrocarbons and pyrethroid insecticides have a relatively low potential to form bound residues. Carbamates, triazines and organophosphates result in significantly higher amounts of bound residues. The large range of residue formation is probably due to the type of soil and the experimental conditions. The fact that agrochemicals accumulate in soil bound residues even when good agricultural practices are applied has increased environmental concern. Therefore, the formation of persistent residues including non-extractable species in soils is a criterium to evaluate the long-term behaviour of agrochemicals in soils (e.g. Domsch 1992).

Apart from pesticides, other anthropogenic chemicals form non-extractable residues. Phenols and anilines can form large bound residue fractions. In the case of chlorophenols and chlorinated biphenyls, the chemically unreactive chlorine substituent reduces residue formation. Chlorine substitution increases the

Table 16.1. Formation of bound pesticide residues in soils (after Calderbank 1989)

Structural type	Bound residue (% of applied)	Parent detected
Herbicides:		
Anilides and ureas	34–90	no
Bipyridyliums	10–90	yes
Nitroanilines	7–85	no
Phenoxyderivatives	28	no
Phosphonate (glycosates)	12–95	yes
Triazines	47–57	yes
Insecticides:		
Carbamates	32–70	yes
Organochlorines	7–25	n.d.
Organophosphates	18–80	yes
Pyrethroids	3–23	no
Fungicides:		
Chlorophenols	45–90	yes
Nitroaromatics	60–90	yes

n.d. = non determined

resistance of biphenyl to biological degradation. This may indicate that metabolically activated metabolites are involved in the process of residue formation.

The chemical reactivity of the parent compound or their metabolic products influences the bound residue formation. One factor controlling the chemical reactivity is suggested to be the abundance of functional groups such as carboxyl, hydroxyl, phosphate, nitro and amino groups which are able to react in condensation processes with natural organic matter. The nitro group of anthropogenic chemicals often requires a reduction to an amino group before a significant residue formation can occur. The amino derivative of certain pesticides contributes to relatively high amounts of residues and causes fast formation rates (Roberts 1984). The covalent binding of aromatic amines results in nitrogen-carbon bonds with humic substances (Thorn et al. 1996). Enzymatic reactions of fungal laccases catalyse the oxidative cross-linking with typical humic structure moieties, and the nucleophilic addition reaction with quinone and other carbonyl groups causes the formation of an amine bond (Thorn et al. 1996; Dawel et al. 1997).

16.2.1 Properties of Soil Components

Soil is a highly variable and complex medium with a large number of possible binding sites for organic chemicals. Moreover, the great variety of soil types differs in sand, silt, clay and humus content. The composition, the texture, the water content and the concentration of non-aqueous phase liquids strongly affect the soil physical properties and influences the soil's ability to interact with xenobiotics. In particular, clay and organic colloids have very large active surface areas and high cation exchange capacity effecting the physical and chemical properties of soils (pH, redox potential, ion concentration). The expanding lattice clays such as montmorillonite and vermiculite provide large external and internal surfaces to interact with organic molecules. They may adsorb organic molecules in their swelling interlayers. The surfaces and cation exchange capacity of non-expanding clays such as illite and kaolinite are, in general, much smaller.

The organic matter fraction is believed to be most responsible for the interactions with hydrophobic organic xenobiotics but clays may also play an important role. In many cases, both soil constituents may be involved (Calderbank 1989; Stevenson 1994). During the past two decades the interaction of pesticides and other organic chemicals with clay minerals has been reviewed several times (White 1976; Theng 1982; Mortland 1986; Zielke et al. 1989). The reactivity and the catalytic properties of clays and metal constituents in soils are discussed in detail by Ruggiero (1998).

The influence of natural organic matter on the behaviour of organic chemicals in soils has different facets. Natural soil organic matter is very complex and exists in many different forms in soil. Organic matter can be divided into living biomass (bacteria, fungi, algae, plant roots, and higher organisms) and non-living organic substances. The latter consists mainly of fragments and decomposition products of the biomass (carbohydrates, proteins, waxes, chitins, and lignines) which rearrange during humification processes forming refractory soil humic substances. Furthermore, soil may receive black carbon in the form of charcoal like material as a result of biomass burning or in form of soot from combustion of fossil fuels. Both carbon sources are difficult to characterise quantitatively in soils and sediments and their role in the environment is a matter of recent research. The interaction of these black carbon fractions with hydrophobic xenobiotics may affect the bioavailability of xenobiotics in soils and sediments (Gustafsson and Gschwend 1998).

Humic substances are the dominant organic carbon species in soil and, therefore, are an important substrate to interact with xenobiotics. Humic substances are separated by their solubility in alkaline and acidic solution into humin, humic acids and fulvic acids fractions. Humic acids are soluble in aqueous alkaline solution. They precipitate upon acidification at pH 2, whereas the fulvic acids remain soluble in acidified water. The humin fraction is, in general, insoluble in alkali, acids, or organic solvents. Humic substances are thought to have a complex macromolecular nature covering a molecular weight range from 1000 to more than 250,000 Daltons (Hedges 1988). Humic substances are continuously synthesised from non-living organic matter entering soils as well as slowly degraded by biological and non-biological processes. They form relatively stable organic fractions in soils that represent, in reality, a continual state of decomposition and synthesis. The continual state can be characterised by its mean residence or turn over time (Stevenson 1994).

The heterogeneity of humic substances does not allow a precise chemical definition to be given of the overall structure, but various conceptual models have been discussed (see for a review Hayes et al. 1989). Humic substances are suggested to be complex soil colloids with a micellar structure (Pauli 1967). The building blocks consist of hydrophobic aromatic and aliphatic nuclei that are linked by certain types of covalent bonds and carry a number of reactive functional groups with a hydrophilic character (fig. 16.1). Both hydrophobic nuclei and hydrophilic functional groups modify the chemical properties of humic substances to interact with xenobiotic substances. More specifically, hydrophobic sites are important for the sorption of hydrophobic compounds. Reactive functional groups may act as suitable binding sites for chemical reactions with xenobiotics.

Natural soil organic matter interacts with clay to form stable clay organic complexes and aggregates. This fact has important consequences on the physical, chemical and biological properties of the soil matrix (Stevenson 1994). In many cases the organic matter content in soils is closely related to the amount of clay, which tends to stabilise organic matter. Organic matter may serve to bridge soil particles together by forming stable aggregates. Because the soil organic matter is closely bound to clays to form clay-organic complexes, the individual contribution of the clay and the organic matter fraction to the interaction with xenobiotics is difficult to assess under field conditions. Organic substances may build coatings on clays and other mineral surfaces and modify their surface properties with respect to sorption of xenobiotics by forming these clay-organic complexes (Calderbank 1989; Stevenson 1994). Small amounts of organic matter may be sufficient to coat large mineral surfaces and effectively prevent soil mineral surfaces from acting as adsorbents for xenobiotics (Menzel and Nelson 1986). Therefore, in many cases, humic substances play a major role in influencing the interactions with hydrophobic pollutants in soils. The presence of reactive functional groups and hydrophobic moieties allows chemical and physical associations with hydrophobic and hydrophilic xenobiotics.

16.2.2 Interaction of Xenobiotic Compounds with Humic Substances

16.2.2.1 Sorption

The binding of xenobiotics with soil organic matter can be classified with respect to two principle mechanisms: sorption and covalent binding. Sorption occurs when the free energy of the sorption reaction is negative and can be driven by entropy and enthalpy-related adsorption forces (Hassett and Banwart 1989). In a first approximation, the negative free energy of the sorption reaction is equivalent to the stability of the formed association.

Sorption is considered to be a reversible process and leads to the formation of a relatively unstable association between xenobiotics and natural organic matter in soils. However, adsorption-desorption hysteresis of xenobiotic compounds during interactions of organic pollutants in soils was observed in some cases and 30–50 % of the adsorbed substances could not be desorbed within several weeks (Kan et al. 1994).

Sorption can be driven by van der Waals forces (1–2 kcal mol^{-1}), hydrogen bonding (2–10 kcal mol^{-1}), dipole-dipole attraction (< 2 kcal mol^{-1}) and electrostatic Coulomb forces (Hassett and Banwart 1989). Moreover, the responsibility for sorption reactions lies with charge-transfer complexes consisting of an electron

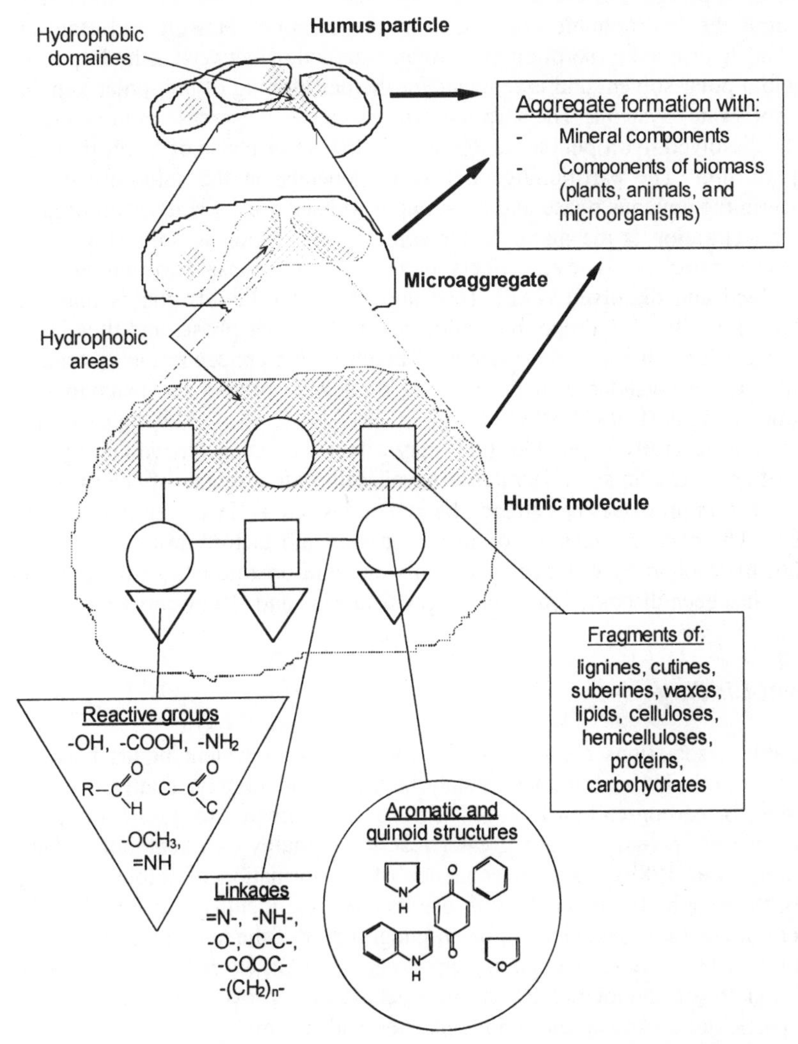

Fig. 16.1. Humic molecules and aggregation model (modified from Kästner 2000), the humic micelles and particles may be encrusted with inorganic components or fragments of biomass.

donor and an electron acceptor molecule with a partial overlap of orbitals and a partial exchange of electron densities. In this category falls hydrogen bonding with an energy attraction ranging from 2 to 10 kcal mol^{-1} and π-bond electron bonds (Hassett and Banwart 1989). In particular, π-bonds *via* overlap of π-electron systems may be relevant for the association of alkenes and aromatic compounds with soil organic matter.

Hydrophobic sorption is the partitioning of non-polar organics out of the polar aqueous phase onto a hydrophobic surface in soil. In this case the primary sorption forces are entropy changes that result from the removal of the solute from the solution. Entropy changes are due to the destruction of the structured water shell surrounding the hydrophobic compound in the solution (Hassett and Banwart 1989). The hydrophobic sorption is a weak interaction between a hydrophobic solute and a polar solvent and is relevant for the partitioning of non-polar xenobiotics in soil-water-systems. The sorption process causes a decrease in the concentration of dissolved hydrophobic xenobiotics in the water phase of a soil to a certain equilibrium. The partitioning coefficient depends on the solubility of the substance in the aqueous phase and on suitable soil surfaces, whereas soil organic matter concentration is the major factor controlling sorption in soils (Pignatello 1989). Furthermore, cross-over effects in liquid-liquid phase partitioning from non-dissolved and dissolved NAPL (non-aqueous-phase-liquid) phases may enhance the solubility of hydrophobic chemicals in the water phase, and thus influence their partition in soil-water systems. Therefore, the concentration of organic matter has to be considered in partitioning models to predict the behaviour of xenobiotics in soils (Karickhoff et al. 1979; Karickhoff 1981). Although sorption processes are generally reversible, they may affect the actual bioavailability and toxicity of chemicals in soils. Limited bioavailability after ageing of chemicals in soils by sorption processes has already been demonstrated (Hatzinger and Alexander 1995). The consequences of sorption processes and factors influencing partitioning of hydrophobic substances between water and organic phases in soils and sediments has been discussed in detail (e.g. Gustafsson and Gschwend 1998).

16.2.2.2 Covalent Binding

More stable associations result from the formation of covalent bonds between xenobiotic molecules and/or their metabolites and soil organic matter, such as ester, ether, or carbon-carbon linkages (fig. 16.2). Compared to physical adsorption, this process reveals a much greater reaction enthalpy (> 7 kcal mol^{-1}) (Hassett and Banwart 1989). As a consequence of the formation of stable chemical linkages, the xenobiotic molecule will lose its chemical identity. Once bound, the xenobiotic fragment is suggested to be an integral part of the humic material (Bollag and Loll 1983). These associations are expected to largely reduce bioavailability and toxicity of xenobiotics by forming persistent non-extractable bound residues of pesticides and other anthropogenic chemicals in soils.

16.3 The Fate of Carbon Derived from Organic Pollutants

Tracing the fate of anthropogenic organic chemicals in soils is one of the major challenges in environmental chemistry. This faces scientists with a number of serious analytical problems according to the transformation of the substance by either biotic or abiotic processes. For example, during the microbial degradation of any organic compounds in soils, many partial degradation and conversion reactions take place, leading to a distribution of carbon in different compartments. Besides metabolisation, mineralisation, and formation of biomass, a certain

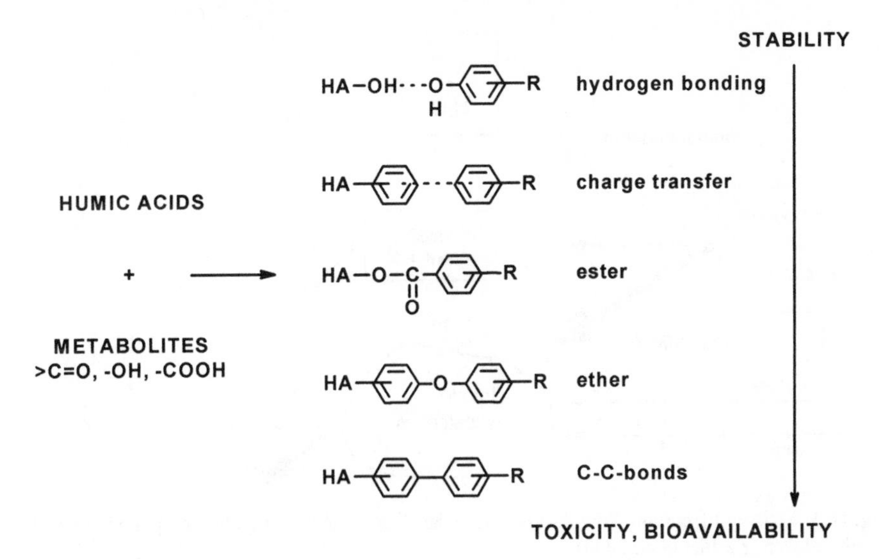

Fig. 16.2. Association of PAH and their metabolites with humic substances (from Richnow et al. 1994)

amount of carbon is generally incorporated into refractory soil organic matter. Humic substances, metabolites and biomass are likewise parent material for subsequent degradation and conversion reactions (fig. 16.3). The stimulation of soil microbial activity by the addition of composts or other organic materials to soils usually leads to enhanced turnover in all parts of the processes. The overall process associated with the degradation of natural compounds is known as humification. Xenobiotic compounds or their metabolites are included in such reaction sequences in an analogous manner.

However, the parent compounds and metabolic products may also react with the organic chemicals as a source of carbon in anabolic processes and may, thus, transform the xenobiotic carbon into biomass. Labelled CO_2 from mineralisation of the compounds may even enter the soil carbonate system and either become available for further abiotic reactions or be used in anabolic pathways by soil organisms ending up in biomass (Kästner et al. 1999). Because of the large number of possible transformation processes, the speciation of the xenobiotic carbon in soils is difficult to predict even in a controlled system. Understanding the carbon budget of the xenobiotic compounds is essential to understand these processes in soils. One possible technique to differentiate among the relevant transformation processes is the use of isotopically labelled model substances, which allows the fate of chemicals to be deciphered on both a bulk and a molecular level.

The general modes of residue formation from ^{14}C-labelled compounds depend on the compounds, the association process, and the binding matrix (table 16.2). These modes are confirmed for residues of xenobiotics and may be valid for the formation of refractory soil organic matter from fragments of the biomass as well. Three possible binding modes with increasing stability and decreasing bioavail-

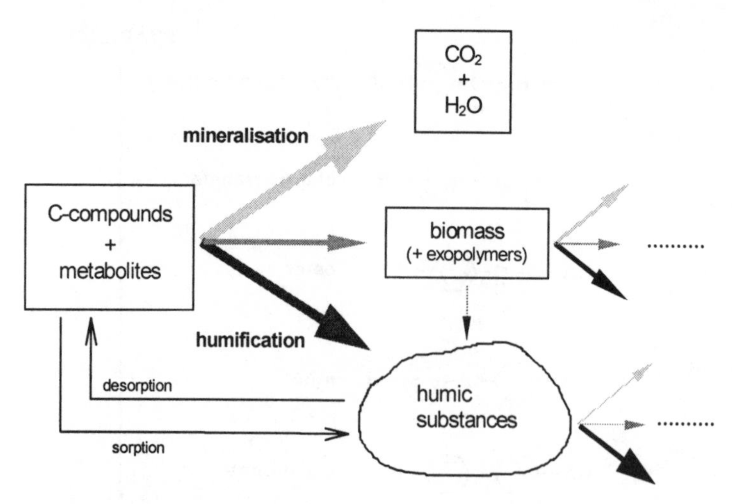

Fig. 16.3. Scheme of carbon flow during microbial degradation of organic compounds in soil (modified from Kästner et al. 2000)

ability can be classified: sorption, physical entrapment into cavities of macromolecules and covalent bonds.

In most cases the incorporation of carbon into the biomass occurs during the main degradation phase in soil. After this phase, a certain amount of residue is formed by the fixation of fragments of dead microbial biomass which incorporates the carbon after the degradation of the parent compound. The residue formation is dependent on the metabolisation of organic compounds to a large extent. With increasing metabolisation of xenobiotic substances in soil, the residue formation becomes similar to humification processes because parts of the xenobiotic carbon are incorporated into the biomass. The formation of residues in soil is not an unusual process. Even after the degradation of natural compounds such as plant material or glucose in soil, up to 35 % of "residues" or refractory soil organic matter are formed from the precursors carbon (Stott et al. 1983a; Baldock et al. 1989). These results are of fundamental significance and can be generalised for the microbial degradation of any organic compounds. Even after the degradation of ^{14}C-glucose, more than 30 % of the radiolabel remained in the soil after 34 d

Table 16.2. Formation of non-extractable residues from ^{14}C-labelled xenobiotic compounds in soil

origin:	modes of association	binding matrix:
parent compound	sorption	xenobiotic matrix (tar oil, coal particles)
metabolites inclusion	inorganic matrix (clay minerals)	(physical entrapment)
biomass/(CO_2)	covalent bonds	organic matrix (humic substances)

and after one year 10 % was still found to be incorporated within the organic matrix of the soil (Baldock et al. 1989; Stott et al. 1983a). 65 % of the initial ^{13}C-signals of glucose in the NMR spectra were still found within the original O-alkyl area, whereas 25 % was shifted to the alkyl and 8 % to the carboxyl areas (Baldock et al. 1989).

16.3.1 Carbon Mass Balances

The classical approach to balance carbon inventories of PAH in soils is the application of ^{14}C-labelled model substances which allows the speciation in soils to be traced during biological and chemical processes. The behaviour of a number of typical anthropogenic contaminants, including PAH, has been investigated using this method. One main parameter necessary to quantify the transformation of the radiolabelled compound is CO_2 evolution, which reflects the ability of the soil microorganisms to mineralise the respective compound (table 16.3).

In recent studies, the formation of bound residues in a soil from Northern Germany (Luvisol, Ah-horizon) spiked with 9-[^{14}C]- and 9-[^{13}C]-anthracene was examined. With mass balance experiments, the concept of applying stable isotope-labelled PAH to trace their transformation in the environment was evaluated and proven (Richnow et al. 1999). Stable isotope-labelled model substances have several advantages compared to classical radiolabelled tracer substances. Radio tracers have continuously lost acceptance in environmental chemistry due to specific safety considerations required in the laboratory. Using radioactive substances is complicated due to issues of waste disposal and strict legislation which prohibits field studies. The description of these experiments and the obtained results is presented in Chapter 17.

The concept to trace the transformation of the ^{13}C-labelled substances was performed in a bioreactor system and is based on carbon isotopic measurements combined with GC-MS analyses. The labelled anthracenes were added to a soil to investigate transformation over time (fig. 16.4). The mineralisation of the ^{13}C-anthracene during biodegradation can be monitored by determining concentration and isotopic composition of the CO_2. The concentration of extractable ^{13}C-anthracene and labelled metabolites was determined by GC and GC-MS measurements. The degree of bound residue formation was calculated from the organic carbon concentration and the isotopic composition of the extracted soil material. For the determination of the natural carbon isotopic composition of the soil, a batch experiment with non-labelled anthracene had to be performed. Both the carbonate fraction and the bound residue fraction from both experiments were analysed for their isotopic composition. The CO_2 concentration and organic carbon content of the soil in the batch experiment with non-labelled anthracene essential data needed to define the natural background concentration of ^{13}C, which is necessary due to the relative high natural abundance of the ^{13}C-isotope (^{12}C: 98.89 %; ^{13}C: 0.11 %). Therefore, during the entire experiment, the ^{13}C-concentration in the CO_2 fraction and bound residue fraction were monitored in the experiment with unlabelled anthracene. To calculate the incorporation of the ^{13}C-label into CO_2 and the bound residue fraction, the ^{13}C-concentration of the are background was subtracted from the respective fraction of the ^{13}C-labelling

Table 16.3. Mineralisation and formation of non-extractable residues during microbial degradation of ^{14}C-labelled PAH in soils and sediment microcosms (modified from Kästner et al. 1999)

^{14}[C]-labelled comp.	Incubation time [days]	CO_2 [%]	Non-extractable residues [%]	Reference
1*-naphthalene	2	51	27	Herbes and Schwall 1978[a]
	56	60 –70	5– 8	Heitkamp et al 1987
	120	56	19	Hosler et al. 1988[a]
	174	≈68	≈20	Eschenbach et al. 1995[a]
9*-phenanthrene	21	38	36	Brodkorb and Legge 1992[a]
9*-anthracene	7	4	23	Herbes and Schwall 1978[a]
	148	25 – 5	35–74	Goodin and Weber 1995[a]
	120	22	65	Hosler et al. 1988[a]
	103	24	42	Kästner et al. 1995
	174	46 –38	40–51	Eschenbach et al. 1995[a]
	176	67 (44)	21 (45)	Kästner et al. 1999
9*-anthraquinone	176	80	10	"
1,2,3,4,4a,5a*-anthracene	159	62	28	"
3*-fluoranthene	117	63	19	Schnöder et al. 1994
4,5,8,10*-pyrene	64	52 –16	13–45	Grosser et al. 1991[a]
	63	27 –48	25–28	Sack and Fritsche 1997
	174	46 –79	22–16	Eschenbach et al. 1995[a]
7*-benzo(a)pyrene	150	48	24	Kanaly et al. 1997
	174	3.5–22	52–48	Eschenbach et al. 1995[a]

[a] soil from contaminated sites

experiment. The concentration of the ^{13}C-label in the extracts is determined based on the concentration of ^{13}C-labelled anthracene by GC and GC-MS. With these data the ^{13}C-balance within the bioreactor was calculated.

The comparison of the ^{13}C- and ^{14}C-labelling experiments revealed similar results. In both experiments CO_2 was formed after a lag phase of about 25 days (fig. 16.4). After this time, anthracene was efficiently removed from the extractable fraction and mineralisation was the main elimination pathway. However, the formation of bound residues started at the beginning of the experiment but was most pronounced in the time interval of intensive mineralisation. Thus, the residue formation was closely associated with mineralisation underlining the important role of microbial activity in the formation process. The highest concentration of residues was observed at day 92. Once formed, the concentration only slightly declined in the time course of the experiment indicating the relatively high stability of this fraction.

The mass balance calculations illustrate that the speciation of PAH during microbial transformation processes in soils can be traced by ^{13}C-labelled model substances. The experiments with ^{13}C- and ^{14}C-labelled PAH demonstrate comparable mass balances and non-extractable residue formation. However, some differences do exist. Conversion of the ^{14}C-labelled tracer into a non-extractable residue fraction was higher (20 %) than in experiments where ^{13}C was used (11 %). At the same time, more ^{13}C (86 %) was transformed into the CO_2-fraction compared to ^{14}C (78 %).

The degradation studies with labelled anthracene demonstrate that two main processes are involved: (I) the mineralisation as measured by the evolution of CO_2, and by (II) the incorporation of the label into soil bound residues. Non-extractable residue formation appears to accompany biodegradation of organic compounds in soils in general, since residues were also formed upon degradation of glucose (Baldock et al. 1989; Stott et al. 1983a). In the case of anthracene, the formation of non-extractable residues is most pronounced at the beginning of the phase of degradation and appears to be closely associated with biodegradation processes. Studies with [^{14}C]-pyrene also show that the formation of residues is predominantly influenced by biotic rather than abiotic processes (Bollag and Loll 1983). The bound residue formation is closely correlated with the microbial degradation of anthracene and, therefore, is most probably a biologically induced process, which mainly occurs in the initial stage of degradation. Relevant amounts of anthracene derived carbon were transformed into the bound residue fraction.

Moreover, the non-extractable residue fraction was found to be rather stable to microbial degradation for the duration of the experiment. Once anthracene is immobilised within non-extractable residues, the bioavailability of residual anthracene fragments decreases by an order of magnitude compared to the parent anthracene. This is demonstrated by the low mineralisation and the slight decrease in residue concentrations in the long-term experiments ($\approx$300 d). Non-extractable residues appear to be relatively resistant to biodegradation over time. Their accessibility to microorganisms and plants is limited (Eschenbach et al. 1998; Kästner et al. 1996; Kästner et al. 1977).

Studies concerning the transformation of PAH in soil indicate that the microbial soil system will most likely respond to the addition of PAH in three distinct phases: First a lag-phase (I) was necessary to develop the ability to degrade anthracene. Phase II is characterised by the formation of bound residues. Phase III is dominated by mineralisation indicating a complete adaptation of the microbial system to anthracene (Richnow et al. 1999; Richnow et al. 2000).

16.3.1.1
Tracing Bound Residue Formation with Stable Isotope Labelled Model Compounds on a Molecular Level

Basically two different concepts have been applied in studies concerning transformation of PAH in soils on a molecular level using stable isotope-labelled model compounds. (I.) A holistic approach using NMR techniques was used to obtain overall structural information on the molecular configuration of a particular labelled atom within the residue containing material. The substitution pattern of the labelled atom is used to track structural information of molecular entities in the macromolecular matrix of residues. (II.) The other concept applies chemical degradation techniques to cleave molecular units from the macromolecular organic matrix of bound residues. Among the chemical degradation products, molecular entities can be identified and characterised with GC-MS and IR-C-GC-MS-techniques (for example Annweiler, see Chapter 17).

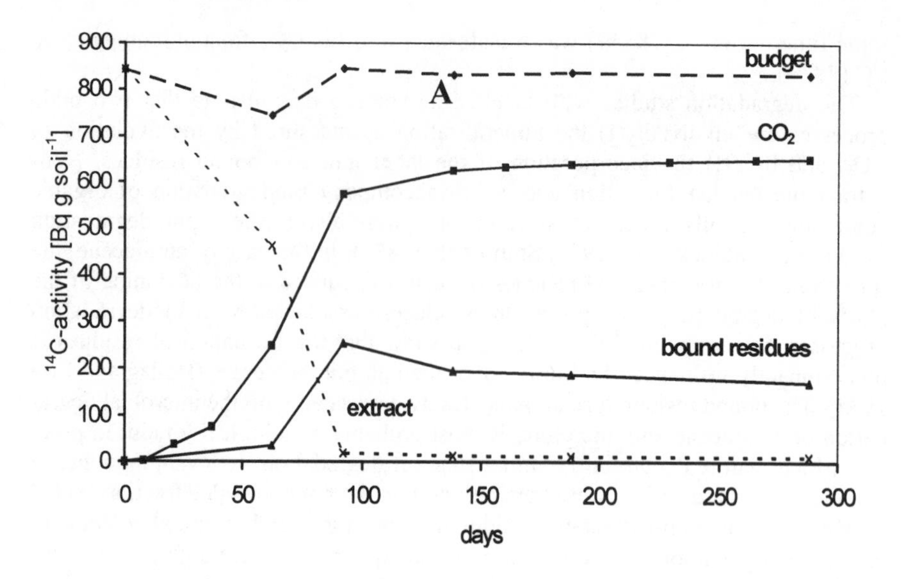

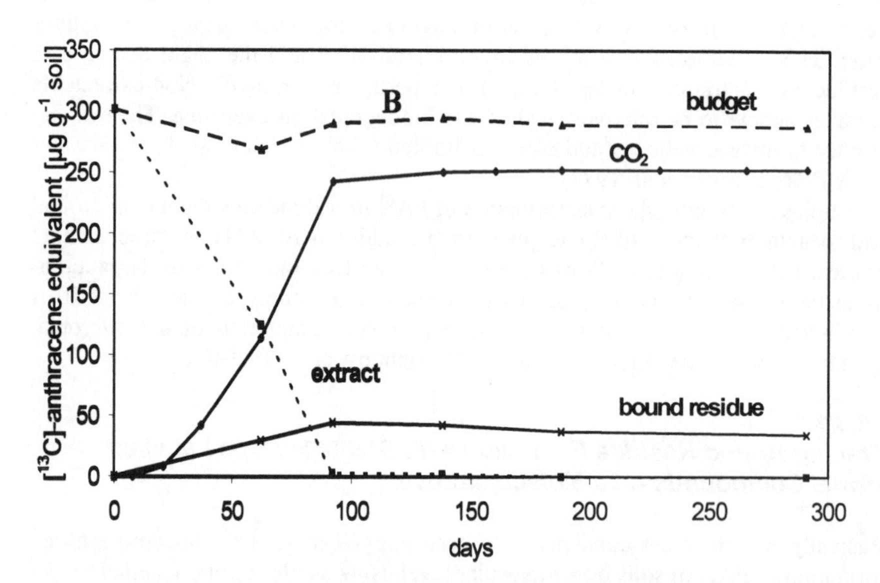

Fig. 16.4. Carbon partitioning during microbial degradation of 9-[^{14}C]- (A) and 9-[^{13}C]- (B) labelled anthracene in soil supplemented with compost (modified after Richnow et al. 1999). (anthracene equivalent = recovery of the labelled ^{13}C-carbon calculated as anthracene equivalents)

NMR spectroscopy is a powerful non destructive method for the structural characterisation of bound residues. Progress in structural studies concerning the interaction of xenobiotic molecules with macromolecular organic matter has been made applying ^{13}C-labelled model compounds to enhance the sensitivity of NMR spectroscopy (Dec et al. 1997; Nanny et al. 1996; Hatcher et al. 1993; Knicker et al. 1999). NMR has also been applied to elucidate residue formation of PAH in soil (Guthrie et al. 1999). The advantage of ^{13}C-NMR spectroscopy in labelling experiments is that the observation of the chemical environment of the label within the macromolecular organic matter is enabled. Changes in the chemical shift indicate the structural arrangement of the ^{13}C-label with adjacent atoms. Hence, it is possible to measure the chemical shift arising from different chemical environments tracing the transformation of the parent labelled PAH within the soil and during the formation of bound residue. Since a sequestered PAH will not change the molecular environment of the ^{13}C-label, this approach may allow examination of covalent and non-covalent interactions of xenobiotic molecules with natural macromolecular organic matter.

Bound residues of the degradation experiment with 9-[^{13}C]-anthracene were examined with NMR to illustrate the transformation during biodegradation in soil. Compared to the control soil (fig. 16.5), the labelling experiment with ^{13}C-anthracene revealed a characteristic pattern with predominant signals around 70 mg kg^{-1} and 170 mg kg^{-1}, which originate from O-alkyl and carbonyl units, respectively (fig. 16.5 A).

Signals in the area of 126 mg kg^{-1} characterise the label at position 9 of 9-[^{13}C]-anthracene with an intact carbon skeleton (fig. 16.5). Their low abundance reflect a minor concentration of intact aromatic units in the bound residue. Sorbed or sequestered anthracene was of minor relevance in the formation of residues in this experiments. Most of the label was transformed into O-alkyl and carbonyl units clearly indicating a structural rearrangement of the position 9 of the former anthracene. Generally, carbonyl units are characteristic structural elements of

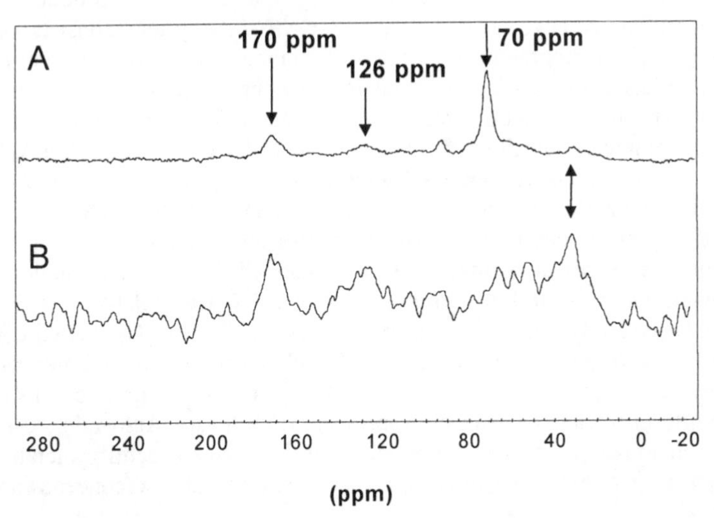

Fig. 16.5. NMR Spectroscopy of ^{13}C-labelled residues originating for a biodegradation experiment of 9-^{13}C-anthracene (A) and a control soil (B). The signals around 30 mg kg^{-1} (arrow) are of comparable intensity.

humic material (fig. 16.5 B) and this may illustrate the transformation of anthracene during biodegradation to humic like bound residues.

16.3.1.2
The Chemical Degradation Approach

The large number of sub-units with different chemical structures cross-linked by various types of chemical bonds complicate structural analysis of residues. The definition of a precise chemical structure for humic macromolecules is difficult today (for review, see Hayes et al. 1989) which may also account for non-extractable bound residues in soil. The chemical degradation approach applies degradation techniques to cleave xenobiotic moieties from the macromolecular matrix and identify the low-molecular weight products with a combination of methods for structural analysis, such as GC-MS and IR-C-GC-MS (Richnow et al. 1996; Richnow et al. 1998). An overview on the application of degradation methods to characterise humic material is discussed by Hayes et al. (1989). Moreover, the application of selective reagents may allow cleavage of specific bonds within the residues and may trace the site and type of linkage of xenobiotic fragments within the natural macromolecular matrix (Richnow et al. 1994). To elucidate structural units of residue materials from labelling experiments with ^{13}C-model compounds chemical degradation studies have been conducted to obtain structural information. The advantage of isotopically labelled model compounds becomes obvious when tracing the precursor product relationship by the means of the isotopic label (see Annweiler, Chapter 17).

For example the incorporation of PAH-related fragments into macromolecular organic matter forming non-extractable residues can be studied by cleaving ester bonds by alkaline hydrolysis. The xenobiotic fragments among the products can be easily identified by IR-C-GC-MS. The IR-C-GC-MS system consists of a GC-unit for the chromatographic separation of complex mixtures. The column effluents are oxidised to CO_2 in a combustion line that is connected to a mass spectrometer. The mass spectrometer monitors continuously the concentration (m/z 44) and isotopic composition (ratio m/z 45/44) of the CO_2 (fig. 16.6 A, B). The relative concentration of single substances is shown in the trace (m/z 44), which is similar to the total ion current (TIC) trace of a conventional GC-MS. The IR-C-GC-MS system is commonly used to analyse the isotopic composition of GC-amenable substances at natural abundance and allow the determination of isotopic signatures at natural occurrence of at least 0.3 ‰ relative the PDB δ-notation. Consequently, this system is extremely sensitive to ^{13}C-labelled compounds and is able to detect metabolites of ^{13}C-anthracene in both complex mixtures and low concentrations. The chemical structure of these xenobiotic units can be subsequently characterised with conventional GC-MS (see Annweiler, Chapter 17). The combination of these techniques can be used to identify metabolic products by the ^{13}C-label, even in complex mixtures and at low concentrations (fig. 16.6).

Structural investigations by GC-MS revealed carboxylic acids as the predominant constituents in extractable carboxylic acid fraction (fig. 16.6). Their isotopic signature ranged between $\delta^{13}C$ -22 and -36 ‰, and thus, the isotopic composition is very similar to those observed in carboxylic acids from bacterial

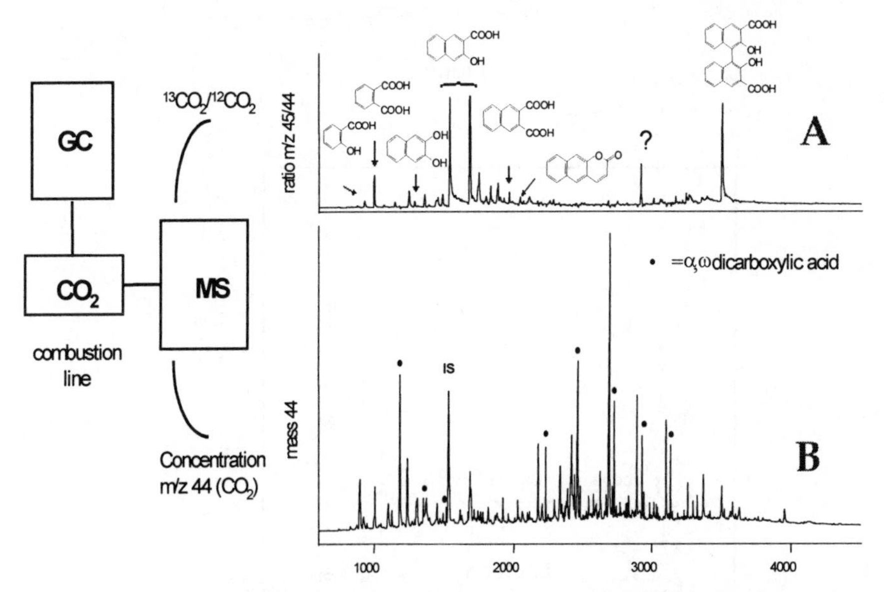

Fig. 16.6. IRM-GC-MS of alkaline hydrolysis products of the bound residue fraction after degradation of 9-^{13}C-anthracene in soil. The upper trace (45/44; A) depicts the variation of mass 45 ($^{13}CO_2$) to mass 44 ($^{12}CO_2$). The lower trace (B) shows that the concentration of $^{12}CO_2$. ^{13}C-labelled metabolites can be easily identified by their characteristic isotope composition even if the absolute concentration is low. (dots = α,ω dicarboxylic acids with an isotopic composition typical for long-chain carboxylic acids in soil).

and plant derived sources. ^{13}C-labelled substances become apparent in the chromatogram of the mass ratio 45/44 which shows very strong signals, although the absolute concentration of the respective compound is relatively low. Extraordinarily high positive signals in the 45/44 trace were recorded for the ^{13}C-labelled metabolites such as, salicylic acid, phthalic acid, 2-hydroxy-3-naphthoic acid and 2,3-naphthalene-dicarboxylic acid. The ^{13}C-label proves the precursor-product relationship to the parent anthracene. Thus, IR-C-GC-MS analysis is a powerful tool to screen ^{13}C-labelled metabolites even in complex mixtures. The distribution of the ^{13}C-label among the various fractions of the degradation experiment allows reconstruction of the pathways of the microbial degradation in the soil (fig. 16.7). For example, among the ^{13}C-labelled compounds are 2-hydroxy-3-naphthoic acid, 1,2-dihydroxynaphthalene, and salicylic acid, which are all typical metabolites of anthracene. These metabolites have been already described in classical oxidative anthracene metabolism in liquid cultures by *Pseudomonas* strains (Evans et al. 1965; Gibson and Subramanian 1984). Similar to results from liquid culture experiments, the appearance of ring-cleavage products such as hydroxynaphthoic acid or salicylic acid may indicate the ability of the soil microbial system to completely mineralise anthracene. Apart from these well known metabolites, products such as phthalic acid and 2,3-naphthalenedicarboxylic acid were determined.

These have not been described as anthracene metabolites before and clearly

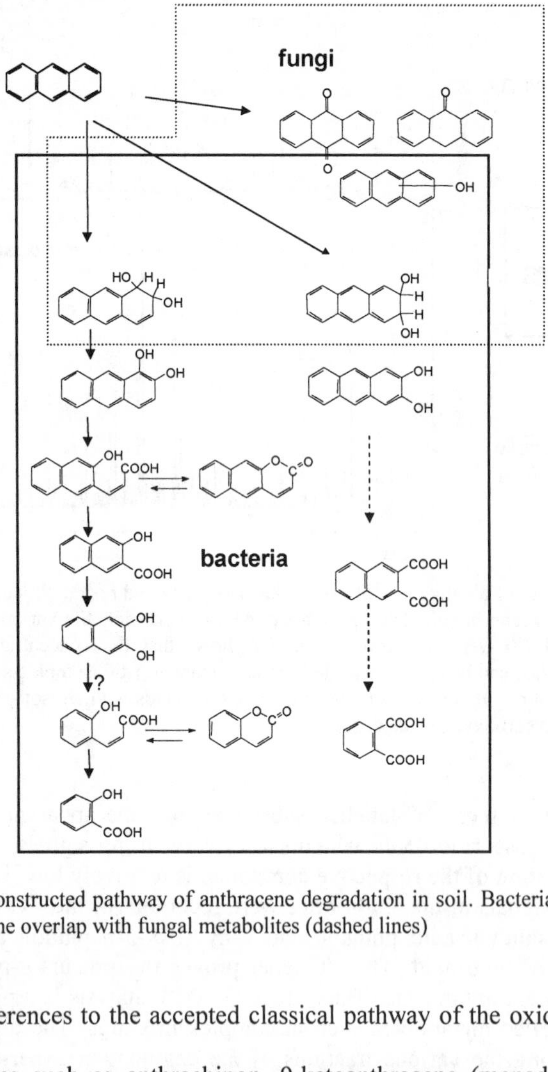

Fig. 16.7. Reconstructed pathway of anthracene degradation in soil. Bacterial metabolites (solid lines) have some overlap with fungal metabolites (dashed lines)

indicate differences to the accepted classical pathway of the oxidative anthracene metabolism.

Metabolites such as anthrachinon, 9-ketoanthracene (menadion) and several hydroxyanthracenes indicate that fungi are also involved in the degradation of anthracene in soil. Moreover, a dimer of hydroxynaphthoic acid was identified which indicates that condensation processes leading to the formation of carbon-carbon bonds are relevant in soils (see fig. 16.6). To our knowledge, high molecular weight metabolites have not been reported in soil systems. Thus, soil microbial systems may produce variants of high-molecular weight condensation or conjugation products. The formation of conjugates is a well known detoxification and excretion strategy of higher organisms and plants (Mulder 1990; Livingston 1993). Microbial high molecular weight metabolites may be functional analogues.

The overall composition of bound anthracene fragments reflects the actual pathway of microbial degradation in soils. In this context, chemical fingerprints of the degradation products are well preserved in the humic material and may reflect former anthropogenic contamination. Using these molecular markers, the biochemical history of the soil may be reconstructed from bound residues even when the parent chemical is removed completely from the low-molecular weight fraction.

16.3.1.3
Types of Chemical Linkages

Only limited information is available regarding types of chemical cross-linkages between xenobiotic fragments and natural organic matter in soil bound residues. Studies have been focused on model experiments with xenobiotics carrying reactive functional groups, such as amino- or hydroxyl functionalities attached to an aromatic ring system. In these experiments, TNT-metabolites containing amino groups and guaiacol dimers representing typical structural units of humic substances were cross-linked by covalent nitrogen bonds with laccases (Dawel et al. 1997). Thorn et al. (1996) present evidence that the binding of aniline to humic substances is a nucleophilic substitution reaction. The presence of quinone moieties is essential for this reaction. Under anoxic conditions, when the quinone group is reduced the reaction is blocked (Weber et al. 1996). Thus, N-containing functional groups may serve as reactive binding sites for bound residue formation.

Oxidoreductase enzymes such as peroxidase, laccase, and tyrosinase are known to oxidise phenolic compounds to aryloxy radicals. These radicals then polymerise to form insoluble humic acid like complexes (Martin and Haider 1980; Filip and Preusse 1985; Sarkar and Bollag 1987; Bollag et al. 1988). The kinetics of enzymatic oxidation of phenols and chlorinated derivatives were studied in various types of soil (Claus and Filip 1990). Enzymatic oxidative cross-coupling of phenols results in ether and carbon-carbon bonds (Bollag 1983). Structural assignments of bonds formed by enzymatic cross-linking reactions of dichlorophenol with humic substances by ^{13}C-NMR give indication of the formation of ethers, esters and alkyl-aryl or aryl-aryl carbon-carbon bonds (Hatcher et al. 1993).

The enzymatic binding with certain types of model compounds, including typical PAH-metabolites catalysed by horseradish peroxidase, has been recognised to form ether and carbon-carbon-bonds (Richnow et al. 1997). In this study PAHs were unreactive but hydroxylated PAH-derivatives representing typical PAH-metabolites could be enzymatically bound to humic material. This indicates that a metabolic activation leading to phenolic derivatives is essential for enzymatic cross-linking reactions by oxidoreductases. Ether-linked PAH-metabolites have been found in bound residues formed in bioremediation experiments (Richnow et al. 1997) and thus indicate bound residue formation to be a biocatalysed process.

An ester linkage of xenobiotic moieties to humic material has been proved applying ^{18}O-labelled NaOH to hydrolyse ester bonds (Richnow et al. 1994; Michaelis et al., 1995). A detailed description of chemical linkages is presented in Chapter 17. During the hydrolytic cleavage with $^{18}OH^-$, the obtained organic acids are labelled with ^{18}O. The amount of labelling can be characterised by GC-MS. Using this method, the ester linkage of PAH-metabolites in humic substances was demonstrated (Richnow et al. 1994). Applying alkaline hydrolysis with $^{18}OH^-$,

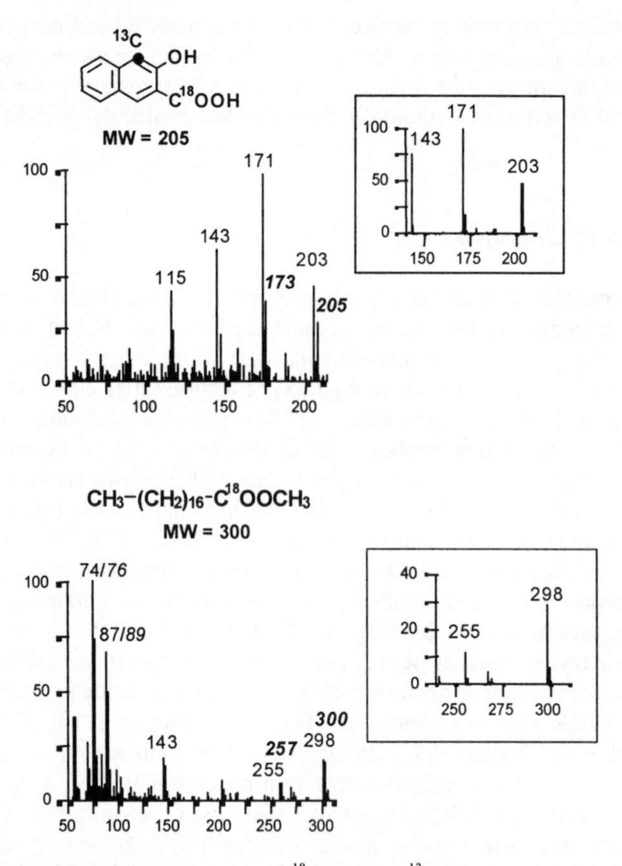

Fig. 16.8. (above) Mass spectrum of ^{18}O-labelled ^{13}C-2-hydroxy-naphthoic acid methyl ester. Mass spectrum of ^{13}C-2-hydroxy-naphthoic acid methyl ester (insert).
(below) Mass spectrum of ^{18}O-labelled octadecanoic acid methyl ester. The insert shows the molecular ion of non-labelled octadecanoic acid. The characteristic shift of 2 amu (figures in italic) indicate the incorporation of ^{18}O.

^{13}C-labelled bound residues yields ^{13}C-labelled metabolites with an additional ^{18}O-label in the carboxyl group. The characteristic shift of 2 amu of the molecular ion proves the incorporation of heavy oxygen into ^{13}C-hydroxynaphthoic acids during the hydrolytic cleavage of the bound residue (fig. 16.8). The mass fragments at m/z 173 and 205 illustrate the incorporation of ^{18}O into the carboxylic group released upon chemical degradation. The ^{18}O-labelled fragments represent the part of hydroxynaphthoic acids that has been linked as non-extractable residue by ester functionality via the carboxylic group. The peak at m/z 203 represents the part of hydroxynaphthoic acids that has not been labelled during alkaline hydrolysis. The non-^{18}O-labelled hydroxynaphthoic acids rather become adsorbed on humic material or ester linked via the hydroxyl group. Consequently, the $^{18}OH^-$ technique allows recognition of former ester bound carboxylic like *n*-octadecanoic acid (stearic acid) shows a very similar fragmentation pattern. The molecular ion m/z 300 represents the ester linked compound and m/z 298 shows the presence of non chemically bound carboxylic acids in the non-extractable organic soil fraction (fig. 16.8). The ratio of the molecular ion intensities may characterise the amount of adsorbed (m/z 298) and chemically bound (m/z 300) species. Thus, about 50 % of the octadecanoic acid in the chemical degradation products have been chemically bound.

Interestingly, the hydroxynaphthoic acid and natural compounds like hexadecanoic acid (palmitic acid) show a similar ^{18}O-labelling and thus a comparable amount of these substances were chemically linked to the organic soil matrix. This may demonstrate that during humification metabolites of xenobiotica undergo similar cross-linking reactions with soil organic matter like substances of natural origin.

16.3.1.4
Formation of PAH-Residues

PAH-transformation studies on a molecular level provide strong evidence that bound residues mainly consist of biological metabolites cross-linked to macromolecular material. Microbial catalysed immobilisation of xenobiotic carbon in soils may take place, although only parts of the structural entities of the total bound residues have been structurally characterised today. Moreover, anthropogenic chemicals without reactive functional groups such as PAH and PCB, require some metabolic activation to form bound residues.

A partial degradation of parent xenobiotics and subsequent incorporation into humic material is very similar to the humification of natural organic matter. Even natural compounds like glucose form residues in soils (Baldock et al. 1989; Stott et al. 1983a). In a preliminary estimation, xenobiotic fragments bound to humic substances, or other natural organic fractions, might behave as binding substrates, depending on the mean residence times of the particular fraction in soil. Clearly, more research addressing the turnover of bound residues in soil is required to substantiate this assumption.

Most of the non-extractable residues, which are formed during microbial degradation of PAH seems to result from the incorporation of PAH-metabolites. When the parent compounds are sequestered, similar residue formation of the different labelled anthracenes should be expected. However, elevated mineralisation rates were observed with 1,2,3,4,4a,5a-[^{14}C]- instead of 9-[^{14}C]-anthracene and higher amounts were incorporated into the residues under similar application conditions (Kästner et al. 1999). This only can be explained by metabolism, since the availability of different C-positions in the anthracene molecule is not equal for the anabolic and catabolic processes of soil microorganisms. Another indication for the formation of residues from metabolites is the fact that large parts of the radioactivity were extractable before relevant amounts of PAH were already degraded. Therefore, the formation of microbial metabolites resulting in reactive functional groups at the PAH-molecule seems to be essential to decrease the extractability. The oxidation of PAH result in phenolic compounds or aryl radicals, which exhibit a high potential for coupling to the organic soil matrix (Bollag and Loll 1983). The extent of bound residue formation depends therefore on the metabolic potential of the soil microflora and the characteristics of the soil. The

Table 16.4. Distribution of ^{14}Carbon after microbial degradation of 9-[^{14}C]-anthracene and in native soil and soil-compost-mixture[a] 174 d) and after incubation under $^{14}CO_2$-atmosphere (modified from Kästner et al. 1999).

	applied radioactivity [%] from 9-[14C]-anthracene		$14CO_2$	
	native soil	soil-compost a	soil-compost a, b	+CHC13[a, b, c]
Σ	98.7	91.9	99.8	98.7
CO_2	43.8	67.2	4.6[e]	11.3e
total soil	54.9	24.6	95.2[f]	87.4
extraction + alk. hydrolysis	9.5	3.9	9.4	44.2
bound residues	45.4	20.7 d	85.8	43.2

[a] soil-compost-mixture (80 %: 20 % dry weight)
[b] 90 d
[c] microflora inhibited after fumigation with $CHCl_3$
[d] amount represents 25.5 Bq g^{-1} of soil
[e] recovered $^{14}CO_2$ (initially applied amount: 24.8 Bq g^{-1} of soil)
[f] represented by 31 % acid labile carbonates, 67 % incorporated into the organic matrix, and < 2 % within clay and silicates

time courses of bound residue formation in the experiments are based on two main processes. Immediately after application of the PAH, a diffusion and sequestration process of the parent compounds into soil pores occurred (Hatzinger and Alexander 1995; Kelsey and Alexander 1997). The following process is the microbial degradation of the compound accompanied by the incorporation of metabolites into the organic soil matrix. Parts of the sequestered portion may desorb and become bioavailable with decreasing concentrations.

In case of PAH, relevant chemical condensation processes include both formation of carbon-carbon bonds, as well as ether and ester bonds. The precise mechanism of ester condensation in soils is still not known. However, natural organic substances such as fatty acids and lignin derived phenolic acids are ester linked in fossil macromolecular organic matter. This implies a naturally occurring condensation process. The formation of ether and carbon-carbon bonds may be rationalised as an enzymatic oxidative cross-linking reaction and thus is a biocatalysed process. Catalytic mineral surfaces in soils may also contribute to the formation of ether and carbon-carbon-bonds (Ruggiero 1998).

A soil-compost-mixture was incubated with $^{14}CO_2$ in the absence of light to investigate the incorporation of $^{14}CO_2$ derived from mineralisation of labelled anthracene as a possible mode of residue formation (Kästner et al. 1999). The total CO_2 concentration applied was equivalent to ≈70 % of mineralised anthracene used in our other degradation experiments. After 90 d of incubation, the major part of the applied $^{14}CO_2$ was found within the matrix of the soil-compost-mixture and the non-extractable activity amounted to 85.8 % (table 16.4). The absolute amounts of residue activity was in the range of residues that were observed during

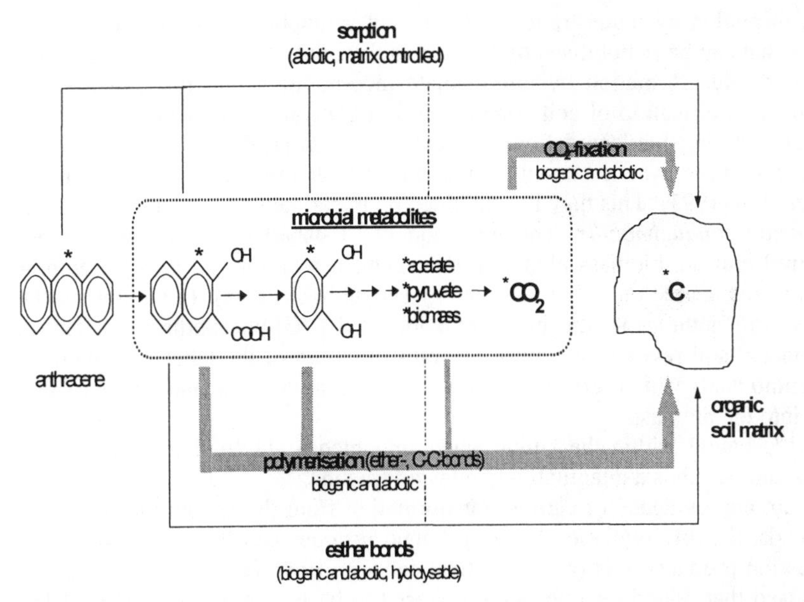

Fig. 16.9. Products from microbial degradation of anthracene and estimated contribution (fat lines) to the formation of non-extractable residues in the organic soil matrix (modified from Kästner et al. 1999)

biodegradation of anthracene. The inhibition of microbial activity by fumigation with $CHCl_3$ (respiration < 0.5 % of the native soil) decreased the immobilisation to 40 % of the applied $^{14}CO_2$ which indicates a microbial contribution to the CO_2 fixation in the soil. Only a part of the CO_2 fixation (31 %) in the native soil mixture was due to the formation of carbonates whereas 67 % were incorporated into the organic matrix. However, the total amount of the $^{14}CO_2$ fixation represents only 4.8 mmol C (≈0.17 %) within a soil carbon matrix of about 2,830 mmol kg^{-1}.

Another indication for the relevance of CO_2 fixation is based on the partitioning of the radiolabel in humines, humic and fulvic acids that were similar to the residues derived from anthracene. The degradation of 9-[^{14}C]-anthracene in the soil-compost-mixture led to a speciation of bound activity in humines (29 %), humic acids (38 %) and in fulvic acids (18 %) (Kästner et al. 1999). In comparison to the degradation of 9-[^{14}C]-anthracene, the experiment with $^{14}CO_2$ showed some striking similarities in the distribution of radioactivity between the different soil fractions.

The processes causing the CO_2 fixation are still unknown and have to be examined in more detail. However, the results give clear evidence that the extent of residue formation by certain PAH-metabolites in the humic matrix is smaller than usually presumed. Based on the presented data, we predict a concept of the mode of residue formation during biodegradation in soil (fig. 16.9). Although no detailed mass balances of the carbon flow via PAH-metabolites, or CO_2 fixation into the residue fraction are available, one may conclude from the transformed amounts that both processes significantly contribute to the residue formation. The residue formation by sequestration of the parent compound or by fixation of metabolites that can be remobilised by hydrolysis does not play a major role. In addition, the residue formation by intermediate incorporation into the biomass and subsequent humification of cell compounds is a general mode, if the carbon was assimilated during biodegradation. However, no radiolabel after degradation of 9-[^{14}C]-anthracene was observed in the biomass in previous soil experiments (Kästner et al. 1995). This may be explained by experiments with pure cultures of *Sphingomonas paucimobilis*. The 9-postion of ^{13}C-labelled anthracene was not transferred into the biomass although the organism was able to grow on anthracene as a sole source of carbon and energy (unpublished data). In contrast to the findings with anthracene, the biodegradation of 1-[^{13}C]-phenanthrene in tar oil contaminated soil revealed a significant transformation of the label into hydrolysable amino acids which proves the anabolic assimilation of phenanthrene related carbon into the biomass.

The ^{13}C bound within the amino acids amounted to 11 % of the total bound residues and was thus a quantitatively relevant fraction.

Without any evidence of carbon transformation from the xenobiotic to natural compounds, the differentiation between bound residues and biogenic residues or humification products is only a theoretical concept. However, it can be generally summarised that bound residues were assessed to be less toxic, less bioavailable and less mobile in comparison to the free parent compounds (Bollag and Loll 1983; Bollag et al. 1988; Park et al. 1988; Wang et al. 1990). Some authors have proposed the use of enhanced formation of non-extractable residues from xenobiotics as a technique to lower the toxic potential and the bioavailability at contaminated sites (Bollag and Loll 1983; Berry and Boyd 1985; Bollag 1992; Verstraete and Devliegher 1996). However, for a critical assessment of such techniques further investigations of the residue remobilisation under environmental "worst case" conditions, long-term stability, and structural assignment are necessary.

16.4 Stability and Metabolism of Bound Residues in Soil

Since only limited information about the stability of the bound residues is available (Nordlohne et al. 1995; Eschenbach et al. 1996; Eschenbach et al. 1998; Kästner et al. 1996; Kästner et al. 1997), the long-term fate and the remobilisation of bound residues has to be examined empirically for each xenobiotic compound in so-called worst case studies. In these studies, extreme physical, chemical or biological situations are simulated in soils in order to predict the remobilisation of xenobiotic carbon from bound residues in soils that may affect environmental quality. For example the long-term behaviour of bound residues according to the natural turnover of organic matter in soils may be simulated by enhancing the biological decay of soil organic matter fractions with a stimulation of the microbial activity (Kästner et al. 1997). The investigation of the remobilisation of bound residues includes:

1. altering the soil texture by physical or chemical treatment grinding, freeze and thaw = simulation of climatic effects,

2. stimulating turnover of soil organic matter in biodegradation experiments with white-rot fungi, and
3. the uptake by plants.

It is well known, that white rot fungi are capable of depolymerizing lignin and humic substances (Kästner and Hofrichter 2001). In some cases, these organisms are able to mineralise these residues to a higher extent than the parent compounds (Haider and Martin 1988).

The soil material for the worst case studies was taken from the degradation experiments with 9-[^{14}C]-labelled anthracene (soil compost material, see table 16.4). After 174 days of incubation, 67.2 % of the applied radioactivity was mineralised, 24.6 % remained non-extractable in the soil, and 4 % of the radioactivity was still extractable with organic solvents. A part of the obtained soil material was ground intensively to break up the soil texture completely (fig. 16.10). Another portion was treated by freeze and thaw techniques five times to simulate climatic effects of several years (fig. 16.11). Both charges of the soil

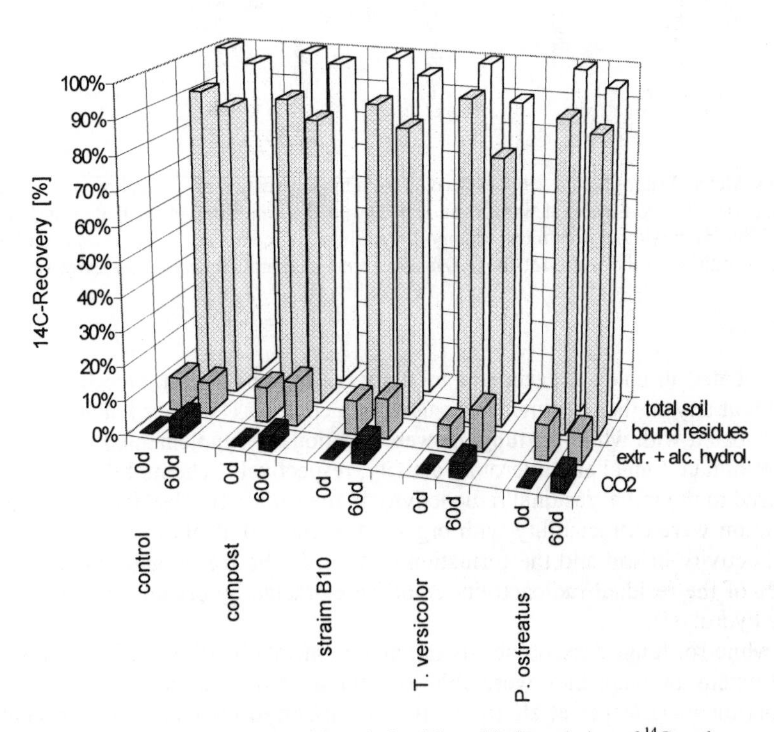

Fig. 16.10. Metabolism of bound residues derived from biodegradation of ^{14}C-anthracene (174 d) in soil material. The soil texture was destroyed by grinding of the material before incubation in batch cultures. (control = untreated soil; compost = addition of compost; strain B10, *T. versicolor* and *P. ostreatus* = addition of white-rot fungi, strain B10, *Trametes versicolor* and *Pleurotus ostreatus*, respectively).

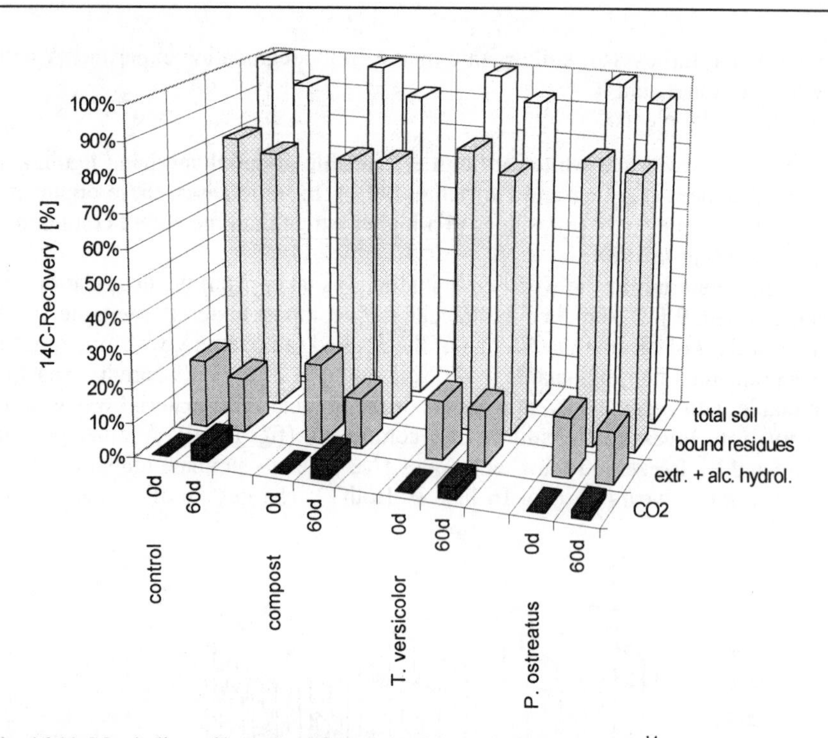

Fig. 16.11. Metabolism of bound residues derived from biodegradation of ^{14}C-anthracene (174 d) in soil material. The soil material was frozen and thawed five times before incubation in batch cultures. (control = without treatment; compost = addition of compost; *T. versicolor* and *P. ostreatus* = addition of the white-rot fungi *Trametes versicolor* and *Pleurotus ostreatus*, respectively)

were incubated in batch experiments to study the remobilisation. One batch culture without any further treatment was used as a control. The other batch cultures were incubated with white-rot fungi or were additionally supplemented with compost to stimulate fungal and microbial activity, respectively. The mobilisation was normalised to the initial residual radioactivity in the soil (= 100 %). Parameters for mobilisation were extractability with organic solvent and alkaline hydrolysis, the residual activity in soil and the formation of CO_2. In the frozen and thawed soil, 15–20 % of the residual radioactivity could be extracted by organic solvents and alkaline hydrolysis.

The white-rot fungi were obviously not able to metabolise the residues to a significant extent although they were able to metabolise the parent PAH in similar soil experiments (Kästner et al. 1997). In comparison to the control experiments with the autochthonous microflora, neither stimulating nor inhibitory effects regar ding mineralisation of bound residues were observed in other soil batches.

Furthermore, supplementing the soil with compost revealed no stimulation of the bound residue mineralisation, although the biodegradation of the parent anthracene was enhanced significantly after addition of compost to the soil in former experiments (Kästner and Mahro 1996). Because no increase of the

Table 16.5. Remobilisation and transfer of radioactivity from bound residues into oats *(Avena sativa)* and cress (*Lepidium sativum*)

	Distribution of radioactivity [%]			
	total soil	extraction + alk. hydrolysis	residues	plants
soil-compost mixture [a]	100	15.8	84.1	–
Control[b]	100.5	16.2	84.3	
+ cress	99.2	14.1	85.1	0.0057
+ oats	99.9	14.6	85.3	n.d
native soil[a]	100	17.3	82.7	–
control[b]	95.9	10.4	89.6	–
+ cress	98.9	8.3	91.7	0.0051
+ oats	97.4	8.2	89.2	0.002

[a] residue age 174 d; residual radioactivity in soil is normalised to 100 %
[b] similar water addition as applied in planted cultures
[c] n.d. = not determined

extractable radioactivity was observed, the residues in the organic soil matrix appear to have a high stability towards microbial remobilisation. In contrast to these findings, the age of residues seems to have a significant effect on the mobilisation. Experiments with residues of lower age showed a significant higher mineralisation of the residues (unpublished data). Therefore, the persistence of the bound residues increase with the age of the residues.

A crucial question in the evaluation of bound residues is the transfer of xenobiotic carbon into plants. An incorporation of xenobiotic fragments into plants may increase their bioavailability by directly entering the food web of higher organisms. In order to investigate the transfer of the radioactivity into higher plants, experiments with soil material containing bound residue were carried out. Inhibition or any other effect on germination and growth of mono- and dicotyle plants (*Avena sativa* = oat; *Lepidium sativum* = cress) was not observed in comparison to batch experiments with the control soil (table 16.5).

Thus, the test system showed no toxic effects on the plants. Only 0.002 % and 0.006 % of the radioactivity in the soil was incorporated into the plants. Similar results of other investigations showed that 0.7–1.0 % of the extractable radioactivity from soils entered tomato plants (Kolb et al. 1996). However, in this study with compost supplemented soil, the main portion of the extractable radioactivity was represented by the parent labelled fluoranthene and primary oxidised metabolites. This may explain the higher incorporation. Results from experiments with pesticide residues lead to the conclusion that small amounts of the residual radioactivity were generally transferred to the plants. The extent of incorporation is mostly below 1 % of the residue radioactivity in soils (Calderbank 1989). These results suggest that during biodegradation, the transfer of radioactivity from residues into plants decreases with the extent of mineralisation of the parent compounds and the intensity of the humification process. These studies demonstrate that the bioavailability of the residues is orders of magnitude lower than the bioavailability of the parent chemicals.

16.5 Conclusions

Progress in understanding the molecular structure of bound residues was made applying ^{13}C-labelled tracer substances. Former xenobiotic structural entities in macromolecular bound residues can be easily recognised by their ^{13}C-signature in low-molecular weight products of chemical degradation experiments. Even though the overall structure is unknown, present data suggest that the main component of bound residues consists of covalently bound metabolites. These metabolites are randomly bound by a variety of chemical linkages within the organic matrix of the soil. The xenobiotic fragments are molecular markers that allow a reconstruction of the biogeochemical soil history in terms of biodegradation processes and impacts from former contamination.

In the case of PAH, bound residues only contain traces of the parent compounds as shown in microbial degradation experiments with ^{14}C-labelled anthracene. The extractability of bound residues by organic solvents and their bioavailability to microorganisms and plants is low, and decreases with age and degree of humification. The cross-links of humified xenobiotics within the macromolecular substances are stable enough to resist degradation by white-rot fungi to a large extent. Consequently, there exists only a minor mobilisation potential of bound residues by biological processes and hence a low probability of a transfer of xenobiotic carbon into plants. Although no significant mobilisation of the radioactivity was observed in worst case studies, a low amount of the residues was mineralised by the metabolic activity of the soil microflora. With increasing degree of metabolism of an anthropogenic compound, the process is similar to the natural humification process. The stability of the residues with respect to mechanical, chemical and microbial access increases with the degree of humification. At the present state, the limited bioavailability of bound residues most likely includes a detoxification of xenobiotics in soils and sediments. However, more research is needed for a final evaluation.

The biogenic formation of residues is a complex process of single reactions. The residues probably are formed coupling the microbial metabolites to the soil matrix, or by the incorporation of labelled carbon into the biomass. This biomass than may be transformed to humic substances after cell death. Residues formed by the latter processes are not bound residues as defined by the IUPAC for pesticides, since the carbon is no longer part of the molecular structure of the parent compound or its related metabolites.

References

Abbott BJ, Gledgehill WE (1971) The extracellular accumulation of metabolic products by hydrocarbon-degrading microorganisms. Adv Appl Microbiol 14: 249–388

Bailey GW, White JL (1964) Review of adsorption and desorption of organic pesticides by soil colloids with implication concerning pesticide bioactivity. J Agric Food Chem 12: 324–332

Baldock JA, Oades JM, Vassallo AM, Wilson MA (1989) Incorporation of uniformly labelled ^{13}C-glucose carbon into the organic fraction of a soil. Carbon balance and CP/MAS-^{13}C-NMR measurements. Soil Biol Biochem 27: 725–746

Berry DF, Boyd SA (1984) Oxidative coupling of phenols and anilines by peroxidase: structure-activity relationships. Soil Sci Soc Am J 48: 565–569

Berry DF, Boyd SA (1985) Decontamination of soil through enhanced formation of bound residues. Environ Sci Technol 19: 1132–1133

Bollag JM (1983) Cross-coupling of humus constituents and xenobiotic substances. In: Aquatic and terrestrial humic material. Christman RF, Gjessing ET (eds) Ann Arbor Publishers, Michigan, USA, pp 127–141

Bollag JM (1992) Decontaminating soil with enzymes. Environ Sci Technol 26: 1876–1881

Bollag JM, Bollag WB (1990) A model for enzymatic binding of pollutants in the soil. Intern J Environ Anal Chem 39: 147–157

Bollag JM, Loll, M J (1983) Incorporation of xenobiotics into soil humus. Experientia 39: 1221–1231

Bollag JM, Shuttleworth KL, Anderson DH (1988) Laccase-mediated detoxification of phenolic compounds. Appl Env Microbiol 54: 3086–3091

Brotkorb TS, Legge RL (1992) Enhanced biodegradation of phenanthrene in oil tar-contaminated soils supplemented with Phanerochaete chrysosporium. Appl Environ Microbiol 58: 3117–3121

Calderbank A (1989) The occurrence and significance of bound pesticide residues in soil. Rev Environ Contam Toxicol 108: 71–103

Cerniglia C E, Heitkamp M A (1989) Microbial metabolism of polycyclic aromatic hydrocarbons (PAH) in the aquatic environment. In: Metabolism of polycyclic aromatic hydrocarbons in the aquatic environment. Varanasi U (ed) CRC Press, Boca Raton, pp 41–68

Claus H, Filip Z (1990) Effects of clays and other solids on the activity of phenoloxidases produced by some fungi and actinomycetes. Soil Biol Biochem 22: 483–488

Craven A (2000) Bound residues of organic compounds in the soil: the significantce of pesticide persistence in soil and water: a European regulatory view. Environmental Pollution 108: 15–18

Dawel G, Kästner M, Michels J, Poppitz W, Günther W, Fritsche W (1997) Structure of a laccase-mediated product of coupling of 2,4-diamino-6-nitrotoluene to guiacol, a model for coupling of 2,4,6 trinitrotoluene metabolites to a humic organic soil matrix. Appl Environ Microbiol 63: 2560–2565

Dec J, Haider K, Benesi A, Rangasvamy V, Schäfer A, Plücken U, Bollag JM (1997) Analysis of soil bound residues of ^{13}C-labelled fungicide Cyprodinil by NMR-spectroscopy. Environ Sci Technol 31: 1128–1135

Domsch KH (1992) Pestizide im Boden – Mikrobieller Abbau und Nebenwirkungen auf Mikroorganismen. VCH Verlagsgesellschaft, Plankstadt, Germany

Eschenbach A, Kästner M, Wienberg R, Mahro B (1995) Microbial PAH degradation in soil material from a contaminated site – mass balance experiments with *Pleurotus ostreatus* and different ^{14}C-PAH. In: Contaminated soil 1995. van den Brink WJ, Bosman R, Arend F (eds) Kluver Academic Publishers, Dodrecht, The Netherlands, pp 377–378

Eschenbach A, Wienberg R, Mahro B (1998) Fate and stability of non-extractable residues of ^{14}C-PAH in contaminated soils under environmental stress conditions. Environ Sci Technol 32: 2585–2590

Eschenbach A, Wienberg R, Mahro B (1996) Entstehung und Langzeitstabilität von nicht extrahierbaren PAK-Rückständen im Boden. In: Biologischer Abbau von polyzyklischen aromatischen Kohlenwasserstoffen. Cuno M (ed) Schriftenreihe – Biologische Abwasserreinigung 7, (SFB 193), Technische Universität Berlin, Germany, pp 63–80

Evans WC, Fernley HN, Griffiths E (1965) Oxidative metabolism of phenanthrene and anthracene by soil pseudomonas, the ring fission mechanism. Biochem J 91: 819–831

Filip Z, Preusse T (1985) Phenoloxidierende Enzyme – ihre Eigenschaften und Wirkungen im Boden. Pedobiologica 28: 133–142

Führ F (1987) Non-extractable pesticide residues in soil. In: Pesticide and biotechnology. Greenhalgh R, Roberts TR (eds) Proc 6th Congr Pestic Chem, IUPAC, Blackwell Scientific Publication, London, UK, pp 381–389

Gibson DT, Subramanian V (1984) Microbial degradation of aromatic hydrocarbons. In: Microbial degradation of organic compounds. Gibson DT (ed) Marcel Dekker, New York, USA, pp 181–252

Goodin JD, Weber MD (1995) Persistence and fate of anthracene and benzo(a)pyrene in municipal sludge treated soil. J Environ Qual 24: 271–278

Grosser RJ, Warshawsky D, Vestal R (1991) Indigenous and enhanced mineralization of pyrene, benzoa]pyrene, and carbazole in soils. Appl Environ Microbiol 57: 3462–3469

Gustafsson Ö, Gschwend P (1998) Phase distributions of hydrophobic chemicals in the aquatic environment: existing partitioning models are unable to predict the dissolved component in several common situations. In: Bioavailability of Organic Xenobiotics in the Environment. Block JC, Baveye Ph, Goncharuk VV (eds). NATO ASI Series 64, Kluwer Academic Publishers, The Netherlands, pp 297–326

Guthrie EA, Bortiatynski JM, van Heemst JDH, Richman JE, Hardy KS, Kovach EM, Hatcher PG (1999) Determination of [^{13}C]-Pyrene Sequestration in Sediment Microcosms Using Flash Pyrolysis-GC-MS and 13-C NMR. Environ Sci Technol 33: 119–125

Haider KM, Martin JP (1988) Mineralization of ^{14}C-labelled humic acids and of humic-acid bond ^{14}C-xenobiotics by Phanerochaete chrysosporium. Soil Biol Biochem 20: 425–429

Hassett JJ, Banwart WL (1989) The sorption of nonpolar organics by soils and sediments. In: Reactions and movement of organic chemicals in soils. Sawhney BL, Brown K (eds) SSSA Spec Publ 22: 31–44

Hatcher PG, Bortiatynski JM, Minard RD, Dec J, Bollag JM (1993) Use of high resolution ^{13}C NMR to examine enzymatic covalent binding of ^{13}C-labelled 2,4-dichlorophenol to humic substances. Environ Sci Technol 27: 2098–2103

Hatzinger PB, Alexander M (1995) Effect of aging of chemicals in soil on their biodegradability and extractability. Environ Sci Technol 29: 537–545

Hayes MHB, MacCarthy P, Malcom RL, Swift RS (eds) (1989) Humic substances II – In: Search of structure. Wiley and Sons, Chichester, UK

Hedges JI (1988) Polymerisation of humic substances in natural environment. In: Humic substances and their role in the environment. Frimmel FH, Christman RF (eds) Wiley and Sons, Chichester, pp 45–48

Heitkamp MA, Freeman JP, Cerniglia CE (1987) Naphtalene biodegradation in environmental microcosms: Estimates of degradation rates and characterization of metabolites. Appl Environ Microbiol 53: 129–136

Herbes SE, Schwall LR (1978) Microbial transformation of polycyclic aromatic hydrocarbons in pristine and petroleum-contaminated sediments. Appl Environ Microbiol 35: 306–316

Hosler KR, Bulman TL, Fowlie PJA (1988) Der Verbleib von Naphtalin, Anthracen und Benz(a)pyren im Boden bei einem für die Behandlung von Raffinerieabfällen genutztem Gelände. In: Altlastensanierung `88. Wolf K, van den Brink WJ, Colon FJ (eds) Kluwer Academic Publisher, Dordrecht Boston London, pp 111–113

Kan AT, Fu G, Thomson MB (1994) Adsorption/Desorption hysteresis in organic pollutant and soil/sediment interaction. Environ Sci Technol 28: 859–867

Kanaly R, Bartha R, Fogel S, Findlay M (1997) Biodegradation of ^{14}C-Benzo[a]pyrene added in crude oil to uncontaminated soil. Appl Environ Microbiol 63: 4511–4515

Karickhoff SW (1981) Semi-empirical estimation of sorption of hydrophobic pollutants on natural sediments and soils. Chemosphere 10: 833–845

Karickhoff SW, Brown DS, Scott TA (1979) Sorption of hydrophobic pollutants on natural sediments. Water Res 13: 241–248

Kästner M, Mahro B (1996) Microbial degradation of polycyclic aromatic hydrocarbons in soils affected by the organic matrix of compost. Appl Microbiol Biotechnol 44: 668–675

Kästner M, Streibich S, Richnow HH, Michaelis W, Fritsche W (1997) Bildung und Schicksal von gebundenen Rückstände aus Umweltschadstoffen im Boden. In: Biologische Sanierung von Rüstungsaltlasten) Berichte zum 3. Statusseminar des Verbundprojektes "Biologische Sanierung von Rüstungsaltlasten" im Umweltbundesamt. *Bundesminister für Bildung und Forschung, Projektträgerschaft Abfallwirtschaft und Altlastensanierung im Umweltbundesamt* (ed) Berlin, 26.–27.2.1997, pp B1–41

Kästner M, Hofrichter M (2001) Biodegradation of humic substances. In: Biopolymers Vol. 1 – Lignin, humic substances and coal. Steinbüchel A, Hofrichter M (eds.) Wiley-VCH, Weinheim, Germany (in press)

Kästner M, Lotter S, Heerenklage J, Breuer-Jammali M, Stegman R, Mahro B (1995) Fate of ^{14}C-labelled anthracene and hexadecane in compost manured soil. Appl Microbiol Biotechnol 43: 1128–1135

Kästner M, Sack U, Streibich S, Beyrer M, Fritsche W (1996) Metabolisierung, Mineralisierung und Humifizierung von PAK durch Pilze. In: Biologischer Abbau von polyzyklischen aromatischen Kohlenwasserstoffen. Cuno M (ed) Schriftenreihe – Biologische Abwasserreinigung 7, (SFB 193) Technische Universität, Berlin, Germany, pp 41–61

Kästner M, Streibich S, Beyrer M, Richnow HH, Fritsche W (1999) Formation of bound residues during microbial degradation of [^{14}C]-anthracene in soil. Appl Environ Microbiol 65: 1834–1842

Kästner M (2000). The "humification" process or the formation of refractory soil organic matter. In: Biotechnology, 2nd. Edition, Vol 11b; Environmental Processes . Rehm HJ, Reed G, Pühler A, Stadler P (eds) Wiley-VCH, Weinheim, Germany, pp 89–125

Kaufman DD (1976) Bound and conjugated pesticide residues. In: Bound and conjugated pesticide residues. Kaufmann DD, Still GG, Paulson GD, Bandal SK (eds) ACS Symposium Series 29, Am Chem Soc Washington DC, USA, pp 1–10

Kelsey JW, Alexander M (1997) Declining bioavailability and inapropriate estimation of risk of persistent compounds. Environ Tox Chem 16: 582–585

Khan SU (1982) Bound pesticide residues in soil and plants. Residue Reviews 84: 1–24

Khan S, Dupont U (1987) Bound pesticide residues and their bioavailability. In: Pesticide Science and Biotechnology. Greenhalgh R, Roberts T (eds) Int Congr Pesticide Chemistry, IUPAC, Blackwell Scientific Publication, London, pp 417–420

Klein W, Scheunert I (1982) Bound pesticide residues in soil, plants and food with particular emphasis on the application of nuclear techniques. In: Agrochemicals: Fate in food and environment. Proc Intern Symp IAEA Vienna, Austria, pp 177–205

Knicker H, Bruns-Nagel D, Drzyzga O, v Löw E, Steinbach K (1999) Characterization of ^{15}N-TNT Residues After an Anaerobic/Aerobic Treatment of Soil/Molasses Mixtures by Solid-State ^{15}N-NMR Spectroscopy. 1. Dertermination and Optimization of Relevant NMR Spectroscopic Parameters. Environ Sci Technol 33: 343–349

Kolb M, Bock C, Harm H (1996) Bioakkumulation und Persistenz organischer Schadstoffe aus Bioabfallkomposten in Pflanzen. In: Neue Techniken der Kompostierung. Stegmann R (ed) Hamburger Berichte Abfallwirtschaft 11. Economica Verlag, Bonn, Germany, pp 345–360

Kovacs MF (1986) Regulatory aspects of bound residues. Residue Reviews 97: 1–17

Livingston D (1993) Biotechnology and pollution monitoring: use of molecular biomarkers in the aquatic environment. J Chem Technol Biotechnol 57: 195–211

Martin JP, Haider K (1980) A comparison of the use of phenolase and peroxidase for the synthesis of model humic acid-type polymers. Soil Sci Soc Am J 44: 983–988

Menzel RE, Nelson JO (1986) Water and soil pollutants. In: Casarett and Dulls Toxicology: The basic science of poisons. Doull JD, Klaassen DD, Amdur MO (eds) Macimillan Pub Co Inc, New York, USA, pp 825–853

Michaelis W, Richnow HH, Seifert R (1995) Chemically bound chlorinated aromatics in humic substances. Naturwissenschaften 82: 139–142

Mortland MM (1986) Mechanisms of adsorption of nonhumic organic species by clays. In: Interaction of soil minerals with natural organics and microbes. Huang PM, Schnitzer M (eds) SSSA Spec Publ 17: 59–75

Mulder GJ (ed) (1990) Conjugation reaction in drug metabolism – an integrated approach. Substrates, co-substrates, enzymes and their interaction in vivo and in vitro. Taylor and Francis, London, UK

Nanny MA, Bortiatynski JM, Tien M, Hatcher PG (1996) Inverstigation of enzymatic alterations of 2,4-dichlorophenol using ^{13}C-nulear magnetic resonance in combination with site specific ^{13}C-labelling: understanding the fate of this pollutant. Environ Tox Chem 15: 1857–1864

Nordlohne L, Eschenbach A, Wienberg R, Mahro B, Kästner M (1995) Versuche in Kleinreaktoren und Batchversuche mit Zudotierung ^{14}C-markierter Schadstoffe. Teilprojekt 4, Wissenschaftliches Untersuchungsprogramm "Veringstraße" (Verbundvorhaben): Sanierungsbegleitende Untersuchung zur Stoffbilanz und zur Metabolitenbildung bei der Durchführung eines Weißfäule-Mietenverfahrens zur Reinigung des PAK-kontaminierten Bodens von dem Schadensfall "Veringstraße 2". Abschlußberichte an die Umweltbehörde der Hansestadt Hamburg, Februar 1995, Germany

Northcott GL, Jones KC (2000) Experimental approaches and analytica techniques for the determining organic compound bound residues in soil and sediment. Environ Poll 108: 19–43.

Park KS, Sims RC, Doucette WJ, Matthews JE (1988) Biological transformation and detoxification of 7,12-dimethyl benz(a)anthracene in soil systems. J Water Poll Control Fed 60: 1822–1825

Pauli FW (1967) Soil Fertility. Adam Hilger, London, UK

Pignatello JJ (1989) Sorption dynamic of organic compounds in soils and sediments. In: Reactions and movement of organic chemicals in soils. Sawhney BL, Brown K (eds) SSSA Spec Publ 22: 45–80

Richnow HH, Annweiler E, Koning M, Lüth J-C, Stegmann R, Garms C, Francke W, Michaelis W (2000) Tracing the transformation of labelled [1-^{13}C]-phenanthrene in a soil bioreactor. Environ Poll 108: 91–101

Richnow HH, Seifert R, Hefter J, Kästner M, Mahro B, Michaelis W (1994) Metabolites of xenobiotica and mineral oil constituents linked to macromolecular organic matter in polluted environments. Org Geochem 22: 671–681

Richnow HH, Eschenbach A, Hefter J, Kästner M, Mahro B, Seifert R, Michaelis W (1996) Bildungsmechanismen von Bound Residues bei der biologischen Behandlung kontaminierter Böden. In: Biologische und chemische Behandlung von PAK-haltigen Böden und Abwässern. Cuno M (ed) Schriftenreihe Biologische Abwasserreinigung 7, TU-Berlin, Berlin 1996, Germany, pp 81–97

Richnow HH, Eschenbach A, Seifert R, Wehrung P, Albrecht P, Michaelis W (1998) The use of ^{13}C-labelled polycyclic aromatic hydrocarbons for the analysis of their transformation in soils. Chemosphere 36: 2211–2224

Richnow HH, Seifert R, Hefter J, Link M, Francke W, Schäfer G, Michaelis W (1997) Organic pollutants associated with macromolecular soil organic matter – a mode of binding. Org Geochem 26: 745–758

Richnow HH, Seifert R, Kästner M, Mahro B, Horsfield B, Tiedgen U, Böhm S, Michaelis W (1995) Rapid screening of PAH-residues in bioremediated soils. Chemosphere 31: 3991–3999

Richnow HH, Eschenbach A, Mahro B, Kästner M, Annweiler E, Seifert R, Michaelis W (1999) The formation of non-extractable soil bound residues – a stable isotope approach. Environ Sci Technol 33: 3761–3767

Roberts TR (1984) Non-extractable pesticide residues in soil and plants. IUPAC Reports on pesticides (17). Pure Appl Chem 56: 945–956

Ruggiero P (1998) Abiotic transformations of organic xenobiotic in soils: A compounding factor in the assessment of bioavailability. In: Bioavailability of Organic Xenobiotics in the Environment. Block JC, Baveye Ph, Goncharuk VV (eds) NATO ASI Series 64, Kluwer Academic Publishers, The Netherlands, pp 159–205

Sack U, Fritsche W (1997) Enhancement of pyrene mineralization in soil by wood-decaying fungi. FEMS Microbiol Ecol 22: 77–83

Sarkar JM, Bollag JM (1987) Inhibitory effect of humic and fulvic acids on oxidoreductases as measured by the coupling of 2,4-dichlorophenol to humic substances. Sci Tot Environ 62: 367–377

Schnöder F, Mittelstaedt W, Führ F (1994) Das Verhalten von Benzo(a)pyren und Fluoranthen in einer Parabraunerde – Lysimeter und Abbaustudien. In: Biologischer Abbau von polyzyklischen aromatischen Kohlenwasserstoffen. Weigert B (ed) Schriftenreihe Biologische Abwasserreinigung 4, SFB 193, TU-Berlin, Germany, pp 217–230

Shen J, Bartha R (1996) Metabolic efficiency and turnover of soil microbial communities in biodegradation tests. Appl Environ Microbiol 62: 2411–2415

Stevenson F-J (1994) Humus chemistry – genesis, composition reactions. John Wiley and Sons, New York, USA

Stott DE, Kassim G, Jarrell WM, Martin JP, Haider K (1983a) Stabilization and incorporation into Biomass of specific plant carbons during biodegradation in soil. Plant and Soil 70: 15–26

Stott DE, Martin JP, Focht DD, Haider K (1983b) Biodegradation, stabilization in humus, and incorporation into soil biomass of 2,4-D and chlorocatechol carbons. Soil Sci Soc Am J 47: 66–70

Theng BKG (1982) Clay activated organic reactions. Dev Sedimentol 35: 197–238

Thorn KA, Pettigrew PJ, Goldenberg WS (1996) Covalent binding of aniline to humic substances. 2. ^{15}N NMR studies of nucleophilic addition reactions. Environ Sci Technol 30: 2764–2775

US Federal Register (1975) Registration of pesticides in the United States – Proposed guidelines. Fed Regis 40, USA, p 123

Verstraete W, Devliegher W (1996) Formation of non-bioavailable organic residues in soil: Perspectives for site remediation. Biodegradation 7: 471–485

Wang X, Yu X, Bartha R (1990) Effect of bioremediation on polycyclic aromatic hydrocarbon residues in soil. Environ Sci Technol 24: 1086–1089

Weber EJ, Spidle DL, Thorn KA (1996) Covalent binding of aniline to humic substances. 1. Kinetic studies. Environ Sci Technol 30: 2755–2763

White JL (1976) Clay-pesticide interaction. In: Bound and conjugated pesticide residues. Kaufmann DD, Still GG, Paulson GD, Bandal SK (eds) ACS Symosiums Series 29: 208–218

Zielke RC, Pinnavaia TJ, Mortland MM (1989) Adsorption reactions of selected organic molecules on clay mineral surfaces. In: Reactions and movement of organic chemicals in soils. Sawhney BL, Brown K (eds) SSSA Spec Publ 22, pp 81–97

17 The Significance of Bound Residues in the Bioremediation Process of PAH Contaminated Sites

E. Annweiler[1], H.H. Richnow[2] and W. Michaelis[1]
[1]Institute of Biogeochemistry and Marine Chemistry, University of Hamburg, Bundesstr. 55, 20146 Hamburg, Germany
[2]UFZ Centre for Environmental Research Leipzig-Halle, Permoserstr. 15, 04318 Leipzig, Germany

17.1 Introduction

The formation of bound residues has intensively been studied in agrochemistry with the main focus on the fate of pesticides in soil. It is assumed that nearly all xenobiotics entering the soil environment build up these residues, and for many classes this has already been proven (Calderbank 1989).

The term *bound residue* is also defined in Chapter 16 and 18. Bound residue is a technical term, operationally defined by the inability of solvents to extract pesticides and their degradation products from soil. Fragments of the parent xenobiotic molecule which through metabolic pathways lead to natural products are in particular excluded by this definition (Roberts 1984; European Union 1994). Therefore, cell biomass constituents like proteins, lipids and carbohydrates, that have been built up from xenobiotic carbon, are not part of the bound residue fraction.

Bound residue formation of pesticides raises two major problems. First, bound pesticides loose their biological activity and a significant amount of the applied pesticide gets lost for pest control. Second, the drastic accumulation of pesticide related components in soils is of great ecological concern as long as questions concerning toxicity, long-term stability or remobilisation of bound residues remain unanswered.

Depending on the pesticide, the soil matrix, and the experimental conditions, the extent of bound residue formation has been shown to be highly variable. Typically, an amount between 20 and 70 % of the applied pesticide is reported to form bound residues (Calderbank 1989). The significance of pesticide incorporation into the soil matrix gave rise to cover bound residue formation by legal regulations. In the legislation of the European Union (1994) the amount of pesticides transformed to non-extractable residues in a soil must not rise above 70 % (Verstraete and Devliegher 1996).

Today, the fate of priority pollutants like polycyclic aromatic hydrocarbons (PAH) in soils is of major concern in bioremediation studies. PAH form non-extractable residues during bioremediation treatment similar to pesticides in soil. Compared to pesticides which comprise a wide range of chemical compound classes, PAH are extremely hydrophobic, chemically stable and relatively inert substances, and are not likely to react with soil material to form bound residues to a significant extent. Thus, it is not the parent PAH but their reactive transformation products that are more suitable substrates which can react with the soil matrix to sequester PAH related carbon in the soil. PAH transformation is mainly due to microbial degradation and the extent of bound residue formation is usually related to microbial activity (Lerch et al. 1997; Guthrie et al. 1998; Niemann et al. 1999; Richnow et al. 1999, 2000) (see also Chapter 16).

The binding reaction is substantially governed by the physical and chemical properties of the xenobiotic compound as well as of the binding capacity of the soil matrix (e.g. humic substances, metal oxides or clays). Furthermore, catalysts like exoenzymes or bivalent cations (iron, manganese) may assist residue formation.

The binding process strongly changes the chemical and biological characteristics of contaminants. A reduced bioavailability of bound xenobiotics has been shown for microorganisms as well as for higher organisms like plants and earthworms (Dec et al. 1997; Chung and Alexander 1998; Guthrie and Pfaender 1998; Tang et al. 1998) (see also Chapter 18). Hence, on the one hand the extent of microbial degradation of bound contaminants is much reduced compared to the extractable contaminant leading to an accumulation of xenobiotic carbon in soil. On the other hand, humus bound chemicals are apparently detoxified because their bioavalability for organisms is drastically reduced compared to extractable chemicals. Obviously, the mobility of bound pollutants decreases significantly because of their immobilisation within the soil, leading to a lower risk of ground water pollution by contaminant leachates from overlying soils (see also Chapter 16).

In this context several authors suggested enhanced bound residue formation as a cost-effective *in situ* remediation technique that may detoxify and immobilise contaminants in the soil (Berry and Boyd 1985; Bollag 1992; Verstraete and Devliegher 1996). The immobilisation approach is in accordance with the modern concept of *Natural Attenuation* that favours natural processes to enhance the restraining capacity of soils and aquifers instead of active remediation techniques (Held 1996; Wienberg 1997). Combined with an intensive monitoring of the contaminated site *Natural Attenuation* represents a passive remediation strategy that may be applied if an *in situ* potential for the mass reduction of contaminants is proved (immobilisation, biodegradation) and an export of contaminants from the site can be excluded.

However, the long-term risk of bound residues is difficult to assess as long as formation processes, chemical structures and the stability of bound chemicals are still not completely understood.

Since bound residues are non-extractable when applying ordinary extraction procedures, they are not suited to most analytical methods and are difficult to characterise and quantify. One way to trace the fate of xenobiotic carbon, even after extensive transformation, is the application of isotopically labelled model substances. Radioactive labelled model compounds are usually applied in closed microcosms to quantify the extent of mineralisation and bound residue formation as bulk fractions. However, the ^{14}C-technique is not appropriate for the structural characterisation of bound residues.

To overcome these limitations, stable isotope labelled ^{13}C- and ^{15}N-substrates have been successfully applied. In combination with NMR spectroscopy, the transfer of the ^{13}C- or ^{15}N-label into specific structural units of the macromolecular matrix was elucidated (Hatcher et al. 1993; Dec at al. 1997; Achtnich et al. 1999; Guthrie et al. 1999). Gas chromatography/mass spectrometry (GC/MS) after chemical degradation reactions associated with isotope analyses allow us to investigate the ^{13}C-transfer into the bulk fractions in combination with detailed structural determinations of the transformation products (Richnow et al. 1998, 1999). We have applied this technique in remediation studies of PAH contaminated soils.

17.2
^{13}C-Tracer Application

As already mentioned in Chapter 16 the use of stable isotope labels provides several advantages. Compared to ^{14}C-tracers no specific safety regulations regarding nuclear radiation are required. Moreover, an application in field studies is possible. Regarding the carbon balance of a system, radioactivity measurements are easily performed with high sensitivity and precision. The ^{13}C-approach does not provide the same sensitivity, but still meets the requirements to quantify the carbon flux of xenobiotics in soils (Richnow et al. 1999).

The carbon balance can be calculated from

- the extractable parent ^{13}C-PAH,
- the ^{13}C-concentration in the produced CO_2, and
- the ^{13}C-concentration in the bound residue fraction.

The high natural abundance of ^{13}C (1.11 % ^{13}C, 98.89 % ^{12}C) requires its careful consideration in quantifications. Therefore, parallel experiments with non-labelled PAH are necessary to monitor the concentration of natural ^{13}C during biodegradation. However, during the course of the experiments with non labelled PAH no isotopic shift ($\pm$ 0.2 ‰) within the CO_2- and the bound residue fraction was observed indicating that the isotopic composition of soil fractions did not change significantly during the experiments and thus did not interfere the calculation of carbon balances. The experimental details and calculations are given in Richnow et al. (1999).

In addition to a complete carbon balance, the ^{13}C approach allows precise structural investigations of the transformation products on a molecular level. Thus, it enables a detailed characterisation of the fate of a contaminant combining multiple objectives:

- establishing a complete ^{13}C-carbon balance (parent PAH, metabolites, CO_2, non-extractable residues),
- studying ageing effects – differentiation of chemically identical structures by the isotopic label,
- determining the molecular structure of bound residues.

17.2.1 Carbon Balance

A closed bioreactor system was constructed to establish a complete ^{13}C-balance (fig. 17.1). The soil material was spiked with a ^{13}C-labelled PAH and introduced into the bioreactor. Oxygen supply was performed continuously depending on the gas pressure controlled by the consumption of oxygen in the reactor. Briefly, pure oxygen was added through a glass vessel containing a NaCl solution (10 %) acidified to pH 2 to prevent diffusional loss of CO_2 out of the bioreactor. A trap filled with 2 M NaOH was connected to the reactor to dissolve microbially produced

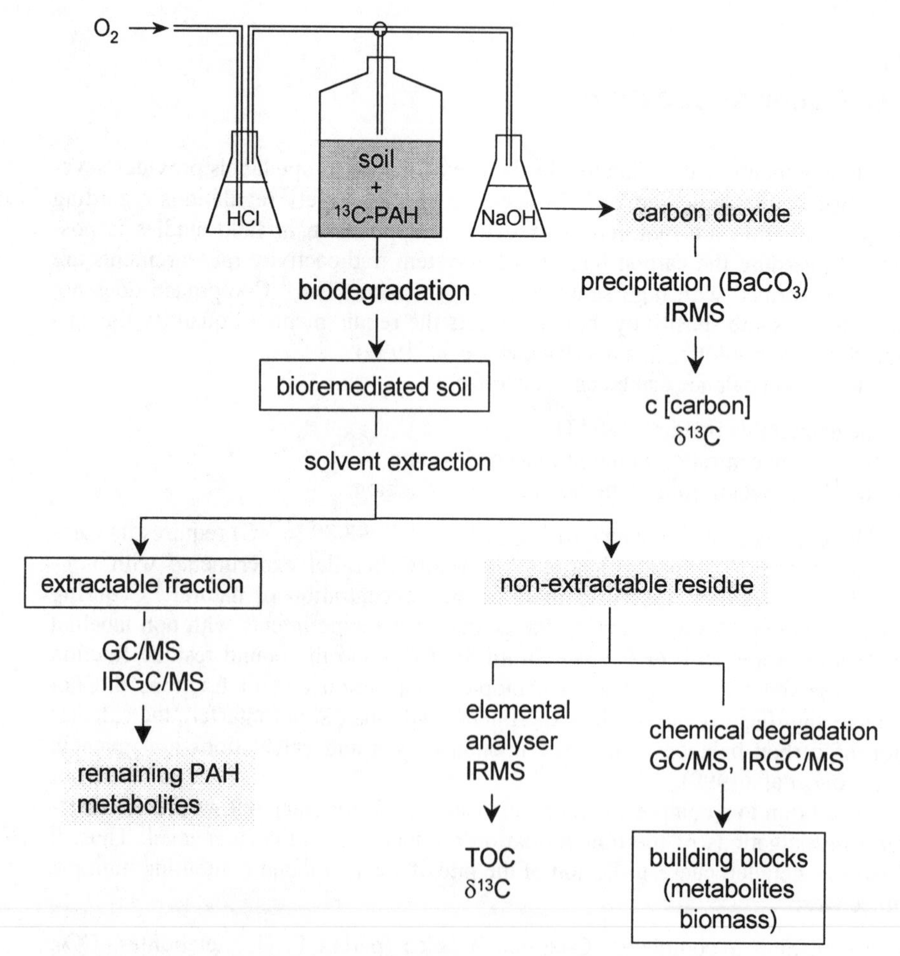

Fig. 17.1. Flow chart of bioreactor experiments
IRMS Isotope Ratio Mass Spectrometry, *IRGC/MS*, Isotope Ratio Monitoring GC/MS, *c* concentration, $\delta^{13}C$ carbon isotope ratio, *TOC* Total Organic Carbon

O_2. The continous trapping of CO_2 maintained a constant flux of oxygen into the reactor. Soil samples were taken periodically, the gas trap was changed within the same time intervals to trace the transformation of ^{13}C-PAH during the time course of the experiment.

With this experimental set-up we investigated the biodegradation of PAH in several soil matrices. Two experiments are discussed below. (I) Degradation of [9-^{13}C]-anthracene in a pristine soil (Ah-horizon, pseudovergleyte Parabraunerde, German systematic, collected near Hamburg) bioaugmented with a water extract of a tar oil contaminated soil to stimulate microbial degradation activity (for a detailed description of the soil treatment before the experiment see Richnow et al. 1999). The experiment was run for 291 days. (II) Degradation of [1-^{13}C]-phenanthrene in a soil that had been heavily contaminated with mineral-oil derived aliphatic hydrocarbons and PAH more than 50 years ago (for a description of the soil treatment before the experiment see Richnow et al. 2000). The duration of the experiment was 467 days.

The carbon balance is deduced from the distribution of the ^{13}C-label between the fractions CO_2, extractable parent PAH/metabolites and bound residues at discrete time intervals during the time course of the experiment. The ^{13}C-speciation among these fractions depicts the degradation capacity of the soil microflora (fig. 17.2).

Extractable PAH – In both studies concentrations of extractable PAH decreased to less than 1 % of the initial amount within the first hundred days of the experiment to reach final concentrations below 1 $\mu g\ g^{-1}$ soil.

CO_2 – The biodegradation experiment with [9-^{13}C]-anthracene in pristine soil inoculated with a water extract of tar oil contaminated material revealed a lag phase of about 25 days. Thereafter, significant amounts of $^{13}CO_2$ were released indicating that the microflora was able to mineralise anthracene. In the study with [1-^{13}C]-phenanthrene in a mineral oil contaminated soil an adaption of microorganisms to PAH degradation could be assumed because of the exposition to these pollutants over decades in the contaminated site. In accordance with this presumption an intense $^{13}CO_2$-formation from ^{13}C-PAH mineralisation was observed within the first 5 days of the experiment. In both studies mineralisation was the major elimination pathway for ^{13}C-PAH in soil. More than 80 % of the applied amount were completely degraded to CO_2 indicating an effective bioremediation process in the bioreactor experiments.

Bound residues – In the two experiments shown bound residue formation was closely related to the biodegradation process; during the phase of high microbial degradation activity a continuous increase in the bound residue concentration was observed. A rough correlation of the course of bound residue formation with the CO_2 evolution was recorded. In the degradation experiment with [9-^{13}C]-anthracene the highest concentration of the bound residue fraction was observed at day 92 (18 % of the applied ^{13}C-anthracene). In the following period only a slight decrease to 14 % of the applied ^{13}C-anthracene occurred. Similar results gave the phenanthrene experiment with the highest concentration of bound

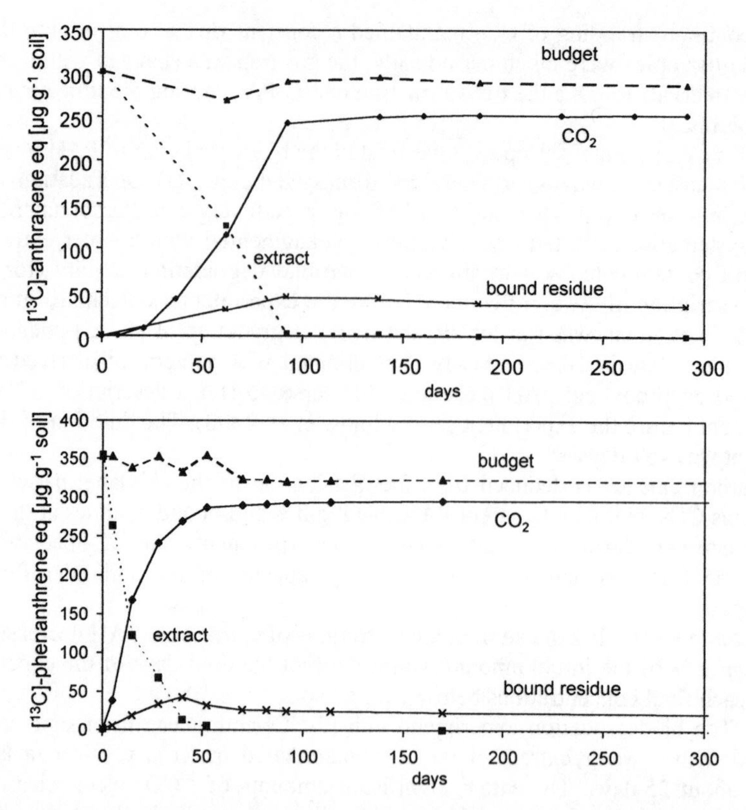

Fig. 17.2. Fate of PAH during the biodegradation experiments. Top, [9-^{13}C]-anthracene in a pristine soil, inoculated with a water extract of a tar oil contaminated soil. Bottom, [1-^{13}C]-phenanthrene in a soil heavily contaminated with mineral-oil derived hydrocarbons more than 50 years ago; for comparison the carbon balance is shown for the first 300 days of the 467 day experiment; eq, equivalents; modified after Richnow et al. 1999, 2000).

residues at day 40 (13 % of the applied ^{13}C-phenanthrene) of which a minor portion was degraded until day 70. Between day 70 and 170 nearly stable amounts of about 8 % were observed suggesting a high stability of bound residues for longer periods.

Previous experiments have shown that after an effective microbial degradation capacity had been established, additional spiking with PAH ([9-^{13}C]-anthracene) did not result in an increase in the bound residue fraction. The additional ^{13}C-anthracene portions were completely mineralised to CO_2, indicating that bound residue formation predominantly occurs during the adaptation phase of the microbial community to PAH degradation (Richnow et al. 1998).

Although different soil materials were used, the degradation efficiency and bound residue formation were comparable in these studies. The development of the bound residue fractions clearly indicated a high stability of this fraction. ^{13}C carbon incorporated into the macromolecular fraction seemed to be barely bioavailable. Thus, toxicity of soil bound residues should strongly decrease. These results present evidence for the suggestion that bound residues show a reduced degradation rate similar to the natural humus turnover (Eschenbach et al. 1998; Niemann et al. 1999).

17.2.2 Impact of Ageing on Bioavailability

As already mentioned in Chapter 16 previous ageing experiments revealed an increasing resistance of chemicals in soils to biodegradation and extraction with time. This was shown by decreasing rates of mineralisation, limited extractability and increasing amounts of PAH resistant to bioremediation (Hatzinger and Alexander 1995; Kelsey and Alexander 1997; Tang et al. 1998).

The application of [1-^{13}C]-phenanthrene in a biodegradation experiment with a tar oil contaminated soil enabled us to compare the biodegradation rate of the freshly added ^{13}C-PAH with the degradation rate of the unlabelled ^{12}C-analogue that had resisted for more than 50 years in the soil matrix. This approach can be used to evaluate bioavailability of aged contaminants in the soil. Prior to incubation in the bioreactor the soil was sieved, pH was adjusted to 6.3 and the water content was set to 50 % of the maximal water-holding capacity. To stimulate microbial activity the soil was supplemented with compost (Richnow et al. 2000). The contaminated soil contained 89 $\mu g\ g^{-1}$ tar oil derived ^{12}C-phenanthrene and was spiked with 354 $\mu g\ g^{-1}$ [1-^{13}C]-phenanthrene.

During the biodegradation experiment the fate of the two phenanthrene species was monitored (fig. 17.3). After 82 days the concentrations of both phenanthrene species had strongly decreased to 0.2 $\mu g\ g^{-1}$ soil (^{12}C) and 0.1 $\mu g\ g^{-1}$ soil (^{13}C), respectively. Both phenanthrene species revealed nearly identical degradation rates, an ageing effect causing reduced bioavailability of the ^{12}C-phenanthrene species was not observed until a residual concentration of less than 1 $\mu g\ g^{-1}$ phenanthrene.

These results contrast previous ageing experiments that reported decreasing rates of naphthalene and phenanthrene biodegradation with increasing duration of ageing (Erickson et al. 1993; Hatzinger and Alexander 1995; Carmicheal et al. 1997; Kelsey and Alexander 1997). Whereas the lack of an ageing effect on the soil-air partitioning of polychlorinated biphenyls was comparable to our observations (Cousins et al. 1998). Bioavailability studies revealed that the concentration of contaminants may govern the degradation rate. Efroymson and Alexander (1995) observed a rapid mineralisation at high phenanthrene concentrations (10 $\mu g\ l^{-1}$ water phase) and slow mineralisation rates at low phenanthrene concentrations (6 $ng\ l^{-1}$ water phase). In the ^{13}C-phenanthrene degradation experiment described here, relatively high contaminant concentrations were applied, and until a residual concentration of $< 1 \mu g\ g^{-1}$ ageing effects were not relevant for the biodegradation of phenanthrene. The results indicate that an ageing effect at high pollutant concentrations, that are typical for tar oil contaminations, can be less pronounced than previously assumed. In our approach an effective degradation of the aged ^{12}C-phenanthrene as well as of the freshly spiked ^{13}C-phenanthrene was achieved by soil treatments such as the adjustment of pH and water content, continuous oxygen supply or amendment of compost.

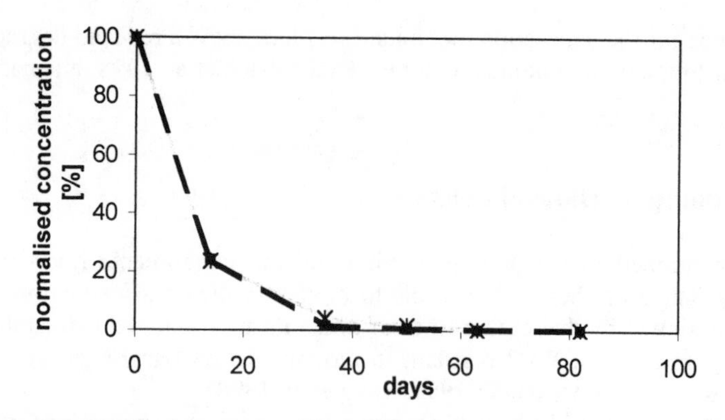

Fig. 17.3. Normalised *concentrations* of the two phenanthrene species during the biodegradation experiment
Grey solid line aged unlabelled ^{12}C-phenanthrene, *dashed black line* freshly added ^{13}C-phenanthrene

However, inconsistent results in variable studies concerning the bioavailability of aged contaminants point out the complexity to predict the fate of xenobiotics in the environment. More investigations with real tar oil contaminated soils are necessary to elucidate the degradation behaviour of aged PAH in soil.

17.3 Molecular Structure of Bound Residues

17.3.1 Formation Process

Two basic reaction mechanisms can be distinguished concerning the binding process: sorption to the soil matrix and chemical reactions leading to the formation of covalent bonds. Sorption generally is considered as a reversible process, – though adsorption-desorption hysteresis of xenobiotic compounds during interactions of organic pollutants in soil was observed. In some cases, 30–50 % of the adsorbed substances could not be desorbed within several weeks (Kan and Tomson 1994) (see also Chapter 16). Partitioning into humic matter or sequestration of the parent chemical and its metabolites in micro- and nanopores is suggested to significantly limit extractability and bioavailability (Hatzinger and Alexander 1995 and references cited therein). For hydrophobic contaminants sorption is supposed to be a partitioning process between the aquatic and solid phase, resulting in an equilibrium that is mainly controlled by the content of organic carbon in soil. Usually, increasing adsorption is observed with increasing organic carbon content. Therefore, organic carbon is taken into account when the sorption coefficient (K_{oc}) is determined (Korte 1992).

As already described in Chapter 16 the second binding type is represented by stable covalent linkages like ester, ether or carbon-carbon bonds. In particular phenolic structures and aromatic amines are known to undergo oxidative coupling to ether and carbon-carbon bonds. This reaction is catalysed either non biologically by clay minerals and metal oxides (Shindo and Huang 1982; Wang at al. 1978) or microbially by phenoloxidases (Bollag 1983; Berry and Boyd 1985; Bollag and Bollag 1990; Hatcher et al. 1993).

Phenolic metabolites are formed during the biodegradation of PAH by fungi, a few bacteria and some cyanobacteria that produce P450-dependent monooxygenases catalysing the oxidation of PAH to arene oxides. The arene oxides can rearrange nonenzymatically to monohydroxy PAH derivatives (Sutherland et al. 1995). Dihydroxy derivatives of PAH (catechol structures) result from dioxygenase catalysed reactions (Cerniglia and Heitkamp 1989). The phenolic metabolites may react with humus constituents or can cross link among themselves.

By enzymatic cross coupling experiments with horseradish peroxidase and H_2O_2 we could demonstrate the oxidative cross-linking capacity of phenolic PAH metabolites with soil humic substances (Richnow et al. 1997). A buffered solution of soil humic substances was spiked with typical PAH metabolites (1-naphthol, 2,3-dihydroxynaphthalene, 1-pyrenol, 9-phenanthrol) and naphthalene. Horseradish peroxidase was repeatedly added and the mixture was activated with H_2O_2. A change of the original light brown colour to dark brown was observed during activation with H_2O_2. After each addition an aliquot of the reaction mixture was extracted and analysed.

A strong concentration decrease of the extractable phenolic substances indicated a significant binding to humic substances, whereas the concentration of naphthalene remained stable (fig. 17.4). No cross-coupling was observed without adding the enzyme. Experiments with 1-naphthol and horseradish peroxidase without humic substances revealed the formation of naphthol-dimers, -oligomers

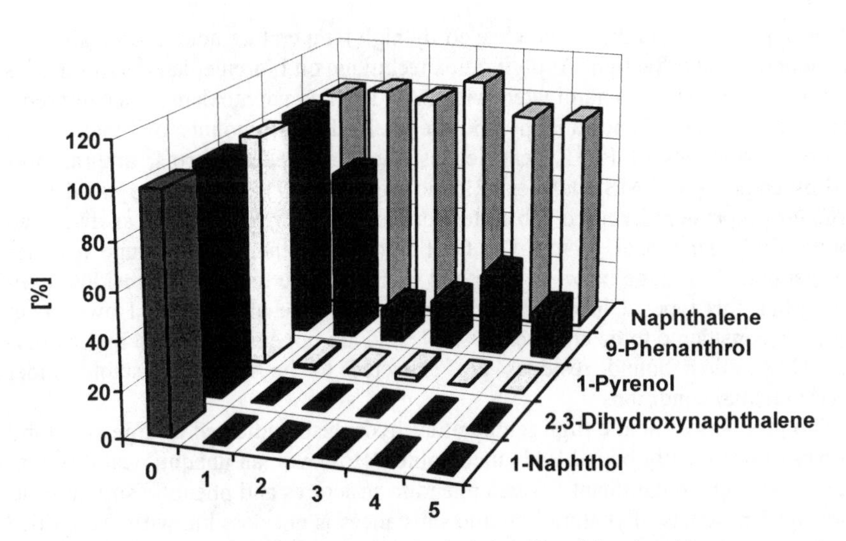

Fig. 17.4. Enzymatic cross coupling of aromatic alcohols with soil humic acids
0 initial analysis before enzyme addition, *1–5* analyses after the repeated addition of the activated enzyme (Richnow et al. 1997)

and -polymers, whereas these polymerisation products lacked if humic substances were present. This clearly indicated that humic substances served as the preferred substrate compared to the naphthol molecules themselves (Richnow et al. 1997).

17.3.2 Structural Characterisation of Building Blocks and their Mode of Binding

The type of binding between xenobiotic and humic material is an important factor governing the chemical characteristics and the long-term stability of bound residues. However, bound residues are not amenable to most structural analytical methods because of their macromolecular structure. For structural investigations NMR-spectroscopy is applied on the bulk macromolecular material (Hatcher et al. 1993; Dec at al. 1997; Achtnich et al. 1999; Guthrie et al. 1999). Detailed information on the molecular level can be obtained by the identification of the macromolecular building blocks after chemical degradation reactions followed by GC/MS-analyses (Hayes et al. 1989; Rullkötter and Michaelis 1989). For characterisation of the bound residue fraction, analytical pyrolysis and selective chemical degradation techniques were applied (Richnow et al. 1994, 1995, 1997, 1998, 2000). Analytical pyrolysis cleaves a variety of covalent bonds non specifically at a reaction temperature usually between 600 and 700 °C (for a review see Saiz-Jimenez 1994). Whereas, chemical degradation reactions allow the cleavage of selective chemical linkages under controlled conditions as discussed below.

17.3.3 Cleavage of Ether Bonds by Acidic High Temperature Hydrolysis

Ether linked compounds can be cleaved by high temperature acid hydrolysis from the bound residue fraction. Applying this technique on bioremediated soil samples that were artificially contaminated with PAH such as naphthalene, phenanthrene, anthracene, fluoranthene, and pyrene we yielded low amounts of ether bound hydroxyderivatives of PAH identified as naphthols, phenanthrols, anthracenols and pyrenols by GC/MS analyses (Richnow et al. 1997). The hydroxylated PAH structures represent favoured substrates for enzymatic cross coupling as discussed above (Sutherland et al. 1995). The total amount of ether linked bound residues was estimated to range between 0.05 and 0.25 % of the total bound residue, indicating that this type of linkage is of small quantitative importance. However, ethers are considered to be chemically stable and are not easily cleaved by hydrolysis. Thus, ether bound residues are expected to be relatively stable under environmental conditions.

The products of acidic high temperature hydrolysis indicated a large structural variety of chemically bound PAH metabolites. However, an unequivocal differentiation between contaminant derived phenolic structures and phenolic structures as hydrolysis products of "natural" humic substances is not possible with the applied method.

To overcome these limitations ^{13}C-labelled PAH were applied in biodegradation experiments. By means of the ^{13}C-label chemical cleavage products of the bound residue fraction can be unequivocally assigned to the ^{13}C-PAH precursor.

Apart from bound residues the same holds true for extractable transformation products.

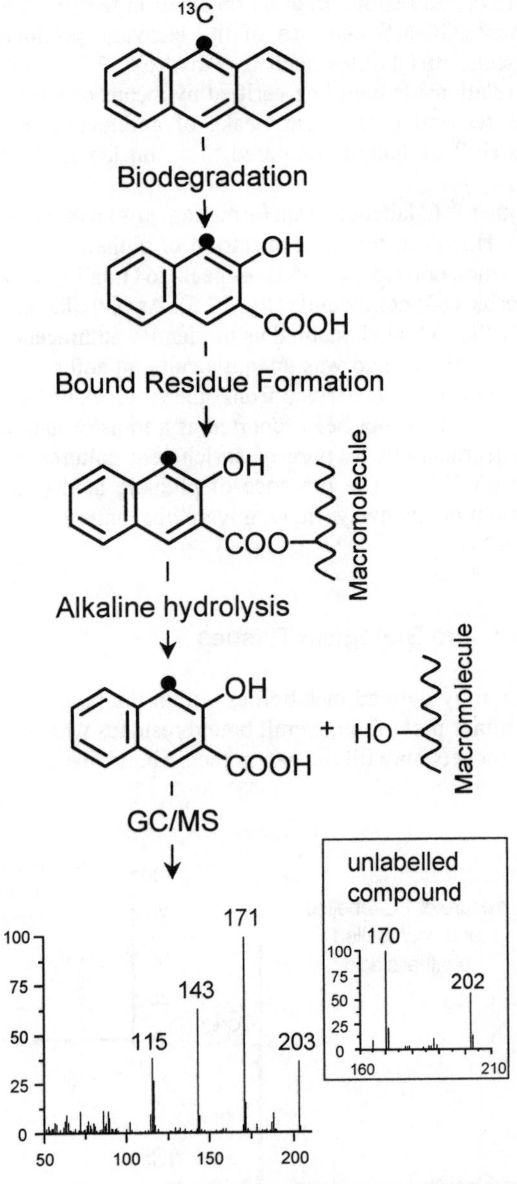

Fig. 17.5. Bound residue formation of ^{13}C-labelled anthracene and chemical degradation traced by GC/MS analysis

17.3.4
Cleavage of Ester Bonds by Alkaline Hydrolysis

Alkaline hydrolysis was applied on the extracted soil material to cleave ester bonds within the bound residue fraction obtained in the [9-^{13}C]-anthracene degradation experiment. GC/MS analyses of the cleavage products revealed typical bacterial anthracene metabolites such as 3-hydroxy-2-napththoic acid. The product-precursor relationship could be verified by means of the ^{13}C-label. The molecular peak and respective fragment peaks of anthracene degradation products showed a mass shift of 1 amu compared to a non labelled reference compound (fig. 17.5).

Moreover, other ^{13}C-labelled transformation products, such as phthalic acid, were idendified. However, the mass spectrum of phthalic acid revealed a mixture of ^{13}C-labelled (molecular peak 195, basepeak 164) and non labelled (molecular peak 194, basepeak 163) compounds (fig. 17.6). As phthalic acid is a common soil component, only the ^{13}C-label enabled us to identify anthracene as its precursor. A minor part of the phthalic acid was unequivocally an anthracene degradation product while the major part was derived from other sources in the soil.

So far phthalic acid has not been reported as a transformation product in bacterial anthracene degradations by pure or enrichment cultures (Cerniglia and Heitkamp 1989; Smith 1990). The presence of phthalic acid might indicate that unexplored degradation pathways may govern the anthracene biodegradation in soils.

17.3.5
Transformation into Biological Tissues

The so far structurally defined metabolites within the bound residue fraction represented only a minor part of the overall bound residues which raising the question – what type of products may fill the gap in the carbon balance? – Microorganisms

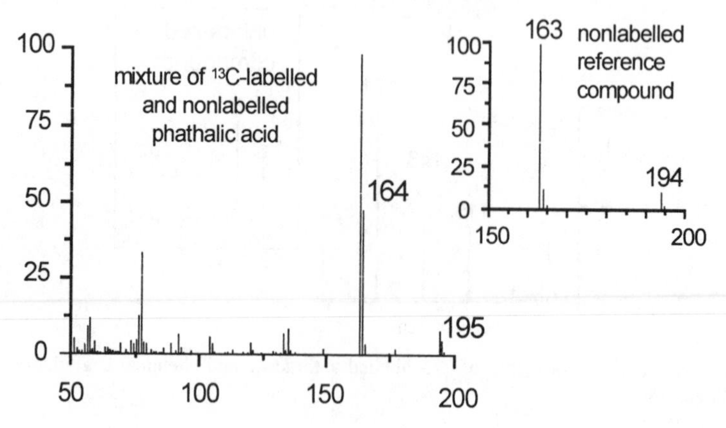

Fig. 17.6. Mass spectrum of phthalic acid from alkaline hydrolysis of a [9-^{13}C]anthracene degradation experiment in comparison with the non labelled reference compound

may use xenobiotic organic substances as a source of carbon for the synthesis of cell biomass transforming xenobiotic to natural products. Although these products may be incorporated into the non-extractable soil fraction after cell death, according to the definition these components that are recycled through metabolic pathways leading to natural products like cell biomass are excluded from the bound residue fraction (Roberts 1984; European Union 1994). This indicates that a precise structural characterisation exceeding bulk analyses is essential for the purpose of the technical definition of non-extractable residues.

Proteinaceous amino acids account for up to 80 % of the cell biomass (Gottschalk 1986). Sørensen (1987) reported the degradation of ^{14}C-labelled barley straw in a sandy soil in a long duration experiment of 20 years. The portion of radioactive carbon in soil amino acids remained almost constant at about 21 % during this period, illustrating the high persistence of transformed organic carbon in natural soil systems.

We applied the ^{13}C-phenanthrene degradation experiment to investigate the transfer of ^{13}C into soil amino acids. After 50 days of incubation a soil sample was hydrolysed and analysed for the amino acid concentration and composition. The isotopic composition of hydrolysable amino acids was investigated by IRGC/MS (Isotope Ratio Monitoring GC/MS).

The isotopic signature of the hydrolysable amino acids resided between –3.1 and –10.4 ‰ (PDB) (table 17.1). A significant enrichment in ^{13}C was found compared to an experiment with non-labelled phenanthrene, wherein the amino acids revealed an isotopic composition between –14 and –26 ‰. The ^{13}C-enrichment clearly demonstrated the transformation of labelled carbon into the soil biomass via anabolic processes.

Quantitatively, the ^{13}C-enrichment in the hydrolysable amino acid fraction corresponded to a significant amount of phenenathrene equivalents. The phenanthrene derived carbon transformed into amino acids accounted for 3.1 $\mu g\ g^{-1}$ phenanthrene equivalents corresponding to 11 % of the total bound residues. The absolute carbon transfer into biomass might be still higher because further cell constituents like lipids, carbohydrates and other biological tissue were not investigated.

Dead biomass may serve as a substrate for other microorganisms, and thus may be transformed several times in soil microbial cycles, unless the xenobiotic carbon may become incorporated into the soil organic matter during humification processes forming a refractory organic fraction in the soil.

The results clearly demonstrate that significant amounts of xenobiotic carbon are transformed by anabolic processes into non hazardous biological components. This process corresponds to a complete detoxification and represents an effective remediation process. Thus, the determination of bound residues without structural assignments is not sufficient for ecotoxicological risk assessments. The classical ^{14}C-approach for soil carbon balances tends to overestimate bound residues because fragments recycled through anabolic pathways to natural products as well as incorporated $^{14}CO_2$ are not excluded from the bulk ^{14}C carbon (Kästner et al. 1999).

Table 17.1. Isotopic signature of the hydrolysable amino acids in a background experiment with non labelled phenanthrene and in the degradation experiment with [1-^{13}C]-phenanthrene (Richnow et al. 2000)

Amino acid	Background (non labelled pheananthrene) δ^{13}C [‰ PDB]*	^{13}C-experiment (day 50) c (amino acid) [µmol g^{-1} soil]	^{13}C-experiment (day 50) δ13C [‰ PDB]*	^{13}C-experiment (day 50) phenanthrene (eq) [ng g^{-1} soil]
Alanine	–17.1	2.38	–6.3	178.5
Glycine	–18.5	3.22	–6.0	186.2
γ-Aminobutyric acid	–17.0	10.00	–6.7	957.9
Valine	–20.6	1.26	–8.0	184.9
Leucine	–24.4	1.65	–9.2	350.4
Isoleucine	–23.9	0.86	–10.4	160.8
Proline	–17.2	1.44	–3.3	231.9
Aspartic acid	–14.1	2.84	–9.3	125.4
Glutamic acid	–14.0	2.50	–3.1	315.2
Phenylalanine	–26.5	0.71	–7.2	286.6
Lysine	–19.9	0.35	–3.3	81.0

*corrected for derivatisation

17.3.6 Conclusion

The application of ^{13}C-labelled xenobiotics revealed to be a valuable tool to characterise the fate of organic contaminants in soil. ^{13}C-PAHs were shown to be suitable for establishing a carbon balance in bioreactors and thus, might substitute ^{14}C-tracers in many approaches. Additionally, stable isotope labels provide several advantages when compared to radioactive tracers. They may be applied without additional safety regulations that are required when working with radioactive compounds. Even field applications like the scale up of laboratory remediation experiments to field studies would be possible. In contrast to radioactive tracers which are usually mixed with the unlabelled analogue to decrease radioactivity, stable isotope labelled compounds can be used as pure substances allowing the differentiation from chemically identical non-labelled structures of different sources. The effect of ageing on bioavailability could be evaluated by spiking labelled PAH to a soil that had previously been contaminated with the unlabelled analogue. The product-precursor relationship is easy to achieve with ^{13}C-labelled compounds for analysis of metabolites or other transformation products.

Bound residue formation seemed to depend strongly on biological activity and was observed during the early phase of PAH biodegradation. Whereas in long-term experiments, the bound residue fraction, once formed, revealed a high stability over time, indicating an incorporation into the natural humification process.

Structural characterisation on a molecular level provided important new information on building mechanisms and stability of bound residues. In particular biologically activated metabolites could be determined as bound residue building blocks. The structural investigations showed that the extent of bound residues is usually overestimated when using the ^{14}C approach without a detailed structural survey. A significant contribution of amino acids to the non-extractable residues has been shown (Richnow et al. 2000).

However, with regard to the fate of pollutants in soil further investigations on structure, long-term stability, and remobilisation potential of non-extractable residues as well as worst case studies are necessary for a final evaluation of an enhanced bound residue formation as a remediation technique (*natural attenuation*).

References

Achtnich C, Fernandes E, Bollag J-M, Knackmuss H-J, Lenke H (1999) Covalent binding of [$^{15}N_3$]-TNT to soil organic matter during a bioremediation process analyzed by ^{15}N NMR spectroscopy. Environ Sci Technol 33: 4448–4456

Bailey GW, White JL (1964) Review of adsorption and desorption of organic pesticides by soil colloids with implication concerning pesticide bioactivity. J Agric Food Chem 12: 324–332

Berry DF, Boyd S (1985) Decontamination of soil through enhanced formation of bound residues. Environ Sci Technol 19: 1132–1133

Bollag JM, Loll MJ (1983) Incorporation of xenobiotics into soil humus. Experientia 39: 1221–1231

Bollag J-M, Bollag WB (1990) A model for enzymatic binding of pollutants in the soil. Intern J Environ Anal Chem 39: 147–157

Bollag J-M (1992) Decontaminating soil with enzymes. Environ Sci Technol 26: 1876–1881

Calderbank A (1989) The occurrence and significance of bound pesticide residues in soil. Rev Environ Contam Toxicol 108: 71–103

Carmichael LM, Christman RF, Pfaender FK (1997) Desorption and mineralization kinetics of phenanthrene and chrysene in contaminated soils. Environ Sci Technol 31: 126–132

Cerniglia CE, Heitkamp MA (1989) Microbial degradation of polycyclic aromatic hydrocarbons (PAH) in the aquatic environment. In: Varanasi M (ed) Metabolism of polycyclic aromatic hydrocarbons in the aquatic environment. CRC Press Inc Boca Raton, Florida, USA, pp 48–68

Chung N, Alexander M (1998) Differences in sequestration and bioavailability of organic compounds aged in dissimilar soils. Environ Sci Technol 32: 855–860

Cousins IT, McLachlan MS, Jones KC (1998) Lack of an aging effect on the soil-air partitioning of polychlorinated biphenyls. Environ Sci Technol 32: 2734–2740

Dec J, Haider K, Benesi A, Rangaswamy V, Schäffer A, Plücken U, Bollag J-M (1997) Analysis of Soil-Bound Residues of ^{13}C-labeled Fungicide Cyprodinil by NMR Spectroscopy. Environ Sci Technol 31: 1128–1135

Efroymson RA, Alexander M (1995) Reduced mineralization of low concentrations of phenanthrene because of sequestering in nonaqueous-phase liquids. Environ Sci Technol 29: 515–521

Erickson DC, Loehr RC, Neuhauser EF (1993) PAH loss during bioremediation of manufactured gas plant site soils. Wat Res 27: 911–919

Eschenbach A, Wienberg R, Mahro B (1998) Fate and stability of nonextractable residues of [^{14}C]-PAH in contaminated soils under environmental stress conditions. Environ Sci Technol 32: 2585–2590

European Union (1994) Council Directive 14/43/EC of 27 July 1994 established Annex VI to Directive 91/454/EEC concerning the placing of plant protection products on the market. Official Jounal of the European Communities 37: L 227/31–L 227/55

Gottschalk G (1986) Bacterial metabolism. Springer Verlag, New York, USA, p 359

Guthrie EA, Pfaender FK (1998) Reduced pyrene bioavailability in microbially active soils. Environ Sci Technol 32: 501–508

Guthrie EA, Bortriatynski JM, van Heemst JDH, Richman JE, Hardy KS, Kovach EM, Hatcher PG (1999) Determination of [^{13}C]-pyrene sequestration in sediment microcosms using flash pyrolysis-GC-MS and ^{13}C NMR. Environ Sci Technol 33: 119–125

Hatcher PG, Bortiatynski JM, Minard RD, Minard RD, Dec J, Bollag J-M (1993) Use of high-resolution ^{13}C NMR to examine the enzymatic covalent binding of ^{13}C-labeled 2,4-dichlorophenol to humic substances. Environ Sci Technol 27: 2098–2103

Hatzinger PB, Alexander M (1995) Effect of aging of chemicals in soil on their biodegradability and extractability. Environ Sci Technol 29: 537–545

Hayes MHB, MacCarthy P, Malcolm RL, Swift RS (1989) Humic substances II – In search of structur. Wiley & Sons, Chichester, UK

Held T (1996) No-action-Variante – Selbstreinigungskraft der Natur nutzen. Altlasten Spektrum 4/96: 176–179

Kästner M, Streibich S, Beyrer M, Richnow HH, Fritsche W (1999) Formation of bound residues during microbial degradation of [^{14}C]anthracene in soil. Appl Environ Microbiol 65: 1834–1842

Kan FG, Tomson MB (1994) Adsorption/desortion hysteresis in organic pollutant and soil/sediment interaction. Environ Sci Technol 28: 859–867

Kelsey JW, Alexander M (1997) Declining bioavailability and inappropriate estimation of risk of persistent compounds. Environ Toxicol Chem 16: 582–585

Korte F (1992) Lehrbuch der Ökologischen Chemie. p 38

Lerch RN, Thurman EM, Kruger EL (1997) Mixed-mode sorption of hydroxylated atrazine degradation products to soil: A mechanism for bound residue. Environ Sci Technol 31: 1539–1546

Niemann JKC, Sims RC, Sims JL, Sorensen DL, Mclean JE, Rice JA (1999) [^{14}C]-Pyrene bound residue evaluation using MIBK fractionation method for creosote-contaminated soil. Environ Sci Technol 33: 776–781

Richnow HH, Seifert R, Hefter J, Kästner M, Mahro B, Michaelis W (1994) Metabolites of xenobiotica and mineral oil constituents linked to macromolecular organic matter in polluted environments. Org Geochem 22: 671–681

Richnow HH, Seifert R, Käster M, Mahro B, Horsfield B, Tiedgen U, Böhm S, Michaelis W (1995) Rapid screening of PAH-residues in bioremediated soils. Chemosphere 31: 3991–3999

Richnow HH, Seifert R, Hefter J, Link M, Francke W, Schäfer G, Michaelis W (1997) Organic pollutants associated with macromolecular soil organic matter – a mode of binding. Org Geochem 26: 745–758

Richnow HH, Eschenbach A, Mahro B, Seifert R, Wehrung P, Albrecht P, Michaelis W (1998) The use of ^{13}C-labelled aromatic hydrocarbons for the analysis of their transformation in soil. Chemosphere 36: 2211–2224

Richnow HH, Eschenbach A, Mahro B, Kästner M, Annweiler E, Seifert R, Michaelis W (1999) Formation of nonextractable soil residues: a stable isotope approach. Environ Sci Technol 33: 3761–3767

Richnow HH, Annweiler E, Koning M, Lüth J-C, Stegmann R, Garms C, Francke W, Michaelis W (2000) Tracing the transformation of stable isotope labelled [1-^{13}C]-phenanthrene in a soil bioreactor. Environ Pollution 108: 91–101

Rullkötter J, Michaelis W (1989) The structure of kerogen and related materials, A review of recent progress and future trends. Org Geochem 16: 829–852

Roberts TR (1984) Non-extractable pesticide residues in soils and plants. Pure Appl Chem 56: 945–956

Saiz-Jimenez C (1994) Analytical pyrolysis of humic substances: pitfalls, limitations, and possible solutions. Environ Sci Technol 28: 1773–1780

Shindo H and Huang PM (1982) Role of Mn(IV) oxide in abiotic fomation of humic substances in the environment. Nature 298: 363

Smith MR (1990) The biodegradation of aromatic hydrocarbons by bacteria. Biodegradation 1: 191–206

Sørensen LH (1987) Organic matter and microbial biomass in a soil incubated in the field for 20 years with ^{14}C-labeld barley straw. Soil Biol Biochem 19: 39–42

Sutherland JB, Rafii F, Khan AA, Cerniglia CE (1995) Mechanisms of polycyclic aromatic hydrocarbon degradation. In: Young LY, Cerniglia CE (eds) Microbial transformation and degradation of toxic organic chemicals. Wiley-Liss, Inc, New York, USA, pp 269–306

Tang J, Carroquino MJ, Robertson BK, Alexander M (1998) Combined effect of sequestration and bioremediation in reducing the bioavailability of polycyclic aromatic hydrocarbons in soil. Environ Sci Technol 32: 3586–3590

Verstraete W, Devliegher W (1996) Formation of non-bioavailable organic residues in soil: Perspectives for site remediation. Biodegradation 7: 471–485

Wang TSC; Li SW, Ferng YL (1978) Catalytic polymerization of phenolic compounds by clay minerals. Soil Sci 126: 15–21

Wienberg R (1997) Nichtstun und beobachten – eine alternative Grundwasser-Sanierungstechnik? Altlasten Spektrum 97/2: 55–59

18 Humification of PAH and TNT During Bioremediation – Evaluation of Long Term Risk and Sustainability

A. Eschenbach[1], H. Mescher[2], R. Wienberg[3], B. Mahro[2]
[1]GKSS Research Centre, Institute of Chemistry, Dept. Environmental Technology, Max-Planck-Straße, 21502 Geesthacht, Germany
[2]Institut für Technischen Umweltschutz, Hochschule Bremen, Neustadtwall 30, 28199 Bremen, Germany
[3]Umwelttechnisches Büro und Labor Dr. R. Wienberg, Gotenstraße 4, 20097 Hamburg, Germany

18.1 Introduction

The depletion of contaminants in soil is not only based on degradation or mineralisation, but also on the fact that a fixation or immobilisation of the xenobiotic substances as bound residues takes place within the soil matrix. This binding of organic contaminants (called a humification process if bound to soil humus) can reduce the bioavailable and analytically detectable part of the xenobiotics. The binding was investigated in detail by the use of ^{14}C-labeled substances for PAH and TNT in the last decades and this immobilisation process has been proposed as a remediation measure. The intentional humification process may be achieved by adjusting bioremediation process parameters such as the supplementation of soil with organic substances (compost etc.) or by changing the incubation conditions (anaerobic and aerobic phases).

The application of this humification strategy however, requires a careful study of the factors which influence the formation and long term fate and stability of these bound residues. The fixation/immobilisation could be accepted as a remediation technique if in the long run, no hazardous substances (original substances or metabolites) can be remobilised in an available form (extractable or even water soluble). This demands a stable covalent binding of the contaminants (and/or their metabolites) to the soil matrix. This paper will give a brief overview of recent investigations on this topic. For completeness also Chapters 15, 16, 17, 18, 20 and 27 should be respected.

18.2
Definition of Bound Residues and Binding Mechanisms

The term "bound residues" is used to distinguish between the portion of a compound that can be extracted from soil, normally by organic solvents without altering the chemical structure of the compound, and the non-extractable portion (Northcott and Jones 2000). Consequently, the amount estimated as bound residues depends on the applied extraction procedure.

The most widely recognised definition of bound residues was given by the IUPAC (Roberts et al. 1984, see also Chapters 16 and 17). A recent and slightly modified definition of bound residues was presented at the DFG-workshop "Pesticide Bound Residues in Soil" as follows (Führ et al. 1998):

> "Bound residues" represent compounds in soil, plant or animal which persist in the matrix in the form of the parent substance or its metabolite(s) after extractions. The extraction method must not substantially change the compounds themselves or the structure of the matrix. The nature of the bond can be clarified in part by matrix-altering extraction methods and sophisticated analytical techniques. To date, for example, covalent ionic and sorptive bonds, as well as entrapments, have been identified in this way. In general the formation of bound residues reduces the bioaccessibility and the bioavailability significantly.

In fact, for most compounds the mechanisms of sorption or binding processes are still under investigation, as the chemical structure of the residues is almost unknown. Until now, nearly all research on bound residues was conducted by the application of radiolabeled compounds, which allows a distinct quantification of the formed bound residues (see also Chapter 17). However, this technique cannot provide any information on the structural identity of the molecule. This method cannot identify different types of binding e.g. the binding or sorption of parent compounds, of major metabolites, or widely degraded transformation products.

Organic contaminants can be adsorbed or bound to soil organic matter in a reversible and a recalcitrant mode, respectively (fig. 18.1). Contaminants can be adsorbed by van der Waals forces, hydrogen bonds, hydrophobic bonds, ionic bonds (electrostatic Coulomb forces), ligand exchange and charge transfer complexes. Since adsorption is considered to be primarily a reversible process, the bulk of the substances remains available and solvent extractable (Bollag et al. 1992; Richnow et al. 1994). However, there is also evidence that adsorbed substances tend to become more resistant to extraction and degradation the longer they interact with the soil matrix (Hatzinger and Alexander 1995; Pignatello 1989).

The covalent binding of xenobiotics to organic matter should result in more persistent associations. However, one should differentiate between easily hydrolysable bounds like ester- and peptide-bounds and more stable covalent bonds like ether- C-C, and C-N-bonds.

Covalent bonds are formed by chemical reactions which require a reactivity of both the xenobiotics and the soil organic matter. The mode of binding depends on the functional groups of the parent compounds or the metabolites which are formed during microbial degradation (Richnow et al. 1994; Kästner 2000). Xeno-

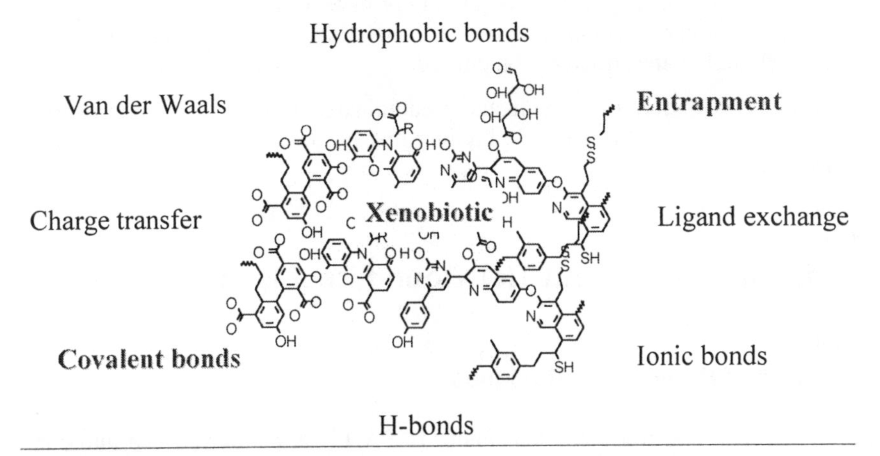

Fig. 18.1. Possible interactions between xenobiotics and humus polymers (modified after: Haider and Schäffer 2000)

biotics without functional groups, require an introduction of reactive groups before covalent binding to organic matter by microbial or chemical degradation occurs. It is presumed that covalently bound xenobiotics become an integral part of the humic soil substances (Mahro and Kästner 1993; Bollag et al. 1992; Bollag 1992) and that they will be released during humification processes at half-lives equivalent to those for humic materials (Northcott and Jones 2000).

The physical entrapment or sequestration of contaminants in macromolecular humus substances is another mechanism of bound residue formation which has recently gained more attention (Haider et al. 1992; Senesi 1993). The sequestration is described as slow partitioning of xenobiotics in micropores of the soil matrix (Alexander 1997). Soil organic matter contains cavities in the nanometer range with reactive inner surfaces or hydrophobic regions, in which the contaminants (e.g. hydrophobic PAH) could be encapsulated and retarded (Pignatello and Xing 1996; Engbrestson and Wandruszka 1994).

A silylation procedure can release parts of these entrapped residues by a derivatisation of the soil organic matter using trimethylchlorosilane. The silylation results in a disaggregation of the humic polymer and a release of entrapped xenobiotics (parent compounds or metabolites) (Dec et al. 1997; Haider et al. 1993). However, the substances could also be released as bound or conjugated parts of humic substances. In that case, the substances were not entrapped, but covalently bound to soil organic matter. To characterise the released products, the silylation should be combined with chromatographic techniques (e.g. thin-layer chromatography, size-exclusion chromatography or HPLC) (Dec et al. 1997) (see also Chapter 16 and 17).

In general, three different contamination and humic matter binding types are involved in bound residue formation:

- adsorptive processes between parent compound and/or metabolites and the soil organic matter, which are primarily reversible,

- covalent binding of parent compounds and/or metabolites,
- entrapment of parent compounds and/or metabolites, which are retained within the matrix unless the structure is modified.

The mode of binding that occurs and predominates depends on several factors, like the compounds themselves, the binding matrix, the time-scale, the microbial activity in soil etc.

18.3 Formation of PAH- and TNT-Bound Residues

18.3.1 Formation of Bound PAH-Residues

It is well known that considerable amounts of PAH may be transformed into non-extractable residues in soil (see Chapter 16 and 17, Schnöder et al. 1994; Kästner et al. 1995; Eschenbach et al. 1998b; Goodin and Webber 1995; Qiu and McFarland 1991; Guthrie and Pfaender 1998). As shown in table 18.1, the fraction of bound residues varied for 2- to 5-ring PAH ranging between approximately 15 and 65 % of applied radioactivity for soil materials, with varying physical and chemical parameters. A clear correlation between the type of PAH and the extent of bound residue formation could not be detected.

Experiments, that were carried out to obtain more information about the characteristics of bound residues, showed that at the beginning of the incubation, almost the total amount of ^{14}C-activity of the bound residues was detected in the NaOH-extractable humic acid fraction. During the incubation period, a significant change in the distribution of ^{14}C-activity within the pool of the soil organic matter was observed. In own studies with biologically active soil, about 80 % of the radiolabel was enriched in the humic fraction. Other experiments have shown similar results (Niemann et al. 1999; Schnöder et al. 1994).

The factors influencing the formation of bound PAH-residues were also recently examined in more detail (Eschenbach 1995; Eschenbach et al. 2000a; Richnow et al. 1998, 1999; Kästner et al. 1999; Guthrie et al. 1999). It was found that the biological activity within the soil material was an essential prerequisite for the formation of bound residues. Although a slight initial formation of bound residues was detected in sterile soil samples, a further increase of bound residues was either not detectable or very small under sterile conditions. Additional experiments with naphtalene, anthracene, pyrene, and benzo(a)pyrene confirmed that in sterile control samples the majority of the ^{14}C-activity remained in an extractable form (Eschenbach 1995; Eschenbach et al. 1995, 1998b; Kästner et al. 1999; Guthrie and Pfaender 1998).

Table 18.1. Extent of non-extractable residue (ner) formation from different ^{14}C-PAH in different soil materials (in % of the initial ^{14}C-activity). (after Eschenbach et al. 2000b)

^{14}C-PAH	soil material	C_{org}	PH	supplement	incubation [days]	Ner [%]	Ref.
naphthalene	contaminated site loamy sand	7.9	7.6	P. ostreatus	194	21	1
anthracene	contaminated site loamy sand	4.6	7.5	*P. ostreatus*	174	38	1
	uncontaminated soil, silty clay	0.9	6.9	none	291	33	2
	contaminated site, loamy sand	6.7	7.2	none	288	51	2
	uncontaminated soil, loamy sand	1.1	4.5	compost	176	21	3
	uncontaminated soil, loamy sand	1.1	4.5	none	176	45	3
	uncontaminated soil, loamy sand	1.1	4.5	compost	103	24	4
	soil	ng	ng	none	120	68	8
pyrene	contaminated site, loamy sand	4.6	7.5	*P. ostreatus*	174	15	1
	contaminated site, loamy sand	7.9	7.6	*P. ostreatus*	194	21	1
	uncontaminated soil, silty clay	0.9	6.9	none	291	56	2
	uncontaminated soil, silty clay	0.9	6.9	compost	291	21	2
	uncontaminated soil	3.4	4.8	none	270	44	7
benzo(a)-pyrene	contaminated site, loamy sand	4.6	7.5	*P. ostreatus*	174	56	1
	contaminated site, loamy sand	7.9	7.6	*P. ostreatus*	194	26	1
	uncontaminated soil, silty clay	0.9	6.9	none	291	21	2
	contaminated soil, loamy sand	6.7	7.2	none	288	12	2
	uncontaminated soil, silty loam	0.5	7.2	*P. chrysosporium*	120	37	5
	uncontaminated soil, loamy silt	1.1	7.7	none	117	20	6

ng: not given
[1] Eschenbach 1995.
[2] Eschenbach et al. 1998b.
[3] Kästner et al. 1999.
[4] Kästner et al. 1995.
[5] Qiu and McFarland 1991.
[6] Schnöder et al. 1994.
[7] Guthrie and Pfaender 1999.
[8] Hosler et al. 1988.

Two processes of binding are discussed in this context to explain the predominantly biological character of bound residue formation. The first mechanism that is assumed to play an important role, is the biologically induced transformation of the binding matrix (soil organic matter). This hypothesis is supported by NMR-studies with ^{13}C-pyrene that indicated a physical entrapment of parent pyrene (non-covalently bound) due to microbiologically mediated diagenetic processes in the humic material (Guthrie et al. 1999). The second biogenic process that may contribute to the binding of PAH in soil is the microbial transformation of PAH into more reactive metabolites. This modification or transformation of the PAH is essential because PAHs themselves do not carry the functional groups necessary for covalent binding. The necessity of a metabolic activation of PAH is also indicated by both; with the observed correlation between the mineralisation and the formation of bound residues (Eschenbach et al. 2000a, 2000b); and by experiments with ^{14}C-labelled bacterial PAH-metabolites (^{14}C-*cis*1,2-dihydro-1,2-dihydroxyanthracene, ^{14}C-4,5-dicarboxy-phenanthrene) where a fast and pronounced binding of these metabolites to the soil matrix was also observed (Eschenbach et al. 1998b). In addition, chemical degradation experiments with bound PAH-residues (hydrolysis with ^{18}O-labelled NaOH and H_2O and with ^{13}C-labelled anthracene) has shown that a covalent binding between anthracene metabolites (hydroxynaphtoic and phthalic acid) and the soil organic matter may occur to some extent (Richnow et al. 1994, 1999).

However, beside these biologically induced binding processes a spontaneous non-biologically derived formation of residues also occur to a minor extent within the first hours of application in both sterile and non-sterile soil samples. This initial binding is independent of biogenic transformations and it is assumed that parts of the parent PAH were bound by sorptive processes as discussed above (Hatzinger and Alexander 1995; Kelsey et al. 1997).

The amount of the extractable PAH fraction could be significantly reduced by the addition of organic supplements like compost (Hupe et al. 1996; Lotter et al. 1993). It was also investigated whether the addition of various supplements like matured compost, bark chips, or forest litter had an impact on the bound residue-formation. However, the experiments showed that these supplements had no positive influence on the extent of non-extractable residues, though occasionally the mineralisation of PAH was enhanced. A specific simulation of the humification process could not be observed. Even the addition of the lignolytic white rot fungus *Pleurotus ostreatus* was not able to enlarge the immobilisation of PAH in our studies, though this fungus is able to excrete exoenzymes which are responsible for unspecific radical reactions and oxidative coupling (Barr and Aust 1994). Therefore previous assumptions that organic supplements might act as an additional binding matrix could not be confirmed by our experiments (Eschenbach et al. 1997, 1998a).

18.3.2 Formation of Bound TNT-Residues

Because TNT persists in a biologically induced mineralisation, due to its electrophilic character and chemical structure, the main depletion process for TNT in soil is the formation of bound residues. TNT can be transformed microbially reduced rather than by oxidation reactions (Rieger and Knackmuss 1995). The main transformation process is the reduction of the nitro groups to amino groups. Under aerobic conditions only partially reduced nitroaromatic compounds were generated (Preuss et al. 1993). A complete co-metabolic reduction of all three nitro groups can occur via aminodinitrotoluene and diaminonitrotoluene to triaminotoluene. A significant mineralisation was only observed in laboratory cultures (Fernando et al. 1990; Scheibner et al. 1997). Nevertheless, it was shown that the extractable TNT-concentration decreases rapidly and nearly to completion in soil during bioremediation processes (e.g. anaerobic/aerobic-treatment) (Lenke et al. 1998; Bruns-Nagel et al. 1998; Breitung et al 1996). It is assumed that reduced TNT-metabolites are covalently bound to soil organic matter and become an integral part of humic substances. An overview of the process and a perspective on a bio-elemination of nitroaromatic compounds in soil is given in reviews (e.g. Lenke et al. 2000).

The application of ^{14}C-labelled TNT has proven that TNT metabolites bind to the soil humic substances, and form of bound residues in soil (Achtnich et al. 1999a; Drzyzga et al. 1998). While the extractable ^{14}C-activity decreased with time, an increasing proportion was found in the humic acids, fulvic acids and the unextractable fraction (humin) (Drzyzga et al. 1998).

Recently, bioremediation processes based on the transformation and binding of TNT metabolites in soil were developed. This is obtained by the addition of oxidisable organic supplements (C-sources), adjustment of anaerobic conditions, the co-metabolic reduction of TNT and binding of these metabolites to soil organic matter. The BMBF-Joint Research Group "Processes for the Bioremediation of Soil" developed three bioremediation processes for humification of TNT, which were applied on the former ammunition plant "Werk Tanne" in Clausthal-Zellerfeld" by the companies Plambeck ContraCon, Umweltschutz Nord and Awia. Awia used white rot fungi for metabolising TNT (see also Chapter 19).

Laboratory experiments with ^{14}C-TNT and ^{15}N-TNT were carried out parallelly to the large scale experiments in Clausthal-Zellerfeld in order to quantify and qualify the binding of TNT metabolites to soil humic substances. Amounts of bound residue fractions determined during the different bioremediation processes are given in table 18.2.

Additional pilot scale experiments (2 m^3), with ^{14}C-TNT and soil from the ammunition plant were carried out in special reactors at the FhIUCT Schmallenberg (Hund-Rinke and Kördel 1999). The declining kinetics in the large scale experiments and the pilot scale experiments were comparable. In both cases the TNT concentration decreased rapidly within the first few days. Like the original TNT concentration, the ^{14}C-TNT concentration decreased in the same way. TNT was transformed almost completely into bound residues.

To characterise the mode of binding of TNT or its metabolites with the soil organic matter, experiments with ^{15}N-labeled TNT and ^{15}N-NMR spectroscopy were carried out. The covalent binding to humic acids was determined by in-vitro experiments with TNT-derivates (aminodinitrotoluenes and diaminonitrotoluenes) and horseradish peroxidase. The major NMR-peaks responded to anilinohydroquinone, anilinoquinone, anilide nitrogen and heterocyclic N-compounds (Thorn 1997).

Table 18.2. Simulation of different bioremediation processes: extent of the formation of non-extractable residues originated from ^{14}C-TNT (after: Eschenbach et al. 2000b)

Bioremediation process	Supplement	TNT concentration [mg kg^{-1}]	Incubation time [d]	Bound residues	Institution
Alternate anaerobic/aerobic process, laboratory experiment	company specific substrate	4500	92	92 %	FhIGB Stuttgart
Dynamic pile process, laboratory experiment	company specific substrate	4100	93	90 %	University of Marburg
Aerobic white rot fungi treatment, laboratory experiment	Straw fungi substrate *S. rugosoannulata*	200	30	86 %	University of Jena
Alternate anaerobic/aerobic process, 2 m^3 scale	company specific substrate	213	120	> 99 %	FhIUCT, Schmallenberg
Dynamic pile process, 2 m^3 scale	company specific substrate	273	170	> 99 %	FhIUCT, Schmallenberg
Anaerobic/aerobic bioreactor process, laboratory experiment	Molasses	590	63	84 %	Drzyzga et al. 1998
Anaerobic/aerobic slurry reactor process, laboratory experiment	glucose and mixed culture of anaerobic microorganisms	350	80	98 %	Achtnich et al. 1999a

Knicker et al. (1999) analysed pyrole, imidazole, indole, chinolone and carbazole by ^{15}N-NMR spectroscopy, in the humic fraction of an anaerobic/aerobically treated soil material. This indicated that TNT is completely transformed and bound covalently to the humic substance. A significant change of the NMR spectra after the silylation of an anaerobic/aerobically treated soil material was observed by Achtnich et al. 1999c). The untreated material showed sharp NMR signals of unreduced free TNT and its metabolites after the application of the TNT and the silylation. After the anaerobic/aerobic treatment a wider NMR spectrum could be observed, which also indicates a covalent binding of TNT metabolites. However, at the end of the incubation nitrogroups were still detectable, indicating that a covalent and multivalent binding did not form in every case (Achtnich et al. 2000).

18.4 Stability of Bound PAH- and TNT-Residues

18.4.1 Strategies to Investigate the Stability of Bound Residues

The mineralisation of soil contaminants is a well accepted remediation strategy as it represents a direct and complete detoxification of the soils. However, the intentional binding or fixation of contaminants (e.g. PAH or TNT) within the soil matrix has to be evaluated carefully in terms of effectiveness and sustainability. It must be guaranteed that the soil does not function as a temporary sink of contaminants which could be remobilised lateron (see also Chapter 16).

The sustainability assessment of the pollutant humification can be investigated using the following strategies (Eschenbach et al. 2000b):

- characterisation of bonds between contaminants (or their metabolites) and soil organic matter by NMR-spectroscopy and stable isotopes,
- investigation of the long-term stability and remobilisation potential of bound residues under environmental stress conditions at a lab-scale,
- investigation of the long-term fate of bound residues in the large-scale under field conditions (e.g. field lysimetres),
- ecotoxicological assessment of soil material with bound residues.

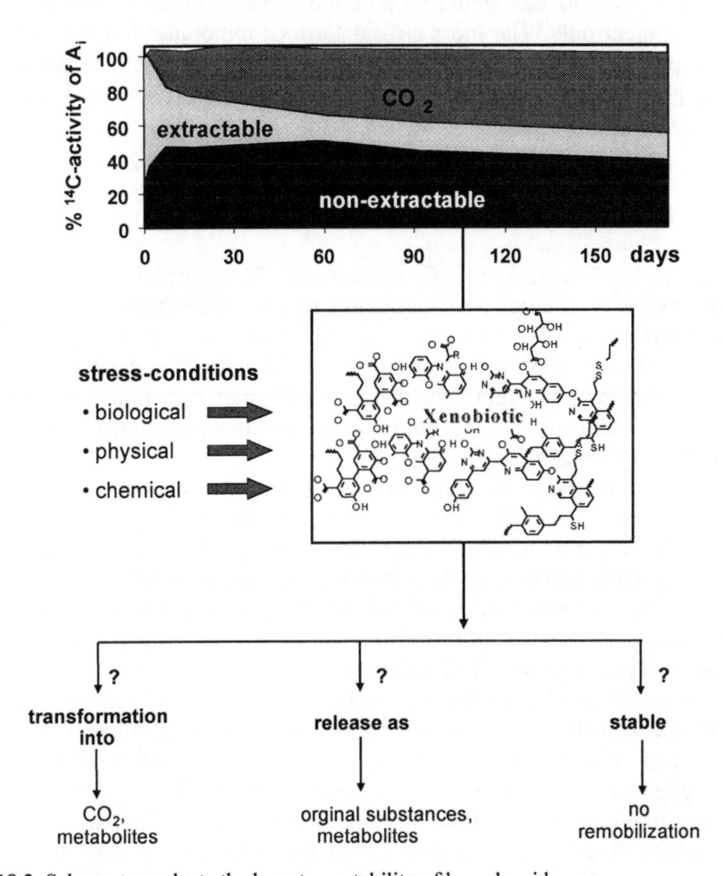

Fig. 18.2. Scheme to evaluate the long-term stability of bound residues

Conservative complementary strategies should be applied in combination whenever possible. A complete assessment program was carried out recently for PAH- and TNT-residues in a BMBF-joint research project (Eschenbach et al. 2000b). Some of these results will be discussed in the following sections

The evaluation of the long-term stability of the bound pollutant residues should also include tests that simulate environmental stress conditions (biological, physical and chemical worst case conditions). An experimental approach for these studies is shown in fig. 18.2.

Soil material with stable fractions of non-extractable or bound residues (e.g. from PAH) is treated under relevant different ecological stress conditions and the fate or the release of the ^{14}C-label is analysed in the different soil compartments (mineralisation in the gas phase, pollutant residues in the water or solvent extractable fraction and, non-extractable fraction in the residual soil matter).

While strong mineralisation would indicate a detoxification of the residues to CO_2 a slight mineralisation rate might also indicate that the bound residues became involved in turnover processes of soil organic matter (Arjmand and Sandermann 1985; Haider and Martin 1988). A critical point would be the release of parent xenobiotics or hazardous degradation products (e.g. PAH-metabolites). The release of parent compounds would be possible if the substances were bound by physical entrapment only. The most critical form of remobilisation would be the transition of previously bound residues into the fraction of extractable substances. Extractable pollutant residues are potentially bioavailable and could therefore cause harm to the environment or to human health.

18.4.2 Stability of Bound PAH-Residues

The stability of PAH-residues of ^{14}C-naphthalene, ^{14}C-anthracene, ^{14}C-pyrene or ^{14}C-benzo(a)pyrene was tested under a variety of environmental stress conditions which are listed in fig. 18.3. The effect of each treatment was analysed in comparison to an untreated control by measuring the ^{14}C-activity in the different fractions (mineralisation, water and solvent extractable part and, non-extractable or bound residues). The analytical procedures of the treatment have been described in more detail by Eschenbach et al. (1998a).

Biological treatment did not result in a mobilisation of bound residues as extractable compounds (Eschenbach et al. 1998a, 2000a; Eschenbach 1995). The addition of microorganisms, which were chosen due to their known ligninolytic and humus degrading activity, or their ability to metabolise PAH, or the addition of isolated enzymes (peroxidases), did not effect the bound residue fraction. The fraction of bound residues decreased slightly in all soil materials. With all PAH-residues being tested, this occurred independently of the addition of the microorganisms or enzymes (Eschenbach et al. 1998a). A rather delayed mineralisation of ^{14}C-activity was detectable with no distinguishable differences in the mineralisation rates among different samples. The same results were obtained, when diverse biological supplements, like bark chips or compost, were added (fig. 18.4). However the most important observations from these experiments was that an increase of the extractable fraction was not observed in any of the experiments during 210 days of incubation. This means that bound residues were not remobilised as extractable parent compounds or even PAH-metabolites. The slight reduction of bound residues was correlated with a slow mineralisation (formation of $^{14}CO_2$).

biological treatments

- impact of isolated and enriched microorganisms (PAH- and humus-degrading)
- impact of isolated oxidative and radicalproducing enzymes
- impact of organic supplements (compost and bark chips)
- impact of litter degrading earthworms
- simulation of a "priming-effect"

physical treatments

- alternating temperatures (freezing/thawing cycles)
- alternating soil water contents (drying/rewetting cycles)
- mechanical disruption of soil aggregates

chemical treatments

- destabilisation of metal-organic complexes
- repeated extraction procedures
- impact of tensides
- impact of radical producing agents
- impact of pH-decrease (simulation of acid rain)

combination of physical and biological treatments

- mechanical disruption of soil aggregates followed by impact of organic supplements, of isolated and enriched microorganisms and of isolated enzymes

Fig. 18.3. Stress conditions to examine the stability of bound residues

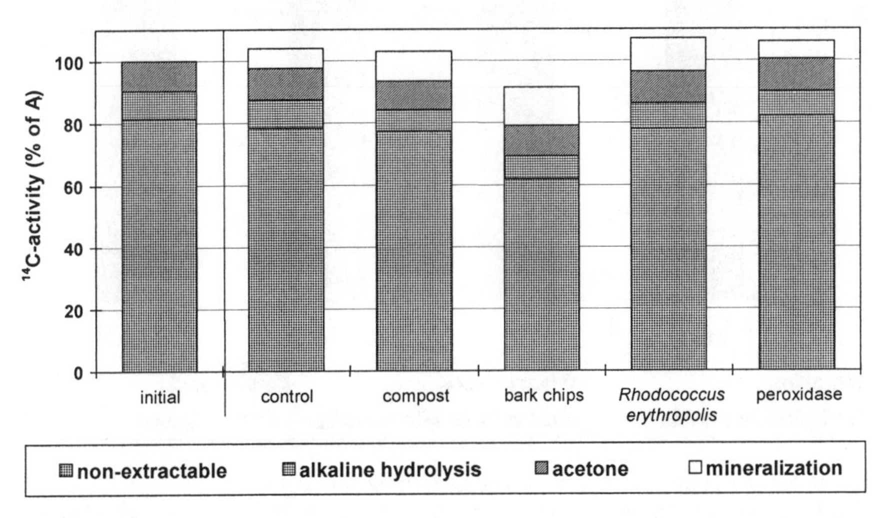

Fig. 18.4. Impact of organic supplements (compost and bark chips), of the humus-degrading micoorganism *R. erythropolis*, and of the enzyme peroxidase on the fate of bound residues from ^{14}C-anthracene

Based on the model that bound PAH-residues are covalently bound or trapped within the humic matrix (Richnow et al. 1994; Engebrestson and Wandruszka 1994; Nanny et al. 1997), this mineralisation of bound residues could be compared to humus turnover-rates in the soil (2 to 5 % per year; data for undisturbed soils under moderate climatic conditions; Saxena and Bartha 1983).

Experiments with artificially produced pollutant/humic-complexes revealed that covalently bound pollutants, like aniline and chlorinated phenols, can be mineralised in approximately the same amount as the humic substances themselves (Arjmand and Sandermann 1985; Haider and Martin 1988). The reduction of the non-extractable ^{14}C-fraction – as it was observed in this study – was slightly higher than the average natural humus turnover rate. This may be due to the optimised incubation conditions, the disturbed nature of the soil samples, the particular soil material or the particular high activity of the autochthonous soil microorganisms. It is possible that bound PAH-residues may have become involved in slow humus turnover and degradation processes.

In another set of remobilisation experiments the influence of earthworms on the stability of bound residues was investigated. The rationale of these experiments was whether earthworms, which play an important role in the destruction and degradation of organic matter (Scheu 1987), might contribute to a presumed remobilisation. During the intestinal passage the ingested material is reduced or

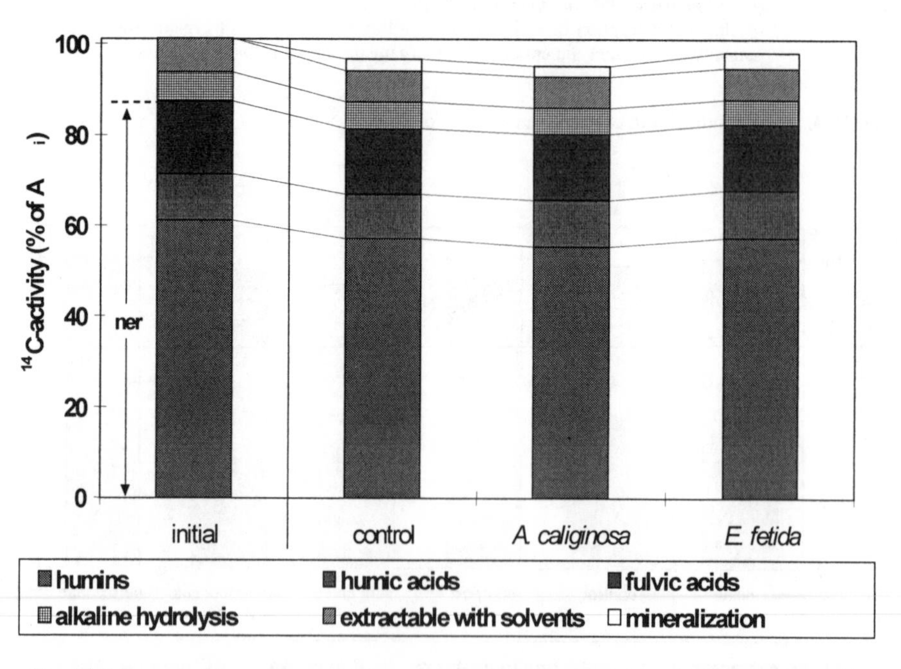

Fig. 18.5. Impact of the earthworms *Aporrectodea caliginosa* and *Eisenia fetida* on the stability of bound residues (ner) from ^{14}C-anthracene. ^{14}C-activity at the beginning and after 42 days of incubation

degraded to small pieces and mixed with mucus, which also leads to an increased microbial activity. Two different worm species were used in the experiments: first the epigeic, typical litter decomposing worm, *Eisenia fetida*, and secondly the endogeic worm, *Aporrectodea caliginosa*, which lives in the mineral soil material. However, as shown in fig. 18.5, the earthworms were unable to mobilise bound PAH-residues within the incubation period.

A slow mineralisation of about 2–3 % of the initial activity was determined during the incubation period. The fraction of bound residues decreased about 5 % but this depletion could not be attributed to earthworm treatment. The incubation of the control without addition of earthworms showed results in the same order of magnitude. The distribution among different humic fractions (humic acids, fulvic acids and humin) had not changed significantly during incubation. Experiments with different soil materials, and with bound residues of different PAH (anthracene, pyrene, benzo(a)pyrene) confirmed these findings.

To investigate the impact of *physical stress treatments* on the release of bound residues, appropriate soil samples were treated by quick and drastic changes, representing climatic changes over several years. This implied repeated freezing and thawing cycles and the adjustment of alternating water contents in the soil. In addition, the soil structure was totally disrupted by mechanical means, either by ultrasonic disintegration or by grinding in a mortar. None of these treatments led to a significant remobilisation of bound PAH-residues. However, a trend of increasing availability of ^{14}C-activity in the water elutable or extractable fraction as obtained. Small amounts of activity were mineralised during the incubation periods (Eschenbach et al. 2000a).

Therefore, the next stress factor tested was the *combination of mechanical and biological treatments*. After the mechanical pre-treatment, the soil samples were supplemented by compost and a specific cocktail of actinomycetes, which had been shown to mineralise PAH (*Sphingomonas, Gordona, Mycobacterium-species*) (Kästner et al. 1994).

As shown in fig. 18.6 the bound residue fraction was not influenced by this combination of drastic treatments either. The amount of bound residues decreased slightly by about 6 %-points during 206 days of incubation. Only the soil sample pre-treated by ultrasonics showed a more distinct reduction of bound residues though this reduction did not cause releases of an extractable fraction, but led to a higher mineralisation.

It can be concluded from the findings that environmental factors such as repeated freezing and thawing – though they may influence the formation of bound residues in soil (Hatzinger and Alexander 1995; Kelsey et al. 1997) – did not destabilise the residues once they were formed. Bound PAH-residues displayed high stability against both, biological and physical mobilisation treatments.

Along with the conduct of the chemical remobilisation treatments, which had the purpose to enhance the desorption and solubility of contaminants, the impact of the treatment was investigated, and was found to preferentially alter the spatial structure of the humic polymers.

Destabilisation of the residues may be achieved using complexifying agents like EDTA. EDTA destabilises naturally occurring metal organic complexes, which are responsible for the formation of chelates and the spatial aggregate structure of humic substances in soils (Senesi 1994).

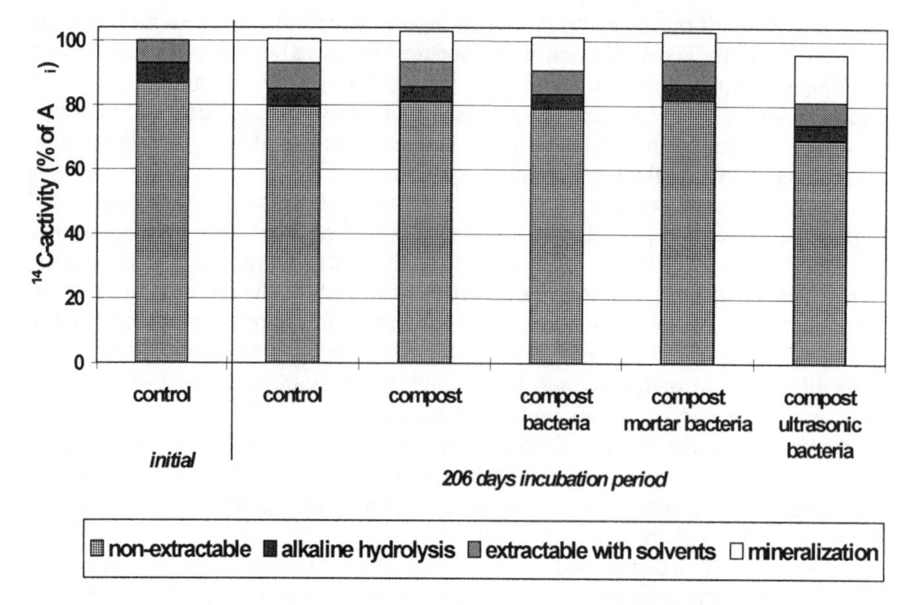

Fig. 18.6. Impact of the combination of mechanical (destroying soil aggregates by grinding or ultrasonic) and biological treatments (addition of compost and PAH-degrading microorganisms) on the fate of bound residues from ^{14}C-anthracene

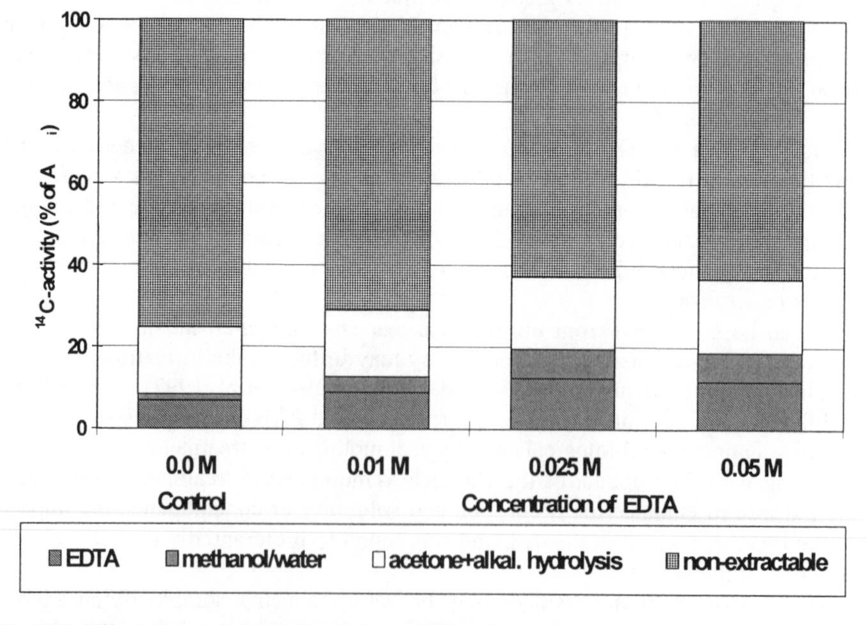

Fig. 18.7. Effect of resolving metal-organic complexes by an extraction with EDTA at different concentrations on the fate of bound residues from ^{14}C-anthracene in soil from a contaminated site (modified after Eschenbach et al. 2000a)

The application of increasing concentrations of EDTA solutions led to an increasing ^{14}C-activity in the EDTA-soil extracts, while only a low ^{14}C- activity was detectable in the water treated control samples (fig. 18.7). In addition, the subsequent extraction with a methanol water mix recovered higher amounts of activity in the treated soil samples than in the untreated control. As shown in fig. 18.7, for a soil sample from a contaminated site, the increasing extractability of ^{14}C-activity at increasing concentrations of the EDTA solution correlated with a considerable reduction of bound residue fraction. The fraction of non-extractable residues was reduced to 12–13 % by treatment with the highest concentration of EDTA (0.05 M). The presented data show that the chemical dissolution of metal organic complexes may lead to an appreciable reduction of the non-extractable residue fraction and to a transfer of ^{14}C-activity in the extractable fractions (extraction by EDTA-solution and methanol/water). The released activity was obviously not caused by physically entrapped PAH or their metabolites, which could have been set free due to the destruction of the spatial structure of the soil organic matter. However, it was shown in additional experiments that the released ^{14}C-activity, was primarily from ^{14}C-atoms that were attached to soluble humic substances. This result supports the assumption that PAH, or rather their metabolites, could be (covalently) bound to humic polymers and thereby become an indistinguishable part of the humic substances.

18.4.3 Stability of Bound TNT-Residues

Similar to the remobilisation experiments with PAH-bound residues, another set of experiments was also carried out to test the stability of bound TNT metabolites in soil (BMBF-Joint Research project "Long-Term Stability and Remobilisation"; Achtnich et al. 2000; Banholczer et al. 2000). These long-term studies were carried out in accordance with the proposed scheme given in fig. 18.2. The results obtained were very similar to those observed with PAH-bound residues. The application of the freeze/thaw cycles (climate simulation) did not lead to a significant remobilisation of ^{14}C-activity from bound TNT-residues.

In long-term elution experiments conducted over 21 months, Achtnich et al. (1999a) observed only 3.8 % of the total bound radioactivity of bound TNT in the leachate. Neither TNT or its reduction products were detected in the eluate by HPLC-analysis. Most of the radioactivity (2.8 %) was eluted in the first 2 months, and was assigned to polar hydrophilic substances, representing dissolved organic matter (e.g. fulvic acids) that contained bound aminotoluene-metabolites or fragments thereof.

Similar to the results being observed with PAH-bound residues, EDTA also affected the remobilisation of ^{14}C-activity of bound TNT-transformation products (fig. 18.8). The EDTA treatment was carried out at three different concentrations (0.01 M, 0.025 M, 0.05 M) and was followed by a successive extraction with water and with methanol. The application of increasing concentrations of EDTA solutions also led to an increase of ^{14}C-activity in the soil extracts. The treatment with 0.05 M EDTA reduced the fraction of bound residues for about 25 % of the total radioactivity. For further characterisation of the remobilised ^{14}C-activity, the soil extracts were examined by HPLC analysis. TNT and its metabolites could not

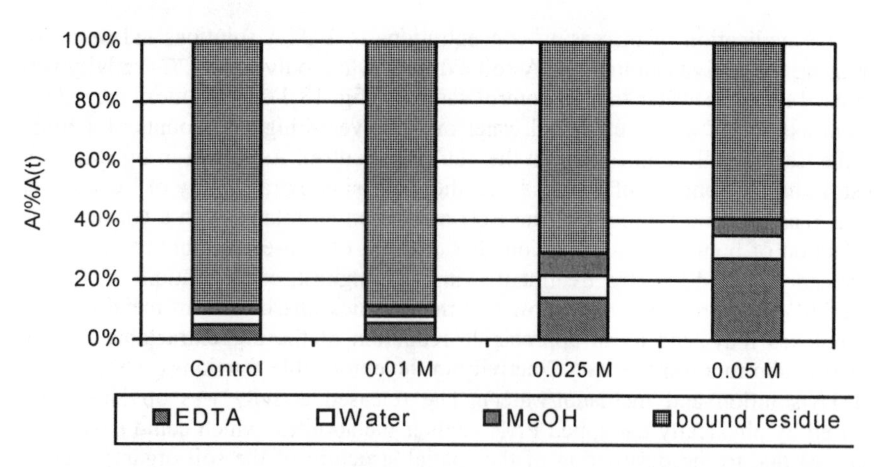

Fig. 18.8. Remobilisation of bound TNT-residues by EDTA extraction. ^{14}C-activity in the fractions (extractable by EDTA, by water or by MeOH and bound residues) in the soil samples treated by different concentrations of EDTA (in % of the total obtained activity; after: Banholzer et al. 2000)

be detected in the extracts. The results indicate that TNT was reduced during the anaerobic/aerobic process and that the released ^{14}C-activity must originate from transformed metabolites being bound to the humic substances (Achtnich et al. 2000; Banholczer et al. 2000).

In another experiment, the impact of continued acid rain was investigated. The impact of acid rain on bound TNT residues was simulated with aqueous solutions of nitric acid and sulfuric acid. The extraction was carried out 10 times in succession. The results showed that up to 15 % of the total radioactivity was eluted by acid rain treatment. The simulation of acid rain released notable quantities of radioactivity similar to the results obtained after the EDTA treatment of the anaerobic/aerobic treated soil material. However, again TNT and its metabolites could not be detected by HPLC analysis. These results confirm the assumption that the released radioactivity must originate from completely reduced TNT metabolites associated with the dissolved humic substances (Achtnich et al. 2000).

18.5 Conclusions

The binding of organic contaminants (e.g. PAH, TNT) to soil organic matter as bound residues has been used as a bioremediation measure; recently at a commercial scale. It was determined in several studies which applied ^{14}C-labelled xenobiotics, that this formation of bound residues occurs to a significant extent for PAH and TNT. 85% to 99% of ^{14}C-TNT was immobilised by humification processes Laboratory experiments indicated, that the residue formation is mainly controlled by the biological activity of the soil, and to some extent by the duration of incubation period. In sterile soil samples nearly the total amount of ^{14}C-label remains extractable, whereas in non-sterile control samples a high degree of bound residues was observed (Eschenbach et al. 1995, 2000a; Guthrie and Pfaender 1998).

This clear correlation of bound-residue formation to biological soil activity can be based on the necessity of;

- a metabolic modification of the soil organic matter for the incorporation of parent compounds or their metabolites in the humic matter fraction (entrapment), or
- a metabolic modification of the parent compound (introduction of functional groups or reduction of nitro-groups for example) for the binding of the compound to soil organic matter (covalent binding).

The first hypothesis was recently proven in studies with ^{13}C-pyrene (Guthrie et al. 1999). However, a covalent binding of derivatives of organic contaminants (PAH and TNT) was confirmed by both, the NMR studies and the chemical derivatisation methods (Achtnich et al. 1999b; Knicker et al. 1999; Richnow et al. 1998; Thorn 1997; Dec et al. 1997).

However, the chemical processes which resulted in the spontaneous occurrence of bound residues at the very beginning of the incubation are not yet understood. It seems possible, that non-covalent adsorptive types of association are involved as well (Eschenbach et al. 2000a; Kästner 2000). The adsorption is assumed to be reversible. However, the stability increases with interaction time.

For risk assessment the stability of bound residues should not only be characterised by the chemical structure of the binding. It is also necessary to determine in detail the long-term stability under environmental stress conditions for each compound.

In general, it was shown by remobilisation tests that the biological formed bound residues of TNT and PAH remained stable in the soil matrix. That does not mean that they remain inert in the soil. However, biological treatments result in a slow mineralisation, which can be compared to humic matter turnover. No remobilisation of parent compounds or major metabolites was detected, even when ^{14}C-activity was released by the treatment (EDTA or acid rain simulation). The further characterisation of the released ^{14}C-activity indicates that it was set free as being bound to humic polymers, which was dissolved by the treatment itself. The treatment process of acid extraction and resolving metal organic complexes resulted in a disaggregation and cleavage of humic material. Entrapped residues (parent compound or main metabolites) are irreversibly bound if the matrix is not modified. However, they would be released when the matrix structure is broken up.

For risk assessment, one should also be aware that dissolved organic matter can act as a carrier for organic contaminants (Deschauer and Kögel-Knabner 1992). On the other hand, the release of ^{14}C-label, which is bound to soluble organic matter molecules actually confirms the assumption of covalent binding.

Different kinds of binding (covalent, adsorptive, entrapped) will be involved in residue formation, to differing extents depending on factors such as the substance itself, the time of interaction, the soil binding matrix, the biological activity of the soil, and the remediation treatment process. However, one has to take into account that binding modes are proven only for small parts of the total residues. Therefore, it is essential to estimate the remobilisation potential and the long-term stability by empirical investigations, case by case, as it has been done for bound PAH-residues and TNT-residues. The results presented show that bound residues of PAH and TNT will probably not cause harmful effects to human health or the environment. The results presented support the concept of using the formation of bound residues as a decontamination strategy.

18.6 Acknowledgments

Parts of this work were supported by grants from "Deutsche Forschungsgemeinschaft (DFG)" within the interdisciplinary research project "Remediation of contaminated soil" (grant SFB 188, project B1) and by grants from "The German Ministry of Education and Research" (BMBF grant 1480937 and 14810925).

References

Achtnich C, Sieglen U, Knackmuss HJ, Lenke H (1999a) Irreversible binding of biologically reduced 2,4,6-Trinitrotoluene to soil. Environ Toxicol Chem 18: 2418–2323

Achtnich C, Fernandes E, Bollag JM, Knackmuss HJ, Lenke H (1999b) Covalent binding of reduced metabolites of ($^{15}N_3$)TNT to soil organic matter during a bioremediation process analyzed by ^{15}N NMR Spectroscopy. Environ Sci Technol 33: 4448–4456

Achtnich C, Lenke H, Knackmuss HJ (1999c) Untersuchungen zur Bindung und Stabilität von festgelegtem reduziertem TNT im Boden nach einer biologischen Anaerob-/ Aerob-Behandlung. In: Umweltbundesamt (ed) Langzeit- und Remobilisierungsverhalten von Schadstoffen. Tagungsband zum Statusseminar des BMBF-Verbundvorhabens "Biologische Verfahren zur Bodensanierung", 22./ 23.10.1998 in Bremen. E1–14

Achtnich C, Peters D, Knackmuss HJ, Lenke H (2000) Alternierender Anaerob-/ Aerob-Prozess: Analyse der Bindunmgsstruktur von metabolisiertem und humifiziertem TNT im Boden. In: Umweltbundesamt (ed) Langzeit- und Remobilisierungsverhalten von Schadstoffen bei der biologischen Bodensanierung. Tagungsband zum Statusseminar des BMBF-Verbundvorhabens "Biologische Verfahren zur Bodensanierung", 22.02.2000 in Bremen, in press

Alexander M (1997) How toxic are toxic chemicals in soil? Environ Sci Technol 29: 2713–2717

Arjmand M, Sandermann H (1985) Mineralization of chloroaniline/lignin conjugates and of free chloroanilines by the white-rot fungus *Phanerochaete chrysosporium*. J Agric Food Chem 33: 1055–1060

Baldock JA, Oades JM, Vassallo AM, Wilson MA (1989) Incorporation of uniformly labelled ^{13}C-glucose carbon into the organic fraction of a soil, Carbon balance and CP/MAS-^{13}C-NMR-measurements. Soil Biol Biochem 27: 725–746

Banholczer A, Fründt J, v Löw E, Bruns-Nagel D, Gemsa D (2000) Dynamisches Beetverfahren: Analyse nicht extrahierbarer TNT-Transformationsprodukte. In: Umweltbundesamt (ed) Langzeit- und Remobilisierungsverhalten von Schadstoffen bei der biologischen Bodensanierung. Tagungsband zum Statusseminar des BMBF-Verbundvorhabens "Biologische Verfahren zur Bodensanierung", 22.02.2000 in Bremen, in press

Barr DP, Aust SD (1994) Mechanisms white rot fungi use to degrade pollutants. Environ Sci Technol 28: 78–87

Bollag JM (1992) Decontaminating soil with enzymes. Environ Sci. Technol 26: 1876–1881

Bollag JM, Myers CJ, Minard RD (1992) Biological and chemical interactions of pesticides with soil organic matter. Sci Total Environ 123/124: 205–217

Breitung J, Bruns-Nagel D, Steinbach K, Kaminski L, Gemsa D, v Löw E (1996) Bioremediation of 2,4,6-trinitrotoluene-contaminated soils by two different aerated compost systems. Appl Microbiol Biotechnol 44: 795–800

Bruns-Nagel D, Drzyzga O, Steinbach K, Schmidt TC, v Löw E, Gorontzy T, Blotevogel KH, Gemsa D (1998) Anaerobic/aerobic composting of 2,4,6-Trinitrotoluene-contaminated soil in a reactor system. Environ Sci Technol 32: 1676–1679

Dec J, Haider K, Schäffer A, Fernandes E, Bollag JM (1997) Use of silylation procedure and ^{13}C-NMR Spectroscopy to charakterize bound and sequestered residues of cyprodinil in soil. Environ Sci Technol 31: 2991–2997

Deschauer H, Kögel-Knabner I (1992) Binding of a herbicide to water-soluble soil humic substances. The Science of the Total Environment 117/118: 393–401

Drzyzga O, Bruns-Nagel D, Gorontzy T, Blotevogel KH, Gemsa D, v Löw E (1998) Incorporation of ^{14}C-labled 2,4,6-Trinitrotoluene metabolites into different soil fraction after anaerobic-aerobic treatment of soil/molasses mixture. Environ Sci Technol 32: 3529–3535

Engebretson RR, Wandruszka R (1994) Microorganization in dissolved humic acids. Environ Sci Technol 28: 1934–1941

Eschenbach A (1995) Einfluß von *Pleurotus ostreatus*, Kompost, *Sphingomonas paucimobilis* und der Kontaminationsdauer auf den Verbleib und Abbau ^{14}C-markierter polyzyklischer Kohlenwasserstoffe (PAK) in Altlastböden. Dissertation, Technische Universität Hamburg-Harburg

Eschenbach A, Kästner M, Wienberg R, Mahro B (1995) Microbial PAH degradation in soil material from a contaminated site – Mass balance experiments with *Pleurotus ostreatus* and different ^{14}C-PAH. In: van den Brink WJ, Bosman R, Arendt F (eds) Contaminated soil '95. Kluver Academic Publ Dodrecht, pp 377–378

Eschenbach A, Wienberg R, Mahro B (1997) Einsatz von Kompost und Rindenmulch bei der biologischen Altlastensanierung PAK-kontaminierter Bodenmaterialien. Mittlgn Dt Bodenkundl Gesellsch 83: 279–282

Eschenbach A, Wienberg R, Mahro B (1998a) Fate and stability of nonextractable residues of [^{14}C]PAH in contaminated soils under environmental stress conditions. Environ Sci Technol 32: 2585–2590

Eschenbach A, Wienberg R, Mahro B (1998b) Untersuchung der Bildung und Langzeitsabiltät von humifizierten PAK in biologisch behandelten Recyclingböden. Abschlußbericht BMBF-Forschungsvorhaben "Biologische Verfahren zur Bodensanierung", Teilprojekt 7, Förder-Nr. 1480937

Eschenbach A, Wienberg R, Mahro B (2000a). Formation, long-term stability and fate of non-extractable ^{14}C-PAH-residues in contaminated soils. In: Wise DL, Trantolo DJ, Cichon EJ Inyang II, Stottmeister U (eds) Remediation of hazardous waste contaminated soils. 2nd Edition, Marcel Dekker, Inc, New York, Chapter 20, pp 429–448

Eschenbach A, Mescher H, Wienberg R, Mahro B (2000b) Humifizierung von Schadstoffen. In: Michels J, Track T, Gehrke U, Sell D (eds) Biologische Verfahren zur Bodensanierung. Grün-weiße Reihe des BMBF, Berlin (in press)

Fernando T, Bumpus JA, Aust SD (1990) Biodegradation of TNT (2,4,6-Trinitrotoluene) by phanerochaete chrysosporium. Appl Environ Microbiol 56: 1666–1671

Führ F, Ophoff H, Burauel P, Wanner U, Haider K (1998) Modification of the definition of bound residues. In: Senate Commission for the Assessment of Chemicals Used in Agriculture (eds) Pesticide bound residues in soil. Workshop, September 3.–4. 1996, Deutsche Forschungsgemeinschaft, Wiley-VCH, Weinheim

Goodin JD, Webber MD (1995) Persistence and fate of anthracene and benzo(a)pyrene in municipal sludge treated soil. J Environ Qual 24: 271–278

Guthrie EA, Pfaender F (1998) Reduced Pyrene Bioavailability in Microbially Active Soils. Environ Sci Technol 32: 501–508

Guthrie EA, Bortiatynski JM, Van Heemst JD, Richman JE, Hardy KS, Kovach EM, Hatcher PG (1999) Determination of ^{13}Cpyrene sequestration in sediment microcosms using flash pyrolysis-GC-MS and ^{13}C NMR. Environ Sci Technol 33: 119–125

Haider K, Martin JP (1988) Mineralization of ^{14}C-labeled humic acids and of humic-acid bond ^{14}C xenobiotics by *Phanerochaete chrysosporium*. Soil Biol Biochem 20: 425–429

Haider K, Schäffer A (2000) Umwandlung und Abbau von Pflanzenschutzmitteln in Böden – Auswirkungen auf die Umwelt. Thieme-Verlag. in press

Haider K, Spiteller M, Reichert K, Fild M (1992) Derivatization of humic compounds: An analytical approach for bound organic residues. Intern J Environ Anal Chem 46: 201–211

Haider K, Spiteller M, Wais A, Fild M (1993) Evaluation of the binding mechanism of anilazine and its metabolites in soil organic matter. Intern J Environ Anal Chem 53: 125

Hatzinger PB, Alexander M (1995) Effect of aging of chemicals in soil on their biodegradability and extractability. Environ Sci Technol 29: 537–545

Hund-Rinke K, Kördel W (2000) Biologische Bodensanierung unter der Lupe. Umwelt 30: 51–53

Hupe K, Lüth JC, Heerenklage J, Stegmann R (1996) Kompost als Mittel zur Reinigung ölkontaminierter Böden. Altlastenspektrum 4: 182–189

Hosler KR, Bulman TL, Fowlie PJA (1988) Der Verbleib von Naphthalin, Anthracen und Benz(a)pyren im Boden bei einem für die Behandlung von Raffinerieabfällen genutztem Gelände. In: Wolf K, van den Brink WJ, Colon FJ (eds) Altlastensanierung '88. Kluwer Academic Publisher, Dordrecht/Boston/London, pp 111–113

Kästner M 2000 "Humification" Process or formation of refractory soil organic matter. In: Rehm HJ, Reed G, Pühler A, Stadler P (eds) Biotechnology, Vol.11 b, Environmental Processes – Soil decontamination, Waste gas treatment, Potable water preparation. Wiley-VCH, Weinheim, pp 90–125

Kästner M, Breuer-Jammali M, Mahro B (1994) Enumeration and characterization of soil microflora from hydrocarbon-contaminated soil sites able to mineralize polycyclic aromatic hydrocarbons (PAH). Appl Environ Biotechnol 41: 267–273

Kästner M, Lotter S, Heerenklage J, Breuer-Jammali M, Stegmann R, Mahro B (1995) Fate of ^{14}C-labeled anthracene and hexadecane in compost-manured soil. Appl Microbiol Biotechnol 43: 1128–1135

Kästner M, Streibich S, Beyrer M, Richnow HH, Fritsche W (1999) Formation of bound residues during microbial degradation of (^{14}C)anthracene in soil. Appl Environ Microbiol 65: 1834–1842

Kelsey JW, Kottler BD, Alexander M (1997) Selective chemical extractants to predict bioavailability of soil-aged organic chemicals. Environ Sci Technol 31: 214–217

Knicker H, Bruns-Nagel D, Drzyzga O, v Löw E, Steinbach K (1999) Characterization of ^{15}N-TNT residues after an anaerobic/aerobic treatment of soil/molasses mixtures by solid-state ^{15}N NMR spectroscopy. Determination and optimization of the relevant NMR spectroscopic parameters. Environ Sci Technol 33: 343–349

Lenke H, Warrelmann J, Daun G, Hund K, Sieglen U, Walter U, Knackmuss HJ (1998) Biological treatment of TNT-contaminated soil, 2. Biologically induced immobilization of the contaminants and full-scale application. Environ Sci Technol 32: 1964–1971

Lenke H, Achtnich C, Knackmuss HJ (2000) Perspectives of bioelimination of polynitroaromatic compounds. In: Spain JC, Hughes JB, Knackmuss HJ (eds) Nitroaromatic compounds and explosives. Lewis Publishers, Boca Raton, pp 91–126

Lotter S, Brumm A, Bundt J, Heerenklage J, Paschke A, Steinhart H, Stegmann R (1993) Carbon balance of a PAH-contaminated soil during biodegradation as a result of the addition of compost. In: Arendt F, Annokkee GJ, Bosman R, van den Brink WJ (eds) Contaminated soils '93. Kluwer Acad Publishers, pp 1235–1245

Mahro B, Kästner M (1993) PAK-Altlasten – Bewertung der mikrobiellen Sanierung. Spektrum der Wisenschaft: 97–100

Nanny MA, Bortiatynski JM, Hatcher PG (1997) Noncovalent interactions between acenaphtenone and dissolved fulvic acid as determined by ^{13}C NMR T1 relaxation measurementes. Environ Sci Technol 31: 530–534

Nieman JKC, Sims RC, Sims JL, Sorensen DL, McLean JE, Rice JA (1999) [^{14}C]Pyrene bound residues evaluation using MIBK fractionation method for creosote-contaminated soil. Environ Sci Technol 33: 776–781

Northcott GL, Jones K (2000) Experimental approaches and analytical techniques for determining organic compound bound residues in soil and sediment. Environmental Pollution 108: 19–43

Pignatello JJ (1989) Sorption dynamics of organic compounds in soils and sediments. In: Sawhney BL, Brown K (eds) Reactions and movement of organic chemicals in soils. 22, Soil Sci Soc Am Inc, Madison, pp 45–80

Pignatello JJ, Xing B (1996) Mechanisms of slow sorption of organic chemicals to natural particles. Environ Sci Technol 30: 1–11

Preuss A, Frimel J, Diekert G (1993) Anaerobic transformation of 2,4,6-trinitrotoluene (TNT). Arch Microbiol 159: 345–353

Qiu X, McFarland MJ (1991) Bound residues formation in PAH contaminated soil composting using *Phanerochaete chrysosporium*. Hazardous Waste and Hazardous Materials 8: 115–126

Richnow HH, Seifert R, Hefter J, Kästner M, Mahro B, Michaelis W (1994) Metabolites of xenobiotica and mineral oil constituents linked to macromolekular organic matter in polluted environments. Adv Org Geochem 22: 671–681

Richnow HH, Eschenbach A, Mahro B, Seifert R, Wehrung P, Albrecht P, Michaelis W (1998) The use of ^{13}C-labelled polycyclic aromatic hydrocarbons for the analysis of their transformation in soil. Chemosphere 36: 2211–2224

Richnow HH, Eschenbach A, Mahro B, Kästner M, Annweiler E, Seifert R, Michaelis W (1999) The formation of nonextractable soil residues – a stable isotope approach. Environ Sci Technol 33: 3761–3767

Rieger PG, Knackmuss HJ (1995) Basic knowledge and perspectives on biodegradation of 2,4,6-Trinitrotoluene and related nitroaromatic compounds in contaminated soil. In: Spain J (ed) Biodegradation of nitroaromatic compounds. Plenum Press, New York, pp 1–18

Roberts TR, Klein W, Still GG, Kearney PC, Drescheer N, Desmoras J, Esser HO, Aharonson N, Vonk JW (1984) Non-extractable pesticide residues in soil and plants. Pure and Applied Chemistry 56: 945–956

Saxena A, Bartha R (1983) Microbial mineralization of humic acid-3,4.dichloroaniline complexes. Soil Biol Biochem 15: 59–62

Scheibner K, Hofrichter M, Herre A, Michels J, Fritsche W (1997) Screening for fungi intensively mineralizing 2,4,6-trinitrotoluene. Appl Microbiol Biotechnol 47: 452–457

Scheu S (1987) The role of substrate feeding earthworms (lumbricidae) for bioturbation in a beechwood soil. Oecologia 72: 192–196

Senesi N (1993) Organic pollutant migration in soils as affected by soil organic matter, Molecular and mechanistic aspects. In: Petruzelli D, Helfferich FG (eds) Migration and Fate of pollutants in soils and subsoils. Springer, Berlin, pp 47–74

Senesi N. (1994) Spectroscopic studies of metal ion-humic substance complexation in soil. In: International Society of Soil Science and Mexican Society of Soil Science, 15th World Congress of soil science. 3A: Commission II: Symposia, July 1994, Acapulco, Mexico, pp 384–402

Schnöder F, Mittelstaedt W, Führ F (1994) Das Verhalten von Benzo(a)pyren und Fluoranthen in einer Parabraunerde – Lysimeter- und Laborstudien. In: TU Berlin (ed) Biologischer Abbau von polycyclischen aromatischen Kohlenwasserstoffen. Schriftenreihe Biologische Abwasserreinigung 4, pp 217–230

Thorn KA (1997) Covalent binding of the reductive degradation products of TNT to humic substances examined by N-15 NMR. Division of Environ Chem Preprints of Extended Abstracts. 37: 305–306

19 Humification as a Remediation Strategy for TNT Contaminations: Applications and Limitations

R. Winterberg
Plambeck ContraCon GmbH, Peter Henlein-Straße 2–4, 27472 Cuxhaven, Germany

19.1 Objectives

Currently, soils from military sites, former ammunition factories and similar sites are simply dumped or remediated by thermal treatment. Low cost alternative microbiological remediation techniques based on humification processes have now been developed for sustainable soil treatment.

19.2 Principle of the *Plambeck ContraCon* Humification Process

The *Plambeck ContraCon* humification technology is based on a two-step process. The first step is a contaminant-transformation under anaerobic conditions, followed by an essential immobilisation step, performed under aerobic conditions. This immobilisation reaction has been described as a humification process (Fritsche 1998; Mahro 2000). This process has been developed using soil, which is contaminated mainly by 2,4,6-trinitrotoluene (TNT).

Due to the fact that TNT is not mineralised under conditions of soil remediation, the addition of organic substrates is essential. After having added these substrates in a concentration range from 15 to 25 % (weight-%, based on dry substance), the organic substrates are utilised by autochthonous microorganisms in the first phase of treatment. Substrate utilisation causes anaerobic conditions due to complete exhaustion of oxygen in the soil pores. Microbial growth conditions in this first step are comparable to agricultural ensilage.

Under these anaerobic conditions, TNT is transformed mainly to monoamino-dinitrolouenes (ADNT) as well as to diamino-mononitrotoluenes (DANT). Triaminotoluene (TAT) may be formed to a minor extent towards the end of the anaerobic phase, but its formation is not an inherent step of the *Plambeck ContraCon* process. Aiming at an optimum homogenisation of the soil/substrate-mixture, the soil is homogenised periodically by mixing or tilling. Nevertheless, the supply of microorganisms with oxygen remains a limiting factor due to the high oxygen consumption rates of facultative anaerobic acid-forming bacteria.

Oxygen limitation is stopped after having obtained a degree of TNT-transformation exceeding about 75 %. In this second phase of soil treatment, intensive mixing is combined with aeration by injecting air into the soil pile. Due to the presence of oxygen, the fermentation products and the remaining original compounds from the organic substrate are mineralised. Comparable to a composting process, the pile temperature raises up to about 65 °C. These aerobic processes are combined with the binding of TNT-transformation products into the soil matrix, thus producing a significant detoxification of the soil material.

19.3 Sequence and Techniques of the *Plambeck ContraCon* Humification Process

Following several years' development, the *Plambeck ContraCon* humification process has been assessed by a full-scale test on the site of the former ammunition factory "Werk Tanne" near Clausthal Zellerfeld (Lower Saxony). The project was part of the joint research project "Processes for the Bioremediation of Soil", which has been initiated and financially supported by the German BMBF (Federal Ministry of Education and Research).

After excavation, the contaminated soil was sieved with a cut-point of 60 mm. Material exceeding a diameter of 60 mm may be optionally crushed and then returned to the soil to be treated. The organic substrate was then added and intensively mixed with the soil material. This soil/substrate-mixture was deposited in triangular piles in a closed tent (fig. 19.1). Soil-mixing was performed with a

Fig. 19.1. Full-scale plant at "Werk Tanne"-site near Clausthal-Zellerfeld (photo: IABG)

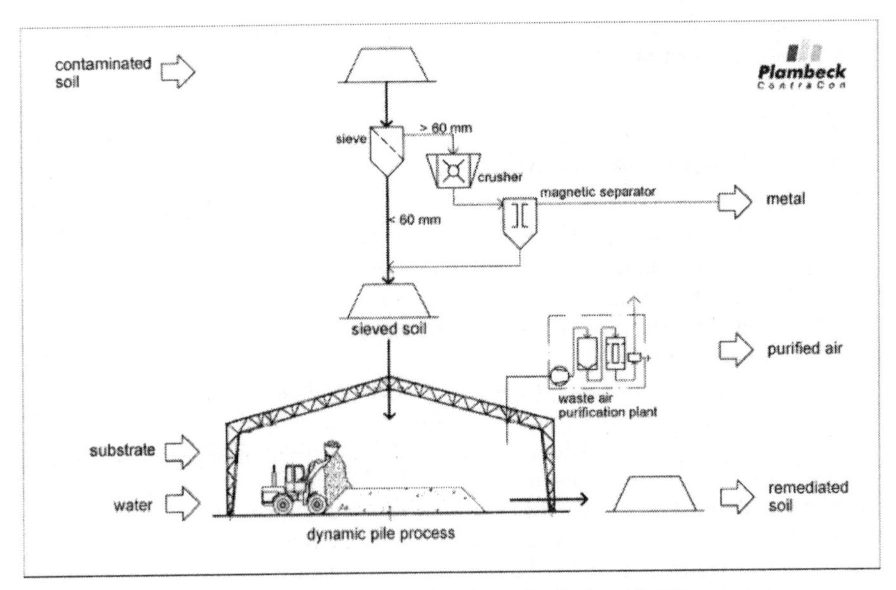

Fig. 19.2. Flow chart of the Plambeck ContraCon dynamic pile humification process

wheel-loader equipped with a special shovel for sieving and mixing soils (fig. 19.2 and fig. 19.3).

The oxygen-supply of the microorganisms was controlled by adjusting the frequency of soil mixing processes, as well as by injection of air into the soil pile during the aerobic phase.

Fig. 19.3. Soil mixing process at "Werk Tanne"-site

The treated soil then was removed and recycled according to the prescriptions of the public authorities.

19.4 Results of Soil Treatment

The full-scale test on the "Werk Tanne"-site showed treatment efficiencies of about 99.9 %. For example, a TNT-concentration of 1160 mg kg^{-1} (dry substance; 917 mg kg^{-1} after substrate-addition) was reduced to 1.3 mg kg^{-1} during a 6 months treatment period. Formation of transformation products that remained in the soil was negligible.

Besides TNT, additional explosives-related contaminants had been detected in the "Werk-Tanne" soil material (i.e. 34.6 mg 2-amino-2,6-dinitrotoluene kg^{-1}; 61.0 mg 4-amino-2,6-dinitrotoluene kg^{-1}; 60.8 mg 4-amino-2,6-dinitrobenzoic acid kg^{-1}; 8.2 mg 1,3,5-trinitrobenzene kg^{-1}). These contaminants were eliminated completely (dinitrotobenzoic acid, trinitrotoluene) or nearly completely, respectively. The elimination of contaminants was confirmed by a significant detoxification of the soil material, as could be shown by different assays (Dahn 2000; Hund-Rinke and Kördel 2000).

Additionally, the irreversible binding of contaminants as well as a sustainable long-term behaviour of the treated soil had to be proven. Corresponding to the formation of substances with high molecular weight, the immobilisation reactions were described as humification processes (Fritsche 1998; Mahro 2000). Drzyzga et al. (1998) and Hund-Rinke and Kördel (2000) proved the immobilisation of ^{14}C-labelled TNT during performance of the *Plambeck ContraCon* process with efficiencies comparable to large degrees of TNT-elimination. A first indication of the formation of substances completely different from the original contamination was obtained by Banholczer et al. (2000) investigating the structures of the formed macromolecules. Hence, these current investigations, performed by ^{15}N-NMR spectroscopy, proved that sequestration (the encasement of the original contaminant) is not responsible for the TNT-elimination. The investigations concerning the important question of sustainability of the remediation process are currently ongoing. The results available to date (Hund-Rinke and Kördel 2000; Mahro 2000) confirm the idea of a sustainable soil treatment by humification.

All these results allowed the treated soil to be refilled at the "Werk Tanne"-site, in accordance with the prescriptions of the public authorities.

19.5 Applications and Limitations of the Humification Process

Due to the full-scale tests, scale-up risks have been eliminated. Thus, a full-scale application of the microbiological treatment of TNT-contaminated soils can be recommended. The proof of the sustainable long term behaviour of the treated soil is the only remaining work to be done. Nevertheless, the available results are in accordance with the idea of a sustainable contaminant humification.

Most of the recent investigations dealt with TNT as the main contaminant. Hence, it has to be shown by feasibility studies, that the obtained results are applicable to other explosives or explosive-related compounds.

Comparable to microbiological mineralisation technologies as well as to established remediation technologies in general, maximum contaminant concentrations have to be defined in order to guarantee reproducible high treatment efficiencies in economically acceptable remediation periods.

19.6 Summary

Soil remediation by microbiological mineralisation is well established and has been applied in numerous projects. Additionally, detoxification by microbiologically mediated immobilisation has been developed in Germany for several years, aiming at recalcitrant contaminants, which are not being mineralised by microorganisms under full-scale remediation conditions. The microbiological technology has been developed as a low cost alternative to relatively expensive thermal treatment technology, which was the only state of the art-technique for the remediation of former ammunition factories and similar sites.

The comprehensive joint research project "Processes for the Bioremediation of Soil" which has been initiated by the German BMBF (Federal Ministry of Education and Research) has enhanced the development of these humification technologies significantly, especially the remediation of 2,4,6-trinitrotoluene (TNT)-contaminated soils. Currently, the humification strategy for TNT-contaminated sites allows soil remediation in a wide range of contaminant concentration with high efficiencies and low costs. Thus, the development has been very successful.

Nevertheless, the principle of humification involves an inherent problem: the characterisation of the humic remediation product is, similar to natural humic substances, very complicated. Thus, the sustainability of the remediation has been discussed intensively. An investigation programme is underway for the assessment of the long-term behaviour of the treated soil. The results available up to now confirm the idea of a sustainable soil treatment by humification.

References

Banholczer A, von Löw E, Fründt J, Bruns-Nagel D (2000) Dynamisches Beetverfahren: Analyse der Bindungsstruktur nicht extrahierbarer TNT-Transformationsprodukte. lecture, 2. Statusseminar zum BMBF-Forschungsverbund "Langzeit- und Remobilisierungsverhalten von Schadstoffen bei der biologischen Bodensanierung", Bremen, Germany, 22.02.2000

Dahn A (2000) Targets and Results of Joint Projekt 6: True-to-Scale-Test of Biological Remediation Processes ar Site "Tanne" near Clausthal-Zellerfeld, State of Lower Saxony, Germany. lecture, Consoil 2000, Satellite Seminar "Abandoned Armament Sites", 21.09.2000

Drzyzga O, Bruns-Nagel D, Gorontzy T, Blotevogel KH, Gemsa D, von Löw E (1998) Incorporation of ^{14}C-Labeled 2,4,6-Trinitrotoluene Metabolites into Different Soil Fractions after Anaerobic and Anaerobic-Aerobic Treatment of Soil/Molasses Mixtures. Environ Sci Technol 32/22: 3529–3535

Fritsche W (1998) Biologische Sanierung von Rüstungsaltlasten. lecture, Statusseminar zum BMBF-Forschungsverbund "Biologische Verfahren zur Bodensanierung", Clausthal-Zellerfeld, Germany, 06.–07.05.1998

Hund-Rinke K, Kördel W (2000) Biologische Bodensanierung – Beurteilung des Langzeitverhaltens und der Remobilisierung festgelegter Schadstoffe (PAK sowie TNT und Metabolite). lecture, 2. Statusseminar zum BMBF-Forschungsverbund "Langzeit- und Remobilisierungsverhalten von Schadstoffen bei der biologischen Bodensanierung", Bremen, Germany, 22.02.2000

Mahro B (2000) Aufgaben und Fragestellung des Statusseminars "Langzeit- und Remobilisierungsverhalten von Schadstoffen im Boden". lecture, 2. Statusseminar zum BMBF-Forschungsverbund "Langzeit- und Remobilisierungsverhalten von Schadstoffen bei der biologischen Bodensanierung", Bremen, Germany, 22.02.2000

Winterberg R, von Löw E, Held T (1998) Dynamisches Mietenverfahren zur Sanierung von Rüstungsaltlasten. TerraTech 3/1998: 39–41

20 Strategies to Improve PAH Bioavailability: Addition of Surfactants, Ozonation and Application of Ultrasound

A. Tiehm, M. Stieber
Water Technology Center, Karlsruher Straße 84, 76139 Karlsruhe, Germany

20.1 Introduction

Hydrophobic compounds, such as polycyclic aromatic hydrocarbons (PAHs), mineral oils, or halogenated chemicals, represent pollutants of high ecotoxicological relevance at many contaminated sites. Most of these pollutants are biodegradable (Cerniglia 1992; van Hylckama Vlieg and Janssen 2000; Wischnak and Müller 2000) but their rate of biodegradation is limited by low bioavailability. In general, microbial uptake and degradation of pollutants predominantly occurs in the aqueous phase. Pollutants present as crystals (Stucki and Alexander 1987; Tiehm 1994) or in nonaqueous phase liquids (Mukherji and Weber Jr. 1998), as well as compounds sorbed by organic or inorganic matter (Guerin and Boyd 1997; Harms and Zehnder 1995), first have to be transferred into the aqueous phase before biodegradation is possible (Mahro 2000). Con-sequently, recently published models taking into account mass transfer and micro-biological parameters correlate well with measured biodegradation kinetics of hydrophobic model compounds (Ghoshal and Luthy 1998; Mulder et al. 1998a).

Mass transfer processes associated with contaminant release into the water phase limit the rate of removal of hydrophobic pollutants rather than the explicit aqueous-phase biodegradation kinetics. In case of PAH-contaminated soil, binding of the hydrocarbons could totally prevent a microbial attack (Erickson et al. 1993; Weissenfels et al. 1992). Slow mass transfer has to be considered the most important factor limiting biotreatment of contaminated sites (Luthy et al. 1994; Tiehm et al. 1997).

It is the objective of this paper to discuss technical options to increase the bioavailability of pollutants such as PAHs. The paper focuses on three different measures: (1) addition of surfactants, (2) treatment with ozone, and (3) application of ultrasound. For each technique, the theoretical background and experimental results are presented. The potential, the limits, and possible drawbacks are discussed.

20.2 Polycyclic Aromatic Hydrocarbons

In our studies, polycyclic aromatic hydrocarbons (PAHs) have been either the main pollutants in contaminated soil, or been used as hydrophobic model pollutants. Because of their ecotoxicological relevance and their frequent occurrence in the environment, the US Environmental Protection Agency (EPA) numbers PAHs among the priority pollutants. Several hundred PAHs are known. In view of the complexity of this group, the EPA has selected 16 non-substituted PAHs as an analytical standard that has been globally accepted. PAHs consist of two or more condensed benzene rings. Their physico-chemical properties vary significantly according to their number of aromatic rings and molecular weight. Table 20.1 gives the water solubility and octanol-water coefficients of selected PAHs. More detailed information can be found in Kästner (2000). Low PAH bio-availability as a limiting factor for biodegradation has been recognised early (Wodzinski and Bertolini 1972). PAHs have been used as model compounds in numerous bioavailability studies (e.g. Guerin and Boyd 1992; Harms and Zehnder 1994; Kotterman et al. 1998; Laha and Luthy 1991; Mihelcic et al. 1993; Mulder et al. 1998 b, Mulder et al. 2000; Ortega-Calvo et al. 1995; Tsomides et al. 1995; Volkering et al. 1992; Wang et al. 1998; Wodzinski and Coyle 1974).

Table 20.1. Physico-chemical properties of PAHs.

PAH	Molecular Weight	Water Solubility[a] [mg L^{-1}]	log Kow[b]
Naphthalene (NAP)	128	30.0	3.37
Acenaphthylene (ACY)	154	3.47	4.33
Acenaphthene (ACE)	152	3.93	4.07
Fluorene (FLU)	166	1.98	4.18
Phenanthrene (PHE)	178	1.29	4.46
Anthracene (ANT)	178	0.07	4.45
Fluoranthene (FLA)	202	0.26	5.33
Pyrene (PYR)	202	0.14	5.32
Benzo[a]anthracene (BaA)	228	0.014	5.61
Chrysene (CHR)	228	0.002	5.61
Benzo[b]fluoranthene (BbF)	252	0.0012	6.57
Benzo[k]fluoranthene (BkF)	252	0.0006	6.84
Benzo[a]pyrene (BaP)	228	0.0038	6.04
Dibenzo[a,h]anthracene (DBA)	278	0.0005	6.75
Benzo[g,h,i]perylene (PER)	276	0.00026	7.23
Indeno[1,2,3-cd]pyrene (IND)	276	0.062	7.66

[a] according to Sims and Overcash (1983)
[b] octanol-water partition coefficient

20.3 Addition of Surfactants

20.3.1 Theory of Surfactant Enhanced Bioavailability

Surfactants consist of a hydrophilic and a hydrophobic moiety thus mediating between immiscible phases. The amphiphilic molecules accumulate at surfaces and interfaces thereby decreasing surface and interfacial tension. Above a specific threshold, known as critical micellar concentration (CMC), surfactant molecules form micellar aggregates in the water phase. The surfactant's hydrophilic parts in the micelles are directed to the water phase and the hydrophobic parts are directed to the micelles core forming a hydrophobic pseudophase inside. Incorporation of hydrophobic compounds in the micelles is termed solubilisation. In equilibrium, the amount of solubilised PAHs linearly depends on the surfactant concentration above the CMC. The solubilisation ratio (SR) describes the mass of solubilised compounds per mass of surfactant and varies considerably for different surfactants (Edwards et al. 1991; Klevens 1950; Myers 1991; Tiehm 1994).

Both the lowering of the surface tension and solubilisation contribute to a facilitated transfer of hydrophobic pollutants into the aqueous phase. The enhanced desorption rate results in enhanced biodegradation. Fig. 20.1 illustrates the mechanism of surfactant enhanced PAH bioavailability.

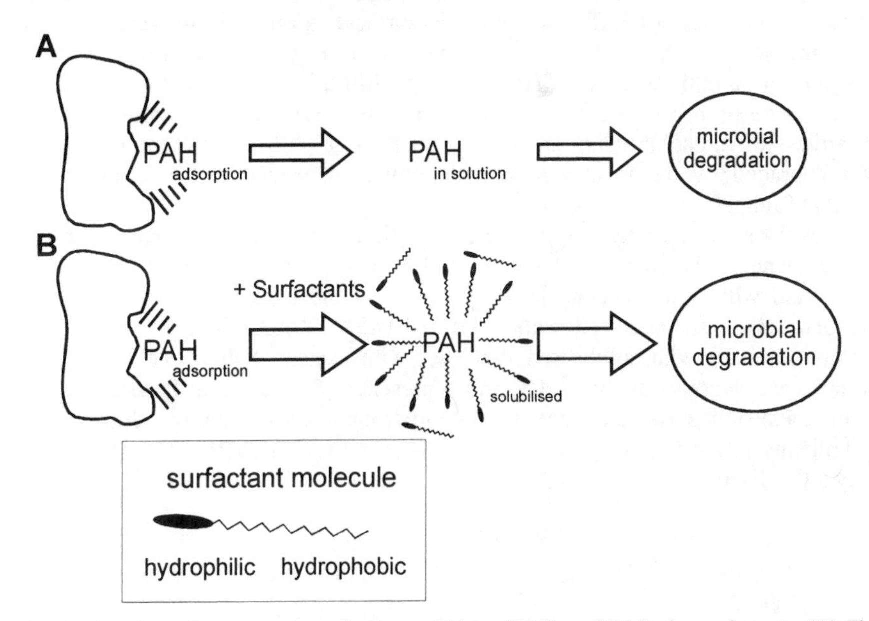

Fig. 20.1. Schematic presentation of enhanced bioavailability of PAHs by surfactants: (A) The mass transfer rate of adsorbed PAHs into the aqueous phase limits the rate of microbial degradation. (B) Enhanced mass transfer and biodegradation of PAHs in the presence of surfactant micelles.

Washing with surfactant solutions has been shown to be effective for the removal of hydrophobic compounds such as hydrocarbons (Barczewski et al. 2000; Gannon et al. 1989; Pennel et al. 1993), polychlorinated biphenyls (Abdul and Gibson 1991) or tetrachloroethene (Fountain et al. 1991) from contaminated soil. However, a surfactant-enhanced mass transfer only results in a successful improvement of bioavailability if the surfactants are non-toxic.

20.3.2 Toxicity of Surfactants

The toxicity of surfactants is one of the key parameters determining their suitability to enhance pollutant biodegradation. It has been demonstrated for several soil bacteria that the toxicity of surfactants is related to their Hydrophil-Lipophil-Balance (HLB) values (Cserhati et al. 1991). The HLB value describes the overall lipophilicity of a surfactant. The lipophilicity decreases with increasing HLB values.

It has been shown that lipophilic surfactants are buried in the lipid layer of liposomes resulting in an increased permeability of the membranes. Hydrophilic surfactants had no effect on membrane permeability (Cserhati et al. 1982; Cserhati et al. 1984). The surfactant "Triton X-100" (HLB 13.5) reduced phenanthrene degradation by a mixed culture (Laha and Luthy 1991) and completely inhibited the sterol transformation of *Mycobacterium fortuitum* (Atrat et al. 1992).

In our study, the relevance of the HLB values was tested for PAH degrading bacteria. The effect of different non-ionic surfactants on PAH degradation was investigated. The surfactants were applied at concentrations resulting in similar amounts of solubilised PAHs (Tiehm 1994). Utilisation of phenanthrene and of fluoranthene by mixed cultures enriched on these PAHs occurred within 7 days, regardless of the solubilising surfactant. Whether solubilised fluorene or pyrene was degraded by the respective mixed cultures depended on the solubilising agents (Table 20.2).

Growth of *Mycobacterium* sp. which is able to utilise acetate, phenanthrene, fluoranthene, and pyrene (Boldrin et al. 1993) was tested with the solubilised PAHs and with 3 mM acetate in the presence of the surfactants. No growth of *Mycobacterium* sp. occurred with "Marlipal 013/90" or "Triton X-102" in the medium, neither with solubilised PAH nor with acetate. Solubilised PAHs and acetate were degraded within 7 days in the presence of the other surfactants.

In conclusion, surfactants that are more hydrophobic and seem to exhibit better solubilising properties (Atrat et al. 1992; Kile and Chiou 1989) are tolerated by only a few bacteria.

Table 20.2. Effect of the Hydrophil-Lipophil-Balance (HLB) of surfactants on the biodegradation of solubilised PAHs by mixed cultures and *Mycobacterium* sp. In the case of *Mycobacterium sp.* degradation of acetate was also tested in the presence of the different surfactants.

Surfactant	Weight SR[a]	Mixed cultures[b]: Degradation of solubilised PAHs				*Mycobacterium*:
	[mg g^{-1}]	FLU	PHE	FLA	PYR	PAHs or Acetate
5 mM Marlipal 013/90 **HLB: 13.3**	34.7	-[c]	+[c]	+	-	-[d]
4 mM Triton X-102 **HLB: 14.6**	26.4	-	+	+	-	-
8 mM Genapol X-150 **HLB: 15.0**	14.1	-	+	+	+	+
3 mM Brij 35 **HLB: 17.0**	22.5	+	+	+	+	+
6 mM Arkopal N-300 **HLB: 17.0**	10.7	+	+	+	+	+
4 mM Sapogenat T-300 **HLB: 17.0**	16.7	+	+	+	+	+

[a] solubilisation ratio determined for phenanthrene
[b] mixed cultures were adapted to the degradation of the respective PAH
[c] +/- : degradation/no degradation of PAH within 7 d of incubation
[d] +/- in case of acetate: increase/no increase in optical density within 7 d

20.3.3 Surfactant Enhanced Bioremediation of PAH Contaminated Soil

A study was undertaken with contaminated soil that was excavated from a former coal gasification site (Tiehm et al. 1997). The concentration of PAHs in the soil ranged between 100 mg kg^{-1} (ACE) and 1,000 mg kg^{-1} (PHE). Bioremediation was studied in laboratory percolator systems at 20 °C (Fig. 20.2). Soil columns had a diameter of 77 mm and contained 2.7 kg of soil. The whole system was operated

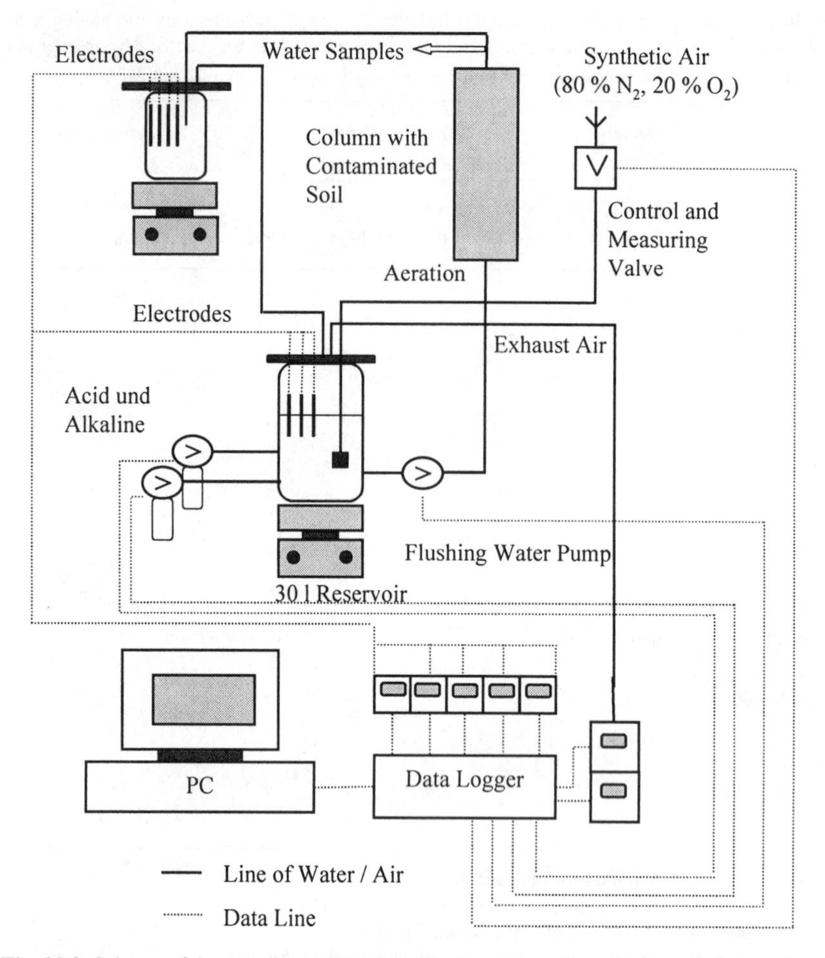

Fig. 20.2. Scheme of the percolator system used in the surfactant studies.

with 27 l of medium that was recirculated with a flow rate of 12 l h^{-1}. O_2 consumption and CO_2 production were measured continuously in the exhaust air. The off-gas was cooled to 2 °C in order to minimise losses of volatile contaminants. The pH was regulated automatically between pH 6.9 and pH 7.1.

In the percolator experiments, the effect of addition of the non-ionic surfactant "Sapogenat T-300" to the recirculating medium on mobilisation and biodegradation of PAHs was investigated. This surfactant was selected due to its low toxicity (Table 20.2). One soil column was operated with a mineral medium. The other column was amended with the surfactant twice in a concentration of 2.5 mM, after 192 and 1150 h of operation.

In the column operated with the mineral medium without surfactant addition the concentration of PAHs in the aqueous phase increased at the beginning of soil flushing. Naphthalene transiently reached a concentration of 7.0 mg L^{-1} after 20 h of percolation, and fluorene and phenanthrene reached concentrations of about 200 µg l^{-1}. The less soluble PAHs anthracene, fluoranthene, and pyrene were detected at concentrations below 50 µg l^{-1}. Due to the growth of PAH-degrading bacteria, the concentration of dissolved PAHs decreased rapidly. In later stages PAH concentrations remained below 30 µg/L. The surface tension decreased only from 71.4 to 69.8 mN/m during the 53 d of incubation indicating that no biosurfactants had been produced.

In the second column, percolation was undertaken with a mineral medium for 8 days. Afterwards, the surfactant Sapogenat T-300 was added twice in a concentration of 2.5 mM. Surface tension was reduced to 43.5 mN m^{-1} after surfactant addition. It was calculated that 2.5 mM surfactant corresponded to 2.3 g l^{-1} Dissolved Organic Carbon (DOC). After a recirculation period of 22 and 48 h, a DOC of 2.1 g l^{-1} was determined in the flushing water. This indicated that about 10 % of the surfactant was sorbed to the soil.

The addition of surfactant facilitated the desorption of PAHs from the contaminated soil. Immediately after surfactant addition, an increase of PAHs in the aqueous phase was observed (Fig. 20.3). The PAHs were solubilised in concentrations several times above their water solubility (Table 20.1). For example, aqueous phase concentrations of fluoranthene and pyrene temporarily were about 25 times and of chrysene about 500 times above their water solubility. Subsequently, concentrations of solubilised PAHs decreased again (Fig. 20.3) due to an increasing microbial PAH degradation as was demonstrated by CO_2 formation (Fig. 20.4). Since surfactant degrading microorganisms were detected first after 350 h, the enhanced microbial activity could be attributed to increased bioavailability of the pollutants.

Our studies (Tiehm and Fritzsche 1995; Tiehm et al. 1997) demonstrate that even more hydrophobic PAHs than used in other studies (Volkering et al. 1995; Guha and Jaffe 1996a; Guha and Jaffe 1996b; Guha et al. 1998) are available in the solubilised state, i.e. although the PAHs were originally enclosed by surfactant molecules. In this context, it should be noted that there is a constant dynamic exchange of aromatic hydrocarbons between surfactant micelles and the water phase (Almgreen et al. 1979). The diffusion out of the micelle′s core into the aqueous medium enables microbial uptake of the PAHs from the water phase.

It has been shown in short term experiments over 100 min that dissolved naphthalene is degraded more rapidly than solubilised naphthalene (Volkering et al. 1995). This finding contributes to a better understanding of the mass transfer processes, but is not critical with respect to a practical surfactant application. Our experiments clearly demonstrate that the bioavailability of PAHs in contaminated soil is significantly enhanced by surfactant addition.

After 350 h, it was impossible to distinguish between CO_2 evolution related to the degradation of PAHs and surfactant. Nevertheless, the slow surfactant degradation did not result in complete consumption of oxygen during 54 d of incubation. After the second addition of 2.5 mM Sapogenat T-300 in the seventh week, microorganisms were adapted to this substrate and rapidly metabolised the surfactant. This was indicated by an increase of surfactant degrading bacteria in the medium, a sharp increase of CO_2 formation, and a decrease of dissolved oxygen to undetectable levels in the percolating aqueous phase.

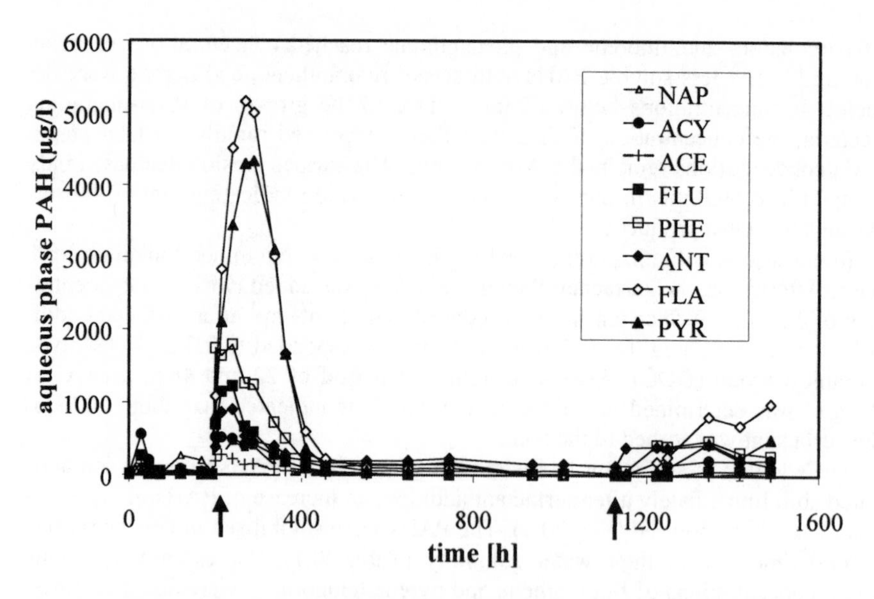

Fig. 20.3. Aqueous phase concentrations of PAHs during the percolation with 2 x 2.5 mM surfactant. Arrows indicate surfactant addition.

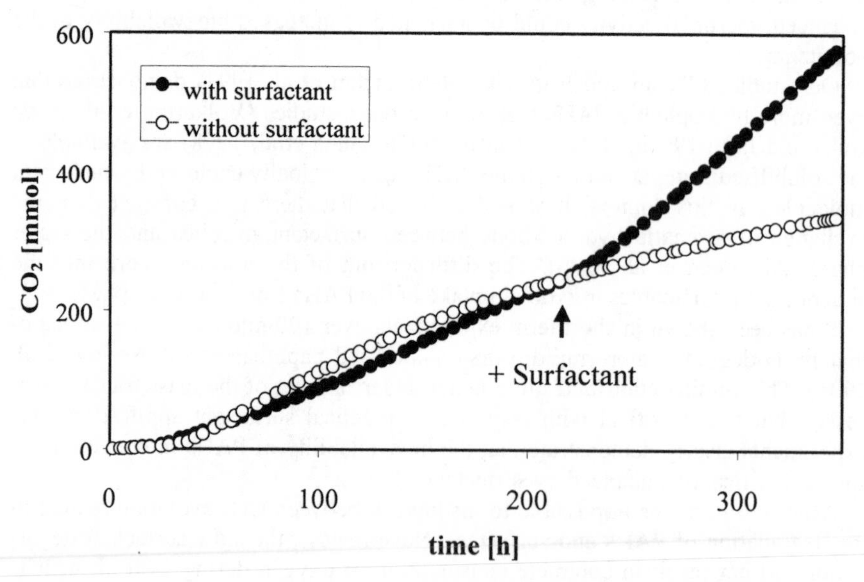

Fig. 20.4. Enhanced formation of CO_2 after addition of Sapogenat T-300. Surfactant degrading bacteria were not detected before 350 h of percolation.

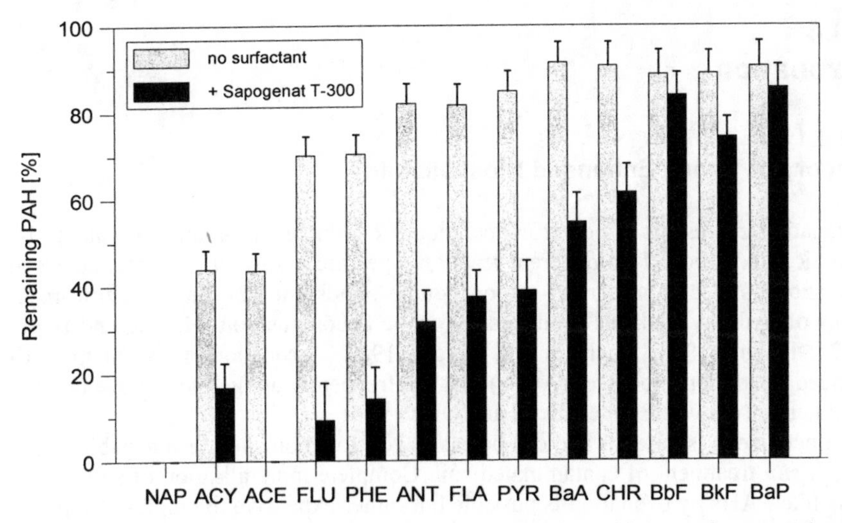

Fig. 20.5. PAHs remaining in the soil after the percolation without surfactant and with 2 x 2.5 mM Sapogenat T-300 (error bars represent standard deviations of five replicates).

Remaining PAH concentrations were significantly reduced by the addition of Sapogenat T-300. Acceleration of PAH degradation was most pronounced for the PAHs with three or four rings (Fig. 20.5). Biodegradation of PAHs in total was doubled as compared to the treatment without surfactant. The lower PAH content of the surfactant-treated soil was in correlation with a reduced toxicity of the soil eluates. In the bioluminescence tests the inhibition by the undiluted eluate was reduced to 41 % and the EC20 increased to 14.7 % (Table 20.3).

In general, surfactant addition is a suitable technique to stimulate PAH mobilisation and biodegradation in contaminated soil. From a microbiologist's point of view, the major obstacle for an *in-situ* application seems to be the biodegradation rate of the surfactant itself. As the microorganisms adapt to the surfactant, the degradation rate increases which may result in oxygen depletion and soil clogging. These drawbacks are less important under on-site conditions.

Table 20.3. Effect of surfactant addition on PAH biodegradation and reduction of eluate toxicity.

	PAH [mg]	Toxicity (bioluminescence test) Original Eluate [%]	EC20[a] [%]
Before Treatment	13.030	91.3	1.4
After Treatment			
Without Surfactant	8.450	69.6	4.7
With Surfactant	3.752	40.9	14.7

[a] EC20: concentration resulting in 20 % inhibition

20.4 Ozonation

20.4.1 Theory of Ozone Enhanced Bioavailability

Ozonation represents a powerful technique to mineralise organic pollutants by chemical oxidation. Two different reaction types are involved: (1) direct attack of the ozone molecule at carbon-carbon double bonds and (2) the unspecific reactions of hydroxyl radicals produced by ozone decomposition (Hoigné and Bader 1975; Hoigné 1988; Staehelin and Hoigné 1982). Ozonation is one of the Advanced Oxidation Processes (AOPs) that are frequently applied to remove organic pollutants from waste water (Scott and Ollis 1995).

Since ozone is gaseous and can be easily generated on site it is amenable to use for *in situ* treatment of contaminated soil. Complete mineralisation of soil pollutants like PAHs by ozonation is possible but rather expensive to employ. Bioremediation is a relative inexpensive but relatively slow process due to limited bioavailability. Therefore, the addition of low doses of ozone to increase pollutant bioavailability seems to be most reasonable. Ozonation of PAHs leads to the formation of partially oxidised reaction products with increased water solubility resulting in an increased mass transfer into the aqueous phase. The mechanism of increasing bioavailability by ozonation is illustrated in Fig. 20.6.

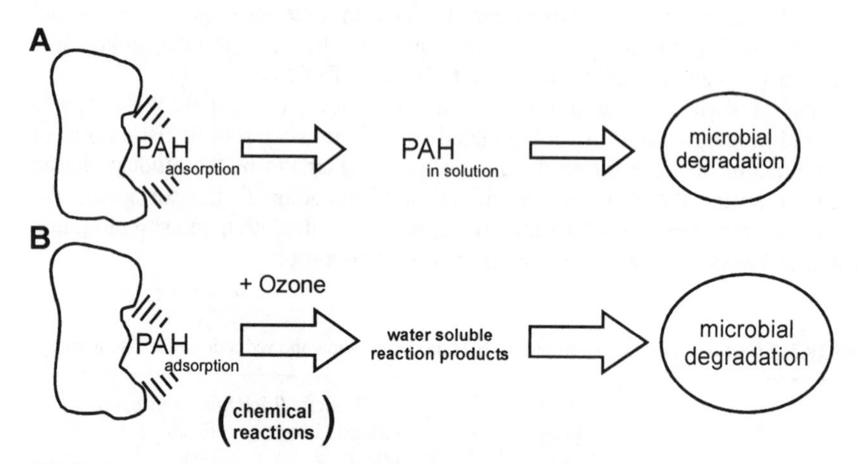

Fig. 20.6. Scheme of ozone-enhanced bioavailability. (A) Mass transfer without ozonation and (B) enhanced mass transfer by chemical transformation of PAHs into more hydrophilic reaction products.

20.4.2
Bioavailability and Biodegradation of Ozoned PAHs

Haeseler et al. (1993) investigated the effect of ozone pre-oxidation on the bioavailability of model PAHs. Suspended crystals of naphthalene and pyrene were ozoned (addition of 0.45 g–3.6 g ozone per g PAH) in stirred batch reactors. After ozonation, the samples rested for two days to achieve complete ozone decomposition. Then the samples were inoculated with mixed cultures of PAH-degrading bacteria and incubated at 20 °C.

In the case of naphthalene, almost complete elimination of the initial compound was obtained after 240 hours of ozonation. The chemical transformation of about 450 mg L^{-1} naphthalene resulted in the formation of about 300 mg L^{-1} dissolved organic compounds (Fig. 20.7). The ozonation time, i.e. the ozone doses applied, significantly affected the biodegradability of the reaction products. The products formed after 30 to 120 min of ozonation were poorly biodegradable: only about 20 % of the DOC was eliminated by biodegradation. In contrast, about 80 % of the products from 240 min ozonation were biodegraded. The dissolved reaction products formed by ozonation resulted in an increased toxicity in the bioluminescence test that was reduced again by the subsequent biological treatment.

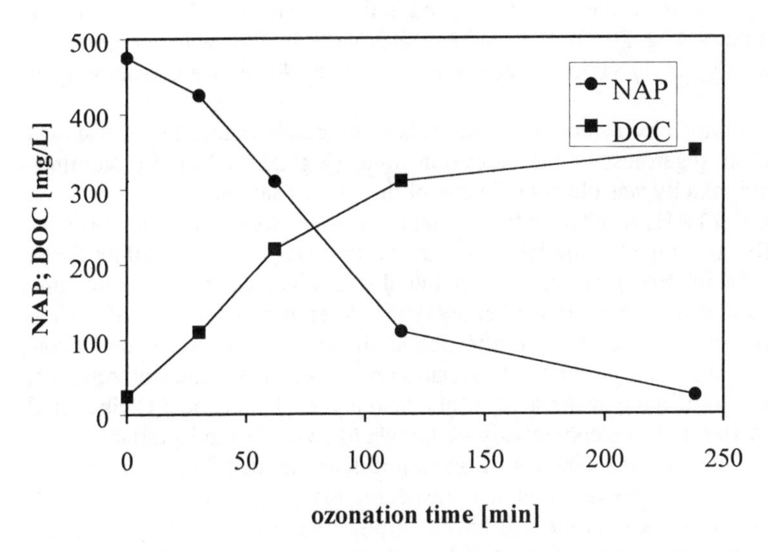

Fig. 20.7. Effect of ozonation on degradation of naphthalene crystals and formation of dissolved organic reaction products.

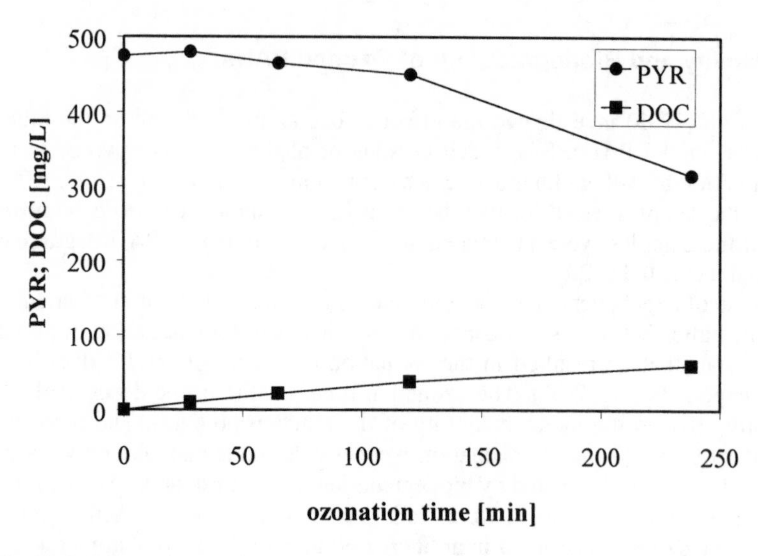

Fig. 20.8. Effect of ozonation on degradation of pyrene crystals and formation of dissolved organic reaction products.

Also ozonation of pyrene crystals resulted in the formation of dissolved organic reaction products (Fig. 20.8). The reaction rates were smaller when compared to naphthalene (Fig. 20.7) which is probably due to the lower water solubility of pyrene.

The DOC formed by pyrene ozonation was significantly reduced by subsequent biodegradation, regardless of the ozonation times (Fig. 20.9). In the bioluminescence test, no toxicity was observed in any of the ozoned samples.

Ozonation of PAHs resulted in the formation of water soluble reaction products in artificially contaminated model soils like quartz sand (Choi et al. 2000; Seibel et al. 1995; Seidel 1995), and in PAH polluted soils obtained from manufactured gas plant sites (Brown et al. 1997; Seidel 1995). A temporary increase of toxicity was observed that decreased again after the application of higher doses of ozone (Kautt 1997). The main products of ozonation of naphthalene, phenanthrene, and pyrene were formic acid, acetic acid, oxalic acid and o-phthalic acid (Seibel et al. 1995; Seidel 1995). These compounds are known to be easily biodegradable.

The self-decomposition kinetics of ozone influence its distribution in the subsurface on account of the very high and unspecific reactivity of the hydroxyl radical. The mineral surfaces in the soil have a strong catalytic reactivity with ozone and significantly affect the ozone half-live. By covering the reactive sites, soil humidity reduces the self-decomposition rate of ozone. With increasing water content of the soil, an increase of the ozone half-life is observed from a few minutes to several hours (Choi et al. 2000; Seibel et al. 1995). Under field conditions, ozone is sufficiently stable to be applied for *in-situ* remediation using air sparging technology.

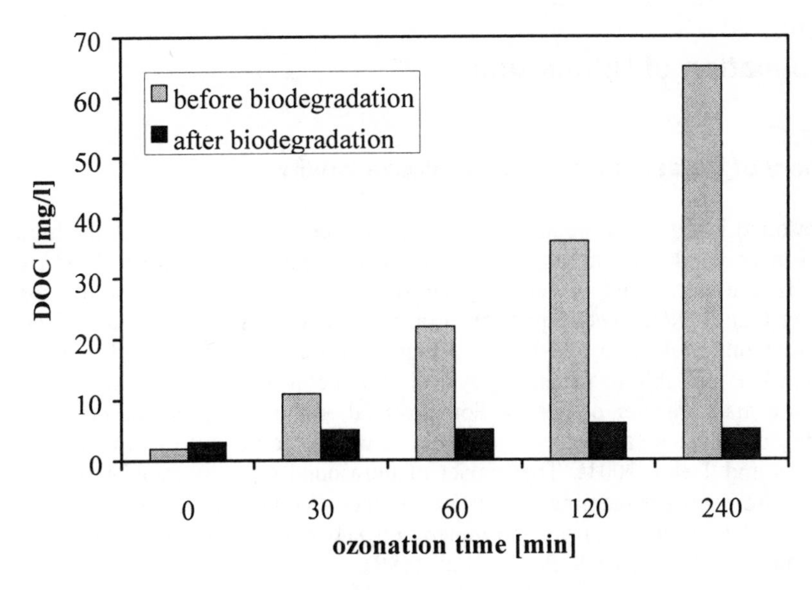

Fig. 20.9. Organic reaction products of ozoned pyrene suspensions before and after biodegradation.

The application of ozone for soil treatment has to consider further reactions with the natural soil matrix. Kautt (1997) showed that low ozone doses can mobilise soil organic matter to a large extent and decrease the pH down to values between 3 and 4. Longer ozonation resulted in increasing pH values and a significant decrease of dissolved organic carbon. Soil acidification and oxidative destruction of humic substances increased the mobilisation of heavy metals into the aqueous phase. These side reactions were less pronounced if only low doses of ozone were applied.

An important factor in integrating chemical oxidation and biodegradation is the impact of an oxidising environment on the microbial population. This is a genuine concern since oxidants have been shown to dramatically reduce biological activity in completely mixed systems with relatively long contact times such as slurry reactors. However, it was reported that ozonation reduced the number of colony forming units only moderately in soil systems (Kautt 1997) and biological activity rebounded quickly (Brown et al. 1997; Seidel 1995). In a field study, elimination of PAHs by the integrated ozonation/biodegradation treatment proved to be a very efficient technique (Brown et al. 1997).

20.5 Application of Ultrasound

20.5.1 Theory of Ultrasound Enhanced Bioavailability

Ultrasound is defined as sound with a frequency beyond that to which the human ear can respond. The normal range of human hearing is between 16 Hz and 18 kHz and ultrasound is generally considered to be sound with a frequency greater than 20 kHz. The impact of ultrasonic waves causes the periodical compression and rarefaction of a liquid. If high acoustic energy is applied to a liquid system it is possible to generate physical and chemical effects that significantly increase mass transfer processes. Sonophysical and/or sonochemical effects can modify dissolved and adsorbed substances as well as particles present in the liquid (Mason and Tiehm 2001). The impact of ultrasound on mass transfer processes results from microstreaming effects at surfaces (Young 1989) and – at higher ultrasound intensities – from the generation and collapse of cavitation bubbles (Mason 1991; Mason 1999; Pétrier et al. 1999).

Cavitation occurs at high acoustic intensities, particularly in the low and mid frequency range (between 20 kHz and 1,000 kHz), when the liquid is broken by sufficiently large negative pressure during rarefaction. Gas bubbles are generated which will grow by taking in gas and vapour from the liquid. They alternate in size according to the acoustic wave action. Finally they collapse violently within a few microseconds in a compression cycle. The phenomenon of sudden implosion of the bubbles is called cavitation (Fig. 20.10).

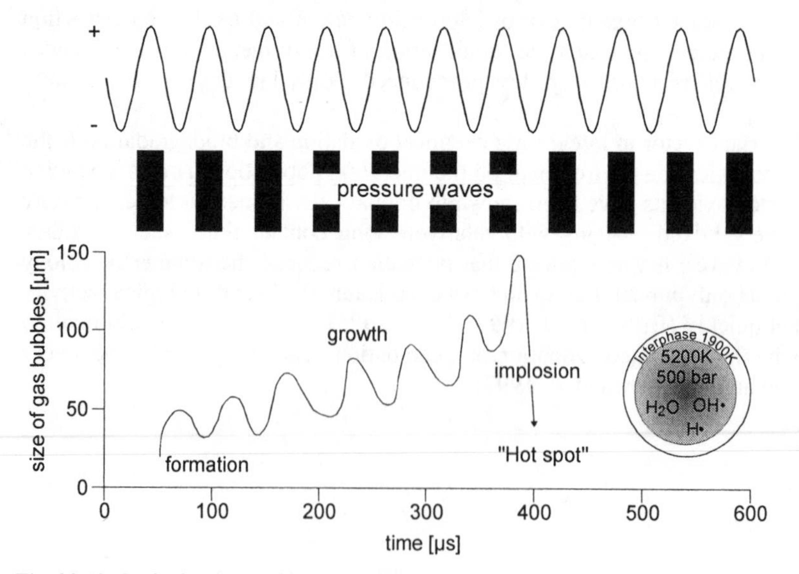

Fig. 20.10. Cavitation induced by ultrasound.

High shear stresses ("jet streams") are produced in the liquid by the collapse of the cavitation bubbles. The temperature and pressure inside the collapsing bubbles rise up to about 5,000 Kelvin and several hundred atmospheres. The extreme conditions inside these hydrophobic "hot spots" lead to the thermal destruction of compounds present in the bubbles and to the generation of the very reactive hydroxyl radicals by water sonolysis. In this way, sonochemical reactions can degrade volatile and hydrophobic pollutants by pyrolytic processes inside the cavitation bubbles. They also degrade non-volatile and more hydrophilic pollutants by subsequent hydroxyl radical reactions in the bulk liquid (Pétrier et al. 1998; Portenlänger 1999; Tiehm et al. 2000).

There are two different effects that might be used for increasing mass transfer of hydrophobic pollutants (Fig. 20.11). Firstly, the ultrasonically induced microcurrents reduce the thickness of diffusion layers surrounding solid surfaces. Due to the reduced diffusion layers, desorption processes are promoted resulting in a faster mass transfer into the liquid. This sonophysical process enhances the bioavailability of the original pollutant (Fig. 20.11 B). In the advent of cavitation, powerful mixing of the liquid and sonochemical processes additionally occur. The sonochemical processes result in the formation of better water soluble reaction products, thus increasing the mass transfer rate and bioavailability (Fig. 20.11 C). Sonochemical treatment often is added to the Advanced Oxidation Processes (AOPs) since hydroxyl radical reactions contribute to the chemical transformation.

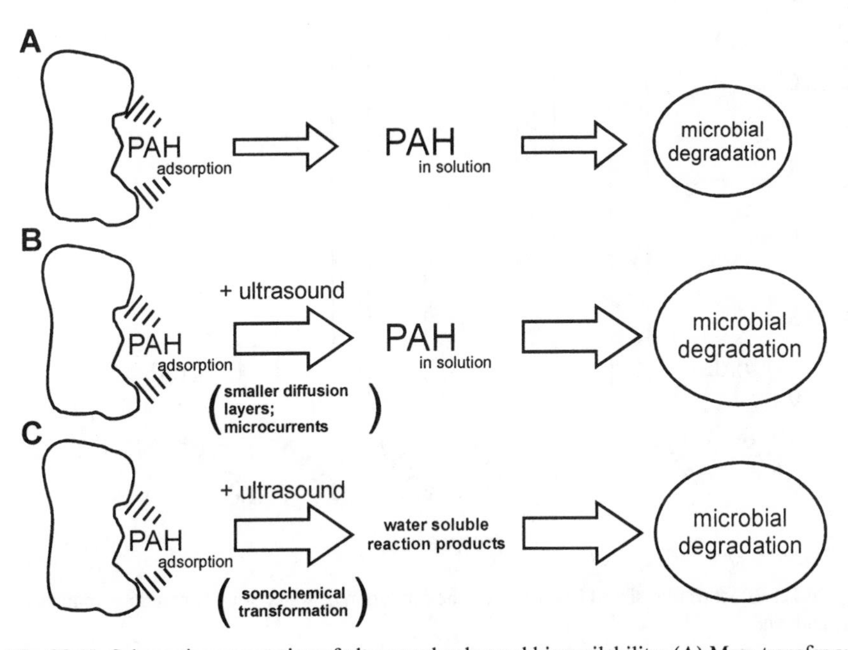

Fig. 20.11. Schematic presentation of ultrasound-enhanced bioavailability: (A) Mass transfer and biodegradation without ultrasound, (B) enhanced bioavailability by sonophysical effects, and (C) enhanced bioavailability by sonochemical effects.

20.5.2
Application of Ultrasound to Enhance PAH Bioavailability

The effect of ultrasound on PAH bioavailability was studied with the model compounds naphthalene and phenanthrene. Sonication was done by disk transducers fixed at the bottom of a cylindrical tube equipped with a stirrer and a thermostated jacket (25 +/- 3 °C). Transducers of similar shape were available at frequencies of 41 kHz, 207 kHz, 360 kHz, 616 kHz, 1,068 kHz, and 3,217 kHz. Identical power input at all frequencies was adjusted by calorimetric measurement.

Phenanthrene crystals were used to study the sonochemical formation of water-soluble reaction products (Tiehm 1999; Tiehm and Neis 1999). After ultrasonic treatment at different frequencies, the Chemical Oxygen Demand (COD) of aqueous phase samples was determined. The maximum solubility of phenanthrene is about 1.2 mg L^{-1} (Table 20.1) corresponding to a maximum aqueous phase COD of about 3.6 mg L^{-1}. The COD increased to > 20 mg L^{-1} in the samples sonicated at 41 kHz, 207 kHz, or 616 kHz (Fig. 20.12). These results are consistent with previous reports that sonochemical effects are most pronounced at frequencies below 1 MHz (Mark et al. 1998; Pétrier et al. 1998; Portenlänger and Heusinger 1997).

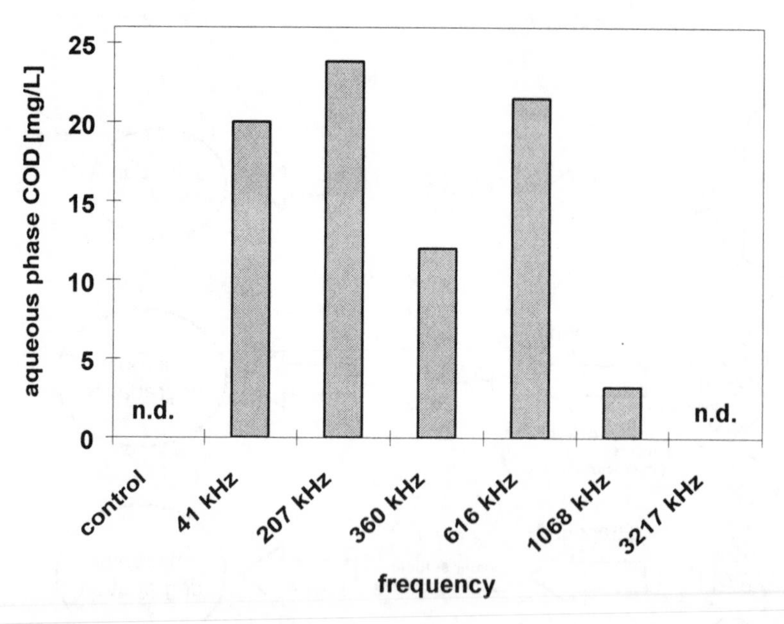

Fig. 20.12. Effect of ultrasound frequency on the formation of water-soluble reaction products of phenanthrene.

Table 20.4. Biodegradability of the sonochemical reaction products of phenanthrene obtained at different ultrasonic frequencies.

Frequency	After Sonication	After Biodegradation	
	COD dissolved [mg L^{-1}]	COD dissolved [mg L^{-1}]	Protein [mg L^{-1}]
Control (no ultrasound)	< d.l.[a]	< d.l.	< d.l.
41 kHz	20.0	2.5	0.85
207 kHz	23.8	6.7	0.81
360 kHz	12.0	3.2	0.46
616 kHz	21.5	3.6	0.93
1068 kHz	3.2	< d.l.	< d.l.
3217 kHz	< d.l.	< d.l.	< d.l.

[a] below detection limit (COD: 1.0 mg L^{-1}; Protein: 0.2 mg L^{-1})

After inoculation, the filtered supernatants were incubated at 20 °C on a rotary shaker. The sonochemical reaction products proved to be biodegradable. A decrease in aqueous phase COD and the formation of biomass was observed (Table 20.4). No inhibiting or toxic products were produced by the sonochemical reactions.

When sonication was done at 3.2 MHz, the dissolution of solid naphthalene and phenanthrene was significantly accelerated by sonophysical processes such as microstreaming. For example, the initial rates of naphthalene dissolution were 0.4 μmol $(\text{min l})^{-1}$ in the unsonicated control and 4.9 μmol $(\text{min L})^{-1}$ with 3.2 MHz ultrasound (Tiehm 1999). Identical absorption spectra of sonicated and control samples indicated that no sonochemical processes occurred at this high frequency (Tiehm et al. 2001b).

Treatment with ultrasound at low frequency (about 20 kHz) is a well-known laboratory method to disrupt microbial cells (Harrison 1991). Recent studies show that the damaging effect of ultrasound on microorganisms decreases with increasing frequency (Hua and Thompson 2000; Tiehm et al. 2001a). Therefore, high frequency ultrasound can be applied simultaneously to biodegradation. Under these conditions, permanent biodegradation of the dissolved pollutants should result in low aqueous phase concentrations, thus maintaining the concentration gradient as a driving force of the ultrasonically accelerated mass transfer. This hypothesis was confirmed by experiments carried out in a fermenter system (Fig. 20.13, Fig. 20.14).

In the fermenter, the stimulation of naphthalene biodegradation by sonication was determined by continuous analysis of the offgas CO_2 by mass spectrometry (Fig. 20.13). Melts of PAH (Mulder et al. 1998a) were fixed in the ultrasound reactor as shown in Fig. 20.14 in order to avoid undefined hydrodynamic conditions due to the pumping of PAH crystals through small tubes. The mineral medium was recirculated in the system (Fig. 20.13). Inoculation was done with mixed cultures pre-adapted to naphthalene degradation.

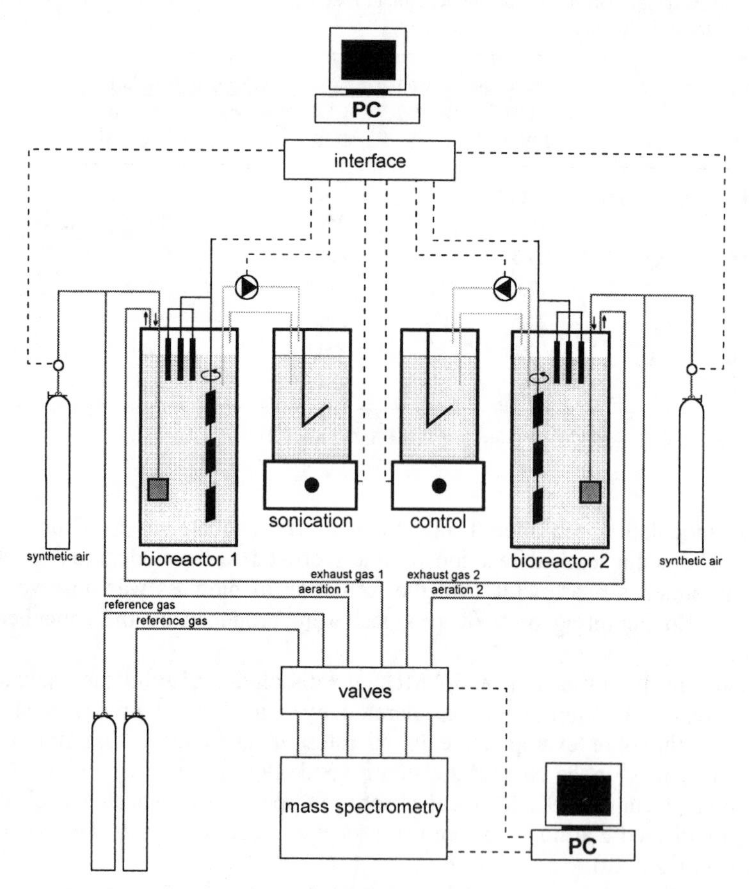

Fig. 20.13. Experimental set-up of the fermenter system used for simultaneous sonication and biodegradation studies.

Application of 3.2 MHz ultrasound simultaneously with biodegradation resulted in an increased formation of CO_2 (Fig. 20.15). Immediately after the start of sonication, the concentration of CO_2 in the fermenter offgas increased. Corresponding to higher CO_2 production, protein concentration and optical density in the fermenter increased.

An even more pronounced enhancement of CO_2 formation was observed at 351 kHz (Fig. 20.16). At this frequency, not only sonophysically reduced diffusion layers but also strong turbulent mixing and sonochemical effects triggered by cavitation contributed to an enhanced mass transfer of solid naphthalene into the aqueous phase. Formation of CO_2 was attributed to biodegradation since sonochemical mineralisation of naphthalene was neglectable and more biomass was produced in the sonicated fermenter. Remarkably, the ratio of biomass produced to CO_2 formed was lower at 351 kHz as compared to

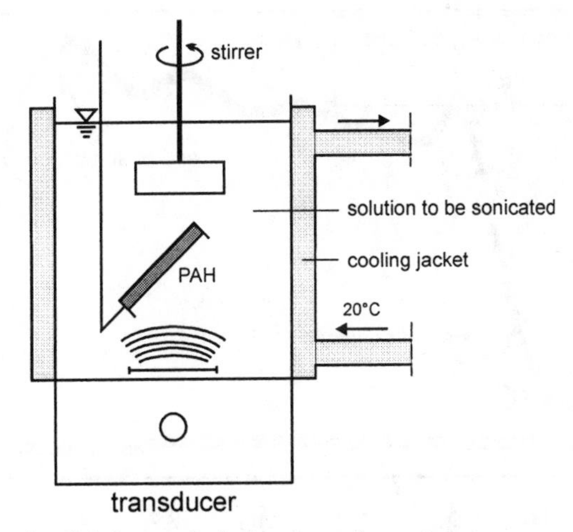

Fig. 20.14. Fixation of naphthalene melts in the ultrasonic reactor.

sonication at 3.2 MHz (Tiehm et al. 2001b). This effect was due to the shear forces produced by ultrasound that disrupted part of the microbial cells (Tiehm et al. 2001a). The released intracellular material was degraded by other microorganisms thus reducing the biomass yield.

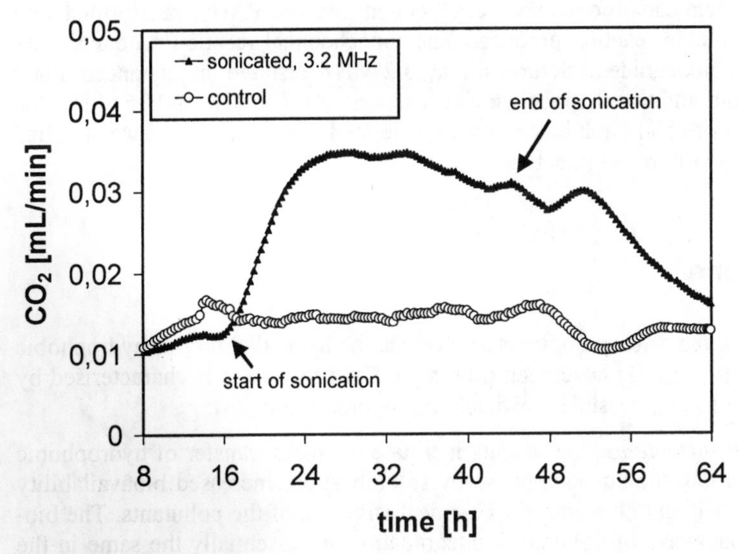

Fig. 20.15. Effect of sonication at 3.2 MHz on the formation of CO_2 during biodegradation of solid naphthalene.

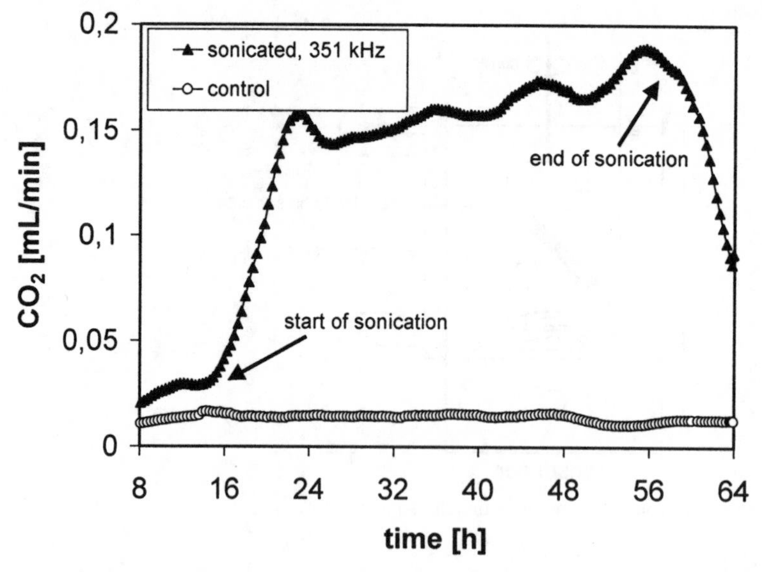

Fig. 20.16. Effect of sonication at 351 kHz on the formation of CO_2 during biodegradation of solid naphthalene.

In conclusion, the application of ultrasound proved to be suitable for improving the bioavailability of hydrophobic environmental pollutants. At frequencies below 1 MHz sonochemical processes were generated and the PAHs transformed into better water-soluble reaction products. The sonochemical reaction products were subsequently biodegraded. Sonication at 3.2 MHz resulted in enhanced mass transfer without additional sonochemical reactions. At 3.2 MHz and 351 kHz, the mineralisation of solid naphthalene was accelerated when ultrasound was applied simultaneously with biodegradation.

20.6 Conclusions

In this paper, three different options to increase the bioavailability of hydrophobic pollutants such as PAHs have been presented. Each technique is characterised by specific strengths and possible drawbacks as summarised below:

1. *Addition of surfactants:* Surfactants facilitate the mass transfer of hydrophobic compounds into the aqueous phase by solubilisation. Increased bioavailability is obtained without changing the chemical structure of the pollutants. The biochemical pathways of pollutant biodegradation are essentially the same in the absence or presence of surfactants. The risk of producing new unknown metabolites is low. When selecting a specific surfactant, its toxicity and biodegradability have to be taken into consideration. A low surfactant toxicity can be expected if the surfactant exhibits a high Hydrophil-Lipophil-Balance (HLB) value. The rate of surfactant biodegradation in the field is difficult to

value. The rate of surfactant biodegradation in the field is difficult to predict since the microbial community will adapt to the substrate. A fast biodegradation would result in oxygen depletion especially under *in-situ* conditions. Under on-site conditions, this possible drawback is considered to be less critical.

2. *Ozonation:* Ozonation increases the bioavailability of hydrophobic pollutants by altering the chemical structure. Partial oxidation results in an enhanced water-solubility. Both the ozone itself and the hydroxyl radicals produced by ozone self-decomposition are involved. The unspecific radical reactions in particular may result in the production of new, probably unknown products. Therefore the possible formation of new toxicologically relevant compounds has to be taken into consideration. On the other hand, ozonation has the potential to even convert pollutants, that completely resist a biological attack, into biodegradable and bioavailable products.
3. *Application of ultrasound:* Two mechanisms contribute to an increased bioavailability by ultrasound: (i) the reduction of the diffusion layers of solid surfaces, and (ii) the high shear forces and sonochemical effects occurring as a consequence of cavitation. By reduction of the diffusion layers, the mass transfer of the original pollutants is increased. The sonochemical effects triggered by cavitation can be used for the sonochemical transformation of hydrophobic pollutants into more water-soluble reaction products. In the case of sonochemical transformation, the benefits and possible drawbacks are the same as already discussed for ozonation. The application of ultrasound in environmental technology still is in its infancy. The main advantages of this technique are: (i) no chemicals are required and (ii) the effects can be easily controlled by adjusting ultrasound frequency and intensity.

In conclusion, all three techniques are suitable for increasing the bioavailability of hydrophobic pollutants. Since these measures are not yet available as standard technology for field application, careful pre-investigations are recommended in each individual case.

References

Abdul AS, Gibson TL (1991) Laboratory studies of surfactant-enhanced washing of polychlorinated biphenyl from sandy material. Environ Sci Technol 25: 665–671

Almgren M, Grieser F, Thomas JK (1979) Dynamics and static aspects of solubilization of neutral arenes in ionic micellar solutions. J Am Chem Soc 101: 279–291

Atrat PG, Koch B, Szekalla B, Hörhold-Schubert C (1992) Application of newly synthesized detergents in the side chain degradation of plant sterols by *Mycobacterium fortuitum*. J Basic Microbiol 32:147–157

Barczewski B, Josef R, Klaas N (2000) Surfactant enhanced extraction of PAH at a contaminated former gas production plant in pilot scale. In: Proc 7th Int FZK/TNO Conf, Contaminated Soil 2000, Vol 2, Thomas Telford Publ, London, UK, pp 975–980

Boldrin B, Tiehm A, Fritzsche C (1993) Degradation of phenanthrene, fluorene, fluoranthene, and pyrene by a *Mycobacterium* spec. Appl Environ Microbiol 59: 1927–1930

Brown RA, Nelson C, Leahy M (1997) Combining oxidation and bioremediation for the treatment of recalcitrant organics. In: Proc 4th Int In Situ On-Site Biorem Symp, Vol 4, Battelle Press, Columbus, Richland, pp 457–462

Cerniglia CE (1992) Biodegradation of polycyclic aromatic hydrocarbons. Biodegradation 3: 351–368

Choi H, Kim J-Y, Lim H, Cho J, Kang J-W, Kim K-S (2000) Oxidation of polycyclic aromatic hydrocarbons by ozone in the presence of sand. In: IWA 1st World Congress 03.–07.07.2000, Paris, France, Vol II pp 477–484

Cserhati T, Szoegyi M, Bordas B (1982) QSAR study on the biological activity on nonyl-phenyl-oxide polymers. Gen Physiol Biophys 1: 225–231

Cserhati T, Szoegyi M, Bordas B, Dobrovolszky A (1984) Structural requirements for the membrane damaging effect of non homologous series of nonionic tensides. Quant Struct-Act Relat 3: 56–59

Cserhati T, Illes M, Nemes I (1991) Effect of non-ionic tensides on the growth of some soil bacteria. Appl Environ Microbiol 35: 115–118

Edwards DA, Luthy RG, Liu Z (1991) Solubilization of polycyclic aromatic hydrocarbons in micellar nonionic surfactant solutions. Environ Sci Technol 25: 127–133

Erickson DC, Loehr RC, Neuhauser EF (1993) PAH loss during bioremediation of manufactured gas plant site soils. Wat Res 27: 911–919

Fountain JC, Klimek A, Beikirch MG, Middleton TM (1991) The use of surfactants for in-situ extraction of organic pollutants from a contaminated aquifer. J Hazardous Materials 28: 295–311

Gannon OK, Bibring P, Raney K, Ward JA, Wilson DJ, Underwood JL, Debelak KA (1989) Soil clean up by in-situ surfactant flushing: III Laboratory results. Sep Sci Technol 24: 1073–1094

Ghoshal S, Luthy RG (1998) Biodegradation kinetics of naphthalene in nonaqueous phase liquid-water mixed batch systems: comparison of model predictions and experimental results. Biotechnol Bioeng 57: 356–366

Guerin WF, Boyd SA (1992) Differential bioavailability of soil-sorbed naphthalene to two bacterial species. Appl Environ Microbiol 54: 1142–1152

Guerin WF, Boyd SA (1997) Bioavailability of naphthalene associated with natural and synthetic sorbents. Wat Res 31: 1504–1512

Guha S, Jaffé PR (1996a) Bioavailability of hydrophobic compounds partitioned into the micellar phase of nonionic surfactants. Environ Sci Technol 30: 1382–1391

Guha S, Jaffé PR (1996b) Biodegradation kinetics of phenanthrene partitioned into the micellar phase of nonionic surfactants. Environ Sci Technol 30: 605–611

Guha S, Jaffé PR, Peters CA (1998) Bioavailability of mixtures of PAHs partitioned into the micellar phase of a nonionic surfactant. Environ Sci Technol 32: 2317–2324

Haeseler F, Stieber M, Werner P, Frimmel FH (1993) Ecotoxicological aspects of chemical pre-oxidation combined with subsequent microbial degradation of polycyclic aromatic hydrocarbons. In: Arendt F, Annokkée GJ, Bosman R, van den Brink WJ (eds) Contaminated Soil '93, Kluwer Academic Publishers, Dordrecht, The Netherlands, pp 1337–1344

Harms H, Zehnder AJB (1994) Influence of substrate diffusion on degradation of dibenzofuran and 3-chlorodibenzofuran by attached and suspended bacteria. Appl Environ Microbiol 60: 2736–2745

Harms H, Zehnder AJB (1995) Bioavailability of sorbed 3-chlorodibenzofuran. Appl Environ Microbiol 61: 27–33

Harrison STL (1991) Bacterial cell disruption: a key unit operation in the recovery of intracellular products. Biotechnol Adv 9: 217–240

Hoigné J, Bader H (1975) Ozonation of water: role of hydroxyl radicals as oxidizing intermediates. Science 109: 782–784

Hoigné J (1988) The chemistry of ozone in water. In: Stucki S (ed) Process technologies for water treatment, Plenum Press, New York, pp 121–143

Hua I, Thompson JE (2000) Inactivation of *Escherichia coli* by sonication at discrete ultrasonic frequencies. Wat Res 34: 3888–3893

Kästner M (2000) Degradation of aromatic and polyaromatic compounds. In: Klein J (ed) Biotechnology Vol 11b, Environmental Processes: Soil Decontamination, Wilex-VCH, Weinheim, pp 211–239

Kaut CC (1997) Über die Auswirkung einer Ozonbehandlung von Boden auf die natürliche organische Bodenmatrix. Dissertation Universität Karlsruhe, Forschungszentrum Karlsruhe GmbH

Kile ED, Chiou CT (1989) Water solubility enhancements of DDT and trichlorobenzene by some surfactants below and above the critical micelle concentration. Environ Sci Technol 23: 832–838

Klevens HB (1950) Solubilization of polycyclic hydrocarbons. J Phys Colloid Chem 54: 283–298

Kotterman MJJ, Rietberg H-J, Hage A, Field JA (1998) Polycyclic aromatic hydrocarbon oxidation by the white-rot fungus *Bjerkandera* sp. strain BOS55 in the presence of nonionic surfactants. Biotechnol Bioeng 57: 220–227

Laha S, Luthy RG (1991) Inhibition of phenanthrene mineralization by nonionic surfactants in soil-water systems. Environ Sci Technol 25: 1920–1930

Luthy RG, Dzombak DA, Peters CA, Raswaswami A, Nakles DV, Nott BR (1994) Remediating tar-contaminated soils at manufactured gas plant sites. Environ Sci Technol 28: 266A–276A

Mahro B (2000) Bioavailability of contaminants. In: Klein J (ed) Biotechnology Vol 11b, Environmental Processes: Soil Decontamination, Wilex-VCH, Weinheim, pp 61–88

Mark G, Tauber A, Laupert R, Schuchmann H-P, Schulz D, Mues A, von Sonntag C (1998) OH-radical formation by ultrasound in aqueous solution – Part II Terephthalate and Fricke dosimetry and the influence of various conditions on the sonolytic field. Ultrasonics Sonochem 5: 41–52

Mason T (1991) Practical sonochemistry: user´s guide to applications in chemistry and chemical engineering. Ellis Horword Ltd, Chichester, UK

Mason T (1999) Ultrasound in environmental protection – an overview. In: Tiehm A, Neis U (eds) TU Hamburg-Harburg Reports on Sanitary Engineering 25: Ultrasound in Environmental Engineering (ISBN 3-930400-23-5), pp 1–9

Mason T, Tiehm A (eds) (2001) Advances in Sonochemistry Vol 6: Ultrasound in Environmental Protection, Elsevier Publishers, Amsterdam, The Netherlands, in press

Mihelcic JR, Lueking DR, Mitzell RJ, Stapleton JM (1993) Bioavailability of sorbed- and separate-phase chemicals. Biodegradation 4: 141–153

Mukherji S, Weber WJ Jr (1998) Mass transfer effects on microbial uptake of naphthalene from complex NAPLs. Biotechnol Bioeng 60: 750–760

Mulder H, Breure AM, Van Andel JG, Grotenhuis JTC, Rulkens WH (1998 a) Influence of hydrodynamic conditions on naphthalene dissolution and subsequent biodegradation. Biotechnol Bioeng 57: 145–154

Mulder H, Breure AM, Van Honschooten D, Grotenhuis JTC, Van Andel JG, Rulkens WH (1998b) Effect of biofilm formation by *Pseudomonas* 8909N on the bioavailability of solid naphthalene. Appl Microbiol Biotechnol 50: 277–283

Mulder H, Breure AM, Van Andel JG, Grotenhuis JTC, Rulkens WH (2000) Effect of mass-transfer limitations on bioavailability of sorbed naphthalene in synthetic model soil matrices. Environ Toxicol Chem 19: 2224–2234

Myers D (1991) Surfaces, interfaces, and colloids; principles and applications. VCH Verlagsgesellschaft, Weinheim, Germany

Ortega-Calvo JJ, Birman I, Alexander M (1995) Effect of varying the rate of partitioning of phenanthrene in nonaqueous-phase liquids on biodegradation in soil slurries. Environ Sci Technol 29: 2222–2225

Pennell KD, Abriola LM, Weber WJ Jr (1993) Surfactant-enhanced solubilization of residual dodecane in soil columns. 1. Experimental investigation. Environ Sci Technol 27: 2332–2340

Pétrier C, Jiang Y, Lamy M-F (1998) Ultrasound and environment: sonochemical destruction of chloraromatic derivatives. Environ Sci Technol 32: 1316–1318

Pétrier C, Jiang Y, Francony A, Lamy MF (1999) Aromatics and chloroaromatics sonochemical degradation: yields and by-products. In: Tiehm A, Neis U (eds) TU Hamburg-Harburg Reports on Sanitary Engineering 25: Ultrasound in Environmental Engineering (ISBN 3-930400-23-5), pp 23–37

Portenlänger G, Heusinger H (1997) The influence of frequency on the mechanical and radical effects for the ultrasonic degradation of dextranes. Ultrasonics Sonochem 4: 127–130

Portenlänger G (1999) Mechanical and radical effects of ultrasound. In: Tiehm A, Neis U (eds) TU Hamburg-Harburg Reports on Sanitary Engineering 25: Ultrasound in Environmental Engineering (ISBN 3-930400-23-5), pp 11–22

Scott JP, Ollis DF (1995) Integration of chemical and biological oxidation processes for water treatment: review and recommendations. Environ Progr 14: 88–103

Seibel F, Stieber M, Werner P, Frimmel FH (1995) Characterization of degradation products of PAH contaminated soil after ozone treatment. SPIE-proceedings of Environmental Monitoring and Harzardous Waste Site Remediation, Vol 2504, pp 86–97

Seidel JP (1995) Über die Anwendung von Ozon zum oxidativen Abbau polycyclischer aromatischer Kohlenwasserstoffe in Böden. Dissertation Universität Karlsruhe, Forschungszentrum Karlsruhe GmbH

Sims RC, Overcash MR (1983) Fate of polynuclear aromatic compounds (PNAs) in soil-plant systems. Residue Rev 88: 1–68

Staehelin J, Hoigné J (1982) Decomposition of ozone in water: rate of initiation by hydroxide ions and hydrogen peroxide. Environ Sci Technol 16: 676–681

Stucki G, Alexander M (1987) Role of dissolution rate and solubility in biodegradation of aromatic compounds. Appl Environ Microbiol 53: 292–297

Tiehm A (1994) Degradation of polycyclic aromatic hydrocarbons in the presence of synthetic surfactants. Appl Environ Microbiol 60: 258–263

Tiehm A, Fritzsche C (1995) Utilization of solubilized and crystalline mixtures of polycyclic aromatic hydrocarbons by a *Mycobacterium* sp. Appl Microbiol Biotechnol 42: 964–968

Tiehm A, Stieber M, Werner P, Frimmel FH (1997) Surfactant-enhanced mobilization and biodegradation of polycyclic hydrocarbons in manufactured gas plant soil. Environ Sci Technol 31: 2570–2576

Tiehm A (1999) Combination of ultrasound and biodegradation: enhanced bioavailability of polycyclic aromatic hydrocarbons. In: Tiehm A, Neis U (eds) TU Hamburg-Harburg Reports on Sanitary Engineering 25: Ultrasound in Environmental Engineering (ISBN 3-930400-23-5), pp 167–180

Tiehm A, Neis U (1999) Ultrasound enhanced biodegradation of phenanthrene. In: Wilhelm AM (ed) Proc 2nd Conf Ultrasound Processing, PROGREP, Toulouse, France, pp 165–170

Tiehm A, Kohnagel I, Neis U (2000) Removal of chlorinated pollutants by a combination of ultrasound and biodegradation. In: IWA 1st World Congress 03.–07.07.2000, Paris, France, Vol I, pp 177–184

Tiehm A, Nickel N, Zellhorn M, Neis U (2001a) Ultrasonic waste activated sludge disintegration for improving anaerobic stabilization. Wat Res: in press

Tiehm A, Ortfeld R, Schmidt O (2001b) Ultrasound-enhanced bioavailability of solid naphthalene: in preparation

Tsomides HJ, Hughes JB, Thomas M, Ward HC (1995) Effect of surfactant addition on phenanthrene biodegradation in sediments. Environ Toxicol Chem 14: 953–959

Young F R (1989) Cavitation. Mc Graw-Hill Book Company, Maidenhead, UK, pp 40–76

Van Hylckama Vlieg JET, Janssen DB (2000) Bacterial degradation of aliphatic hydrocarbons. In: Klein J (ed) Biotechnology Vol 11b, Environmental Processes: Soil Decontamination, Wilex-VCH, Weinheim, Germany, pp 193–209

Volkering F, Breure AM, Sterkenburg A, Van Andel JG (1992) Microbial degradation of polycyclic aromatic hydrocarbons: effect of substrate availability on bacterial growth kinetics. Appl Microbiol Biotechnol 36: 548–552

Volkering F, Breure AM, Van Andel JG, Rulkens WH (1995) Influence of nonionic surfactants on bioavailability and biodegradation of polycyclic aromatic hydrocarbons. Appl Environ Microbiol 61: 1699–1705

Wang J-M, Marlowe EM, Miller-Maier RM, Brusseau ML (1998) Cyclodextrin-enhanced biodegradation of phenanthrene. Environ Sci Technol 32: 1907–1912

Weissenfels WD, Klewer H-J, Langhoff J (1992) Adsorption of polycyclic aromatic hydrocarbons (PAH's) by soil particles: influence on biodegradability and biotoxicity. Appl Microbiol Biotechnol 36: 689–696
Wischnak C, Müller R (2000) Degradation of chlorinated compounds. In: Klein J (ed) Biotechnology Vol 11b, Environmental Processes: Soil Decontamination, Wilex-VCH, Weinheim, Germany, pp 241–271
Wodzinski RS, Bertolini D (1972) Physical state in which naphthalene and dibenzyl are utilized by bacteria. Appl Microbiol 23: 1077–1081
Wodzinski RS, Coyle JE (1974) Physical state of phenanthrene for utilization by bacteria. Appl Microbiol 27: 1081–1084

21 Bioaugmentation: Advantages and Problems Using Micro Organisms with Special Abilities in Soil Decontamination

R. Müller[1] and B. Mahro[2]
[1]Department of Biotechnology II, Technical University of Hamburg-Harburg, Denickestraße 15, 21071 Hamburg, Germany
[2]Institute for Environmental Technology, Hochschule Bremen, Neustadtwall 30, 28199 Bremen, Germany

21.1 Introduction

The use of micro organisms with specific properties in the environment is a practice which has been known for about a century. Examples can be found in agriculture, where *Rhizobia* have been used to enhance nitrogen fixation at the roots of leguminoses or where *Bacillus thuringiensis* has been used to fight insect pests (Elsas et al. 1991). The inoculation of contaminated soils with bacteria, that have the ability to degrade environmental contaminants, a process called *bioaugmentation*, is a relatively new technique (Trevors et al. 1994). At present it is still unclear which factors determine the success of such a treatment. This situation prompted us to investigate which prerequisites are required to make bioaugmentation a successful tool in bioremediation of contaminated soils.

The knowledge on degradation pathways in the microbial degradation of xenobiotics is mainly derived from studies with isolated pure microbial strains in liquid culture in the laboratory. Many micro organisms are known that can degrade environmental pollutants in the laboratory in liquid culture experiments (Müller 1999). However, it is still unclear which factors determine the ability of micro organisms to establish their degradative potential in soil. In this chapter we shall use the term "*specialists*" for these bacteria with the ability to degrade special compounds.

The use of these specialists in bioremediation is a matter of debate. Some people suggest using these specialists only in special cases like in fresh oil spills where adaptation of the autochthonous flora has not yet occurred. In this case the ability of the specialists to survive in the given environment has to be checked in the laboratory (Sprenger and Ebner 1990).

In contrast, Portier et al. (1988) stress the importance of specialists in the treatment of old contaminated soils. They suggest that the degradative potential of a site could be enhanced by the exchange of genetic information between the specialists and the autochthonous flora. Since it is difficult to assess the degradative potential in a given soil, the use of specialists is always considered to be useful when, under favourable environmental conditions, no degradation occurs or when complex mixtures of pollutants have to be degraded rapidly (Brunsbach 1993). Schwefer (1988) doubts the usefulness of specialists when the autochthonous microflora is stimulated sufficiently. He doubts the survival of exogenously added micro organisms in a given environment.

21.1.1
Successful Examples of Bioaugmentation in Soil

In several cases the use of specialists led to a significant increase in degradation rates or to a more efficient degradation. Huntjens et al. (1988) increased the dechlorination rate for α-hexachlorohexane four-fold by the addition of *Pseudomonas versicularis* compared to a non inoculated control. Viney and Bewley (1990) obtained 30 % degradation of chlorinated biphenlys when they added a mixture of *Corynebacterium sp.*, *Alcaligenes eutrophus*, *Pseudomonas putida* and *Phanerochaete chrysosporium* compared to 13 % in 14 weeks in the non inoculated sample. Valo and Salkinoja-Salonen (1986) report the acceleration of chlorphenol degradation by the addition of chlorophenol degrading bacteria. Edgehill and Finn (1983) reduced the pentachlorophenol in a contaminated soil by the addition of *Arthrobacter sp.* by 85 % within 12 days compared to 30 % without addition. Middeldorp et al. (1990) achieved complete mineralisation of pentachlorophenol by inoculation with *Rhodococcus chlorophenolicus,* while in the control without inoculation pentachloroanisols and other chlorinated compounds were formed. Barbeau et al. (1997) used activated soil biomass and PCP acclimated flocculent biomass to degrade pentachlorophenol in a sandy soil. Barles et al. (1979) describe increased degradation of parathion in soil in the laboratory and in the field when a mixed culture containing the parathion degrading *Pseudomonas stutzeri* was added. Newcombe and Crowley (1999) used repeated additions of atrazine-degrading bacteria to decontaminate atrazine-contaminated soil. Similar results were obtained in the degradation of 3-chloroanilin by *Pseudomonas* strains in soil slurries (Brunsbach 1993) and in the degradation of 1,2,4-trichlorophenol by *Pseudomonas sp.* P51 in a contaminated soil column (Meer et al. 1987) or in the degradation of 2,4,6-trichlorphenol by *Alcaligenes eutrophus* in contaminated soil microcosms (Andreoni et al. 1998). The addition of specialists and nutrients in an off site field trial also led to an improved decrease in hydrocarbon content compared to samples with the addition of nutrients or activated sludge, or without addition (Eiermann and Peter 1990).

Often the degradation of a pollutant only starts after the addition of specialists. This was the case in the degradation of 3-chlorobenzoate by *Pseudomonas alcaligenes* (Focht and Shelton 1987), of 2,4-dichlorophenol by *Pseudomonas sp.* (Goldstein et al. 1985), of *o*-xylene and dichlorobenzene by *Pseudomonas sp.* (Oldenhuis et al. 1989) or in the degradation of pentachlorophenol by *Rhodococcus chlorophenolicus* (Salkinoja-Salonen et al. 1990; Middeldorp et al. 1990). Chen et al. (1998) report the conversion of polychlorinated biphenyls after inoculation with an adapted mixed culture from a contaminated sediment, which did not occur without inoculation. Klibane et al. (1983) report the successful degradation of the herbicide 2,4,5-trichlorophenol in soil by *Pseudomonas cepacia* and Unterman et al. (1988) report that only after the addition of specialists was the degradation of a complex mixture of polychlorinated biphenyls initiated by the removal of the higher chlorinated compounds. This then allowed the autochthonous flora to remove the remaining compounds.

21.1.2
Examples where Bioaugmentation Was not Successful

In contrast to the above mentioned examples there are several reports in literature, where the addition of specialists showed no effect in soils or where the specialists were even inferior to the existing flora. Valo and Salkinoja-Salonen (1986) found a higher pentachlorophenol mineralisation rate in an insterile soil sample with compost (30–40 % in 40 d) compared to a sterile sample with *Rhodococcus chlorophenolicus* (10 % in 40 d). There was also no increase in pentachlorophenol degradation by the addition of an adapted mixed culture isolated from a soil contaminated with pentachlorophenol. Focht and Brunner (1985) obtained the same degradation rates in soil contaminated with lower chlorinated biphenyls with or without the addition of *Acinetobacter sp.* P6. Comparisons of the degradation of different PCB-isomers by specialists in liquid cultures and soils showed that *Pseudomonas sp.* LB400 did not metabolise 4,4-chlorinated isomers in soil, which it degraded in liquid culture, while *Corynebacterium sp.* MB1 retained this ability (Unterman et al. 1988). Macrae and Alexander (1965) observed that a *Flavobacterium sp.* which degraded 4-(2,4-dichlorophenoxy)-butyrate in liquid culture did not degrade this compound in soil. Similar results were obtained by Anderson et al. (1970) in the degradation of DDT by the fungus *Mucor alternans* and by Lehtomaki and Niemela (1975) with mineral oil degrading micro organisms.

21.2
Aim of our Study

The previous chapters show, that on one hand several reports suggest that bioaugmentation could be a valuable tool in bioremediation. On the other hand several cases were reported in which bioaugmentation was not successful and bioaugmentation was not employed successfully on a commercialscale. Therefore, we want to investigate which reasons determine the success of bioaugmentation. We wanted to know whether micro organisms behave differently in soil and in liquid culture, and we investigated how different parameters influence the activity of specialists in soil.

21.3
Strains and Soils Used in this Study

In order to solve the problems mentioned above we used several model systems. As substrates we chose 4-chlorobenzoate, the different chlorophenols from mono-to pentachlorophenol and the polycyclic aromatic compounds anthracene, phenanthrene and pyrene. Then we chose the micro organisms with the ability to degrade these compounds. For the degradation of 4-chlorobenzoate we used the gram-negative *Pseudomonas sp.* CBS3 and the gram-positive strains *Arthrobacter sp.* SU, *Nocardia sp.* CBS2 and *Nocardia sp.* FL5g (Klages and Lingens 1979, 1980; Löffler 1993; Ruisinger et al. 1976). The 4-chlorphenol degrading strain *Arthrobacter sp.* HEID was isolated from garden soil, the *Pseudomonas putida* PaW85 (pEST 4011) was a gift from Dr. Allan Nurk (Talin, Estland) and was originally isolated with 2,4-dichlorophenoxy acetic acid. This strain was also able to degrade the 2,4-dichlorophenol. *Azotobacter sp.* GP1 was a gift from Dr. Jürgen Eberspächer (Stuttgart, Germany) (Li et al. 1991). This strain was isolated with 2,4,6-trichlorophenol as substrate. The strain was later reclassified as *Ralstonia eutropha. Arthrobacter sp.* ATCC 33790 (Stanlake and Finn 1982) and degraded pentachlorophenol; it was later reclassified, this time as *Sphingmonas chlorophenolicus.* The strains *Mycobacterium sp.* VF1 and *Sphingomonas paucimobilis* BA2 degrade phenathrene, anthracene and pyren and were isolated from a contaminated soil in Hamburg (Kästner et al. 1994). The organisms are listed in table 21.1.

These micro organisms were then added to different soils. Unless mentioned otherwise we used the Ah-horizon of a soil from Hamburg, Germany. The Euro-soils were used as reference soils because they represent the agricultural soils within the European community and because they had been selected for the testing of chemicals as reference soils (Kuhnt et al. 1991). They have a high variability of the relevant parameters like particle size distribution, organic carbon content, C/N ratio, pH, iron and aluminium content and cation exchange capacity. These parameters are listed in table 21.2. The soil termed Veringstraße was derived from a site in Hamburg contaminated with polycyclic aromatic compounds and with tar oil. The Euro-soils were delivered dry and sieved to 2 mm and sterilised by γ-radiation. The other soil materials were delivered humid and insterile. All soils except from Veringstraße were uncontaminated, and had to be contaminated at the beginning of the experiment.

Table 21.1. Strains used in this study

Bacteria used	Substrate degraded in liquid culture
Pseudomonas sp. CBS3 (DSM-Nr. 6613)	4-chlorobenzoate
Nocardia sp. CBS2	4-chlorobenzoate
Nocardia sp. FL5g	4-chlorobenzoate
Arthrobacter sp. SU (DSM-Nr. 6612)	4-chlorobentoate
Arthrobacter sp. HEID	4-chlorpophenol
Pseudomonas putida PaW85 (pEST 4011)	2,4-dichlorophenol
Azotobacter sp. GP1 (*Ralstonia eutropha*)	2,4,6-trichlorophenol
Arthrobacter sp. ATCC 33790 (*Sphingomonas chlorophenolicus*)	2,3,4,6-tetrachlorophenol pentachlorophenol
Sphingomonas paucimobilis BA2	anthracene, phenanthrene
Mycobacterium sp. VF1	anthracene, phenanthrene, pyrene

Table 21.2. Characterisation of the soil materials used in this study

Name	Horizon	pH	C_{org}-content [%]	C/N	Cation exchange capacity [mval (100 g)$^{-1}$]	Water holding capacity [%]	Origin
Ah	AhAp*	4.5	1.1	14.0	3.0	31	Germany, Hamburg**
Bt	IISwBt SdSWBt*	4.5	0.1		5.4	29	Germany, Hamburg**
Klei	aMAp Ap Go Gro*	4.7	2.3	11.0	34.7	67	Germany, Hamburg**
Klei Torf	nHFor nH*	3.7	11.0	20.0	41.8	55	Germany, Hamburg**
Euro-Soil 1	A	5.1	1.3	7.6	29.9	58	Italy, Aliminusa***
Euro-Soil 2	A	7.4	3.7	18.5	28.3	62	Greece, Souli***
Euro-Soil 4	A	6.5	1.5	9.7	17.5	54	France, Normandie***
Euro-Soil 5	A	3.2	9.2	30.8	32.7	27	Germany, Lauenburg***
Euro-Soil 6	B	7.2	0.2	12.5	11.4	42	France, Normandie ***
Veringstraße		6.2	4.1				Germany, Hamburg, Veringstr. **/contaminated site

* determined by University Hamburg, Institute for Soil Science
** obtained from University Hamburg, Institute for Soil Science
*** obtained from Joint Research Centre, Ispra Establishment, 21020 Ispra (Varese), Italy

21.4 Experimental Set-up

In our experiments with liquid cultures we used the following procedure:

- Growth of the preculture.
- Inoculation of 50 ml of mineral salts medium with 1 ml of the preculture.
- Incubation under the desired conditions.
- Sampling (3x1 ml) from t_0 to t_{end}.
- Determination of optical density in the sample.
- Centrifugation and determination of the contaminant by HPLC. In case of chlorinated compounds the chloride release was determined by ion chromatography.

In the experiments with soil the following procedure was used:

- Conditioning of the soils (sieved and dried).
- Contamination with the contaminant.

- Homogenisation.
- Inoculation with bacteria.
- Homogenisation.
- Incubation under the desired conditions.
- Sampling (3x1 ml) from t_0 to t_{end}.
- Storage of the samples at –18 °C until they were extracted.
- Extraction of the samples.
- Determination of the remaining contaminant by HPLC. In case of chlorinated compounds the chloride release was determined by ion chromatography.

At the beginning and the end of the experiments the water content, the water holding capacity and the pH of the soil were determined.

21.5 Factors that Influence the Performance of Bacteria in Soil

When we introduced the different xenobiotics into the different soils and added the micro organisms that degrade these compounds, the bacteria successfully removed the contaminant from soil within a couple of days in all cases. Fig. 21.1 shows as an example the degradation of 2,4,6-trichlorophenol by *Ralstonia eutropha* in Ah-soil. As can be seen, within 5 days the trichlorophenol was completely degraded and a corresponding amount of chloride was formed indicating complete mineralisation. When after two weeks another 3 mM of the trichlorophenol was added, this was removed within two days. In the control also a slight decrease in

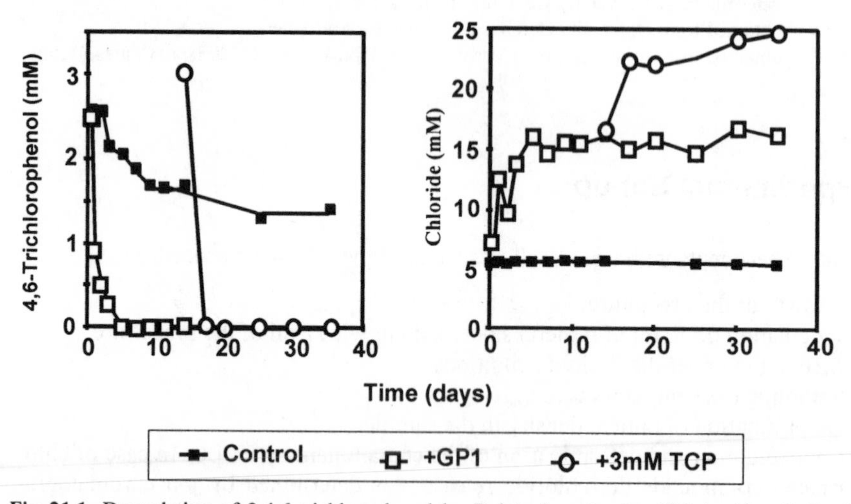

Fig. 21.1. Degradation of 2,4,6-trichlorophenol by Ralstonia eutropha GP1 in Ah-soil. In the control no bacteria were added.

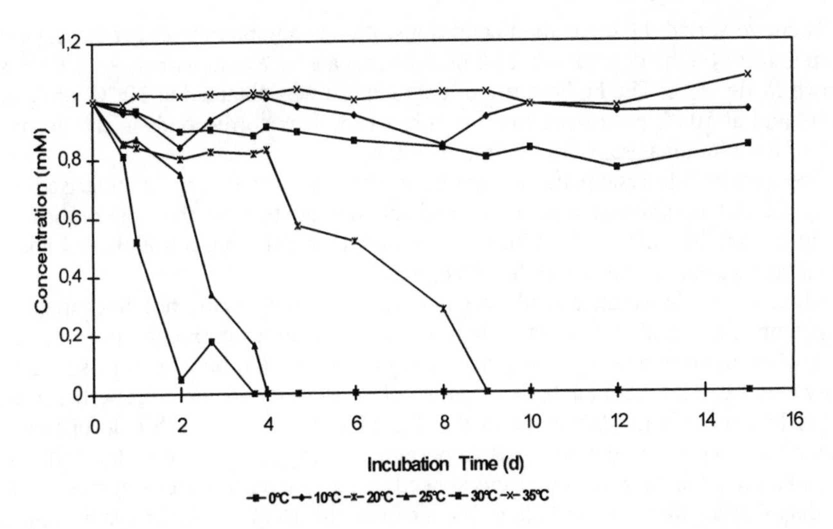

Fig. 21.2. Degradation of 4-chlorobenzoate by *Pseudomonas sp.* CBS3 in soil at different temperatures

trichlorophenol concentration was seen, however no chloride was released, indicating that no mineralisation had occurred and that this decrease was either due to evaporation or to analytical problems with strongly sorbed chlorophenol. Similar results were obtained with all compounds and strains.

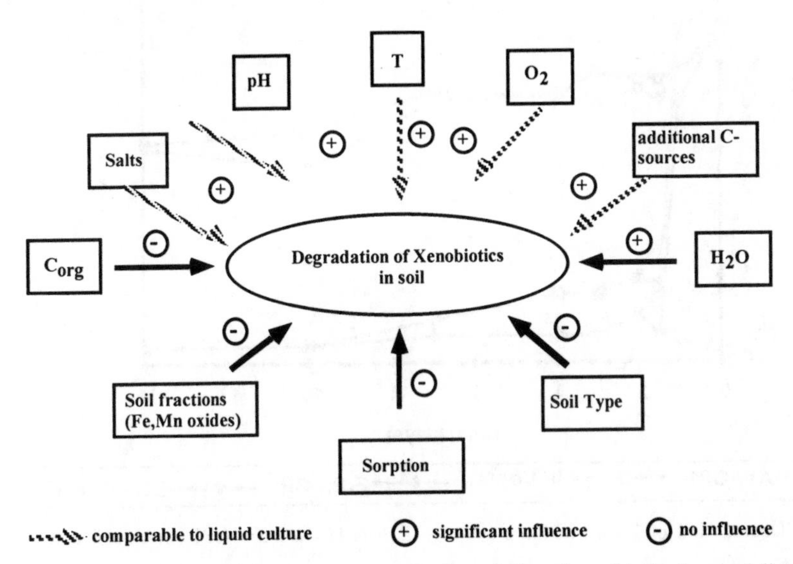

Fig. 21.3. Effect of different parameters on the degradation of xenobiotics by specialists in soil

Then we varied the incubation conditions. As an example, the effect of varying temperatures on the degradation of 4-chlorobenzoate by *Pseudomonas sp.* CBS3 is shown in fig. 21.2. The highest rate of degradation was obtained at 30 °C while at 36 °C and at 10 °C no degradation was observed. This corresponds to the behaviour of Pseudomonas *sp.* CBS3 in liquid culture.

The results of systematic variation of the parameters are summarised in fig. 21.3. All parameters that influenced the degradation of the xenobiotics in liquid culture like pH, temperature, oxygen content, salt concentration and additional carbon sources had a similar effect in soil.

All soil specific parameters like organic carbon content, water holding capacity, soil composition and soil structure had no effect. As an example, the degradation of 2,4,6-trichlorophenol by *Ralstonia eutropha* in Ah-soil and in Euro-soil 2 is shown in fig. 21.4. Although these soils differ strongly in their properties (see table 21.2) there is no difference in the degradation rates of the trichlorophenol. The water content of the soil was important. As expected in very dry soils no degradation occurred since the bacteria need water to live. In water saturated soils the degradation was also very slow. We attribute this to the limited oxygen supply under these conditions.

As expected, the size of the inoculum had an influence on the initial degradation rate. The more bacteria were added, the faster the degradation started. When different strains were available for the degradation like in the case of 4-chlorobenzoate, no difference was seen in the performance between the different strains.

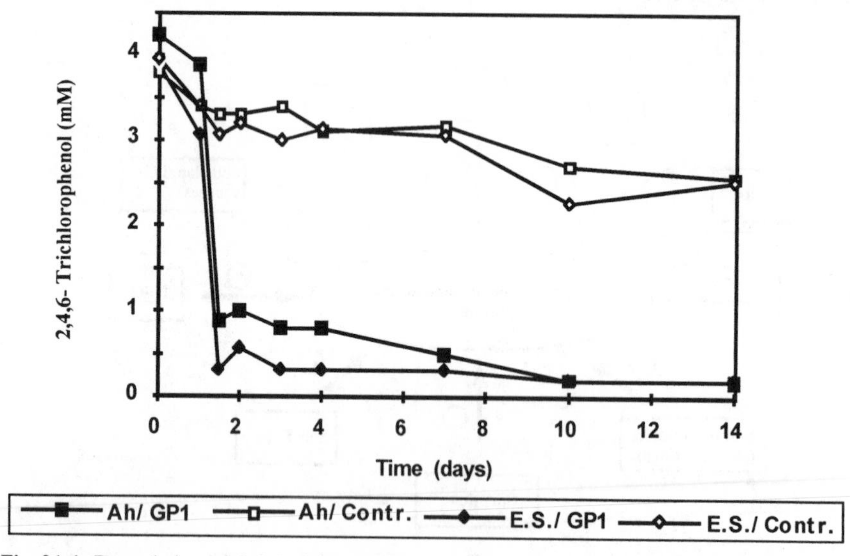

Fig. 21.4. Degradation of 2,4,6-trichlorophenol by *Ralstonia eutropha* GP1 in Ah-soil and in Euro-Soil 2

We could not confirm the observation that immobilised bacteria were better than free cells (Briglia et al. 1990). With 4-chlorobenzoate and with trichlorophenol as substrate we observed in both cases the same degradation rates. Surprisingly the autochthonous flora had no effect on degradation. This is demonstrated in fig. 21.5 in the degradation of 4-chlorobenzoate in sterile and non sterile soil. Survival of bacteria in the environment is often discussed as one reason for the failure of bioaugmentation. However, there are several reports in the literature, which show that after a rapid initial decrease in cell number, the cells remain detectable by modern molecular biological techniques even after years in soil (Ripp et al. 2000). In some cases the transfer of degradation pathways located on plasmids to endogenous bacteria may be an alternative to maintain a certain ability in soil (Top et al. 1998). However, in the degradation of pollutants after bioaugmentation this factor seems not to be important. When we added the bacteria to contaminated soil in all cases the pollutants were completely degraded within a few days.

From these results we conclude that the behaviour of bacteria in freshly contaminated soil is identical to that in liquid culture. We found no specific factors in such soils, which inhibit the bacteria in their ability to degrade pollutants. The pollutants were always degraded, provided the conditions in soil were in accordance with the requirements of the bacteria as determined in liquid culture.

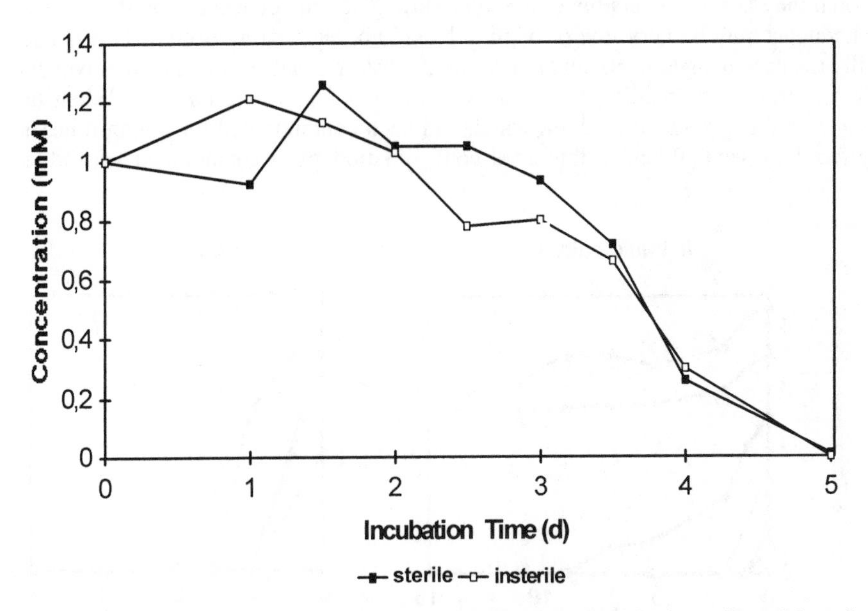

Fig. 21.5. Degradation of 4-chlorobenzoate by *Pseudomonas sp.* CBS3 in sterile and insterile Ah-soil

21.6
Reasons for the Failure of Bioaugmentation

21.6.1
Sorption, Diffusion and Bioavailability

From the results described above there seems to be no reason for a possible failure in bioaugmentation. The bacteria worked in soil like in liquid culture. Nevertheless, differences were observed between liquid cultures and soil. However, it was not the bacteria that behaved differently, but the pollutants. In soil these substances may be sorbed to surfaces or enter pores that are not available to microbes or may be entrapped into the organic soil matter. In some cases this may have a positive effect on degradation of pollutants. Fig. 21.6 shows, that in soil much higher concentrations are tolerated by bacteria than in liquid culture. In soil *Sphingomonas chlorophenolicus* degraded 5 mM of 2,3,4,6-tetrachlorophenol within 36 hours whereas in liquid culture 3.5 mM of the same substance was toxic. This can only be explained by the fact that in soil the effective concentration of the pollutant is lower through reversible removal of the compound by the effects described above.

Sorption, entrapment or diffusion into small pores becomes a serious problem, when they are not reversible or are very slow. Then the concentration the bacteria encounter may be very low or negligible and no degradation occurs. This is usually the case in aged contaminations. Fig. 21.7 shows that freshly added polycyclic hydrocarbons are readily degraded by the added *Mycobacterium* VF. When the same strain was added to a soil which had been contaminated for a long time, no effect was seen. When to this aged contamination fresh contaminant was added

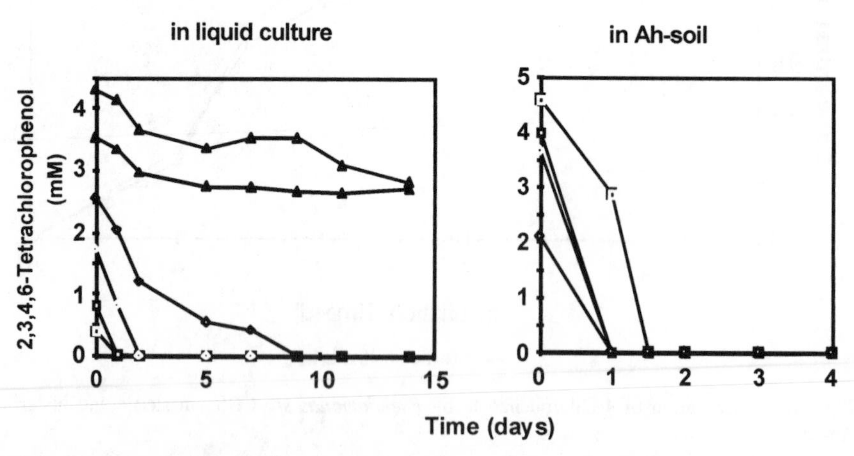

Fig. 21.6. Degradation of different concentrations of 2,3,4,6-tetrachlorophenol by *Sphingomonas chlorophenolicus* in liquid culture and in Ah-soil

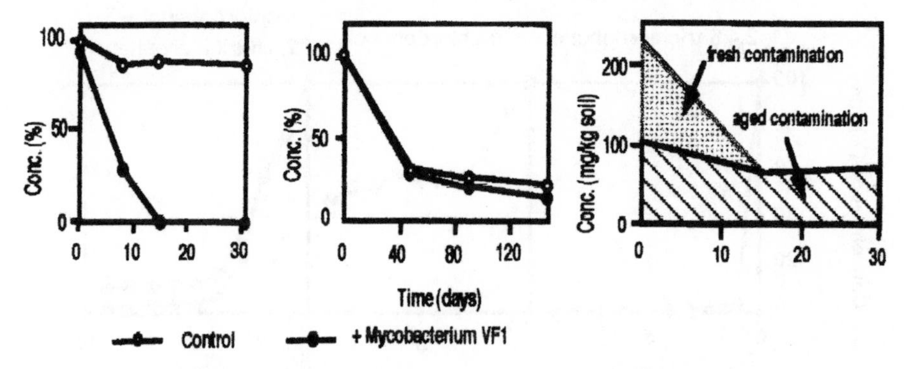

Fig. 21.7. Degradation of polycyclic aromatic compounds by Mycobacterium VF1 in freshly contaminated soil (left) in an aged contamination (middle) and in an aged contamination with freshly added contaminants (right). Note the different time scales.

only the portion that corresponded to the fresh contamination was removed. Since the aged and the fresh contaminant are chemically identical, this behaviour can only be explained by the fact that in the aged contamination the contaminant is no longer bioavailable. In this case the bacteria cannot degrade the contaminant.

One must assume that normally in aged contaminations the bioavailable fraction has been degraded and only the non bioavailable fraction is remaining. In this case it is of no use to add bacteria, because the bacteria are not the limiting factor. Here one must try to find measures to improve bioavailability instead. The lack of bioavailability is certainly the major reason, when bioaugmentation is not successful. It is a matter of debate, whether the non bioavailable pollutants are of concern, because if they are not bioavailable they are also not toxic and do not pose a threat. The problems associated with bioavailability are discussed in several other chapters of this book.

21.6.2 Synergistic and Antagonistic Effects in Mixtures of Pollutants

In contaminated soils very often the pollution is not caused by a single contaminant but by complex mixtures. Therefore the ability to degrade a certain pollutant may not be sufficient to ensure the success of bioaugmentation, but the bacteria must be able to degrade the pollutant in the presence of other contaminants. In the literature there are only very few examples, where the influence of mixtures on the degradation of pollutants was studied even in liquid culture. Therefore, we used the system described above to investigate on a systematic basis, how the presence of other substances influences the performance of a specialist. The results obtained show that the answer to this question is very complex. Depending on the substrates and the strains used, very different results were obtained. The principle possibilities shall be discussed in this chapter.

The first possibility is that a compound, which is not degraded by a bacterium, can be degraded when another substance is present. For example, 4-chlorophenol

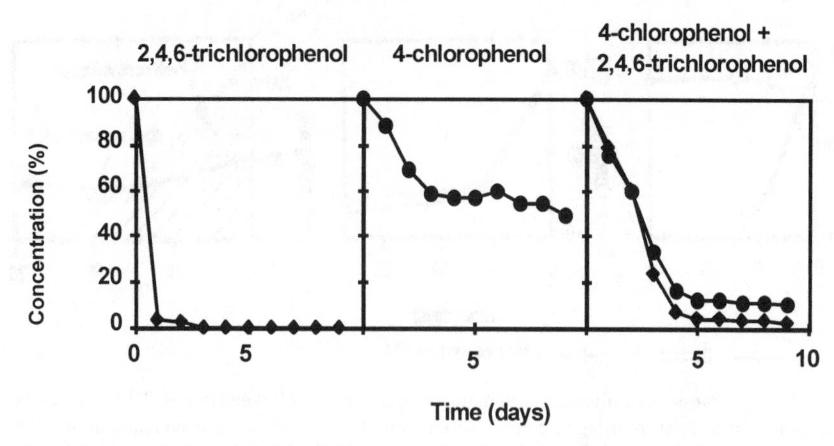

Fig. 21.8. Degradation of 2,4,6-trtichlorophenol and 4-chlorophenol by *Ralstonia eutropha* as single substances and in the mixture

is not degraded by *Ralstonia eutropha*, when it is the only substrate. However, the 4-chlorophenol is also degraded when 2,4,6-trichlorophenol is present in the medium, the substrate on which this strain was isolated. This can be explained by the fact, that the trichlorophenol induces the enzymes for its degradation. These enzymes can then also attack the 4-chlorophenol (fig. 21.8). This phenomenon is termed co-metabolism.

Another very common phenomenon is shown in fig. 21.9. The degradation of 2,4-dichlorophenol by *Pseudomonas putida* PaW85 (pEST 4011) is completely inhibited in the presence of 2,4,6-trichlorophenol. This may be explained by the fact that the trichlorophenol is toxic for this strain.

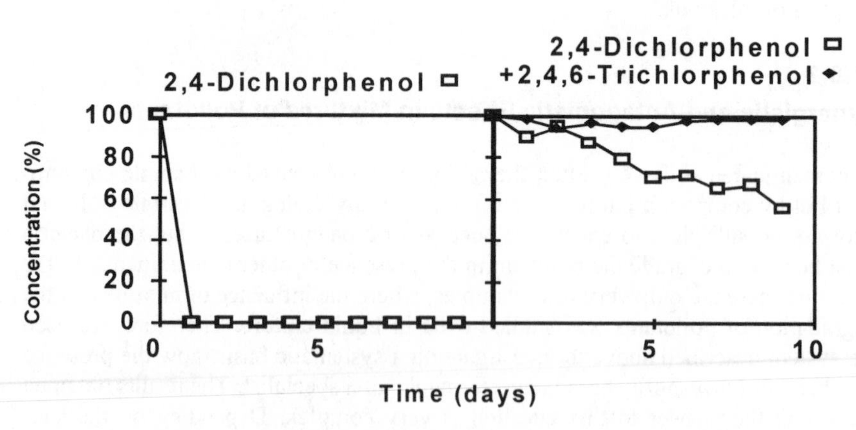

Fig. 21.9. Inhibition of 2,4-dichlorophenol degradation by *Pseudomonas putida* PaW85(pEST4011) in the presence of 2,4,6-trichlorophenol

But the situation may also be more complicated as shown in fig. 21.10. *Sphingomonas chlorophenolicus* degrades its substrate 2,3,4,6-tetrachlorophenol in the presence of 4-chlorophenol. *Arthrobacter sp.* HEID, the strain that normally degrades the 4-chlorophenol, cannot grow on this mixture because it is inhibited by the tetrachlorophenol. When *Arthrobacter sp.* HEID and *Sphingomonas chlorophenolicus* are added together to the mixture neither the 4-chlorophenol nor the tetrachlorophenol is degraded. This can only be explained by the assumption that *Arthrobacter sp.* HEID, although it does not grow in the mixture, converts a tiny fraction of one of the chlorphenols to a metabolite toxic to *Sphingomonas*.

Table 21.3 summarises the results we got with the mixtures of the different chlorophenols and the bacteria that should theoretically degrade these mixtures. As can be seen, in complex mixtures normally no degradation at all occurred, although all the bacteria necessary for the degradation were present. This may be an explanation, why in such complicated situations where many pollutants are present in one site bioaugmentation is not successful. A lot of research is needed to understand at least part of all the processes that can occur under such circumstances.

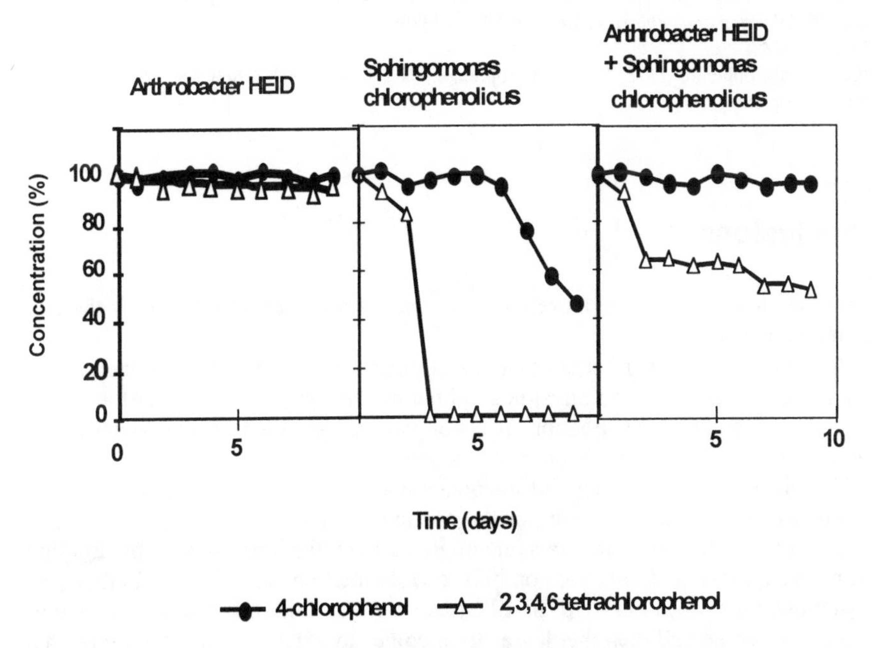

Fig. 21.10. Degradationof a mixture of 4-chlorophenol and 2,3,4,6-tetrachlorophenol by *Arthrobacter* HEID, *Sphingomonas chlorophenolicus* and by the mixture of the two strains

Table 21.3. Degradation of mixtures of chlorophenols by mixtures of degrading bacteria. Without antagonistic effects theoretically all chlorophenols should be degraded.

Combination of degrading micro organisms	Combination of chlorophenols	Degradation of the single phenols in the mixture			
		MCP	**DCP**	**TCP**	**TeCP**
HEID+EST	MCP+DCP	++++	+++++		
HEID+GP1	MCP+TCP	+++		+++	
HEID+SP	MCP+TeCP	-			⊕
EST+GP1	DCP+TCP		++++	++++	
EST+SP	DCP+TeCP		-		-
GP1+SP	TCP+TeCP			+++	+++
HEID+EST+GP1	MCP+DCP+TCP	+	++++	+	
HEID+EST+SP	MCP+DCP+TeCP	-	-		-
HEID+GP1+SP	MCP+TCP+TeCP	-		-	-
EST+GP1+SP	DCP+TCP+TeCP		-	-	-
HEID+EST+GP1+SP	MCP+DCP+TCP+TeCP	-	-	-	-

Strains:
HEID Arthrobacter sp., HEID, EST Ps. putida PaW85 (pEST4011), *GP1 Ralstonia eutropha* GP1, *SP Sphingomonas chlorophenolicus* ATCC 33790
Chlorophenols:
MCP 4-Chlorophenol, *DCP* 2,4-Dichlorophenol, *TCP* 2,4,6-Trichlorophenol, *TeCP* 2,3,4,6-Tetrachlorophenol

21.7 Conclusions

From the survey of the literature and from our experiments we can draw the following conclusions:

Bacteria do, in freshly contaminated soil, the same things they do in the laboratory in liquid culture. If the physiological parameters are favourable for the bacteria, then the bioavailable fraction of a pollutant is degraded by the bacteria that possess the ability to degrade these compounds.

The big problem in biological decontamination of soil is bioavailability. The organisms can only degrade the portion of the pollutant that is available to them. Especially in aged contaminations one might assume, that the bioavailable fraction has been degraded and only the non bioavailable fraction is remaining. In this case bioaugmentation does not help. In all cases where an initial degradation of a pollutant is observed and then the degradation comes to a halt, one must assume, that not the bacteria are limiting and therefore in these cases bioaugmentation is not useful. In such cases measures have to be taken that increase bioavailability. This may explain why in many cases in the field, bioaugmentation was not successful.

Bioaugmentation is only successful when the pollutants are available, but the autochthonous microflora did not have time to adapt to the pollutant. This is certainly the case in fresh contaminations e.g. after accidents. Here bioaugmentation can certainly speed up degradation considerably. In cases where the physiological conditions, have not been suitable for biodegradation after the change in conditions bioaugmentation certainly helps to stimulate the degradation in the initial stages of remediation.

In cases where despite the bioavailability of pollutants and of the correct physiological conditions no degradation occurs, antagonistic effects of mixtures of contaminants have to be considered. However, in this area a lot of research is still needed in order to understand and to solve these problems.

References

Anderson JPE, Lichtenstein, EP Whittingham WF (1970) Effect of *Mucor alternans* on the persistence of DDT and dieldrin in culture and in soil. J Econ Entomol 63: 1595–1599

Andreoni V, Baggi G, Colombo M, Cavalca L, Zangrossi M, Bernasconi S (1998) Degradation of 2,4,6-trichlorophenol by a specialized organism and by indigenous soil microflora: bioaugmentation and self-remediability for soil restoration. Lett Appl Microbiol 27: 86–92

Barbeau C, Deschenes L, Karamanev D, Comeau Y, Samson R (1997) Bioremediation of pentachlorophenol-contaminated soil by bioaugmentation using activated soil. Appl Microbiol Biotechnol 48: 745–752

Barles R, Daughton WCG, Hsieh DPH (1979) Accelerated parathion degradation in soil inoculated with acclimated bacteria under field conditions. Arch Environ Contam Toxicol 8: 647–660

Briglia M, Nurmiaho-Lassila EL, Vallini G, Salkinoja-Salonen M (1990) The survival of the pentachlorophenol-degrading *Rhodococcus chlorophenolicus* PCP-1 and *Flavobacterium sp* in natural soil. Biodegradation 1: 273–281

Brunsbach FR (1993) Abbau von Chloraromaten in Bodenschlämmen durch Laborstämme und Bodenpopulationen. Dissertation, Wuppertal: BUGH, Fachbereich Naturwissenschaften II

Chen M, Hong CS, Bush B, Rhee G-Y (1988) Anaerobic biodegradation of polychlorinated biphenyls by bacteria from Hudson river sediments. Ecotoxicol Environ Saf 16: 95–105

Edgehill RO, Finn RK (1983) Microbial treatment of soil to remove pentachlorophenol. Appl Environ Microbiol 45: 1122–1125

Eiermann DR, Peter J (1990) Wirksamkeit des Einsatzes von schadstoffabbauenden Mikroorganismen anhand eines Off-site Feldversuchs. In: Thomé-Kozmiensky KJ (ed) Altlasten. 3 EF, pp 469–477

Elsas JD, van Hekman W, Overbeek LS, van Smitt E (1991) Problems and prospectives of the application of genetically engeneered micro organisms to soil. Trends Soil Sci 1: 373–392

Focht DD, Brunner W (1985) Kinetics of biphenyl and polychlorinated biphenyl metabolism in soil. Appl Environ Microbiol 50(4): 1058–1063

Focht DD, Shelton D (1987) Growth kinetics of *Pseudomonas alcaligenes* C-O relative to inoculation and 3-chlorobenzoate metabolism in soil. Appl Environ Microbiol 53(8): 1846–1849

Goldstein RM, Mallory LM, Alexander M (1985) Reasons for possible failure of inoculation to enhance biodegredation. Appl Environ Microbiol 50: 977–983

Huntjens JLM, Brouwer W, Grobben K, Jansma O, Scheffer F, Zehnder AJB (1988) Biodegradation of α-hexacyclohexane by a bacterium isolated from polluted soil. In: BMFT/UBA (ed) Zweiter Internat TNO/BMFT-Kongreß über Altlastensanierung, Hamburg 11–15 April: Altlastensanierung ´88. Dordrecht: Kluwer Academic Publishers 1, pp 747–753

Kästner M, Breuer-Jammali M, Mahro B (1994) Enumeration and characterisation of the soil microflora from hydrocarbon contaminated soil sites able to mineralize polycyclic aromatic hydrocarbons (PAH). Appl Microbiol Biotechnol 41: 267–273

Kilbane JJ, Chatterjee DK, Chakrabarty AM (1983) Detoxification of 2,4,5-trichlorophenoxyacetic acid from contaminated soil by *Pseudomonas cepacia.* Appl Environ Microbiol 45: 1697–1700

Klages U, Lingens F (1979) Degradation of 4-chlorobenzoic acid by a *Nocardia* species. FEMS Microbiol Lett 6: 201–203

Klages U, Lingens F (1980) Degradation of 4-chlorobenzoic acid by a *Pseudomonas sp.* Zbl Bakt Hyg I Abt Orig C1: 215–223

Kuhnt G, Härtling T, Schmotz W, Vetter L (1991) Auswahl von Referenzböden für die Chemikalienprüfung im EG-Bereich, Umweltforschungsplan des Bundesministers für Umwelt Naturschutz und Reaktorsicherheit, Forschungsbericht 10602058

Lehtomäki M, Niemelä S (1975) Improving microbial degradation of oil in soil. Ambio 4: 126–129

Li DY, Eberspächer J, Wagner B, Kutzner J, Lingens F (1991) Degradation of 2,4,6-trichlorophenol by Azotobacter sp GP1. Appl Environ Microbiol 57: 1920–1928

Löffler F (1993) Untersuchungen zum Abbau und zur Funktion der 4-Chlorbenzoat-Dehalogenase aus *Pseudomonas sp* CBS3. Dissertation, Universität Hohenheim, Fachbereich Biologie

Macrae IC, Alexander M (1965) Microbial degradation of selected herbicides in soil. J Agric Food Chem 13: 72–76

Meer JR, van der Roelofsen W, Schraa G, Zehnder AJB (1987) Degradation of low concentrations of dichlorobenzenes and 1,2,4-trichlorobenzene by *Pseudomonas sp* strain P51 in non sterile soil columns FEMS. Microbiol Ecol 45: 333–341

Middeldorp PJM, Briglia M, Salkinoja-Salonen M (1990) Biodegradation of pentachlorophenol in natural soil by inoculated *Rhodococcus chlorophenolicus.* Microbial Ecology 20: 123–139

Müller R (1999) Grundlagen des mikrobellen Schadstoffabbaus. VDI-Berichte 1478: 391–396

Newcombe DA, Crowley DE (1999) Bioremediation of atrazine-contaminated soil by repeated applications of atrazine-degrading bacteria. Appl Microbiol Biotechnol 51: 877–882

Oldenhuis R, Koijk L, Lammas A, Jannssen J, Witholt B (1989) Degradation of chlorinated and non chlorinated aromatic solvents in soil suspensions by pure bacterial cultures. Appl Microbiol Biotechnol 30: 211–217

Portier R, Bianchini M, Fojisaki K, Henry C, McMillin D (1988) Comparison of effective toxicant biotransformation by autochthonous micro organisms and commercially available cultures in the situ reclamation of abandoned industrial sites. In: Schriftenreihe Verein WaBoLu 80, Internationales Fachgespräch, Langen 11–12 Mai 1987 Stuttgart: Fischer, pp 273–292

Ripp S, Nivens DE, Werner C, Sayler G (2000) Bioluminescent most-probable-number monitoring of a gentically engineered bacterium during a long-term contained field release. Appl Microbiol Biotechnol 53: 736–741

Ruisinger S, Klages U, Lingens F (1976) Abbau der 4-Chlorbenzoesäure durch eine *Arthrobacter*-Spezies. Arch Microbiol 110: 253–256

Salkinoja-Salonen M, Middeldorp P, Briglia M, Valo R, Häggblom M, McBain A (1990) Clean-up of old industrial sites. In: Kamely D, Chakrabarty A, Omenn GS (eds) Advances in applied biotechnology series. Houston, Gulf Publishing Company 4, pp 347–367

Schwefer R-J (1988) Stand der Entwicklung der biologischen in-situ-Sanierungsverfahren, dargestellt an Beispielen aus Europa und USA. In: BMFT/UBA (ed) Zweiter Internat TNO/BMFT-Kongreß über Altlastensanierung, Hamburg 11–15 April: Altlastensanierung ′88. Dordrecht: Kluwer Academic Publishers 1, pp 697–705

Sprenger B, Ebner HG (1990) Ermittlung der biotechnischen Sanierbarkeit kontaminierter Böden. In: Arendt F, Hinsenveld M, van den Brink WJ (eds) Altlastensanierung `90, dritter internationaler KfK/TNO-Kongreß über Altlastensanierung. Dordrecht: Kluwer Academic Publishers, pp 1063–1070

Stanlake GJ, Finn RK (1982) Isolation and characterization of a pentachlorophenol-degrading bacterium. Appl Environ Microbiol 44: 1421–1427

Top EM, Van Daele P, De Saeyer N, Forney LJ (1998) Enhancement of 2,4-dichlorophenoxyacetic acid (2,4-D) degradation in soil by dissemination of catabolic plasmids. Antonie van Leeuwenhoek 73: 87–94

Trevors JT, Kuikman P, van Elsas JD (1994) Review article: release of bacteria into soil: cell numbers and distribution. J Microbiol Methods 19: 247–259

Unterman R et al. (1988) Biological approaches for polychlorinated biphenyl degradation. In: Omenn GS (ed) Environmental biotechnology. London, Plenum Press

Valo R, Salkinoja-Salonen M (1986) Bioreclamation of chlorophenol-contaminated soil by composting. Appl Microbiol Biotechnol 25: 68–75

Viney I, Bewley RJF (1990) Preliminary studies on the development of a microbiological treatment for polychlorinated biphenyls. Arch Environ Contam Toxicol 19: 789–796

22 Optimisation of Microbial Soil Treatment

K. Hupe[1], M. Koning[2], J.-C. Lüth[2], J. Heerenklage[2], R. Stegmann[2]
[1]Consultants for Waste Management, Prof. R. Stegmann and Partner, Nartenstraße 4a, 21079 Hamburg, Germany
[2]Technical University Hamburg-Harburg, Department of Waste Management, Harburger Schloβstraße 37, 21079 Hamburg, Germany

22.1 Introduction

Biological treatment of contaminated soil is the most common method of soil treatment during the redevelopment of old sites (ITVA 1997). Biopiles in particular are frequently used due to the simple handling involved and process realisation. The significance of the bioremediation of old sites depends on the reutilisation of the soil material to be treated. The objective should be an increase in the amount of decontaminated soil for ecological and economic reutilisation e.g. as a culture medium in landscape architecture.

In spite of the widespread establishment of biological remediation methods, there is still a whole string of unsolved problems. First of all, it is necessary to examine the degradation processes in detail in order to optimise them afterwards, taking into consideration the knowledge gained, and to initiate their translation into technical remediation practice by scale enlargement.

In the practice of remediation there appears to be a tendency towards "minimal expenditure". To an increasing degree, contaminated sites are less often redeveloped and more often secured because of the costs that arise. This means that redevelopment companies have to offer more economical methods. Therefore, the technical requirements for the treatment of "simple" damages, such as mineral-oil damage, have to become lower. This means that an increase for example, in treatment capacity per unit area is necessary as far as biopile methods are concerned. This can be realised e.g. by the application of high biopiles (biopile height 4–5 m). First of all, the degradation processes in the biopile body have to be investigated more closely for a further optimisation of the biopile methods, with the aim of accelerating degradation.

22.2 Preliminary Tests in Test Systems

For a general understanding of biological conversion processes, fundamental investigations into the application of test systems and laboratory reactors are necessary. When conceiving biological redevelopment methods, preliminary tests at laboratory scale should be geared towards answering questions concerning microbiological, chemical-physical and process-related geological conditions. The most important problem – and at the same time the most difficult – is to evaluate to what extent the soil can be bioremediated. The laboratory results of investigations into the microbial degradation of relevant pollutants in soil have to be transferable into practice and should, therefore, provide data concerning degradation such as degradation rates, achievable final concentrations, and obtained intermediate and final products. The choice of which complex test system to use should be made according to the initial questions.

Owing to the complexity of the "soil" system, the rate of degradation is often decisively determined by chemical/physical parameters (such as transport or sorption processes) rather than real microbiological degradability. Bioavailability is of great importance and primarily depends on the type of soil matrix – that is, on its material composition and particle-size distribution – and also on the duration of exposure to contamination. A preliminary decision concerning the application of a biological method is made by means of a bioavailability test. When the results are positive, laboratory investigations have to be carried out in order to optimise the biodegradation and to ascertain to what extent the physical and chemical soil parameters influence biodegradation (DECHEMA 1991).

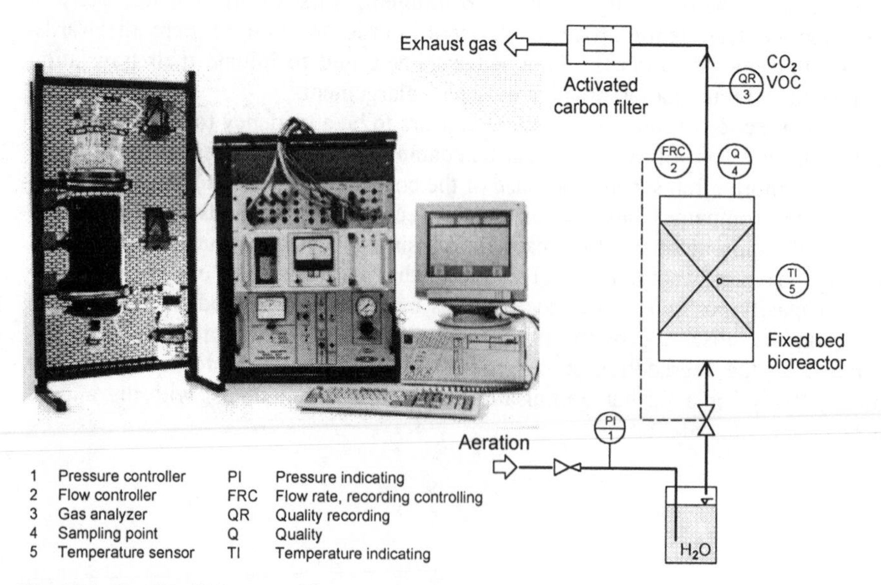

Fig. 22.1. Fixed-bed laboratory bioreactor system

Table 22.1. Gas-measuring instruments and equipment

Parameter	Method, instrument, equipment
CO2 – on-line	IR: ADC DB2E
VOC – on-line	FID: Bernath-Atomic – total carbon analyser 3002
Measurand pick-up	Regulation program: WorkBench PC 2.0 on MS-DOS

VOC volatile organic carbons, *IR* Infrared spectrophotometer, *ADC* Analytical Development Company, *FID* Flame ionisation detector

The fixed-bed laboratory bioreactor system shown in fig. 22.1 is used for the simulation of the conditions in biopiles or in large-scale reactors. The system was used for the investigations concerning the optimisation of the milieu and operating conditions. There is a sieve in the lower part of the reactors onto which the soil material (0.5–1 kg dry-weight test portion per litre reactor volume) is delivered. Different sampling devices are installed at different heights on the walls of the reactors. For sufficient aeration, the volumetric rate of flow is adjusted in general to 1 [litre $kg^{-1}{}_{dry\ weight\text{-}soil\ material}$ h^{-1}]. To guarantee a constant water content in the soil material, the gas is led through a wash bottle filled with water before entering the reactor. At the waste-gas side, CO_2 concentrations are measured on-line by means of an infrared spectrophotometer (IR), and the off-gassing of volatile contamination is measured as total carbon (VOC: volatile organic carbon) by a flame ionisation detector. For the on-line detection of the volume flow, an inductive flowmeter is installed on the waste-gas side.

The on-line controlling and recording of the aeration rate (mass flowmeter) as well as of the measured values of the exhaust air from the bioreactors (up to 20 setups in parallel) was carried out by means of a PC (controlling and recording program: WorkBench PC 2.0 on MS-DOS). The single reactors can be connected to the serialised measuring instruments via a solenoid-valve block which is controlled by a PC. Manual checks can be carried out by taking gas samples from the reactor with a syringe through a septum, and injecting the samples into a gas chromatograph (GC). Detailed information about the gas-measuring instruments and equipment used can be seen in table 22.1 (see also: Hupe 1998 or Hupe et al. 1998a).

22.3 Optimisation of Treatment Conditions

For economic soil remediation, a significant enhancement of the biological degradation of the pollutant is necessary. This can be realised by an optimisation of the milieu and operating conditions and by process engineering. In general, optimisation has to be carried out for every single contaminated soil/site to be treated. In practice, investigations of this kind are usually carried out only to a limited extent due to financial reasons and time constraints. The authors highlight the importance of these pre-investigation tests.

Within the framework of the optimisation tests, the influence of the following parameters were intensively examined:

- water content,
- temperature,
- oxygen content,
- compost content,
- dynamic treatment.

Operating parameters as well as parameters that significantly influence the milieu conditions, and therefore also biological degradation (e.g. pH value, nutrient ratio) were varied. The results of own investigations are described in the following chapters.

22.3.1 Water Content

The optimal water content for the static treatment (no mixing) of oil-contaminated soil material is in the range of 35–65 % of the maximum water-holding capacity, as shown for example, by investigations into the long-term influence of the water content on pollutant conversion in biopiles. Concerning the treatment in mixed systems – such as tube or other mixer reactors – the optimal range shifts towards the lower water contents. This is necessary in order to avoid the formation of agglomerates and pellets. A range of 35–55 % of the maximum water-holding capacity is recommended, as described by Koning et al. (2000).

22.3.2 Temperature

The optimal temperature can vary greatly, depending on the volatility and the solubility of the pollutants to be treated. For mineral oil contamination, the conversion output in the contaminated soil can be significantly improved by a temperature increase from 10 to 30 °C (see also Koning et al. 2000, Chapter 26). Year round biological remediation strongly depends on temperature, especially in colder climatic zones with temperatures below 10 °C in the winter months. In these cases, heat insulation can be, at most, a basic measure. The use of waste heat from other processes (e.g. composting of biowaste or municipal solid waste) or from eventually available block-heating power plants (of a disposal site or an anaerobic treatment plant for example) can be efficient particularly in central soil treatment plants. In general, a treatment temperature below 10–15 °C should be avoided. With increasing treatment temperature, the biodegradation of mineral oil in contaminated soils can be accelerated.

22.3.3 Oxygen

To avoid anaerobic conditions, the oxygen supply is of great importance. Oxygen supply can be effected in a passive/diffusive or in an active way. According to extensive investigations into the influence of the oxygen content on pollutant conversion, the oxygen content in the soil body should not drop below 1 vol. %. This means that periodic oxygen measurements in the soil or in the biopile body at different depths are indispensable.

The influence of oxygen on the biodegradation of diesel fuel in unsaturated soil/compost mixtures at 30 °C was analysed over a period of 7 weeks. Oxygen concentrations varying between 0 % and 80 vol. % O_2 were investigated, with concentrations below 5 vol. % O_2 being examined more closely.

A slightly loamy sand (sieved to particle size ≤ 2 mm) was artificially contaminated with 1 % (% of dry weight of soil) diesel fuel. Mature biocompost (sieved to ≤ 4 mm; German standard degree of maturity: V) was used as soil supplement. The water contents of the soil/compost mixtures were adjusted with distilled water to 55 % and 60 % of the maximum water-holding capacity (WC_{max}).

Aeration of the soil materials was carried out by means of synthetic air (O_2/N_2 mixtures). While the aeration rate for the test series OX1 was adjusted to a fixed value of 2 l h^{-1} for the whole 48 hour test period, the aeration rates in test series OX2 and OX3 were regulated based on oxygen requirements predicted from preliminary tests. In these tests, oxygen consumption was determined, and then converted into an aeration rate for each test series based on the required oxygen concentration using a safety factor of 3. The test set-ups are described in table 22.2.

As a concluding examination of the influence of the oxygen content on the pollutant's turnover, the mineralisation rates of test series OX1, OX2 and OX3 were compared. The cumulative mineralisation of the different test series was compared after 7 days and after 7 weeks (at the end of the test period) to the oxygen content and was evaluated analogously using the Monod kinetics formula (equation 22.1).

Table 22.2. Parameters of test series OX1-3

Test series	OX1	OX2	OX3
Soil material [g dwt]	1420	1045	800
Diesel fuel [g (kg dwt)$^{-1}$]	10	10	10
Compost [g dwt]	360	105	80
Soil/compost ratio	4:1	10:1	10:1
pH	6.9	6.7	6.7
$^{W}C_{max}$ [g H_2O (100 g dwt)$^{-1}$]	55.9	49.2	45.6
Water content [% of $^{W}C_{max}$]	60	55	55
Temperature [°C]	30	30	30
Oxygen content [vol. %]	0; 5; 10; 21; 40; 80	1; 2; 3; 4; 5; 21	0; 0.1; 0.2; 1; 2; 5; 21
Gas analysis	discontinuous	on line	on line
Flow rate [l h^{-1}]	2	varying #	varying #
Test period [d]	48	49	7

dwt dry weight, WC_{max} maximum water-holding capacity, # depending on microbial activity

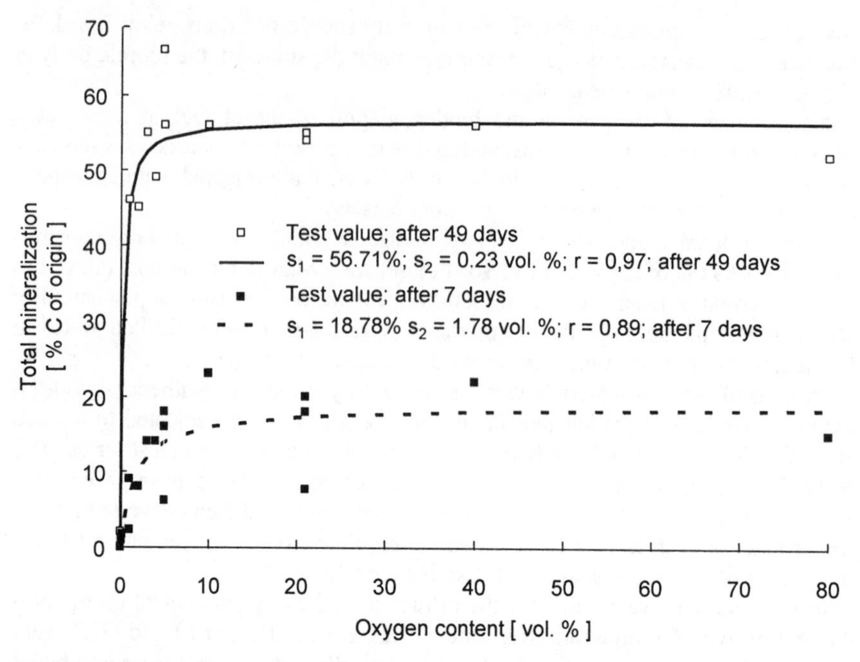

Fig. 22.2. Influence of the oxygen content on the total mineralisation (s_1 and s_2 analogous equation 22.1)

$$\Delta C_{co_2} = \frac{s_1 \cdot C_{o_2}}{s_2 + C_{o_2}} \tag{22.1}$$

with:

ΔC_{CO2} cumulative mineralisation [% C of origin]
C_{O2} oxygen content in the aeration stream [vol. % O_2]
s_1, s_2 rate constants [% of original C], [vol. % O_2] respectively

Fig. 22.2 shows the influence of the oxygen content on the total mineralisation. For the evaluation of the oxygen influence, the total mineralisation after 7 days and at the end of the test was determined. The two curves were calculated using equation 22.1. This description takes into consideration further data concerning the oxygen content, obtained from the test series OX3, following OX1 and OX2.

The results showed that the total mineralisation for oxygen concentrations between 0 and 80 % in the phase of the maximum degradation rate after 7 days, can be described by the equation

$$\Delta C_{CO2} = 18.78\,\% \cdot C_{O2} / (1.779 \text{ vol. } \% + C_{O2}).$$

During the whole test period, the mineralisation rates followed the equation

$$\Delta C_{CO2} = 56.71\,\% \cdot C_{O2}/(0.23 \text{ vol. } \% + C_{O2}).$$

When using a passive oxygen supply, the soil material is converted in the soil body depending on the microbial activity and on the oxygen content. As far as the height of the biopile is concerned, no limit of the pollutant conversion could be measured during investigations of biopiles of 1.5 m height after a 9-week treatment period, using passive aeration and without moving of the biopiles. Thus, this kind of aeration seems to be sufficient for a normal biopile height (Hupe et al. 1998). However, especially in high biopiles (4–5 m height) and during the treatment of cohesive soil material, a limit due to an insufficient diffusive oxygen supply is to be expected. In this case, active aeration via drainage systems, or intermittent aeration via pneumatic aeration lances could be applied. The aeration intervals for intermittent aeration could be controlled via the oxygen content in the biopile body. Furthermore, it can also be useful to add structural material to increase the air-void volume for an improved air exchange, depending on the soil structure. Inert material with a small portion of organic material or material which can be separated from the soil material via mechanical processes after the treatment, should be applied in order not to limit the reutilisation possibilities of the decontaminated soil material. The effect of using organic structural material like compost strongly depends on the quantity added.

22.3.4 pH Value

The pH value is of special importance, as most of the microorganisms that degrade mineral oil prefer a slightly acid to neutral milieu. To obtain a pH value of 6–7.5, it is generally necessary to raise the pH value of acidic soil materials. In slightly acidic soils with low buffer capacity and low organic content, this can be achieved by the addition of a pH-neutral material with high buffer capacity, such as mature compost (rotting degree V). On the other hand, liming is necessary in acidic soils with high organic content and high buffer capacity. By the use of calcium carbonate, the pH value can be adjusted to about 7–8. Beyond that, calcium oxide as well as calcium hydroxide can be used when:

- the pH value needs to be raised,
- an increased addition of carbonates could disturb the measurement of the mineralisation,
- a rise in temperature via the exothermic reaction of calcium oxide in the soil is specifically desired,
- oxidation effects are desired.

22.3.5 Nutrients

The C:N ratio seems to have a lasting influence on the biodegradation of mineral oil contaminated soils. Oil contamination has the effect of increasing the C:N ratio in the soil material, which has a negative effect on microbial activity. A reduction of the C:N ratio to less than 20:1 is useful. Furthermore, own investigations on the biodegradation of mineral oil showed that the C:P ratio should be reduced to less than 50:1. This can be achieved by the addition of nutrient solutions, mineral fertilizer, or compost.

22.3.6 Structural Material

Aggregates are added for different reasons. The additive can be applied as structural material for better aeration, as a supplier of pollutant-utilising microorganisms, as a nutrient supplier, as a pH buffer or buffer for the water balance, or can be used specifically for humification. Within the scope of our investigations, mature compost with rotting degree V showed the best characteristics for an increase in pollutant conversion in recently contaminated soil material, poor in organic matter (Hupe et al. 1996).

The investigations showed that biological degradation of contaminated soils could be enhanced by compost addition. The biological degradation, which depended on the addition of compost of different degrees of maturity and the significant parameters of the compost, was investigated.

Several investigations have shown that, depending on the history of the contaminated soil, the degradation of organic contaminants may be enhanced by adding biocompost. Investigations on the biodegradation of PAHs have shown that biocompost addition can primarily enhance the degradation of the bioavailable PAHs in freshly contaminated soil materials. However, no significant effect of compost addition on the degradation of PAHs in an old contamination (Eschenbach 1995) or on the degradation of 4-chloro-benzoate could be detected (Breidenbach 1995).

In own investigations it was observed that the degradation rate of oil contaminants increased with increasing compost content (from 10 to 33 % dry weight of soil). However, for remediation in practice, a compost content of 33 % (dry weight of soil) is too high. Thus the compost content should be reduced to < 10 %.

Furthermore, these effects were investigated in several respirometer and bioreactor test series using compost at different degrees of maturity. Compost has the following applications, derived from the biocompost tests, that may be of importance:

- as a bulking agent to improve aeration particularly in cohesive soil materials,
- as a supplier of a great variety of microorganisms (Kästner et al. 1995),
- as an agent to improve the pH buffer capacity and water storage capacity after compost addition,
- as a structural material to reduce pellet formation when cohesive soil materials are treated in mixing reactors ("dynamic treatment"),
- as a source of nutrients and trace components (particularly nitrogen and phosphorus); compost as a depot fertilizer,
- as an agent to incorporate the contaminant into the organic matrix of the compost (incorporation of the contaminants in the soil/humus matrix; Lotter et al. 1993).

Moreover, ecotoxicity could be reduced in oil contaminated soils by means of compost addition. This was measured using bacterial activity as well as algae and plant toxicity tests (Ahlf et al. 1993).

The amount of compost that can be added is limited by the increase in volume and the increased space requirement for treatment, and by the increased quantity of material to be reutilised. Furthermore, reutilisation of the decontaminated soil material is already considerably limited by the quantity of added organic material. Thus, there is a necessity to optimise the kind and quantity of the additives, depending on the type of soil and pollutant as well as on the reutilisation strategies. Details concerning the maximum organic content for an unlimited reutilisation are not yet available. Thus, the application of additives, which are poor in organic material and/or are inert, is of great importance. As far as the selection of suitable material is concerned, there is still a need for research.

22.3.7 Homogenisation

One of the main problems in the specific treatment of contaminated soil material is the considerable inhomogeneity of the contaminated material. Different aggregates are available for homogenisation. For the homogenisation of screened soil material, the use of mixer systems could be suitable. To avoid pellet formation in these systems, de-agglomerating units, such as rapidly rotating cutter heads, are available. With an increasing treatment period in these mixer systems there is also a tendency towards pellet formation – depending on the water content of the soil and on the size of the fine grain component (Hupe 1998). To avoid this pellet formation, the improvement of the microbial pollutant conversion in soil-solid reactors by de-agglomeration is being investigated, for example at the Technical University of Braunschweig, Germany within the scope of a DFG project (Scholz et al. 1996). Mixers are not only used for homogenisation and for an eventual de-agglomeration, but also for the breaking-up (aeration) and addition of aggregates before and during the biological treatment.

22.3.8 Application of Dynamic Reactors

Based on experience so far, dynamic dry soil treatment does not seem to be useful in practice for biological remediation. Dynamic treatment in slurry reactors seems to be more promising, but according to a cost-benefit analysis, it is to be expected that these dynamic reactors will only be applied to special problems, in particular highly contaminated single charges (such as residues derived from soil washing).

Even though the application of dynamic bioreactors has not been particularly economical, mixers can be useful nonetheless for the mechanical pretreatment, prior to biological treatment, of soil material (in the solid phase), providing relatively short treatment times. Mixers with de-agglomerating devices, such as a plowshare mixer with an integrated quickly rotating cutter head, are suitable for de-agglomeration, for homogenisation and for the homogeneous addition of structural material and nutrients, and for moistening. Nevertheless, it has been ascertained for this treatment, that the water content has to be set to the optimal level below 50 % of the maximum water-holding capacity. To achieve this, the soil material either has to be dried thermally or the water content has to be decreased by the addition of "dry" structural material with a high water-holding capacity. The quantity of additives is especially limited by the reutilisation of the decontaminated soil material.

22.4 Summary and Outlook

Biological soil-purification methods are state of the art for the decontamination of mineral oil contaminated soil. They are successfully applied in situ as well as ex situ. An acceleration in the biodegradation processes of the pollutants in the soil and an increase in the conversion rate can be effected by an optimisation of the milieu and operating conditions.

By optimising the milieu conditions as well as the chemical and physical properties of the contaminated soil, decontamination can be accelerated and carried out to a higher degree. This is especially true for recently polluted soil materials with a high bioavailability of the contaminant due to the optimisation of temperature, water content, oxygen content, pH value, nutrient supply, and process-related operating conditions (such as separation of disturbing material, homogenisation, de-agglomeration, addition of aggregates etc.) (see table 22.3).

Table 22.3. Advisable milieu and operating conditions for the treatment of TPH contaminated soils

Parameter	"Optimum" (advice for practical treatment)
Water	static treatment: 35–65 % WC_{max} dynamic treatment: 35–55 % WC_{max}
Temperature	dependent on the kind of contamination increased oil conversion by maintaining temperature in the range of 10–30 °C (in cold seasons/regions: heat insulation, heat coupling)
Oxygen	> 1 vol. % in the soil (structural material can be added for improved oxygen supply; active aeration via drainage system, pneumatic aeration lances, movement)
pH value	slightly acid to neutral (pH 6–7.5) (lime addition increases of the pH value for acid soil with a high buffer capacity; compost addition increases pH and buffering for a slightly acid soil with a low buffer capacity)
Nutrients	C:N:P ratio about 100:8:2 (C:N < 20; C:P ≈ 50) (nutrient solutions, fertilizers or compost can be added to reduce the C:N ratio
Sieving/crushing	separation of foreign and coarse material (> 20 mm)
Mixing	for preparation before starting the treatment or treatment accompanying for homogenisation, breaking-up (aeration), de-agglomeration
Addition of aggregates	inoculation material: biologically purified soil mature biocompost inert structural material for improved aeration (additive quantity: depending on the used additives and on the reutilisation of the purified soil)

WC_{max} maximum water-holding capacity, *C:N:P* carbon : nitrogen : phosphorus

There are a number of possibilities to increase the bioavailability of organic pollutants in soil materials where there is a lack of bioavailability:

- reduction of the diffusion paths by corn or aggregate crushing or by deagglomeration,
- heating in order to increase the diffusion coefficient or to increase the solubility,
- acceleration of biotransformation by mixing of the soil (Harms 1996),
- degradation of the bioavailable pollutant fractions and observation/securing of the remaining material (Harms 1996),
- increase of the mobility or of the solubility of the pollutants by addition of tensides/solutizers,
- combination of chemical oxidation and biological conversion of the preoxidised components (Stegmann et al. 1996).

In all cases, it is strongly recommended to make adequate pre-investigation tests due to uncertainties in the prediction of the actual biodegradibility of specific contaminated soils (see Chapter 41).

References

Ahlf W, Gunkel J, Rönnpagel K (1993) Toxikologische Bewertung von Sanierungen. In: Stegmann R (ed) Bodenreinigung. Hamburger Berichte 6, Economica Verlag, Bonn, Germany, pp 275–286

Breidenbach C (1995) Der Abbau von 4-Chlorbenzoat durch *Pseudomonas* sp CBS3 im Boden und in Flüssigkultur. Thesis, Technical University of Hamburg-Harburg, Germany

DECHEMA (1991) Einsatzmöglichkeiten und Grenzen mikrobiologischer Verfahren zur Bodensanierung. 1. Bericht des Interdisziplinären Arbeitskreises der DECHEMA "Umweltbiotechnologie – Boden", DECHEMA, Frankfurt a M, Germany

Eschenbach A (1995) Einfluß von *Pleurotus ostreatus*, Kompost, *Sphingomonas paucimobilis* und der Kontaminationsdauer auf den Verbleib und Abbau ^{14}C-markierter polyzyklischer aromatischer Kohlenwasserstoffe (PAK) in Altlastböden. Thesis, Technical University of Hamburg-Harburg, Germany

Harms H (1996) Bioverfügbarkeit organischer Schadstoffe – der Einfluß von Sorption und Massentransfer. In: Stegmann R (ed) Neue Techniken der Bodenreinigung. Hamburger Berichte 10, Economica Verlag. Bonn (distribution: Verlag Abfall aktuell, Stuttgart), Germany, pp 257–268

Hupe K (1998) Optimierung der mikrobiellen Reinigung mineralölkontaminierter Böden in statischen und durchmischten Systemen. In: Stegmann R (ed) Hamburger Berichte 15. Verlag Abfall aktuell, Stuttgart, Germany

Hupe K, Heerenklage J, Woyczechowski H, Bollow S, Stegmann R (1998) Influence of oxygen on the degradation of diesel fuel in soil bioreactors. Acta Biotechnol 18 (2): 109–122

Hupe K, Lüth J-C, Heerenklage J, Stegmann R (1998 a) Test systems for balancing and optimizing the biodegradation of contaminated soils: A German perspective. In: Sikdar SK, Irvine RL (eds) Fundamentals and Applications, vol 1, Bioremediation: Principles and Practice. Technomic Publishing Co, Inc Lancaster, Basel, Switzerland, pp 665–692

Hupe K, Lüth J-C, Heerenklage J, Stegmann R (1996): Enhancement of the biological degradation of soils contaminated with oil by the addition of compost. Acta Biotechnol 16 (1): 19–30

ITVA (1997) ITVA-Preisspiegel zur Dekontamination von Böden – ITVA-Fachausschuß H1 "Technologien und Verfahren". Altlasten Spektrum 5 (5): 248

Kästner M, Lotter S, Heerenklage J, Breuer-Jammeli M, Stegmann R, Mahro B (1995) Fate of ^{14}C-labeled Anthracene and Hexadecane in compost manured soil. Applied Microbiology and Biotechnology 43: 1128–1135

Koning M, Scholz J, Hupe K, Schwedes J, Stegmann R (2000) Einfluss des Sauerstoffgehaltes auf die Umsetzung von MKW im Boden, Teil 2: Sauerstoffeintrag in Sanierungsmieten. TerraTech 9 (1): 64–67

Lotter S, Brumm A, Bundt J, Heerenklage J, Paschke A, Steinhart H, Stegmann R (1993) Carbon balance of a PAH-contaminated soil during biodegradation as a result of the addition of compost. In: Arendt F, Annokkeé GJ, Bosman R, van den Brink WJ (eds) Contaminated Soil'93, vol II. Kluwer Academic Publishers, Dordrecht, The Netherlands, pp 1235–1246

Scholz J, Müller BG, Schwedes J, Deckwer W-D (1996) Verbesserung der mikrobiellen Bodenreinigung in Feststoffreaktoren durch Desagglomeration. In: Stegmann R (ed) Neue Techniken der Bodenreinigung, Hamburger Berichte 10. Economica Verlag, Bonn (distribution: Verlag Abfall aktuell, Stuttgart), Germany, pp 411–422

Stegmann R, Förstner U, Hupe K, Reimers C (1996) Sonderforschungsbereich (SFB 188) der DFG "Reinigung kontaminierter Böden". In: Stegmann R (ed) Neue Techniken der Bodenreinigung, Hamburger Berichte 10. Economica Verlag, Bonn (distribution: Verlag Abfall aktuell, Stuttgart), Germany, pp 463–510

23 Carbon Balance and Modelling of Oil Degradation in Soil Bioreactors

S. Lotter[1], J. Heerenklage[2], R. Stegmann[2]
[1]Bundesverband der deutschen Zementindustrie e.V., Immenhof, 22087 Hamburg, Germany
[2]Technical University Hamburg-Harburg, Department of Waste Management, Harburger Schloßstraße 37, 21079 Hamburg, Germany

23.1 Introduction

Biological treatment is a useful tool for mineral oil contaminations as they are easily biodegradable. Numerous publications about oil degradation are available and many remediations have been carried out. However, to date, there is little information about carbon mass balances and carbon turnover of oil in soil.

The aim of this study is to quantify the carbon pathways during biological remediation processes. A method has been developed to estimate the carbon balance of oil in soil bioreactors. Futhermore, a model of carbon turnover of organic matter in soil has been modified and adapted to fit the results in reactor experiments. The constants of the model (rate constants and proportions leaving the carbon compartments) have been optimised on the basis of the test results.

23.2 Experimental Methods and Calculation of the Carbon Balance

The procedure for balancing carbon has been developed through the artificial contamination of soil with diesel oil. The balance parameters are hydrocarbon (HC) content, CO_2-evolution (mineralisation), biomass (BIO), and TOC (total organic carbon) in the exhaust air (volatilisation).

The test conditions in the sealed aerated soil bioreactors are listed in table 23.1. To enhance the oil decomposition, compost is added to the soil (Stegmann et al. 1991). Moreover an increased interaction of organic soil matter and contaminants was to be expected due to the addition of biologically active organic matter.

Three reactors with a contaminated soil/compost mixture and two uncontaminated control reactors were examined. One reactor was chosen to represent the time-course of hydrocarbon content, CO_2-evolution, and biomass. The carbon-balance was calculated from the mean values of the contaminated and control

Table 23.1. Experimental conditions in the soil bioreactors

Parameter	*Specification*
Material:	
soil type	parabrown soil undergoing pseudogleyisation
soil horizon	Ah Ap
soil texture	low loamy sand, 78 % sand, 16 % silt, 6 % clay
compost	organic yard and house waste, maturity 8 months
moisture	60 % of moisture-holding capacity
mixture ration	soil:compost = 4:1
Contamination:	
substance	diesel fuel
concentration	1 % of soil dry matter
HC-content	8,000 mg HC $(kg\ dm)^{-1}$ (soil/compost-mixture)
Reactors:	
volume reactor	3 l
aeration rate	2 l h^{-1}
temperature	30 °C

Table 23.2. Summarisation of the analytical methods

Parameter	*abbr.*	*Methods of analysis*
hydrocarbons	HC	a modification of Deutsche Einheitsverfahren (DIN 38409 H18) (extraction with tri-chloro-tri-flouro-ethan in saelted glass flasks by ultrasound)
biomass	BIO	SIR-method (substrate-induced-respiration)
CO_2 exhaust air	CO_2	gas chromatography (thermal conductivity detection)
TOC exhaust air	TOC	gas chromatography (flame ionisation detection)

reactors. The analytical methods used to measure the balance parameters are summarised in table 23.2.

Regression analysis and cubic spline approximation are computed to compensate random fluctuations and to calculate values for each day. Thereafter, the carbon content for each parameter is calculated. The method of biomass estimation gives mg C biomass per kg dm (dm: dry matter). The carbon content of the oil was determined by elementary analysis to be 86.1 %. We assume that this remained unchanged over the whole test period.

The differences between the oil-amended treatment and the pure soil/compost-mixture control is computed for each parameter. It is assumed that these differences result from the oil addition. The sum of the measured balance parameters is derived at any time from HC + BIO + cumulative TOC. The parameter values and the sum are related to the amount of oil carbon added.

Biomass estimation using the substrate-induced respiration (SIR) method (Anderson and Domsch 1978) is complicated by the fact that oil is a relatively available substrate. Therefore, comparative research was carried out to determine the reaction of an oil-consuming microflora to a glucose addition. The results show that it is possible, with certain qualifications (i.e., the reactor should not be in the phase of maximum oil degradation), to estimate the biomass in oil contaminated soils by the SIR-method (Fricke 1992).

The oil carbon turnover is derived from the differences between the contaminated treatment and the uncontaminated control. The effect of the oil addition an the basal metabolism of organic matter was investigated using ^{14}C-labelled wheat straw (data not shown). There was no change in the $^{14}CO_2$-evolution to be observed after the oil addition.

23.3 Results

In fig. 23.1, the HC-content (mean values ± deviation from three samples) is plotted as a function of time. During the test period from day 2 to 53, the decrease in HC-content can be described by an exponential function.

The course of the CO_2-concentrations and the biomass are illustrated in fig. 23.2 and 23.3. The same parameters in the control runs changed slightly over time. However, after a short time, CO_2- and biomass concentrations were surged in the contaminated reactors. The same values for the contaminated and the control again aligned with each other after 40 days. The volatilisation of oil, described by the TOC-content of the exhaust air, decreased rapidly and exponentially after the experiment had begun. In this case, the values of the contaminated and uncontaminated tests corresponded after 14 days.

The oil carbon balance, calculated from the difference values (contaminated

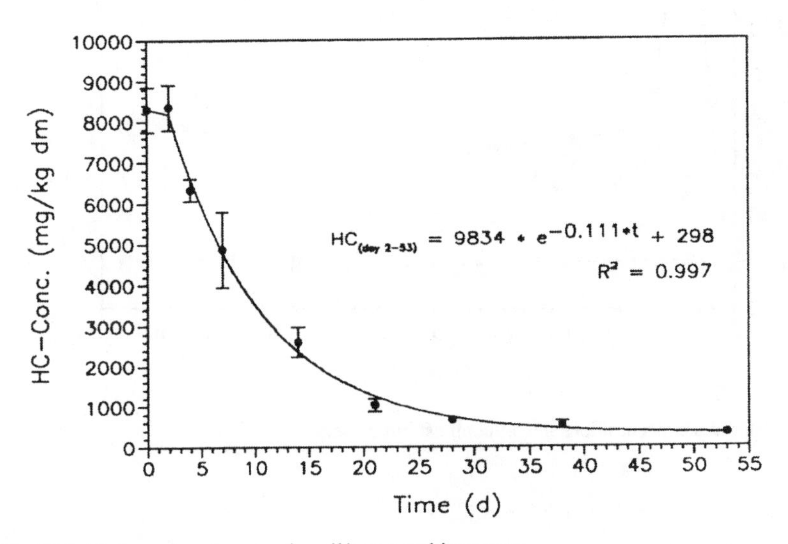

Fig. 23.1. Hydrocarbon decrease in soil/compost bioreactors

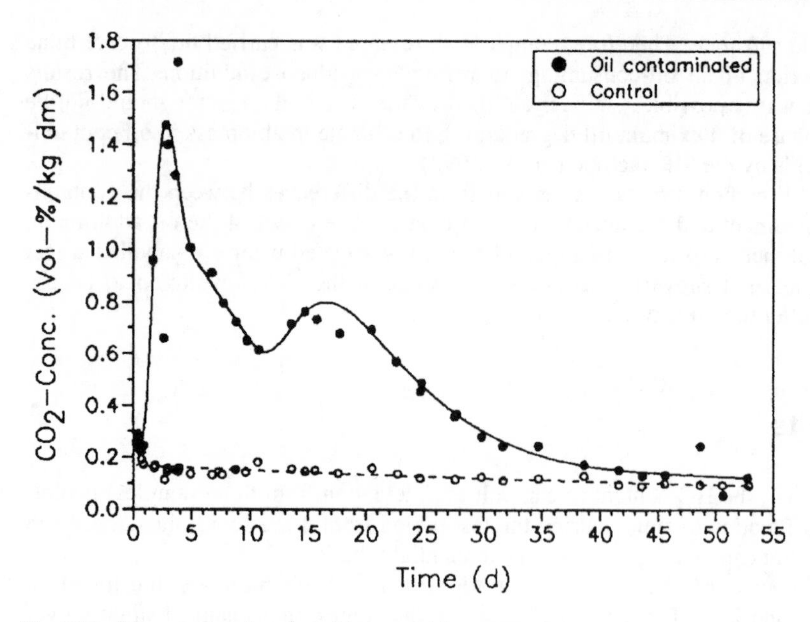

Fig. 23.2. CO_2-concentrations in soil/compost bioreactors

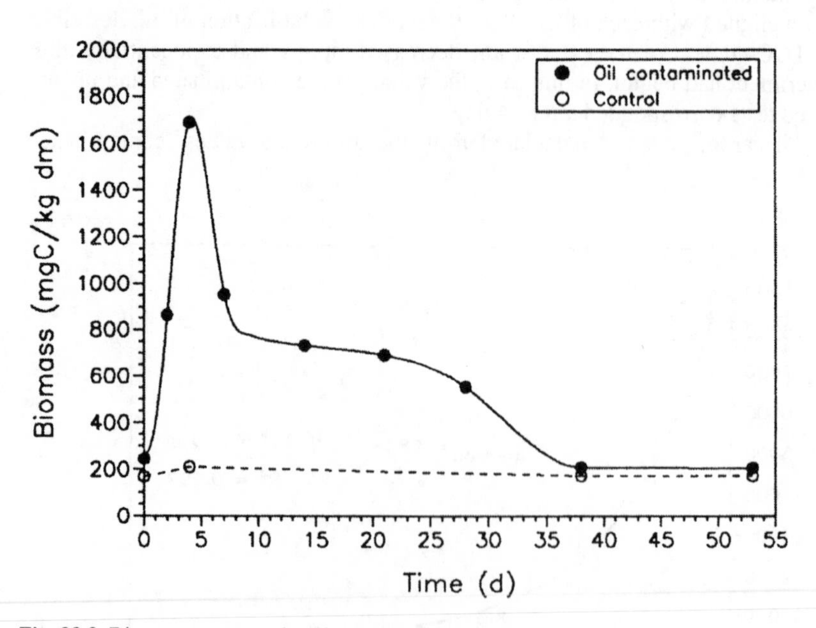

Fig. 23.3. Biomass concentrations in soil/compost bioreactors

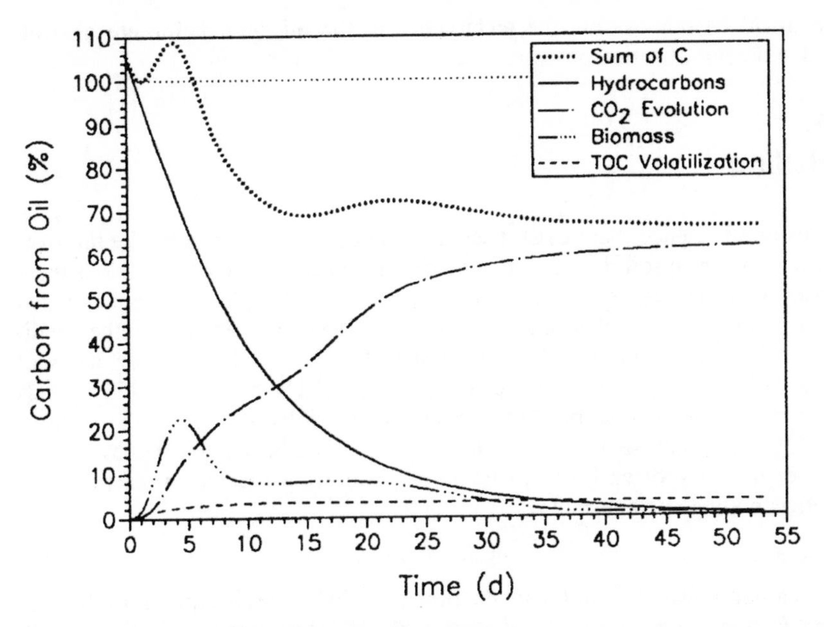

Fig. 23.4. Carbon balance in soil/compost bioreactors

minus uncontaminated) is graphed in fig. 23.4. At the end of the experiment on day 53, the HC- and BIO-contents amounted to less than 1 %. The mineralisation quantity (CO_2) was 61 % and the volatilisation (TOC) was 3.3 %. The derivative of all curves tended towards zero. A balance gap of 35 % was observed. The scatter between the three reactors set in parallel was for the parameters HC, BIO, and TOC normally < 1 %. For the sensitive parameter CO_2, it varied within ± 5 % at the end of the testing.

23.4 Discussion

At the end of the test period, the oil degradation processes had progressed to the extend that the soil system seemed to be in an equilibrium. Hydrocarbons were no longer detectable by the H18 method. The volatilisation process was finished; the mineralisation and the biomass levels in the contaminated reactors were the same as the control reactors. A possible overestimation, due to the SIR method, of the biomass during the highest oil degradation phase is seen in the peak of the sum of carbon curve between the 3rd and 5th test day. The balance gap at 35 % can be explained by the formation of bound residues. The form of the bound residues remains open (humification; sorption processes with the organic matter, the clay minerals or the clay-humus-complex). The formation of bound residues from hydrocarbons was shown in another test series with ^{14}C-labelled hexadecane (Heerenklage et al. 1992). How well they can be distinguished from the original soil humus is still unclear. It is quite conceivable that besides the biomass produced in oil decomposition, the metabolites of the oil degradation are also involved in the humification process.

23.5 Modelling

In agricultural science there exist a series of models which describe the turnover of organic matter in soil. Due to the good degradability of oil contaminations there is reason to believe that similar turnover processes take place in contaminated soils. A model, based an that of Jenkinson et al. (1981), was adapted to the conditions existing during biological soil treatment. In our model, 4 compartments of organic carbon in soil are assumed: hydrocarbon (HC), biomass (BIO), humus (HUM), and a sorbed fraction (SOR). Furthermore, volatilisation (GAS) is introduced (fig. 23.5). Further material on this modelling has been published by Lotter.

The degradation of each compartment (C_i) follows a first order reaction with the corresponding rate constant r_i:

$$dC_i/dt = - r_i * C_i \tag{23.1}$$

The carbon released from the compartments redistributes according to the proportion factors (f_i) of the compartments BIO, HUM and mineralisation (CO_2). Somewhat simplified, these proportion factors are assumed to be equal for all decay processes.

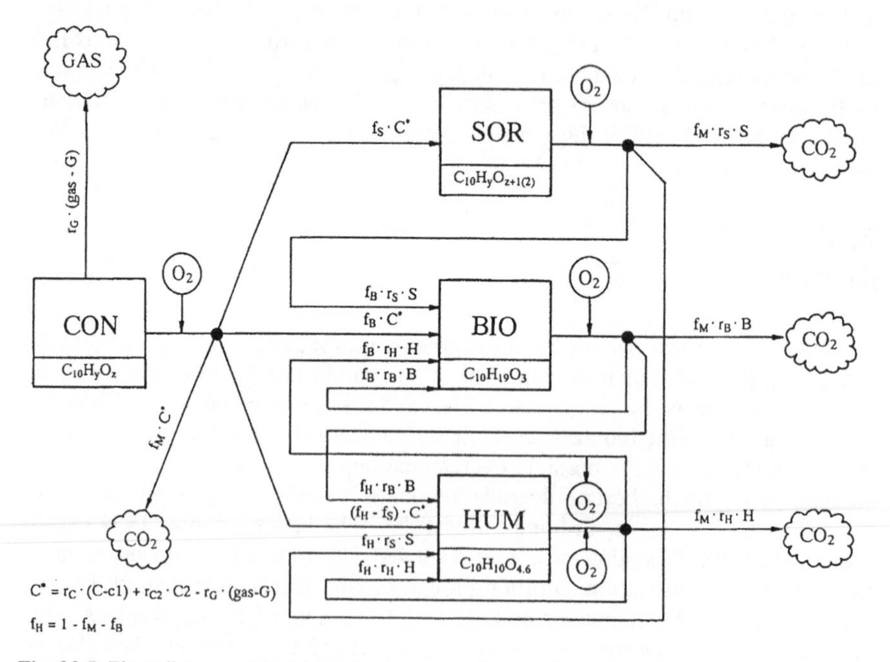

Fig. 23.5. Flow-diagram of the model for turnover of contaminants in soil

The compartment HC is only subject to degradation; the compartment SOR only receives C-Input from HC. A series of influence factors (temperature, moisture, plant growth), that have to be taken into consideration under field conditions, can be set at 1 in the bioreactor tests.

The model inputs are the HC decrease and volatilisation via the corresponding rate constants. BIO, HUM, SOR, and CO_2 are then computed by the model according to the time interval of one day. The model includes 6 unknown constants (f_{CO2}, f_{Bio}, f_{SOR}, r_{BIO}, r_{HUM}, r_{SOR}), which were chosen to best fit the measured CO_2 cumulative curve and biomass concentration. An optimising algorithm is used according to Townsend and Zarak (1983). The result of the modelling of the bioreactor tests is presented in fig. 23.6.

In this modelling, the compartment SOR is not taken into account, as no better fit would be achieved. The constants amount to:

$$F_{CO2} = 0.15,\ f_{Bio} = 0.74,\ _{rBio} = 1\ (1\ d^{-1}),\ r_{HUM} = 0.01\ (1\ d^{-1}) \qquad (23.2)$$

The relatively high proportion factor, f_{Bio}, reflects the considerable turnover of carbon into the biomass. The high degradation rate, r_{Bio}, points to the rapid turnover of the biomass. The degradation rates in the turnover of ^{14}C-labelled plant material, published by Jenkinson et al. (1987), are determined to be lower by a factor of 40 to 70 compared to those reported here. The proportion factor in this tudy for the mineralisation is 0.78.

In fig. 23.7, the measured CO_2-evolution and biomass concentration are compared with the results of the modelling. Agreement between model and data is

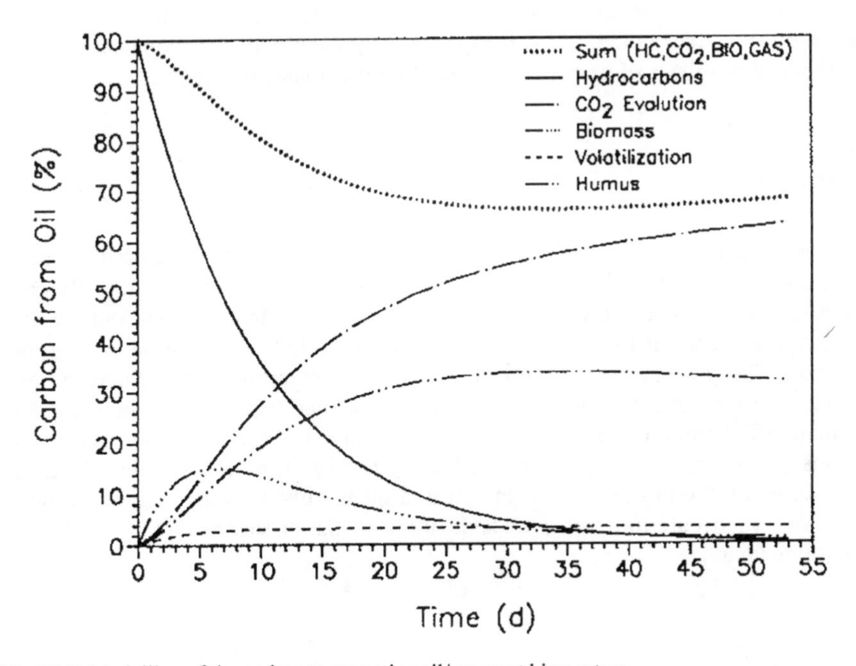

Fig. 23.6. Modelling of the carbon turnover in soil/compost bioreactors

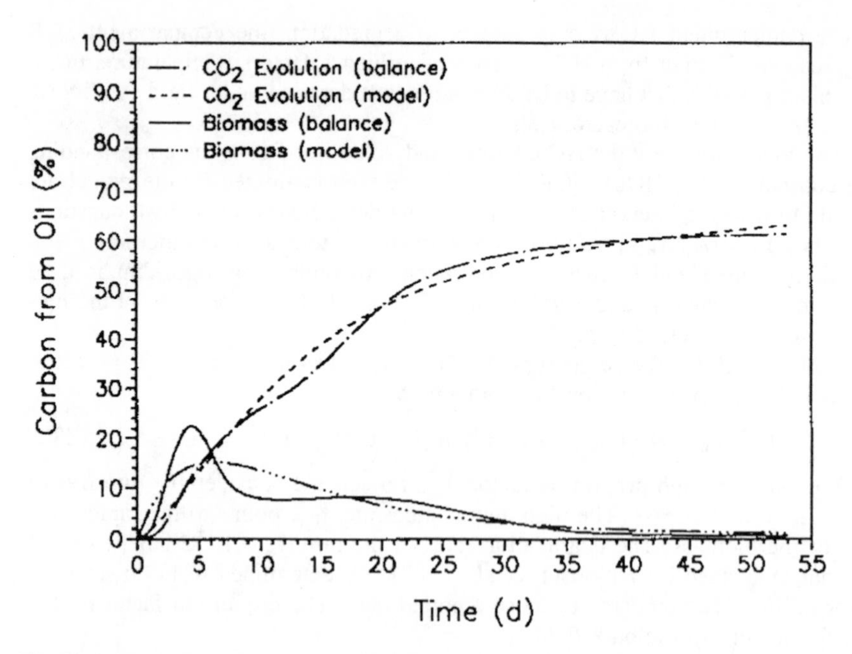

Fig. 23.7. Comparison of the measured and the modelled data: cumulative CO_2-evolution and the biomass concentration

rather good. However, the peak of the biomass concentration after 5 days is not represented in the model. This divergence may eventually be explained by the methodical problems of biomass determination previously mentioned.

23.6 Summary

The presented method for balancing the oil turnover in soil bioreactors considers the parameters hydrocarbons, biomass, CO_2-evolution, and volatilisation. By calculating the difference between the data of the contaminated samples and the uncontaminated control for each parameter, it is possible to calculate an oil carbon balance. The processes of oil turnover in the soil/compost mixture may be assumed to have finished after 53 test days. A balance gap of 35 % suggests the formation of bound residues or humus. The modelling of the oil decomposition by means of a simple model yielded satisfactory results. The model cannot describe with certainty the turnover of oil in soil in a quantitatively exact manner; it may, however, contribute to a better understanding of the processes that occur during soil remediation.

23.7 Abbreviations

CON,C	Contamination
SOR,S	Sorbed fraction
BIO,B	Biomass
HUM,H	Humus
f_i	Proportion factor of the compartments
r_i	Degradation rate of a first order reaction of the compartments
r_c	Maximum degradation rate
c1	Residual concentration
C2	Second contamination fraction
GAS,G	Volatilisation
gas	Maximum percentage of volatilisation
O_2	Oxygen
CO_2	Carbon dioxide

References

Anderson JPE, Domsch KH (1978) A physiological method for the quantitative measurement of microbial biomass in soils. Soil Biol Biochem 10: 215–221

Fricke K (1992) Untersuchungen zur Bestimmung der Biomasse in kontaminierten Böden mit der SIR-Methode (Estimation of biomass in contaminated soils using the SIR method). Internal report, unpublished, SFB 188, TU Hamburg-Harburg, Germany

Kästner M, Lotter S, Heerenklage J, Breuer-Jammeli M, Stegmann R, Mahro B (1992) Fate of ^{14}C-labeled anthracene and hexadecane in compost-manured soil. Appl Microbiol Biotechnol (1995) 43: 1128–1135

Jenkinson DS, Ladd JN, Rayner JH (1981) Microbial biomass in soil: Measurement and tumover. Soil Biochemistry. Paul EA, Ladd JN (eds) Dekker, New York, pp 415–471

Jenkinson DS, Hart PBS, Rayner JH, Parry LC (1987) Modelling the turnover of organic matter in long-term experiments at Rothamsted. INTECOL Bulletin 15: 1–8

Lotter S (1995) Kohlenstoffbilanzierung und Kohlenstoffumsetzung während der biologischen Bodensanierung – Untersuchungen in beschlossenen Testsystemen (Carbon balance and carbon turnover during biological soil remediation – investigations with sealed test systems). PhD thesis, TU Hamburg-Harburg Germany

Stegmann R, Lotter S, Heerenklage J (1991) Biological treatment of oil-contaminated soils in bioreactors. On-site bioreclamation – processes for xenobiotic and hydrocarbon treatment. Hinchee RE, Olfenbuttel RF (eds) Butterworth-Heinemann, Stonehame, UK, pp 188–208

Townsend MA, Zarak CE (1983) Accelerating flexible polyhedron searches for nonlinear minimization. J Mechanisms, Transmissions and Automation in Design 105: 196–200

24 Degradation of Xenobiotics at Elevated Temperatures

H. Feitkenhauer, S. Hebenbrock, U. Deppe, H. Märkl, G. Antranikian
Technical University Hamburg-Harburg, Institutes of Technical Microbiology and Bioprocess and Bioengineering, Denickestraße 15, 21073 Hamburg, Germany

24.1 Introduction

In the last decade, attention has been paid to the development of new strategies for the biodegradation of pollutants such as aliphatic and aromatic hydrocarbons. These compounds are usually found in the wastewaters of petrochemical plants or in contaminated soils. The bioremediation of such sites has been investigated using pure and mixed microbial cultures under mesophilic conditions (Liu 1985; Cuno 1996). Due to the limited bioavailability of these compounds at ambient temperatures, their biodegradation is rather limited (Kästner et al. 1993; Wilson and Jones 1992). A promising approach to improve the bioavailability of these hydrophobic compounds without using additional chemicals is to develop biological processes at elevated temperatures. Although the treatment of large amounts of polluted water or soils at temperatures above 60 °C requires additional energy, there are, however, many advantages that make such an approach very attractive.

The physicochemical parameters favour biodegradation at high temperatures. Under these conditions, the solubility and diffusion coefficients of organic compounds in water increase and the viscosity decreases. As a consequence, it becomes feasible in some cases to apply a biological process instead of incineration. Furthermore, many industrial effluents are discharged at temperatures between 50 and 130 °C, with temperatures frequently exceeding 80 °C (Hamer et al. 1989). From an energetic point of view, the treatment of such effluents would be inexpensive. If the feed for the bioreactor is at ambient temperature, the energy required to run the process at a higher temperature will be compensated by lower costs due to a shorter treatment time. In addition to this, the temperature in aerated bioreactors increases upon oxidation of the substrates (Pöpel 1970). Finally, the risk of releasing pathogenic microorganisms from such treatment plants will be dramatically reduced, because no pathogens have been found to date that grow at temperatures above 60 °C. In the past however, little attention has been paid to the degradation of pollutants at elevated temperatures (Hamer et al. 1989; Mutzel et al. 1996; Nazina et al. 1993; Sorkoh et al. 1993). In this chapter, we will present results on the ability of thermophilic bacteria to degrade pollutants at high temperatures. Furthermore, the limiting factors and the potential for application of this fascinating group of microbes will be demonstrated.

24.2 Extremophiles for Bioremediation

Extremophiles are organisms that can survive and grow optimally under extreme conditions. Although there are no strict definitions of a "normal" environment, there is a general view on what most of the important physical and chemical factors available under these conditions are. A normal environment has a temperature between 10 and 40 °C, a pH from 5.0 to 8.0, a pressure of 1 atmosphere and a salinity corresponding to the salinity of freshwater or seawater (up to 30 g l^{-1}) (Kristjansson and Hreggvidsson 1995). Many parts of the world, such as geothermal environments, polar regions, acid and alkaline springs and the cold pressurised depths of the oceans, are considered to be extreme. As conditions become more and more demanding, extreme environments become exclusively populated by microorganisms belonging to the bacterial and archaeal domains (prokaryotes). It is very likely that higher organisms are unable to survive under extreme conditions due to their cellular complexity and compartmentation. The realisation that extreme environments harbour different kinds of prokaryote lineage has resulted in a complete reassessment of our concept of microbial evolution and has given considerable impetus to extremophile research (Horikoshi and Grant 1998).

There are several biotic factors that determine the growth of extremophiles, but the most important ones are the nutrition and energy sources that an organism can utilise, the availability of these sources, and the biochemical mechanism that the organism possesses. Metabolic processes and specific biological functions of these microorganisms are mediated by enzymes and proteins that work under extreme conditions. The steadily increasing scientific interest has been aroused not only by the fact that microorganisms survive and grow in some of the harshest environments on earth, but also because they provide a valuable resource of unique enzymes that are able to metabolise hydrocarbons such as xenobiotics.

Microorganisms that are adapted to grow optimally at high temperatures (60–108 °C) have been isolated from high-temperature terrestrial and marine habitats (Stetter 1998; Stetter 1996; Kristjansson and Hreggvidsson 1995; Tunnicliffe 1991). The most common biotopes are vulcanically and geothermally heated hydrothermal vent systems such as solfataric fields, neutral hot springs and submarine hot vents. Submarine hydrothermal systems are situated in shallow and abyssal depths. They consist of hot fumaroles, springs, sediments and deep-sea vents with temperatures up to 400 °C ("black smokers"). Shallow marine hydrothermal systems are located at the beaches of Vulcano, Naples and Ischia (all Italy), Sao Miguel (Azores) and Djibouti (Africa). Examples of deep-sea hydrothermal systems are the Guaymas Basin (depth 1500 m; Canganella et al. 1998), the East Pacific Rise (depth 2500 m; Jeanthon et al. 1999), both off the coast of Mexico, the Mid-Atlantic Ridge (depth 3700 m; Jeanthon et al. 1998) and the Okinawa Trough (depth 1400 m; Gonzáles et al. 1998). Because of their ability to convert volcanic gases and sulfur compounds at high temperatures, hyperthermophilic communities living in such hydrothermal vents are expected to play an important role in marine ecological, geochemical and volcanic processes (Huber et al. 1990). Microorganisms capable of growing optimally at temperatures between

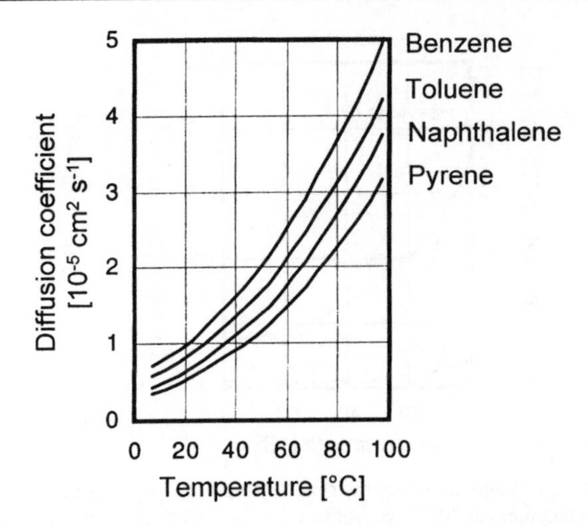

Fig. 24.1. Temperature dependence of the diffusion coefficient of selected monoaromatic compounds (benzene, toluene) and polycyclic aromatic hydrocarbons (PAH) (naphthalene, pyrene) in water (Yaws et al. 1995)

60 and 80 °C are designated as extreme thermophiles. These microorganisms are widely distributed among the genera *Bacillus*, *Thermus*, *Clostridium*, *Thermoanaerobacter*, *Fervidobacterium*, *Thermotoga* and *Aquifex*. Members of the genera *Bacillus* and *Thermus* are aerobic and are able to oxidise a variety of carbohydrates and proteins. Shallow as well as deep-sea hydrothermal systems harbour members of hyperthermophiles such as *Igneococcus*, *Thermococcus*, *Methanococcus*, *Archaeoglobus, Pyrodictium*, *Pyrolobus*, *Pyrobaculum*, *Pyrococcus* and *Thermotoga*, which grow anaerobically between 80 and 108 °C. Most of these organisms gain energy by fermentation of peptides, amino acids and sugars. Nothing, however, is known on their ability to utilise xenobiotics. Recently discovered biotopes of extremophiles and hyperthermophiles are deep, geothermally heated oil reservoirs some 3,500 m below the bed of the North Sea and the permafrost soil of North Alaska (Lien et al. 1998; Stetter et al. 1993).

24.3 Influence of Temperature on the Aerobic Biodegradation Process; Physicochemical Considerations

Since aliphatic and aromatic hydrocarbons can be efficiently metabolised aerobically, the influence of elevated temperature on the transport process and on bioavailability of these compounds under aerobic conditions was considered. The most important parameters are the following: diffusion coefficient, solubility of hydrophobic pollutants in water, the viscosity of water and oil, and the melting points of various compounds. Diffusion processes play an important role since small diffusion pores and boundary layers of particles can be limiting factors for

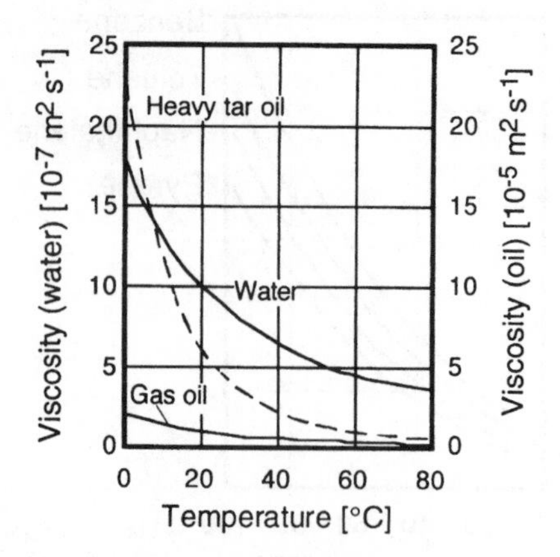

Fig. 24.2. The influence of temperature on the relative solubility of PAHs in water. The solubility of the individual compound at 20 °C was set to 1 (IUPAC solubility data series. 1989)

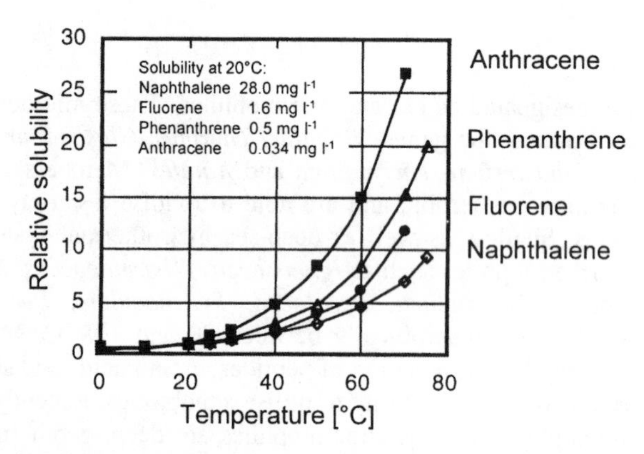

Fig. 24.3. The viscosity of heavy tar oil, gas oil (Bohl 1991) and water (Brockmann 1986) as a function of temperature

bioremediation. As shown in fig. 24.1, the diffusion coefficient D of pollutants in water increases when the temperature is raised.

Furthermore, the mass transfer coefficient is also influenced by temperature and can be assumed to be proportional to $D^{0.5}$ (Brauer 1971). The biodegradation rate of certain compounds is usually determined by its concentration in the aqueous phase. In most kinetic models, the growth rate of microorganisms is dependent on the dissolved concentration of the compound in water (Moser 1989). As shown in fig. 24.2, the mass transfer rate of crystalline pollutants in an aqueous phase strongly depends on the solubility of the pollutant.

It is also evident that the viscosity of both the medium and the substrate influences several transport parameters, e.g. the mass transfer coefficient and the sinking velocity of suspended particles in the bioreactor. As shown in fig. 24.3, the

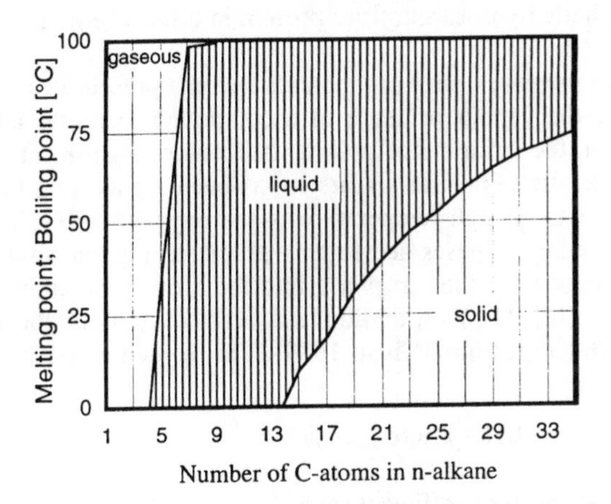

Fig. 24.4. The temperature range of the solid, liquid and gaseous state of *n*-alkanes (Griesbaum 1989)

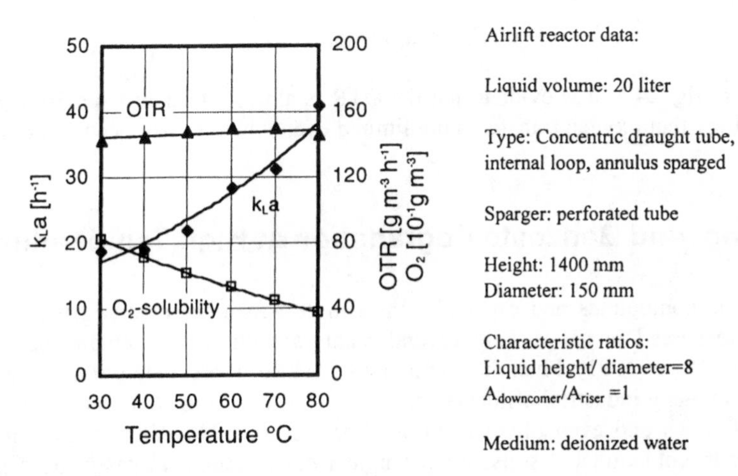

Airlift reactor data:

Liquid volume: 20 liter

Type: Concentric draught tube, internal loop, annulus sparged

Sparger: perforated tube

Height: 1400 mm
Diameter: 150 mm

Characteristic ratios:
Liquid height/ diameter=8
$A_{downcomer}/A_{riser}$ =1

Medium: deionized water

Fig. 24.5. Maximal oxygen transfer rates (OTR), k_La and oxygen solubility at temperatures between 30 and 80 °C at a superficial gas (air) velocity in the riser of 1.02 cm s^{-1} (0.25 vvm) as measured by a dynamic method (Wolf 1994). (Oxygen solubility data: Mettler-Toledo 1996)

viscosity of the medium and the oily pollutants decreases with increasing temperature. The steepest decrease is observed between 0 °C and 30 °C. Consequently, droplets can be formed more easily and higher specific interfacial areas of the oily pollutant can be formed.

Last but not least, the rate of biodegradation of an organic compound changes significantly when the compound melts. This is of special interest for long chain aliphatic hydrocarbons that are hardly soluble in water. Under mesophilic conditions, long chain aliphatic hydrocarbons are present in a solid form and are hardly attacked by microbes (fig. 24.4).

As the temperature increases, such compounds become present in a liquid form while simultaneously the vapour pressure of a compound increases (Reid et al. 1977). For this reason, the stripping of pollutants becomes significant, especially when the system is aerated. For many years it has been argued that the limiting factor for aerobic growth at high temperature is the oxygen supply. In fact, the solubility of oxygen and other gases decreases with increasing temperature. However, the oxygen transfer from the gas phase into the growth medium is faster at elevated temperatures (fig. 24.5). In addition, in aerated systems, the interfacial area (a) increases with temperature (Chisti 1989). The oxygen transfer rate (OTR) is calculated as follows:

$$OTR = k_L a\,(c_{l,max} - c_l) \qquad (24.1)$$

k_L	mass transfer coefficient [m s^{-1}]
$c_{L,max}$	maximal solubility of oxygen at the given temperature [kg m^{-3}]
c_L	actual oxygen concentration [kg m^{-3}]

The maximum value of the oxygen transfer rate OTR is reached when no oxygen is dissolved in the medium:

$$OTR_{max} = k_L a\, c_{l,max} \qquad (24.2)$$

From fig. 24.5, it is evident that the OTR is almost constant from 30 to 80 °C, indicating that oxygen transfer is not limited even when growth occurs at 80 °C.

24.4 Phenol and Benzoate Degradation at High Temperatures

Phenolic compounds and especially the simplest member of this group, phenol, are widespread in nature, and several plants are known to contain these compounds (Bayer 1991). Phenolic compounds are also widespread pollutants and are found in many industrial effluents such as wastewater from coal processing plants, oil refineries and even olive oil mills (Borja et al. 1996; Vijayaraghavan et al. 1995). In contaminated soils, for example from former coal gasification plants, phenols are found at high concentrations. Even small concentrations of phenol or other phenolic compounds can inhibit microbial growth. Moreover, phenol-containing wastewater may be difficult to treat because phenolic compounds can slow down or even inhibit the degradation of other contaminants, such as fats in olive oil mill effluents (Borja et al. 1996). Therefore, phenol degradation can be a prerequisite if mixed contamination has to be treated.

The ability to degrade phenol and other phenolic compounds is widespread in mesophilic microorganisms. *Pseudomonas* sp. have been used for phenol degradation studies (Allsop et al. 1992). For several reasons, the treatment of mixed contamination at high temperatures is an alternative to mesophilic treatment. For example, hot, highly contaminated streams in refineries can be treated before being discharged to a larger wastewater treatment facility. Soil bioremediation at

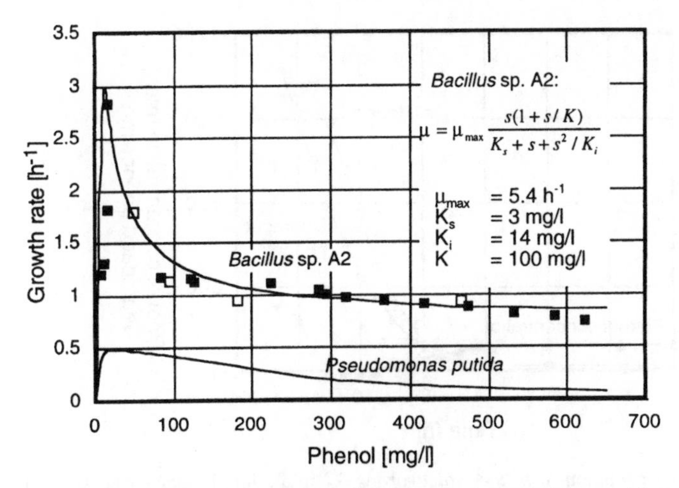

Fig. 24.6. Kinetics of the phenol degradation by *B. thermoleovorans* A2 at 65 °C and pH 6 in a 2-l bioreactor at 800 rpm. A dynamic method with increasing phenol content in a fed-batch fermentation (■) was compared to a method using the growth rate during the early exponential growth phase in batch fermentations at different phenol start concentrations (□). The line (-) represents the calculated curve obtained using the equation shown and the parameters given. The growth rate of *B. thermoleovorans* A2 was compared to the mesophilic microorganism *Pseudomonas putida* (Yang and Humphrey 1975) (μ = specific growth rate; K_i = inhibition constant; K_s = constant in Monod-type equation; s = substrate concentration).

high temperatures is another promising process. Little information however is available on the degradation of phenols at elevated temperatures (Adams and Ribbons 1989). The degradation kinetics of thermophilic phenol-degrading microorganisms have not been investigated, and studies for medium and fermen tation requirements have not been published. For the *Bacillus thermoleovorans* strain A2, which was recently isolated and used in this study, the optimal growth temperature was found to be 65 °C; where phenol serves as the sole carbon and energy source (Duffner et al. 1997; Mutzel et al. 1996). This strain was also able to utilise starch, xylan and olive oil, but not 2-nitrophenol, 4-nitrophenol, catechol, naphthalene, fluoranthene, benzthiazole, xylene, pectin, pullulan or cellulose. The efficiency of phenol degradation is highly dependent on the phenol concentration in the effluents and it is reduced due to the inhibitory effects of phenolic compounds and their metabolites.

Several methods have been developed to determine the dependence of the growth rate of *B. thermoleovorans* A2 on the substrate concentration. A dynamic method (Götz 1994) was used to determine the values displayed in fig. 24.6. During a fed-batch fermentation, the phenol concentration was continuously raised and phenol concentrations and cell densities were determined at short intervals. The values shown in fig. 24.6 were found to be in good agreement with values measured by a method where the growth rate during the early exponential growth phase at a certain phenol concentration was determined in independent batch fermentations (47, 94, 188, 470 mg l^{-1} phenol). The sharp decrease of the growth rate at relatively low substrate concentrations, and the nearly constant growth rate in

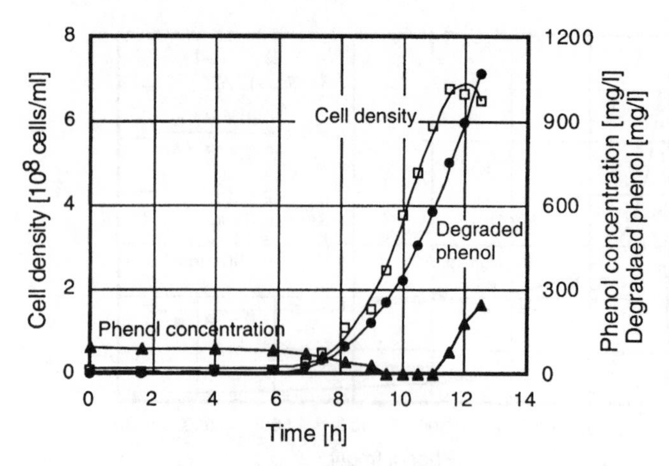

Fig. 24.7. Fed-batch fermentation of *B. thermoleovorans* A2 in a 2 litre bioreactor at 65 °C and pH 6 with a stirrer speed of 1,400 rpm using phenol as the sole carbon and energy source. The fermentation process was started at a low phenol concentration (94 mg l^{-1}) to avoid inhibiting concentrations of phenol. After 7.5 h of growth, a pump was activated to deliver a stock solution of phenol continuously to the bioreactor (to maintain a phenol concentration of 1 mM). The feed of 0.1 mol l^{-1} stock solution was increased incrementally after every measurement from 5.5 ml h^{-1} to 946 ml h^{-1} after 11 h, according to the concentration of phenol detected in the bioreactor.

the range of 150–650 mg l^{-1} could not be described adequately by the Haldane equation. Another equation described in fig. 24.6 was used to fit the growth data.

From fig. 24.6 it is obvious that the newly isolated phenol degrader is very efficient compared to *Pseudomonas putida.* During a fed-batch fermentation, the highest biomass specific phenol turnover was reached after 10 h (fig. 24.7). During the exponential growth phase, phenol degradation was almost proportional to cell growth , but large quantities of phenol were still consumed during the stationary growth phase. At the end of the exponential growth phase and in the stationary growth phase, an increasingly yellowish colour of the fermentation broth was observed. The spectrum of the supernatant was found to be identical to the spectrum of 2–hydroxymuconic acid semialdehyde (HMSA), a common metabolite detected during mesophilic phenol degradation.

The continuous fermentation experiments shown in fig. 24.8 served as a model for the operation of a continuous bioremediation process or an activated sludge wastewater treatment. Due to the instability of the process (inhibitory growth kinetics), the derivation of kinetics constants from continuous one-stage fermentation processes is problematic. Both the reservoir concentration and the dilution rate were changed during continuous fermentations. A period equivalent to at least 5 hydraulic residence times was necessary before steady state conditions could be reached.

Continuous fermentation was also performed with immobilised cells. At dilution rates of up to 0.8 h^{-1}, the phenol concentration in the effluent was close to zero

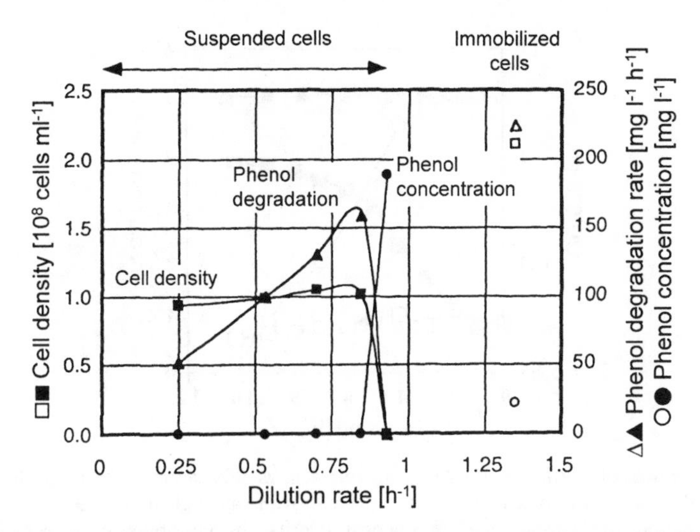

Fig. 24.8. Continuous fermentation of *B. thermoleovorans* A2 in a 2-litre bioreactor at 800 rpm, pH 6.0 and 65 °C (●, ■, ▲). The reservoir concentration of phenol was 188 mg l^{-1} and the reported values were measured in the reactor (effluent). The reported data of cell densities were combined from 2 independent sets of experiments. The maximal dilution rate (0.9 h^{-1}) was determined in a wash-out experiment. The same continuous fermentation was performed with cells immobilised on Aquacel carriers (O,□,Δ,) as described by Feitkenhauer (1998). The experiment was started at a dilution rate of 0.5 h^{-1} and then set to 1.3 h^{-1}. Cell densities were recalculated from protein measurements. The degradation of phenol dropped almost to zero when the dilution rate was increased to 1.6 h^{-1}.

(fig. 24.8). Degradation rates of phenol up to 7 g_{phenol} l^{-1} d^{-1} were determined in continuous fermentations with a feed concentration of 660 mg l^{-1} phenol. The immobilisation of cells allowed the dilution rate to be further increased to 1.3 h^{-1}; 5 g_{phenol} l^{-1} d^{-1} were degraded at a feed concentration of 188 mg l^{-1}. The experiment with carrier material was started at a dilution rate of 0.5 h^{-1} and then set to 1.3 h^{-1}. The degradation of phenol dropped to almost zero, when the dilution rate was increased to 1.6 h^{-1}.

Another thermophilic strain *Bacillus* sp. H2M was isolated from hot springs of the Azores (Iceland) and was found to be able to degrade benzoate as the sole carbon source. The optimum temperature ranges between 55 and 65 °C at pH 6.5. Growth experiments have shown that benzoate (5 mM) can be rapidly utilised within 8 h of cultivation at 60 °C at an agitation speed of 800 rpm (fig. 24.9). The organism utilises acetate, glucose and 3-hydroxybenzoate as sole carbon source. 2- and 4-hydroxybenzoate, 3- and 4-chlorobenzoate, and methoxybenzoates were not used as substrates for growth.

One key enzyme responsible for benzoate utilisation, catechol-2,3-dioxygenase, was detected in *Bacillus* sp. H2M and found to be optimally active at 60 °C with an activity range between 30 and 70 °C; the pH optimum ranges from 7 to 8. The half-life of the enzyme at 60 °C is 30 min. In the enzymatic assays (Hebenbrock 1998), the substrate catechol was converted to a yellow product with an increased, new maximum of absorption at 375 nm. This indicates the formation of

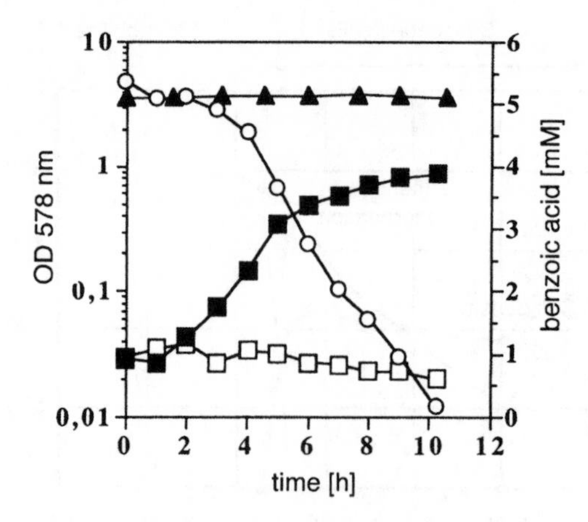

Fig. 24.9. Growth of *Bacillus* sp. H2M on benzoic acid as sole carbon and energy source. The shake flasks were incubated at 60 °C, pH 6.5 with an agitation speed of 160 rpm (O = degradation of benzoic acid, ■ = optical density in the presence of benzoic acid, ▲ = benzoic acid concentration in uninoculated cultures, □ = optical density in the absence of carbon source).

2-hydroxymuconic acid semialdehyde, which is known as a common metabolite of the mesophilic degradation of benzoic acid via a *meta*-cleavage.

24.5 Utilisation of Naphthalene

The degradation of polycyclic aromatic hydrocarbons (PAH) was under intense investigation in the last two decades (Cerniglia 1984; Bryniok 1997). One of the most investigated compounds is naphthalene, which is often the PAH present at high concentrations in, for example, tar oil contamination. Concentrations of up to 5,769 $mg_{naphthalene}$ (kg dry matter)$^{-1}$ were reported for a creosote production site (Wilson and Jones 1993). Several strains have been isolated that are capable of degrading naphthalene (Cerniglia 1992). They include the following species: *Acinteobacter calcoaceticus*, *Alcaligenes denitrificans*, *Mycobacterium* sp., *Pseudomonas* sp., *Pseudomonas putida*, *Pseudomonas fluorescens*, *Pseudomonas vesicularis*, *Pseudomonas cepacis*, *Pseudomonas testosteroni*, *Rhodococcus* sp., *Corynebacterium renale*, *Moraxella* sp., *Streptomyces* sp. and *Bacillus cereus*. Furthermore, several species of fungi, cyanobacteria and algae oxidise naphthalene. The reports on naphthalene degradation by thermophilic microorganisms are rare. Sorkoh et al. (1993) reported on the utilisation of naphthalene by *Bacillus stearothermophilus*.

A number of aerobic and anaerobic thermophilic microorganisms were tested for their ability to degrade naphthalene as the sole carbon and energy source. The substrate (1–2 mM) was added to a minimal medium (Hebenbrock 1998) and incubated between 60 and 70 °C. None of the aerobic thermophiles tested (*Bacillus thermoleovorans*, *Thermoplasma acidophilum*, *Metallosphaera sedula*) or the

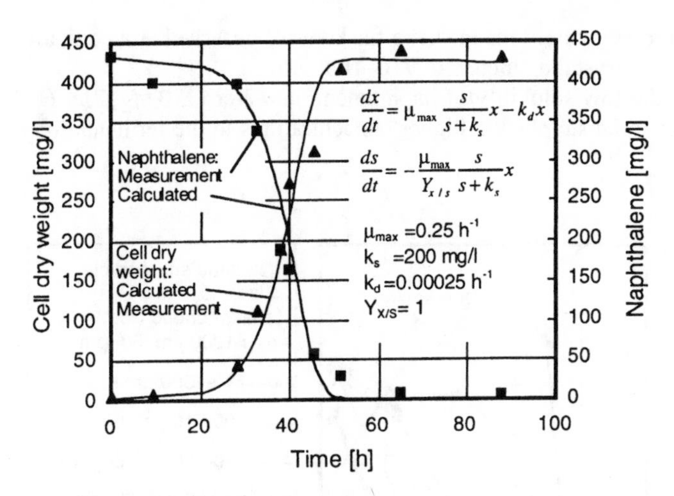

Fig. 24.10. Degradation of unsized, crystalline naphthalene by *Bacillus thermoleovorans* Hamburg 2 in 100 ml shake flasks at a shaker speed of 150 rpm, 60 °C and an initial pH value of 6.5. Flasks were sealed using PTFE lined screw caps to prevent any loss of substrate. Each point represents the extraction of one shake flask (K_d = constant for microbial decay; $Y_{x/s}$ = yield coefficient, g cell dry weight (g substrate)$^{-1}$; x = cell density).

anaerobes (*Thermococcus stetteri, Thermoanaerobacter ethanolicus, Desulfotomaculum thermobenzoicum, Thermotoga thermarum*) were able to metabolise naphthalene. Therefore, attempts were made to isolate naphthalene degraders from an oil refinery and a compost heap in Bitterfeld, Germany. Using naphthalene as the sole carbon source, a thermophilic *Bacillus* strain that grows optimally at 60 °C and pH 6.5 was isolated. This strain was identified as *Bacillus thermoleovorans* Hamburg 2. The growth temperature is above that of the naphthalene degrading strains that have already been characterised and described in the literature (Grund et al. 1992; Garcia-Valdes et al. 1988; Cerniglia and Gibson 1972). In addition to naphthalene, benzol and toluol (1 and 2 mM) were utilised as sole carbon sources. On the other hand, no growth was observed with phenol, dibenzothiophene, fluoranthene, pyrene and eicosane. Growth experiments have shown that more than 400 mg l^{-1} of naphthalene was utilised within 50 h of aerobic cultivation. The degradation of naphthalene was found to proceed parallel to the formation of cell dry mass (fig. 24.10).

The growth rate of *B. thermoleovorans* Hamburg 2 in the early growth phase was determined to be 0.11 h^{-1}, the yield coefficient was 1.03 g cdw (g naphthalene)$^{-1}$. Further experiments were conducted in order to find out whether the naphthalene crystal size has any effect on the growth rate. Under vigorous shaking conditions (rotary shaker, 200 rpm, baffled shake flasks) all particle sizes yielded similar maximum growth rates of about 0.13 h^{-1}. However, under moderate shaking conditions (60 rpm) the influence of the crystal size was evident (fig. 24.11).

Growth rates were about 0.03 h^{-1} for crystals with particle size larger than 1,000 μm, and they increased to 0.08 h^{-1} for crystals with a size of 63–180 μm.

The best growth conditions in shake flasks were observed at an agitation speed of 150 rpm with crystal diameters of 710–1000 μm.

Due to the low solubility of naphthalene in water (130 mg l^{-1} at 60 °C), naphthalene has to be supplied at higher concentrations to the fermentation medium to

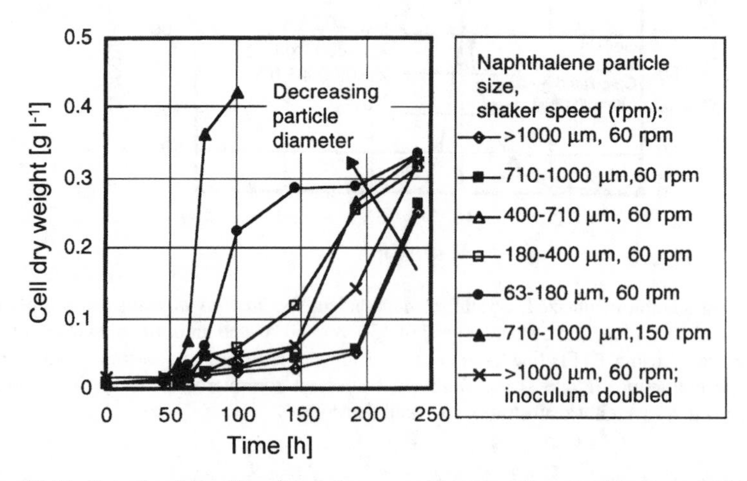

Fig. 24.11. Growth of *Bacillus thermoleovorans* Hamburg 2 on naphthalene at 60 °C with an initial pH value of 6.5 in baffled shake flasks. 1 g l^{-1} naphthalene was supplied in the form of crystals of the indicated particle size classes. The shaker speed was set to 60 rpm, while a control experiment was performed at 150 rpm. No growth was obtained in non agitated cultures within a two month period. One shake flask was inoculated with double the amount of pre-culture at 60 rpm.

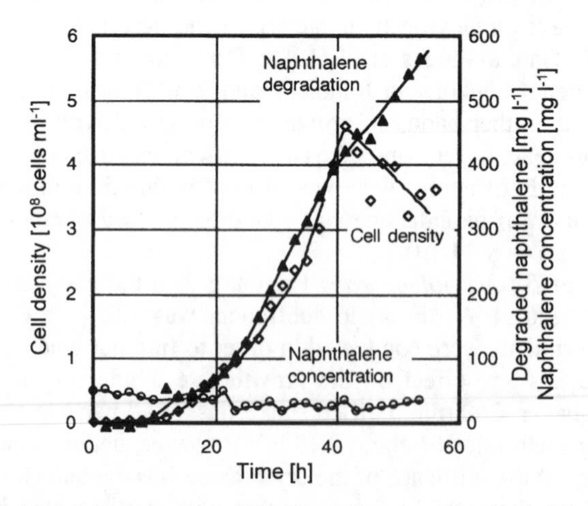

Fig. 24.12. Fed-batch fermentation of *B. thermoleovorans* Hamburg 2 in a 7-litre bioreactor at 60 °C, pH 6.5, at a stirrer speed of 1200 rpm and an aeration rate of 0.03 vvm. Naphthalene was determined every second hour by GC-FID and was added after each sampling to keep the naphthalene concentration at 100 mg l^{-1}.

reach high cell densities. On the other hand, high levels of crystalline naphthalene in the growth medium are inhibitory, and therefore a large surplus of crystalline naphthalene should be avoided. Hence, a fed-batch fermentation mode was developed so that every second hour, the amount of crystalline naphthalene necessary to reach 100 mg l^{-1} in water could be supplied (fig. 24.12).

After growth of *Bacillus thermoleovorans* Hamburg 2 on naphthalene, one of the key enzymes (1,2-dihydroxy-naphthalene dioxygenase) was characterised. Enzymatic activity was measured between 40 and 55 °C with an optimum at 60 °C, which is also the temperature required for optimal growth of the organism. The pH range for enzymatic activity is narrow (between 7 and 8.5). The enzyme did not lose activity after 6 h of incubation at 60 °C; the half-life at 80 °C was 2.5 h.

Furthermore, experiments with [1-^{13}C] labelled naphthalene have shown that almost 29 % of naphthalene was converted to CO_2, 2 % was detected in fatty acids and more than 60 % in biomass. Apart from typical metabolites of naphthalene degradation known from mesophiles, intermediates such as 2,3-dihydroxynaphthalene, 2-carboxycinnamic acid, and phthalic and benzoic acid were identified. These compounds indicate that the naphthalene degradation pathway of the thermophilic *B. thermoleovorans* Hamburg 2 differs from the known pathways found for mesophilic bacteria (Annweiler et al. 2000).

24.6 Aliphatic and Polycyclic Aromatic Hydrocarbons

Earlier studies have focused on the production of single cell proteins from hydrocarbons, but the increase of oil prices stopped these efforts (Dellweg 1987). The hydrocarbon degradation studies were often performed with mesophilic microorganisms and environmental issues in general were considered (Cuno 1996; Wilson and Jones 1993).

The ability to degrade hydrocarbons is widespread among mesophilic microorganisms (Kästner et al. 1993). The following genera have been shown to be able to degrade alkanes: *Achromobacter*, *Acinetobacter*, *Actinomyces*, *Aeromonas*, *Alcaligenes*, *Arthrobacter*, *Bacillus*, *Benecka*, *Brevibacterium*, *Corynebacterium*, *Flavobacterium*, *Micromonospora*, *Mycobacterium*, *Nocardia*, *Pseudomonas* and *Vibrio* and 14 genera of yeasts (Briton 1984). The ability of organisms to degrade polycyclic aromatic hydrocarbons is more limited (Kästner et al. 1993). However, several microorganisms have been described that oxidise phenanthrene or pyrene (among other PAH compounds). Phenanthrene degradation has been described for *Aeromonas*, *Alcaligenes*, *Arthrobacter*, *Beijerinckia*, *Micrococcus*, *Mycobacterium*, *Pseudomonas*, *Rhodococcus*, *Vibrio*, *Nocardia*, *Flavobacterium*, *Streptomyces* and *Acinetobacter*. Pyrene oxidation has been described for *Alcaligenes*, *Mycobacterium* and *Rhodococcus* species (Cerniglia 1992).

Although some thermophilic alkane-degrading microorganisms have already been isolated (Matales et al. 1967), studies on the degradation kinetics using thermophilic microorganisms are not available yet. The obligate thermophilic alkane degrading microorganisms known to date are *Thermoleophilum album* (Zarilla and Perry 1984) and *T. minutum* (Perry 1985), the facultative *n*-alkane degraders are

Thermomicrobium fosteri (Phillips and Perry 1976), *Bacillus stearothermophilus* (Sorkoh et al. 1993) and *B. thermoleovorans* (Zarilla and Perry 1987). Few members of the genus *Thermus* were reported to utilise alkanes at elevated temperatures. Such microorganisms have been detected in compost piles and recently isolated. Different mechanisms for the uptake of aliphatic hydrocarbons have been proposed. Due to the low water solubility of long chain aliphatic hydrocarbons, transport through the water phase in a dissolved state was ruled out (Cuno 1996; Singer and Finnerty 1984). Some commonly discussed hydrocarbon uptake mechanisms by bacteria include transport of small hydrocarbon droplets (micelles) into the cells, or uptake of hydrocarbons by contact of the cells with a large hydrocarbon phase. Emulsifiers seem to play an important role in the transport of hydrocarbons, and they have been identified from a variety of microorganisms. These emulsifiers include fatty acids, lipoproteins, rhamnolipids and polysaccharide-fatty acid complexes (Hommel 1990).

With the aim of isolating thermophilic aliphatic and aromatic hydrocarbon degraders, an aerobic strain was isolated from hot compost piles at 65 °C and identified as *Thermus brockii* Hamburg. Growth of this extreme thermophile was observed at 70 °C with the substrates *1*-chlorohexadecane, *1*-alkenes (C_{12} and C_{20}), and alkanes of a chain length ranging from 8 to 32 carbon atoms. Weak growth was found on tetradecanoic acid, phenanthrene and pyrene. Pyrene (100 mg l^{-1}) dissolved in heptamethylnonane (HMN) (1 g l^{-1}) was degraded more efficiently

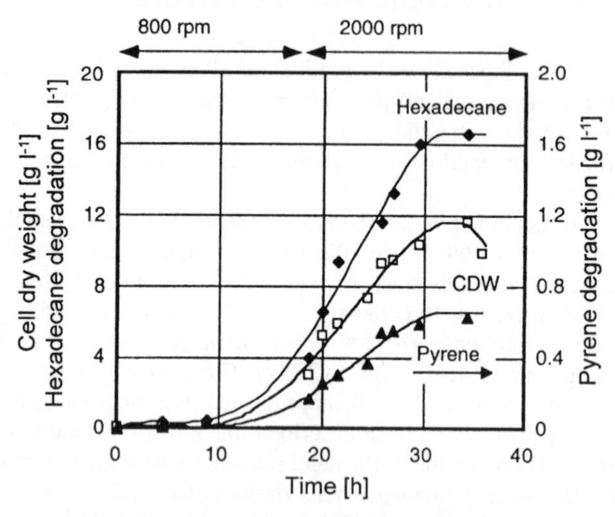

Fig. 24.13. Fed-batch fermentation of *T. brockii* Hamburg at 70 °C and pH 6.25 using hexadecane plus pyrene as the substrates. The process was started at a stirrer speed of 800 rpm in the presence of 800 mg l^{-1} hexadecane and 40 mg l^{-1} pyrene. After 8.75 h, 20 h, and 27 h the following amounts of hexadecane and pyrene were added: after 8.75 h, 4,800 mg l^{-1} hexadecane and 200 mg l^{-1} pyrene; after 20 h, 15,400 mg l^{-1} hexadecane and 440 mg l^{-1} pyrene; after 27 h, 2,000 mg l^{-1} hexadecane and 200 mg l^{-1} pyrene. The airflow was increased from 0.05 vvm to 0.5 vvm during the course of the fermentation to keep the oxygen level above 10 % air saturation. (CDW = cell dry weight).

compared to crystalline pyrene or pyrene solubilised with a synthetic emulsifier (ET 5) (Feitkenhauer 1998). Under optimal conditions, 34 mg l^{-1} pyrene was utilised after 10 days of incubation at 70 °C. On the other hand, α,ω-diclordecane, eicosanoic acid, benzene, naphthalene and diesel oil were not utilised.

The ability of *T. brockii* Hamburg to utilise pyrene however was only stable for up to 3 cell transfers. Organic supplements or vitamins were not essential for the growth of *T. brockii* Hamburg, nor did they influence the degradation rates of hexadecane. The maximal growth rate was calculated to be $\mu = 0.13\ h^{-1}$. The growth rate of *T. brockii* Hamburg was not changed when the complex medium, described by the DSMZ (DSMZ-Catalogue 1993), was also used.

As shown in fig. 24.13, *T. brockii* Hamburg was able to utilise 16 g l^{-1} of hexadecane and 0.5 g l^{-1} of pyrene after 30 h of aerobic cultivation at 70 °C. The residual concentration at the end of the co-metabolic turnover of the aromatic compound was 6.4 g l^{-1} hexadecane and 0.24 g l^{-1} pyrene.

Hence, this strain possesses very interesting features for wastewater treatment and bioremediation under thermophilic conditions.

24.7 Application of Mixed Cultures for Bioremediation

The feasibility of bioremediation processes at elevated temperature was examined using a consortium of eleven thermophilic bacteria in a 1 l stirred bioreactor at 65 °C. The consortium contained the following thermophilic bacterial strains: *Bacillus thermoleovorans* Hamburg 1, *Bacillus thermoleovorans* Hamburg 2, *Bacillus* sp. K, *Bacillus* sp. H2M, *Bacillus* sp. D5, *Bacillus* sp. M1, *Bacillus* sp. M2, *Bacillus* sp. M3, *Bacillus* sp. M4, *B. thermoleovorans* A2 and *Thermus brockii* Hamburg. To increase the biological activity, cells were immobilised on Aquacel carrier material (10 mg l^{-1}) (Feitkenhauer 1998). In order to run experiments under defined conditions, the soil material was contaminated with the following components: 1.5 g each of tetradecane, heptadecane, eicosane, tetracosane, dotriacontane, naphthalene, flourene, phenanthrene, pyrene, 1.0 g fluoranthene, 10 ml squalane. These components were dissolved in 90 g ethylacetate and 38 g hexane and subsequently sprayed onto 750 g of soil and mixed for 1 h. Afterwards, the solvent was removed by evaporation for 16 h at room temperature and the soil material was mixed again for 15 minutes and stored for two months at 4 °C.

As shown in fig. 24.14, around 9 $g_{hydrocarbons}$ (kg soil)$^{-1}$ was utilised after 14 days of operation using the immobilised mixed culture. The reduction of naphthalene concentration under these conditions was mainly due to stripping. Fluoranthene and pyrene were reduced by more than 75 %. Without immobilisation, the same mixed culture degraded only 1 $g_{hydrocarbon}$ (kg soil)$^{-1}$ within 14 days.

In order to evaluate the thermophilic consortium described above, further experiments were performed on soil material with real contamination (contamination age 50 years). The following hydrocarbons were identified: benzo[a]anthracene, pyrene, chrysene, fluoranthene, cyclopentaphenanthrene, anthracene, fluorene, phenanthrene, acenaphthene, tetradecane, pentadecane, hexadecane, heptadecane, nonadecane, eicosane, heneicosane, docosane, tricosane, tetracosane,

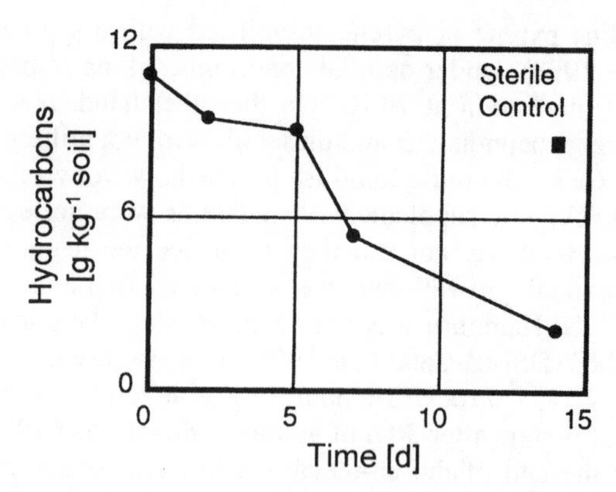

Fig. 24.14. Reduction of the hydrocarbon content at 65 °C and pH 6.5 in soil material spiked with model oil. The mixed culture (Feitkenhauer 1998) immobilised on Aquacel carrier material (10g l^{-1}) was used in a 1 litre stirred bioreactor (1400 rpm, aeration 0.01–0.03 vvm). The initial content of hydrocarbons (11 g kg_{soil}^{-1}) was reduced to about 2 g kg_{soil}^{-1} in the inoculated sample, but only to about 9 g kg_{soil}^{-1} in the sterile control.

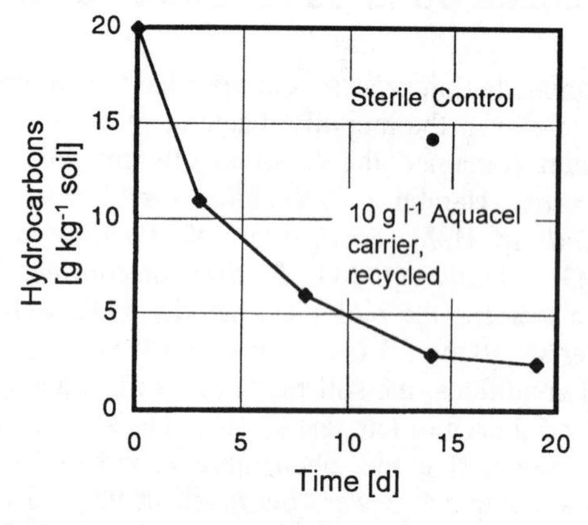

Fig. 24.15. Bioremediation of the tank storage site material by a thermophilic mixed culture immobilised on 10 g l^{-1} Aquacel regenerated cellulose carrier material. 90 % of the Aquacel carrier material was recycled from the previous experiment and 10 % fresh carrier material was added. Experiments were performed with 20 % w/v soil in a 1 litre bioreactor, 1400 rpm, 65 °C, pH 6.5, aeration rate 0.01–0.03vvm.

pentacosane, hexacosane, and heptacosane. As shown in fig. 24.15, the bioremediation was found to be very efficient. About 80 % of the hydrocarbons (20 g kg_{soil}^{-1}) were degraded within 19 days of operation.

The experiments were designed to resemble the conditions of a possible treatment plant. 90 % of the biomass (immobilised on carrier material) was recycled and 10 % was replaced by Aquacel carrier immobilised with fresh cells of the mixed culture. As shown in fig. 24.15, hydrocarbons were not degraded completely, and a residual contaminant level of 12.5 % remained in soil after 20 days of treatment. The GC-MSD analysis showed that none of the compounds originally detected could be removed to 100 %.

It could be demonstrated that at high temperatures, the physico-chemical parameters are improved. While it was possible to obtain high growth and degradation rates in liquid cultures, it was not possible to obtain the same high growth and degradation rates in the soil slurry. More research is necessary to optimise the slurry process at high temperatures, but the feasibility of the biodegradation of aliphatic and aromatic hydrocarbons at elevated temperatures could be clearly demonstrated.

24.8 Abbreviations

a	Specific surface area $[m^2\ m^{-3}]$
c_l	Concentration in aqueous phase
D	Diffusion coefficient $[m^2\ s^{-1}]$
k_d	Constant for microbial decay
K_i	Inhibition constant
k_L	Mass transfer coefficient $[m\ s^{-1}]$
K_s value	Constant in Monod-type equation
HMN	Heptamethylnonane
HMSA	2-hydroxymuconic acid semialdehyde
OTR	Oxygen transfer rate
PAH	Polycyclic aromatic hydrocarbon
S	Substrate concentration
X	Cell density
$Y_{x/s}$	Yield coefficient, g cell dry weight $(g\ substrate)^{-1}$
μ	Specific growth rate
μ_{max}	Constant in Monod-type equation

References

Adams D, Ribbons DW (1989) The metabolism of aromatic compounds by thermophilic bacteria. Appl Biochem Biotechnol 17: 231–244

Allsop PJ, Chisti Y, Moo-Young M, Sullivan GR (1992) Dynamics of phenol degradation by *Pseudomonas putida*. Biotechnol Bioeng 41: 572–580

Annweiler E, Richnow HH, Antranikian G, Hebenbrock S, Garms C, Franke S, Francke W, Michaelis W (2000) Naphthalene degradation and incorporation of naphthalene derived carbon into biomass by the thermophile *Bacillus thermoleovorans*. Appl Environ Microbiol 66: 518–523

Beffa T, Blanc M, Lyon PF, Vogt G, Marchiani M, Fischer JL, Aragano M (1996) Isolation of *Thermus* strains from hot composts (60–80 °C). Appl Environ Microbiol 62: 1723–1727

Behrend JB (1993) Biologisch-chemische Behandlung von Prozeßwässern der Bodenbehandlung sowie von kontaminierten Grundwässern. Fortschrittsberichte VDI, VDI-Verlag, Düsseldorf, Germany

Bohl W (1991) Technische Strömungslehre. 9. ed, Vogel Verlag, Würzburg, Germany, pp 28–289

Borja R, Alba J, Banks CJ (1996) Impact of the main phenolic compounds of olive mill wastewater on the kinetic of acetoclastic methanogenesis. Process Biochem 32: 121–133
Brauer H (1971) Stoffaustausch. Verlag Sauerläder, Aarau, Germany, pp 197–295
Briton LN (1984) Microbial degradation of aliphatic hydrocarbons. In: Gibson DT (ed) Microbial degradation of organic compounds. Marcel Decker, New York, USA, pp 89–129
Brockmann HJ (1986) Thermodynamik. In: Beitz W, Küttner KH (eds): Dubbel-Taschenbuch für den Maschinenbau. 15. ed, Springer Verlag, Berlin, Germany, p 1356
Bryniok D (1997) PAK-Abbau und Freisetzung von Metaboliten. In: Knorr C, von Schell T (eds) Mikrobieller Schadstoffabbau. Vieweg Verlag, Braunschweig, Germany, pp 118–129
Canganella F, Jones WJ, Gambacorta A, Antranikian G (1998) *Thermococcus guaymasensis* sp. nov. and *Thermococcus aggregans* sp. nov, two novel thermophilic archaea isolated from the Guaymas Basin hydrothermal vent site. Int J Syst Bacteriol 48: 1181–1185
Cerniglia CE, Gibson DT (1977) Metabolism of naphthalene by *Cuninghamella elegans*. Appl Environ Microbiol 34: 363–370
Cerniglia CE (1984) Microbial metabolism of polycyclic hydrocarbons. Adv Appl Microbiol 30: 31–71
Cerniglia CE (1992) Biodegradation of polycyclic aromatic hydrocarbons. Biodegradation 3: 351–368
Chisti MY (1989) Airlift bioreactors. Elsevier Science Publishers, New York, USA
Cuno M (1996) Kinetische Untersuchungen zum biologischen Abbau von Mineralölen und polyzyklischen aromatischen Kohlenwasserstoffen. Dissertation TU-Berlin, Germany
Dellweg H (1987) Biotechnologie. VCH Verlag, Weinheim, Germany, pp 180–185
DSMZ (1993) Catalogue of strains. Deutsche Sammlung von Mikroorganismen und Zellkulturen, Braunschweig, Germany
Duffner FM, Reinscheid UM, Bauer MP, Mutzel A, Müller R (1997) Strain differentiation and taxonomic characterization of a thermophilic group of phenol degrading *Bacilli*. J Appl System Microbiol 20: 602–611
Eaton RW, Chapman PJ (1992) Bacterial metabolism of naphthalene: construction and use of recombinant bacteria to study ring cleavage of 1,2-dihydroxynaphthalene and subsequent reactions. J Bacteriol 174: 7542–7554
Einsele A (1983) Biomass from higher n-alkanes. In: Rehm HJ, Reed G (eds) Biotechnology. Vol 3, VCH Verlag, Weinheim, Germany, pp 44–81
Feitkenhauer H, Hebenbrock S, Terstegen L, Schnicke S, Schöb T, Meyer M, Müller B, Antranikian G, Märkl H (1996) Bodenreinigung mit thermophilen Mikroorganismen. In: Stegmann (ed) Neue Techniken der Bodenreinigung. Economia Verlag, Bonn, Germany, pp 361–372
Feitkenhauer H, Hebenbrock S, Antranikian G, Märkl H (1997) Behandlung von organisch kontaminierten Fluiden aus der Bodenreinigung mit Hilfe extrem thermophiler Mikroorganismen. In: Reinigung kontaminierter Böden, Abschlußbericht des SFB 188 der DFG, Hamburg, Germany
Feitkenhauer H (1998) Biodegradation of aliphatic and aromatic hydrocarbons at high temperatures: Kinetics and applications. Dissertation TU Hamburg-Harburg, Germany
Garcia-Valdes E, Cozar E, Rotger R, Lalucat J, Ursig J (1988) New naphthalene-degrading marine *Pseudomonas* strains. Appl Environ Microbiol 54: 2478–2485
Grund E, Denecke B, Eichenlaub R (1992) Naphthalene degradation via salicylate and gentisate by *Rhodococcus* sp. strain B4. Appl Environ Microbiol 58: 1874–1877
Götz P (1994) Ermittlung einer mikrobiellen Reaktionskinetik für inhibitorische Substrate im Fed-Batch-Prozeß. In: Schmauder HP (ed) Methoden der Biotechnologie. Gustav Fischer Verlag, Jena, Germany, pp 96–102
Gonzáles JM, Masuchi Y, Robb FT, Ammerman JW, Maeder DL, Yanagibayashi M, Tamaoka J, Kato C (1998) *Pyrococcus horikoshii* sp. nov., a hyperthermophilic archaeon isolated from a hydrothermal vent at the Okinawa Trough. Extremophiles 2: 123–130
Griesbaum K (1989) Hydrocarbons. In: Elvers B, Hawkins S, Ravenscroft M, Schulz G (eds) Ullman`s encyclopedia of industrial chemistry. Volume A 13, 5. Ed, VCH Verlag Weinheim, Germany, pp 227–275

Hamer G, Al-Awadhi N, Egli T (1989) Biodegradation of petrochemical industry pollutants at elevated temperatures. DECHEMA Biotechnology Conference, Vol 3, pp 823–827

Hamzah RY, Al-Baharna BS (1994) Catechol ring cleavage in *Pseudomonas cepacia*: the simultaneous induction of *ortho* and *meta* pathways. Appl Microbiol Biotechnol 41: 250–256

Hebenbrock S (1998) Isolierung und Charakterisierung von thermophilen Mikroorganismen und deren Untersuchungen zum mikrobiellen Abbau von Naphthalin und Benzoesäure. Dissertation TU Hamburg-Harburg, Germany

Hommel RK (1990) Formation and physiological role of biosurfactants produced by hydrocarbon-utilising microorganisms. Biodegradation 1: 107–119

Huber R, Stoffers P, Cheminee JL, Richnow HH, Stetter KO (1990) Hyperthermophilic archaebacteria within the crater and open-sea plume of erupting MacDonald Seamount. Nature 345: 179–182

IUPAC Solubility data series (1989) Vol 89: Hydrocarbons with water and seawater, Part II. International union of Pure and Applied Chemistry, Pergamon Press

Jeanthon C, L'Haridon S, Reysenbach AL, Corre E, Vernet M, Messner P, Sleytr UB, Prieur D (1999) *Methanococcus vulcanius* sp. nov., a novel hyperthermophilic methanogen isolated from East Pacific Rise, and identification of *Methanococcus* sp. DSM 4213T as *Methanococcus fervens* sp. nov. Int J Syst Bacteriol 49: 583–589

Jeanthon C, L'Haridon S, Reysenbach AL, Vernet M, Messner P, Sleytr UB, Prieur D (1998) *Methanococcus infernus* sp. nov., a novel hyperthermophilic lithotrophic methanogen isolated from a deep-sea hydrothermal vent. Int J Syst Bacteriol 48: 913–919

Kästner M, Mahro B, Wienberg R (1993) Biologischer Schadstoffabbau in Böden unter besonderer Berücksichtigung der polyzyclischen aromatischen Kohlenwasserstoffe. Hamburger Berichte Band 5, Economia Verlag, Bonn, Germany

Kelley I, Freeman JP, Cerniglia CE (1990) Identification of metabolites from degradation of naphthalene by a *Mycobacterium* sp. Biodegradation 1: 283–290

Kristjansson JK, Hreggvidsson GO (1995) Ecology and habitats of extremophiles. World J Microbiol Biotech 11: 17–25

Lien T, Madsen M, Rainey FA, Birkeland NK (1998) *Petrotoga mobilis* sp. nov., from a North Sea oil-production well. Int Syst Bacteriol 48: 1007–1013

Liu D (1985) Biodegradation of petroleum via fermentation. In: Cheremisinoff PN, Ouellette RP (eds): Biotechnology. Technomic Publishing, Lancaster, UK, pp 103–134

Matales RI, Baruah JN, Tannenbaum SR (1967) Growth of a thermophilic bacterium on hydrocarbons: A new source of single cell proteins. Science 157: 1322–1323

Mettler-Toledo (1996) Oxygen transmitter. Company publication

Moser A (1989) Bioprocess technology. Springer Verlag, New York-Wien, pp 197–295

Mutzel A, Reinscheid UM, Antranikian G, Müller R (1996) Isolation and characterization of a thermophilic *Bacillus* strain, that degrades phenol and cresols as sole carbon and energy source at 70 °C. Appl Microbiol Biotechnol 46: 593–596

Nazina TN, Ivanova AE, Mityushina LL, Belyaev SS (1993) Thermophilic hydrocarbon oxidizing bacteria from oil strata. Microbiology 62: 359-365

Perry JJ (1985) Isolation and characterization of thermophilic, hydrocarbon utilizing bacteria. Adv Aquat Microbiol 3: 109–139

Philips WE, Perry JJ (1976) *Thermomicrobium fosteri* sp. nov., a hydrocarbon utilizing obligate thermophile. Intern J Sys Bacteriol 26: 220–225

Pöpel F (1970) Selbsterwärmung bei der aeroben Reinigung hochkonzentrierter Substrate mit Hilfe von Umwälzlüftern. Landtechnische Forschung 18: 140–142

Reid RC, Prausnitz JM, Sherwood TK (1977) The properties of gases and liquids. McGraw-Hill, New York, USA

Singer ME, Finnerty WR (1984) Microbial metabolism of straight chain and branched alkanes. In: Atlas RM (ed) Petroleum microbiology. Macmillian Publishing company, New York, USA, pp 1–59

Sorkoh A, Ibrahim AS, Ghanoum MA, Radwan SS (1993) High-temperature hydrocarbon degradation by *Bacillus stearothermophilus* from oil polluted kuwaiti desert. Appl Microbiol Biotechnol 39: 123–126

Stetter KO (1998) Hyperthermophiles: Isolation, classification and Properties. In: Horikoshi K, Grant WD (eds.) Extremophiles - Microbial Life in Extreme Environments. Wiley-Liss, New York, USA, pp 1–24

Stetter KO (1996) Hyperthermophilic procaryotes. FEMS Microbiol Rev 18: 149-158

Stetter KO, Huber R, Blöchl E, Kurr M, Eden RD, Fiedler M, Cash H, Vance I (1993) Hyperthermophilic archaea are thriving in deep North Sea and Alaskan oil reservoirs. Nature 365: 743–745

Vijayaraghavan S, Srinivasaraghavan T, Musti S, Kar S, Swaminathan T, Baradarajan A (1995) Biodegradation of phenol by *Arthrobacter* and modelling of growth kinetics. Bioprocess Eng 12: 227–229

Williams PA, Sayers JR (1994) The evolution of pathways for aromatic hydrocarbon oxidation in Pseudomonas. Biodegradation 5: 195–217

Wilson SC, Jones KC (1993) Bioremediation of soil contaminated with polynuclear aromatic hydrocarbons (PAH): A review. Envir Pollu 81: 229–249

Wolf KH (1994) Bestimmung von Sauerstofftransportparametern in Wasser. In: Schmauder HP (ed) Methoden der Biotechnologie. Gustav Fischer Verlag, Jena, Germany, pp 102–112

Yang RD, Humphrey AE (1975) Dynamic and steady state studies of phenol biodegradation in pure and mixed cultures. Biotechnol Bioeng 17: 1211–1235

Yaws CL, Nijhawan S, Li KY (1995) Diffusion coefficients in water. In: Yaws C (ed) Handbook of transport property data. Gulf Publishing Company, pp 141–168

Zarilla K, Perry JJ (1987) *Bacillus thermoleovorans*, sp. nov., a species of obligately thermophilic hydrocarbon utilizing endospore forming bacteria. Syst Appl Microbiol 9: 258–264

Zarilla K, Perry JJ (1984) *Thermoleophilum album* gen. nov. and sp. nov., a bacterium obligate for thermophily and n-alkane substrates. Arch Microbiol 137: 286–290

25 Influence of Oil-Contamination on N-Mineralisation in Soils

A. Kiene, G. Miehlich, A. Gröngröft
Institute for Soil Science, University of Hamburg, Allende-Platz 2, 20146 Hamburg, Germany

25.1 Introduction

The mineralisation of soil organic matter depends on the activity of fungi and bacteria. The different organisms degrade and transform the organic matter by various processes resulting in an availability of nutrients as part natural nutrient cycles. The activity of microorganisms, and thus the intensity of mineralisation, is strongly related to the physicochemical environment (pH, temperature, water content a.o.) and the existence of appropriate substrate and energy sources. As the natural degradation of organic matter includes various hydrocarbons, microorganisms are used to purify soils and wastewater on a technical scale. Investigations of these techniques mostly focus on the degree of mineralisation caused by the activity of heterotrophic microorganisms neglecting those autotrophic organisms (e.g. nitrifying bacteria) that are not directly involved in mineralisation. The strongly increased activity of heterotrophic microorganisms in oil contaminated soil may result in a considerable change of environmental conditions for the anaerobic autotrophic nitrifying bacteria, especially with regard to oxygen supply and the availability of NH_3 as electron donator for the first step of nitrification.

Toxic or inhibitory effects of pesticides (Harden et al. 1993), heavy metals (Joergensen 1993) and essential oils (Janssen 1989) on heterotrophic microorganisms have been reported as well as the toxicity of crude oil for plants (Jittler-Strahlendorf and Neugebohrn 1988). In a detailed literature review McGill et al. (1981) reported on the effects of addition of hydrocarbons to soils resulting in an increase of microbial activity of soils, toxicity for plants and inhibition of nitrification. Nitrification is reduced to about 50 % by addition of 0.4 % of crude oil to soil and totally stopped by 1 %.

Our investigations thus focus on the following questions:

- What is the influence of the addition of different kinds of oil to soil on the activity of heterotrophic microorganisms?
- How does intensity and duration of oil contamination influence mineralisation?
- How does the soil environment change esp. in regard to oxygen supply and ammonification?
- Is there a direct or indirect effect on nitrification?

25.2 Materials and Methods

The soils used in this study were sampled at investigation sites of the Collaborative Research Centre 188. The investigations were carried out with two topsoil samples (A, K). Samples were sieved moist to < 2 mm (A) or < 4 mm (K), homogenised and stored at 4 °C with constant moisture until further investigation. For the measurement of pH-values soil samples were air-dried. For the analysis of total C- and N-contents the samples were oven-dried at 105 °C.

Three types of oils were used to contaminate the soil samples: an artificial mixture of hydrocarbons (A, hexadecane 45 %, pristane 20 %, *cis*-decalin 10 %, *n*-nonylbenzene 16.8 %, 2-methylnaphthalene 6.45 %, dibenzothiophene 1.5 %, anthracene 0.2 %, perylene 0.05 %), a fuel oil (F) and a lubricating oil (L). The characteristics of the oils are summarised in table 25.1.

The contamination process followed four steps:

- the sieved samples were spread out,
- the oil was equally added to an amount of 0.1, 1, 5 and 10 % (dry weight),
- the samples were intensively homogenised and
- distilled water was added to 60 % of the maximum water holding capacity (WHC_{max}).

Standardised incubation tests were carried out to analyse ammonification and nitrification (Schinner et al. 1991; Alef 1991). An aliquot of the contaminated material was weighed into an Erlenmeyer bottle so that each bottle contained 10 g of dry matter without oil. All tests were performed with 2 replicates. The bottles were loosely closed with aluminium caps and stored in an incubator for 7, 14 and 28 days at 25 °C. The water content was controlled weekly andany evaporated water was replenished. Following incubation the samples were extracted using 100 ml of 1 % K_2SO_4-solution according to Alef (1991). The extract was filtered and the concentration of NH_4-N, NO_3-N and NO_2-N was analysed using standard methods. The uncontaminated reference sample was frozen for seven days and afterwards extracted and analysed as the contaminated samples.

50 g (K) or 100 g (A) dry matter of contaminated soil samples were weighed into laboratory bottles, closed with a screw cap with septum and stored at 25 °C as well to test the respiration activity of microorganisms. At varying intervals (about twice a week) gas samples were taken from the headspace of the bottle and analysed for CO_2, O_2 and N_2 using a gas chromatograph. Initial gas concentrations were determined by sampling immediately at the beginning of incubation. Some samples had to be ventilated during incubation as the concentration of O_2 decreased to < 10 %.

Aliquots of the contaminated samples were used to quantify NH_4- and NO_2-oxidants with the most probable number (MPN) method. In contrast to Alef (1991), 1 g of dry matter was suspended in 5 ml water and for each dilution step 0.5 ml were added to 4.5 ml of solution. All steps were performed with three tubes.

Table 25.1. Properties of oils

	Artificial oil	Fuel oil	Lubricating oil
origin	artificial mixture	Shell	Shell SAE 10W-40
density at 20 °C (kg m^{-3})	809	843	893
boiling point (°C)	not detected	190–360	> 350
kinematic Viscosity at 20 °C (m^2 s^{-1})		$3.5 \cdot 10^{-6}$	$2.4 \cdot 10^{-4}$
surface tension (N m^{-1})		$33.3 \cdot 10^{-3}$	not detected

25.3 Results

25.3.1 Properties of Soil Materials

Table 25.3 summarises the properties of the tested soil materials. Both samples do not contain any inorganic carbon and are slightly acid. They differ in grain size distribution, cation exchange capacity, total organic carbon content and C/N-ratio. Based on the lower C/N-ratio and the higher content of OC, it was suspected that the microbial activity of sample K would exceed the activity of sample. A substantially. The pH-values of both samples are significantly lower than the optimum for nitrifying bacteria, which is reported as pH 7 to 8.

25.3.2 Activity of Heterotrophic Microorganisms

The production of carbon dioxide or the consumption of oxygen can be used as a measure for the activity of heterotrophic microorganisms. In both cases it has to be taken into account that autotrophic microorganisms may consume CO_2 or O_2 as well.

Table 25.2. Laboratory methods

pH	in soil: solution ratio of 1:5 with A. de-ion. and 0.01 M $CaCl_2$ with pH-glass electrode
water content	water lost upon drying to constant mass at 105 °C per unit mass of dry soil
total C- and N-content	combustion at 970 °C and determination of gases in an elementary analyser (CHNO-Rapid, Heraeus)
NH_4-N, NO_2-N, NO_3-N	photometric

Table 25.3. Properties of soil material

Parameter	Sample A	Sample K
soil type [FAO]	Haplic Alisol	Eutric Gleysol
depth [cm]	0–0.1 m	0–0.6 m
clay [%]	6	55
silt [%]	15	42
sand [%]	79	3
water content [% dry weight]	7.7	23.4
Water holding capacity [% dry weight]	30.3	61.8
pH [in H_2O]	5.4	5.4
pH [in $CaCl_2$]	4.3	4.6
total organic carbon [%]	1.36	2.96
total nitrogen [%]	0.074	0.262
C/N-ratio	18	11
CEC [$mmol_c$ kg^{-1}]	30	347

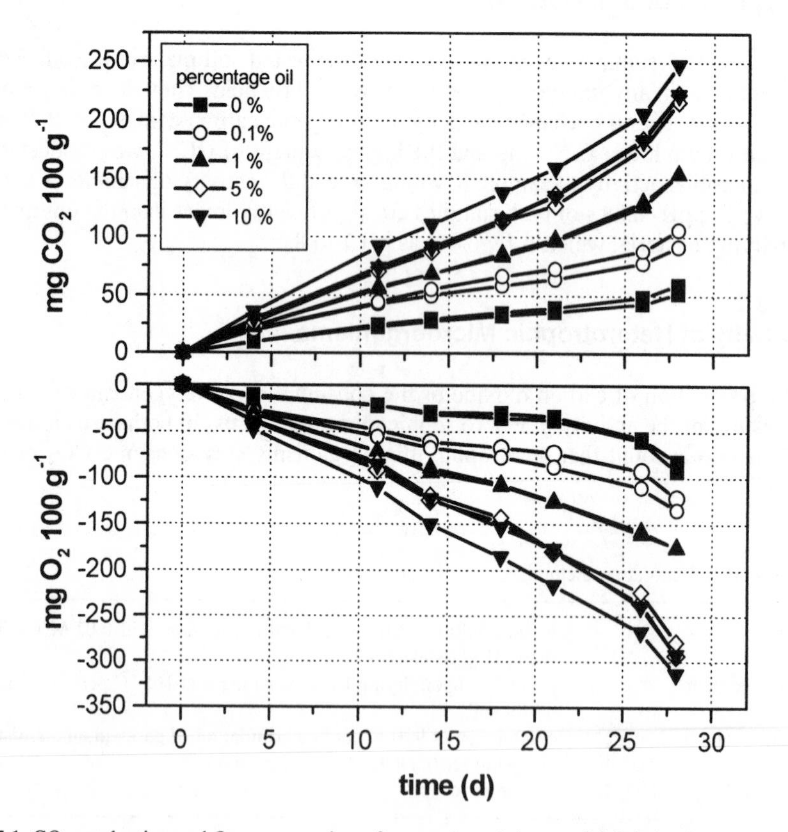

Fig. 25.1. CO_2-production and O_2-consumption of contaminated A-material (lubricating oil)

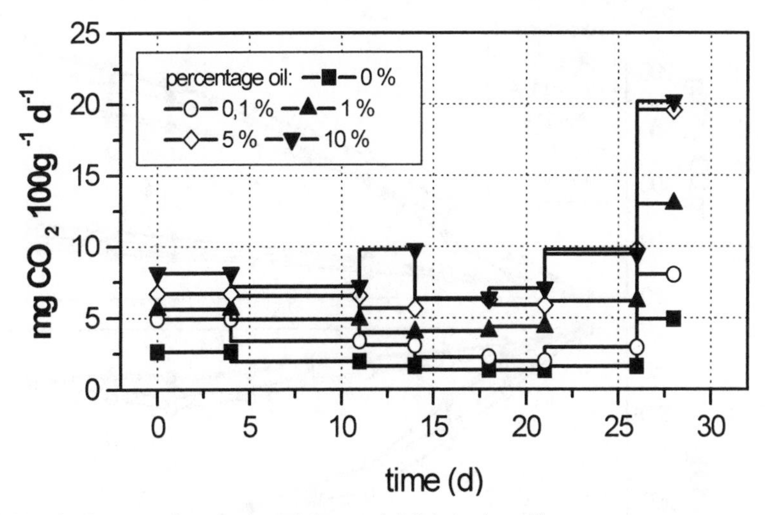

Fig. 25.2. Respiration rate of contaminated A-material (lubricating oil)

The production of CO_2 and the consumption of O_2 with time by the contaminated *A-material* are shown in fig. 25.1 for the case of lubricating oil contamination. In most cases the replicates are quite similar whereas the contamination with the other oil types leads to more variability amongst the replicates. There is a general increase in the activity of heterotrophic microorganisms with increasing oil contamination.

For the zero-contamination variant the daily CO_2-production rate (fig. 25.2) decreases until the 21st day. After this time a small increase could be detected (values of the 28th day are uncertain). The contamination variants with lubricating oil react similarly with time whereas the higher contaminated variants with the other oil types show a different temporal response in comparison to the zero-variant.

The addition of 5 % or 10 % of artificial oil or of 1 %, 5 % or 10 % of fuel oil leads to a strong increase of CO_2-production rates until the 11th day. As in the zero-variants, a reduction of production rates could be observed from the 11th to the 21st day. Afterwards the CO_2-production rate is nearly constant.

The influence of oil contamination on the microbial activity of the K-material is comparable to that on the A-material (fig. 25.3, 25.4). However, the increase of microbial activity with the intensity of contamination is not as linear as in the A-material. The addition of 0.1 % of oil increases the microbial activity only slightly for all oil types (29–121 % uncontaminated variant), whereas 1 % of oil results in a significant increase (290–490 %). With further contamination (5 %, 10 %) the CO_2-production rate remains nearly constant.

The CO_2-production rate in the uncontaminated K-material (fig. 25.4) constantly decreases with time. With the addition of 1 % of oil or more, the microbial activity increases significantly until the 10th day, followed by a period of

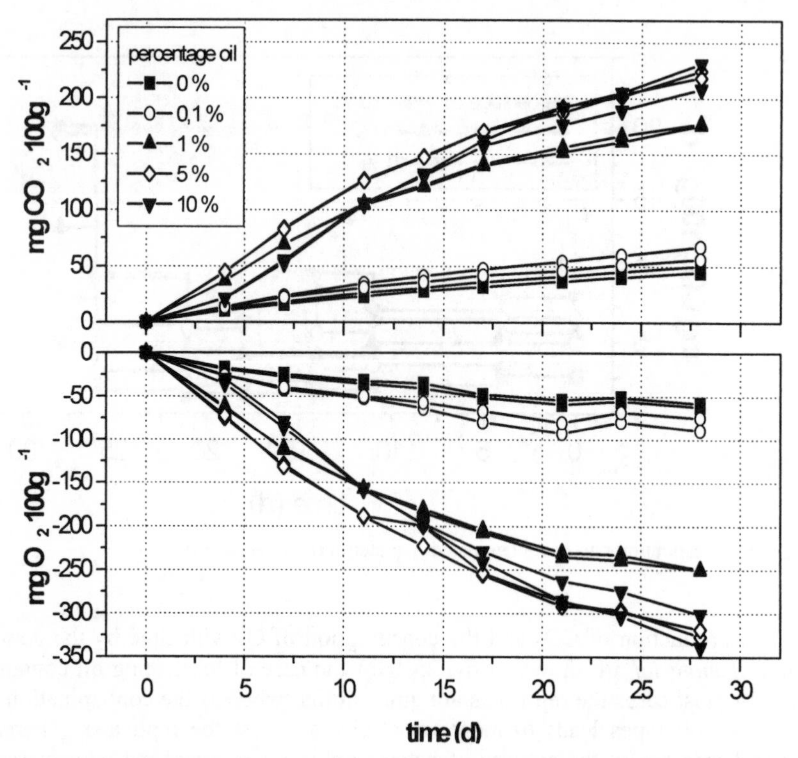

Fig. 25.3. CO_2-production and O_2-consumption of contaminated K-material (lubricating oil)

decreasing activity. At the end of the incubation period, CO_2-production and O_2-consumption are clearly related to the intensity of oil contamination.

25.3.3 N-Mineralisation

25.3.3.1 A-Material

The mineralisation of nitrogen-containing organic compounds results in the production of NH_3 or NH_4^+, which further serve as substrate for nitrification. As the oils tested in this study are hydrocarbons lacking C-N-compounds, it was possible to derive the intensity of initial mineralisation of soil organic matter (SOM) from the extractable NH_4-contents in comparison to the uncontaminated samples. Theoretical values for total N-production have been calculated from the amount of CO_2 produced and the C/N-ratio of the SOM. These were then compared to measured data (sum of NH_4-, NO_2- and NO_3-N, fig. 25.5).

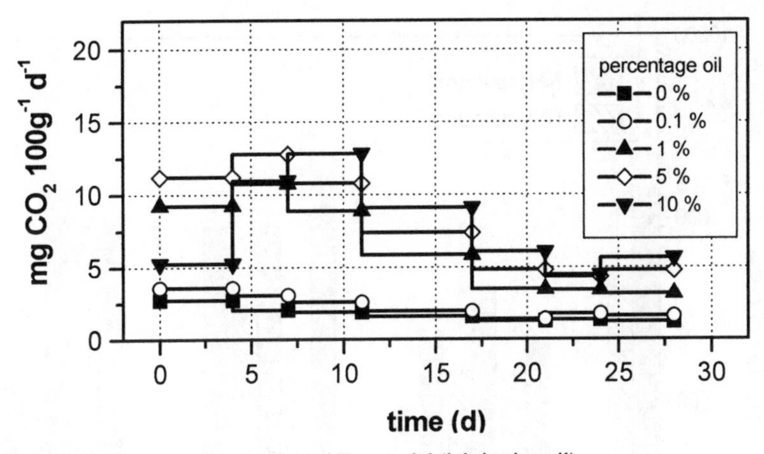

Fig. 25.4. Respiration rate of contaminated K-material (lubricating oil)

All contaminated samples show a strong difference between theoretical and measured values of N-production indicating an intensive mineralisation of oil. The heterotrophic microorganisms of the tested soil samples prefer to mineralise the added lubricating oil even when the contamination is as small as 0.1 %. By the

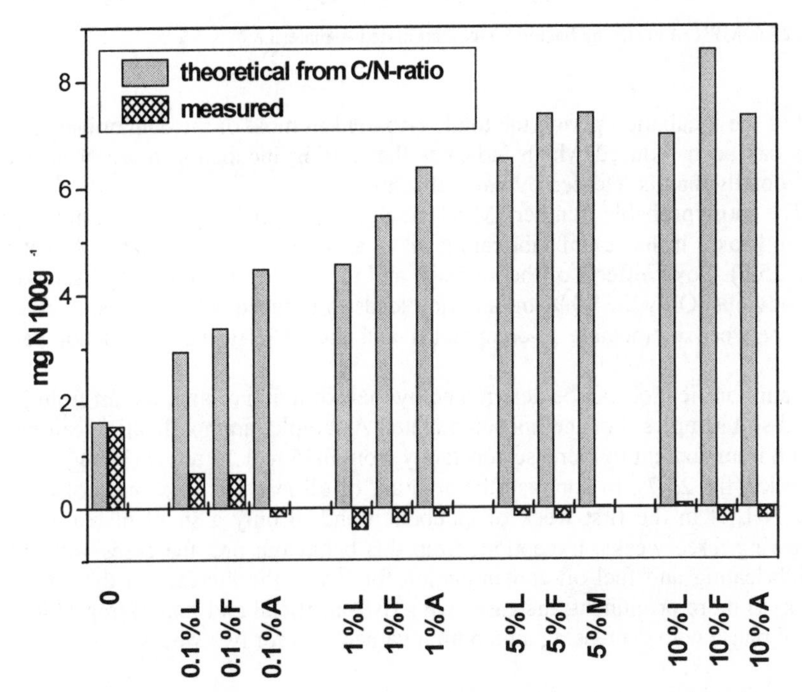

Fig. 25.5. Comparison of theoretical N-production and measured data in A-material in 28 days of incubation

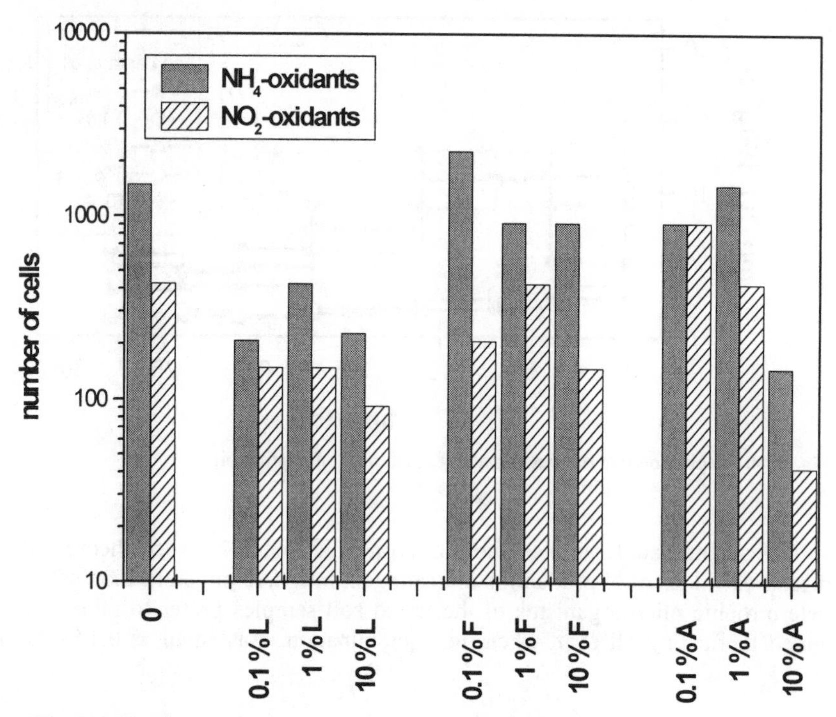

Fig. 25.6. MPN of nitrifying bacteria in contaminated A-material

end of the incubation period the total N-content in most of the contaminated samples has been reduced which indicates that during incubation more N is bound microbially than is released by mineralisation.

The most probable number (MPN) analysis for nitrifying bacteria indicates a general toxic influence of lubricating oil in all cases of contaminated A-material (fig. 25.6). Toxic effects of the artificial and the fuel oil on nitrifying bacteria are not obvious. Only the 10 % oil addition leads to reduced MPN values. A correlation between extractable N-compounds and the MPN-values could not be detected.

Net ammonification can be determined by the extraction of ammonium from incubated soil samples. For the uncontaminated A-sample, ammonification causes the ammonium content to increase constantly from 0.35 to 1.15 mg N $(100\ g)^{-1}$ within 4 weeks (fig. 25.7). In contrast, the addition of oil leads to a decrease of extractable NH_4-N in the first week of incubation and to only a small increase in the following three weeks. Exceptions from this behaviour are; the cases with 0.1 % of lubricating and fuel oil contamination for which the increase in the last three weeks is more prominent; the cases with 10 % artificial and fuel oil for which the initial phase with decreasing ammonium values lasts for two weeks.

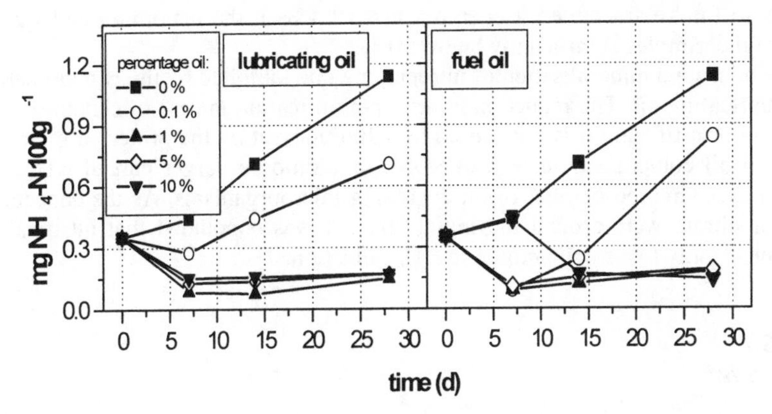

Fig. 25.7. Extractable ammonium in contaminated A-material (lubricating and fuel oil)

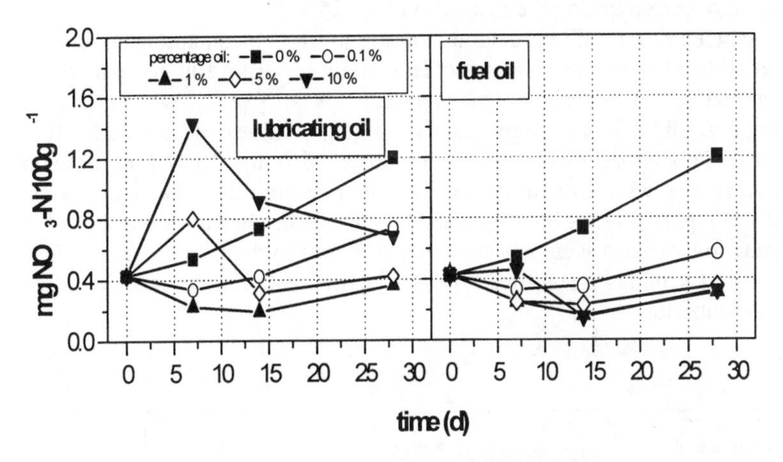

Fig. 25.8. Extractable nitrate in contaminated A-material (lubricating and fuel oil)

Extractable nitrate contents (net nitrification) of uncontaminated A-Material linearly increased to a maximum threefold value at the end of the incubation period. In contrast, the addition of oil resulted in more dynamic behaviour of mobile nitrate. For the fuel and the artificial oil contamination a reduction of extractable nitrate to half of the initial value was observed soon after the start of the incubation, followed by a gradual increase to values close to initial (fig. 25.8). In the case of stronger lubricating oil contamination (5 and 10 %) a significant increase of extractable nitrate content occurred at the beginning of the incubation.

In all samples extractable nitrite was present above the detection limit. However, reliable conclusions on nitrite production or consumption could not be drawn as the differences between the two replicates were considerable. Nevertheless, these cases were higher than the uncontaminated sample results. The proportion of

NO_2-N to total extractable nitrogen reached 12.4 % in the maximum and was in most of the samples significantly below 10 %.

The microbial mineralisation of nitrogen may be inhibited by the contamination with lubricating oil. The reduction in net ammonification in case of contamination with fuel or artificial oil is not caused by inhibition but by the preferred mineralisation of oil components instead of SOM. In addition, a larger part of NH_4-N is fixed in an increased number of oil-degrading microorganisms. As the concentrations of nitrate were around the initial level, it was concluded that nitrification contributes only to a small amount to NH_4 reduction.

25.3.3.2
K-Material

In contrast to the A-material, a steady decline of the extractable ammonium contents in the oil-contaminated, as well as in the uncontaminated samples of the K-material occurred during the incubation (fig. 25.9).

The concentration of extractable nitrate in the uncontaminated clayey K-material showed a much stronger increase of about 30 fold in comparison to the uncontaminated A-material. The increase was from low initial values (0.07 mg N $(100\ g)^{-1}$) to significantly higher values (2.33 mg N $(100\ g)^{-1}$, fig. 25.10). This strong increase could be observed in the 0.1 % fuel oil contamination variant as well. Contamination with lubricating oil of the same low intensity (0.1 %) partly reduced nitrification. The final nitrate values of all other contamination variants were not markedly different to their initial values. In some contaminated samples a significant peak of extractable nitrate occurred after one week of incubation.

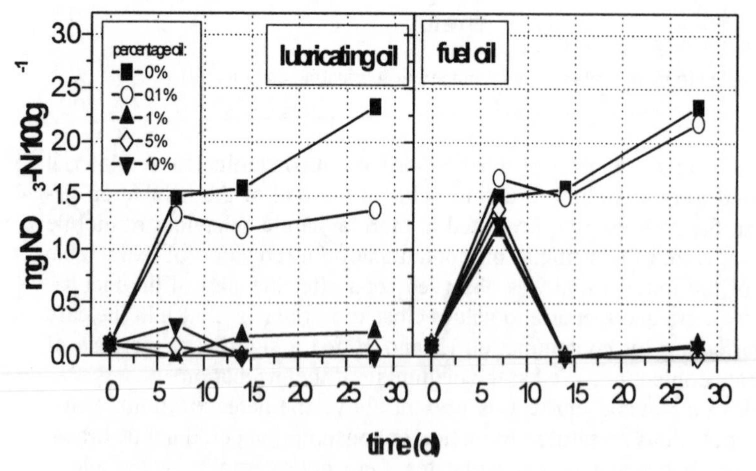

Fig. 25.9. Extractable ammonium in contaminated K-material (lubricating and fuel oil)

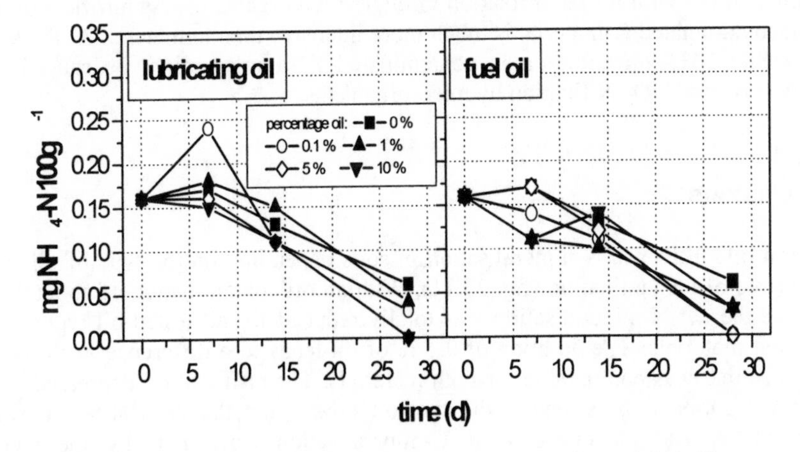

Fig. 25.10. Extractable nitrate in contaminated K-material (lubricating and fuel oil)

No extractable nitrite could be detected (detection limit = 0.015 mg N $(100\ g)^{-1}$) in all contaminated and uncontaminated samples of K-material.

As in the A-material, the reduced net ammonification due to the addition of oil to K-material can be explained by a preferred mineralisation of oil instead of SOM

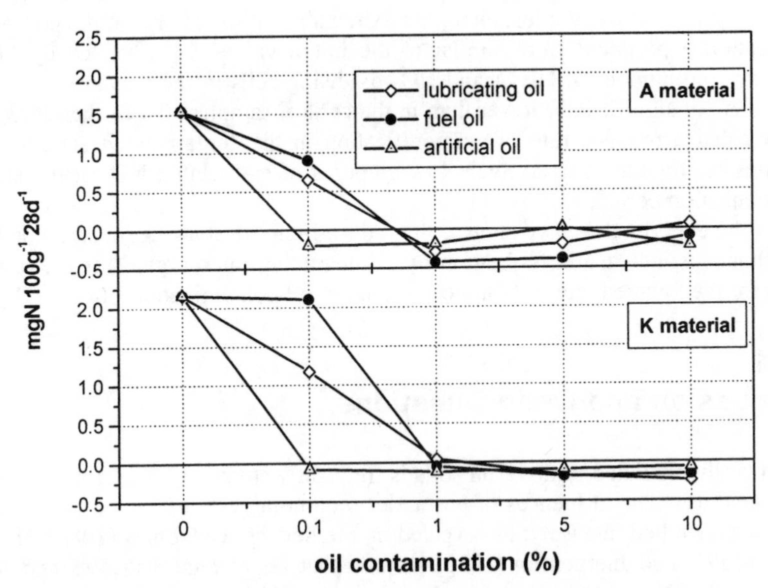

Fig. 25.11. Influence of oil contamination on the difference of extractable nitrogen after 28 days of incubation

in combination with nitrogen fixation through microorganisms. As nitrification in the uncontaminated K-material is much more intensive than in material A, the low extractable NH_4-concentration in contaminated K-samples might be caused by nitrification and NO_3-N fixation by microorganisms.

25.3.4
Conclusions

Mineral oils influence N-mineralisation in soils depending on the type of oil and the degree of contamination (fig. 25.11). With an oil contamination of about 1 % or more the net N-mineralisation was totally reduced for all tested oil types and both soil materials. The addition of 0.1 % of oil leads to a difference in oil type influence that was comparable for both materials. The artificial oil depressed nett N-mineralisation most strongly, the fuel oil lowest and the lubricating oil was intermediate. As the intensity of C-mineralisation (measured by the CO_2-production) was highest in the artificial oil, it can be assumed that the different degradability of the oils tested is responsible for different extents of nitrogen fixation by heterothrophic bacteria. Also, the soil materials A and K tested in this study showed different rates of N-mineralisation. After 28 days of incubation the amounts of extractable NH_4^+ and NO_3^- in uncontaminated A-material have tripled compared to the initial values. In the uncontaminated K-material, however, the NH_4^+-content decreased whereas the NO_3-content showed a strong increase.

Theoretical and measured values of N-mineralisation are in good agreement in both uncontaminated A- and K-materials. This is not true, however, for the contaminated A- and K-samples. Here, the extractable NO_3-contents at the end of the incubation experiment were similar to the initial values. As revealed by MPN-analysis, the number of NH_4^+- and NO_2-oxidising cells in the contaminated samples was not significantly lower than in the control samples. It was therefore concluded that a reduced rate of ammonification in the contaminated samples was responsible for the comparatively low amounts of extractable NO_3 at the end of the incubation period.

As the experiments have shown, the reduced rate of SOM degradation and the simultaneous increase in N-fixation by oil-degrading microorganisms are responsible for the lowered rate of N-mineralisation in oil-contaminated soils.

25.4
Discussion and Open Questions

Microbially mediated turnover in soils is frequently studied in the light of agricultural land use. The influences of pesticides on non-target organisms, in particular, have been studied intensely, as revealed in a review by Malkomes (1985). Harden et al. (1993) and Juergensen (1993) showed that for certain pesticides and some heavy metals the tested substances lowered the efficiency of substrate utilisation by microorganisms. Substrates were increasingly metabolised to yield CO_2 while less was incorporated into the microbial biomass. Domsch (1985) assigned 5 % of the reversible and 6 % of the persistent critical effects of pesticides on microorganisms as being due to influences on the nitrification process. Nitrifying microorganisms are partly considered as target organisms of nitrification inhibitors (e.g. Amberger and Gutser 1986). However, evidence on the influence of these substances can not be related to other substances. Studies by Kralova et al. (1992) examined the influence of the redox potential on ammonification and nitrification. It was suspected that the accumulation of ammonium may be the result of low redox potentials (high water content). In the study presented here the influence of the redox potential on N-mineralisation is negligible as the water content of the samples was adjusted to 60 % WHC_{max}, in accordance with various authors (Schinner et al. 1991; Klein 1992). In a study conducted on copper polluted soils, Minnich and McBride (1986) could not detect any inhibitory effects of copper on N-mineralisation. Domsch (1985), however, attributed inhibitory effects on the nitrification process to elevated concentrations of Cu, Zn, Ni, Ag and Hg as well as to residues of oil. Kandeler et al. (1994) found a lower rate of N-mineralisation (NH_4-N, determined in an anaerobic incubation experiment) with respect to microbial biomass (C_{mic}) in a degraded chernozem contaminated with fuel oil, than measured in the control. With respect to soil dry matter, this soil and a lime-containing chernozem showed an increased rate of N-mineralisation in comparison to uncontaminated material. These results are contradictory to those presented in this study.

The study presented here did not include the analysis of microbial biomass (C_{mic} and N_{mic}) and cell numbers of heterotrophic microorganisms. It was therefore not possible to conclusively assess the share of both rate of SOM degradation and N-fixation in the lowered rate of N-mineralisation of the oil-contaminated samples. Open questions also concern the type and extent of toxic effects of mineral oils and their components on nitrifying microorganisms, as well as the long-term dynamics of N-mineralisation in oil-contaminated materials, even after the degradation of the contaminant.

References

Alef K (1991) Methodenhandbuch Bodenmikrobiologie. ecomed Verlagsgesellschaft mbH, Landsberg/Lech, Germany

Amberger A, Gutser R (1986) Wirkung und Einsatz von Nitrifikationshemmstoffen in Kombination mit Mineraldüngern. Bayerisches Landwirtschaftliches Jahrbuch, BLV Verlagsgesellschaft mbH, München, 64. Jahrg, Heft 6: 717–726

Domsch KH (1985) Funktionen und Belastbarkeit des Bodens aus der Sicht der Bodenmikrobiologie. Verlag Kohlhammer GmbH, Stuttgart, Mainz, Germany

Harden T, Joergensen RG, Meyer B, Wolters V (1993) Mineralization of straw and formation of soil microbial biomass in a soil treated with Simazine and Dinoterb. Soil Biol Biochem 25, 9: 1273–1276

Janssen AM (1989) Antimicrobial activities of essential oils. – A pharmacognostical study – Dissertation, Rijksuniversiteit te Leiden

Jittler-Strahlendorff M, Neugebohrn L (1988) Untersuchungen zum Einfluß schwerer Verölungen des Deichvorlandes auf die Möglichkeit und Durchführbarkeit schneller Rekultivierungen von Pflanzenbeständen mittels unterschiedlicher Verfahren. Universität Hamburg, Institut für angewandte Botanik, Germany

Joergensen RG (1993) Die Messung der mikrobiellen Biomasse mit der $CHCl_3$-Fumigations-Extraktions-Methode in kontaminierten Böden. ECOINFORMA 2: 385–396

Kandeler E, Pennerstorfer C, Bauer E, Braun R (1994) Mikrobiologische Beurteilung biologischer Bodendekontaminationsverfahren im Modellversuch. Z Pflanzenernähr Bodenk 157: 345–350

Klein J (ed) (1992) Labormethoden zur Beurteilung der biologischen Bodensanierung, 2. Bericht des interdisziplinären Arbeitskreises "Umweltbiotechnologie – Boden". DECHEMA, Frankfurt a M, Germany

Kralova M, Masscheleyn PH, Patrick jr WH (1992) Redox potential as an indicator of electron availability for microbial activity and nitrogen transformation in aerobic soil. Zentralbl Mikrobiol 147: 388–399

Malkomes HP (1985) Einflüsse von Pflanzenschutzmitteln auf Bodenmikroorganismen und ihre Leistungen. In: Pflanzenschutzmittel im Boden: Eintrag und Bilanzierung – Dynamik im Boden – Besondere Stoffe – Einflüsse auf Bodenflora und Bodenfauna. Berichte über Landwirtschaft, Sonderheft 198, Verlag P Parey, Hamburg, Berlin, Germany

McGill WB, Rowell MJ, Westlake DWS (1981) Biochemistry, Ecology, and Microbiology of Petroleum Components in Soil. In: Paul EA, Ladd JN (eds) Soil Biochemistry, Vol 5. Marcel Dekker Inc, New York, Basel

Minnich MM, McBride MB (1986) Effect of copper activity on carbon and nitrogen mineralization in field-aged copper-enriched soils. Plant and Soil 91: 231–240

Schinner F, Öhlinger R, Kandeler E (1991) Bodenbiologische Arbeitsmethoden. Springer Verlag, Berlin, Germany

26 Development and Application of an Oxygen-Controlled High-Pressure Aeration System for the Treatment of TPH-Contaminated Soils in High Biopiles (a Case Study)

M. Koning, I. Cohrs, R. Stegmann
Technical University Hamburg-Harburg, Department of Waste Management, Harburger Schloβstraße 37, 21079 Hamburg, Germany

26.1 Introduction

Petroleum releases to the environment can cause safety hazards, ecological harm and adverse human health effects, and therefore the treatment of petroleum contaminated sites is necessary (Weismann et al. 1998). Today, the treatment of soils contaminated by petroleum products is one of the most frequently occurring cases in soil remediation, and many different techniques are available to purify petroleum contaminated soils (e.g. Koning et al. 2000a). Within the biological remediation processes in particular, the biopile process has been established as one of the most effective and competitive technologies for the treatment of petroleum contaminated soils. Nevertheless, in order to increase the economic efficiency of the process further developments are crucial (Schulz-Berendt 2000).

Lab scale investigations have shown that sufficient oxygen supply to the TPH (Total Petroleum Hydrocarbons) degrading microorganisms is one of the major requirements for an efficient TPH degradation (Hupe et al. 1999) (see Chapter 22). Since the passive oxygen penetration into biopiles is often limited to depths between 0.5 and 2.0 m (e.g. Huesemann et al. 1995; Benazon et al. 1995) active ventilation measures become necessary if TPH contaminated soils are treated in higher biopiles. In order to realise an economical treatment of TPH-contaminated soils in biopiles with heights of more than 3 meters, an oxygen-controlled high-pressure aeration system for biopiles was developed and tested in a two-year project under real life conditions. Comparative investigations between aerated and unaerated biopiles of different heights (1.50 m and 3.20 m), as well as lab scale investigations in bioreactors were conducted to describe the milieu changes within high biopiles, and to characterise their impact on the biological TPH degradation (Koning et al. 2000b).

26.1.1
Material and Methods

26.1.1.1
Soil Material

The soil material used for the investigations was a diesel fuel contaminated silty sand. The contamination of the soil resulted from a railway accident where diesel fuel drained off a leak in a tank wagon, infiltrated into the soil and partially burned off. After the fire-fighting operations, the soil was excavated and brought to an off-site treatment facility. Here, the soil was mechanically pre-treated. Water, nutrients and substrates (compost, straw, bark ~30 wt.-%) were added to improve the soil characteristics and to loosen up the soil structure. Table 26.1 gives an overview of the soil characteristics after the pre-treatment and shows treatment targets that have to be reached in order to reuse the soil under defined technical securing measures (LAGA 1994). Examination of the data showed that only the TPH-content exceeded the regulatory requirements and therefore had to be reduced.

26.1.1.2
Pilot Plant

In order to describe the biological degradation of TPH in biopiles, comparative investigations between aerated and unaerated biopiles of different heights (1.50 m and 3.20 m) were conducted. The aeration of the biopiles was performed by an oxygen-controlled high-pressure venting system developed by the working group (fig. 26.1) (Koning et al. 2000b). In regular intervals, gas samples were taken automatically from different depths of the biopiles (0.20 m, 1.20 m, 2.20 m, 3.20 m), dried and analysed online regarding their O_2, CO_2 and CH_4 contents. The results were used to generate gas profiles of the biopiles, and at the same time were used to control the aeration system. If the oxygen concentration dropped at one measuring point below 2 vol.-%, the biopile was automatically aerated with a pulse of 7 m^3 air. The oxygen concentration limit of 2 vol.-% for the start of the forced aeration was based on lab scale investigations on the influence of the oxygen content on the biological degradation of TPH in soils, and is slightly above the limiting concentration of 1 vol.-% (Hupe et al. 1999). In addition to the online gas monitoring, gas samples were taken in regular intervals and analysed for volatile organic carbon (VOC) and trace gases. Soil samples were taken in parallel and the TPH-content as well as the biomass concentration was determined. Finally, the temperatures at the soil gas sampling tubes, the water contents and the pH-values of the soil were measured in order to describe milieu conditions within the biopiles.

Table 26.1. Parameters, analytical methods and results of the initial soil examination

Parameter	Analytical method	Soil[a]	LAGA Z2[j]	unit
TPH[b]	DIN 38409 H18 (IR-Spectroscopy)	8000 –14000	1000	[mg (kg dm)$^{-1}$]
PAH[c]	EPA Methode 610 (GC-MSD)	2.1 – 4.5	20	[mg (kg dm)$^{-1}$]
As	PE AAS 5500, FIAS 400	0.97– 2.77	150	[mg (kg dm)$^{-1}$]
Cd	AAS PE-Z3030	0.16– 0.22	10	[mg (kg dm)$^{-1}$]
Cr	AAS PE-SIMAA 6000	10.7 – 19.0	600	[mg (kg dm)$^{-1}$]
Cu	PE-Plasma 2000 OES, ICP	4.4 – 10.9	600	[mg (kg dm)$^{-1}$]
Hg	AAS PE-FIMS, Amalgam	0.01– 0.03	10	[mg (kg dm)$^{-1}$]
Mn	PE-Plasma 2000 OES, ICP	81 – 281	–	[mg (kg dm)$^{-1}$]
Ni	AAS PE-SIMAA 6000	5.1 – 13.7	600	[mg (kg dm)$^{-1}$]
Pb	PE-Plasma 2000 OES, ICP	6.4 – 17.3	1000	[mg (kg dm)$^{-1}$]
Zn	PE-Plasma 2000 OES, ICP	21.4 – 49.7	1500	[mg (kg dm)$^{-1}$]
TOC[d]	DIN EN 14848	27966 –30143	–	[mg C (kg dm)$^{-1}$]
TKN[e]	VDLUFA A.2.2.1	1834 – 2391	–	[mg (kg dm)$^{-1}$]
P	PE-Plasma 2000 OES, ICP	564 – 1237	–	[mg C (kg dm)$^{-1}$]
biomass	SIR-method (Anderson and Domsch 1978) Sapromat D12, Voith	5441 – 6419	–	[mg C (kg dm)$^{-1}$]
C_{org}	calculated: C_{org} = TOC – C-MKW – biom.	12074 –15019	–	[mg C (kg dm)$^{-1}$]
CO	IR-Spectroscopy (ADC DB 2E)	–	–	[mg C (kg dm)$^{-1}$]
VOC[f]	FID C-Analyser 3002, Bernath Atomic	–	–	[mg C (kg dm)$^{-1}$]
water content	DIN 18121 T2	28.08	–	[mg H_2O (100g dm)$^{-1}$]
WC[g]	(Alef 1991)	43.11	–	[mg H_2O (100g dm)$^{-1}$]
Wp[h]	DIN 18122 T1	26.52	–	
particle fines <63	DIN 4188 T1 sieving	18	–	[%]
pH-value	VDLUFA A.5.1.1	7.0	–	[–]
conductivity	DIN 38404 C8	201	–	µS cm^{-1}

[a] Soil: silty sand with additives (compost, straw, bark ~30 %-wt)
[b] total petroleum hydrocarbons
[c] polycyclic aromatic hydrocarbons
[d] total organic carbon
[e] total Kjehldahl nitrogen
[f] volatile organic carbon
[g] maximum water-holding capacity
[h] plastic limit (Atterberg 1911)
[I] dry matter
[j] treatment targets that allow a reuse of the soil under defined technical measures in Germany (Länderarbeitsgemeinschaft Abfall 20/1)

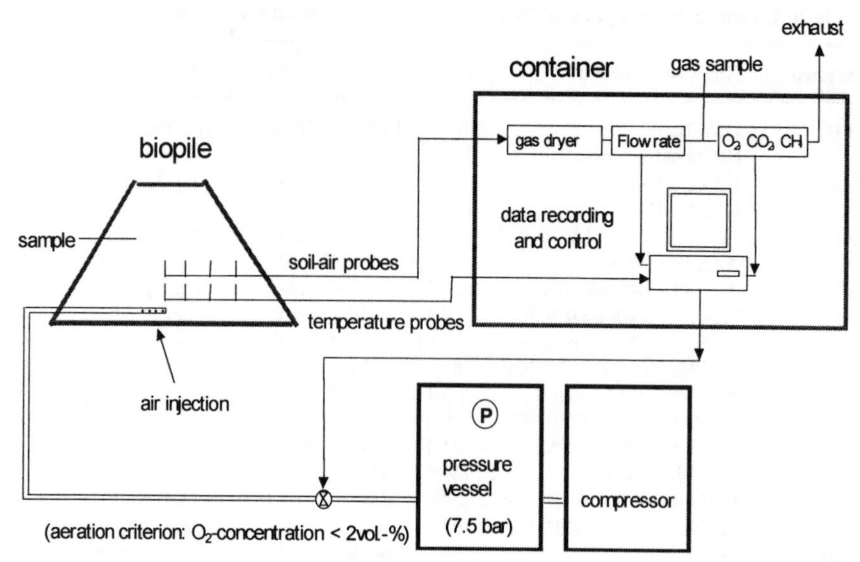

Fig. 26.1. Oxygen-controlled high-pressure venting system for biopiles

1.1.1.3
Laboratory Test Systems

In order to balance the TPH degradation process, carbon balances were simultaneously determined in lab-scale bioreactors (fixed-bed reactors (fig. 26.2)). Those test systems were forced aerated closed bioreactors with a volume of 6 litres (Hupe et al. 1998) (see also Chapter 22).

1.1.1.4
Soil Sampling, Preparation, Storage and Analysis

The soil sampling within the biopile was performed accordingly to the guidelines of the GDCh "Sampling strategy for the clean-up control in ex-situ bioremediation" (GDCh 1994). Soil samples were taken at different sampling points and different depths of the biopile (0–1 m and 1–2 m). At each sampling date, 3 soil samples of approximately 4 kg were taken at different sampling spots and mixed into a composite sample. After a homogenisation of this composite sample, two samples of 1 litre volume were separated, stored at -20 °C, and analysed later on. For quality control, multiple determinations, interlaboratory tests and statistical evaluations were performed (Kraß et al. 1999). Table 26.1 shows the analysed parameters as well as the analytical methods used.

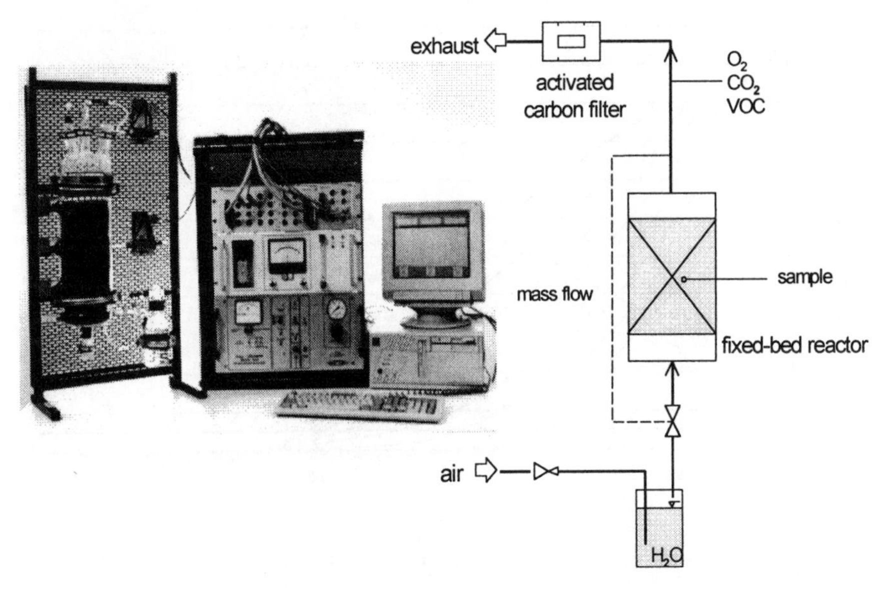

Fig. 26.2. Fixed-bed reactors

26.2 Results

The investigations show that the passive aeration of biopiles was limited by the water content, soil texture, compaction ratio, and biological activity of the soil. Fig. 26.3 shows the development of the oxygen concentrations within an unaerated high biopile (height: 3.20 m) during 320 days of bioremediation. A few hours after the setup, the oxygen within the centre of the biopile was depleted due to the respiration activity of the microorganisms. Only in the areas near the surface (depths < 1.20 m) did the oxygen concentration remain above 1 vol.-%.

With progressing treatment time, an increased drying of the biopile surface could be observed (fig. 26.4). In consequence, the outside air migrated deeper into the biopile and the oxygen concentrations slowly increased. Due to this behaviour, zones of optimal milieu conditions and high biological activity developed and moved from the biopile surface towards the direction of the biopile centre. Within the small biopile (height: 1.50 m) these zones of high biological activity reached the inner centre of the biopile and a continuous TPH-degradation could be generated within the whole biopile. Within the high biopile (height: 3.20 m) the oxygen concentrations which allow an unrestricted TPH-degradation (> 1 vol.-%) were limited to depths below two metres (fig. 26.3).

Consequently, anaerobic zones with a significant methane production developed in the centre of the unaerated high biopile. In these zones, no TPH degradation could be observed (fig. 26.10). Instead, an anaerobic degradation of the organic additives (compost, straw, bark ~30 wt.-%) took place and led to the noticeable methane production (fig. 26.5).

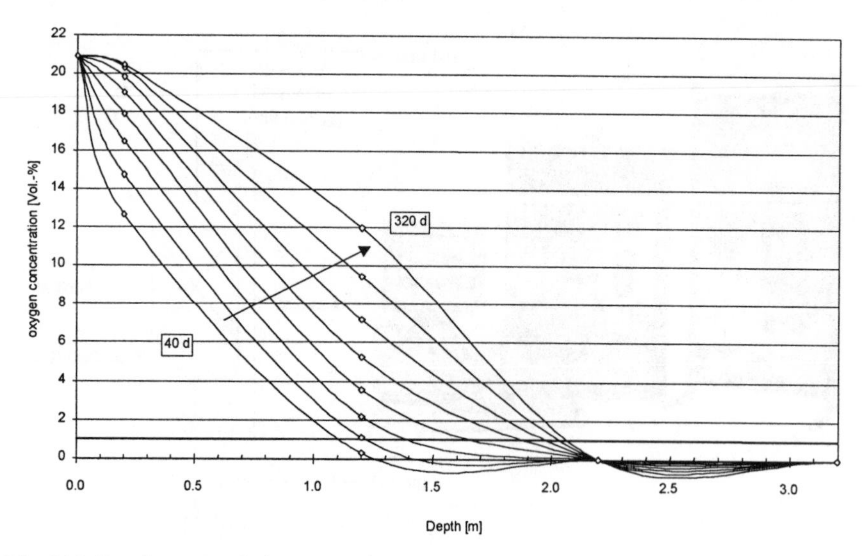

Fig. 26.3. Development of the oxygen concentration in the unaerated high biopile (height: 3.20 m)

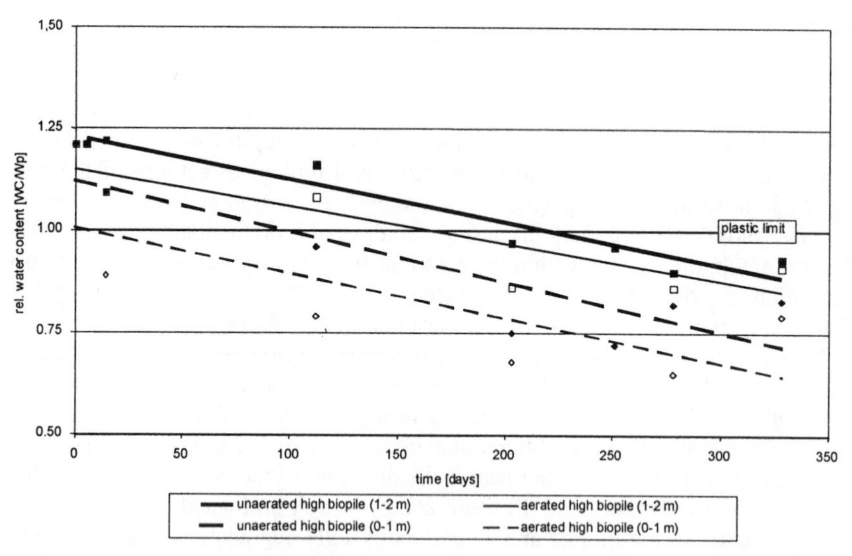

Fig. 26.4. Development of the water content within the aerated and unaerated high biopiles (heights: 3.20 m)

At the beginning of the aeration process oxygen concentrations below 2 vol.-% were observed every 2–3 hours, while methane could not be detected anymore.
The oxygen-controlled high-pressure aeration of the aerated biopiles was started at day 82. If the oxygen concentration dropped at one measuring point below

2 vol.-%, the biopiles were automatically aerated with a pulse of 7 m^3 of air which spreads into the soil within approximately 2 minutes (fig. 26.6).

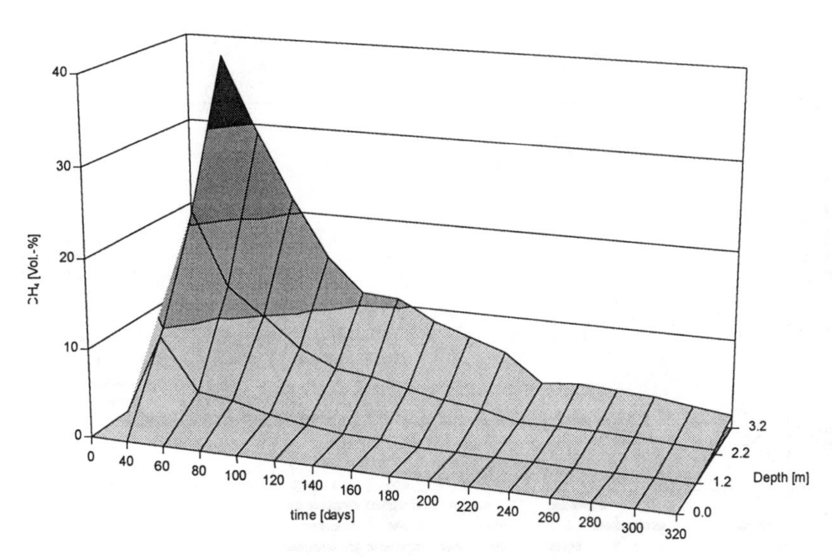

Fig. 26.5. Development of the methane concentration within the unaerated high biopile (height: 3.2 m)

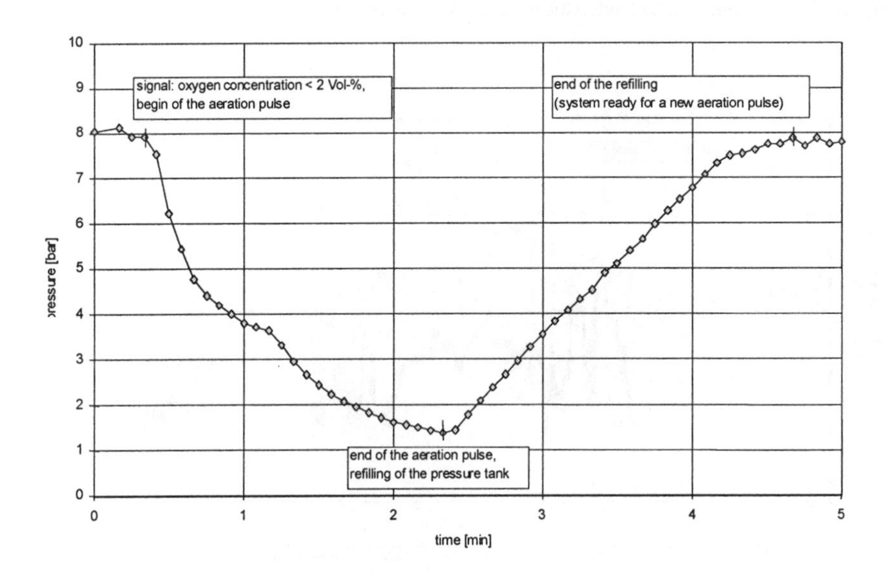

Fig 26.6. Temporal course of one aeration pulse

With progressing aeration of the biopiles and decreasing biological activity time periods between the single aeration pulses increased up to 35 hours. Finally, after 233 days of aeration (day 315) the oxygen-controlled high pressure venting was not requested anymore (fig. 26.7).

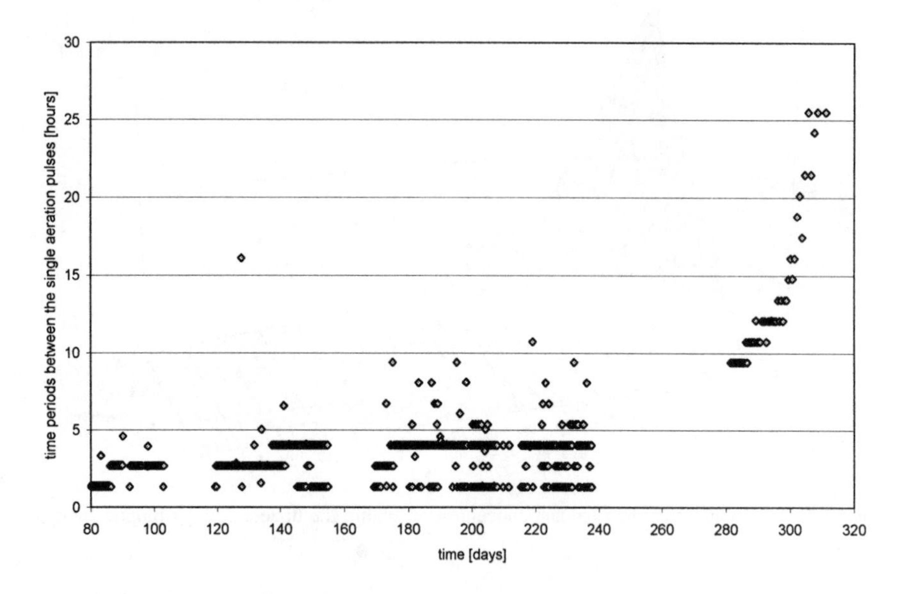

Fig. 26.7. Development of the intervals between aeration pulses

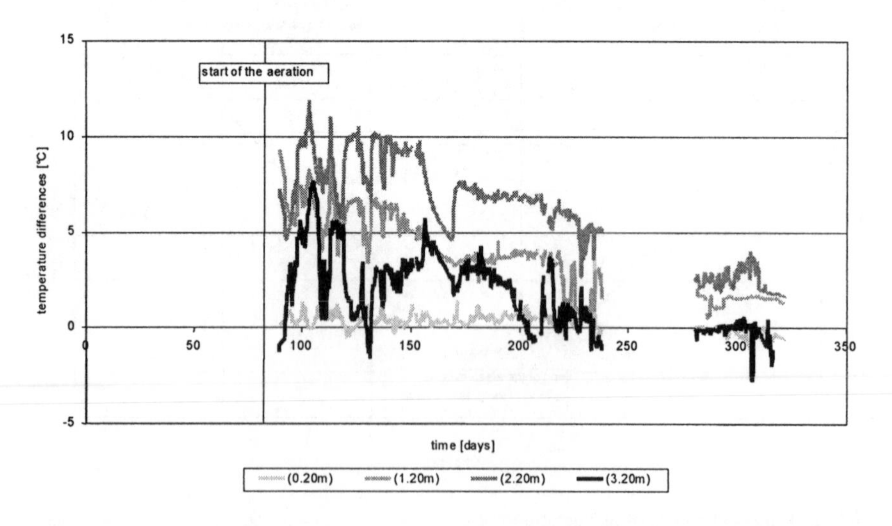

Fig. 26.8. Temperature differences between the aerated and the unaerated high biopile (height 3.2 m)

How the aeration stimulated the microbiological activity can be observed by the development of the soil temperature. Comparing the temperature development within an aerated and an unaerated high biopile of identical composition, temperature differences of up to 12 °C could be observed (fig. 26.8).

Thereby, maximum temperatures of 40 °C were reached within the aerated high biopile (fig. 26.9). Since the temperature of the infiltrated air amounted to an increase of approximately 1 °C, it was clear that the heating of the aerated biopiles was due to an increased microbiology activity. The assumption, that the active aeration and the resulting heating of the biopile will lead to an early drying out of the soil material was not observed. Comparing the water contents within the aerated and the unaerated high biopile, the influence of the impulse aeration on the water content was comparatively small. During 320 days of treatment the water content stayed within a range of 70–125 % of the plastic limit (wp) (fig. 26.4).

The influence of the impulse aeration on the TPH degradation is shown in fig. 26.10. In the passively aerated areas near the surface (depth: 0–1 m) of the aerated as well as of the unaerated biopile, a continuous degradation of TPH started at day 26. A TPH degradation within the inner areas (depth: 1–2 m) could only be observed within the aerated biopile (starting at day 84).

In contrast, the TPH contents within the unaerated biopile did not start to decrease until day 251. This means, that the time necessary to induce a TPH degradation at a depth of 1–2 m could be reduced by approximately 170 days of treatment.

The lab-scale investigations carried out simultaneously in the fixed-bed bioreactors (FBR) (fig. 26.2) confirm the results of Hupe et al. (1999), and show that without any oxygen supply no recognisable TPH reduction could be achieved within 80 days of treatment (fig. 26.11).

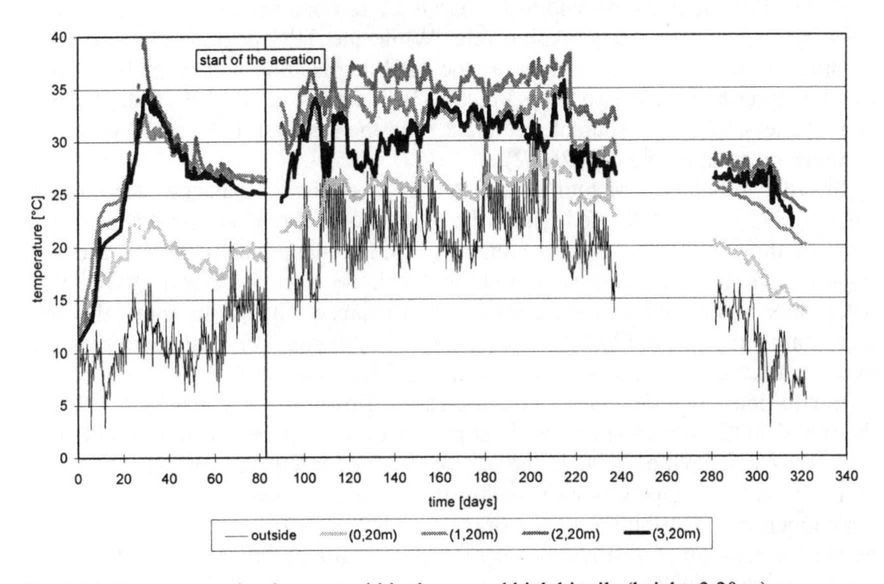

Fig. 26.9. Temperature development within the aerated high biopile (height: 3.20 m)

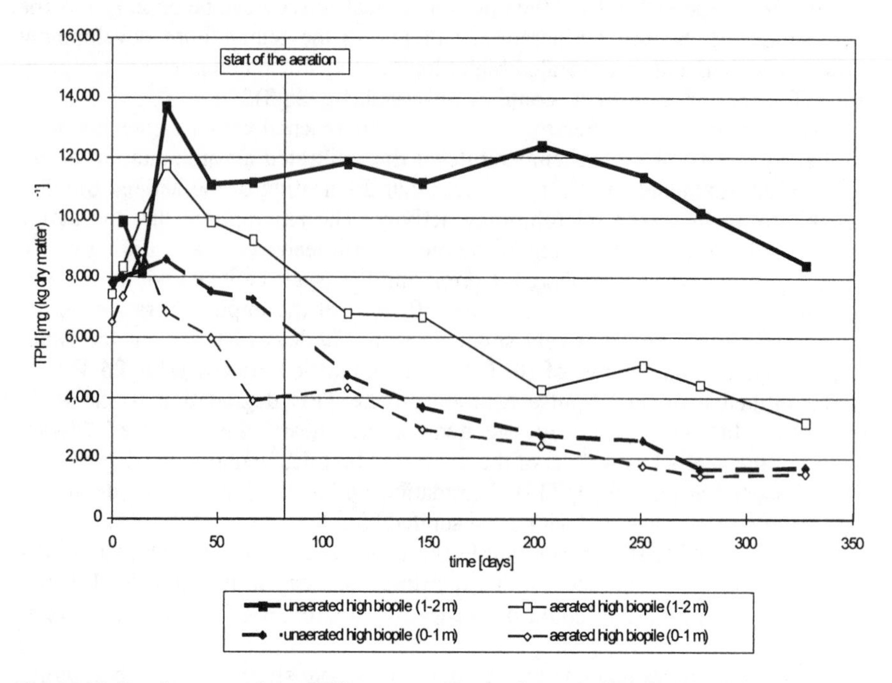

Fig. 26.10. Development of the TPH concentration in the aerated and unaerated high biopiles (height: 3.20 m)

Compared to the TPH degradation within the passively aerated parts of the biopiles, the TPH degradation within the aerated fixed-bed reactors showed a different kinetic and higher degradation rate. While the TPH degradation within the biopiles followed a first order kinetic, the TPH degradation within the bioreactors can be described as a two step kinetic (Williamson et al. 1997) (table 26.2). Nevertheless, after a sufficient treatment time the residual TPH contents in both systems are comparable (fig. 26.12).

In order to balance the biological TPH turnover, carbon balances according to Lotter (1995) were performed within the bioreactors. Therefore, the different fractions of the initial total carbon content (TOC), contamination (TPH), biomass, organic carbon (C_{org}), carbon dioxide and volatile organic carbon (VOC) were determined and combined for the total carbon balance. The results show, that with increasing temperature (20 °C → 30 °C → 40 °C) higher TPH degradation as well as mineralisation rates could be achieved (fig. 26.13 and table 26.3).

In addition, the VOC output also increased. However, only less than 1.9 % of the initial TOC content (this equals approximately 3.5 % to the initial TPH content) was carried out as VOC in the gas phase as a consequence of the active aeration. The balance gaps observed within the bioreactor experiments are due to an inhomogeneous distribution of the organic additives (30 % compost, bark, straw) within the soil, which resulted in fluctuating amounts of the organic carbon (C_{org}) and therefore of the TOC contents.

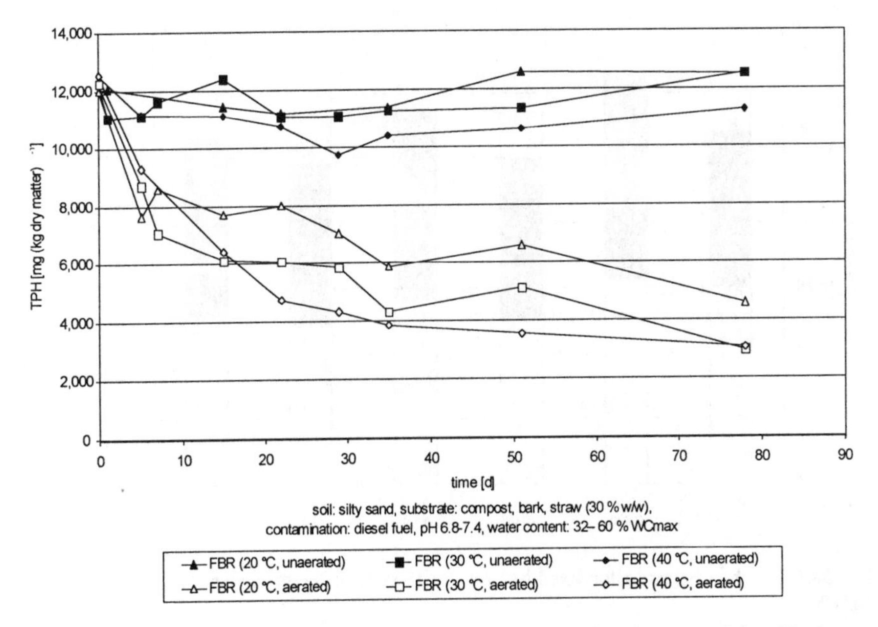

Fig. 26.11. Development of the TPH concentration in the aerated and unaerated fixed-bed reactors (FBR)

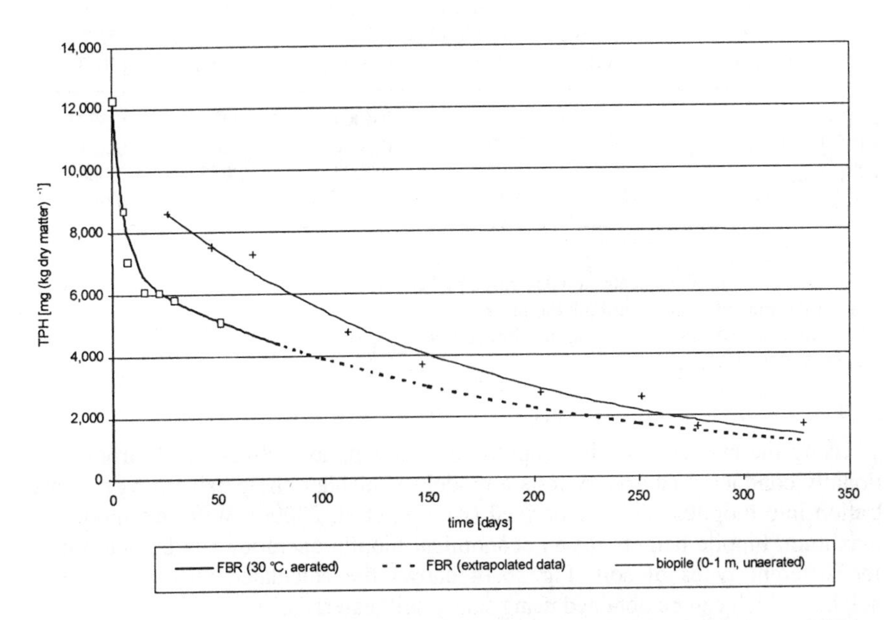

Fig. 26.12. Development of the TPH concentration within an aerated fixed-bed reactor (FBR) at 30 °C and within the passively aerated part of a high biopile

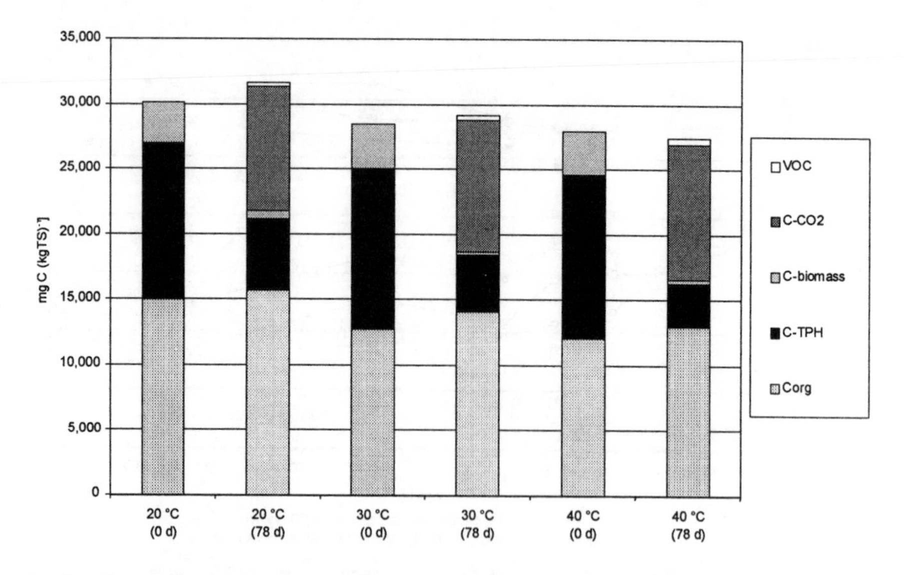

Fig. 26.13. Carbon balances for fixed-bed reactor tests at different temperatures (20 °C; 30 °C; 40 °C)

Table 26.2. TPH degradation kinetic parameters

	unaerated high biopile (0–1 m)	aerated high biopile (0–1 m)	unaerated high biopile (1–2 m)	aerated FBR[a] (20 °C)	aerated FBR[a] (30 °C)	aerated FBR[a] (40 °C)
F[b] [–]	–	–	–	0.4001	0.4478	0.5721
k1[c] [–]	0.0062	0.0063	0.0046	0.1036	0.1900	0.2313
k2[d] [–]	–	–	–	0.0034	0.0055	0.0062
r[e] [–]	0.98	0.94	0.96	0.96	0.99	0.98

[a] fixed-bed reactor
[b] fraction of the initial concentration described by fast released kinetics
[c] first order rate constant describing the fast release
[d] first order rate constant describing the slow release
[e] regression coefficient

Using the results from the biopile investigations as well as results from additionally conducted laboratory tests a model to calculate the passive oxygen penetration into biopiles could be derived (Koning et al. 2000c). With this model the maximum biopile heights for an economical biopile operation can be determined for different types of soil. Fig. 26.14 shows the calculated maximum biopile heights, which can be obtained using sandy soil materials.

It is stated that with sandy soil materials, biopile heights of up to several meters theoretically can be passively aerated. However, with rising water content and increasing respiration activity the maximum biopile height which can be passively aerated clearly drops.

Table 26.3. Carbon balances for fixed-bed reactor tests at 20 °C, 30 °C and 40 °C

Parameter	20 °C (day 0)	20 °C (day 78)	30 °C (day 0)	30 °C (day 78)	40 °C (day 0)	40 °C (day 78)
VOC[a] [mg C (kg dm)$^{-1}$]	0	246	0	294	0	439
C-CO_2[b] [mg C (kg dm)$^{-1}$]	0	9581	0	10172	0	10455
C-biomass[c] [mg C (kg dm)$^{-1}$]	3110	607	3442	183	3337	269
C-TPH[d] [mg C (kg dm)$^{-1}$]	12014	5530	12268	4411	12555	3312
C_{org}[e] [mg C (kg dm)$^{-1}$]	15019	15687	12791	14091	12074	12985
Sum [mg C (kg dm)$^{-1}$]	30143	31651	28501	29151	27966	27460

[a] volatile organic carbon
[b] carbon dioxide carbon
[c] biomass carbon
[d] total petroleum hydrocarbon carbon
[e] organic carbon (without biomass, without contamination)

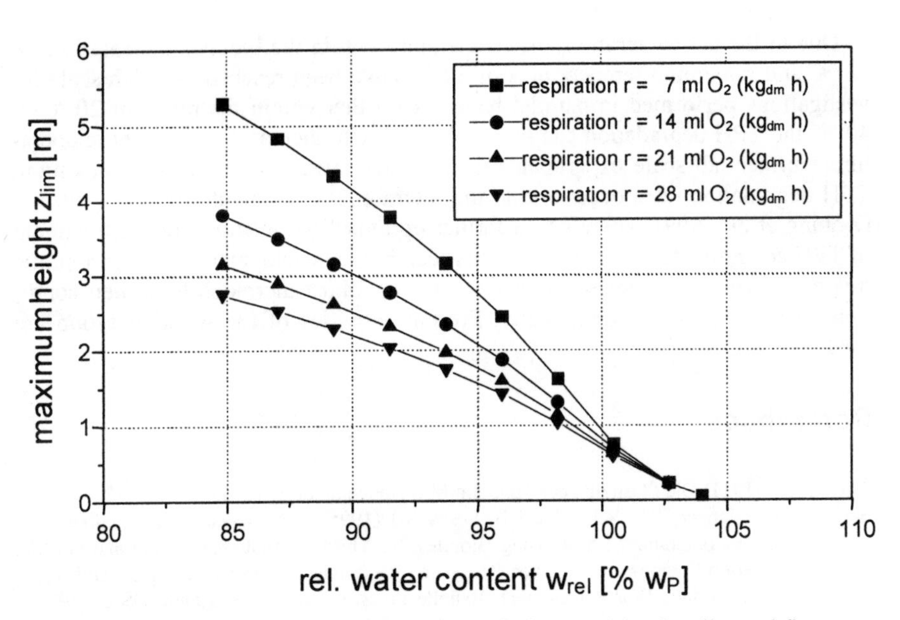

Fig. 26.14. Maximum biopile heights which can be passively aerated (sandy soil material)

Thus, the adjustment of the water content within the biopiles has important implications (Scholz et al. 1999). In order to obtain an optimal biopile operation, the water content should be adjusted below the plastic limit (wp) (~95 % wp). With this water content a maximum biological activity can be ensured; higher water contents mean an impairment of the oxygen penetration is likely (Koning et al. 2000c).

26.3 Summary and Outlook

The rate of the biological TPH degradation in biopiles is mainly influenced by the milieu conditions within the soil. Thereby, the oxygen supply of the TPH degrading microorganisms is of great importance. When the oxygen concentration within the soil drops below 1 vol.-%, the TPH degradation rate will be reduced to a degree that the biopile process becomes inefficient (Hupe et al. 1999). In order to ensure an unrestricted TPH degradation, oxygen concentrations above 1 vol.-% must be guaranteed within all areas of the biopiles. Without an active aeration the oxygen transfer into biopiles is determined by diffusion and usually limited to depths below two meters. Biopiles with heights of more than 3 m have to be actively aerated. One possible solution to aerate high biopiles economically is given by the presented oxygen-controlled high-pressure venting system. The results of the investigation program showed that with the help of the venting system, degradation-limiting oxygen contents < 1 vol.-% could be avoided and a continuous TPH degradation in biopiles with heights of 3.20 m could be implemented (k_1 = 0.0046–0.0063 d^{-1}) (table 26.3).

Due to the active aeration, the temperature within the biopiles increased up to 12 °C and maximum temperatures of 40 °C have been reached. The lab scale investigations performed in parallel have shown that within the range of 20 °C to 40 °C the TPH degradation can be accelerated with increasing treatment temperature. Further lab scale experiments using slurry systems indicate that maximum TPH degradation rates occur within the thermophilic temperature range (~60 °C) (Koning et al. 1999). Therefore, a further optimisation potential for the treatment of TPH contaminated soils in biopiles is seen in the development of a thermophilic treatment process. A prerequisite for this is to adjust thermophilic milieu conditions within biopiles economically. Present activities of the working group are focussed on this topic.

References

Atterberg A (1911) Die Plastizität der Tone. Int Mitt Bodenkunde 1: 10–43

Benazon B, Belanger DW, Scheurlen DB, Lesky MJ (1995) Bioremediation of Ethylbenzene- and Styrene-Contaminated Soil Using Biopiles. In: Third International In-Situ and On-Site Bioreclamation Symposium. Hinchee RE, Skeen RS, Sayles GD (eds) Biological Unit Processes for Hazardous Waste Treatment. Battelle Press, Columbus, Richland, USA 3 (9), pp 179–191

GDCh (1994) Leitfaden – Erfolgskontrolle bei der Bodenreinigung. Arbeitskreis "Bodenchemie und Bodenökologie" der Fachgruppe "Umweltchemie und Ökotoxikologie", GDCh-Monographien, Bd 4, Frankfurt, Germany

Huesemann M (1995) Application of passive aeration to improve the cost effectiveness of bioremediation for petroleum contaminated soils. Battelle Press, Columbus, Richland, USA

Hupe K, Koning M, Lüth J-C, Heerenklage J, Stegmann R (1998) Einsatz von Testsystemen zur bilanzierenden Untersuchung der biologischen Schadstoffumsetzung im Boden. Altlastenspektrum 7: 360–366

Hupe K, Heerenklage J, Stegmann R (1999) Influence of oxygen on the degradation of TPH-contaminated soils. In: Bioreactors and Ex-Situ Biological Treatment Technologies. Alleman BC, Leeson A (eds) Proceedings of the Fifth International In-Situ and On-Site Bioremediation Symposium, April 19–22, San Diego, California. Battelle Press, Columbus, Richland, USA, 5(5), pp 31–36

Koning M, Lüth J-C, Reifenstuhl R, Hintze H, Feitkenhauer H, Stegmann R (1999) Combined chemical and biological treatment of mixed-contaminated soils in slurry reactors. In: Bioreactors and Ex Situ Biological Treatment Technologies. Alleman BC, Leeson A (eds) Proceedings of the Fifth International In Situ and On-Site Bioremediation Symposium, April 19–22, San Diego, California. Battelle Press, Columbus, Richland, USA, 5(5), pp. 25–30

Koning M, Hupe K, Stegmann R (2000 a) Thermal Processes, Scrubbing/Extraction, Bioremediation and Disposal. In: Biotechnology 11b. Rehm HJ, Reed G, Pühler A, Stadler P (eds) Environmental Processes II – Soil Decontamination. Klein J (ed) Wiley-VCH Verlag GmbH, Weinheim, New York, Chichester, Brisbane, Singapore, Toronto, pp 305–317

Koning M, Braukmeier J, Cohrs I, Lüth J-C, Stegmann R, Viebranz N, Schulz-Berendt V (2000 b) Untersuchungen zur Optimierung der biologischen Bodenbehandlung im Mietenverfahren im Hinblick auf eine gezielte Wiederverwertung des zu behandelnden Bodenmaterials. Abschlußbericht zu dem durch die Deutsche Bundesstiftung Umwelt geförderten Gemeinschaftsprojekt. (AZ 09815)

Koning M, Scholz J, Hupe K, Schwedes J, Stegmann R (2000 c) Einfluß des Sauerstoffgehaltes auf die Umsetzung von MKW im Boden – Teil 2: Sauerstoffeintrag in Sanierungsmieten. TerraTech 1: 64–67

Kraß JD, Mathes K, Schulz-Berendt V (1999) Scale up of biological remediation processes: Evaluating the quality of laboratory derived prognoses for the degradation of petroleum hydrocarbons in clamps. Proceedings of the Fifth International Conference on Ecotoxicology and Environmental Safety, Secotox 99, GSF-National Research Center, Neuherberg/Munich, Germany, March 15–17, PK6

LAGA (1994) Anforderungen an die stoffliche Verwertung von mineralischen Reststoffen/Abfällen. Mitteilung der Länderarbeitsgemeinschaft Abfall 20/1

Lotter S (1995) Biologische Bodenreinigung – Untersuchungen zur Kohlenstoffumsetzung, Kohlenstoffbilanzierung und Prozeßmodellierung. In: Hamburger Berichte 9. Stegmann R (ed) Economica Verlag GmbH, Bonn, Germany

Scholz J, Schwedes J, Müller BG, Deckwer WD (1999) Conditions and effects of microbial soil remediation in solid-state bioreactors. In: Bioreactors and Ex Situ Biological Treatment Technologies. Alleman BC, Leeson A (eds) Proceedings of the Fifth International In Situ and On-Site Bioremediation Symposium, April 19–22, San Diego, California. Battelle Press, Columbus, Richland, USA, 5(5), 31–36

Schulz-Berendt V (2000) Bioremediation with heap technique. In: Biotechnology 11b. Rehm HJ, Reed G, Pühler A, Stadler P (eds) Environmental Processes II – Soil Decontamination. Klein J (ed) Wiley-VCH Verlag GmbH, Weinheim, New York, Chichester, Brisbane, Singapore, Toronto, pp 320–328

Weismann et al. (1998) Analysis of Petroleum Hydrocarbons in Environmental Media. Total Petroleum Hydrocarbon Criteria Working Group Series, Volume 1

Williamson DG, Loehr RC, Kimura Y (1997) Measuring release and biodegradation kinetics of aged hydrocarbons from soils. In: Alleman BC, Leeson A (eds) In Situ and On-Site Bioremediation: Volume 5, Proceedings of the Fourth International In Situ and On-Site Bioremediation Symposium, April 28–May 1, New Orleans, Battelle Press, Columbus, Richland, USA, 4 (5), pp 605–610

Physical Treatment

27 Remediation of Soils by Washing Processes – an Historical Overview

M. Wilichowski
Department of Mechanical Engineering/Process and Environmental Engineering, Hochschule Wismar, University of Technology, Business and Design, Philipp-Müller-Straße, 23952 Wismar, Germany

27.1 Introduction

In the early Seventies the protection of soil was included in the environmental program of the government of the Federal Republic of Germany (Umweltprogramm der Bundesregierung 1971) as one of the major goals for the environmental policy. Nevertheless, the enormous relevance of the contaminated sites problem was not perceived in its total extent until the registration of contaminated sites began in the early Eighties. By 1993 more than 139,000 presumably contaminated sites were registered in Germany, including about 86,000 abandoned waste disposal sites and about 53,000 closed industrial sites (Franzius 1995).

Due to the risk of pollution of both groundwater and air caused by contaminated soils, many different remediation methods have been developed in the last twenty years. Nowadays soil washing is a widely used technique for the highly efficient remediation of soils contaminated with both heavy metals and organic pollutants.

27.2 Basic Principles of Soil Washing

Most of the main process steps in soil washing plants were not developed especially for the treatment of contaminated soils, but rather represent conventional unit operations for the treatment of solid raw materials. However, the process sequence of soil washing processes was adapted to the special requirements which result from the physicochemical properties of contaminated soils.

27.2.1
Particle Size-Dependent Distribution of Pollutants

Soil washing as a wet-mechanical pollutant extraction is usually used for the treatment of soils in which the pollutants are accumulated in the fine fraction of the solid matter.

In most cases the contaminants, both organic pollutants and heavy metals, are attached to the surfaces of soil particles. Therefore fine particles in particular with a very high specific surface, usually have a significantly higher content of pollutants than the coarse particles. The specific surface area S_V is dependent on the particle diameter d_p (spherical particles are assumed) as given by the equation (27.1).

$$S_V = \frac{6}{d_p} \tag{27.1}$$

The functional relation in accordance with this equation is shown in fig. 27.1.

The figure demonstrates the significant increase of the specific surface area with decreasing particle diameter. Due to this correlation the contribution of the coarse fraction (particles with diameter above approximately 100 µm) to the total surface area of soils is minor even for coarse-grained soils. As a result more than 90 % of the pollutants can mostly be found in the silt and clay fraction of contaminated soils.

Additionally the accumulation of contaminants in the fine fraction is enlarged by humic matter. It coats the surfaces of fine particles and is a strong bonding agent for both organic pollutants and heavy metals.

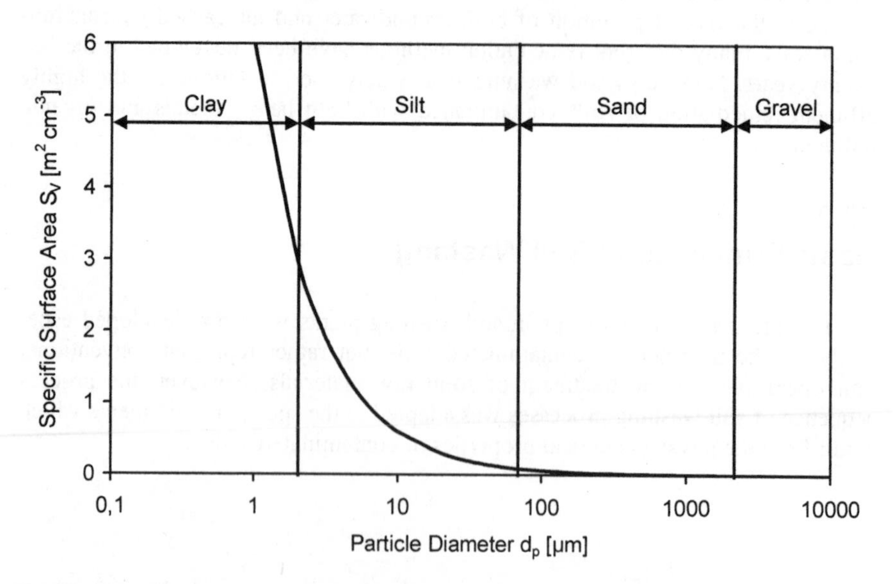

Fig. 27.1. Specific surface area of spherical particles dependent on the particle diameter

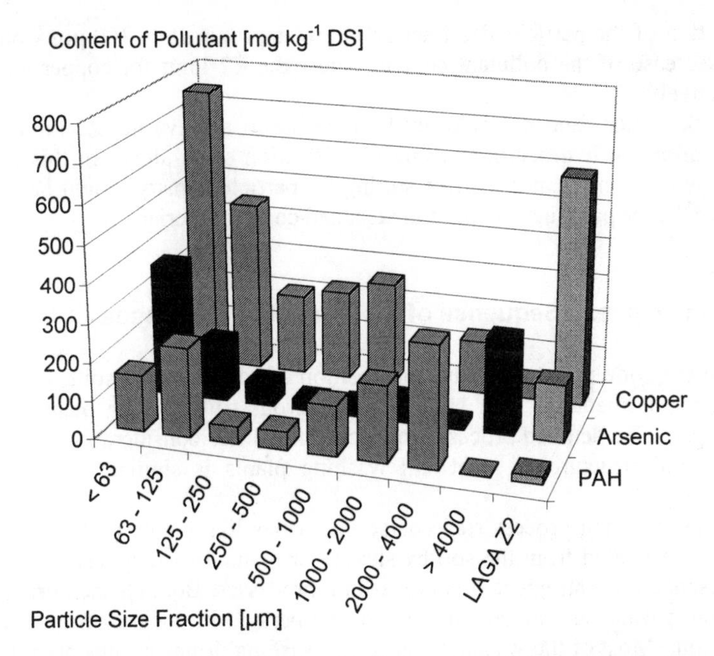

Fig. 27.2. Particle size-dependent distribution of pollutants of a soil from a chemical works site (DS = dry substance, LAGA Z2 = upper limit for reutilisation of cleaned material according to the guideline of the Länderarbeitsgemeinschaft Abfall)

The particle size-dependent contents of some pollutants – copper, arsenic and polycyclic aromatic hydrocarbons (PAHs) – are shown in fig. 27.2, which is an example of a soil from a chemical works site. The bars to the right indicate the pollutant specific limits Z2 according to the German guideline *Anforderungen an die stoffliche Verwertung von mineralischen Reststoffen/Abfällen* of the *Länderarbeitsgemeinschaft Abfall* (LAGA 1994). Provided that cleaned soils are below these limits, they can be reutilised for instance in the construction industry, for noise protection dams or road building.

Concerning copper and arsenic, a distinct accumulation of these pollutants can be observed in the fine fraction of the soil (particle size fraction below 125 µm). The high specific surface area of fine particles can obviously be a factor to account for the high contamination grade of the fine fraction. On the other hand the distribution of PAHs does not seem to be determined by the particle size alone, because in the coarse fraction, high contents of PAHs could be found as well. Further analytical investigations resulted in the conclusion that the PAHs were not only attached to the particle surfaces but also form tar particulates with diameters of several hundred µm which are responsible for the increasing content of PAHs in the coarse fraction.

Based on the particle size-dependent distribution of pollutants first inferences for the process sequences of soil washing plants can be made. On the one hand classification processes are necessary to separate the highly contaminated fine fraction from the less contaminated coarse one. For the example presented, the separation of the particle size fraction below approximately 100 μm would result in a decrease of the pollutant content below the Z2-limit for copper and arsenic respectively.

On the other hand it would not be possible to achieve the Z2-limit for polycyclic aromatic hydrocarbons using classification steps alone. In this case additional separation techniques, like sorting by particle density, might be applied to remove the tar particulates out of the cleaned coarse material.

27.2.2
General Process Sequence of Soil Washing Processes

Due to the wide variation in the composition of contaminated soils, and the associated requirement for their treatment, numerous soil washing plants have been developed. The detailed process sequences may differ considerably. Nevertheless, the general structure of most soil washing plants is similar and is shown in fig. 27.3.

During the first process step coarse impurities like wood, rubbish, metals and foils are separated from the soil by sieving or manual sorting. The following steprepresents the centrepiece of soil washing processes. Both cleaned process water and mains-water are added and the soil is liberated by means of mechanical energy input. Most of the washing plants use washing drums or attritors for the wet-liberation. Until the high-pressure washing process of *NORDAC* (Nord-

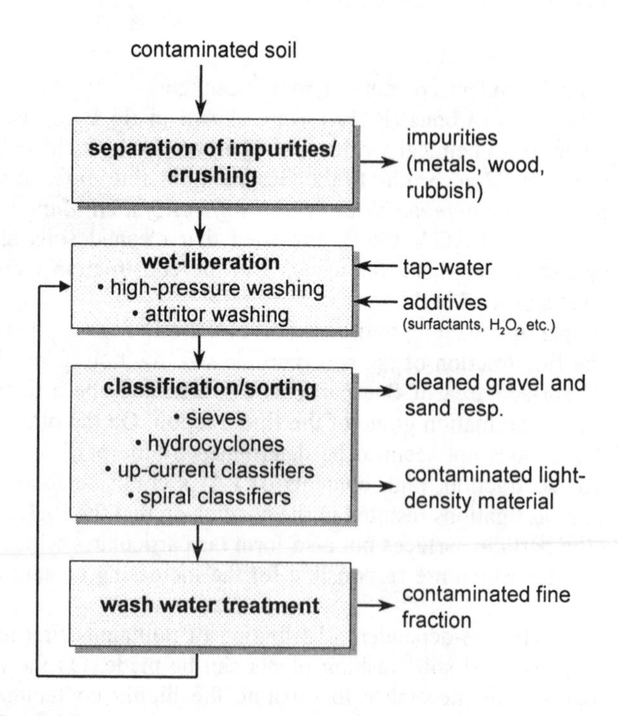

Fig. 27.3. General flow sheet of soil washing processes

deutsches Altlastensanierungszentrum 1991) closed down in 1999, a high-pressure steel jet tube, in which the soil material was treated with water under a pressure up to 500 bars, was used.

However, each of these different devices serve to put high mechanical energy into the soil material to detach the surface-bound pollutants from the coarse soil particles. The detachment of contaminant layers may be supported by means of additives (surfactants, oxidising agents, acids etc.). In subsequent steps of washing procedures the cleaned soil fractions, gravel and sand in particular, are separated from the wash water which still contains the non-treatable fine-grained soil. The separation of cleaned materials is carried out in different classification and sorting apparatus (sieves, hydrocyclones, up-current classifiers, spiral classifiers and jigging machines for instance).

The wash water is treated afterwards by process-specific operations to separate both the fine particles and the dissolved pollutants from the water phase. For that purpose, different techniques like precipitation and flocculation of solid matter, filtration as well as adsorption on activated carbon if necessary, are used. The cleaned wash water can be recycled in the washing process again. The highly contaminated fine fraction has to be disposed of in a suitable manner (e.g. deposition on disposal sites or treatment in incineration plants). Because there is a closed wash water circuit, the washing plants do not have to discharge any waste water outside the plant

27.2.3 Improvement of the Efficiency of Washing Processes

The basic principle of washing processes, as mentioned above, consists of the separation of the highly contaminated particles from particles with a low pollutant content. Therefore washing processes are suitable for the treatment of a broad pollutant spectrum of both organic pollutants and heavy metals. Nevertheless, the possibilities for the improvement of the cleaning efficiency of soil washing processes are limited, especially during the actual operation of the plants. In table 27.1 some procedures for the improvement of the cleaning efficiency are listed.

The primary objective of all procedures, mentioned above, is the reduction of the residual pollutant content in the cleaned soil materials. Simultaneously, attention has to be paid to the yield of the cleaned matter. Besides the reduction of the residual level of contamination, each measure should result in an increase of the yield of cleaned material so that the additional treatment costs can be compensated by the saving in costs for the disposal of contaminated residues.

Table 27.1. Procedures for the improvement of soil washing efficiency

Procedure	Effect
Decreasing of solid mass flow rate	• Increasing of the residence time of solids and therefore increasing of the specific energy input
Addition of additives	• Chemical destruction of pollutants (e.g. by oxidising agents) • Increasing of solubility of pollutants (e.g. by acids, surfactants)
Change of the cut size of classifiers (mostly not practicable without constructive changes)	• Adjustment of the cut size on the particle size-dependent distribution of the pollutants
Increasing of the separation efficiency (only with constructive changes practicable e.g. by arrangement of classifiers in cascades)	• Reducing of the residual pollutant content in cleaned soil materials

27.3 Practice of Soil Washing

Together with the registration of presumably contaminated sites, German enterprises began enthusiastically to develop and construct soil washing plants. Due to the high number of contaminated sites which might endanger the environment, a high demand on soil treatment facilities was predicted for decades (SRU 1990).

At first small mobile plants designed in modular systems were constructed for on-site operation. Later stationary washing plants were constructed to which contaminated soils have to be transported after excavation (off-site plants).

27.3.1 Mobile Soil Washing Plants

The first projects of soil remediation projects were performed with mobile washing plants because of a crucial advantage compared to stationary plants. In accordance with the German regulations, the *Bundes-Immissionsschutzgesetz* (BImSchG 2000) and the *Vierte Bundes-Immissionsschutzverordnung* (4. BImSchV 1999) in particular, the running of mobile soil washing plants was not required to be approved, provided the total operation time did not exceed 12 months (some years before, the time was only 6 months).

Additionally, the measures for the prevention of harmful impacts on the environmental compartments (air, water systems, soil) could be adapted to the local requirements. For example, the sealing of the ground around the washing plants, which should protect the underground from percolation of contaminated leaking water, had to fulfil higher demands for the treatment of soils contaminated with organochlorine substances, than those for oil contaminated soils.

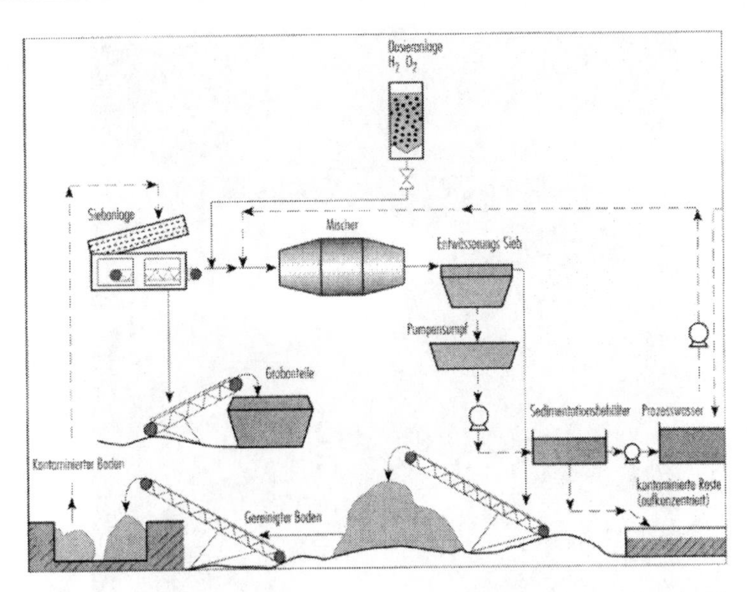

Fig. 27.4. Flow sheet of the *FERMENTA-TEC* soil washing process (reproduced with permission, copyright Peter Neumann Baugrunduntersuchungen GmbH 2000)

However, a considerable disadvantage of mobile soil washing plants has to be taken into account, namely that the German regulations only permit the treatment of soils derived from the local contaminated site. No other soils from other contaminated sites are allowed to be treated in mobile plants. Nevertheless, mobile washing plants proved to be a fairly successful solution.

As an example, the flow sheet of the mobile soil washing plant *FERMENTA-TEC* (Peter Neumann Baugrunduntersuchungen GmbH 2000) is shown in fig. 27.4. The arrangement of the washing plant can be seen in the photography in fig. 27.5.

The first step of the *FERMENTA-TEC* process consists of a screen to separate non-treatable coarse impurities. Subsequently, the soil is conveyed into mixing drums where wash water and some specific additives (e.g. hydrogen peroxide and caustic soda) are added. An oxidation of organic pollutants begins in which temperatures up to 60 °C may occur. After the chemical reaction, the soil is washed and the water phase is separated with the help of a drainage screen. The cleaned soil can then be refilled into the ground provided that the residual pollutant content corresponds to the limits given by the local authorities.

The wash water still contains residuals of organic pollutants, as well as heavy metals and fine particles. The wash water enters sedimentation tanks where yeast and citric acid, if necessary, are added to induce a hydrolytic, oxidative, and reductive disintegration of the substrates. At the same time, non-degradable substances like heavy metals are adsorbed by the yeast. Afterwards all the solid matter is precipitated by means of lime-milk and can be removed out of the wash water by gravity separation in the sedimentation tanks. The cleaned wash water can be led back into the process water circuit. Finally, the highly contaminated

Fig. 27.5. *FERMENTA-TEC* soil washing plant in operation on a contaminated site (reproduced with permission, copyright Peter Neumann Baugrunduntersuchungen GmbH 2000)

residues have to be treated in a suitable manner (e.g. by disposal or incineration). The application of the *FERMENTA-TEC* process has been demonstrated for the treatment of a broad spectrum of contaminated soils in many projects.

27.3.2
Stationary Soil Washing Plants

For the last ten years, greater numbers of stationary soil washing plants were established in different locations in Germany. Even though the running of stationary soil washing plants has to be approved by the responsible authorities in accordance to the German regulations, as mentioned in Chapter 27.3.1, considerable advantages prevail. The higher expenditures for the technical infrastructure of a stationary soil treatment plant compared to the mobile ones can be justified as described below.

On the one hand, extensive measures have to be carried out to prevent harmful emissions (e.g. ground sealing systems, exhaust air decontamination, roofed in storage areas, noise protection). On the other hand, the running of stationary plants is permitted for an unlimited period. However, the most important advantage is that soils from all contaminated sites may be treated in stationary plants. Because of the broad spectrum of soils which are supposed to be treated, the process sequence of stationary soil washing plants is reasonably complex.

As an example, the flow sheet of a stationary soil washing plant is shown in fig. 27.6. It is operated at the soil remediation centre Wittenberge as a joint ven-

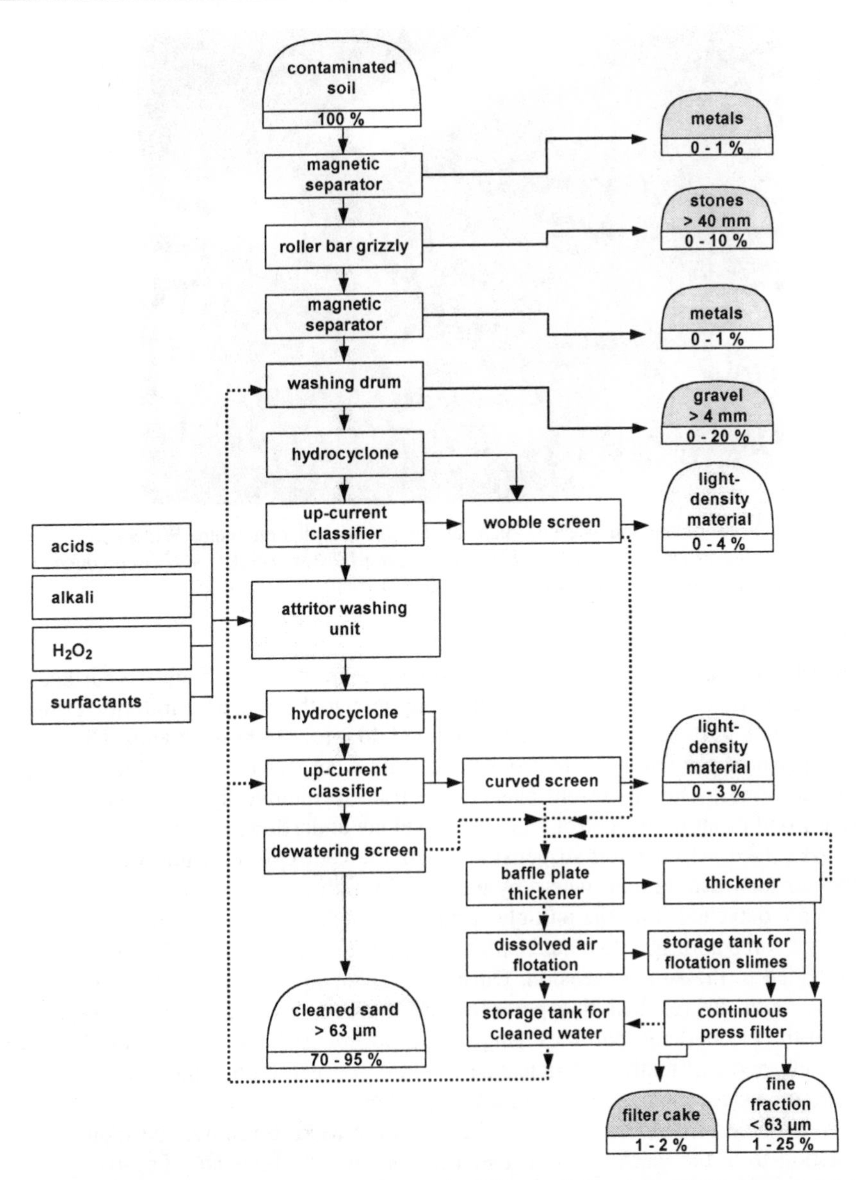

Fig. 27.6. Flow sheet of the soil washing plant at the soil remediation centre Wittenberge, a joint venture of *Eggers Umwelttechnik GmbH* and *Preussag Wassertechnik GmbH* (Reproduced with permission, copyright Preussag Wassertechnik GmbH 2000)

ture by the companies Eggers Umwelttechnik GmbH and Preussag Wassertechnik GmbH.

The dry treatment section of the process consists of two magnetic separators for the removal of ferromagnetic metals, and a roller bar grizzly to separate stones with diameters above 40 mm. The first wet-liberation of the soil takes place in the

Fig. 27.7. View of the soil washing plant at the soil remediation centre Wittenberge, a joint venture of *Eggers Umwelttechnik GmbH* and *Preussag Wassertechnik GmbH* (Reproduced with permission, copyright Preussag Wassertechnik GmbH 2000)

subsequent washing drum, where the pollutants are detached from the surfaces of coarse particles. By means of a vibrating screen the material initially cleaned, gravel with particle diameters between 4 and 40 mm, can be separated. The undersize particles are fed into a hydrocyclone, with a cut size of about 63 μm, to separate the fines. The up-current classifier is used to remove light-density materials like wood or tar particles from the hydrocyclone underflow.

As a next step the soil suspension enters the centrepiece of the washing plant, the attritor washing unit, where by means of a high energy input, layers of pollutants are detached from the particle surfaces. In order to improve the cleaning efficiency some additives like acids, alkaline, hydrogen peroxide, and surfactants can be added to the soil suspension. During the subsequent classification and sorting steps, both the residual fine particles and the detached pollutants are separated from the cleaned soil particles using a hydrocyclone and an up-current classifier. The cleaned soil matter (particle size fraction between 0.063 and 4 mm) is dewatered on a screen and conveyed to a storage area.

A wobble screen and two curved screens are used to remove the light-density fraction from the wash water, which still contains the fines and dissolved pollutants (heavy metals in particular). The solid matter is separated in a baffle plate thickener, whereas heavy metals are removed after precipitation with caustic soda by means of dissolved air flotation. After further thickening both the sludge from the baffle plate thickener and the flotation slime are dewatered on a continuous press filter. The filter cake now contains all the pollutants and has to be disposed of.

The annual throughput of the plant amounts to approximately 40,000 tons. In addition to soils contaminated with heavy metals, soils with different organic pollutants (e.g. mineral oils, polycyclic aromatic hydrocarbons, and polychlorinated biphenyls) can also be treated. A view of the washing plant at its location in Wittenberge is given in fig. 27.7.

27.3.3 Reutilisation of Cleaned Materials – Disposal of Residues

When more extensive soil remediation started in Germany, no legitimate guidelines for the reutilisation of treated soils were available. The local authorities had to set their own standards for the limits of residual pollutant contents in cleaned soil materials. It took several years before some states of the Federal Republic of Germany enacted regulations regarding this, e.g. *Altlasten-Leitfaden für die Behandlung von Altablagerungen und kontaminierten Standorten in Bayern* (N.N. 1991), *Prüfwerte und Verfahrensregeln für kontaminierte Böden* for Hamburg (Schuldt 1990), *Vorläufige Leitwerte für die Sanierung von Grundwasser- und Bodenkontaminationen aus Sicht des Grundwasserschutzes – Mineralölkohlenwasserstoffe,* enacted by the environmental department of the city of Hamburg (N.N. 1990), and *Berliner Liste* (Barkowski et al. 1993).

Nowadays the guideline *Anforderungen an die stoffliche Verwertung von mineralischen Reststoffen/Abfällen*, elaborated by the *Länderarbeitsgemeinschaft Abfall* (LAGA 1994), has become an obligatory regulation for all states of the Federal Republic of Germany. This guideline identifies four different quality classes of soil materials (Z0, Z1.1, Z1.2, and Z2), depending on their residual content of pollutants. In accordance with the quality classes, specific procedures are required to reutilise the soil materials. For instance there are few restrictions for reutilisation of cleaned materials with residual pollutant contents corresponding to the Z0-limits, except for their use for sensitive areas like filling material for children's playgrounds.

On the other hand materials, which fulfil only the Z2-limits, have to be refilled underground with a distance of at least one meter to the highest ground water table, and to then be covered with a watertight layer. Generally, treated soil materials can be utilised for recultivation measures in the mining industry, in earthwork and street construction and also for noise protection dams.

The highly contaminated residues resulting from soil washing have to be disposed in accordance with the regulations for the disposal of hazardous wastes. Mostly the residues are deposited on disposal sites or burnt in waste incineration plants. Nevertheless, depending on their composition, the residues might for instance be used as a substitute raw material in the cement industry.

27.3.4 Chronology of Soil Washing in Germany

In the middle of the Eighties German industry was expecting a growing market for soil remediation techniques, because of the high number of contaminated sites to be treated in order to prevent further impacts on the ecosystem. So, many enterprises started to construct and operate soil washing plants designed in both mobile and stationary modes. Some of the milestones of the development of the soil washing market in Germany are listed in table 27.2.

Table 27.2 Some milestones on the development of the soil washing market in Germany

Year	Status of Soil Washing Activities
1986	• First mobile soil washing plants in operation, e.g. — Harbauer GmbH, Berlin (former site of the Pintsch-Company), — afu GmbH, Berlin.
1991	• First stationary soil washing plant[a] — NORDAC GmbH, Hamburg.
1993	• 11 stationary soil washing plants in operation[b].
1995	• 22 stationary soil washing plants in operation with an annual capacity of about 1.2 mio. tons[c]. • Other 17 plants with a total capacity of 850,000 t a^{-1} in construction and in planning respectively[c].
1998	• Last new stationary soil washing plant for the present — AB Umwelttechnik GmbH, Biburg.
since 1999	• Concentration of soil remediation market by mergers and collaborations respectively. • First close down of soil washing centres, e.g. — close down of the NORDAC soil washing plant, Hamburg, in the end of 1999.

[a] Norddeutsches Altlastensanierungszentrum 1991.
[b] Kielhuber and Schmitz 1993.
[c] Schmitz and Laun 1995.

In the middle of the Nineties the height of the this development has been reached, with 22 stationary soil washing plants operating in Germany, with an annual capacity of about 1.2 M t. At the same time, it became clear that the quantities of soils being supplied to treatment did not increase to the same extent that treatment capacities did. Due to the fact that the initiator of soil contamination can usually not be held responsible for remediation measures, the public (states and municipalities respectively) have to pay for the treatment of contaminated soils. However, because of the high costs for soil remediation and the financial positions of municipalities, a crucially smaller quantity of contaminated soil than expected has been supplied for treatment.

Other means of reutilisation of contaminated soils are still allowed, for example the use as packing material in the mining industry, or as material for the covering of disposal sites. As a result the rivalry in the market became harder, and as a consequence the costs for soil treatment dropped significantly. Nevertheless, in 1995 the average load on all soil washing plants amounted to only 50 % of their capacity (Schmitz and Laun 1995).

In recent years only a few new washing plants were put into operation. In 1999 even one of the most acknowledged soil washing centres, the *NORDAC* plant in Hamburg, was closed down. Since then, a rising tendency of forming monopolies by take-overs and mergers respectively can be observed. Nowadays only a few enterprises which combine different treatment techniques (e.g. biodegradation, thermal treatment and soil washing) are on the market for soil remediation. Nevertheless, the state of the market seems static and no further development of soil washing techniques is expected.

27.4 Costs of Soil Washing

The development of the German market for soil washing is linked tightly to the costs of soil washing processes. The limited amount of contaminated soils supplied to treatment resulted in a severe drop in the cost of soil washing.

Since 1995 the *Bundesanstalt für Materialforschung und -prüfung (BAM),* on behalf of the *Ingenieurtechnischer Verband Altlasten e.V. (ITVA)*, has been compiling annual price lists which contain the average prices for the remedation of three different kinds of soil (various percentages of fines and different contents of mineral oils and PAHs, respectively) by different treatment techniques (ITVA Preisspiegel 1996–1999). The price lists result from annual questionnaires to German enterprises dealing with soil remediation.

In fig. 27.8 the price trends for the soils "type 1" (< 63 μm 30 %, PAHs 2000 mg kg^{-1}, mineral oil 1000 mg kg^{-1}) and "type 3" (< 63 μm 15 %, no PAHs, mineral oil 5000 mg kg^{-1}) are shown for soil washing only.

After a significant price drop between 1995 and 1998, the prices for soil washing for each kind of soil seem to have become more stable at a fairly low level of about 50 % of the initial prices. The reasons for the significant initial higher price level should not be attributed to exaggerated profit margins; they are connected

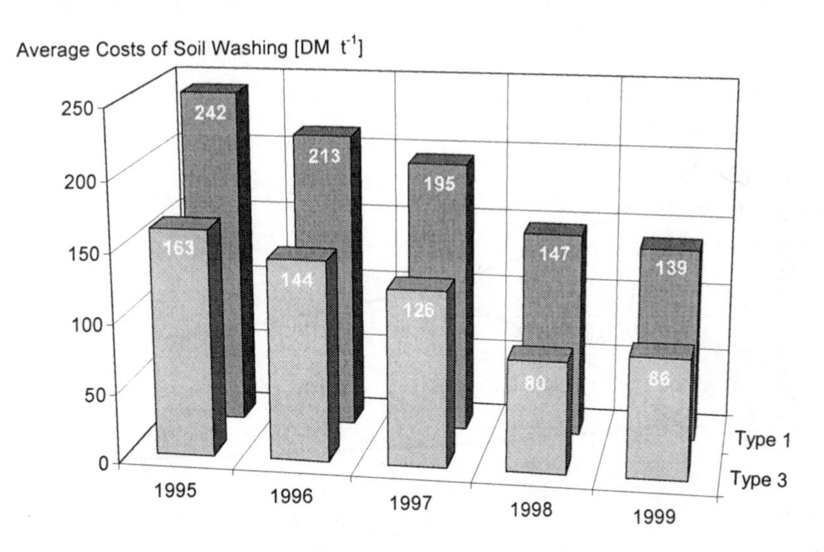

Fig. 27.8. Price levels for soil washing of two different kinds of soil between 1995 and 1999 (ITVA-Preisspiegel 1996–1999)

directly to the cost structure of soil washing.

The costs of soil washing processes are composed of the plant costs, the disposal of residues and the costs for the reutilisation of cleaned materials. Though the cleaned soil materials (gravel and sand) are valuable substitute raw materials in the construction industry, the reutilisation has to be paid by the soil washing enterprises (from approximately 5 up to 25 DM t^{-1}). The costs for the disposal of the residues may be assumed to be 60 up to 250 DM t^{-1} depending on the physico-chemical properties (the disposal costs have also decreased significantly also in recent years). The last cost factor is the fixed costs for operation; consisting of the capital expenditure, personnel expenses and the costs for environmental protection measures. The fixed costs predominantly depend on the throughput of the plant.

The portions of the total costs for soil washing is dependent on the percentage of fines in the soil, as shown in fig. 27.9. The calculation of the cost structure was carried out with costs for disposal of residues, fixed costs and costs for reutilisation of cleaned matter assumed to be 100 DM t^{-1}, 50 DM t^{-1} and 8 DM t^{-1} respectively.

Fig. 27.9 demonstrates clearly the high influence on cost of the percentage of fines forming the residual fraction after soil washing. If the percentage of fines exceeds approximately 30 mass-% , the disposal costs become the main cost factor for soil washing.

The high influence of the percentage of fines on the total washing costs is shown in fig. 27.10. The calculation is based on the same data assumed in fig. 27.9. Fig. 27.10 shows the cost for the washing of a soil containing about 30 % of fines amount to about 100 DM t^{-1}, if the disposal of residues could be achieved for 100 DM t^{-1}. However, the total washing cost increases rapidly up to approximately 170 DM t^{-1}, if 250 DM t^{-1} has to be paid for the disposal of fines.

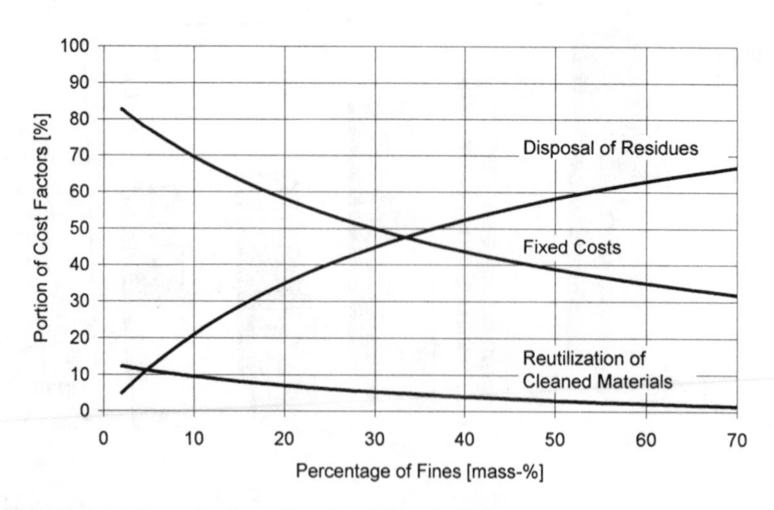

Fig. 27.9. Cost structure of soil washing depending on the percentage of fines (calculated on the basis of the following data: residual water content of residues = 40 mass-%, disposal of residues = 100 DM t^{-1}, fixed costs = 50 DM t^{-1}, reutilisation of cleaned matter = 8 DM t^{-1})

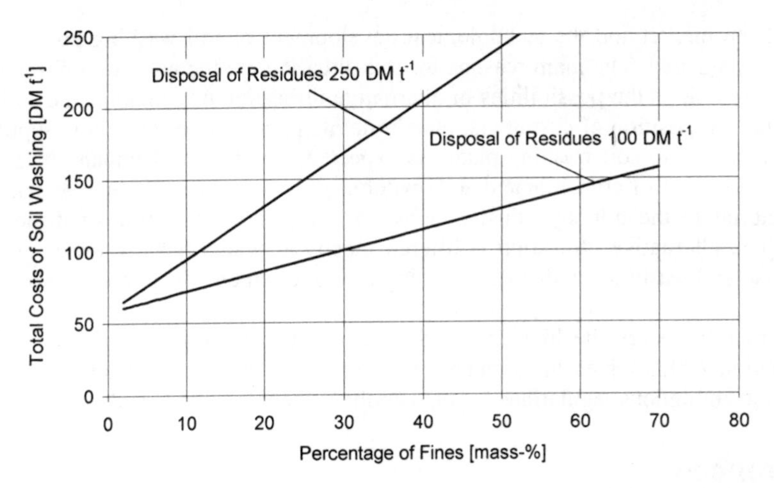

Fig. 27.10. Total costs of soil washing depending on the percentage of fines, calculated for two different cost levels for the disposal of residues (calculated on the basis of following data: residual water content of residues = 40 mass-%, fixed costs = 50 DM t^{-1}, reutilisation of cleaned matter = 8 DM t^{-1})

Regarding the price trends, mentioned in fig. 27.8, nowadays soils can only be acquired for soil washing if both the prices for the disposal of residues and the fixed costs are on a reasonably low level. Due to this fact, it becomes understandable that soil washing plants can only operate profitably if the capital expenditure of the washing plant has already been depreciated. These circumstances may be held responsible for the actual development on the German soil washing market, the closing down of soil remediation centres on the one hand, and the tendency of merging and collaboration on the other hand.

27.5 Summary

Soil washing has proved to be a successful technique for soil remediation for the last 15 years. In soil washing processes the pollutants are not attacked directly. A separation of highly contaminated particles from the less contaminated ones takes place, where conventional separation and sorting techniques are brought into operation. Therefore, soil washing processes are suitable for the treatment of soils contaminated with both organic pollutants and heavy metals. The cleaning efficiency of soil washing processes can be increased to a very high level using high energy inputs in the wet-liberation step on the one hand, and the combination of different classification and sorting techniques on the other hand. Nevertheless, in soil washing processes a fairly high portion of the soil remains as a highly contaminated fine fraction and has to be disposed of.

In spite of the high number of contaminated sites in Germany, the initial hope that soil washing would find a considerable market for German enterprises has not been realised. After a fairly strong boom in the beginning of the Nineties, both the state of the market and the technological development of soil washing processes has now stagnated. The main reasons for this situation are the costs of soil remediation, as well as the possibilities of alternative utilisation of contaminated soils. The financial position of customers, mostly municipalities, does not allow supply of as much soil to soil washing plants as expected. In addition, German law still allows utilisation of contaminated soils without previous treatment (e.g. as packing material in the mining industry or as material for the covering of disposal sites). The alternative utilisation is offered mostly at a significantly lower price level than soil washing, so that soil washing cannot often be now operated profitably.

In order to secure the high standard, and to support further development of remediation technologies in Germany, the legislation needs to be improved to overcome ambiguous regulations for the handling of contaminated soils.

References

Barkowski D, Günther P, Hinz E, Röchert R (1993) Zusammenstellung von Handlungswerten für die Medien Feststoffe, Wasser und Gas. In: Franzius V, Stegmann R, Wolf K, Brandt E (eds) Handbuch der Altlastensanierung, 16th delivery, Chapter 4.1.8. R v Decker's Verlag, G Schenck, Heidelberg, Germany

BImSchG (2000) Gesetz zum Schutz vor schädlichen Umwelteinwirkungen durch Luftverunreinigungen, Geräusche, Erschütterungen und ähnliche Vorgänge (BImSchG) in der Fassung der Bekanntmachung vom 14. Mai 1990 (BGBl I Nr 23 vom 22.05.1990, p 880) zuletzt geändert am 3. Mai 2000 durch Artikel 9 des Zweiten Gesetzes zur Erleichterung der Verwaltungsreform in den Ländern (2. Zuständigkeitslockerungsgesetz). BGBL I Nr 20 vom 10.05.2000, p 6324

BImSchV (1999) Vierte Verordnung zur Durchführung des Bundes-Immissionsschutzgesetzes (4. BImSchV) in der Fassung der Bekanntmachung vom 14. März 1997 (BGBl I Nr 17 vom 20.03.1997, p 504) zuletzt geändert am 23. Februar 1999 durch Artikel 3 der Verordnung zur Änderung der Siebzehnten, der Neunten und der Vierten Verordnung zur Durchführung des Bundes-Immissionsschutzgesetzes. BGBl I Nr 8 vom 26.02.1999, p 186

Franzius V (1995) Remediation Methods. In: Ullmann's Encyclopedia of Industrial Chemistry, Vol B7, Chapter 7.1, VCH Verlagsgesellschaft, Weinheim, New York, Basel, Camebridge, pp 695–707

ITVA-Preisspiegel zur Dekontamination von Boden
altlastenspektrum 5 (1996) 5: 248
altlastenspektrum 6 (1997) 4: 191–192
altlastenspektrum 7 (1998) 5: 310–311
altlastenspektrum 5 (1999) pp 311–312

Kielhuber G, Schmitz HJ (1993) Bodenbehandlungszentren: Die Jagd nach dem Boden hat begonnen. TerraTech 3/1993: 46–57

LAGA (1994) Anforderungen an die stoffliche Verwertung von mineralischen Reststoffen/Abfällen Hamburg. 1.3.1994 und 7.9.1994, Mitteilung der Länderarbeitsgemeinschaft Abfall (LAGA) 20/1 und 20/2, Erich Schmidt Verlag, Berlin, Germany

N.N. (1990) Vorläufige Leitwerte für die Sanierung von Grundwasser- und Bodenkontaminationen aus Sicht des Grundwasserschutzes – Mineralölkohlenwasserstoffe – Stand September 1990. Freie und Hansestadt Hamburg, Umweltbehörde, Amt für Umweltschutz, Gewässer und Bodenschutz, Senatsdrucksache

N.N. (1991) Altlasten-Leitfaden für die Behandlung von Altablagerungen und kontaminierten Standorten in Bayern. eds Bayrisches Staatsministerium für Landesentwicklung und Umweltfragen, Druckerei Lehmann, München

Norddeutsches Altlastensanierungszentrum (1991) Errichtung und Betrieb eines Bodenrecyclingzentrums. Praxis der Altlastensanierung, Band 2, Economica Verlag GmbH, Bonn

Peter Neumann Baugrunduntersuchungen GmbH (2000) FERMENTA-TEC – Das Bodenreinigungsverfahren. Brochure of Peter Neumann Baugrunduntersuchungen GmbH, Eckernförde, Germany

Preussag Wassertechnik GmbH (2000) Brochure "Bodensanierung" and personal information given by Mr Dickhoff, Preussag Wassertechnik GmbH, Niederlassung Zwingenberg, Zwingenberg, Germany

Schmitz HJ, Laun M (1995) Die Jagd nach dem Boden geht weiter. TerraTech 3/1995: 34–47

Schuldt M (1990) Prüfwerte und Verfahrensregeln für kontaminierte Böden. Umweltbehörde Hamburg. In: Arendt F, Hinsenveld M, van den Brink WJ (eds) Altlastensanierung '90, Vol II, Kluwer Academic Publishers, Dordrecht, Boston, London, pp 235–236

SRU (Der Rat von Sachverständigen für Umweltfragen) (1990) Altlasten. Sondergutachten Dezember 1989, Moetzler-Poeschel, Stuttgart, Germany

Umweltprogramm der Bundesregierung (1971) Bundestags-Drucksache 6/2710

28 Ways to Improve the Efficiency of Soil Washing

J. Werther, O. Malerius, J. Schmidt
Chemical Engineering I, Technical University Hamburg-Harburg, Denickestr.15, 21073 Hamburg, Germany

28.1 Introduction

The physical treatment with respect to soil washing is an established technique for cleaning contaminated soils. Its basic principle is the removal of the fine particle fraction from the soil material, in which the contaminants are predominantly concentrated. The goal of the separation process of the fine fraction is to achieve a lowly contaminated coarse fraction that is economically reusable at a yield which should be as high as possible. Industrial soil washing plants basically consist of a wet liberation step and a classification unit. The wet liberation step, which is carried out by application of mechanical energy through impact, is used to transfer the contaminants from the coarse into the fine particle fraction, and to concentrate the contaminant in the fine part of the solids input. The highly contaminated fines are removed in the classification step by means of screens and hydrocyclones, for example.

The cleaning principle of collecting the contaminants in the fine particle fraction does not work efficiently for silty or loamy soils, because the yield of cleaned solids mass is too low. Further problems arise for contaminants that are not displaceable to such an extent that the concentration of the contaminant in the coarse particle fraction meets the required threshold limits (Tittel et al. 1995; Neeße et al. 1991). In these cases, additional process steps are required to achieve both a lower contaminant concentration in the cleaned fraction, and a higher yield of cleaned soil, by concentrating the contaminants in a very narrow particle fraction. These further treatment steps require either a sharper separation at low particle sizes (e.g. cut sizes below 20 µm), or a process which separates according to particle properties other than size. Furthermore, the liberation step should be as intense as necessary to enhance the change of the contaminant distribution.

In the research work carried out at TUHH's Chemical Engineering Department I in the 12 years of SFB, several treatment methods were investigated to improve the efficiency of soil washing. Attrition and high pressure water jet liberation was investigated to improve the change in the contaminant distribution. Flotation as a method that separates particles according to different surface properties was tested for the hydrocyclone underflow to improve the mass yield of the soil washing process. More recently, the hydrocyclone overflow was processed with two different classification methods which work in the particle range below 20 µm, namely the hydro deflector wheel classifier and the centrifugal fluidised bed counter-current classifier.

28.2 Wet Liberation

The wet liberation step is carried out to remove contaminants and highly contaminated fines from large particles and in this way to concentrate the contamination in the fines fraction. Three mechanisms predominantly cause the shifting process into the fine fraction, namely abrasion, which means that contaminants and fine particles are washed off from larger particles; destruction of agglomerates and disintegration of particles. Different methods can be applied to realise the liberation step technically: hydraulic transport for dredged sludges, impact stress in high pressure water jets, attrition in rotating drums and attrition with agitators for soils.

An attrition drum is used on the industrial scale in the LURGI-DECONTERRA® process (Hankel et al. 1992) to concentrate the contaminants in the fine particle fraction. This attrition drum consists of a horizontal tube with special lining and inserts. The contaminant distributions of soils polluted with mineral oils were altered effectively by treating the soil according to this process. After the treatment, the contaminants were found to be concentrated in the fines fraction below 80 µm.

In the authors' group, wet liberation was carried out in an attrition cell, and with a high pressure jet liberation unit for soils contaminated with heavy metals. It was investigated whether the shifting of the contaminant distribution exceeds the shifting of the size distribution of the solids material.

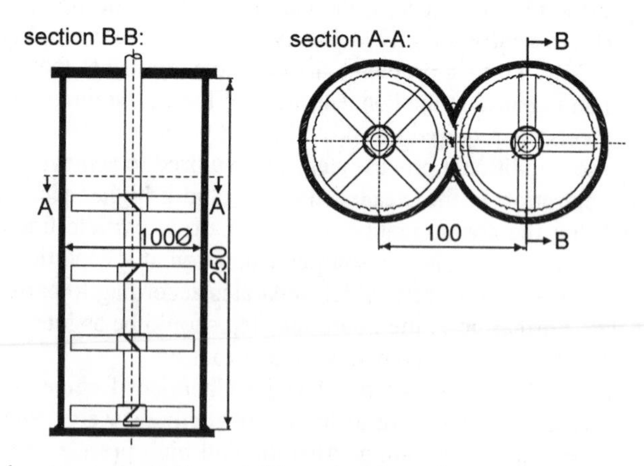

Fig. 28.1. Laboratory scale attrition cell (dimensions in mm)

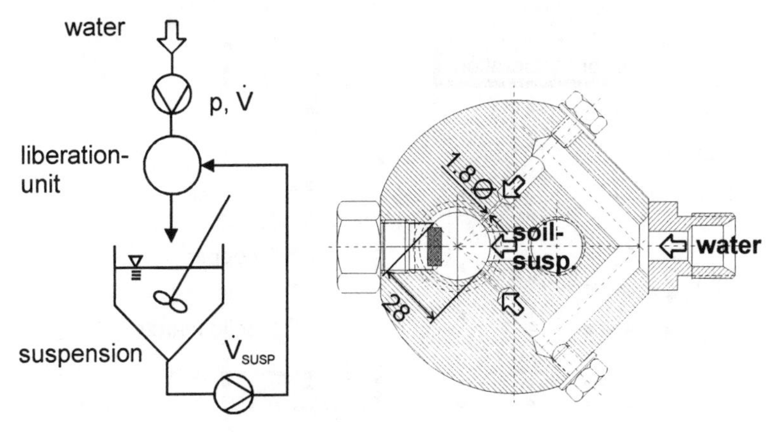

Fig. 28.2. High pressure liberation unit (dimensions in mm)

28.2.1
Experimental Facilities

The wet liberation step was investigated for five different soils contaminated with heavy metals. An attrition cell (fig. 28.1) was compared with a high pressure pump liberation unit (fig. 28.2) for varied input of energy. The attrition cell consists of two stirrers counter-rotating in a steel vessel to induce shear stress. The heart of the high pressure liberation unit consists of two high pressure jets which are focused into the suspension to accelerate it against a deflector plate.

28.2.2
Results of the Wet Liberation Investigation

Fig. 28.3 depicts cumulative mass particle and contaminant distributions before and after wet liberation in the attrition cell. After the wet liberation step, about 90 wt-% of the contaminants are concentrated in the particle fraction below 32 μm which contains only 27 % of the total mass of the solids. The shifting of the contaminants obviously dominates the shifting of the solids which indicates a successful washing operation.

Further experiments show that the extent of the shifting depends on the energy input and the particular way of energy dissipation. The experiments indicate that the higher the energy amount introduced into the suspension, the higher is the extent of the shifting of contaminants into the fines fraction. In addition to this, the way of dissipation affects the particle mass shifted. It was found that for equal values of the energy input, attrition works more efficiently than high pressure jet liberation. An explanation could be that the residence time of the particles in the attrition unit is longer than in the high pressure liberation unit (Venghaus et al. 1997).

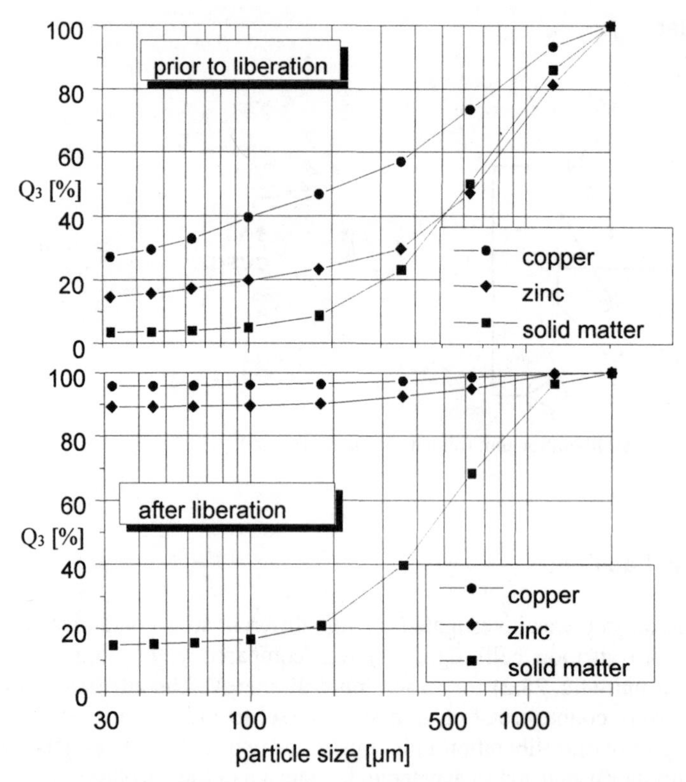

Fig. 28.3. Particle and contaminants mass distributions prior to and after wet liberation in the attrition cell (soil AS, duration of attrition 240 min, Q_3 is the cumulative mass distribution)

28.3 Flotation

28.3.1 State of the Art

Flotation processes were originally developed in the mineral processing industry for the separation of valuable minerals from tailings (Arbiter 1985). In this field great experience with different flotation agents has been gained for the flotation of metal ores (Yarar 1988). The use of flotation in soil washing is seldom discussed in the scientific literature. Pearl and Wood (1993) mention the possibility of improving soil washing for soils contaminated with mineral oils by means of flotation. They report flotation experiments carried out on a soil excavated from a former coking plant. They achieved a mineral oil removal of 95.9 wt-% with a remaining mineral oil concentration of 1400 mg (kg dwt soil)$^{-1}$ in the cleaned material, which was 68.1 wt-% of the original soil mass. The objective of the research work carried out in SFB 188 was to improve the efficiency of the soil washing process for soils contaminated either with mineral oil or with heavy metals by means of flotation.

28.3.2 Some Fundamentals of Flotation

The basic principle of flotation is the separation of particles according to their different surface properties. Hydrophobic particles are caught by rising air bubbles whereas hydrophilic particles remain in the pulp. The flotated particles are removed at the surface of the suspension in a froth which is stabilised by the addition of frothing agents (Kihlstedt 1974). The wet tability of the particles can be altered by adding conditioning agents to enhance the effectiveness of the flotation process. "Collectors" are added to render the particles, which are intended to be removed from the pulp, with a hydrophobic surface. "Depressants" enhance the wettability of particles that shall remain in the pulp. The chemical nature of particle surfaces can be altered using "modifying agents" in order to promote or inhibit the collector action (Abdo and Darwish 1991).

In the field of physical soil washing processes, the use of flotation is limited by the particle size of the material. Particles exceeding a certain particle size cannot be transported upwards by rising air bubbles, and particles in the micron range disturb the flotation process by unselectively adsorbing the flotation reagents due to the large specific surface area of the particles. The flotation step will therefore be used for the soil material, after both the coarse and the finest fraction have been removed by screens, and hydrocyclones respectively. In the present work, the underflow of a hydrocyclone cascade (fig. 28.4) has been used for the flotation experiments. Both mineral oil-containing soils and soils contaminated with heavy metals were treated.

The following characteristic quantities are commonly used to describe the cleaning results achieved by flotation: The solids removal β_S is that portion of the solids in the pulp which is entrained into the flotate fraction:

$$\beta_S = \dot{m}_{fl} / \dot{m}_f \tag{28.1}$$

where $\dot{m}_{fl}$ denotes the solids mass flow rate in the flotate, and $\dot{m}_f$ denotes the solids mass flow rate in the feed. The contaminants removal β_C is defined in the respective way:

$$\beta_C = \dot{m}_{C,fl} / \dot{m}_{C,f} \tag{28.2}$$

where $\dot{m}_{C,fl}$ describes the contaminant mass flow rate in the flotate and $\dot{m}_{C,f}$ describes the contaminant mass flow rate in the feed.

The concentration ratio α is the ratio of the contaminant concentration c_{fl} in the flotate, to the contaminant concentration in the feed c_f. It is equal to the ratio of the contaminant removal β_C, to the solids removal β_S.

$$\alpha = c_{fl} / c_f = \beta_C / \beta_S \tag{28.3}$$

The concentration ratio α is significant for the selectivity of the flotation process. For a selective separation and entrainment of the contaminants in the flotate, α must be larger than 1, and should be as large as possible. Finally, the decontamination ratio r_d describes the decrease of contaminants concentration from the feed material to the cleaned tailings:

$$r_d = (c_f - c_t)/c_f \tag{28.4}$$

where the contaminants concentration in the cleaned tailings is denoted by c_t. The theoretical maximum decontamination ratio $r_d = 1$ should be approached as much as possible. A decontamination ratio above 0 indicates a cleaning of the soil material and an enrichment of the contaminant in the flotate.

28.3.3 Application of the Flotation Principle to Solids Contaminated with Mineral Oil

28.3.3.1 Experimental

Fig. 28.4 shows the laboratory-scale soil washing facility of the Technical University Hamburg-Harburg (TUHH). It consists of a wet liberation step, a screening unit and a hydrocyclone cascade. The flotation is additionally used to clean the hydrocyclone underflow. Different soil materials with mineral oil concentrations

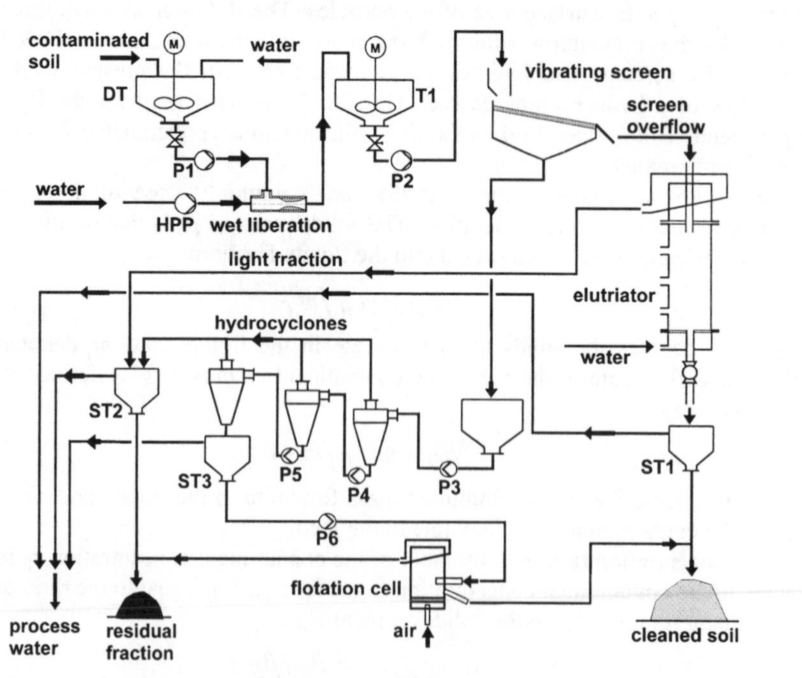

Fig. 28.4. Soil washing test plant at TUHH (DT: dispersing tank; HPP: high pressure pump; P1–P6: pumps; T1, T2: tanks; ST1, ST2, ST3: sedimentation tanks)

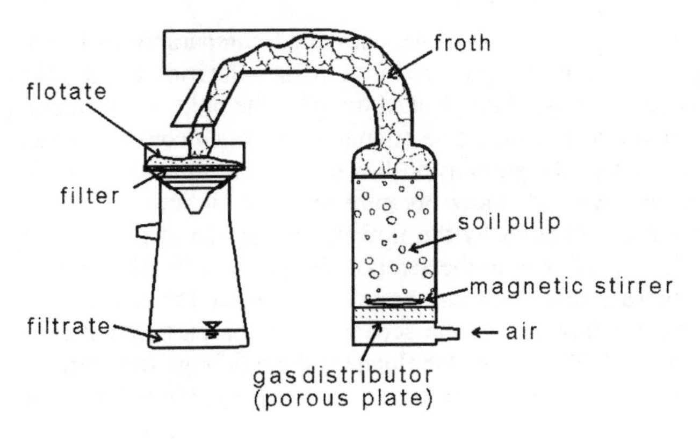

Fig. 28.5. Laboratory-scale flotation cell

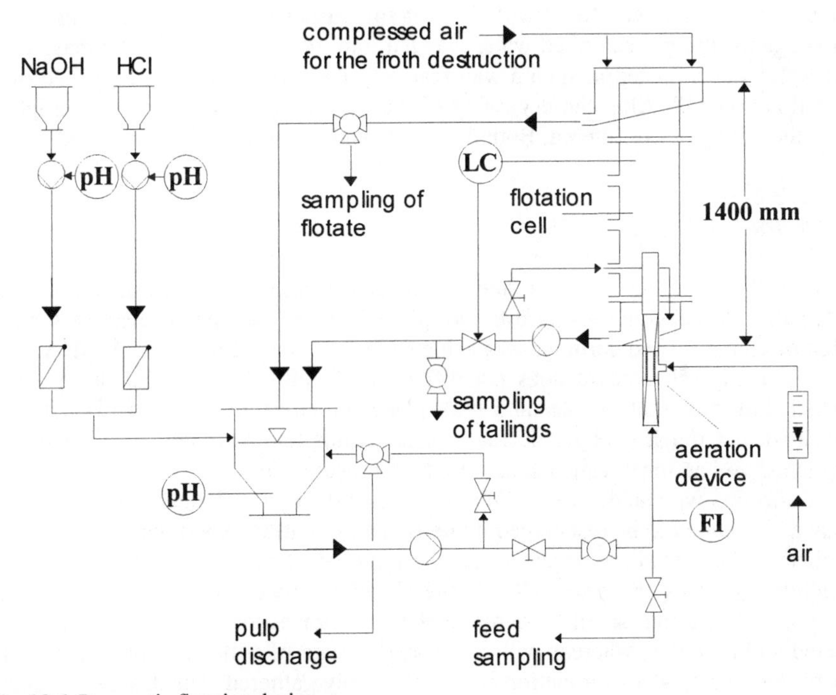

Fig. 28.6. Pneumatic flotation device

between 10,000 and 100,000 mg (kg dwt soil $)^{-1}$ and different particle size distributions were treated in this facility.

The flotation experiments were carried out with three different flotation units. A modified gas washing bottle was used as a laboratory-scale flotation cell for screening experiments (fig. 28.5). Pilot-scale flotation experiments were carried out in a pneumatic flotation facility (fig. 28.6) and a pressure flotation unit.

The pneumatic flotation unit was designed comparably to technical flotation units. The suspension was prepared in a vessel, flotation agents were added and the pH was adjusted by adding NaOH or HCl. The pulp was then pumped into the flotation cell via the pneumatic aeration device, which consisted of a porous metal pipe, through which the suspension flowed into the flotation cell. The air outside the porous pipe was set under pressure to force air into the pulp. The air was sheared into small bubbles by the flowing suspension at the inner surface of the pipe. The flow conditions in the aeration device were highly turbulent in order to attain an intense contact between the soil particles and the air bubbles, whereas in the subsequent calming area, the separation of the particle-bubble agglomerates from the pulp took place. Both the flotate and the tailings were returned to the feed vessel. Samples were taken from both the tailings and the flotate through sampling valves.

The pressure flotation unit was run batchwise, and consisted of a flotation cell and a pressure vessel, inside of which the water was enriched with air. After the flotation cell had been filled with 2.8 l of soil suspension conditioned with flotation agents, the gas-saturated water was fed into the flotation cell. The mass flow of water was adjusted in such a way that only the forming froth was spilled over the flotation cell. After the degassing of the water, the flotation cell was emptied and the tailings were filtered. Both the flotate and the tailings were analysed.

28.3.3.2 Results

One of the most important parameters characterising the flotation medium is the pH value. A screening test at different pH values showed that a selective separation of contaminated particles was achievable only for pH below 4. At pH values above 4, the oil removal does not differ significantly from the solids removal, which indicates that no cleaning took place. Flotation media with pH below 4 resulted in a much stronger decrease of the solids removal than the oil removal, which shows an improving selectivity of the process.

Collectors were added to the pulp to enhance the hydrophobia of those particles that are intended to be discharged in the flotate. Flotation experiments at increased collector concentrations showed that the removal of both soil and oil increased slightly, whereas the selectivity of the flotation process became worse. These results indicate that small amounts of the collector adsorb onto highly contaminated soil particles, whereas with increasing dosage of collectors, more and more particles with low contamination are unselectively gathered. The dosage of collector should therefore not exceed a certain amount that is necessary for the sufficient removal of the oil, because higher concentrations will lead to both a lower selectivity of the process, and a higher loss of clean soil material (Wilichowski et al. 1998).

A comparison of pressure flotation and pneumatic flotation showed better cleaning results for pneumatic flotation. Not only twice the decontamination ratio of pressure flotation was achieved, but also the selectivity ($\alpha = 2.07$) was higher than with pressure flotation ($\alpha = 1.5$) (Wilichowski et al. 1995).

Fig. 28.7 shows a mass balance of silty sand cleaned in a conventional soil washing process with an additional flotation step for the hydrocyclone underflow.

Without the flotation unit, 90 % of the feed solid matter is found in the cleaned fraction with an oil content of 1,907 mg (kg dwt soil)$^{-1}$. Such an oil concentration is too high for the "cleaned" fraction to be acceptable. However, with an additional flotation step, the oil concentration in the coarse fraction comes down to 467 mg (kg dwt soil)$^{-1}$ while a solids recovery of 85 % is still achieved. This example illustrates that with an additional flotation step, the decontamination ratio can be increased significantly, while the solids loss remains low (Wilichowski et al. 1998).

In summary, the integration of the flotation step into the soil washing test facility leads to a significantly lower contaminant concentration in the cleaned soil, whereas the solids loss is very low.

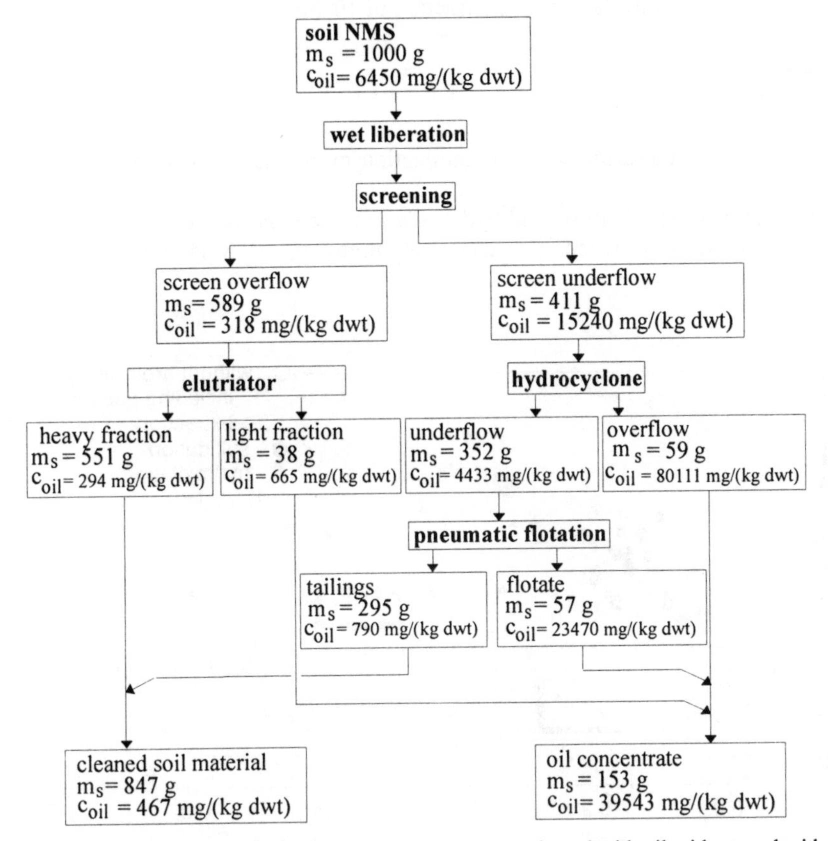

Fig. 28.7. Mass balance of the washing of a silty soil contaminated with oil without and with an additional flotation of the hydrocyclone underflow

28.3.4
Application of the Flotation Principle to Soils Contaminated with Heavy Metals

28.3.4.1
Experimental

Different soils contaminated with heavy metals (e.g. copper, zinc) were treated in the soil washing test plant at TUHH. Laboratory-scale flotation experiments were carried out to investigate various flotation agents and conditions. The results were used to calculate the possible decontamination ratio and solids recovery of the whole soil washing process. The results presented below were gained for a soil contaminated with 3200 mg (kg dwt soil)$^{-1}$ zinc. The soil was charaterised by a sand fraction of 80 wt-% and a clay fraction of 10 wt-%.

28.3.4.2
Results

Fig. 28.8 shows the achieved decontamination ratio r_d as a function of the solids removal β_S.

The application of sulfide and hydroxide specific collectors during the flotation was dissatisfactory when no chemical pretreatment was practised (these

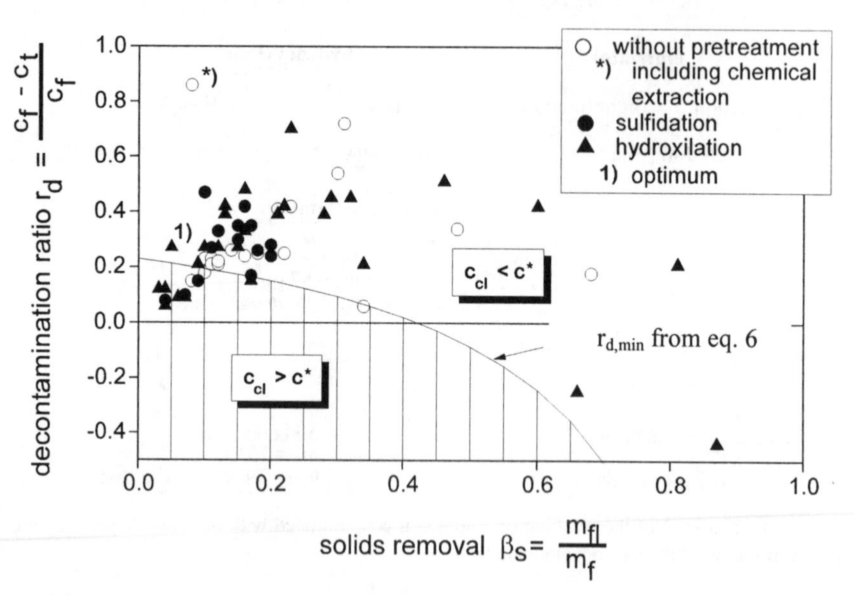

Fig. 28.8. Flotation results of a zinc contaminated soil residue (c_{so} = 379 mg (kg dwt soil)$^{-1}$, c_{hu} = 3,179 mg (kg dwt soil)$^{-1}$, m_{so} = 419 g, m_{hu} = 451 g, c^* = 1,500 mg (kg $_{dwt}$ soil)$^{-1}$ from LAGA Z.2) with and without chemical pretreatment, determination of the optimal flotation conditions

experiments are depicted as open dots in fig. 28.8). Either the decontamination ratio was too low or the solids loss was too high. The conclusion that the zinc exists in different chemical compounds in the soil and not in one particular form, as in an ore, was confirmed by a sequential extraction (Förstner and Calmano 1982). The zinc was found to be exchangeable (12.8 wt-%), carbonatic (32.5 wt-%) and easily reducible (51.5 wt-%), respectively.

During further experiments, the soil was therefore chemically pretreated prior to the flotation experiment (Eberius 1989). The soil was either hydroxylated or sulfidated to achieve a uniform chemical bonding of the zinc, thus allowing a better flotation with the suitable agents. Fig. 28.8 shows that chemical pretreatment can improve the flotation results significantly. By sulfidation or hydroxylation prior to the flotation of the soil, a higher decontamination ratio can be achieved with the same solids removal, or the same decontamination ratio may be obtained with a smaller solids removal in the flotate, respectively.

In practice, the flotation step must improve the overall cleaning result to such an extent that the soil flotation becomes economically feasible. The minimum decontamination ratio $r_{d,min}$ that must be achieved in the flotation step to meet the legal requirements indicated by a certain limiting concentration c^* in the cleaned material, may for the washing process depicted in fig. 28.9 be calculated from mass balances for the flotation, and for the whole soil washing process (Venghaus et al. 1998),

$$r_d \geq r_{d,\min} \tag{28.5}$$

with

$$r_{d,\min} \overset{!}{=} 1 - \frac{c^*}{c_{hu}} - \frac{m_{so} \cdot \left(c^* - c_{so}\right)}{m_{hu} \cdot c_{hu}} \cdot \frac{1}{1-\beta_s} \tag{28.6}$$

In equation 28.6, c_{so} is the contaminant concentration of the sieve overflow, and c_{hu} denotes the contaminant concentration of the hydrocyclone underflow. m_{so} and m_{hu} are the respective masses of soil.

Equation (28.6) allows the calculation of the minimum decontamination ratio $r_{d,min}$ from the solids removal β_s, depending on the limiting contamination concentration c^* in the cleaned fraction. The boundary has been plotted together with the experimental results in fig. 28.8. Experimental points and the respective flotation operating conditions above the curve lead to zinc concentrations c_{cl} in the cleaned soil below the limiting concentration c^*, which was in this case the Z2 limit of the German LAGA list. The cost-effective optimum among these "good" flotation results is the one that shows the smallest soil removal β_s and thus the largest yield of cleaned material. The mass balance of the soil washing test plant under these conditions is shown in fig. 28.9

Without the flotation step, 87 wt-% of the feed is found in a "cleaned" fraction consisting of the sieve overflow and the hydrocyclone underflow, however the zinc concentration exceeds the LAGA Z2 limit. With the additional flotation step, the zinc concentration in the cleaned soil fraction, at1,328 mg (kg dwt soil)$^{-1}$, meets the LAGA Z2 limit of 2,000 mg Zn (kg dwt soil)$^{-1}$, and 85 wt-% of the soil material can thus be reused. This example shows that the flotation process can be tailored down to achieve cleaning results close the concentration limit by choosing

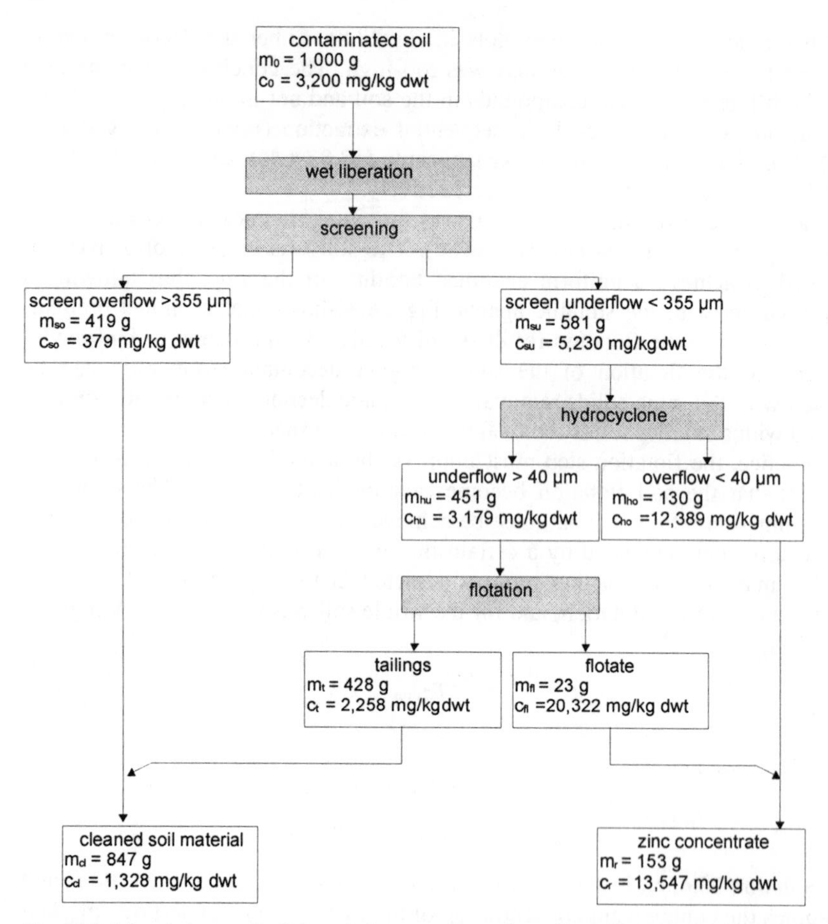

Fig. 28.9. Mass balance of solids and zinc over the entire soil washing process including a flotation step

proper flotation adjustments, with the objective of yielding a reusable mass that is as large as possible (Venghaus et al. 1998).

In summary, the flotation of soils contaminated with heavy metals can be tailored down to the appropriate flotation conditions to enhance the cleaning efficiency of the soil washing process significantly.

28.4 Increasing the Efficiency of Soil Washing by Improved Classification in the Fines Range

As has been described in the introduction, a general idea in soil washing is the inhomogeneous distribution of the contaminants over the particle size range of contaminated soil. This idea is based on the assumption that the contamination is located on the surface of the particles. This will certainly be valid in the case of contamination with mineral oils. The relationship between surface area and volume, for the case of a sphere with diameter d, is given by

$$\frac{\text{surface area sphere}}{\text{volume sphere}} = \frac{\pi \cdot d^2}{\pi / 6 \cdot d^3} = \frac{6}{d} \tag{28.7}$$

It can be seen that for decreasing particle diameters, the ratio of surface to volume increases. This means that for contaminants located on the surface, the relative contamination compared to the particles volume increases. The result is an accumulation of the contamination in the fine particle range. The basic concept of physical soil cleaning is therefore to separate the contaminated soil into a coarse fraction, with no or very low contamination, and a highly contaminated fines fraction to be further processed. All physical soil cleaning plants utilise therefore some sort of classification unit in the process.

In the soil washing plant at TUHH, a sieve is used to remove the sand fraction with particles of size larger than 315 μm. For further classification, a hydrocyclone is used. The basic function of a hydrocyclone is to separate solids of different sedimentation velocities in the centrifugal field. The suspension is fed into the apparatus tangentially, which causes a swirling flow pattern. Due to the centrifugal forces, the particles tend to settle at the circumference of the apparatus while the radial flow to the center of the apparatus induces a drag force acting on the particles. Depending on the particles' sedimentation velocity, which is influenced by size, shape and density, either the drag force or the centrifugal force dominate, and thus particles either report to the fines fraction or the coarse fraction. Since the hydrocyclone does not require any moving parts, it is extremely reliable and combines low operating costs with high relative throughput for its volume. Despite these benefits, hydrocyclones have a major drawback when used for separation in the fines range. In order to obtain the required high centrifugal forces to allow settling of fines, the cyclone geometry has to be rather small to ensure high flow velocities. Besides a much higher tendency of these small apparatuses to plug up, the characterisation of the separation for such a hydrocyclone reveals a poor separation performance in the fines range.

Such a separation is usually characterised by the separation efficiency curve T(x), also called the Tromp curve (Leschonski 1988), which represents the separation grade as a function of particle size. T(x) describes the mass of particles of size x reporting to the coarse fraction, which is related to the mass of particles of this size in the feed stream.

Fig. 28.10 shows the typical parameters of a Tromp curve and the effect of separation sharpness on the composition and total mass of the obtained fractions of contaminated soil. The upper chart shows the mass density distribution q_3 of the feed, the fines and the coarse fraction. The fines and coarse particles' distributions have been multiplied with their respective mass fractions, such that the mass fraction of fines f and the mass fraction of coarse particles g can be read from the area below the individual distribution curves, if the area below the feed particle size distribution curve is set to one. Since the separation efficiency for a particle size class x is given by

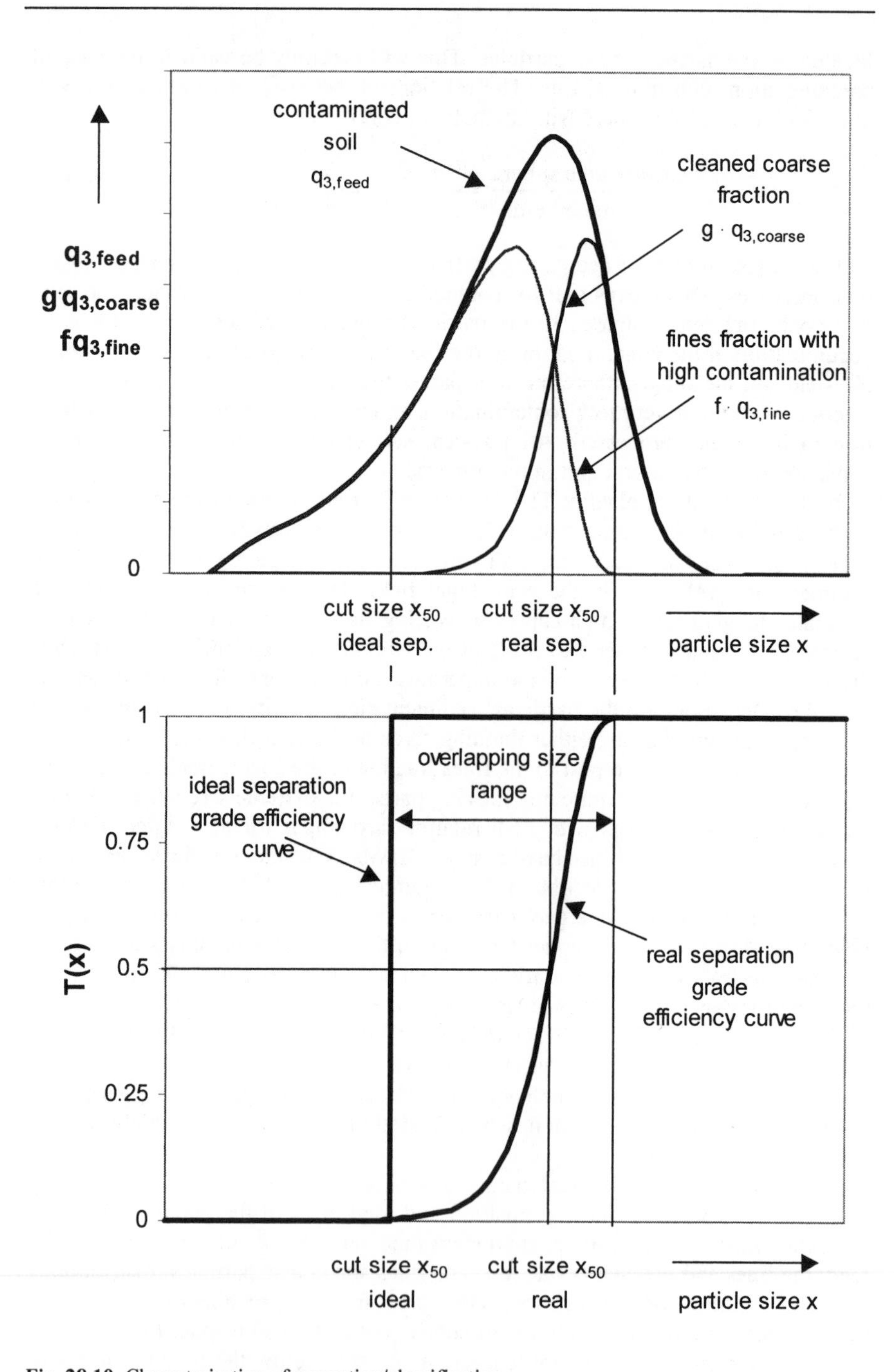

Fig. 28.10. Characterisation of separation/classification

$$T(x) = \frac{g \cdot q_{3,coarse}}{q_{3,feed}} \tag{28.8}$$

The fractional separation efficiency can be easily read from this plot and the Tromp curve shown also in fig. 28.10 may be constructed accordingly.

It can be seen that for a certain range of particle sizes, the separation is not definite, i.e. the particles only have a certain chance of getting into the fines or the coarse fraction, respectively. The lower chart of fig. 28.10 shows the Tromp curves or separation efficiencies for an ideal and a non-ideal separation. It can be seen that an ideal separation characterised by the ideal Tromp curve results in two distinct fractions, with no overlapping size ranges. This means that all particles smaller than a certain size report to the fines fraction, while all bigger particles report to the coarse fraction. The size that represents this decisive boundary is generally called 'cut size' and is an important parameter for characterisation of a separation. The wider the range of overlapping sizes, the less efficient is the separation. A non-ideal separation is generally characterised by an S-shaped Tromp curve. The cut size in this case is defined as the particle size at which the Tromp curve reaches the 50 % value.

The sharpness of separation is important for the cleaning of contaminated soil, as the misled fines with their high contamination would increase the contamination level of the clean coarse fraction. Depending on the sharpness of the separation, the cut size has to be set in practice more or less far in the coarse region to avoid highly contaminated fines reporting to the coarse stream. This means that in turn a large amount of clean coarse particles reports to the highly contaminated fines fraction, thus increasing the total mass and the required processing capacity and costs for further treatment of the fines.

A typical separation efficiency curve, obtained for a hydrocyclone of the soil

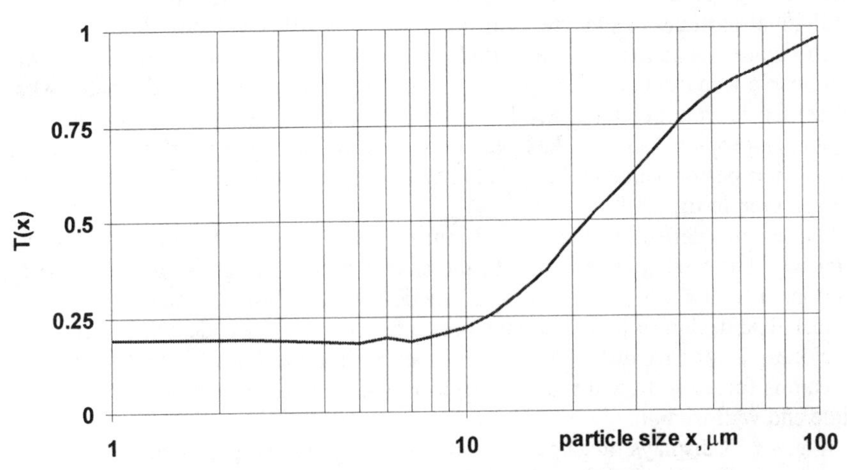

Fig. 28.11. Tromp curve for hydrocyclone in soil washing plant at TUHH

washing plant at the TUHH, can be used to explain the drawback of hydrocyclones for applications in soil washing:

It can be seen that this Tromp curve indicates a rather low sharpness of separation. The hydrocyclone shows a wide range of overlapping sizes between 5 and 150 µm which, according to the effects described before, leads to a poor cleaning efficiency of the process with respect to contaminated soil. The reasons for this low performance are low retention times of the particles inside the apparatus due to the high throughput required to achieve the high velocities, as well as friction effects and shortcut streams inside the apparatus. A typical attempt to improve the separation performance of hydrocyclones therefore involves setting up a hydrocyclone cascade. This means that additional hydrocyclones are installed in sequence in order to post-classify the overflow or underflow, respectively, of the previous cyclone, and thus enhance the total separation sharpness. This can improve the overall separation performance to a limited extent, but in general, still does not lead to sufficient results with respect to the special demands arising from the contamination distribution in contaminated soils.

For the field of cleaning of contaminated soil, it is therefore highly desirable to develop solid-liquid classification devices which allow a sharper separation in the fines range in conjunction with sufficient throughput and low operating costs.

28.4.1 Hydro Deflector Wheel Classifier

A frequently used classification apparatus for particle classification in gas streams in the fines range is the deflector wheel classifier. Here, the solids loaded flow is led through a rotating deflector wheel from the outside to the axis of the apparatus. While the centrifugal force acting on the particles, caused by the rotation of the deflector wheel, forces the particles to settle at the circumference of the apparatus, the drag force of the gas stream points to the center of the wheel. Depending on the rotating speed, particle size and density, and hence the settling velocity of the individual particles, either the drag force or the settling velocity dominates, and such decides which fraction a particle will report to. The same principle is also applicable to particle classification in liquid flows. Such a hydro deflector wheel classifier is commercially available from Hosokawa Alpine, Augsburg. Their model Hydroplex® AHP 63 has been evaluated in the present work with respect to separation performance and applicability to soil washing units. The basic design can be seen in fig. 28.12.

A ceramic deflector wheel of 63 mm diameter is located inside a cylindrical housing. The feed flow is split into the coarse stream, which is sucked off at the bottom near the wall, and the fines fraction which flows through the deflector wheel. The design of this apparatus prevents some problems described for hydrocyclones such as friction and short cut streams into the fines fraction as the fines stream is forced to flow through the centrifugal force field, regardless of retention time and wall friction.

Fig. 28.13 displays the separation efficiency against particle size for classification of limestone with the hydro deflector wheel classifier at 7000 rpm and in a hydrocyclone with 42 mm diameter. It can be seen that the Tromp curve of the hydro deflector wheel classifier shows a sharper separation than the hydrocyclone

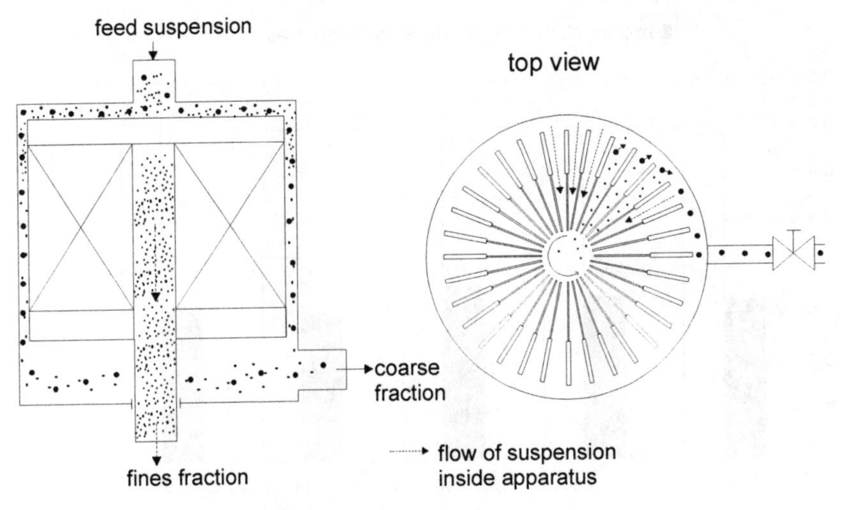

Fig. 28.12. Design of a hydro deflector wheel classifier

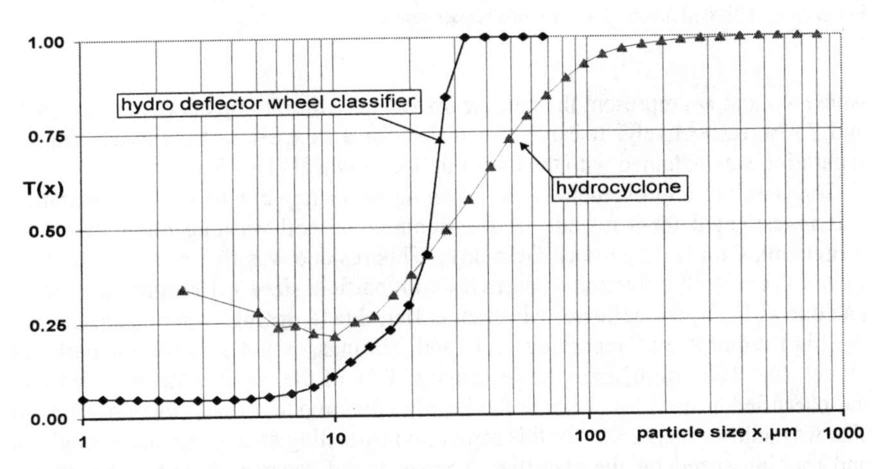

Fig. 28.13. Comparison of Tromp curves for classification of limestone with hydro deflector wheel classifier (at 7000 rpm) and hydrocyclone (42 mm diameter)

at a comparable cut size of about 25 µm. The sharpness of separation was compared in the present work for use in further experiments with different cut sizes, to verify the results of the experiment, depicted in fig. 28.13.

Fig. 28.14 shows these separation sharpnesses for the hydrocyclone and Hydroplex® for different cut sizes. The separation sharpness is defined here as

$$\kappa = \frac{x_{25}}{x_{75}} \tag{28.9}$$

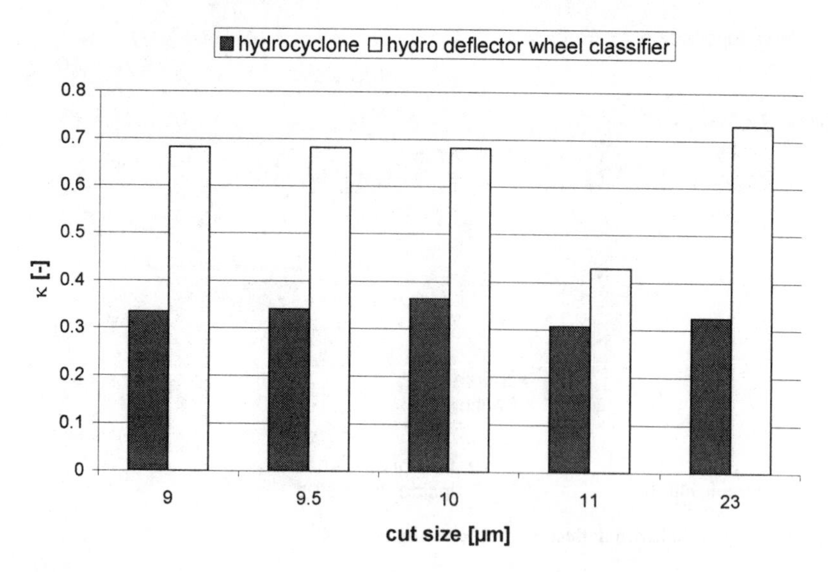

Fig. 28.14. Comparison of separation sharpness for deflector wheel classifier and hydrocyclone in the case of classification of a limestone suspension

where x_{25} and x_{75} represent the particle sizes where the Tromp curve reaches 25 % and 75 %, respectively. It can be seen that for all experiments, a much sharper separation was obtained with the hydro deflector wheel classifier.

The possible improvement of the cleaning efficiency due to sharper separation was investigated for a residue from a commercial soil washing plant (AB Umweltechnik GmbH, Lägerdorf, Germany). This residue was first classified using a sieve to remove the fraction of particles with particle sizes > 100 μm. This has to be done as the hydro deflector wheel classifier, due to the high rotating speeds and the high requirements regarding seals and bearings, is not suitable for particles bigger than 100 μm in size. The remaining 70 % of the residue was then fed into the classifier in order to separate the highly contaminated fines fraction. The cut size was adjusted to 22 μm for this separation by setting an appropriate throughput and rotating speed for the classifier. The separated "coarse" fraction of particles between 22 μm and 100 μm, after hydro deflector wheel classification, was then added to the coarse fraction from the sieving process. As depicted in fig. 28.15, it was possible to reduce the mass of highly contaminated fines through sieving and subsequent classification in the Hydroplex® AHP 63 by almost 50 %. This would consequently mean a saving of about 50 % for a commercial soil washing plant if the contaminated residue would have to be sent to a special disposal site.

Even though the Hydroplex® AHP 63 by Hosokawa Alpine has proven to be able to give good separation results, and is capable of improving the cleaning efficiency of soil washing processes with standard classification devices such as hydrocyclones, the bypass of fines into the coarse fraction is still a major draw back for its application to soil washing processes. Therefore, additional concepts have been investigated with respect to their applicability to solid-liquid separation in the fines range.

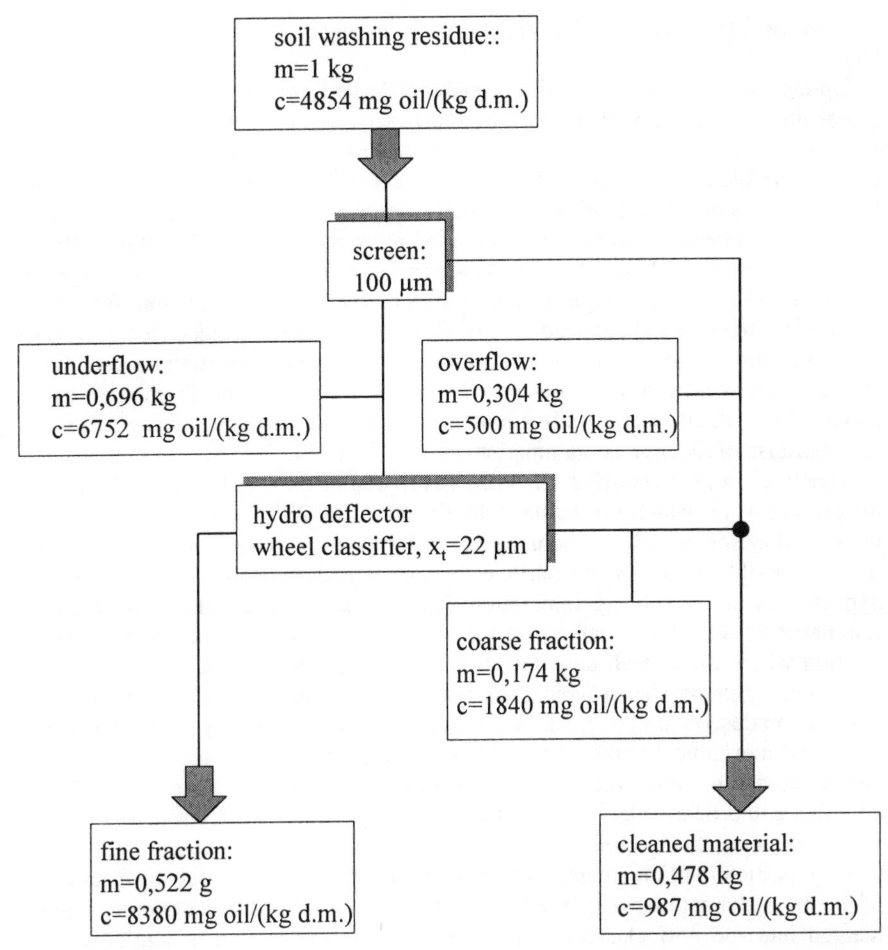

Fig. 28.15. Balance of screening and hydro deflector wheel classifier separation for soil washing residue (residue provided by AB Umwelttechnik GmbH, Lägerdorf, Germany)

28.4.2 Centrifugal Fluidised Bed Counter-Current Classifier

Sharp separations of solid-liquid suspensions on the technical scale have frequently been achieved in up-current classifiers. Here, the suspension flows from the bottom of the classifier to the top against gravity, while the particles settle by the gravitational force in the counter-current direction. A major drawback for this kind of separation apparatus is that, for small particle sizes, the up-current classifier shows a rapid increase in the required cross-sectional area due to the low settling velocities of the small particles, which makes a low velocity of the liquid flow mandatory. A simple way to increase the settling velocity of particles, and thus to keep the required cross-sectional area small, is to change from gravitational settling to settling in the centrifugal force field, as it is done in centrifuges and several other classification apparatus including, for example, the hydrocyclone.

Up-current classification in the centrifugal force field therefore seems to be a promising alternative with respect to sharp separation combined with low cut sizes. First attempts in this area have been made by Colon, van Heuven and van der Laan (Colon et al. 1970), who report on centrifugal elutriation of particles in liquid suspension. A test unit of a counter-current centrifugal classifier at laboratory scale has been developed and examined more recently by Priesemann (1994) and Timmermann (1998). Through classification of quartz sand and limestone, it was found that the centrifugal counter current classification is suitable for solid-liquid classification and can lead to very sharp classification results. However, due to the method of the coarse fraction removal, the results of Priesemann and Timmermann show a bypass of fines into the coarse fraction in the Tromp curves of about 10 %, which is especially counterproductive for cleaning of contaminated soils because of the high contamination level in the fines.

Therefore, a new classifier has been developed at the TUHH in the course of the present work, which is schematically drawn in fig. 28.16. It consists of a rotating classification chamber, where the liquid flows towards the center of a rotating disc. A possible cause for fine particles bypassing a classification process, besides bypass streams, could be agglomeration. The agglomerates of highly contaminated fine particles would then act as one bigger particle, and thus report to the coarse fraction where the overall contamination would rapidly increase. In order to destroy such agglomerates, a typical soil washing plant has a wet liberation or attrition unit as described above, where the agglomerates are destroyed through shear stress and mechanical impact. During the soil washing process, agglomerates may be formed again, which then would deteriorate the separation performance. Such an effect could be avoided if the particles are subjected to shear stress and impact during the separation as well.

High particle-particle interactions, causing shear load and impact stress, are one of the main characteristics of fluidised beds. A fluidised bed was therefore implemented into the new classifier which, besides the destruction of agglomerates, facilitates the removal of the contaminants from the particles' surfaces through the induced shear stress. The suspension is therefore fed into the unit homogeneously over the circumference of the apparatus. A classification water flow is added through a frit which acts as a distributor plate. The flow direction of the water and the suspension points to the center of the rotating apparatus, creating a fluidised bed on the distributor plate, with the particles tending to settle to the frit in a direction counter to the liquid flow. The fines settling slower than the counter current liquid flow are transported to an overflow weir, while the coarse particles accumulate in the fluidised bed and are removed from the system through valves at the circumference of the apparatus. As the settling velocity strongly depends on the centrifugal force, which itself strongly depends on the radius, the superficial velocity of the liquid flow has to be adjusted, depending on the radius, to keep the cut size constant. Therefore, geometric wall profiles are adopted to adjust the cross-sectional area, and thus the velocity, for a constant volumetric flow rate to compensate for the change in centrifugal force.

The major advantages of this design, compared to the hydro deflector wheel

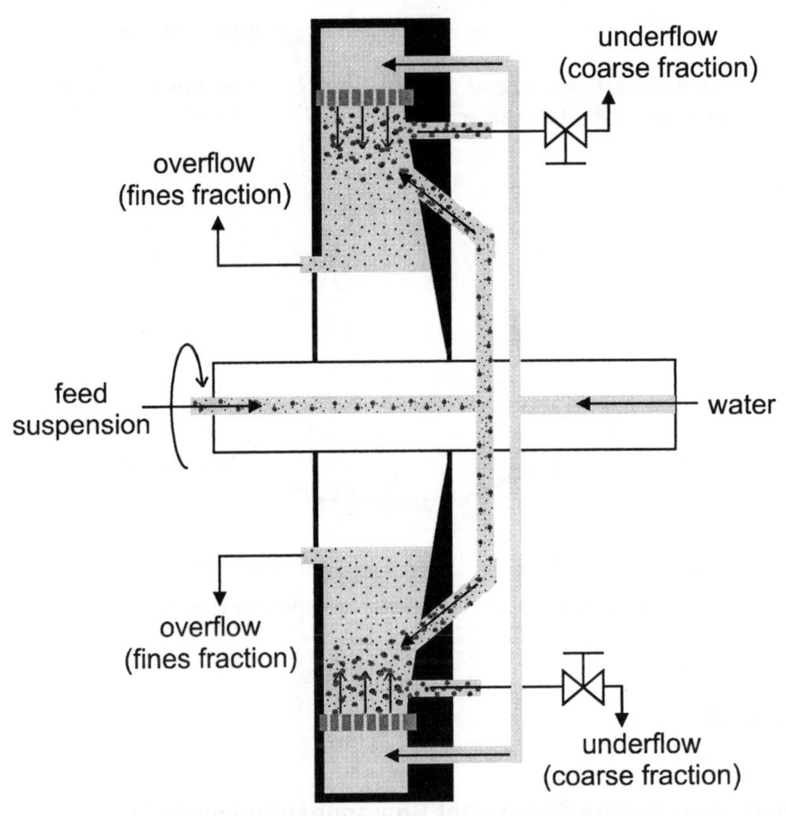

Fig. 28.16. Schematic design and principle of the centrifugal fluidised bed, counter-current classifier

classifier or hydrocyclone, are the total absence of bypass streams, the flexibility to adjust its cut size by changing rotating speed, suspension feed flow rate, classification water flow rate or coarse fraction removal rate, and the positive effect of the fluidised bed in destroying agglomerates and further removing contamination from the particle surfaces. As can be seen in fig. 28.17, very sharp separations can be achieved with this apparatus. For the given outer diameter of 1000 mm and cylinder height of 70 mm, throughputs of up to 4 $m^3 h^{-1}$ are technically obtainable during continuous operation. At a comparably sharp separation, the centrifugal fluidised bed counter-current classifier does not show the undesirable bypass of fines into the coarse fraction as observed for the hydro deflector wheel classifier and hydrocyclone.

Fig. 28.17 shows the Tromp curves of different separation devices. It can be seen that the centrifugal fluidised bed counter-current classifier allows a sharp classification (characterised by a narrow range of overlapping sizes) in the size range of 10 µm and below. This allows not only a sharp separation, but also the fractionation of a given particle collection. Fractionation of particle collections has multiple applications in industry and analytical investigations. The possibilities with respect to fractionation have been investigated in an additional experiment

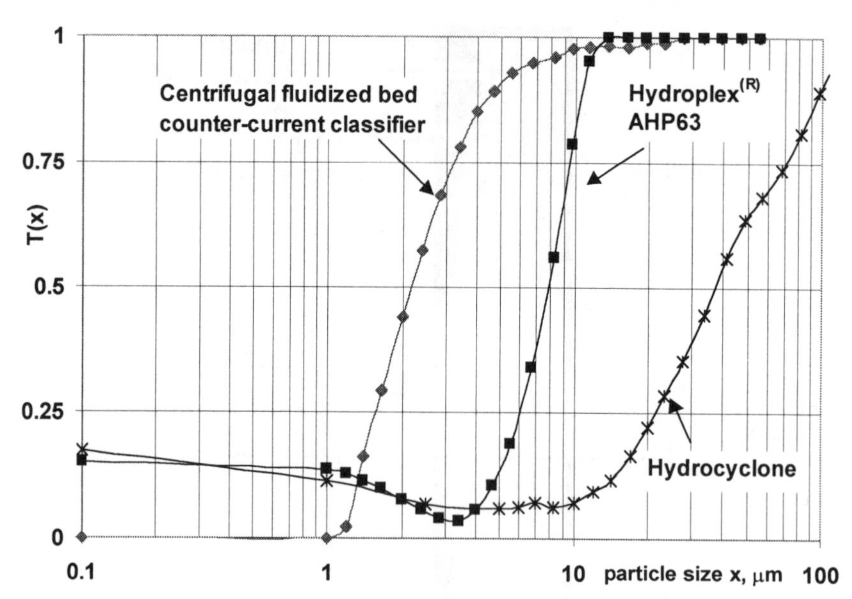

Fig. 28.17. Comparison of separation results for different classifiers for classification of a limestone suspension

with this classifier.

28.4.3 Analysis of Particle Size Dependent Contamination Levels in the Fines Range Using the Centrifugal Fluidised Bed Counter-Current Classifier

In order to verify the assumption of the presence of a disproportionately high contamination in the fines range of contaminated soil particles, a classification and an analysis was conducted for the particle size range between 1 μm and 25 μm. A soil contaminated with mineral oil was first classified in a hydrocyclone. The remaining fines fraction with sizes up to 100 μm was then fractionated into distinct size classes and these fractions were analysed with respect to their mineral oil contamination. The centrifugal fluidised bed counter-current classifier was used for the classification. In order to fractionate the contaminated particles into the different size ranges, six different cut sizes were set by variation of the rotating speed, the feed flow rate and the flow rate of the classification water. After these classifications, the individual contamination level of each fraction was determined with the Horiba H18 infrared spectrometer.

The particle size distribution and the mineral oil balance obtained can be seen in fig. 28.18.

The cumulative mass distribution of particle sizes and the fractional mineral oil contamination is depicted as a function of the particle size. The plot reveals an inhomogeneous distribution of the contamination over the particle size distribution. The observed concentration profile matches the results expected from the

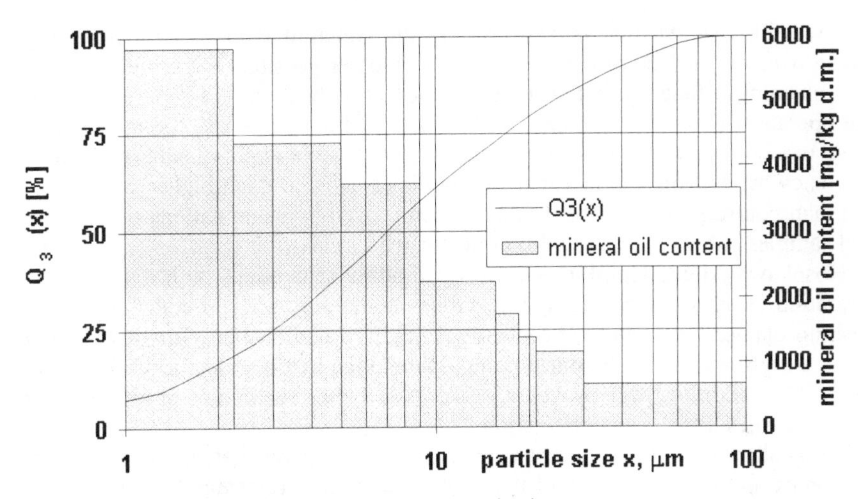

Fig. 28.18. Particle size dependent mineral oil contamination

relation of surface to volume of spherical particles, derived from equation 28.6 above. It is obvious that for the larger particles, particularly above 10 μm, the contamination level is so low that a separation at this cut size would lead to a significant increase of the yield of cleaned material without significantly increasing its overall content of contamination. Further calculations (not shown here) reveal that for the present example, a sharp separation would allow the reduction of the mass of the remaining highly contaminated fines fraction by approximately 50 %, and thus reduce costs of the further processing of the residue.

28.5 Summary and Conclusions: How Can the Efficiency of Soil Washing Be Improved?

Generally, the physical treatment of contaminated soil is efficient and reliable, but for contaminated soil with an extremely high fraction of fines, the standard treatment can lead to a rather big residual fraction, with a high degree of contamination. Depending on the soil's particle size distribution and the corresponding distribution of contamination over particle size, a further treatment of the residue may increase the cleaning success of the overall soil washing plant with respect to decontamination ratio and mass fraction of cleaned soil, related to feed mass.

In this chapter, several methods for further treatment of soil washing residues from commercial soil washing plants have been described. Flotation has been shown generally to be capable of selectively removing highly contaminated soil fractions. The selectivity of the process depends on the nature of the soil material, the nature of the contaminants and the concentration of the contaminant. The optimum flotation environment (type and concentration of the flotation agents, pH-value, chemical pre-treatment for heavy metal contaminated soils) must be found from experiments carried out on the respective contaminated soil.

It was also shown that in cases where the contamination is homogeneously spread over the surface of the soil particles, another possible way to increase the total cleaning efficiency is the integration of classification and separation units that operate with smaller cut sizes in the fines fraction than those commonly used in today's commercial soil washing plants. The possible improvement in cleaning efficiency arises from the fact that fine particles show a much higher contamination than coarse particles. This has also been proven by examinations of fractionated samples shown here. A hydro deflector wheel classifier (Hydroplex® AHP63 by Hosokawa Alpine GmbH) was shown to allow separations with much higher separation sharpness than the hydrocyclones commonly used in commercial soil washing plants. The examined centrifugal fluidised bed counter current classifier shows a high separation sharpness comparable to the hydro deflector wheel classifier and, in addition, was proven to give even better separation results without bypass of fines into the coarse fraction.

It was shown, for the different contaminated soils investigated, that a total reduction of up to 50 % in mass of the highly contaminated residue from the washing process should be possible through application of the investigated new treatment methods and devices. This would mean a substantial increase in the overall cleaning efficiency of such a plant.

28.6 Symbols

c	$[mg\ (kg_{dwt\ soil})^{-1}]$	contaminant concentration
c*	$[mg\ (kg_{dwt\ soil})^{-1}]$	limiting value of the contaminant concentration in the cleaned soil
d	[m]	diameter of sphere
f	[-]	mass fraction of fines after classification
g	[-]	mass fraction of coarse particles after classification
m	[g]	solids mass
mC	[g]	contaminant mass
Q3	[%]	cumulative particle distribution by mass
rd	[-]	decontamination ratio
T(x)	[-]	separation efficiency depending on particle size
x	[μm]	particle size

28.6.1 Greek symbols

α	[-]	concentration ratio
β_C	[-]	contaminant removal
β_s	[-]	solids removal
κ	[-]	separation sharpness ($x_{25}\ x_{75}^{-1}$)

28.6.2 Subscripts

0	untreated soil
cl	cleaned fraction

f	feed
fl	flotate
ho/hu	hydrocyclone overflow/underflow
oil	mineral oil
r	highly contaminated residual fraction
s	solids
so/su	sieve overflow/underflow
t	tailings

References

Abdo MSE, Darwish, AM (1991) Effect of some operating variables on the flotation of aluminium particles. Chem Eng Tech 14: 119–121

Arbiter N (1985) SME Mineral Processing Handbook. Section 5, Society of Mining Engineers, New York

Colon F, van Heuven J, van der Laan H (1970) Centrifugal Elutriation of Particles in Liquid Suspension, Particle Size Analysis. Society of analytical chemistry, Bradford, pp 42–52

Eberius E (1989) Verfahren zur Dekontaminierung schlammartiger Sedimente. European Patent 0332 985

Förstner U, Calmano W (1982) Bindungsformen von Schwermetallen in Baggerschlämmen. Vom Wasser 59: 83–92,

Hankel D, Rosenstock F, Biehler G (1992) The Effect of Attrition in the LURGI-DECONTERRA® Process for Decontamination of Soil. Aufbereitungs-Technik 33: No 5

Kihlstedt G (1974) Flotation. In: Ullmanns Enzyklopädie der Technischen Chemie, Band 2. Verlag Chemie GmbH, Weinheim, pp 110–142

Leschonski K (1988) Particle Size Analysis and Characterisation of a Classification Process. In: Gerhartz W (ed)Ullmann's Encyclopedia of Industrial Chemistry. Vol B2: Unit Operations I, 5th edition, VCH, Weinheim, pp 2-1–2-33

Pearl M, Wood P (1993) Trennprozesse für die Behandlung verunreinigter Böden. In: Arendt F, Bosman R, van den Brink WJ (eds) Altlastensanierung '93, Kluwer Academic Publishers, Dordrecht, Boston, London, pp 1325–1333

Priesemann C (1994) Naßklassierung in einer Aufstromzentrifuge. PhD Thesis, TU Clausthal

Timmermann D (1998) Kontinuierliche Naßklassierung in einer Aufstromzentrifuge im Feinstkornbereich. PhD Thesis, TU Clausthal

Venghaus T, Werther J (1998) Flotation of a zinc-contaminated soil. Advances in Environmental Research 2(1): 77–99

Venghaus T (1997) Aufbereitungsverfahren für schwermetallkonatminierte Böden. Arbeits- und Ergebnisbericht (1995–1997) des SFB 188, Reinigung kontaminierter Böden, Technische Universität Hamburg-Harburg

Wilichowski M, Venghaus T, Werther J (1998) Flotation as a process step in the treatment of soils contaminated with mineral oils. Env Technol 79: 801–810

Wilichowski M (1995) Aufbereitung mineralölkontaminierter Böden durch Bodenwäsche und Flotation. PhD Thesis, Shaker, Aachen

Yarar B (1988) Flotation. In: Gerhartz W (ed) Ullmann's Encyclopedia of Industrial Chemistry. Vol B2: Unit Operations I, 5th edition, VCH, Weinheim, pp 23-1–23-30

29 New Developments in Soil Washing Technology

Th. Neeße
Department of Environmental Process Engineering and Recycling, University of Erlangen-Nürnberg, Paul-Gordan-Straße 3, 91052 Erlangen, Germany

29.1 Introduction

In Germany we can now look back on more than a decade of experience in soil washing based on the state of the art of mineral processing techniques (Neeße and Grohs 1990a; Neeße and Grohs 1990b; Neeße and Grohs 1991a; Neeße and Grohs 1991b; Wilichowski and Werther 1996). In recent years washing technology has changed remarkably. The cost of soil cleaning 10 years ago is now no longer acceptable. As a result of competition with biotechnology, in situ-technologies and natural attenuation, the tonnages sent for soil washing have been reduced dramatically. This is a challenge that can be met only by a new generation of soil washing plants, which is the subject of this paper.

29.2 Process Flow Sheet of a Soil Decontamination Plant

Industrial soil washing processes to meet the specific requirements of the contamination to be removed can be developed on the basis of laboratory tests (Feil et al. 1994; Feil 1997; Neeße et al. 1997). The process stages of a soil decontamination unit are shown schematically in fig. 29.1. The configuration of individual process stages can vary greatly, depending on the material and the objective of decontamination.

The actual washing process is preceded by a dry soil preparation. This includes storage, feed regulation and pre-crushing of the material. Of decisive importance for the washing effect is the subsequent wet liberation, the aim of which is to disperse the agglomerated fine particles. In this process, soil is suspended in water by the application of mechanical energy. If required, liberation may be performed in two stages, when the sand particles are subjected to a second treatment in the previously thickened suspension. The relevant apparatus are encapsulated and equipped with air exhaust systems to allow highly volatile components to be removed. The exhaust air is passed through activated carbon filters and cleaned.

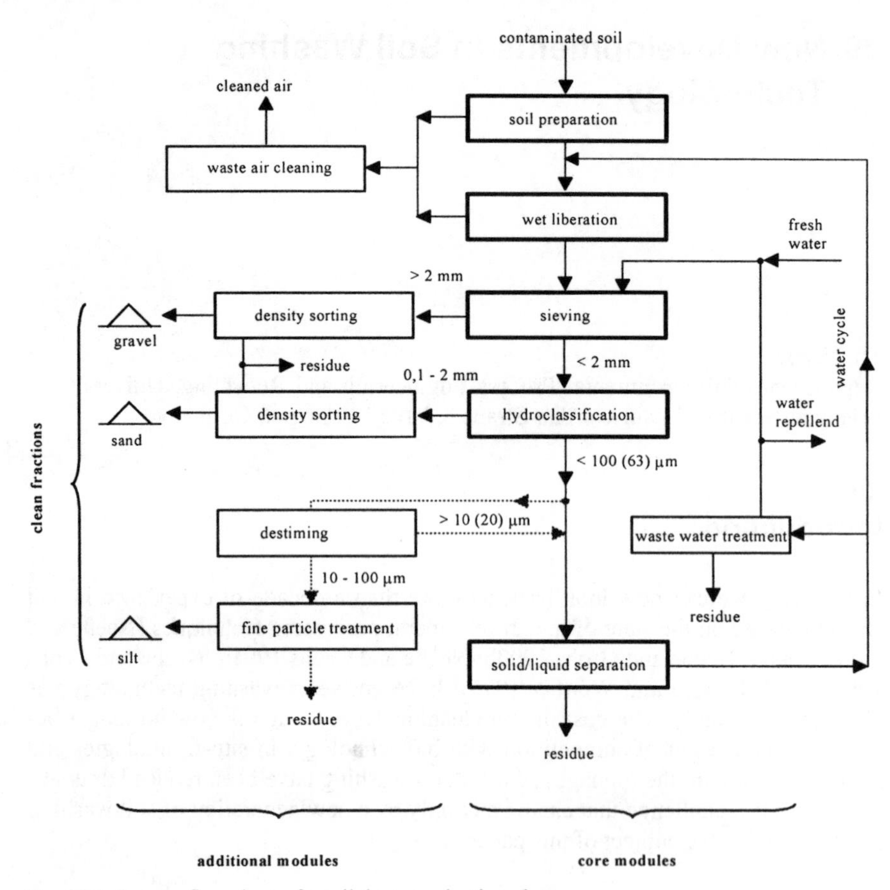

Fig. 29.1. Process flow sheet of a soil decontamination plant

The main principle of soil washing is a selective classification of fines, which are highly contaminated due to the large specific surface. In stepwise fashion the soil is sized on sieves and hydrocyclones, where cleaned gravel and sand can be separated.

The remaining suspension with fines < 63 μm is subjected to a solid/liquid phase separation process resulting in a cohesive contaminated waste product. The process water is returned to the wash water circuit. These processes are basic to the core modules of every soil washing plant.

For special cases auxiliary modules can be added. In these, density sorting for the elimination of highly contaminated organic matter is often carried out.

Further, alternative processes for the cleaning of fines < 100 μm can be considered: flotation (Wilichowski 1994), leaching processes (Kramer and Koch 1992), high gradient magnetic separation and microbiological processes.

29.3
First Generation of Soil Washing Plants

The first generation plant in Germany was characteristically:

- Stationary centralised plants with a relatively high throughput of up to 30–70 t h^{-1}.
- Expensive cleaning of fines and waste water treatment.
- Negative water balance and consequently indirect discharge of a more or less large amount of cleaned waste water.
- Operational costs of up to 200–400 DM t^{-1}, depending on the level/type of contamination.

Advantages of the washing technologies are:

- Soft technology using natural principles of wet self-cleaning of the soil with a moderate energy requirement.
- High cleaning efficiency, as can be seen from table 29.1.
- Low operational costs for the coarse soil fractions.
- High level of technology based on the state of the art of modern mineral processing.

Disadvantages are:

- No soil remediation, but production of a low-cost building material (gravel and sand).
- Highly contaminated residue of the fine fractions.
- High operational costs for cleaning of fine fractions and waste water treatment.

Table 29.1. Data on soil washing plants

Process Parameters	Value
Throughput [t h^{-1}]	
• low	< 10
• moderate	10 –30
• high	30 –70
soil recovery [%]	75 –90
cleaning efficiency [%]	85 –95
recirculation water flow [m³ t^{-1}]	3 – 4
fresh water demand [m³ h^{-1}]	0.1– 0.3
specific energy consumption [kWh t^{-1}]	7 –35
reagent demand [kg t^{-1}] (flocculants, surfactants)	0.3– 0.8

29.4
New Generation of Soil Washing Plants

The second-generation plants have been developed mainly because of the need to reduce operating costs. This can be accomplished by the following measures:

- Use of mobile or semimobile plants in modular form with low or moderate throughput of 10–20 t h^{-1}.
- Treatment of contaminated soil and other mineral waste.
- Reduction of the core cleaning modules to those needed for gravity processes, classification and density sorting.
- Application of costly processes for fine fraction cleaning, such as leaching, and flotation only in special cases.
- High intensity attrition as a special development for soil washing.
- Use of more sophisticated waste water treatment if possible with a positive water balance (no waste water discharge).

A typical flow sheet of a modern soil washing plant, offered by AKW Apparate + Verfahren GmbH & Co. KG in Hirschau/Oberpfalz (Germany), can be seen in fig. 29.2.

29.4.1
Wash Water Circuit

For a soil washing plant to function correctly, the logistics of water flow are of prime importance. Some variants of the flow circuit are presented in fig. 29.3. Fig. 29.3a, b demonstrate two traditional variants with negative water balance (waste water discharge). Fig. 3c shows the modern configuration with process water treatment integrated into solid/liquid separation. The feed to waste water treatment is a fine sludge with a solids content of approx. 50–100 g l^{-1}. In solid/liquid separation, the objective pursued is to bind contaminations largely to the solid phase. One requirement for a positive water balance without waste water discharge is a suffcient proportion of fines in the soil. However, a fines fraction > 20 % threatens the profitability of soil processing because the recovery of washed soil will be too low. When the proportion of fine sludge is too low, more sludge can be produced artificially by means of hydroxide precipitation or addition of betonite.

The addition of Fe (III) or Al (III) leads to the formation of voluminous hydroxide sludges which, for example, can bind heavy metal ions through adsorption. For de-emulsification of hydrocarbon contaminants, organic emulsion splitting agents are used. This is achieved with polymers having molar masses of between 50,000 and 500,000, which have a flocculation effect on the solid phase and a coagulating effect on the oil phase. By integration of flocculation and adsorption, a highly contaminated sludge mixture of organic and mineral substances is formed, which is dewatered in the solid/liquid separation process stage to obtain a solid product. The sludge mixture is either taken to a special disposal site or subjected to a special process (e.g. thermal, biological). In this process stage, process water in which the dissolved and emulsified contaminants are reduced is recirculated.

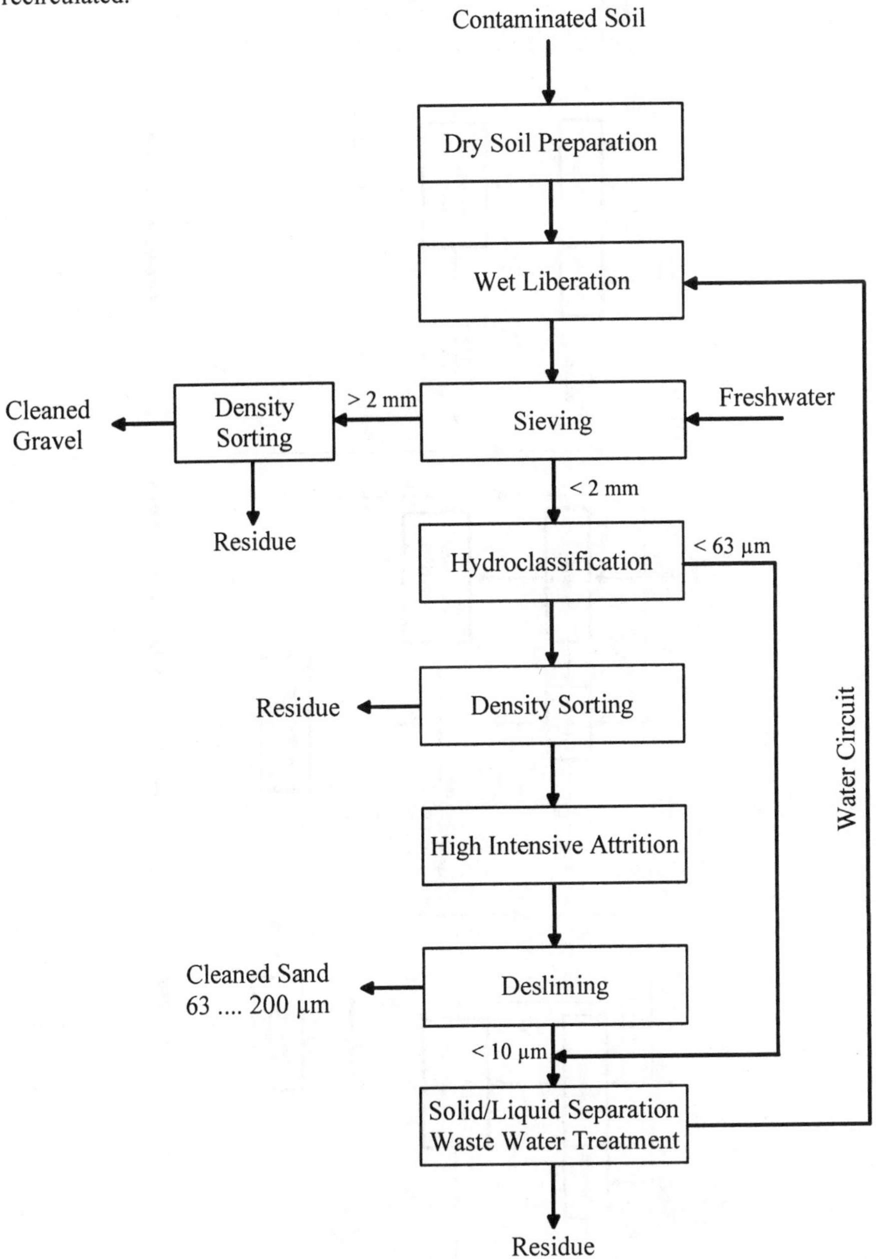

Fig. 29.2. New flow sheet of a soil washing plant

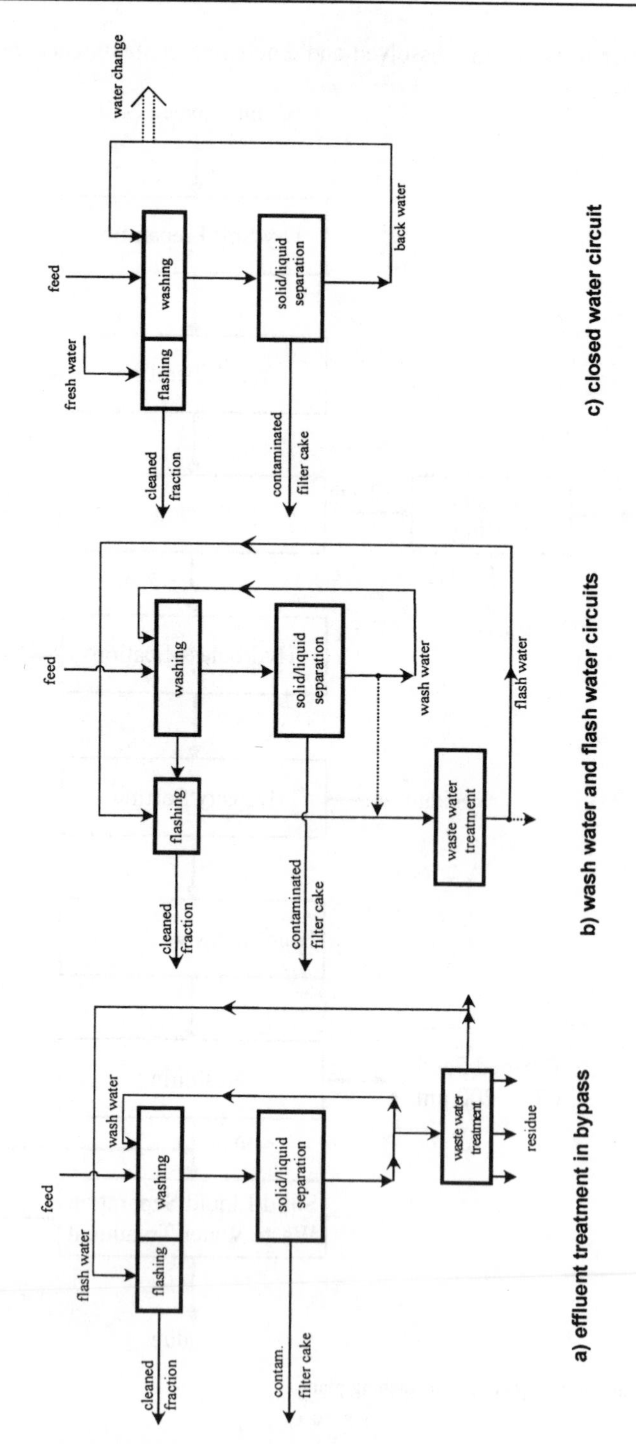

Fig. 29.3. Flow circuit variations in soil washing plants

29.4.2
High Intensity Attrition – a New Process for Fine Particle Cleaning

One result of research and development in soil washing is a new variant of mechanical fine particle cleaning that takes account of the need to reduce costs – high intensity attrition (Tiefel et al. 1999; Schricker 1999; Information by AKW Apparate + Verfahren 2000).

With today's state-of-the-art processing technology, conventional attrition is performed relatively simply by agitation of the material in a thickened suspension. Practical experience with available attrition systems, however, has shown that they are often unable to maintain the necessary parameters and thus to guarantee maximum cleaning efficiency.

A common reason is that the processing conditions necessary for an intensive attrition effect, that is, intensive particle contact, vary during the course of the processing cycle. In this connection, the solids concentration in the suspension fed to the attrition unit must be regarded as a determining factor for efficient processing. An adequate attrition effect can only be ensured if the solids content in the suspension is so high that the free mobility of the individual particles can be largely restricted.

High intensity attrition uses solid concentrations > 40 vol. % and higher mechanical energy input, for producing a higher percentage of attrited fines < 10 µm of up to 10–15 %. The production of fines by attrition of an oily sand can be seen in fig. 29.4.

In the industrial scale application of attrition units, high solids concentrations of 40 % are hardly ever reached, vary widely, or cannot be maintained permanently. When the concentration falls below the pre-set level, which corresponds to a defined solids content, additional coarse material is fed to the first attrition cell. The coarse material consists of part of the > 0.5 mm fraction, which is separated from the abraded material and then recirculated to the feed. Both the recirculated coarse material fraction, and the material fed into the attrition cell, are monitored and controlled by a central SPC unit.

The flowsheet for this new processing stage is shown in fig. 29.5, as exemplified by cleaning contaminated sands.

The suspension is fed (1) directly from the upstream processing installations, e.g. via hydrocyclones, into the attrition unit (2), where additional reagents (8) can be added if required. A torque sensor mounted to the drive shaft of the agitator determines the actual suspension viscosity and passes the signal to the central control system. If the actual value falls markedly below the set value, coarse material is added from the storage (4). The coarse material is obtained by screening at 0.5 mm and dewatering (3) of the particle fraction after attrition. Depending on the filling level in the storage container, controlled diversion of the coarse material flow to the container takes place. The main part can be discharged directly as cleaned material fraction.

In a hydrocyclone (5), the fine fraction formed during attrition is separated from the < 0.5 mm fraction. Depending on the specified processing aims and the cleaning values required, a cut size between 20 and 80 µm may be necessary. The fine fraction (6), which contains the contaminants in concentrated form, is directed to the process water or sludge preparation, where flocculation and sedimentation

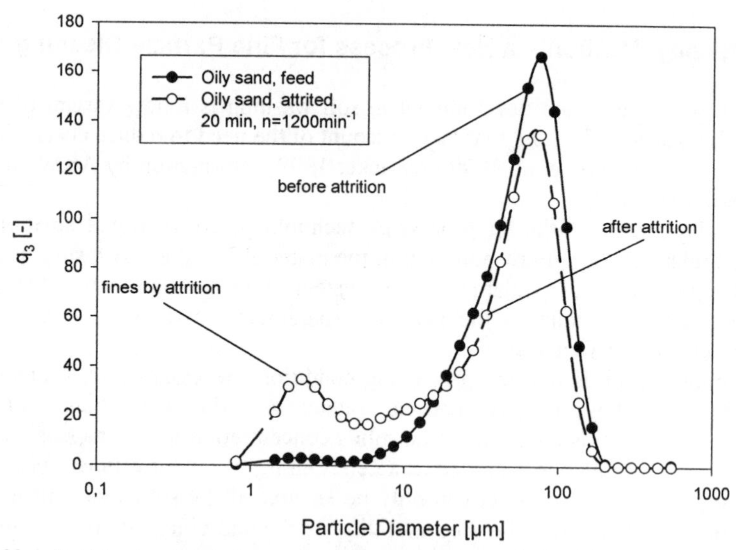

Fig. 29.4. Particle size distributions of an oily sand before and after high intensity attrition (Information of AKW Apparate + Verfahren 2000)

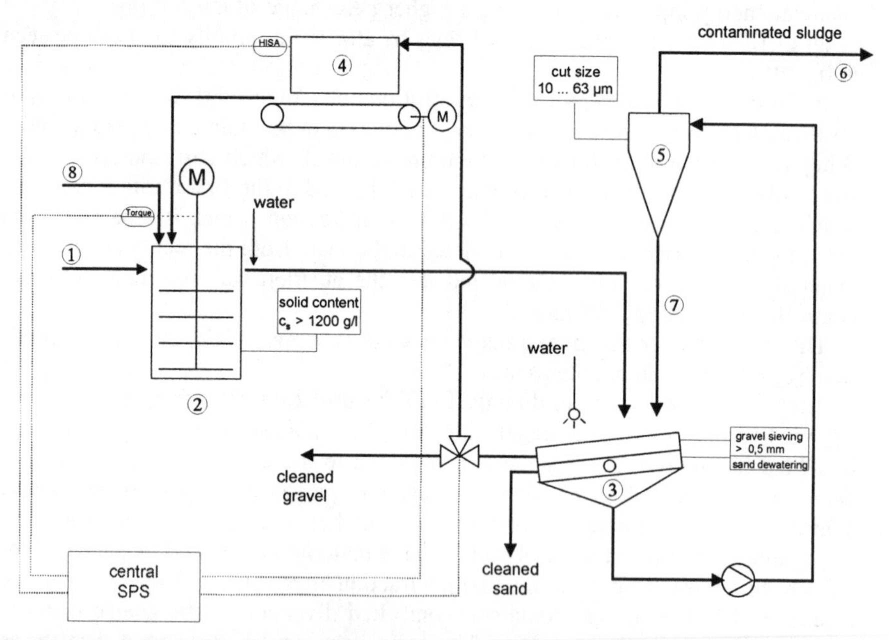

Fig. 29.5. Flow sheet of the high-performance attrition (Tiefel et al. 1999)

take place to produce a product suitable for dumping. From the underflow of the hydro-cyclone (7), the cleaned sand fraction flows on to a dewatering screen and is then stockpiled by means of belt conveyors.

The actual attrition process is carried out in conventional attrition units, no additional alterations are required for high-performance attrition (Mitteilung der AKW Apparate + Verfahren 2000). A particular advantage of this latter is that the entire plant equipment required – with the exception of the measurement and control system – consists of conventional, commercially available units, so that existing plants can be upgraded cost effectively.

29.5 Cleaning Example

In order to study the efficiency of high-performance attrition in comparison with conventional attrition for cleaning different feed materials, and to determine the achievable contaminant concentrations, tests were conducted in a special test rig.

High-performance attrition can be used, for example, for cleaning soils and soil-like materials contaminated with mineral oil hydrocarbons. A typical oily sand with particles ranging between 63 μm and 2 mm and a pollutant concentration of 1845 mg mineral oil (kg dry solids)$^{-1}$ was cleaned by means of high-performance attrition to 300 mg kg^{-1} in the > 63 μm fraction. This corresponds to a cleaning efficiency of 84 %. In comparison, cleaning efficiency with conventional attrition treatment reached just 840 mg (kg dry solids)$^{-1}$. The fine fraction < 63 μm formed during high-performance attrition accounted for 9.5 mass.-%. For high-performance attrition, part of the cleaned coarse material > 1 mm was recirculated to the material feed and added to the attrition process (mass ratio of fine to coarse material 1:1). In this way a constant solids concentration of 1200 g l^{-1} (corresponding to a solids content of 45 vol.-% for quartz) could be maintained in the attrition process.

Table 29.2. Process conditions and cleaning results for high-performance attrition of a 63 μm–2 mm sand contaminated with mineral oil (Tiefel et al. 1999)

Process Parameters	Value
Specific energy input [kWh t^{-1}]	20.0
Attrition time [s]	280
Primary feedback ratio [-] in the closed circuit system	1.0
Quantity of fine particles < 63 μm after high-performance attrition [%]	9.5
MHC concentration > 63 μm before high-performance attrition [mg kg^{-1}]	1845
MHC concentration > 63 μm – after conventional attrition [mg kg^{-1}]	840
MHC concentration > 63 μm after high-performance attrition [mg kg^{-1}]	300
Decontamination efficiency [%]	84

29.6 Summary

A new generation of soil washing plants has been developed mainly because of the need to reduce operating costs. This can be accomplished by the following measures:

- Reduction of the core cleaning modules to those needed for gravity processes, classification and density sorting, and attrition.
- Application of costly processes for fine fraction cleaning, such as leaching, and flotation only in special cases.

High intensity attrition uses solid concentrations > 40 vol.-% and higher mechanical energy input for producing a higher percentage of attrited fines < 10 μm up to 10–15 %.

The stabilisation of a high-solids concentration by controlled recirculation of endogenous coarse material ensured the cleaning of fractions that could otherwise not be cleaned sufficiently, using conventional attrition techniques.

References

Feil A, Wetzel H, Neeße T (1994) Charakterisierung der Sanierbarkeit kontaminierter Böden. Aufbereitungs-Technik 3: 56–60

Feil A (1997) Untersuchungen zur Aufbereitbarkeit kontaminierter Böden. PhD-Thesis, Universität Erlangen-Nürnberg, Germany

Kramer U, Koch P (1992) Zur Anwendung von Laugungstechnologien bei der Dekontamination von Boden. Neue Bergbautechn 22 1: 13–17

Mitteilung der AKW Apparate + Verfahren GmbH und Co KG, Hirschau (2000) Hochleistungsattrition unter Produktionsbedingungen. Aufbereitungs-Technik 41 4: 195

Neeße T, Grohs H (1990a) Naßmechanische Aufbereitung kontaminierter Böden. Aufbereitungs-Technik 31: 563–569

Neeße T, Grohs H (1990b) Die Aufbereitungstechnik des Bodenwesens. Aufbereitungs-Technik 31: 656–662

Neeße T, Grohs H (1991a) Waschen und Klassieren kontaminierter Böden. Aufbereitungs-Technik 32: 72–77

Neeße T, Grohs H (1991b) Fest/flüssig-Trennung für die Entsorgung in Bodenwaschanlagen. Aufbereitungs-Technik 32: 294–302

Neeße T, Feil AU, Schricker B (1997) Washability curves for contaminated soil. Proc XX. Int Mineral Processing Congr, Aachen, Germany, 21–26 Sept

Schricker B (1999) Intensivierung der Attrition bei der physikalisch-chemischen Sanierung kontaminierter mineralischer Abfälle. PhD-Thesis, Universität Erlangen-Nürnberg, Germany

Tiefel H, Schricker B, Neeße T (1999) Hochleistungsattrition zur mechanischen Reinigung von mineralischen Roh- und Reststoffen. Technische. Aufbereitungs-Technik 40 4: 160–164

Wilichowski M, Werther J (1996) Mathematische Modellierung von Naßaufschluß, Siebung und Hydrozyklontrennung bei der physikalischen Reinigung mineralöl-kontaminierter Böden. Aufbereitungs-Technik 37: 87–96

Wilichowski M (1994) Aufbereitung mineralölkontaminierter Böden durch Bodenwäsche und Flotation. PhD-Thesis, TechnischeUniversität Hamburg-Harburg, Germany

30 Clean-Up and Assessment of Metal Contaminated Soils*

W. Calmano, S. Mangold, H. Stichnothe and J. Thöming
Technical University Hamburg-Harburg, Environmental Science and Technology, Eissendorfer Str. 40, 21073 Hamburg, Germany

30.1 Introduction

Heavy metal contamination of soils is still an unsolved problem although metals are associated with human life and have been used for thousands of years. Mining activities in particular have led to the spreading of large amounts of heavy metals in the environment. In many industrial processes (e.g., in the plating industry, accumulator production, chlorine-alkali-electrolysis, pesticide production), metals and their compounds are used, produced and subjected to different finishing processes, and then applied in various fields of human life. When introduced into soils, heavy metal compounds are hazardous pollutants because they are not biodegradable, toxic at relatively low concentrations, and they may be mobilised under changing physical-chemical conditions like redox potential or pH. Soils have a limited capacity to accumulate pollutants, and if this retention capacity is exceeded, the environment, (e.g. ground and surface water, plants and livestock) is likely to be at further risk.

Wherever heavy metals have been handled in the past, large areas with unacceptably high metal concentrations, as well as "hot spots" (small amounts of soil bearing heavy metal compounds at concentrations up to several 10 % of dry weight) can be found. For example, at many accident sites contaminated with metal-containing substances, often only small areas are affected, in some cases only a few cubic meters of soil. An immediate excavation of the material offers the chance to restrict the extent of the damage and spreading of the pollutants. For such cases in particular, sophisticated clean up techniques are needed to treat excavated soil.

The present state of development of soil remediation technology is neither sophisticated nor economic enough to allow the treatment of such small amounts of soil with existing decontamination plants. Hence, there is an urgent need for development of clean up methods which are able to treat on site small amounts of soil contaminated with different metals and metal compounds to the required criteria.

30.2 Treatment Techniques for Metal Polluted Soils: An Overview

The large variety of fine grained solids in a soil contaminated with heavy metals requires clean up techniques which can be adapted to the specific, substantial properties of the materials. For example, using thermal methods, a large spectrum of organic pollutants – including mercury – can be treated. Physical washing can only be applied if there are mixtures of particles consisting of distinct fractions with different heavy metal affinities. This may be the case for excavated soil where heavy metals have been introduced in dissolved form, and thus are mainly sorbed onto particle surfaces and bound in the fine grained fraction.

Treatment methods for contaminated soils are usually divided into *on/off site* and *in situ* techniques (fig. 30.1).

Mechanical procedures are suitable only as pre-treatment techniques. With high pressure soil washing, in particular of sandy soils, a relatively clean coarse grained fraction can be separated, leaving a strongly contaminated fine fraction. This procedure may distinctly reduce the amount of material, but the fine grained fraction has to be further treated or disposed of. Also, a considerable amount of mercury can be separated mechanically if the soil contains very high amounts of this metal in its elemental form. Another method commonly applied to mercury polluted soils involves transporting the contaminant by *thermal desorption* into the gas phase. This method represents a dilution step, and requires expensive waste air conditioning and involves high energy consumption and high investment costs.

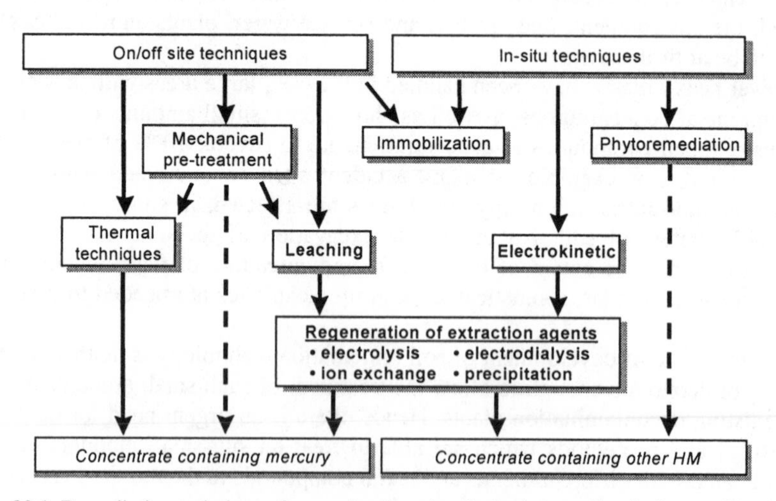

Fig. 30.1. Remediation techniques for metal polluted soils (hydrometallurgical operations in bold letters)

Immobilisation comprises in a wider sense all techniques and measures preventing the spread of heavy metals, i.e. the securing of the toxic substances. In a more narrow sense, the term "immobilisation" stands for both the different processes *solidification* and *stabilisation*. By the installation of barriers (e.g. sealing walls underground or sealing surface layers), the flow-in and flow-out of contaminated water or toxic vapours can be prevented. If it is impossible to seal the polluted ground, or if a natural impermeable underground layer is missing, the seepage water can be intercepted by geohydrological isolation, for example by a system of wells. Apart from these physical, in-situ measures, chemical immobilisation processes are also used, in which the contaminants are fixed in as stabile as possible compounds. Such chemical measures are also partly applicable in-situ.

Hydrometallurgical treatment is characterised by the application of separation processes in which dissolved metal compounds are processed. To get an overview of the techniques available, it is useful to distinguish between different groups of elements. The main group consists of lead, zinc, copper and cadmium. For hot spots containing these metals, there are known hydrometallurgical methods that can be applied: for excavated soil, leaching is a convenient technique (Thöming and Calmano 1995), and electrokinetic extraction is suitable for in-situ applications (Hansen et al. 1997). During these operations, metal loaded liquids are produced; for their treatment, precipitation, electrolytic deposition or ion exchange are proposed. Before treating excavated soil, a conventional soil washing procedure might be used as a pre-concentration step.

The use of hydrometallurgical techniques in mineral processing is widespread and well known. Their advantage is that they offer the possibility of separating the mobile metal fractions from the soil matrix, and producing metal concentrates. However, contrary to ores, soils have a much more complex structure and a high variety of metal binding sites which restrict the applicability of these techniques. The complex nature of the binding forms of the heavy metals within the solid soil matrix renders a specific separation difficult. As a consequence, wet physical-chemical techniques have been rarely used in soil remediation in the past, even for the treatment of hot spots. The main difficulty was to optimise the key operation, i.e. the leaching. Thus, a systematic method is needed for assessing the applicability and the limitations of such processes for a soil clean up. Furthermore, it was assumed that a systematic method would provide a couple of hints on how to best perform the leaching. Fig. 30.2 shows a hydrometallurgical process for the treatment of metal contaminated soils.

Particular attention should be given to mercury because of its special physical and chemical properties as well as its high toxicity – especially for methyl mercury. Excavated soil containing mercury is usually treated by thermal techniques (Hempel and Thöming 1999), however, leaching and electrokinetic methods are also tested at laboratory and pilot scale (Thöming et al. 1996; Wasay et al. 1995; Cox et al. 1996).

Extraction processes work on the basis of enhanced solubility of the extracted compounds in the solvent. According to requirements, desorbing, complexing, oxidising, reducing, acid or alkaline solvents can be used for the purpose of separation. The selection of a certain procedure depends on the metal binding forms on the solids. Fig. 30.3 shows an example of different mercury species bound to the surface of a soil particle, and various extraction mechanisms.

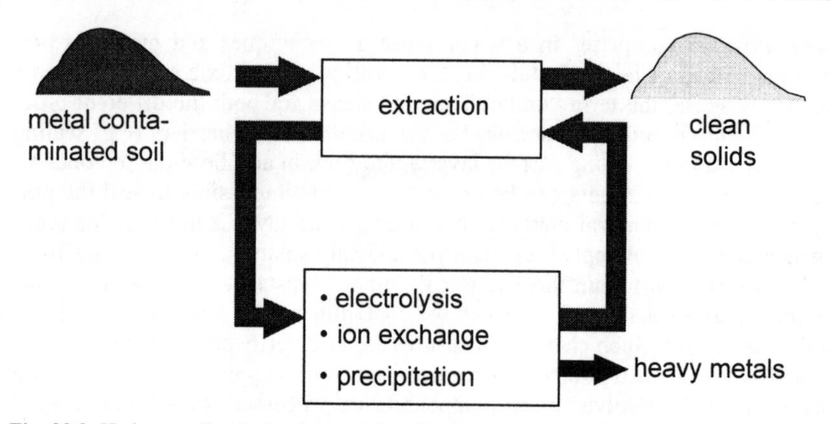

Fig. 30.2. Hydrometallurgical unit operations for the treatment of metal contaminated soils

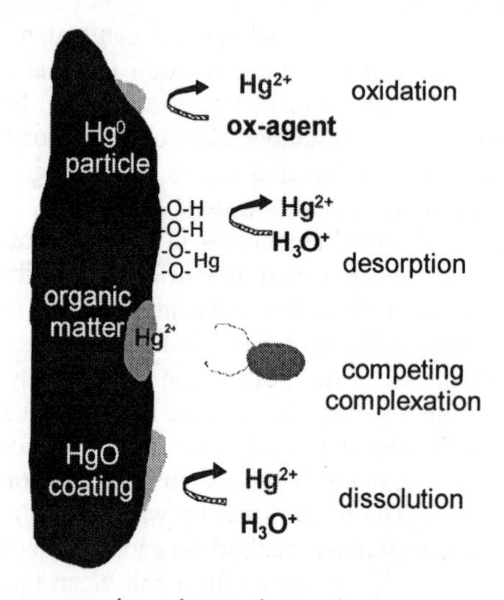

Fig. 30.3. Solid bound mercury species and extraction mechanisms

It should be stressed that the thermodynamical equilibria controlling these reactions are dependent on each other. Thus, it is possible that the mere desorption of surface bound metal species may lead to a following precipitation of metal compounds, i.e. they are only transferred from one solid phase to another. Additionally, re-adsorption and re-complexation processes of dissolved metal species take place. Since all these processes have different reaction kinetics, an optimisation of the treatment process is necessary.

In *electrokinetic processes,* the contaminated soil is treated between rows of electrodes. The electrical field induces a transport of substances by:

- migration of ions (electromigration),
- transport of pore water (electroosmosis),

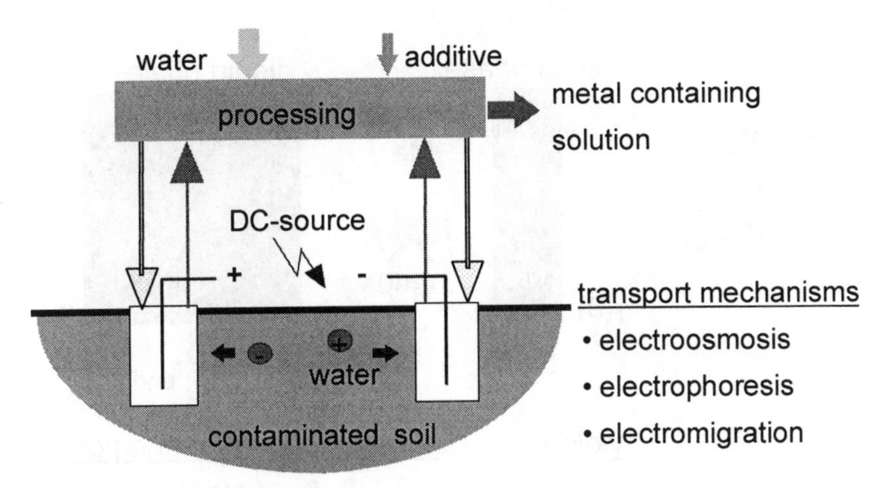

Fig. 30.4. Electrokinetic process

- transport of charged particles in pores (electrophoresis).

In principal, such methods can also be used in situ. Fig. 30.4 shows the principle of an electrokinetic process.

Finally, there remains the problem of the *treatment of the extracts*. The dissolved metals can be removed from the solvents or soil solution by sulfidic precipitation or iron cementation, but these processes produce large amounts of precipitation sludge. A separation by activated carbon or ion exchange is also possible, but these methods have the added disadvantage of production of waste materials and concentrates which have to be treated or disposed of.

Concentrates containing dissolved metals can also be treated electrochemically but a weak point of this process is that decreasing metal concentrations require an exponential amount of energy for further reduction. This can be avoided by means of a circuit process, which allows not only a minimisation of sewage water, but also high circuit concentrations of dissolved metals in the range of 30–60 mg l^{-1}, where treatment at such concentrations is easy. All applied separation processes have to compete with the high solubility of metal complexes formed during a successful extraction step.

30.3 Factors Affecting the Extraction of Heavy Metals from Soils

The solid/liquid extraction is a process that involves dissolving metals from a solid matrix using a solvent. A very important factor is the solid/solution ratio σ.

The bigger the value of σ, the "thicker" the suspension, and according to this fact the maximum attainable concentration of metals in the extract decreases at the same time. The effect of this parameter on the extraction result decisively depends on the selectivity of the solvent and the metal binding forms on the solids.

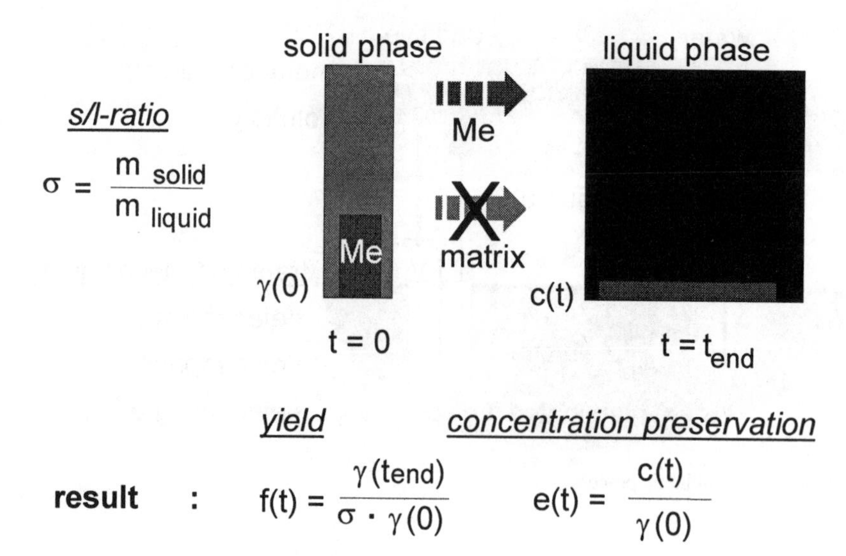

Fig. 30.5. Factors controlling extraction processes

The extraction grade can be described by certain parameters which depend on the various presentation forms of the data. The yield f(t) specifies which part of the contaminant has been extracted from a soil with an original content $\gamma(0)$, and the residue $\gamma(t_{end})$ is the part of a contaminant remaining in the soil. The concentration preservation e(t) expresses what percentage of the original metal content in the soil $\gamma(0)$ is found in the liquid phase as extract concentration c(t).

The extraction is successful if the selectivity α is high, i.e., if dissolution of the matrix ω has been avoided. Matrix dissolution ω is defined as:

$$\omega = \frac{1}{m_{soli,before}} \cdot \left[m_{soli,before} - \left(m_{soli,after} + \sum_i f_i c_i m_{soli,before} 10^{-8} \right) \right] \quad \text{[mass-\%]} \tag{30.1}$$

$$\alpha = \frac{1}{\omega} \sum_i f_i \tag{30.2}$$

The selectivity of the extraction is high if α is much bigger than 1, i.e. much heavy metal has been dissolved and at the same time only a very small part of the soil matrix has been dissolved.

The *pH value* is a very important factor for the extraction of heavy metals because it strongly affects the metal-solid interactions, and thus both mobilisation and immobilisation of the metals and their compounds. With decreasing pH, the activity of hydrogen ions increases exponentially, and the solubility of the metals also increases compared to their binding strength on the solids. Furthermore, at very high pH values, metals (e.g. copper or zinc) can be mobilised because of the formation of easily soluble metal hydroxo complexes and dissolution of humic bound metals. For the extraction of mercury, high pH values are of less importance because only small amounts are dissolved under these conditions.

The most important mechanisms of metal extraction at low pH conditions are displacement of metals from their binding places at the solid surfaces and the acid dissolution of metal precipitates, like metal carbonates, hydroxides, oxyhydrates, and oxides. This acid induced mobilisation affects the various metals differently, and depends on the binding forms of the metals and the content of organic substances.

The *redox potential* is of importance in the case when the metal compound is available as a poorly soluble, reduced or oxidised species. Under such conditions a transformation into a more soluble form is necessary by increasing or decreasing the redox potential. In this instance, the redox potential is the measurable mixed potential of the solvent-solid mixture. In many electrolyte systems it is coupled with the pH value.

The *duration* of the extraction decisively depends on the binding forms of the metals. For old deposits showing a deep diffusion of metals compounds into intermediate layers of clay minerals or other soil phases, slow diffusion processes determine the reaction rates. In such cases, electrokinetic processes could have a favourable effect because they may accelerate the transport. On the other side, if quick physical-chemical reactions play a decisive role, a stirring extraction can achieve equilibrium even after a few minutes. Due to such equilibria, a multi step extraction could be necessary, which may be achieved for example, by column leaching. Fig. 30.6 shows the principle of a column leaching unit and modelling by a stirring vat cascade.

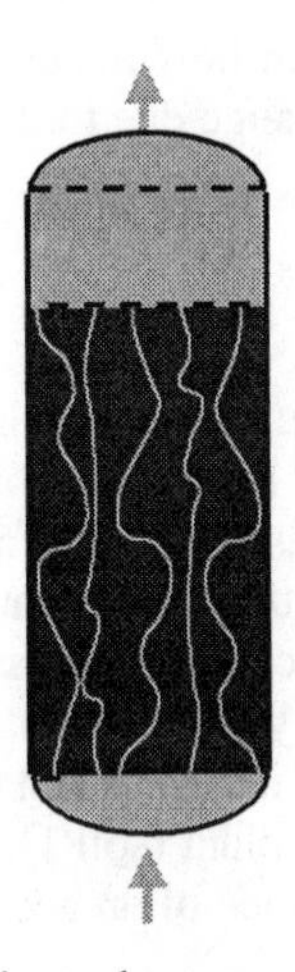

Fig. 30.6. Principle of column leaching and process modelling

30.4 Process Optimisation by Applying Heavy Metal Speciation

Heavy metal speciation entails analytical methods that produce information about the real binding forms of heavy metals in soils. Electron microscopy studies combined with micro probe or X-ray diffraction analyses can provide initial hints of the distribution of metal compounds in the sample. A quantitative instrumental method was described by Manceau et al. (1996) and Welter et al. (1999). They applied X-ray absorption fine structure spectroscopy (XAFS) to analyse metal species in soil samples. The advantages are that the analytical procedure hardly interferes with the matrix or changes the chemical status of the metal species. The produced XAFS-spectra contain information about the original binding form of metals in different soil components like carbonates or oxides. Taking this information as input data for thermodynamic equilibrium calculations, conclusions can be drawn about how the system might behave if the chemical conditions change (mainly the pH or the complexing anion of a leaching agent). One has to bear in mind that such results describe thermodynamic constraints, and there are, in the real process, further constraints due to the kinetics of the reactions.

Contrary to the *instrumental* speciation, an *operational* method like sequential extraction defines "solid bound metal associations" which are not chemical species but fractions that characterise the distribution of a metal in the heterogeneous soil system. These fractions are operationally defined and contain different chemical species. The procedures of operational speciation cause reactions that might change the matrix or the chemical status of the chemical species. Thus, it is a method which provides information on how the system behaves when the chemical conditions change in the soil or change due to the treatment process.

30.4.1 Applications

There are two main tasks in developing and optimising wet soil clean-up processes: to identify the best conditions for the leaching step in terms of high selectivity and efficiency, and to keep the used liquid in a closed loop for reuse, thereby avoiding waste water. To what extent these tasks can be accomplished is shown by example of two hydrometallurgical circuit processes that make use of organic acids and sodium chloride, respectively.

The four soil samples used for this investigation originate from different industrial sites (table 30.1): an accumulator plant (soil 1), a chlorine alkali plant (soil 2), a Brazilian gold mine (soil 3) and a place of an accident involving elemental mercury (soil 4).

The total metal content of the solid samples was determined by aqua regia digestion followed by analyses with atomic absorption spectroscopy; in the case of mercury the cold vapor technique was used after reduction with $SnCl_2$. Particle sizing was achieved by sieving (250 µm) and by laserspectrometric measurement (63 µm) with a Sympatec Helos Vektra. Organic carbon was measured as loss of ignited fraction by heating dry samples at 550 °C for one hour. To measure the

Table 30.1. Characterisation of the soil samples

	Soil 1	Soil 2	Soil 3	Soil 4
Total mercury [mg kg^{-1}]	n.m.	210	6.6	1,170
Total lead [mg kg^{-1}]	10,850	n.m.	n.m.	n.m.
Content of organic carbon [%]	3.6	2.1	0	0.5
Acid neutralising capacity [mol kg^{-1}]	2.2	n.m.	n.m.	0.03
Amount of particles < 63 μm [%]	49	16	8	5

n.m. not measured

acid neutralising capacity, the consumption of nitric acid was recorded that was necessary to maintain a constant pH of 4 for 24 hours.

As the instrumental speciation method for *lead,* XAFS (X-ray absorption fine structure spectroscopy) was used. XAFS-spectra were measured at the HASYLAB at DESY (Hamburg, Germany) at experimental station X1 and A1. All spectra were measured in the transmission mode, using an elemental metal foil as standard for energy calibration. The spectra of the PbL_3 edge were recorded between 12,800 eV and 13,900 eV, covering the pre-edge, edge and post-edge region of the PbL_3 absorption edge. The first steps of the evaluation procedure were subtraction of the background, normalisation of the spectra to an edge jump of 100 %, and the energy recalibration. Linear combinations of spectra of 17 reference compounds were used to identify the species (Welter et al. 1999).

As the operational speciation method to characterise the lead distribution in soil 1, a sequential extraction scheme was applied which was developed by Zeien and Brümmer (1989) and modified according to Thöming and Calmano (1998). This method divides the distribution of a heavy metal in a soil sample into seven operationally defined fractions.

For the operational speciation of *mercury,* the contents of elemental mercury and calomel (Hg_2Cl_2) were determined by a thermal release analysis. To evaporate only the elemental mercury, the samples were heated at 100 °C, and to evaporate also the calomel, the samples were heated at 150 °C (Windmöller et al. 1996),

Table 30.2. Used sequential extraction as described by Zeien and Brümmer (1989) and Thöming and Calmano (1998)

Step No.	Defined phase	Extractant(s)
1	exchangeable	1.0 mol l^{-1} NH_4NO_3
2	easily soluble	1.0 mol l^{-1} NH_4OAc
3	easily reducible	0.1 mol l^{-1} NH_2OH-HCl
4	EDTA-extractable	0.025 mol l^{-1} NH_4-EDTA
5	moderately reducible	0.2 mol l^{-1} NH_4-oxalate
6	strongly reducible	0.1 mol l^{-1} ascorbic acid 0.2 mol l^{-1} NH_4-oxalate
7	residual	aqua regia

each for 3 hours. Afterwards, the amount of the remaining mercury in each sample was analysed according to the procedure for the measurement of the total mercury content. The amount of elemental mercury was calculated as the difference between the total mercury content and the amount remaining after the 100 °C heating. The amount of calomel then resulted from the total mercury content subtracted by the amount remaining after the 150 °C heating, plus the amount remaining after the 100 °C heating.

To investigate the leaching procedure, three agents were used: acetate and citrate for treating the lead-containing soil 1, and a sodium chloride solution which was electrolytically activated for treating the mercury-containing soils 2, 3 and 4. Both acetate and citrate were applied in single leaching steps, or as a cascade of up to four cross-flow leaching steps followed by rinsing with ammonium acetate. The latter agent used in the rinsing is the same as that used in the second step of the sequential extraction. The remaining solids were digested.

The sodium chloride solution for the mercury leaching was activated electrolytically, pumping it through an undivided electrolysis cell for 20 minutes at $i = 1\ A\ dm^{-2}$. The Plexiglas cell contained a stainless steel cathode and a titanium anode plated with ruthenium oxide.

Thermodynamical equilibrium calculations were undertaken using PHREEQC (available under http://h2o.usgs.gov/software/phreeqc.html) or HSC of Outokumpu, Pori, Finland.

30.4.2 Lead Contaminated Soil

The lead contaminated soil sample 1 was analysed via XAFS and sequential extraction before and after a single leaching step with acetic acid. In fig. 30.7, the XAFS-investigations show that the main chemical species of the original sample are Pb-carbonate (cerussite), Pb-sulfate (anglesite) and Pb associated with humic acid. The operational speciation with the sequential extraction method shows that 10 % of the total amount of lead was exchangeably bound in the original soil sample whereas the main part of the lead was equally distributed in the fractions 2 to 4 that contain about 85 % of the total Pb-content.

The clean-up procedure by single leaching with acetic acid caused a release of lead in all operational fractions except the first one. In fig. 30.7 it can also be seen that there is a shift of lead from the less mobile fractions to the exchangeable fraction. Despite the acidic extraction at pH 2, there is a large amount of weakly bound lead remaining. A better understanding of this phenomenon can be obtained from a comparison of the operationally defined species with the chemical distribution that was analysed instrumentally with XAFS.

The results of the instrumental speciation with XAFS, performed after the leaching, corroborate with the described findings and give additional information about what happened during the leaching. Half of the anglesite was removed and most of the cerussite was dissolved. The reduction of the pH led to a dissolution of the cerussite. The lead bound to humic substances disappeared and a new compound was produced: lead bound to clay.

Thermodynamical equilibrium calculations were undertaken for a system that contained 1 M ammonium acetate, 14.5 mM Pb, 14.5 mM SO_4^{2-} and 14.5 mM

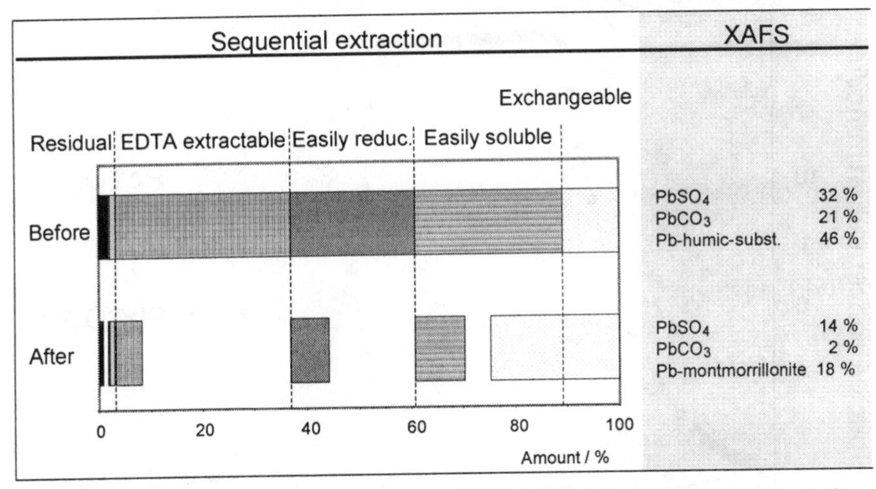

Fig. 30.7. Results of an operational and instrumental speciation of Pb in soil 1 before and after a single leaching step with acetic acid at pH 2

CO_3^{2-}. As fig. 30.8 shows, anglesite is stable below pH 3.5. This means that if the soil contains anglesite, it cannot be removed by acetic acid. However, anglesite becomes dissolved at a pH above 3.5. This is due to the complexing strength of acetate that increases at higher pH. This finding was verified via batch experiments by Stichnothe et al. (1999). Cerussite, on the contrary, is stable under alkaline conditions even in acetate solution, but it becomes dissolved at pH values lower than 5.8.

Similar behaviour of the lead species anglesite and cerussite was calculated for a 0.05 M citrate solution (Stichnothe et al. 1999). For citrate as the complexing anion, the pH-window of a total lead dissolution was determined to be between pH 3.6 and 5.7. But what was the difference between using citrate or acetate? If an acetate solution of pH 5 was used, the extraction yield of a single leaching step was about 40 % (see fig. 30.7, sum of the exchangeable and easily soluble fraction) while at pH 2 the leaching rate was much higher: about 70 % of the lead was leached in a single step (fig. 30.9, step 1). After a fourfold acidic leaching in a cross-flow manner, the cumulative yield became a bit higher: 85 % was reached. However, 15 % of lead was remaining and this amount was easily soluble. It could be extracted using ammonium acetate as a rinsing agent. These numbers are referred to a 100 % label which was calculated as the sum of all leaching and rinsing steps, plus a final extraction with concentrated nitric acid (determination of the residual amount). In opposition to the acetate leaching, citrate was successfully used at pH 5: in the fourfold extraction, 98 % of the lead was leached directly and a last rinsing step kept without effect.

In fig. 30.7, the effect of an acetic acid extraction has been demonstrated for lead. It can be seen that, despite the very acidic extraction, there is a large amount of weakly bound lead remaining. A better understanding of this phenomenon can

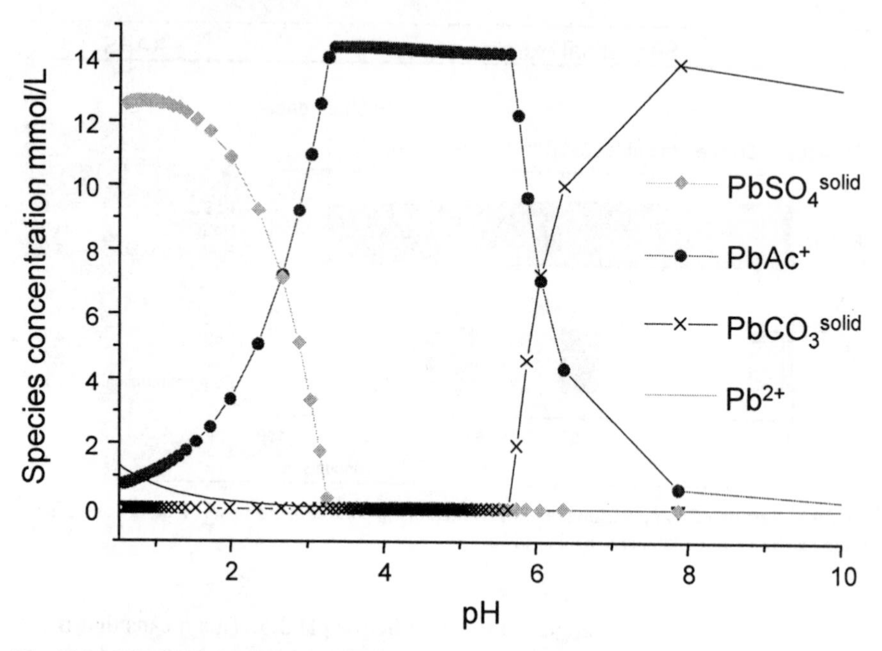

Fig. 30.8. Calculated distribution of Pb-species suspended or dissolved in an acetate solution containing also sulfate and carbonate (after Stichnothe et al. 1999)

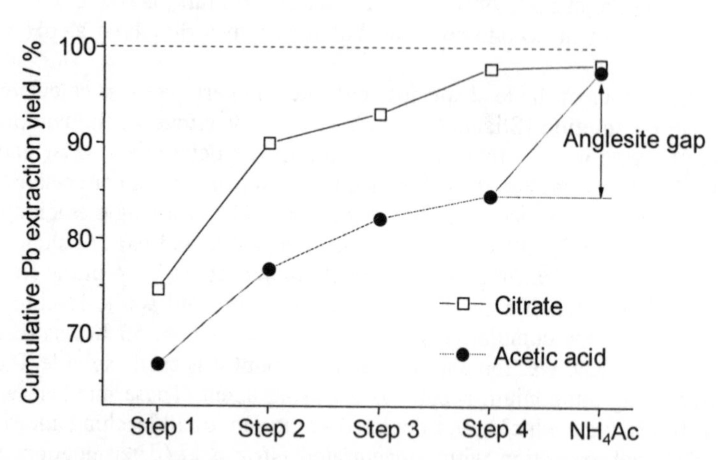

Fig. 30.9. Cumulative amount of lead leached from soil 1 in a series of four cross-flow steps (citrate: pH = 5.5, acetic acid: pH = 2) followed by a rinsing with 1 M ammonium acetate, pH = 5

be obtained from a comparison of the operationally defined species with the chemical distribution that was analysed instrumentally with XAFS. The acetic acid completely extracted the amount of lead bound to humic substances. Furthermore, it dissolved nearly all the cerussite ($PbCO_3$), and to a large extent, the

$PbSO_4$. However, during the leaching, a re-adsorption occurred with 18 % of the total amount of lead being adsorbed by a clay (montmorrillonite) after leaching; in other words, a shift occurred from one soil fraction to another. This was due to solid-liquid (metastable) equilibria, which always determine extraction processes.

How do the three chemical species spread over the operationally defined fractions after the leaching? Here the equilibrium calculations gave an interesting clue: for conditions equal to those of the second fraction of the sequential extraction (acetate solution, pH 5), anglesite ($PbSO_4$) could be dissolved, but not under acidic conditions. This fact indicates that, to a great extent, it was anglesite that was found in the first and the second fraction after the acidic leaching. This means that the availability of anglesite for an acetate dissolution increased. It might be described by a shift of anglesite from fractions with a lower availability to those with a higher one. Furthermore, these findings indicate that the mobile lead fraction should be removable by a set of steps: a leaching with acetic acid followed by a rinsing step that is capable of extracting the exchangeable fraction. This was actually shown afterwards (fig. 30.9). The rinsing step prevented a remarkable quantity of weakly bound lead (25 % of the original lead content) from remaining in the treated soil. As an appropriate pH-window for this rinsing, fig. 30.8 suggests a pH somewhere between 4 and 6.

These results were verified with pure $PbSO_4$ in previous batch experiments (Stichnothe et al. 1999): a 1 M acetate solution dissolved 250 times more anglesite than 10 % acetic acid. This explains the results of the sequential extraction after the treatment with acetic acid, and its limitation as an extraction solvent. Half of the anglesite was removed during the leaching, and it can be assumed that this happened at the beginning of the leaching, when the pH was still high at the surface of the soil particles, due to the acid neutralising buffer capacity of the soil. Then, during the leaching process, even at the particle surface the pH decreased, and no more dissolution of $PbSO_4$ occurred. This limited the decontamination of lead in this soil with acetic acid.

If, instead of the acetate combination, citrate is used, the range of the pH can be much higher (pH about 5 instead of 2) to reach a similar leaching yield. This is due to the complexing strength of citrate, which is much higher at moderate pH values than the complexing strength of acetate.

30.4.3 Mercury Contaminated Soils

When thinking about an appropriate leaching agent for mercury, two peculiarities of mercury have to be considered. The first is that in soil, mercury might appear in elemental form, as is the case for the three soils used for this investigation (table 30.3). To leach elemental mercury, it has to be transformed to a soluble species. Thus, as a first special quality, the leaching agent has shown an oxidising character.

The second peculiarity of mercury is its high affinity to organic matter, like humic acids, if the mercury species are in the divalent state (Xu and Alard 1991). To overcome this, a strong complexing anion is needed to keep the oxidised mercury in the leachate. Chlorocomplexes of mercury are known for their high stability constants, high enough to compete successfully with organic matter in a

Table 30.3. Content of total mercury and operationally defined Hg species in soils 2, 3 and 4 (amounts and standard deviations given in mg kg^{-1})

Sample	Hg_{tot}	Hg(0)	Hg(I)
Soil 2	210 ± 11.7	5.7 ± 5.6	27 ± 21.8
Soil 3	6.6 ± 0.34	1.8 ± 0.28	2.3 ± 0.56
Soil 4	1170 ± 16	990 ± 11	72 ± 4

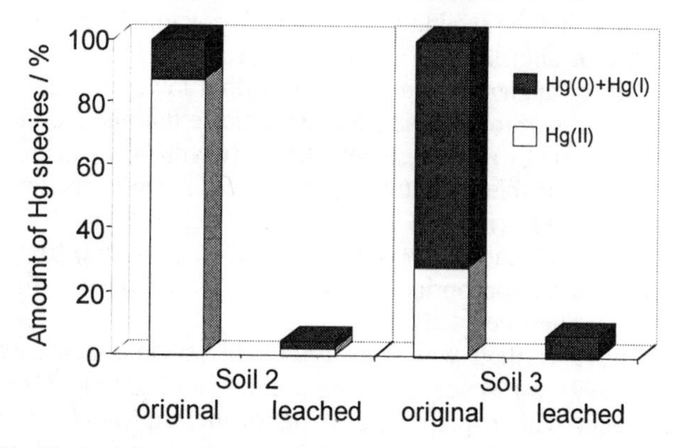

Fig. 30.10. Distribution of operational mercury species in soils 2 and 3 before and after a single leaching step using an NaCl/HOCl-solution

leaching step (fig. 30.10). Furthermore, the bulk of the formed mercurychloro-complexes are negatively charged, and thus they cannot become re-adsorbed by organic matter.

Such an oxidising chloridic leaching agent can be produced electrochemically in-line in a circuit process developed by Thöming et al. (1999). Electroleaching offers a cheaper alternative and is a hydrometallurgical process combining wet extraction and electrolytic preparation of the leachate. There are two advantages over retorting processes. Because of far larger thermodynamic separation factors, the capital costs are much lower. Moreover, air pollution problems are avoided. Fig. 30.11 shows the principle of the electroleaching process.

In a column (sandy soil) or stirred tank (loamy soil), the solid is leached by an oxidising and complexing agent, which is a sodium chloride solution with low concentration hypochlorous acid. After leaching, the leachate is treated electrolytically to deposit the metals in the elemental form cathodically, and to regenerate the leaching agent anodically for re-circulation. The leaching mechanism is based on the oxidation of mercury and complexation of mercury ions by chloride ions, forming a tetra-chloro-mercurate-complex, for instance. The speciation of mercury is controlled by the main process parameters, such as pH, redox potential and element concentrations. How leaching efficiency and electrolytic mercury deposition are both dominated by the mercury species in solution, is illustrated.

During electrolysis, the redox potential of the chloride solution increases dramatically due to the production of hypochlorous acid, HClO (equation 30.3).

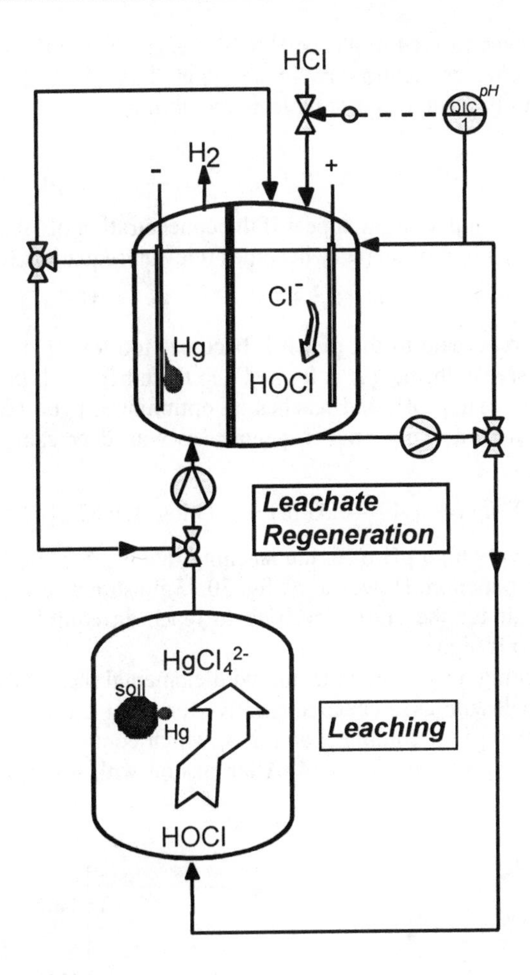

Fig. 30.11. Principle of the electroleaching process

Anodic reaction 1 $\quad H_2O - 2\,e^- + Cl^- \quad \rightarrow \quad ClO^- + 2\,H^+ \quad$ (30.3)

This reaction competes with the oxidation of water (equation 30.4) but it is favoured if titanium electrodes are used that are plated with ruthenium oxide.

Anodic reaction 2 $\quad H_2O - 2\,e^- \quad \rightarrow \quad \frac{1}{2}\,O_2 + 2\,H^+ \quad$ (30.4)

The pH, however, is not affected by this competition. In both cases, two protons are produced anodically (given that a moderate current density i is used, e.g. $i = 0.3\ A\ dm^{-2}$). These protons become neutralised by the hydroxide ions that are produced cathodically:

Cathodic reaction 1 $\quad 2\,H_2O + 2\,e^- \quad \rightarrow \quad H_2 + 2\,OH^- \quad$ (30.5)

As a result of equation 30.3, the solution is capable of readily oxidising Hg(0) and Hg(I) compounds in soils to Hg(II). At low chloride concentrations, the mercurychlorocomplexes formed might be $HgCl_2^{aq}$, as calculated thermodynamically. Whereas, at chloride concentrations higher than 0.5 mol l^{-1}, the extracted Hg-species form mainly strong and soluble tetrachloromercurate complexes (equation 30.6).

$$Hg + HOCl + 3\ Cl^- + H^+ \quad \rightarrow \quad HgCl_4^{2-} + H_2O \qquad (30.6)$$

A slight decrease in pH might appear if the concentration of leached mercury is high. Then the deposition of mercury becomes relevant (equation 30.7).

$$\text{Cathodic reaction 2} \qquad HgCl_4^{2-} + 2\ e^- \quad \rightarrow \quad Hg^0 + 4\ Cl^- \qquad (30.7)$$

Attention has to be paid to the pH: if it becomes too low (pH 4 and less), poisonous and corrosive chlorine gas is formed. On the other hand, the oxidation rate increases with decreasing pH, and reaches an optimum at pH 4 (fig. 30.12). This rate is linked to the redox potential that increases with decreasing pH, following equation 30.8.

$$E^0/mV = 1.49 - 0.03\ pH + 0.03 \log([HOCl]/[Cl^-]) \qquad (30.8)$$

At a comparatively high pH of 6, the leaching rate is poor but the procedure is free of chlorine production. However, as fig. 30.13 illustrates, it is a time consuming process to wait for the Hg(0) leaching to reach an equilibrium state, if the solid/liquid ratio σ is high.

Instead of leaching a soil contaminated with elemental mercury in a cascade of several cross-flow leachings, a cheap alternative for a fast and efficient clean-up is reverse flow leaching in a column. For soil 4, which contained mainly elemental mercury, a result is given in fig. 30.14. After rinsing with water, the total Hg re-

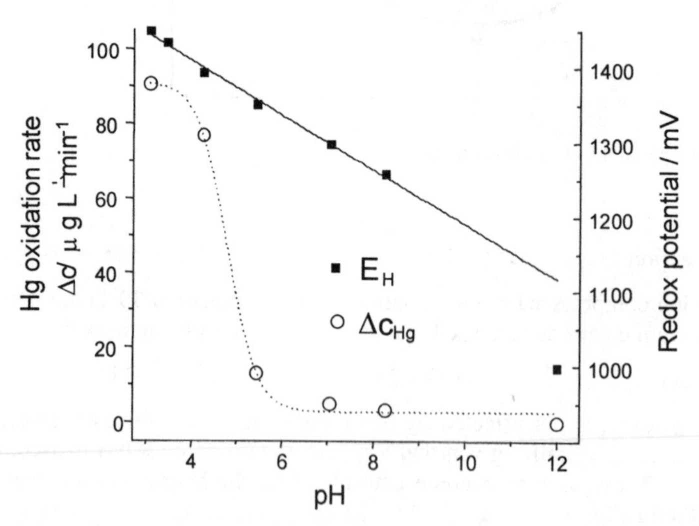

Fig. 30.12. Oxidation rate of elemental mercury in an electrochemically activated sodium chloride solution, with the straight line representing the redox potential, calculated with equation 30.8

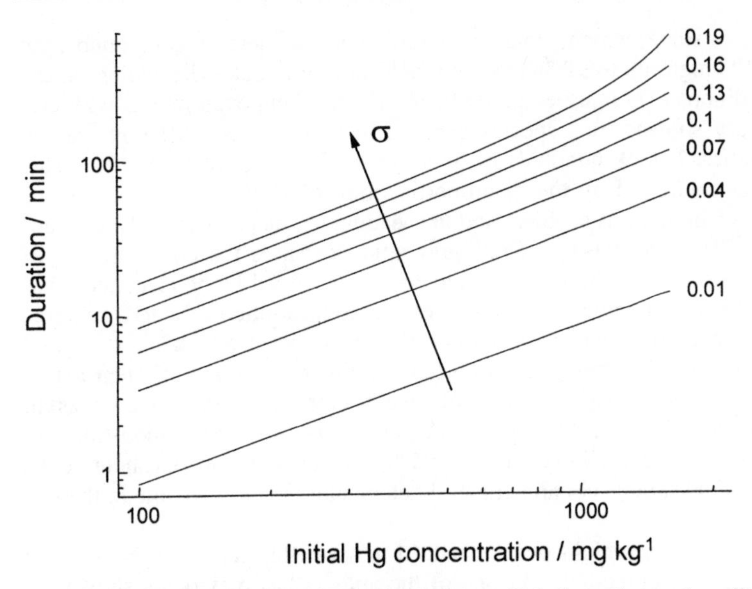

Fig. 30.13. Time to reach a total Hg(0) dissolution at pH 6, depending on the solid/liquid ratio σ (in kg l^{-1}) and the initial Hg concentration of the soil

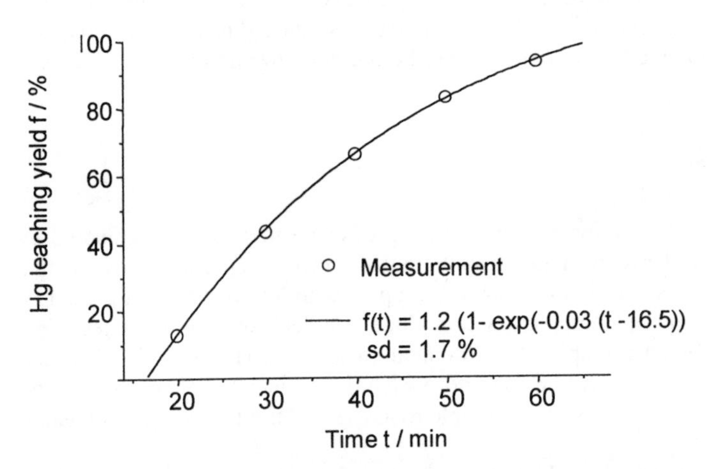

Fig. 30.14. Leaching of 1 kg of soil 4 in a column percolated by reverse flow, with 48 mL min^{-1} of a 1 M NaCl solution kept in a circuit, and regenerated permanently by electrolysis

moval rate was 99.6 %.

In contrast to the leaching of lead, which is determined by the solid/liquid equilibria of Pb(II) species, the leaching of mercury is redox controlled. Since redox reactions are much slower than complex reactions, attention is mainly paid to the oxidation kinetics of elemental and monovalent mercury. Thus, the speciation here is of less importance, and used only to quantify the amount of mercury that has to be oxidised.

In the case of soil 2, which contained mainly oxidised mercury compounds, the leaching rate is high: up to 98 % leaching yield in five minutes depending on the leaching conditions (Thöming et al. 1999). As the leaching experiments with elemental mercury showed, this species behaves in a completely different way in terms of kinetics. It took hundreds of hours to reach an equilibrium stage. This equilibrium is controlled by the disproportionation of Hg(0), Hg(I) and Hg(II). The key to overcoming this problem and to speeding up the process, is the concentration of Hg(II): a decreasing solid/liquid ratio decreases the concentration of Hg(II) and thus increases the leaching rate, as demonstrated in the calculated diagram fig. 30.13. On the other hand, a decreasing concentration of Hg(II) requires more technical equipment and raises costs. What might be a way out?

Instead of greatly decreasing the solid/liquid ratio σ, a fast and efficient leaching might be obtained using a cascade of several cross-flow leachings, or, as an alternative, a leaching in a column percolated by reverse flow. As demonstrated in fig. 30.14, there was a leaching yield of 95 % for soil 4 within one hour, even though the original sample contained mainly elemental mercury, visible to the eye as small droplets.

Generally, a problem might occur if HOCl is used: this causes an increase of chlorinated organic compounds. As one of the authors has previously shown for soils 2 and 4, this depended on the initial concentration of these compounds, as well as on the leaching parameters (Thöming and Franke 1998). In soil 2, a total of 25 chlorinated organic compounds was detected by GC/MS analyses. After leaching, the total amount of chloro-compounds was reduced in soil 2, whereas for soil 4, these compounds were found neither before, nor after treatment.

30.5 Conclusions

Leaching is a possible option for the treatment of excavated soil contaminated by heavy metals. In order to evaluate this potential, a protocol is proposed involving a schedule of combined speciation methods, coupled with leaching tests (fig. 30.15). The suitability of the sequential extraction alone as a decision tool is limited. This could be shown by an example of a Pb-contaminated soil. However, in combination with the instrumental XAFS-method, which determines the chemical binding form of heavy metals in soils, it is possible to assess different lixiviants and leaching parameters.

Detailed knowledge of binding forms is most important for hot spots, where metal species form their own solid phases, e. g. as precipitated salts or as a liquid phase (if mercury is present). Only a sophisticated knowledge of the chemical state of these phases enables a systematic search for appropriate leaching parameters to be made – in other words, to optimise the leaching conditions.

It was shown that, on the one hand, leaching processes can lower total amounts of heavy metals by up to 99 %, but on the other hand, a total separation of mobile fractions cannot be guaranteed. This limits the applicability of these processes, and a great effort has to be made to control remaining amounts of mobile metals. XAFS-results have shown that a separation of just the mobile metal fractions from the soil is not practicable due to equilibrium adjustments which happen during

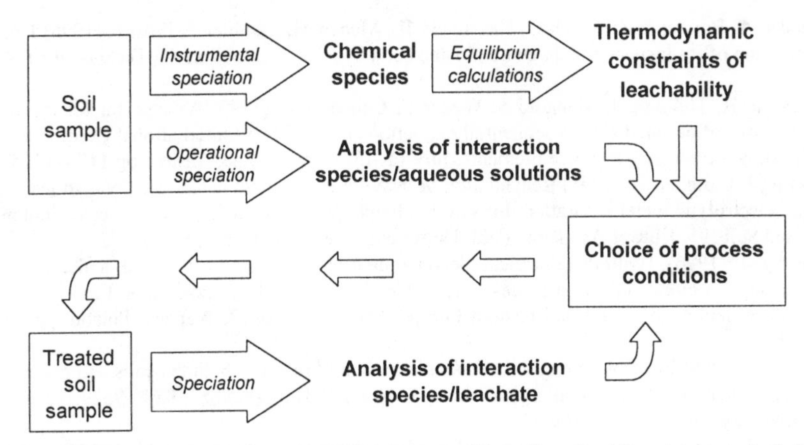

Fig. 30.15. Systematic method for identifying best conditions for a leaching step as part of clean-up techniques for metal contaminated soils

extraction processes.

The kinetic behaviour during extraction has to be investigated experimentally. Although there are sophisticated "Surface Complex Formation" models describing solid-liquid equilibria, treatment results can hardly be predicted. These models assume the metal bindings being surface controlled, but this is not the case in hot spots. Especially if soils are highly contaminated, it can be assumed that particles of precipitated metals are forming their own soil-phases. In extraction processes, these phases behave differently to surface bound metals and could lead to "mobile" metal compounds remaining.

Soil 1, which was highly polluted with lead, could be leached efficiently with acetate solution if a sequence of fourfold acidic steps and a final rinsing under weak acidic conditions was applied. If citrate is used instead of acetate, the whole leaching can be performed at moderate pH conditions (pH about 5). This is advantageous because the soil matrix is damaged to a lesser extent.

Also, three mercury-contaminated soils could be cleaned hydrometallurgically in a circuit process, if sodium chloride is used, and oxidising conditions are chosen. The integrating step of this process is the electrolysis: the mercury loaded aqueous stream can be regenerated in one step that combines the cathodic deposition of mercury and the anodic re-oxidation of the chloride.

References

Cox CD, Shoesmith MA, Ghosh MM (1996) Electrokinetic remediation of mercury-contaminated soils using iodine/iodide lixiviant. Environ Sci Technol 30(6): 1933–1938

Hansen HK, Ottosen LM, Kliem BK, Villumsen A (1997) Electrodialytic remediation of soils polluted with Cu, Cr, Hg, Pb and Zn. J Chem Tech Biotechnol 70: 67–73

Hempel M, Thöming J (1999) Remediation techniques for Hg-contaminated sites with emphasis on Europe. In: Ebinghaus R, Turner RR, de Lacerda LD, Vasiliev O, Salomons W (eds) Mercury contaminated sites. Characterization, risk assessment and remediation. Springer Environmental Science, Berlin, Germany, pp 114–130

Manceau A, Boisset M-C, Sarret G, Hazemann JL, Mench M, Cambier P, Prost R (1996) Determination of lead: speciation in contaminated soils by EXAFS. Environ Sci Technol 30: 1540–1552

Stichnothe H, Thöming J, Mangold S, Welter E, Calmano W (1999) Assessment for clean-up processes of polluted soil by sequential extraction and XAFS. 5th International Conference on the Biogeochemistry of Trace Elements. July 11–15, 1999, Vienna, Austria, pp 1174–1175

Thöming J, Calmano W (1995) Remediation of heavy metal contaminated soils by acid leaching and electrolytic metal separation. In: van den Brink WJ, Bosman R, Arendt F (eds) Contaminated Soil '95. Kluwer Academic Publ, Dordrecht, The Netherlands, pp 895–902

Thöming J (1996) Hydrometallurgical decontamination of mercury polluted soils: electroleaching – a mobile clean-up process. Proc Third International Symposium on Environmental Contamination in Central and Eastern Europe, September 10–13, Warsaw, Poland, pp 907–909

Thöming J (1998) Electroleaching for a decontamination of mercury polluted soils and residues – development of a hydrometallurgical circuit process. Ph D thesis, ISSN 0344-9629, Technical University Hamburg-Harburg, Germany

Thöming J, Calmano W (1998) Applicability of single and sequential extractions for assessing the potential mobility of heavy metals in highly polluted soils. Acta hydrochim hydrobiol 26: 338–343

Thöming J, Franke S (1998) Decontamination of mercury polluted soils by electroleaching: assessment of the ecological clean-up benefit. Proc Sixth International FZK/TNO Conference on Contaminated Soil, May 17–21, 1998, Thomas Telford Publ, London, Great Britain, pp 1123–1124

Thöming J, Stichnothe H, Calmano W (1998) Clean-up of heavy metal polluted soils: applicability and limitations of hydrometallurgical techniques. Proc Sixth International FZK/TNO Conference on Contaminated Soil, May 17–21, 1998, Thomas Telford Publ London, Great Britain, pp 1093–1094

Thöming J (1999) Modeling of a redox-activated extraction of mercury polluted soil for a process integration. 5th International Conference on the Biogeochemistry of Trace Elements, July 11–15, 1999, Vienna, Austria, pp 814–815

Thöming J, Sobral L, Wilken R-D (1999) Electroleaching: a mobile clean-up process for mercury contaminated materials. In: Ebinghaus R, Turner RR, de Lacerda LD, Vasiliev O, Salomons W (eds) Mercury Contaminated Sites. Characterisation, Risk-Assessment and Remediation. Springer Environmental Science, Berlin, Germany, pp 442–455

Wasay SA, Arnfalk P, Tokunaga S (1995) Remediation of a soil polluted by mercury with acidic potassium iodide. J Haz Mat 44: 93–102

Welter E, Calmano W, Mangold S, Tröger L (1999) Chemical speciation of heavy metals in soils by use of XAFS spectroscopy and electron microscopical techniques. Fresenius J Anal Chem 364: 238–244

Windmöller C, Wilken R-D, Jardim WF (1996) Mercury speciation in contaminated soils by thermal release analysis. Water Air Soil Pollution 89: 399–416

Xu H, Allard B (1991) Effects of a fulvic acid on the speciation and mobility of mercury in aqueous solutions. Water, Air and Soil Poll 56: 709–717

Zeien H, Brümmer GW (1989) Chemische Extraktionen zur Bestimmung von Schwermetallbindungsformen in Böden. Mitt Dtsch Bodenkd Ges 59: 505–510

* Parts of this chapter are a modified extract from our paper "Hydrometallurgical approaches to soil remediation – process optimization applying heavy metal speciation" published in Land Contamination & Reclamation 8(1): 19–32 (2000). We thank EPP Publications for the permission to use figures from the above paper in this book.

31 Supercritical Water and Supercritical Carbon Dioxide for Cleaning of Soil Material

G. Brunner[1], B. Misch[1], A. Firus[2], K. Nowak[3]
[1] Technical University of Hamburg-Harburg, Department of Thermal Process Engineering, Eißendorfer Str. 38, 21073 Hamburg, Germany
[2] Bayer AG, ZT-TE, 51368 Leverkusen, Germany
[3] Bremer Sonderabfall Beratungsgesellschaft mbH, 28195 Bremen, Germany

31.1 Introduction: Soil and Scope of the Work

Thermal treatment of contaminated soil has gained great interest. If biological methods have no sufficient cleaning effect, the soil material must be deposited or incinerated. Thermal treatment of contaminated soil with supercritical water (T_C = 374 °C, P_C = 22.1 MPa), in contrast to incineration, leads to clean soil material without creating nitrogen oxides and sulfur dioxide. Depending on the type of contamination, reaction products are CO_2, H_2O, inorganic acids and highly volatile hydrocarbons. The process is carried out at elevated pressures (25 MPa) and, compared to incineration, at moderate temperatures (375–600 °C). Residence times are short (< 120 s) and reaction products can be controlled by conditions of state relating to the oxidation (Brunner 1994).

A supercritical fluid can be regarded as a dense gas. The physico-chemical properties are between those of a liquid and a gas. Near the critical point, the density of the fluid can be varied easily over a wide range by altering the applied pressure. In the same way, the viscosity, solvent power and dielectric constant change their values significantly. The increasing solvent properties of the supercritical water, with respect to organics, lead to complete miscibility, which is the essential effect for the extraction of soil with supercritical water. The produced oil-in-water emulsion can be further treated by supercritical water oxidation. The lack of surface tension and the complete miscibility of the reactants – water, organic contamination and oxygen – lead to effective oxidation.

Research on extraction of organic compounds from soil material with sub- and supercritical water had not been published before our investigation started in 1989. However, extraction with supercritical carbon dioxide had been published, and the influence of water as an entrainer is well documented (Michel 1992; Schleußinger 1996). Supercritical water oxidation of effluents has been intensively investigated. Comprehensive reviews have been written by, amongst others, Mishra et al. (1995) and Savage et al. (1995).

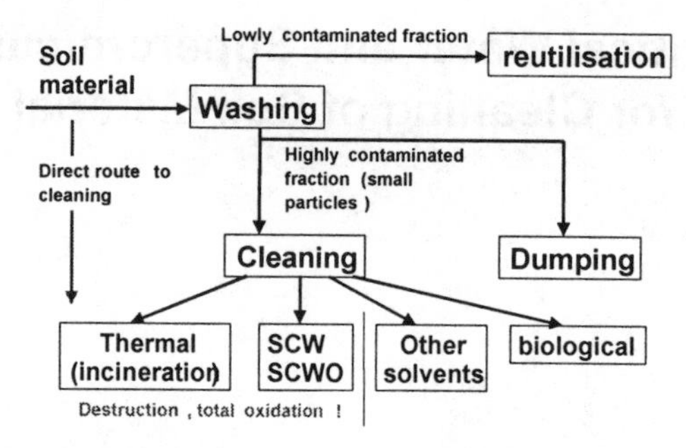

Fig. 31.1. Cleaning processes for soil and soil materials. SCW – supercritical water, SCWO – supercritical water oxidation.

31.2 Properties of Water near the Critical Point

It is basic knowledge that the physico-chemical properties of water and its solvent power for organic substances, salts and gases are necessary for a supercritical water destruction process. P,v,T and transport properties are that of a supercritical gas: Density is relatively high – up to liquid-like densities – but varies strongly with slight changes in pressure and temperature. Viscosity is of the order of a normal gas and the diffusion coefficient is at least one order of magnitude higher than that of a liquid.

The unusually high dielectricity constant of water of about 80 Debye under ambient conditions decreases with increasing temperature, but remains high in the critical and supercritical region. The ion product of water (10–14 at ambient conditions) decreases rapidly at temperatures in the critical region if pressure is lower

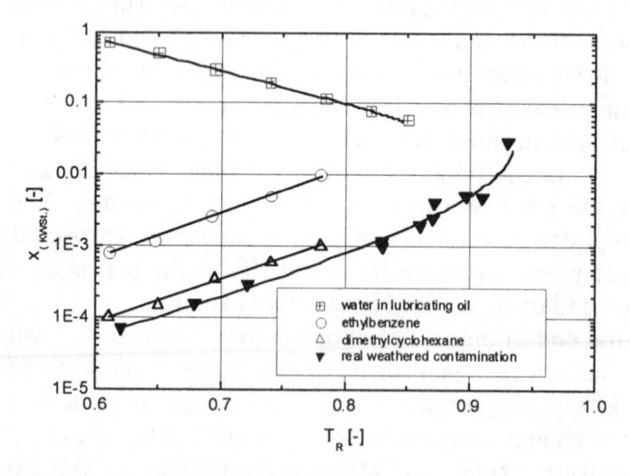

Fig. 31.2. Solubility of a real contamination (extracted from various soil materials) in water

than the vapour pressure. Yet at supercritical pressures (> 22 MPa), the decrease of the ion product is less dramatic, and ionic properties of water are partially maintained in the supercritical region.

Solubility of gases in water is high in the critical region. For conditions of state, which are of interest for destruction processes with supercritical water, gases like O_2, N_2, NH_3, CO_2, and CO are completely miscible with water. The solvent power of water decreases for inorganic compounds in the critical region. Furthermore, it is drastically reduced in the region of 450 °C. Organic compounds, on the other hand, are readily dissolved by water in the critical and supercritical region up to total miscibility (fig. 31.2). However, one must bear in mind that these statements are valid for binary or quasi-binary systems only. Phase equilibria of ternary and multicomponent systems of water, organic compounds, inorganic salts and common gases may deviate from binary behaviour in such a way that a destruction-of-waste process is affected.

31.3 Extraction (Removal of Contaminants from Soil Material)

31.3.1 Semi-Continuous Extraction of Organic Contamination from Soil Material with Water

Cleaning processes for soil material, such as washing processes or burning, face some problems. Soil material contaminated with halogenated compounds, or soils that contain a high percentage of fine particles ($d_P < 63$ μm) cannot be treated. Furthermore, high emissions, high residual concentrations, and large quantities of highly contaminated sludge are problems of the currently applied processes.

Table 31.1. Investigated soil materials

	1	2	3	4
Type of contamination	diesel artificial	diesel/lubricating oil; weathered	diesel/lubricating oil; weathered	diesel/lubricating oil; weathered
Amount of contamination [mg hc (kg dry solid material)$^{-1}$]	24,418	136,587	20,231	35,066
Age [years]	0.5	≈ 20	≈ 2	≈ 45
Type of soil material	weak loamy sand	strong clayey loam	sand	strong clayey sand
Particle mean diameter (sphere of equivalent volume) [mm]	182	18	211	213
Pretreatment	–	sieved at 710 mm	from washing process	sieved at 710 mm

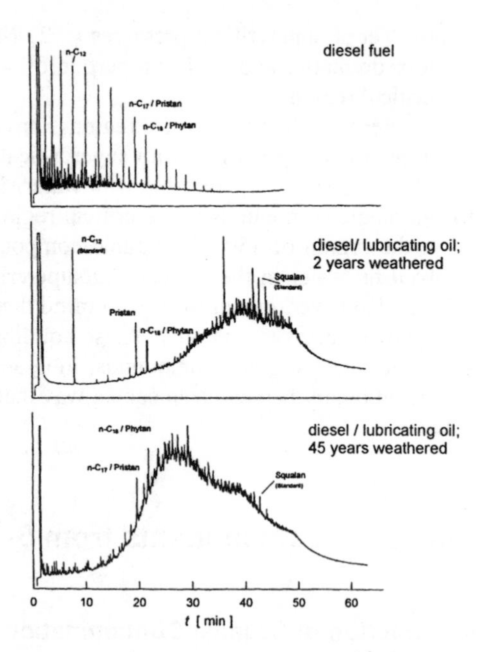

Fig. 31.3. Organic contamination of weathered soil materials (see table 31.1)

Cleaning of soil with supercritical water is an alternative that avoids some of the above-mentioned problems. In our case, the goal was to clean soil material and to modify the contaminating compounds to such a degree that they become biologically degradable, or to destroy them with supercritical water oxidation.

Different types of organic contamination are illustrated by the chromatograms in fig. 31.3. Some more data are given in table 31.1.

Materials with a high content of weathered compounds (≈ 100,000 mg $(kg\ ds)^{-1}$ of hydrocarbons; ds = dry substance) and with a high content of fine mineral particles (up to 85 % with $d_P < 63$ μm) can be cleaned effectively (see table 31.2). Remaining contaminants are present at concentrations below 100 mg kg^{-1}. During treatment with supercritical water, an effluent stable emulsion of oil in water is produced, with concentrations of about 1000–2000 mg l^{-1} TOC, which can be treated with microorganisms with limited success, or totally destroyed by supercritical water oxidation (see below).

The separation of contaminants from soil material (extraction) was carried out in a fixed-bed reactor (fig. 31.4). The solvent passed through a fixed bed of soil material (460 cm^3, e). Whenever water was applied, reaction experiments were carried out downstream in an 88 cm^3 pipe (o), which could be connected in series to the autoclave or operated separately. Oxygen (synthetic air) for the reaction was added externally by water displacement (g). Carbon dioxide was used in some experiments as an alternative solvent only for the extraction. In these cases, the whole process gas was discharged at (r). The applied pressure was always 25 Mpa, and a single fluid phase could be maintained. Temperature and flow rate of water were varied between 200 and 450 °C, and 1 and 2.5 l h^{-1} respectively. The organic content (table 31.1) of the soil material was analysed by Soxhlet-extraction with cyclohexane/acetone as solvent, and the hydrocarbon-content of the aqueous solutions (r) was analysed by gas chromatography and infrared spectroscopy.

Artificial contamination (table 31.1, column 1) could be removed from the soil 4 to 6 times faster than weathered contamination (table 31.1, columns 2, 3, 4). This effect is remarkable at very high degrees of extraction (E [(mohc-mhc) mohc^{-1}] > 0.99). Most of the contamination is removed independently of the age of the contamination, therefore, it is supposed that the hydrocarbons exist as a separate phase between the particles. Only a small amount is adsorbed onto the soil material, and the solvent ratio has no significant effect on the

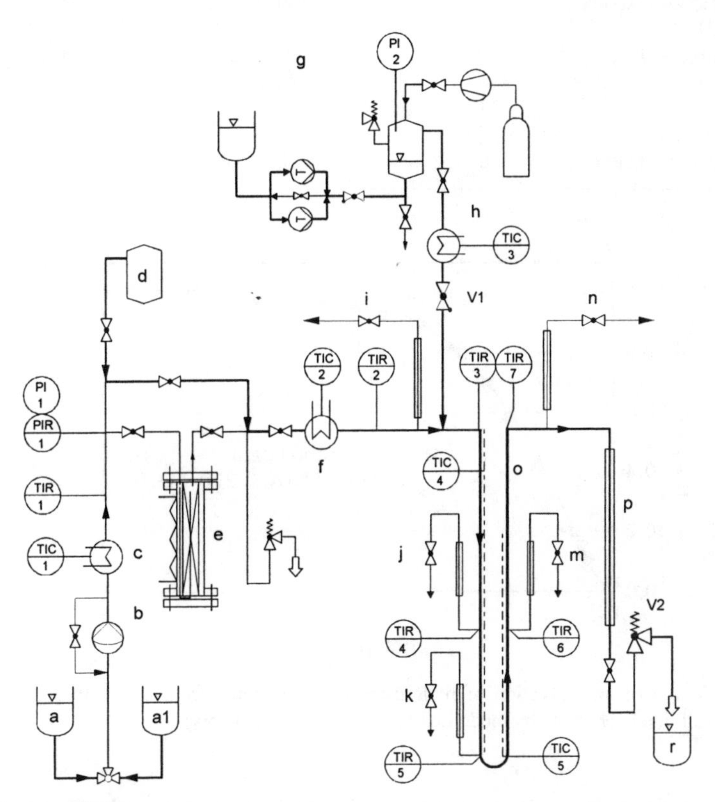

[a, a1] Water supply
[b] Membrane pump
[c] Preheater
[d] Buffer autoclave
[e] Extraction autoclave, fixed bed (soil material)
[f] Double-pipe heat exchanger
[g] Synthetic air supply
[i, j, k, m, n] Sample system
[o] Reactor
[p] Double-pipe heat exchanger
[r] Process water (CO_2, H_2O, oil)

Fig. 31.4. Simplified flow scheme for extraction and reaction experiments

Table 31.2. Extraction results of real contaminated soil materials (adapted from Nowak 1996)

Soil material from:	Petrol Station	Industrial Site	Old Barrel Depot
Sauter diameter [μm]	30	7	14
Contamination	lubricating oil aged 1 year	lubricating oil aged > 20 years	PAH aged > 45 years
Initial concentration [mg hydrocarbons (kg dry substance)$^{-1}$]	20,200	103,500	11,050
Extraction			
Final concentration [mg hydrocarbons (kg dry substance)$^{-1}$]	< 20	< 100	< 10
Amount extracted [%]	> 99.9	> 99.9	> 99.9
Temperature [°C]	390	382	392
Pressure [MPa]	24	24	25
Solvent/soil ratio [kg water (kg dry matter)$^{-1}$]	6	6	12

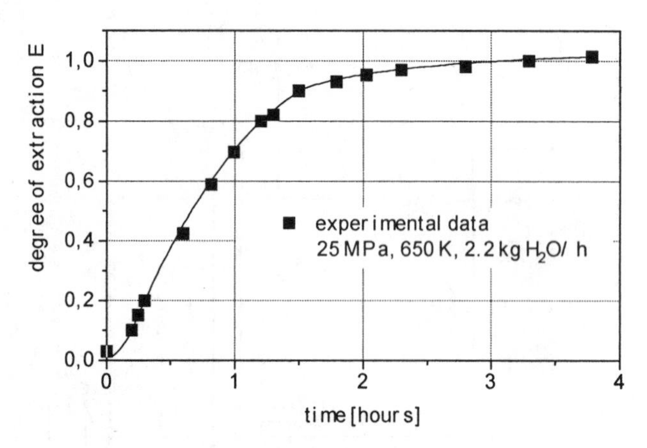

Fig. 31.5. Course of an extraction of hydrocarbon contaminants from soil with supercritical water. The full line is obtained by modelling the extraction (see below)

total degree of extraction. Phase behaviour is not the only reason for the excellent extraction results for water under supercritical conditions. At a temperature of 450 °C (25 MPa), the hydrocarbon contamination is already completely miscible with water, yet the extraction rate is higher at 380 °C (25 MPa). The increase of the diffusion coefficient at supercritical water conditions is supposed to be the reason for this. During the extraction with supercritical water, the pore volume of the soil material increases. The effect is greatest in the soil material with the highest content of fine particles (table 31.2, column 2). In this case, the contamination diffused into the mesopores. Helium-pycnometer measurements showed that the density of the solid material is lower after an extraction with supercritical water.

Extraction experiments with non-contaminated soil material demonstrated that natural organic (humic) substances had been removed. Clay minerals are not extracted due to their low solubility in supercritical water.

Fig. 31.5 shows the relative amount of extracted contamination over the course of an extraction experiment. The extraction can be described by a mathematical model predicting extraction results for different operating conditions and soil materials (Nowak 1996; Firus and Brunner 1996; Firus 1996). Modelling is discussed in context with continuous extraction below.

31.3.2 Semi-Continuous Extraction of Organic Contamination from Soil Material with CO_2

Experiments with supercritical CO_2 as a solvent were carried out for comparison. For analytical purposes, CO_2 is regularly used, in combination with modifiers, for analysing soil material. Literature results raised expectations that contaminated soil material could be cleaned with supercritical carbon dioxide. A review of the literature can be found in the thesis of Michel (1992). Our results showed that supercritical water is a much better agent than supercritical carbon dioxide (fig. 31.6) for extracting weathered contamination from soil (Firus 1996; Firus et al. 1997). The total degree of extraction with supercritical carbon dioxide is low, $E_{CO2} \approx 0.1 \rightarrow 0.2$, while the total degree of extraction using supercritical water as the solvent is very high, $E_{H2O} > 0.99$. The solubility of the hydrocarbons in supercritical CO_2 is very low (e.g. at 60 °C, 25 MPa, the solubility of naphthalene in CO_2 is: 0.1 [mole mole^{-1}]). Even an increase of the mass flow of CO_2 to8 $g_{CO2}\ g_{ds}^{-1}$ had no significant effect (fig. 31.6, 31.7 and 31.8). The total degree of extractionthat could be achieved remained at $E_{CO2} \approx 0.20$. The addition of water as a

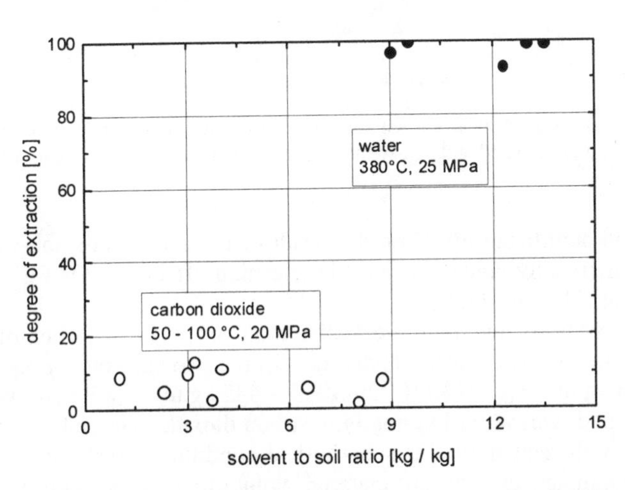

Fig. 31.6. Extraction results (total degree of extraction) for a weathered soil material. Comparison of supercritical water and carbon dioxide as solvents

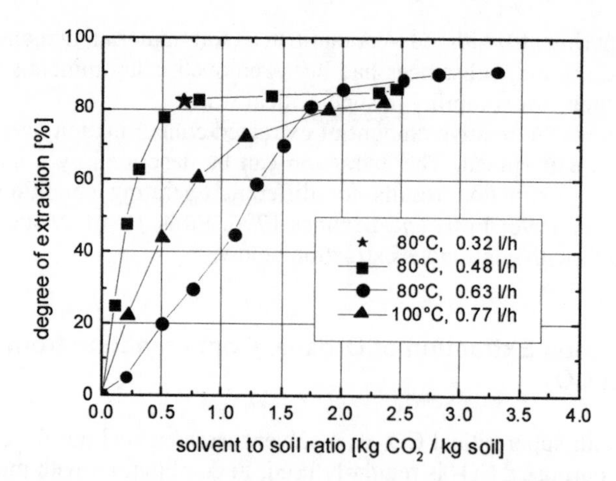

Fig. 31.7. Extraction for a non-weathered soil material with supercritical carbon dioxide at 20 MPa. The soil material was artificially contaminated with diesel oil.

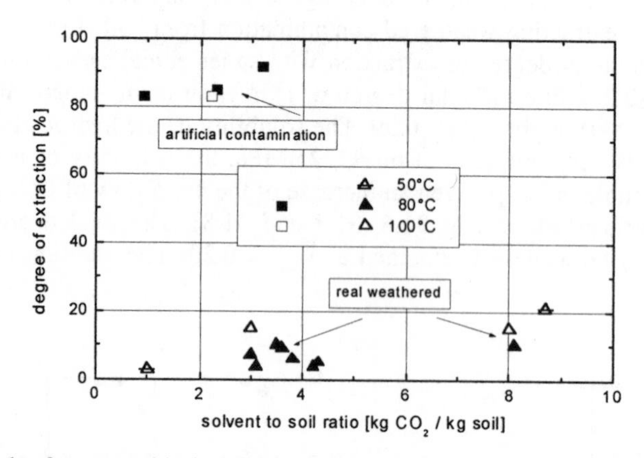

Fig. 31.8. Results for supercritical carbon dioxide extraction (total degree of extraction) at 20 MPa carried out on weathered and non-weathered (artificially contaminated) soil material

modifier had no significant effect on the result of extraction. For non-weathered soil material (artificially contaminated), the cleaning effect of CO_2 was found to be sufficient (fig. 31.7 and 31.8).

Supercritical water is much more effective than supercritical carbon dioxide at cleaning soil material. At comparable densities of the two solvents – ρ_{H2O} (380 °C, 25 MPa) = 445 kg m^{-3}, ρ_{CO2} (100 °C, 20 MPa) = 482 kg m^{-3} – soil material can be effectively cleaned with water but not with carbon dioxide. The polar structure of water and the high temperature at supercritical conditions break any bonds between the contaminants and the soil material. Solubility does not limit the extraction and a high diffusivity favours rapid desorption.

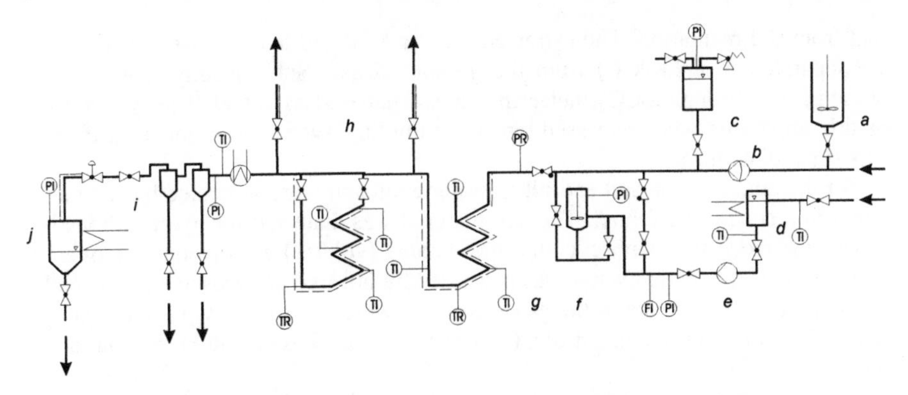

a soil suspension; b membrane pump; c buffer autoclave; d CO_2 condenser; e CO_2 pump; f stirring autoclave (V = 200mL); g fluidized bed heate: extraction zone 1: V = 135mL, extraction zone 2: V = 115mL; h sample system; i cyclones; j process water

Fig. 31.9. Flow scheme of continuous extraction apparatus

31.3.3 Continuous Extraction of Solid Soil Material

A semi-batch extraction with supercritical water is able to remove all of the initial hydrocarbon contamination from a soil material, even after extended weathering. The residence time of the semi-batch extraction was about 6 hours. Although extraction results are excellent, extraction times are too long. In order to improve productivity, the extraction process must be operated continuously.

A new apparatus was built for continuous extraction experiments on contaminated soil material with supercritical water. Fig. 31.9 shows a flow scheme of the apparatus. For a homogeneous charge, an electromagnetic stirrer is placed in the feed vessel (a). A static mixer in the vessel prevents the suspension from performing a tornado. Soil-in-water suspensions (0–10 wt-%) are pumped into the piping by a conventional membrane pump (b). The suction and discharge valves of the pump each use a two-ball configuration. The whole apparatus is protected against excessive pressure (> 35 MPa) by means of a valve connected to the nitrogen filled buffer (c). Different extraction plant were built with different lengths of extraction piping, where all piping was made of stainless steel o.d. 0.25 inch, i.d. 0.125 inch. The apparatus used for the mixed contaminated soil had an extraction volume of 250 ml; all other results were achieved using a plant with a total extraction volume of 88 ml, parted into a preheating section (38 ml) and an extraction zone (50 ml). The extraction tube of the configuration shown above is heated by 2 fluidised sand bed heaters, which can be regulated separately. The temperature of each heater is recorded as well as the system pressure. System pressure can be visually checked at two points. After the extraction process, the suspension is cooled down. Through a pneumatically operated check valve, which is regulated by a pressure control system, the suspension is allowed to expand to ambient pressure. For proper operation, the ceramic spindle of the check valve requires a pressure of 0.6 MPa (compressed air). An abrupt opening of the valve prevents the seat from sedimentation. The extracted soil material and the oil-in-water emulsion are separated in a flask (j) from the gaseous phase. Subsequently, the gaseous phase passes through a CO_2-meter and an evacuated glass vessel. The cyclones (i) of this apparatus were bypassed and the sampling system was not used during experiments with soil.

For the analysis, the soil and oil-in-water emulsion were separated by sedimentation in a separating funnel. The soil material was analysed for hydrocarbons by supersonic extraction with cyclohexane/acetone (vol 1:1) as solvent. The overall oil content of the loaded water was measured by infrared spectroscopy, and single hydrocarbon components in the gaseous phase were measured by gas chromatography. The volumetric content of CO_2 in the gaseous phase could be measured in-line by a CO_2-meter.

The soil material used for the experiments, a clayey loam with an equivalent particle diameter of 10 µm, was taken from a former paint factory. It was weathered for more than 20 years and contained 19 wt-% hydrocarbons (37 % long alcanes, 34 % monoaromatic, 16 % diaromatic, 12 % polyaromatic hydrocarbons). Only the agglomerate fraction (hydrocarbons and sand) smaller than 355 µm was used. Demineralised water was applied as the supercritical solvent. With the apparatus described above, concentrations of soil material in water up to 10 wt-% were investigated under operating conditions of 250–400 °C at 24 MPa. Representative results are shown in fig. 31.10.

Fig. 31.10 shows semi-batch and continuous extraction results as a function of solvent to soil ratio. Compared to the semi-batch extraction, residence times can be reduced remarkably by continuous operation. In the continuously operated apparatus, the degree of extraction can be brought up to 100 % with a residence time of only 28 s. With semi-batch extraction, the same result is only yielded after 6 h of extraction. If the extraction is operated continuously, high solvent to soil ratios are needed for acceptable extraction results. Only at concentrations below 0.5 wt-% soil in water (ie. over 200 kg_{water} $(kg_{soil\ material})^{-1}$) can more than 92 %

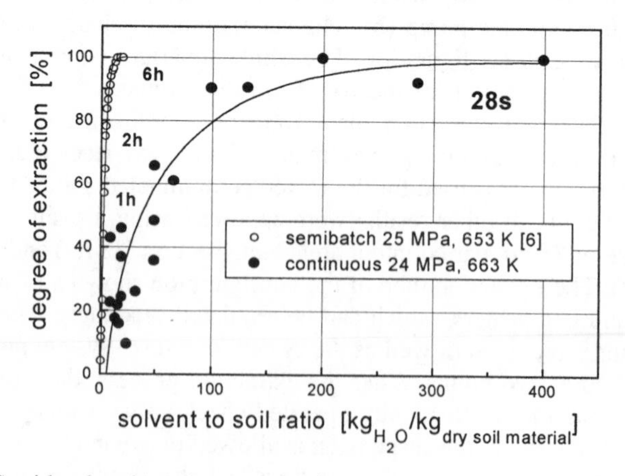

Fig. 31.10. Semi-batch and continuous extraction results versus solvent to soil ratio

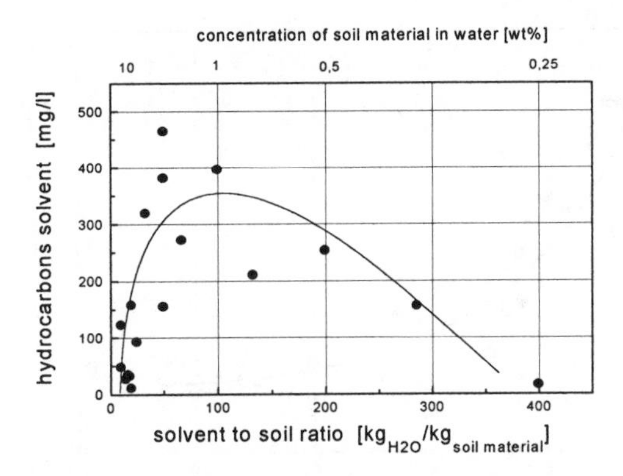

Fig. 31.11. Concentration of hydrocarbons in the fluid aqueous phase versus concentration of soil material in water (solvent to soil ratio) at 390 °C and 24 MPa

of the contamination be extracted in the tubular reactor of given length in one run. Due to extraction in the con-current mode, solvent to soil ratios must be high for a high concentration gradient.

Fig. 31.11 shows the concentration of hydrocarbons in the fluid solvent phase at the outlet of the extractor and its relation to the solvent-to-soil ratio. From the data, one may conclude that the hydrocarbon concentration in water reaches a maximum at a particular solvent to soil ratio. At low solvent to soil ratios, only a small amount of hydrocarbons could be extracted. This amount rises with increasing solvent to soil ratio, enhancing mass transfer, and therefore also increasing the loading of the solvent phase. In the region above 50 kg_{water} $(kg_{soil\ material})^{-1}$, it is assumed that all of the contamination is extracted from the soil material, but the amount of soil material in water decreases as does the amount of extractable contamination.

Experiments with temperatures ranging between 320 °C and 385 °C at a pressure of 24 MPa, for a constant concentration of soil in water of 1 wt-% and a residence time of 45 s, show a remarkable increase in the degree of extraction with increasing temperature. Although hydrocarbon contaminants are already miscible with water at 350 °C and 25 MPa, the best extraction results were obtained at 377 °C and 24 MPa. At a constant concentration of 1 wt-% of soil in water, the extraction result could be improved by about 8 % by extending the residence time from 28 s to 45 s.

Due to thermal decomposition and hydrolytic reactions, carbon dioxide and methane could be detected in the gas phase, particularly above 387 °C at 24 MPa. The measured amount of CO_2 in the gas phase rises linearly with increasing amount of solids in suspension. At low solvent to soil ratios (< 50 kg_{water} $(kg_{soil\ material})^{-1}$), the degree of extraction is below 50 %. For total cleaning, the soil material must then be charged again. Longer residence times, up to 60 s, without regeneration of the solvent, do not lead to better extraction results. Experiments with a suspension of 2 wt-% of soil material showed that after three runs of extraction, the degree of extraction could be improved to 94 %. The total time of extraction was less than 90 s.

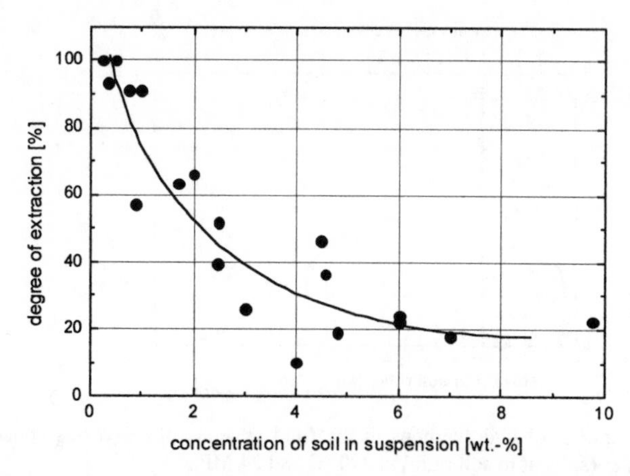

Fig. 31.12. Degree of extraction and its dependence on the amount of soil material in the solvent (solvent to soil ratio). This diagram corresponds to the left branch of fig. 31.12, i.e. up to solvent-to-soil ratios of ≈ 100. T = 390 °C, P = 24 MPa, residence time: 28 s

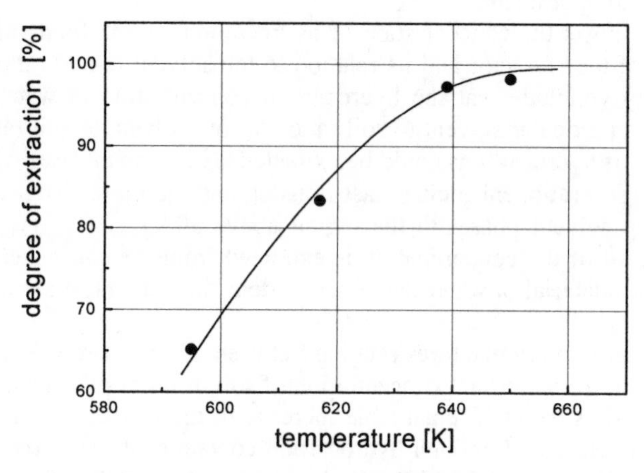

Fig. 31.13. Extraction results versus temperature at 24 MPa, with residence time 45 s; and 1 wt-% soil in water

31.3.4 Extraction of Mixed Contaminations of Heavy Metals and Organic Compounds with Water and Carbon Dioxide

31.3.4.1 Soil Materials

Two different types of soil material were investigated. Both contained heavy metals and hydrocarbons. The soil materials were taken from the highly contaminated effluent streams of a mechanical cleaning of soil material. The materials are labeled "GKSS" and "Lägerdorf". In addition to the contaminating hydrocarbons, interest was focussed on lead, copper and antimony for the material "GKSS", and lead and zinc for "Lägerdorf". The composition of the materials, with respect to the heavy metals, is given in tables 31.3 and 31.4 for the material "GKSS", and in tables 31.5 and 31.6 for the material "Lägerdorf". More details are presented in Misch and Brunner (2000) and Misch (2001).

Table 31.3. Composition of soil material "GKSS". Analysis by Umweltanalytik Kiel

	Method	Unit	Result
Total dry matter	DIN 38 414 S 2	%	31.8
Lead	DIN 38 406 E 22	mg kg^{-1}	4200
Copper	DIN 38 406 E 22	mg kg^{-1}	300
Antimony	DIN 38 406 E 22	mg kg^{-1}	3500
Hydrocarbons	DIN 38 409 H18	mg kg^{-1}	39000

Table 31.4. Composition of soil material "GKSS". Analysis after Zeien and Brümmer (Misch and Brunner 2000; Misch 2001)

Fraction	Lead [%]	Copper [%]	Antimony [%]
Mobil	15.3	3.6	3.3
Easily extractable	22.7	8.6	1.9
Occluded in Mn-Oxides	17.4	8.7	1.7
Organic bonded	27.6	0	5.3
occluded in amorphous Fe-Oxides	6.6	15.1	18.3
occluded in crystalline Fe-Oxides	4.0	0	33.2
Residually bonded	6.2	64.0	39.3

Table 31.5. Composition of material "Lägerdorf". Analysis by Umweltanalytik Kiel

	Method	Unit	Result
Total dry matter	DIN 38 414 S 2	%	47.0
Lead	DIN 38 406 E 22	mg kg^{-1}	360
Zinc	DIN 38 406 E 22	mg kg^{-1}	420
Hydrocarbons	DIN 38 409 H18	mg kg^{-1}	43000

Table 31.6. Composition of material Lägerdorf. Analysis after Zeien and Brümmer, and BCR (Misch and Brunner 2000; Misch 2001)

Zeien and Brümmer		*BCR*	
Fraction	Lead [%]	Fraction	Lead [%]
Mobile	0.9	Soluble in acetic acid	4.7
Easily extractable	7.5	Can be reduced	73.3
Occluded in Mn-Oxides	14.0	Can be oxidised	9.0
Organically bonded	34.6	Aqua regia	13.3
Occluded in amorphous Fe-Oxides	27.1		
Occluded in crystalline Fe-Oxides	3.8		
Residually bonded	11.2		

31.3.4.2 Mechanism for Mobilising Heavy Metals

Heavy metals are fixed to the soil material by different bonding mechanisms. Therefore, various methods can lead to mobilisation. The most obvious parameter is the pH-value of the environment of the soil material. In a neutral environment, heavy metals are least mobile. At pH-values higher than neutral, a mobilisation of humic compounds can lead to an increased mobilisation of heavy metals. At pH-values lower than neutral, a dissolution of hydroxides, carbonates, and oxides occurs. This mechanism is used in several investigations of soil-cleaning with acid treatment. This mechanism was also used for cleaning-experiments on the mixed contaminations characterised above. Differently to other investigations, the dissolution of supercritical carbon dioxide in water was used to lower the pH-value. The mechanisms are:

Dissolution of hydroxides: $Me(OH)_2 + 2\,H^+ \leftrightarrow Me^{2+} + 2\,H_2O$

Dissolution of carbonates: $MeCO_3 + 2\,H^+ \leftrightarrow Me^{2+} + H_2O + CO_2$

Dissolution of oxides: $MeO + H_2O \leftrightarrow Me(OH)_2$

$$Me(OH)_2 + 2\,H^+ \leftrightarrow Me^{2+} + 2\,H_2O$$

Competition for adsorption sites at the surface:

$$MeX + 2\,H^+ \leftrightarrow Me^{2+} + H_2X$$

Carbon dioxide dissolves in water at high pressures to relatively high degrees (Wiebe 1941; Franck 1961). The pH-value decreases with increasing pressure, as has been shown (Toews et al. 1995; Härtelet al. 1980). For temperatures up to 300 °C and pressures up to 25 MPa, as were used in this investigation, no data were available. The pH-value was therefore calculated using data for the 1st ionisation constant (Read 1975) and equilibrium data (Takenouchi 1964) from the

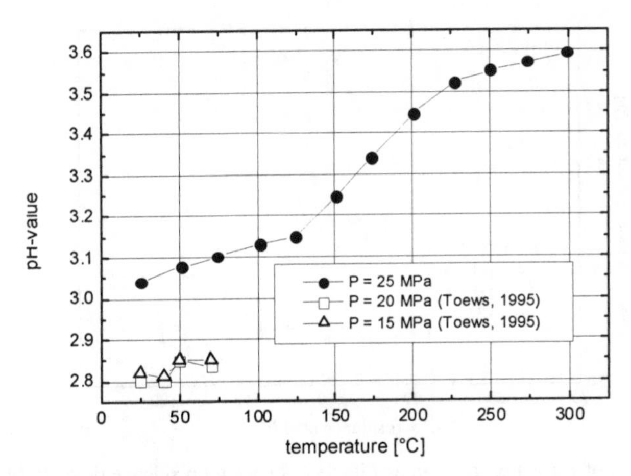

Fig. 31.14. Calculated pH-value of water for carbon dioxide saturated solutions at a pressure of 25 MPa. Equilibrium data from Wiebe (1941), Takenouchi (1964)

literature, according to the following equations:

$$K_S = \frac{\left[H_3O^+\right] \cdot \left[A^-\right]}{\left[HA\right]} \tag{31.1}$$

or rearranged:

$$\left[H_3O^+\right] = K_S \cdot \frac{\left[HA\right]}{\left[A^-\right]} \tag{31.2}$$

The resulting calculated pH-value is shown in fig. 31.14.

Acetic acid was applied in some experiments to enhance the extraction efficiency. By using acetic acid, it was easily possible to control the pH-value. Acetic acid dissociates according to:

$CH_3COOH \longleftrightarrow H_3O^+ + CH_3COO^-$ (pKs = 4.75)

Experiments with acidic mobilisation have been carried out only in continuous extraction experiments. The results are therefore presented in context with the extraction results of hydrocarbon contaminated soil materials in the next section.

A parameter investigation was carried out for the extraction of mixed contaminations. The degree of extraction was determined, as usual, with respect to the initial concentration. Since the heavy metals concentration, determined by analysis of the same sample in different analysing laboratories, deviates considerably, the highest determined value was taken for calculating the degree of extraction. This may have led to values which are lower than the actual results. In this context, only some examples for the material "GKSS" can be presented. All heavy metal extractions were achieved by mixing two liquid streams: soil in water and carbon dioxide. Measurements of pH-values inside the pressurised and heated system of the water-rich phase were not taken. Results were obtained by analysing the solid particles after collecting and filtering.

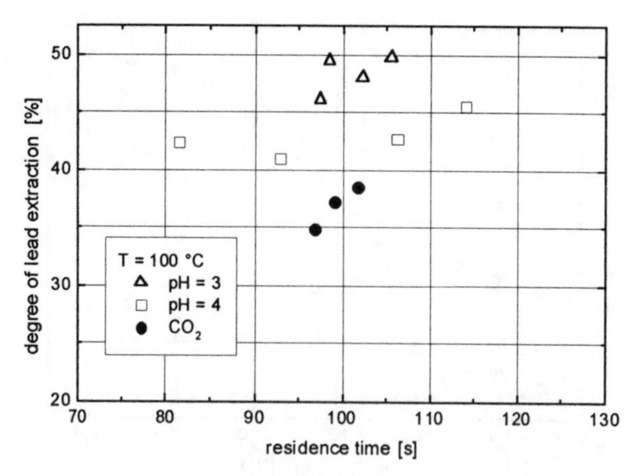

Fig. 31.15. Degree of extraction for lead and its dependence on residence time. 25 MPa, 100 °C; Material "GKSS"

In fig. 31.15, the extraction yield of lead from soil is shown as a function of residence time. Different pH-values of the feed stream at a temperature of T = 100 °C are mentioned. The highest extraction yield is achieved with pH = 3, adding acetic acid. The required residence time to obtain this result is about 100 s. With decreasing acidic power at constant residence time, the amount of remaining lead increases. Compared to the calculated pH-value of carbonic acid in fig. 31.14, the results shown in fig. 31.15 support the theory that a steady state is not reached. With the acidic power of the added carbon dioxide, not even a pH-value of 4 is obtained. Summing up all experiments carried out with the aim of lead removal, extraction results for temperatures equal or near to T = 100 °C were most satisfying, yielding about 50 %. Compared to the composition determined after Zeien and Brümmer (given in table 31.4), the fractions *mobile, easily extractable, and occluded in Mn-oxides* should have been removed from the soil. Different comparisons on how the composition changes for different extraction parameters (temperature) during the process, are described elsewhere (Misch 2001).

Further investigations were carried out concerning the organic content of the soil material. The ideal cleaning process provides the principles of a simultaneous removal of both heavy metals and hydrocarbons. There are only few investigations carried out so far that deal with mixed contamination, although such contamination is a material from soil washing facilities. It was in our interest to give initial hints on how to handle such material, providing a reduction in both heavy metal content and organic pollution. In fig. 31.16, the extraction results of the organic compounds of the soil are shown.

As a function of residence time, the extraction yield concerning the hydrocarbons is shown. With increasing temperature, organics can be almost completely removed. It is remarkable that there is a cleaning effect at very low temperatures. Compared to the extraction results of earlier experiments with soils containing only organic contaminants, the required residence time is higher.

The degree of extraction of antimony is presented in fig. 31.17. Extraction yields of 80 % can be obtained. As a function of the soil in solvent concentration,

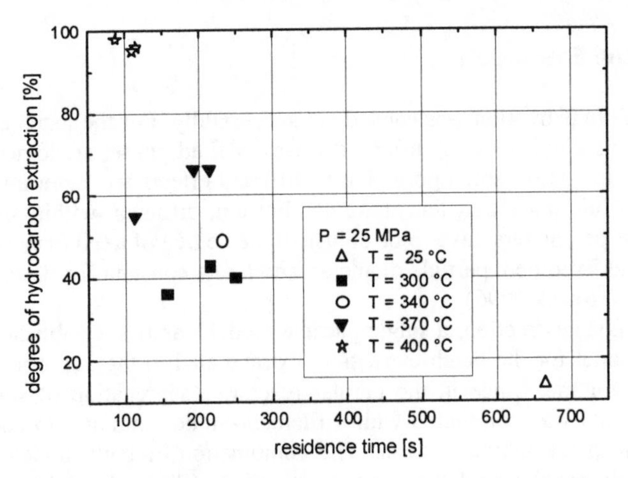

Fig. 31.16. Degree of extraction for hydrocarbons and its dependence on residence time. 25 MPa, material "GKSS"

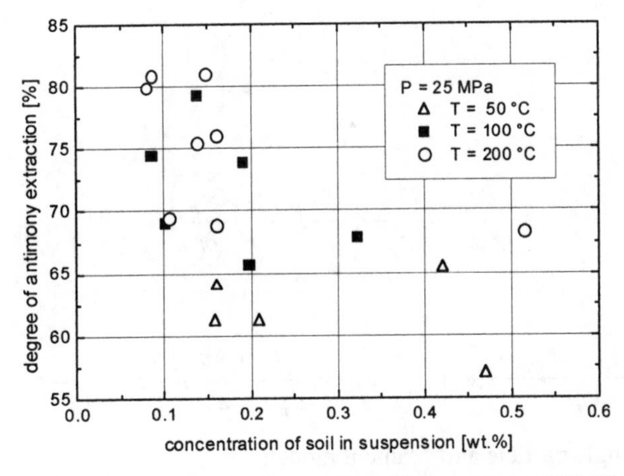

Fig. 31.17. Degree of extraction of antimony. P = 25 MPa, material "GKSS"

the extraction result decreases, though the total concentration is very small at high extraction rates. In contradiction to experiments on lead extraction, the results at higher temperatures are comparable to those at lower temperatures (T = 50 °C). This high extraction result is surprising according to the composition given in table 31.4. Referring to those data, a notable amount of the residually bonded fraction must also have been removed during extraction.

The extraction of mixed contaminants clearly is more difficult than that of hydrocarbons (Misch 2001). Cleaning with water and dissolved carbon dioxide is possible. However, better analyses and more experiments are necessary to evaluate the efficiency. A precise knowledge of the metal speciation and binding conditions is advantageous.

31.3.5
Modelling the Extraction

Modelling of the extraction was carried out successfully. For the batch extraction, the fixed bed is divided into 3 different phases: a fluid phase, agglomerates consisting of a solid phase, and a pore phase. Different effects are taken into account such as fluid-fluid mass transfer, phase equilibrium, diffusion within and outside of pores. Four parameters have to be adjusted: the degree of axial dispersion, contact part of the fixed bed, particle diameter, and a rate constant for desorption and solubilisation (Nowak 1996).

For continuous extraction, a single particle model was first established. This is a modified model for the batch extraction, extrapolated to the appropriate values for the void volume. Since in the tubular reactor, a segregation of solids takes place, this model was completed with a filament-model, taking into account the sedimentation in the tubular reactor. The combination of both models provides excellent results for the modelling of the extractions (Firus 1996). The following equations were established:

1. Desorption:

$$\frac{\partial c_s}{\partial t} = -k_{de} \cdot c_s \cdot \left(1 - \frac{c_p}{c_p^{\max}}\right) \tag{31.3}$$

2. Diffusion:

$$\frac{\partial c_p}{\partial t} = \frac{D_{eff}}{\varepsilon_p} \cdot \left(\frac{\partial^2 c_p}{\partial r^2} + \frac{2}{r} \cdot \frac{\partial c_p}{\partial r}\right) \tag{31.4}$$

3. Convection:

$$\frac{\partial c_f}{\partial t} = \frac{\beta \cdot A_{Part} \cdot V_{Part}}{V_f} \cdot \left(c_p\big|_{r=R} - c_f\right) = \frac{6\beta}{d_p} \cdot \frac{(1-\varepsilon_f)}{\varepsilon_f} \cdot \left(c_p\big|_{r=R} - c_f\right) \tag{31.5}$$

Combination of single particle and filament model:

$$E_{comb} = \frac{\varphi \cdot E_{single} + (\Phi - \varphi) \cdot E_{filament}}{\Phi} \tag{31.6}$$

The parameter φ, which represents the ratio of single particle extraction to extraction of agglomerates or clouds of particles, is considered as an adjustable parameter since only few data exist on real extraction experiments. The efficiency of the modelling is shown in fig. 31.18 for a weathered material (hydrocarbon contamination only).

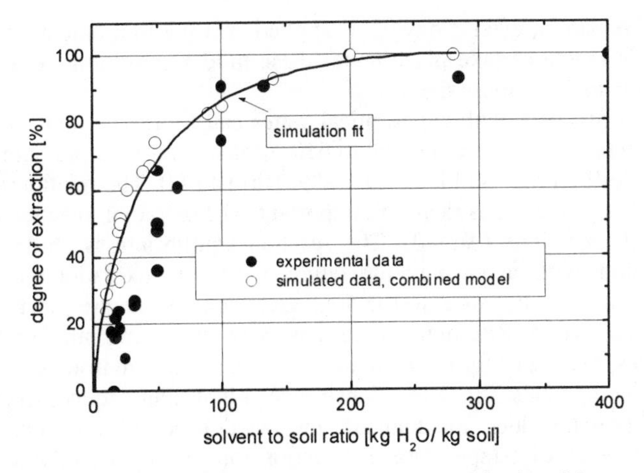

Fig. 31.18. Modelling of the continuous extraction with the combined model

31.4 Destruction of Organic Compounds in the Effluent from the Cleaning Process: Supercritical Water Oxidation

Organic waste can be destroyed by burning, biological treatment or oxidation in an atmosphere of water. Oxidation in supercritical water is fast and can provide total oxidation of the organic compounds. Products are confined to a controllable space. In spite of these advantages, supercritical water oxidation is the least developed technique compared to other methods of destruction. In this investigation, supercritical water oxidation was investigated as a method for the destruction of extraction-effluent compounds, in order to improve the cleaning and disposal processes used in extraction.

Oxidising agents like air, oxygen or hydrogen peroxide, can initiate and promote oxidation reactions. For the cleaning of soil material, synthetic air was first used. In later experiments, oxygen supplied by in-line high-pressure electrolysis was used as an oxidising agent. No catalysts were used. At supercritical conditions for water, operating conditions are limited at the lower end by the critical constants of water: 374 °C and 22 MPa. The upper limit of supercritical water oxidation is somewhat arbitrary at about 640 °C and 25 MPa.

Under conditions of supercritical water oxidation, organic compounds can be decomposed to the end products carbon dioxide and water. Other elements in the compounds, like sulfur, chlorine or phosphorus, react with the corresponding mineral acids. Organic nitrogen preferentially forms ammonia at temperatures below 500 °C. At temperatures above 600 °C, nitrogen (N_2) and nitrous oxide (N_2O) are formed. For total decomposition, sufficiently high temperatures of more than 600 °C and pressures around 23 MPa must be insured, in order to enable residence times of less than 1 minute. Sufficient oxygen must also be available. Milder conditions in supercritical water and similar short residence times do not lead to total oxidation, even if oxygen is applied. Yet the first oxidation reactions and hydrolytic reactions take place under these milder conditions, enabling biological systems to decompose the rest.

Destruction of waste with supercritical water can be carried out over a wide range of temperatures, from about 400 to 650 °C at supercritical pressures (Barner et al. 1982; Modell 1982; Killilea et al. 1989; Hirth 1992). The relationship of the reactions with temperature is therefore important. The oxidation reactions in water are exothermic reactions ($\Delta h < 0$). The reaction equilibrium is shifted towards educt compounds with increasing temperature. Yet for the oxidation reactions and temperatures involved in waste destruction, educt compounds are practically non-existent at equilibrium. Reactions proceed with an increasing number of moles. High pressures, necessary for maintaining the supercritical environment, therefore favour educt compounds. Since reactions proceed in a dilute aqueous system, the influence of pressure does not dominate the reaction equilibrium. On the other hand, the influence of temperature on reaction rate is important. According to experimental data, the reaction rate can be enhanced, so that the yield is raised by a factor of 102–103 by increasing temperature by 100 °C. Destruction of chemical compounds can be modelled globally using a first order reaction, with which experimental data can be sufficiently represented.

After the extraction of contaminants from the soil material, an oil-in-water emulsion is produced with tiny droplets of the contaminants. The mean diameter of the droplets is d = 1.7 µm. Reaction experiments on the effluent produced by the extraction of a weathered, contaminated soil material showed that at supercritical (380 °C) and subcritical (200 °C) conditions, the total destruction of the contaminants to H_2O and CO_2 could be achieved (table 31.7).

Experiments on a model oil (polycyclic, heterocyclic, aromatic and aliphatic compounds) were carried out at concentrations of 0.6 wt-% and 5 wt-% oil relative to water. Independently of the residence time, the dependence of the maximum degradation attainable on temperature was investigated (fig. 31.19). One major problem of supercritical water oxidation is corrosion. If the water is supercritical, both halogenic compounds and oxygen are present simultaneously, and unacceptably high corrosion rates have been observed for all metallic materials. Therefore, composition of the feed dominates the choice of the construction material for the reactor and at least the first heat exchanger, where supercritical water conditions are maintained.

A destruction process with supercritical water enables the construction of a small processing unit due to the short reaction times needed. Residence times may

Table 31.7. Reaction results at 25 MPa, 88 cm^3 reactor

Temperature [°C]	Loading [$g_{hc}\ l_{H2O}^{-1}$]	Residence time [s]	O_2 excess [mol O_2 mol $O_{2,min}^{-1}$]	Degree of conversion U [$(m_{ohc}-m_{hc})\ m_{hc}^{-1}$]
290	6.7	135	1.68	0.963
330	0.7	107	26	0.999998
385	3.0	68	6.99	0.9376

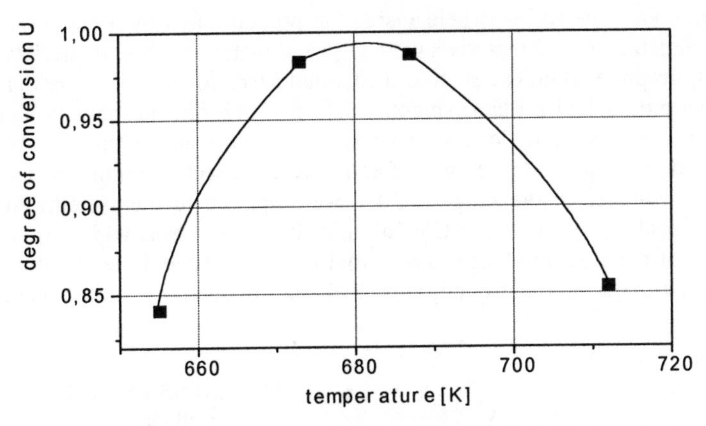

Fig. 31.19. Reaction data with model oil, 25 MPa, 0.6 wt-% oil relative to water

be very short, especially if only biological degradation is intended. Such a small waste destruction unit can be designed to meet the requirements of the byproducts of a process, and may well be integrated into the production plant.

Beside soil material, the effluent stream of a conventional soil washing process can be treated effectively with supercritical water, as well as the effluent streams of dumping sites. The effluent of the supercritical water reactor is sterile.

31.5 Supercritical Water and In-line Produced Oxygen

For small process units, the supply of oxygen at pressures > 20 MPa needs great effort. Therefore, the idea of directly producing oxygen by in-line electrolysis, just at or before the place where it is needed, was conceived (Brunner 1993) and investigated during this research.

The experimental part of the high-pressure electrolysis is as follows: the main components of the plant are a membrane pump, a pressure check valve, and an electrically heated tubular system (Misch and Brunner 1999; Misch et al. 2000). After passing the heated components, the fluid was fed through the electrolysis cell and cooled before depressurising. All piping was made of stainless steel 1.4404. The soil analyses yielded no or an extremely low halogen content, thus the same was assumed for the extract. Therefore no corrosion or corrosion-caused failure of the material has been detected so far, and nor is it to be expected. The temperature in the preheating system was measured by NiCrNi-thermocouples and recorded. The reference temperature indicated in the results is the temperature of the fluid before entering the cell. Two-ball configuration pump valves were used so that fibrous feed or small particles would not cause failure of the whole experiment.

The body of the electrolysis cell (fig. 31.20) is a 1" high-pressure pipe, with high-pressure connections for coupling the spark plug. For the two electrodes, the pipe was used as the cathode, and a noble metal rod ($d_a = 2$ mm, $l = 200$ mm) was used as the anode. The rod was held inside the pressure area by a custom-made spark plug. Satisfactory rod materials were gold and platinum. Other materials like Inconel 600, graphite, stainless steel and titanium were found not to be durable enough. The anodes had a plain, smooth surface which did not change significantly during use. The feed flows first vertically onto the anode surface. At this point the anode is assumed to be most effective with respect to oxidation capability. Different voltages in the range 0–30 V were applied to the system. At the beginning of each experiment, a KOH-solution (0.1 mol l^{-1}) was added to enable the transport of the electric charge. The initial conductivity and the resulting pH were measured under ambient conditions. Effluent conductivity and effluent pH

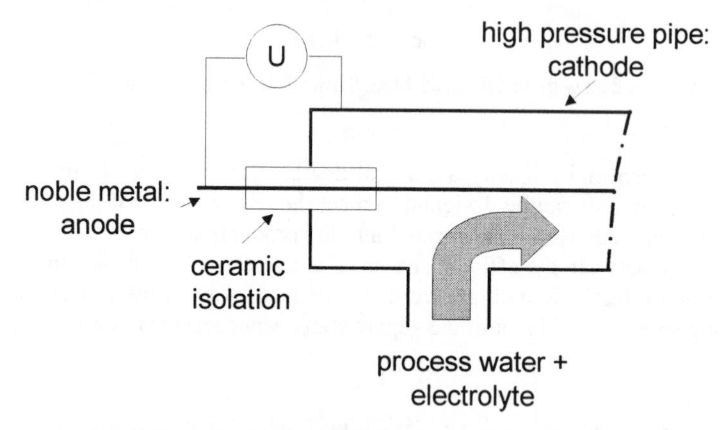

Fig. 31.20. Schematic drawing of electrolysis cell

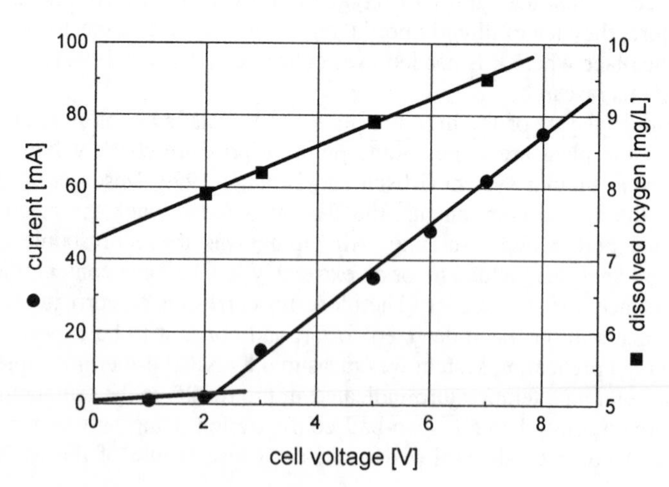

Fig. 31.21. Current flow and dissolved oxygen as a function of cell voltage; 23.2 MPa, 17 °C, κ = 4.33 mS cm^{-1}, V'a = 3.1 l h^{-1}

were not examined. The electrolysis cell is not heated directly, but the temperature inside is maintained by preheating of the tube. Ohmic heating inside the cell at the maximum voltage and a conductivity of 1.6 mS cm^{-1} was in the range of 10–20 °C. This is assumed to be small compared to absolute temperatures of up to 400 °C.

Experiments were all carried out in the pressure range of 23 to 25 MPa (Misch Brunner 1999; Misch et al. 2000). The variable parameters are the organic load of the feed stream, the feed conductivity and/or pH, the reaction temperature, the residence time, and the surface area of the anode. The yield of destruction was determined by IR-spectroscopy corresponding to method DIN 38409 H18. The reference was the characteristic peak of the asymmetric valence oscillation of the methyl group at 2930 cm^{-1}. Total organic carbon (TOC) was determined with a Dohrmann TOC-analyser.

In fig. 31.21, the resulting current and the amount of dissolved oxygen are shown as a function of cell voltage. Both the current and the oxygen amount are linearly dependent on cell voltage. With increasing cell voltage, the current is low but rises suddenly after overcoming the decomposition voltage. The real decomposition voltage was not measured potentiometrically during the experiments. The decomposition voltage is a characteristic of the decomposed molecule and therefore a function of state parameters. There are two competing effects influencing the change in the decomposition voltage under high pressure and high temperature. With increasing pressure the dependency decreases. With increasing temperature, the decomposition voltage decreases (Ogata et al. 1988). The overall effect caused by changes in experimental conditions is so insignificant, that it can be neglected in this context.

To optimise cell operation conditions, experiments with soil extracts were carried out. These extracts were obtained from semi-batch experiments and then by

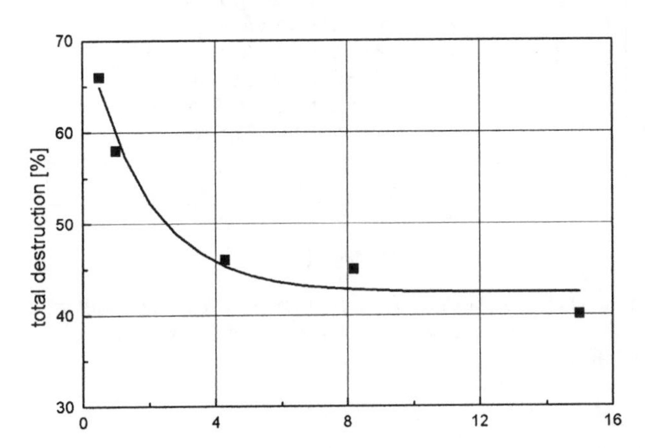

Fig. 31.22. Dependence of total destruction of soil extract on conductivity. 25 MPa, 300 °C, U = 30 V

continuous con-current extraction of weathered contaminated soil. This soil, taken from a former paint factory, exhibited a hydrocarbon contamination. It had been weathered for over 20 years and was washed and sieved down to 355 µm before extraction. The amount of contamination removed from the soil was determined, without paying attention to destruction pathways or particular compounds. For the soil extracts, we assumed the autooxidative processes to be of minor importance, because the electrolysis temperature is equal to or less than the oxidation temperature. As a function of conductivity, the destruction yield of the soil extract is shown in fig. 31.22.
The diagram shows that with increasing conductivity, the efficiency of the destruction decreases. This can be explained by a hindrance of the reaction, which is caused by a restriction of ion mobility due to increasing ionic concentration in the water. From this information, it was concluded that for this kind of extracted contamination, it is advantageous to use the highest possible temperature and voltage, at low conductivity.

In order to prove the success of combining the extraction and oxidation processes into one step, the electrolysis cell was linked to the extraction reactor. Thus, extraction and oxidation of the contaminated soil were also studied in one step. Fig. 31.23 shows the degree of destruction as a function of cell voltage.

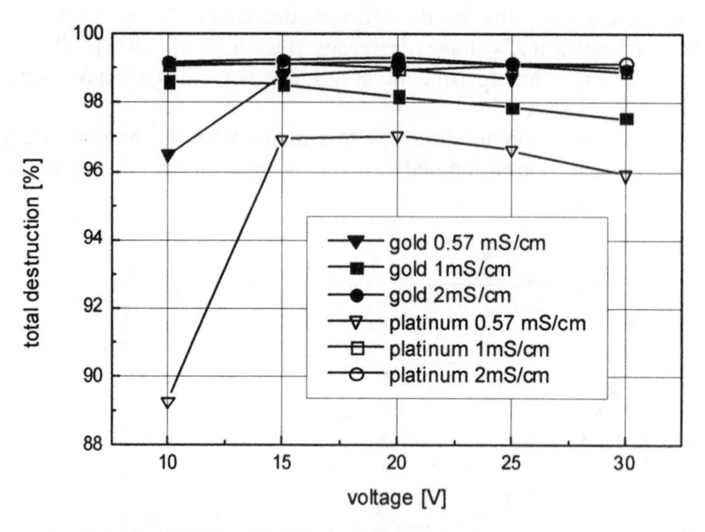

Fig. 31.23. Total destruction of soil extract by supercritical water oxidation with in-line produced oxygen by electrolysis as a function of voltage; coupled extraction and electrolysis; P = 24.1 MPa, T = 340 °C

31.6 Conclusions

Supercritical water is an excellent solvent for organic compounds. Furthermore, it has outstanding properties as a reaction medium because of the complete solubility of reactants such as oxygen and hydrocarbons. A new apparatus was developed for continuous extraction of contaminated soil material at high pressure (25 MPa) and high temperature (390 °C). The extraction of hydrocarbon contaminants from the soil material could be realised with supercritical water under parallel flow. Within a residence time of only 30 s, suspensions of less than 0.5 wt-% solid matter could be cleaned to 100 %. The extraction can be carried out as a multistage process, so that more highly concentrated suspensions can be totally cleaned in more than one extraction cycle. A unit for the separation of solid matter and aqueous phase must be developed.

The continuous extraction of hydrocarbon from soil material can be followed by supercritical water oxidation. Supercritical water extraction and the following supercritical water oxidation are alternative cleaning processes providing up to 100 % extraction. Therefore, a second new apparatus was developed which produces oxygen under high pressure by means of electrolysis. Regarding the example of the alkaline water electrolysis, a high pressure electrolysis cell was invented that works at pressures up to 25 MPa and temperatures up to 350 °C. When determining the influence of destruction on the organic load of the waste stream, the electrical conductivity and the cell voltage were considered as well as pressure and temperature. With respect to the cleaning of contaminated soil, different hydrocarbon loaded streams were fed into the system. Concerning biological waste streams, the destruction yield of wool scouring wastewater was investigated. Another promising field of use is the destruction of aqueous waste streams on ships. For this, the destruction yield of bilge and grey water was investigated. The costs of a destruction process with supercritical water are still not known. However, it is probable that the costs are similar to those of a wet air oxidation process, with the higher cost per unit volume of the reactor and heat exchanger roughly balancing out with the lower volume required. A destruction process with supercritical water enables the construct of a small processing unit due to the short reaction times needed. Residence times may be very short, especially if only biological degradation is intended. Such a small waste destruction unit can be easily designed to meet the requirements of the byproducts of a process, and may well be integrated into the production plant.

31.7 Summary

Cleaning of soil materials contaminated with hydrocarbons was effectively carried out with supercritical water in different processing modes. In the semi-continuous fixed-bed extraction, a variety of weathered and heavily contaminated soil materials was completely cleaned. Supercritical carbon dioxide as an alternative solvent for the contaminants is only effective for fresh hydrocarbon contamination.

Weathered soil materials cannot be treated by supercritical carbon dioxide. A continuous extraction with supercritical water in a tubular reactor can be carried out con-currently. Extraction time needed for cleaning could be reduced drastically, from 6 h for the semi-continuous extraction, to less than 1 minute for the continuous extraction. At a temperature of 380 °C and a residence time of 45 s, a 98 % degree of cleaning could be achieved. However, solvent to feed ratio must be kept high, greater than 100, corresponding to about 1 wt-% of soil material in the aqueous feed suspension. Alternatively, a longer tubular reactor (increased residence time) or multiple treatment may be applied. Mixed contaminations consisting of heavy metals and hydrocarbons can be cleaned with water at temperatures between 250 and 350 °C and a pressure of 25 MPa, if supercritical carbon dioxide is dissolved in the liquid phase. Evaluation of the results of this cleaning process is hampered by diverse analytical results. Fines (soil material with particle diameters < 63 µm) can be treated without additional problems and with the same success.

For the destruction of the extracted contaminants, the effluents were treated biologically. In weathered soil, the contaminants were only partially degradable biologically. Supercritical water oxidation proved to be a good alternative. Oxidation of the hydrocarbons in the extract in supercritical water is fast and can lead to total oxidation of the organic compounds. Products are confined to a controllable space. The necessary oxygen can be supplied in-line using high-pressure electrolysis of water. The hydrocarbons removed from the soil materials are effectively oxidised in supercritical water by the in-line electrolytically-supplied oxygen.

References

Barner HE, Huang CY, Johnson T, Jacobs G, Martch MA, Killilea WR (1982) Journal of Hazardous Materials 31: 1–17

Brunner G (1993) Verfahren zur Oxidation von organischen und anorganischen Stoffen mit Sauerstoff in wässeriger Umgebung. Patentschrift DE 41 32 915 C1

Brunner G (1994) Extraction and Destruction of Waste with Supercritical Water. In: Kiran E, Levelt Sengers JMH (eds) Supercritical Fluids. Kluwer Academic Publishers, pp 697–705

Firus A (1996) Reinigung von Bodenmaterial durch Extraktion und Reaktion mit überkritischem Wasser und Kohlendioxid. Thesis, Technische Universität Hamburg-Harburg

Firus A, Brunner G (1995) Hydrolytic and Thermolytic Clean-Up of Contaminated Soil with Supercritical Water. In: van den Brink WJ, Bosman R, Arendt F (eds), Contaminated Soil '95. Kluwer Academic Publishers, p 1023–1028

Firus A, Brunner G (1996) Continuous Extraction of Contaminated Soil with Supercritical Water. In: von Rohr PR, Trepp C (eds) Process Technology Proceedings, vol 12, p 179–184

Firus A, Weber W, Brunner G (1997) Supercritical carbon dioxide for the removal of hydrocarbons from contaminated soil. Sep Sci & Techn 32: 1403–1414

Fischer J, Hofmann H, Luft G, Wendt H (1980) Fundamental investigations and electrochemical engineering aspects concerning an advanced concept for alkaline water electrolysis. AIChE Journal 26: 794

Franck EU (1961) Überkritisches Wasser als elektrolytisches Lösungsmittel. Angewandte Chemie 10: 309–322

Härtel G, Peter S, Tunn W (1980) Über die Acidität wäßriger Lösungen der Chloride des Natriums, Calciums und Magnesiums in Anwesenheit von Kohlendioxid bei hohen Drücken

Hirth T (1992) Pyrolyse, Hydrolyse, und Oxidation kohlenstoffhaltiger Verbindungen in überkritischem Wasser bei Drücken bis 1000 bar. Thesis, Universität Karlsruhe

Killilea W, Swallow K, Malinowski K, Staszak C (1989) The modar process for the destruction of hazardous organic wastes – Field test of a pilot scale unit. Waste Management 9: 19–26

Michel S (1992) Grundlagenuntersuchung zur Extraktion von polyzyklischen aromatischen Kohlenwasserstoffen aus kontaminierten Böden mit überkritischem Kohlendioxid. Thesis, Universität Dortmund

Misch B (2001) Kontinuierliche Extraktion von Feststoffen und Reinigung mischkontaminierten Bodenmaterials mit überkritischen Fluiden. Thesis, Technische Universität Hamburg-Harburg

Misch B, Brunner G (1999) An alternative method of oxidizing aqueous waste in supercritical water, Oxygen supply by means of electrolysis. In: Steam, water and Hydrothermal Systems, Proc 13th Int Conf On Properties of Water and Steam. Toronto, Canada

Misch B, Brunner G (2000) Deutsche Forschungsgemeinschaft (eds) Sonderforschungsbereich 188, TP A10, Final report 1998–1999–2000

Misch B, Firus A, Brunner G (2000) An alternative method of oxidizing aqueous waste. Journal of Supercritical fluids 17: 227–237

Mishra VS, Mahajani VV, Joshi JB (1995) Wet Air Oxidation. Industrial & Engineering Chemistry Research 34: 2–48

Modell M (1982) Processing Methods for the Oxidation of Organics in Supercritical Water. United States Patent B1 4338199

Nowak K (1996) Reinigung kontaminierter Bodenmaterialien mit überkritischem Wasser. Thesis, Technische Universität Hamburg-Harburg, 1995, Printed by Shaker, Aachen

Ogata Y, Yasuda M, Hine F (1988) Effects of the operating pressure on the performance of water electrolysis cells at elevated temperatures. J of the Electrochemical Soc: Electrochemical Science and Technology: 2976

Read AJ (1975) The first ionization constant of carbonic acid from 25 to 250 °C and to 2000 bar. J of solution chemistry 4, 1: 52–70

Savage PE, Gopalan S, Mizan TI, Martino CJ, Brock EE (1995) Reactions at Supercritical Conditions: Applications and Fundamentals. AIChE Journal 41: 1723–1778

Schleußinger A (1996) Einfluß von Schleppmitteln auf die Hochdruckextraktion am Beispiel der Bodensanierung. Thesis, Universität Dortmund

Takenouchi S, Kennedy G (1964) The binary system H_2O – CO_2 at high temperatures and pressures. American J of Science 262,pp 1055–1074

Toews K, Shroll R, Wai CM (1995) pH-Defining Equilibrium between Water and Supercritical CO_2. Influence on SFE of Organics and Metal Chelates. Anal Chem 67 (22), pp 4040–4043

Wiebe R (1941) The binary system carbon dioxide – water under pressure. Chemical Reviews 29: 475–481

32 Application of the Physico-Chemical Desorption Technology for Soil Decontamination

B. Niemeyer
GKSS National Research Centre, Institute for Physical and Chemical Analysis, Max-Planck-Strasse, 21502 Geesthacht, Germany
University of Federal Armed Forces, Institute of Thermodynamics, Holstenhofweg 85, 22043 Hamburg, Germany

32.1 Actual Situation and Motivation

The last decade was characterised by intensive efforts to gather information on contaminated sites (such as location, chemical species and scale). For example, in Germany 240,000 to 300,000 relevant sites were identified (Franzius et al. 1995; Umweltbundesamt 2000). The increasing knowledge of the potential risks to mankind and the natural environment has stimulated profound investigations to understand the mechanisms and the procedure related with soil remediation and to develop new processes for decontamination. The direct application of the research results was funded both in the USA and in Europe (U.S. Environmental Protection Agency 1996; Rulkens 2000; Stegmann 2000).

Although numerous techniques are available, further research and development is required to discern the *complex system* of SOIL (including the kind of solid, the structure of particle and agglomerates, the natural organic matrix like humic substance etc.) and *CONTAMINANTS* (the type and interaction of hazardous components, as well as with soil and the biological system). The investigation of modern techniques concentrate on the minimisation of energy input into the processes and reduction of residues to be deposited leaving a cleaning process (see Chapter 28, Wiesner 1995). Further investigation of soil purification procedures are addressed by the German Council of Advisors on Environmental Questions (examples are given in section 470 in: Rat von Sachverständigen für Umweltfragen 1995).

Modern technologically powerful and cost effective techniques are required to gain an advantage in this dynamic economic area with large budgets available world-wide. For example, in the USA payments for site remediation under the Superfund and Resource Conservation and Recovery Act (RCRA) are estimated to be $ 750 billion (Wilson and Clarke 1994). Thus even well-established techniques are being investigated further in order to raise economical efficiency eventually.

32.2 Classification of the Soil Cleaning Processes

A variety of activities are involved in the management of soil decontamination ranging from sampling, analyses, evaluation of technological and economical features, risk assessment of hazardous compounds and during decontamination process, permissions and legal regulations and treatment of the soil. The process is one of the central aspects, influencing all the other aforementioned items. Consequently this contribution gives a survey of the processes applicable and focuses on those that are physico-chemical, especially desorption techniques. Due to practical considerations, biotechnological processes including natural attenuation are not addressed, although their importance in practice is significant.

32.2.1 Physico-Chemical Remediation Technologies

The large group of physico-chemical processes may be ordered into the technologies shown in fig. 32.1 as suggested by Rat von Sachverständigen für Umweltfragen (1995). Generally contaminated soil will be treated by one of these techniques if human health is threatened, and a fast solution of remediation is required e.g. to maintain an ongoing construction on the site, respectively.

Some of the techniques are well-established for large-scale remediation such as *chemical conversion* which comprises chemically oxidising as well as reductive processes, e.g. incineration or pyrolysis which separate and convert contaminants simultaneously. High investment and operating expenditures (energy consumption) lead to elevated expenses.

The other presented technologies need additional treatment of the eliminated matter. Mostly it is fed in a concentrated manner into an incineration plant. *Soil washing* is, in fact, an extraction technique using aqueous extractants. During this procedure the contaminant load is dislocated to the fine-grained particles. The cleaned coarse fractions are separated from the fine-grained matter by means of-mechanical procedures (see Chapter 29). A special development is the usage of solvents like CO_2, and water under supercritical conditions for the removal of contaminants (see Chapter 31). *Desorption techniques* are based on the vaporisation of mainly hydrocarbons. A more detailed description will be given later. Both technologies are widespread and are employed in numerous large-scale applications. *Electrokinetic processes* exploit the mobility of charged species in an electrical field. These techniques are being investigated at a small scale (Thöming 1998).

The plants may be operated on site (mobile technical system) or as a stationary facilities (to which contaminated matter is transported). Main criteria for the selection of this option are legal regulation for plant permission, quantity of soil to be processed on site or within the region and energy costs for transport. This topic is discussed further by O'Brien and Rouleau (1995).

As desorption techniques have been proved their technological and economical effectiveness (U.S. Environmental Protection Agency 1999) further focus is put on them.

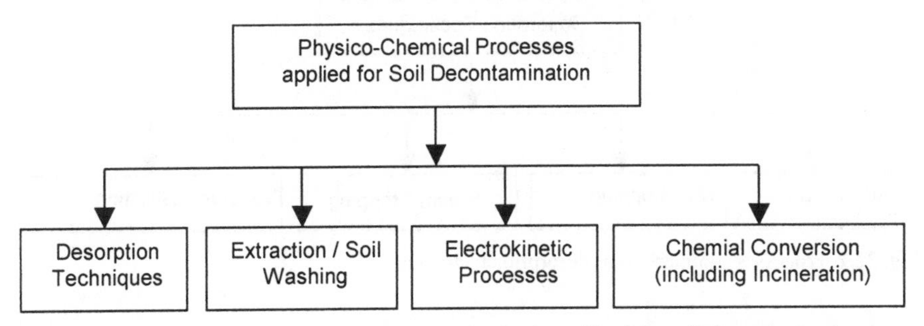

Fig. 32.1. Classification of physico-chemical technologies utilised for soil decontamination

32.2.2
Desorption Techniques and their Arrangement

Thermal desorption techniques are divided into low-temperature (360 K to 600 K) and high-temperature (600 K to 830 K) processes according to the soil temperature at the treatment. The different techniques summarised under *desorption techniques* are presented in fig. 32.2.

These techniques are especially effective if low temperatures are sufficient to release contaminants to gain the required cleaning results. This is mostly applicable to the removal of mineral hydrocarbons (MHC) and other volatile or semi-volatile organic compounds (VOC, SVOC). Moreover, if they can be used in situ the process economy will increase further. This demands presumptions of the area to be purified, such as medium permeability of the underground, minimum slope, removal of larger obstacles as underground wires etc. (U.S. Environmental Protection Agency 1991).

Ex situ treatment is not limited by geological restrictions and the time necessitated for the purification is much shorter than for in situ processes. In addition, even crucial chemical species like polycyclic aromatic hydrocarbons (PAH) or even heavy metals like mercury were separated successfully (Wietstock et al. 1997; Sullivan 1999).

Air suction and *air stripping* are normally employed in situ. These techniques remove gaseous components of the area. Air is sucked through the soil gasifying moisture and contaminants. For air stripping technique air is injected into the ground mostly at elevated temperature improving the technical performance. The gasified components have to be eliminated from the off-gases.

The application of *steam stripping* enforces energy input and transfer to the contaminants improving the separation results with the consequence of higher operating expenditures. Different modifications of these techniques have been utilised both in situ as well as ex situ. The recently developed steam stripping process focuses on the treatment of fine-grained particles enabling the decontamination of sludges and residues from washing processes with their especially high contamination load (Höhne et al. 2000a).

Vacuum distillation is a combined measure of the above described processes whereby a reduced system pressure is additionally applied (Lord et al. 1990).

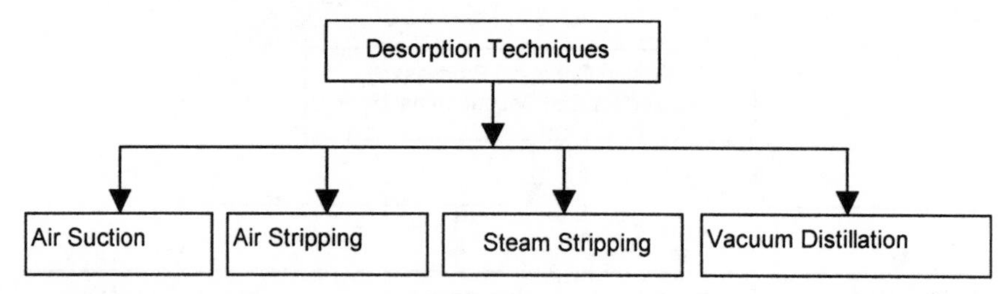

Fig. 32.2. Processes summarised as desorption techniques

32.3 Principle and Theoretical Background of the Desorption Technology

Desorption technology utilises physico-chemical parameters to remove contaminants from solids by volatilisation. After their elimination hydrocarbons may be discharged, and reused for energy conversion processes in concentrated manner, respectively. These processes are flexible due to the availability of several separation parameters to be applicable. Thus a wide scale of different systems of soils and chemical compounds have been treated technically effective (fulfilment of the legal limitations) under economical conditions.

32.3.1 Separation Principles

The main aim of all is to transfer contaminants from a liquid phase or from an adsorbed state mainly by steam-distillation and by desorption effects into a gaseous phase (see Chapter 33). The basis of the successful remediation is the understanding of the interactions of the species involved on microscale, as well as the limiting transport effects such as diffusion (Connaughton 1993; El-Shoubary et al. 1994; Rodriguez-Maroto et al. 1995; Yuan and Udell 1993).

The separation parameters used are elevated *temperature*, reduced system *pressure* and *stripping effect*. The simultaneous employment of these parameters enables an effective processing.

The energy input during processing for delivering phase transfer enthalpy for contaminant volatilisation normally results in *temperature* increase which consequently lowers viscosity. Experience from laboratory to industrial scale proves that temperature is the *most significant process parameter*. It affects the technological course of remediation as well as the costs, both to a high degree. As temperature is a transport-bound parameter, diffusion effects have to be considered. In situ operation has to provide a homogeneous distribution of the gaseous phase removed from the site. Diffusion effects into and out of the particles and their agglomerates may be of significant order lengthening the time needed for a successful remediation or even lead to an unsuccessful processing which does not meet the residual limitations. Mainly during ex situ processing mechanical forces and drying de-agglomerate and thus reduce diffusion effects which, improves purification performance.

The gaseous medium (air/steam) introduced, or even generated by evaporation of moisture present around spills, dilutes the contaminants in the gaseous phase serving as the *stripping effect*. In the end it lowers the partial pressure and thus the boiling point of these species leading to their enhanced release into the gas phase.

The effectiveness of the process is connected with the time required for the successful treatment. The *residence time* may vary to a wide extend ranging from one day for in situ processes (Iben et al. 1996) down to some seconds for a newly developed ex situ technique (Höhne 2000).

The influence of the *pressure* on purification is not significant. More important is the reduced pressure for the prevention of fugitive hazardous emissions and thus safety considerations for plant personnel.

The *ratio* of the *sweep gas* (carrying the volatised components) *to* the *volatised contaminants* affects the separation efficiency only if it is rather low. Under process conditions this would be not of importance.

The parameters themselves are mutually interconnected. For example, the higher the temperature the shorter the residence time needing a shorter time for energy input. PAH, in comparison to MHC, need a higher process temperature and coarse particles, such as sand, may be easier and thus faster to be cleaned than fine-grained particles as silt or clay. In the end socio-economic conditions, such as total expenditure for the remediation, time and space available for the technical procedure and legal limitations also influence the decision on the process selected and its operating conditions.

32.3.2 Plant Design and Operation

For both treatment modes of in situ and ex situ operation the technical designs are basically different.

For *in situ* remediation the focus is put on the wells for the introduction, and collection of the sweep gas, respectively. Furthermore gas distribution in the ground is of highest importance to the final success. For a comprehensive treatise refer to Bellandi (1995).

Alternatively *ex situ* processing may be carried out in continuous or discontinuous operation (Wietstock et al. 1997). Generally the plant may be divided into the following subsystems (Anderson 1993; Anonymus 1998).

32.3.2.1 Pre-Treatment

Several aspects have to be considered for the feed in ex situ operating plants as a wide range of organic species in various media have to be treated. Suitable moisture conditions often result in dewatering of the material to be processed (Sullivan 1999). A recently developed technique even treats suspensions (Wietstock et al. 1998). Additional particle sizing and blending of soil fractions should provide consistent conditions for the process. These activities are cost intensive (energy consumption, analyses of the chemical species) but needed if widely varying feed has to be processed.

32.3.2.2
Separation Unit

The fairly homogenated feed is transferred into the directly or indirectly fired thermal process unit, which is often a rotary or conveyor system. Depending on the abrasive behaviour of the feed, protective measures against material damage have to be undertaken.

32.3.2.3
Post-Treatment of Effluent Streams

The *cleaned solid* has to be removed from the fluid phase, re-wetted and filled back on site, reused or disposed off.

In conventional plants a comprehensive *off-gas treatment* is implemented:

- elimination of hydrocarbons requires integration of oxidising units (afterburner, catalyst application etc.), adsorption systems (active carbon), and condensers for organic recovery and for thermal use, respectively;
- particulate and acid gas separation by the utilisation of cyclones, backhouse filters, scrubbers and active carbon adsorbents are common techniques.

Water purification mainly removes fine solids, traces of organic and heavy metals by well known precipitation and adsorption processes.

32.3.2.4
Management of Decontamination Procedures

The *analyses* of incoming and leaving soil as well as other effluent streams (off-gas, water) have to confirm their quality with respect to legal obligations. Frequently automated analysis devices are in operation which additionally deliver data supporting *process control* and *quality assurance. Safety requirements* are of importance, primarily for plant personnel and also for the direct environment around the plant. The reduced system pressure in the plant prevents contamination loss and thus fugitive emissions. Under certain processing conditions explosion prevention has to be considered at the plant design stage, as well as during operation.

32.3.3
Modelling

Numerous models have been developed parallelly to the extensive laboratory and field studies on remediation techniques. For in situ remediation, models are available describing physico-chemical pathways and transport phenomena for steam stripping of organic species in porous media (van der Ham and Brouwers 1998). The importance of transport phenomena has also been evaluated by modelling an ex situ steam stripping process carried out by Niemeyer and Luo in 1999. The transient behaviour of the separation unit of a low-temperature process is also described (Luo and Niemeyer 1999).

32.4 Economical Aspects

The economical evaluation of soil treatment has to consider operational expenses including capital expenditures for the investment as well as long term costs mostly related with prevention methods or biotechnological treatment.

In general the legal limitations on residues on soil after treatment and effluents from the facilities (fine-grained solids to be disposed off, off-gas and waste water) primarily influence the overall economics of the soil decontamination. They mainly consist of costs for:

- solid features (including the level of humic substances and moisture present),
- kind and concentration of contaminants (volatility),
- transport of soil (especially for off site processing),
- personnel,
- capital refund (reflecting plant investment, including extra expenditures for vendors equipment limitations, interest rate and permission management).

The higher local flexibility of plants (mobile equipment) results in elevated specific treatment prices in comparison to stationary set-ups while reducing the payments for transport.

Operational expenses are widely governed by energy consumption as fuel input becomes increasingly expensive. Low-temperature techniques allow the use of energy from other processes normally emitted at a low temperature level (e.g. less than 450 K).

The United States based literature intensively focuses on economical features and even presents numbers for specific costs (U.S. Environmental Protection Agency 1995a). Using theses numbers as indicative values one should take into account that greater economical changes may have occurred after their publication, such as decrease in plant related investment costs, increase in energy prices etc.

The U.S. Environmental Protection Agency (1995b) provides data for soil decontamination by low-temperature treatment ranging from $ 40 to $ 100 per ton of average soil and for fine-grained matter from $ 100 to 300 per ton soil (cited in Sullivan 1999).

In order to estimate the scale of the soil to be processed in a mobile plant, applicable data are published by Anderson (1993). For MHC contamination of small volumes (around 1,000 ton) expenses range from $ 90 to $ 130 per ton soil and for a 10,000 ton MHC-decontamination $ 40 to $70 per ton are estimated. For more crucial contamination containing PAH, prices range from $ 300–600 per ton (for 1,000 ton site) and from $ 150–200 per ton for the treatment of larger volumes (10,000 ton) in a mobile plant.

The influence of economics on the treatment to be selected should not be underestimated, offering the possibility to apply modern techniques on an industrial scale.

32.5 Advantages and Limitations of Desorption Techniques

One main advantage of *desorption* techniques applied for soil decontamination is the reliable removal of hazardous components and consequently the protection of the environment from further pollution. The processing expenses have to be calculated against long-term expenditures for monitoring connected with prevention methods, and natural attenuation, respectively.

The powerful efficiency of these techniques concerning hydrocarbon removal including crucial compounds like PAH and chlorinated species, is comparable to chemical conversion. For both technologies the insufficient separation of heavy metals is regarded as a limiting factor, especially if mixed contamination of hydrocarbons and heavy metals has to be treated.

The recycling of contaminants (preferably in higher concentrations) for energy conversion purposes is regarded as advantage, finally contributing to eco-efficiency of these processes as well as to their economical performance.

In general all low temperature desorption techniques widely preserve soil fertility without demanding additional effort for recultivation (Urban et al. 1998).

Due to the few, and relatively simple, process units employed for in situ operations the processing costs are quite low. Even the decontamination of soils under buildings and other facilities are enabled; the main objections to in situ operations are the presumptions necessitated for the application as discussed above.

Economical performance for ex situ processing increases by indirectly fired separation units. Due to the neglectable off-gas volume, gas treatment expenditure, and therefore overall payments notably diminish.

The technological effectiveness of the treatment will depend on site-specific and chemical-specific parameters. Consequently each technology has specific degrees of effectiveness over the full spectrum of organic species (U.S. Environmental Protection Agency 1995b). Pitfalls of regularly effective techniques (U.S. Environmental Protection Agency 1995c) reflect illustratively the experience that the complex system is not yet fully understood. Consequently pilot scale plants for testing and evaluating technical and economical features are essential within the course of decontamination procedures.

32.6 Future Perspective

The flexible response to the varying remediation demands concerning the variety of contaminants to be removed and origin of soil, costs, legal limitations etc. is the combination of different methods. This includes prevention and decontamination treatment as well as the consecutive application of different techniques. These measures may deliver a higher operational safety against process failure and may provide a better economical performance (Page 1997; Stegmann 2000).

Further research and development focus on hybrid processes which allow the combination of several unit operations into one processing step. The U.S. Environmental Protection Agency has reported on the simultaneous separation of VOC and heavy metals (1995a).

A further important innovation is the transfer of the widespread knowledge of soil processing to production residues cleaning for recycle valuables, either for the market or for the feed of the production process itself. As an example, the processing of drilling sludge as residues from crude oil exploration showed a reduction from 225,000 mg MHC $(\text{kg SM})^{-1}$) to less than 100 mg MHC per kg SM^{-1} at a process temperature of 500 K (Höhne et al. 2000b). This is a purification performance of 99.9 %; the processed matter are useable without restrictions.

32.7 Notation

MHC	mineral oil hydrocarbons
PAH	polycyclic aromatic hydrocarbons
SM	dry solid matter
SVOC	semi-volatile organic compounds
VOC	volatile organic compounds

References

Anderson WC (1993) Innovative Site Remediation Technology: Thermal Desorption. Volume 6, American Academy of Environmental Engineers, Annapolis, MD, USA, Springer, Berlin, ISBN 3-540-59066-8

Anonymus (1998) Verfahrensbeschreibung Desorptionsanlage, Verfahrensschema Desorptionsanlage. Fa Leonhard Weiss GmbH & Co, Umwelttechnik, Crailsheim

Bellandi R (1995) Innovative Technologies for Hazardous Waste Remediation. Van Nostrand Reinhold, New York, USA

Connaughton DF (1993) Description of Time-Varying Desorption Kinetics: Release of Naphthalene from Contaminated Soils. Environmental Science and Technology 27: 2397-2403

Franzius V (1995) Altlastsituation und Finanzierungsbedarf in den neuen Bundesländern. In: Franzius V, Wolf K, Brandt E (eds) Handbuch der Altlastsanierung. CF Müller Verlag, Hüttig GmbH Heidelberg, ISBN 3-8114-9700-6, contribution no 1843

El-Shoubary Y, et al. (1994) Desorption of Contaminants from Different Host Matrices. Environmental Progress 13: 93-144

Höhne J, Wilichowski M, Niemeyer B (2000a) Decontamination of Residues from Soil Washing Process by Steam Stripping. 7th International FZK/TNO Congress on Contaminated Soil, ConSoil 2000, Leipzig, 18–22 September 2000, Vol. 2, Telford, London, ISBN 07277 29543, pp 1163 f

Höhne J, Eschenbach A, Niemeyer B (2000b) Minimization of Waste by Sludge Treatment Applying the Steam Stripping Process. In: Poon CS, Lei PCK (eds), Waste Management: The Challenge for Asian Cities – Search for a Sustainable Future, Proceedings of the ISWA International Symposium & Exhibition on Waste Management in Asian Cities, Hong Kong, 23–26 October 2000, Vol. 2, ISBN 962-367-303-5 pp 292-297

Höhne J (2000) Untersuchungen zur Reinigung von feinkörnigen, mit organischen Schadstoffen beladenen Bodenmaterialien durch gekoppelten Einsatz von Wasserdampfdestillation und Desorption in einer Labor- und einer Technikumsanlage. Thesis, Universität der Bundeswehr Hamburg (in preparation)

Iben IET, et al. (1996) Thermal Blanket for In-Situ Remediation of Surficial Contamination: A Pilot Test. Environmental Science and Technology 30: 3144–3154

Lord AE Jr, Hullings DE, Koerner RM, Brugger JE (1990) Vacuum-Assisted Steam Stripping to Remove Pollutants from Contaminated Soil: A Laboratory Study. In: Proceedings of 16th Annual Hazardous Waste Research Symposium, Cincinnati, OH, April 3–5 1990, EPA/600/9-90/037

Luo X, Niemeyer B (1999) Modelling and Simulation of Transient Transport Processes Using Axial Dispersion Model. In: Keil F, Voß H, Werther J (eds) Scientific Computing in Chemical Engineering II. Springer, Berlin, ISBN 3-540-65848-3, pp 167–174

Niemeyer B, Luo X (1999) A Diffusion Model for Desorption of Hazardous Components from Solids. Proceedings of the International Colloquium on Modelling of Material Processing, Riga, Latvia, May 28–29 1999, pp 140–145

O′ Brien JF, Rouleau JF (1995) Mobile Thermal Desorption. Military Engineer 87: 55–56

Page GW (1997) Contaminated Sites and Environmental Cleanup: International Approaches to Prevention, Remediation, and Reuse. Academic Press, San Diego, CA, USA

Rat von Sachverständigen für Umweltfragen (1995) Altlast II, Sondergutachten Februar 1995. Metzler-Poeschel, Stuttgart, ISBN 3-8246-0367-5

Rodriguez-Maroto JM, Gomez-Lahoz C, Wilson DJ (1995) Removal of Semivolatiles from Soil by Steam Stripping. IV. Effects of Adsorption/Desorption Kinetics. Separation Science and Technology 30: 2659–2678

Rulkens, WH (2000) An Overview of Soil Treatment Research in the Netherlands. In: Stegmann R (ed) Bodenreinigung – 12 Jahre Forschung und Entwicklung. Technische Universität Hamburg-Harburg, 29–31 März 2000, pp 9–11

Stegmann, R (2000) Bodenreinigung – 12 Jahre Forschung und Entwicklung. Abschlußveranstaltung des Sonderforschungsbereiches 188 der Deutschen Forschungsgemeinschaft REINIGUNG KONTAMINIERTER BÖDEN. Technische Universität Hamburg-Harburg, 29–31 März 2000

Sullivan TP (1999) Thermal Desorption: The Basics. Chemical Engineering Progress 10: 49–56

Thöming J (1998) Elektrolaugung zur Dekontamination quecksilberbelasteter Böden und Reststoffe – Entwicklung eines hydrometallurgischen Kreislaufprozesses. Thesis, Technische Universität Hamburg-Harburg

Umweltbundesamt (2000) Bundesweite Übersicht zur Altlasterfassung. Internet: http://www.umweltdaten.de/altlast/web1/deutsch/1_6.htm

Urban AI, Friedel M, Kugler P, Schmeisky H (1998) Bodenqualität und Rekultivierbarkeit nach einer Altlastsanierung mit einer Niedertemperatur-Desorptionsanlage. Wasser & Boden 50: 15–18

US Environmental Protection Agency (1991) In Situ Steam/Hot-Air Stripping Technology Novaterra. Inc EPA/540/A-5-90/008, Center for Environmental Research Information, Cincinnati, OH, USA

US Environmental Protection Agency (1995a) Process for the Treatment of Volatile Organic Carbon and Heavy-Metal-Contaminated Soil. EPA/540/F-95/509, National Risk Management Research Laboratory, Cincinnati, OH, USA

US Environmental Protection Agency (1995b) Remediation Technologies Screening Matrix and Reference Guide. 2nd ed, EPA/542/B-94/013, Office of Solid Waste and Emergency Response, Washington, DC, USA

US Environmental Protection Agency (1995c) In Situ Steam Enhanced Recovery Process. EPA/540/R-94/510a, National Risk Management Research Laboratory, Cincinnati, OH, USA

US Environmental Protection Agency (1996) Superfund Innovative Technology Evaluation Program: Technology Profiles. 9th ed, EPA/540/R-97/502, Office of Research and Development, Washington, DC, USA

US Environmental Protection Agency (1999) Treatment Technologies for Site Cleanup: Annual Status Report. 9th ed, EPA/542/R-99/001, Washington, DC, USA

Van der Ham AGJ, Brouwers HJH (1998) Modelling and Experimental Investigation of Transient, Nonequilibrium Mass Transfer During Steam Stripping of a Nonaqueous Phase Liquid in Unsaturated Porous Media. Water Resources Research 34: 47–54

Wietstock P, Wrage B, Luther G, Niemeyer B, Gronholz C (1997) Process for Separation of Volatile Pollutants from Contaminated Materials, Soil, Wood, Sludge and Sediments. In:

Calmano W, Roeters P (eds), International Conference on Contaminated Sediments (ICCS). Volume 1, Rotterdam, The Netherlands, 7–11 September 1997, pp 444–451
Wietstock P, Wilichwoski M, Niemeyer B (1998) Entfernung von PAK aus feinkörnigen Böden durch Niedertemperatur-Desorption. Chemie-Ingenieur-Technik 70: 1167–1168
Wiesner J (1995) Fachtreffen Umwelttechnik. Chemie-Ingenieur-Technik 67: 1533–1534
Wilson DJ, Clarke AN (1994) Hazardous Waste Site Soil Remediation: Theory and Application of Innovative Technologies. Marcel Dekker, New York, USA
Yuan ZG, Udell KS (1993) Steam Distillation of a Single Component Hydrocarbon Liquid in Porous Media. International Journal of Heat Transfer 38: 1965–1976

33 Elimination of Hazardous Components from Fine-Grained Particles and Sludges by the Application of the Steam Stripping Process

J. Höhne[1], B. Niemeyer[1,2]
[1] GKSS Research Centre, Max-Planck-Straße, 21502 Geesthacht, Germany
[2] University of Federal Armed Forces, Institute of Thermodynamics, Holstenhofweg 85, 22043 Hamburg, Germany

33.1 Introduction

In recent years the prevention and management of waste have become an important issue of environmental protection. In the production of raw materials, industrial processing, and soil washing processes large amounts of fine-grained solids and sludges (dp < 100 μm) are produced. Due to the lack of reprocessing plants working economically these materials are currently being dumped. It is estimated that in Germany 180,000 t of fine-grained residues are generated annually, solely by soil washing processes.

The removal of hydrocarbons from solid surfaces is a complex issue. Different contaminants summarised as mineral hydrocarbons (MHC), and polycyclic aromatic hydrocarbons (PAH), as well as their chlorinated derivatives may interact with various types of soil matrices containing diverse contents of humic substances.

Thermal desorption techniques has been proven as powerful separation processes (Anderson 1993). The Steam Stripping Process has been developed for the decontamination of fine-grained particles from a wide range of natural origins. The main aim of the process is to transfer the contaminants from a liquid phase, as well as from solid surface-bindings, into the gaseous phase. In this manner, the decontamination of fine-grained particles is carried out. After the separation of hydrocarbon contaminants from the off-gas stream they are discharged in a concentrated manner. The effective decontamination with a cleaning performance of more than 99 mass % is accomplished by applying different separation conditions such as elevated temperatures and reduced system pressures simultaneously.

33.2 Principles of the Process

The main parameters which influence physico-chemical response due to Steam Stripping are temperature and pressure. The basic operations used are *steam distillation* and *desorption*, which will be discussed in detail.

33.2.1 Steam Distillation

Organic contaminants, e.g. MHC and PAH, which generally show slight solubility in water, build up a second distinct liquid phase in the water. PAH which have not been adsorbed can be dissolved in the organic oil phase (Eschenbach 1995). The contact of two liquids, e.g. oil/water, which show little mutual solubilities, leads to a wide miscibility gap which is the basis of the steam distillation. Within the miscibility gap every substance vapourises without influence of the composition of the liquid phase at its respective saturation vapour pressure as a pure component at the set temperature. For a compound whose melting point is higher than the stripping temperature, e.g. for most PAH, the vapour pressure can be viewed as one of a hypothetical subcooled liquid (Hwang et al. 1992). Application of Dalton's law,

$$p = \Sigma p^*_i \tag{33.1}$$

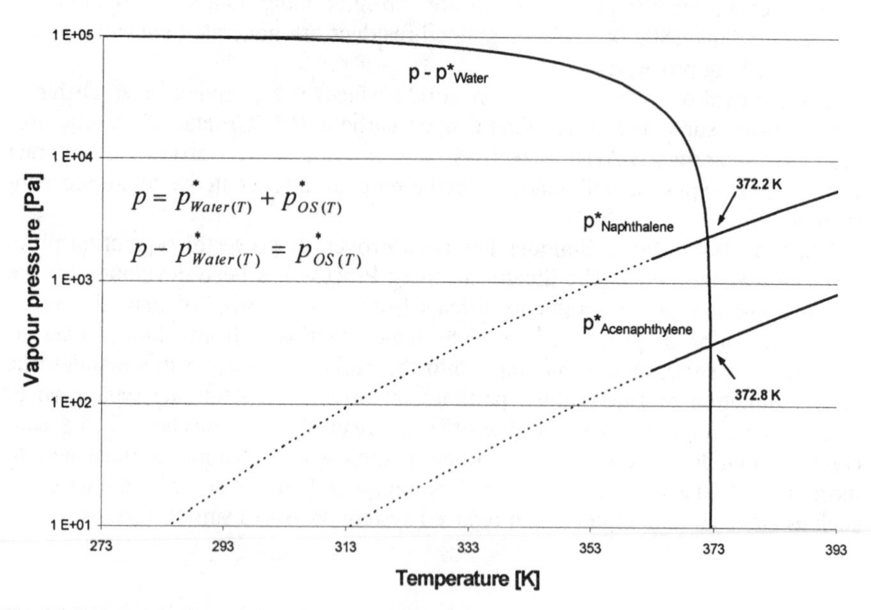

Fig. 33.1. Determination of the common boiling-points of binary mixtures (water, PAH) according to the principles of steam distillation at a system pressure of $1 \cdot 10^5$ Pa

demonstrates that the saturation vapour pressures p^*_i of the components i are lower than the total system pressure. This results in a common boiling point, which is lower than the boiling temperatures of the pure components i at the system pressure (Mersmann 1980). In fig. 33.1 examples of the determination of the common boiling points of binary mixtures (i.e. water-naphthalene and water-acenaphthylene) are presented. As described the resulting boiling points are less than the boiling points of the pure substances ($T_{b,Water} = 373$ K, $T_{b,Naphthalene} = 491$ K, $T_{b,Acenaphthylene} = 538$ K at a pressure of $1.0 \cdot 10^5$ Pa). Necessary data of chemical compounds can be attained from different literature, e.g. Gmehling et al. (1981), Landolt-Börnstein (1999).

Hudel (1995) stated that in steam distillation a rise in pressure will increase the removal of PAH. Additionally the lower the molecular weight of the contaminant, the higher the amount of the organic component transferred into gaseous phase by steam distillation. Following the vapourisation of water, and thus the removal of contaminants by steam distillation, those contaminants, which have not been removed, undergo a "dry desorption".

33.2.2 Desorption

Non-ionic organic compounds are preferentially adsorbed at soil by the soil organic matter. Consequently the sorption is expected to occur rather via non-specific mechanisms as hydrophobic interactions, hydrogen bonding and van der Waals interactions (Schlebaum 1999). The hydrophobic interactions in particular are characterised by relatively low bonding forces. Nevertheless, the soil organic matter is a complex system of molecules with a varying hydrophobicity and size. This causes widely different bonding features like affinity and sorption capacity. During the desorption process organic contaminants are transferred into the steam phase. Seewald and Jüntgen (1977) have shown that low amounts of contaminants may be strongly bound, and therefore the respective desorption only takes place at higher temperatures. For soils with increasing contaminant loads further adsorption areas with lower bonding energy levels are occupied which can already be removed at lower temperatures. Therefore, a rise in process temperature causes the removal of adsorbed contaminants with rising bonding forces. Only high temperatures allow an improvement in contaminant-removal to the lowest levels. At a fixed residence time, an improved desorption kinetic (e.g. by high temperatures) raises the amount of substances desorbed during a designed process.

The influence of pressure causes two contrary effects in the Steam Stripping Process upon steam distillation and desorption. A rise in pressure will increase the removal of PAH by steam distillation. In contrast, the sorption equilibrium is enforced at a lower system pressure.

33.3 Process Design and Operational Conditions

33.3.1 Process Design

The investigations into the process development were carried out in a small scale laboratory unit for the determination of equilibrium data of different solid-contaminant systems and different process conditions (variation of temperature, pressure, water content in the sludge, etc.).

On the basis of these results experimental investigations were carried out into the continuous processing of the pilot scaled Steam Stripping Process for fine-grained particles and sludges. In this plant contaminated matter can be treated under a wide range of separation parameters at a maximum throughput of 5 kg SM h^{-1}. The process temperature can be varied up to 610 K and the pressure ranges from $0.5 \cdot 10^5$ Pa to $1.0 \cdot 10^5$ Pa. The treatment of fine-grained dispersed particles requires that they must be free-flowing, and have suitable spraying features. The content of dry solid matter (SM) ranges from < 20 mass % (sediment) up to 60 mass % (drilling sludge) to provide the necessary spraying features within the nozzle used (fig. 33.2). The flow sheet of the plant used for the investigations into the Steam Stripping Process (Wietstock et al. 1997) is presented in fig. 33.3. The plant may be divided into four main parts, which are described below.

33.3.1.1 Feeding Device

The fine-grained matter is introduced into the plant as an aqueous suspension which is stored in the feed container (C1). An even dispersion of the sludge suspension is achieved by means of an appropriate stirrer preventing sedimentation

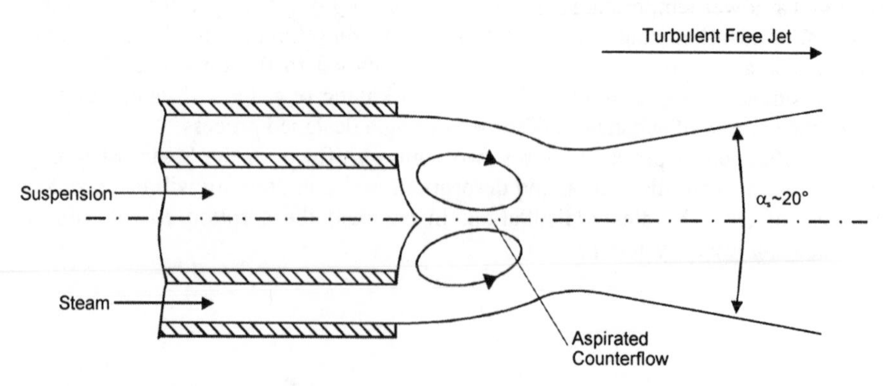

Fig. 33.2. Scheme of the two-fluid nozzle of the Steam Stripping Process

and guaranteeing a nearly homogenous solid-loaded feed at the entrance of the reactor. The eccentric screw pump (P1) feeds the contaminated suspension into the specially designed bi-fluid nozzle. Additionally, superheated steam from the thermal energy supply is introduced into the nozzle. It serves as the energy transport medium as well as the stripping medium for the process.

33.3.1.2 Decontamination Unit

The decontamination takes place in the tube reactor (R1). At the beginning of the reactor a three-phase-system (contaminated solid – liquid – gas) leaves the nozzle. As the spraying of the suspension occurs polydisperse jets are generated containing a high amount of drops consisting of multiple diameters, velocities and outset directions. The turbulent flow of the fluids in the spray reaches an effective heat, mass and impulse exchange between drops and steam. Thus the water contained in the sludge-drops spontaneously vapourises and contaminants are removed by steam distillation. Only two phases (solid and gas) are present after the vapourisation of the water or drying of the spray-drops. During the horizontal pneumatic transport through the tube reactor, particle-contaminant agglomerates are split up and thus the particle surfaces are widely exposed and the desorption process takes place (Höhne and Niemeyer 1999).

At the end of the reactor the flow containing the solid and gaseous phases is postprocessed for the recovery of decontaminated solids, contaminants and cleaning effluents such as water.

33.3.1.3 Postprocessing

Solid recovery: After leaving the tube reactor the decontaminated particles are directed into a cyclone (C2) which separates the cleaned solid particles from the steam and directs them into a collecting hopper (C3).

Gas treatment: The steam containing the gaseous contaminants is led from the cyclone to a quench cooler (condensation chamber, C4), where it will undergo condensation. The condensed water-contaminant mixture is separated in a liquid-liquid separation device (C6), which allows disposal off the hydrocarbon contaminants in concentrated manner.

33.3.1.4 Process Support

Thermal energy supply: The required thermal energy for the process is delivered by the supply units (H1 to H3).

Vacuum generation: The reduced pressure in the system is generated and delivered by the vacuum pump (P2).

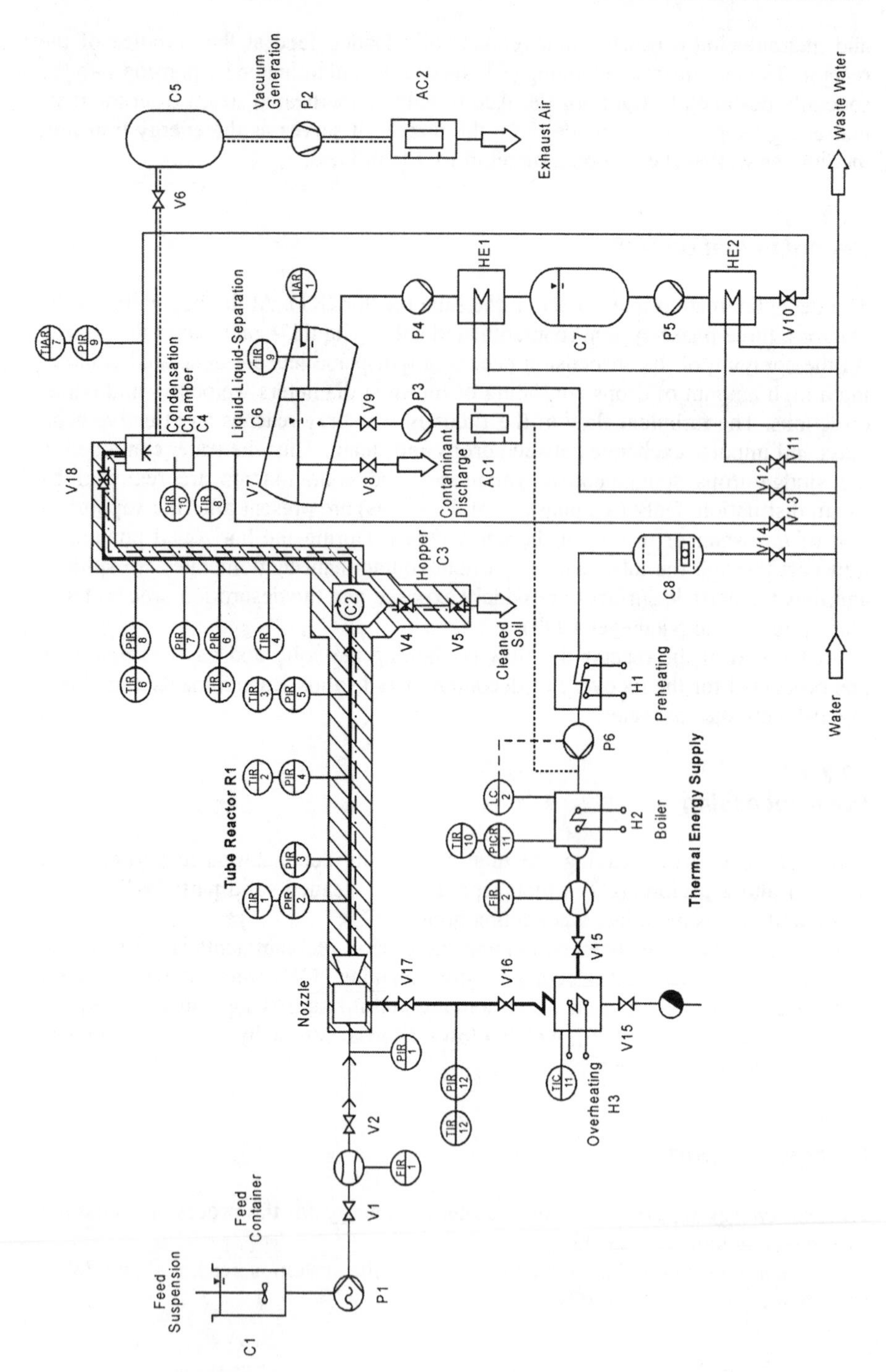

Fig. 33.3. Flow sheet of the pilot plant of the Steam Stripping Process

Table 33.1. Feed-sludges treated in the Steam Stripping Process

Sludge	SM [mass %]	q_{MHC} [mg MHC (kg SM)$^{-1}$]	q_{PAH} [mg PAH (kg SM)$^{-1}$]	TOC [mass %]
Sludge "A"	40	943	4,800	17
Sludge "B"	35	6,892	3,760	11

33.3.2
Operational Conditions

To establish the parameters that influence the process and to evaluate the cleaning results, two residues resulting from the soil washing process have been applied. As seen from table 33.1, both sludges have nearly similar contents of solid matter.

While the PAH-content (analysed according to the standards set by the U.S. Environmental Protection Agency 1979) at 4,800 mg PAH (kg SM)$^{-1}$ "A" and 3,760 mg PAH (kg SM)$^{-1}$ "B" are comparatively similar, both sludges differentiate distinctly in MHC-content as well as in the total organic carbon (TOC)-content. Residue sludges obtained from the soil washing process have normally been treated in a suspension at a level of 35–40 mass % SM. Furthermore the investigated process is very rapid (residence time of the solid particles within the reactor is less than 1 second).

33.4
Experimental Results of Separation

For the process development investigations into the parameters related with the *processing* of decontamination and with the conditions of the *feed* introduced into the process, were carried out. For all experiments sludges from industrial scaled soil washing processes were applied.

33.4.1
Process Parameters

The content of dry matter in the suspension to be processed is not crucial under the presumption specified in the section "Operational Conditions".

For the investigations sludge "A" carrying a medium MHC load, a high PAH load and a high TOC content of 17 % was applied. It is a typical residue sludge from the treatment of soils originating from the premises of former gas works. The results of the treatment of sludge "A" are presented in fig. 33.4. The remaining load of the treated material is shown as a function of the process temperature with system pressure varied from $0.5\ 10^5$ Pa to $1.0\ 10^5$ Pa.

For both MHC as well as PAH the remaining load decreases exponentially with an increasing temperature. By raising the process temperature from 443 K to 593 K at a system pressure of $1.0\ 10^5$ Pa a reduction of the remaining PAH-load from 420 mg PAH (kg SM)$^{-1}$ to < 50 mg PAH (kg SM)$^{-1}$ could be achieved.

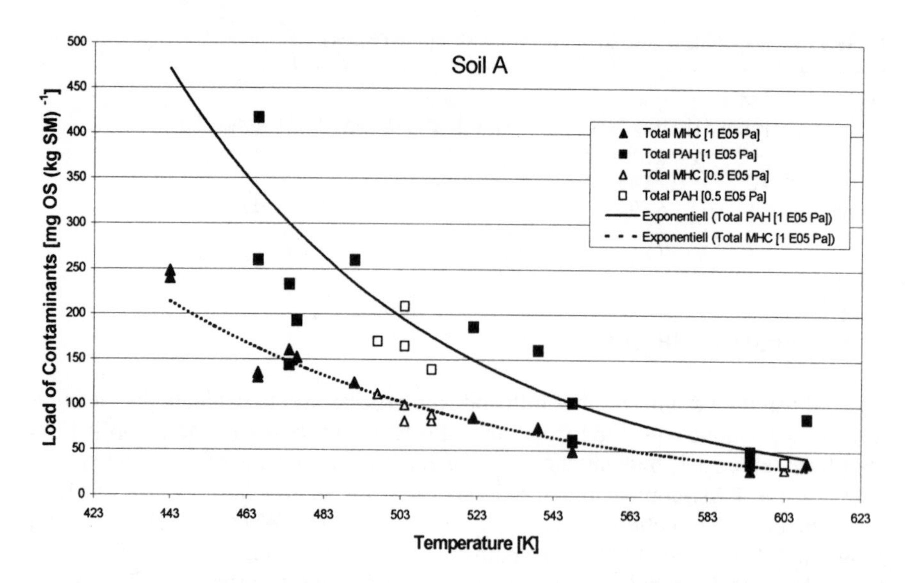

Fig. 33.4. Remaining load of MHC and PAH from sludge "A" in dependence on pressure and process temperature

The degree of separation of MHC from the soil surface is due to the lower initial contaminant load and is consequently not as spectacular as the separation of PAH. For MHC there is only a slight remaining charge which can already be achieved at low process temperatures. In fig. 33.4 the two curves for MHC and PAH prove an asymptotic response indicating a remaining load of approximately 30 mg OS $(kg\ SM)^{-1}$ at a temperature of 603 K, so that an additional increase in temperature will not significantly alter the cleaning performance.

According to the experimental results, the pressure seems to have a minor effect on the degree of contaminant elimination. The results obtained at a reduced system pressure of 0.5 10^5 Pa are within the reproducibility range of the results gained at 1 10^5 Pa. The reproducibility of the values is influenced by the experimental and analytical procedures. The analyses of a contaminated soil, which is a complex system of mineral matter and different natural organic compounds (e.g. humic substances) itself, show problems in sampling and preparation for analyses in particular. Thus analytical values of a treated soil may differ.

33.4.2 Feed Parameters

The soil matrix and *the contaminants* influence the separation process the most.

The soil matrix exhibits a large variety in chemical composition and also, therefore, in solid characteristics. The mineral particles of a soil can be "coated" with polymer organic matrices (Wilichowski 1994).

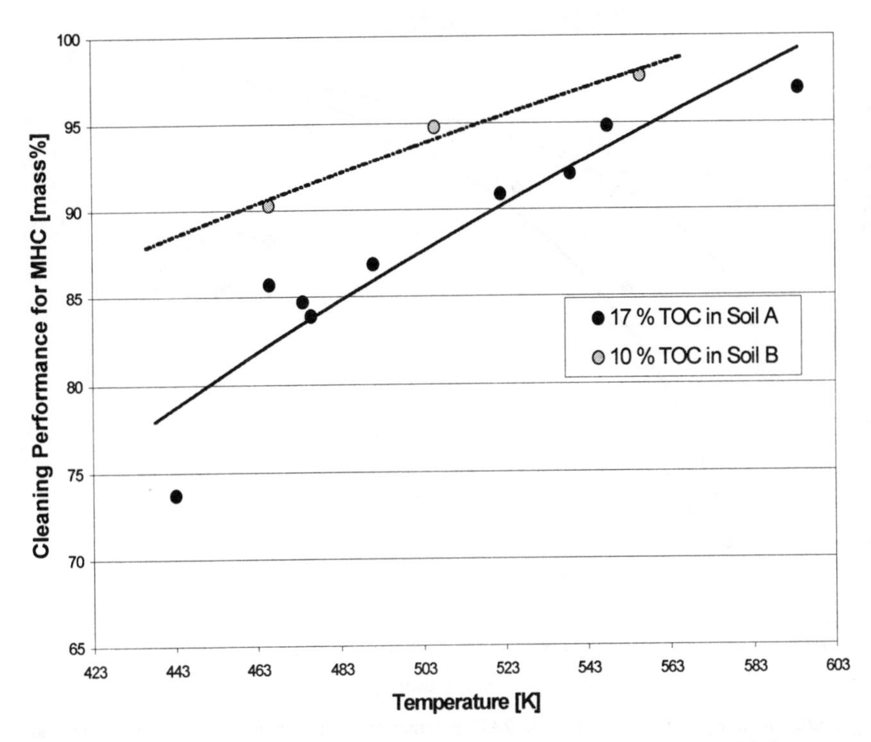

Fig. 33.5. Cleaning performance for the MHC-elimination from sludge "A" and "B" in dependence on the TOC-content and the process temperature (at $1.0 \cdot 10^5$ Pa)

For soils (soil sludges) containing more than 1 mass % of organic matrix, e.g. humic substances, it is predicted that the soil features are dominated by the characteristics of the humic substances (Hassett et al. 1989; Chiou 1989). For an overall characterisation of a soil the content of the total organic carbon (TOC), which embraces the humic substances, is an important property. In various experiments, Means et al. (1980) found out that the amount of adsorbed contaminants correlates with the TOC-content of a soil.

The cleaning performance of the MHC separation for the above mentioned sludges "A" and "B" is expressed as:

$$\eta = (q_{OS,0} - q_{OS,transferred}) * 100/q_{OS,0} \qquad (33.2)$$

The load of remaining contaminants after treatment is inverse related to the cleaning performance. In fig. 33.5 the graphs take an asymptotic course which clearly approaches a value close to 100 %. The sludges differ from each other mainly in their content of total organic carbon (TOC). The higher the TOC-content the more difficult it is to achieve the desired cleaning performance at the same temperature (Sullivan 1999). Henceforth, the TOC-content seems to be connected with the adsorption capacity (Dzombak and Luthy 1984).

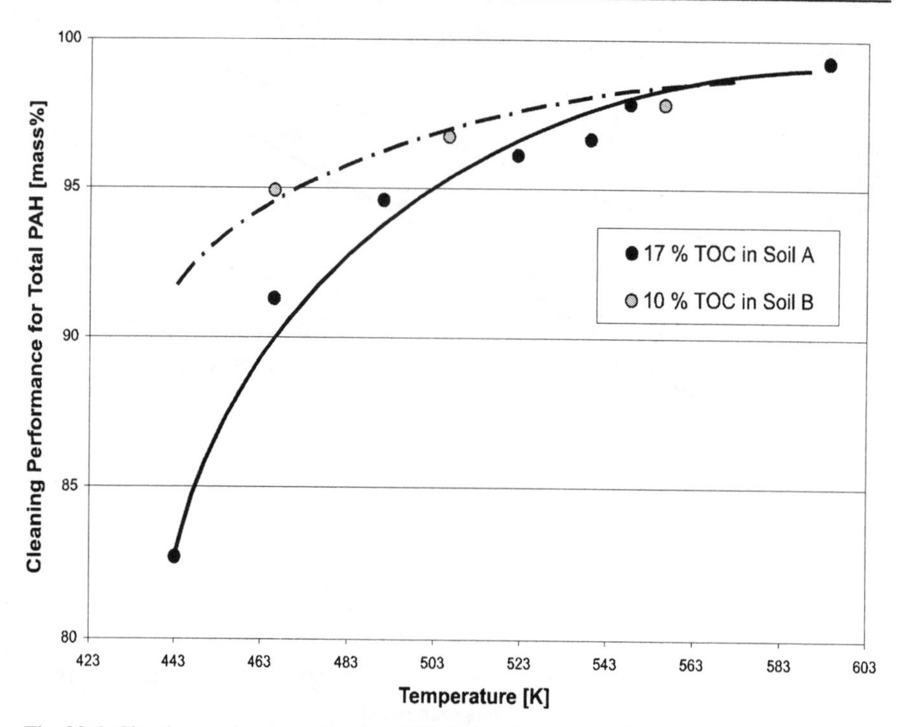

Fig. 33.6. Cleaning performance for the PAH-elimination from sludge "A" and "B" in dependence on the TOC-content and the process temperature (at $1.0 \cdot 10^5$ Pa)

In order to reach a comparable cleaning performance for sludges with different TOC-contents, the process temperature, and therefore the input of heat energy into the process must be adapted. As seen from fig. 33.6 an increase in TOC from 10 mass % to 17 mass %, for instance, requires a temperature increase from 465 K to approximately 500 K for obtaining a similar cleaning performance for PAH removal.

The relation between the cleaning performance and the TOC-content, as indicated above for MHC contaminants, can also be found in the cleaning of PAH-contaminated soil-sludges (fig. 33.6). A cleaning performance of more than 98 % for both sludges "A" and "B", can already be achieved for the PAH removal at temperatures of approximately 550 K. Similar results were obtained with other residues from soil washing processes (Niemeyer et al. 2000).

The contaminants: A contamination frequently consists of a wide range of single compounds, especially if they are of a hydrocarbon nature. One way of characterisation of such mixtures is the use of cumulative parameters like "Total-MHC" or "Total-PAH" as used above. For a more detailed view on the PAH-decontamination patterns the individual PAH, instead of the "Total PAH", have to be examined. In fig. 33.7 the separation efficiency for the 16 PAH selected reference components (U.S. Environmental Protection Agency 1979) are depicted. The load of contaminated matter before cleaning (feed) is compared with the load leaving the process (after treatment).

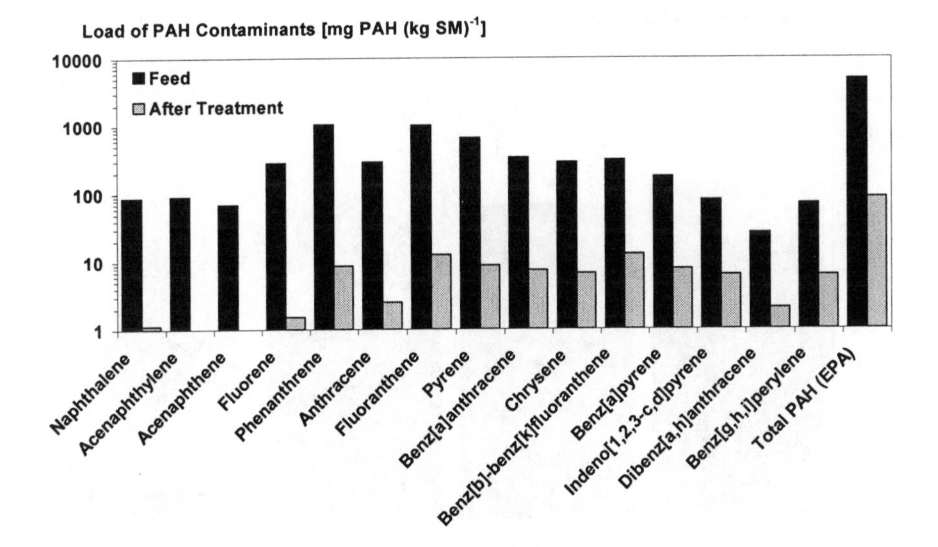

Fig. 33.7. Remaining load of selected PAH contaminants from sludge "A" processed at 609 K, $1.0 \cdot 10^5$ Pa

Attention has to be paid to the logarithmic scale of the ordinate used because of the large differences between the load of contaminants in the feed and that of the reprocessed material. Low condensed ring systems like naphthalene or acenaphthene are separated by more than 99 %. Higher condensed systems, like benz[a]pyrene have a slightly lower separation efficiency. The total load reduction of the 16 PAH is in the range of 98 % (Wietstock et al. 1998).

The slight increase in the amount of residual contaminants found at the particle surface clearly corresponds with an increase in molecular weight and also the boiling points of the contaminants. For example acenaphthene (boiling point: $T_b = 552$ K) can be eliminated by 99.4 %; benz[a]pyrene has a much higher boiling point of $T_b = 768$ K and thus it is more difficult to be eliminated. Nevertheless it can be removed by the Steam Stripping Process to 95.5 %. The separation performance of the process and the PAH boiling points are illustrated in fig. 33.8.

33.4.3 Kinetic of Contaminant Vapourisation

In order to design a technical process based on the physio-chemical behaviour of the contaminant-solid-system it is essential to understand and describe its separation kinetics. Desorption processes can be classified as a reaction of first order ($z = 1$), whereby the energy E_A describes the activation of desorption (Seewald and Jüntgen 1977).

$$\frac{dn_{OS}}{dt} = k \cdot n_{OS,0}^{z=1} \cdot \exp\left(\frac{-E_A}{R \cdot T}\right) \tag{33.3}$$

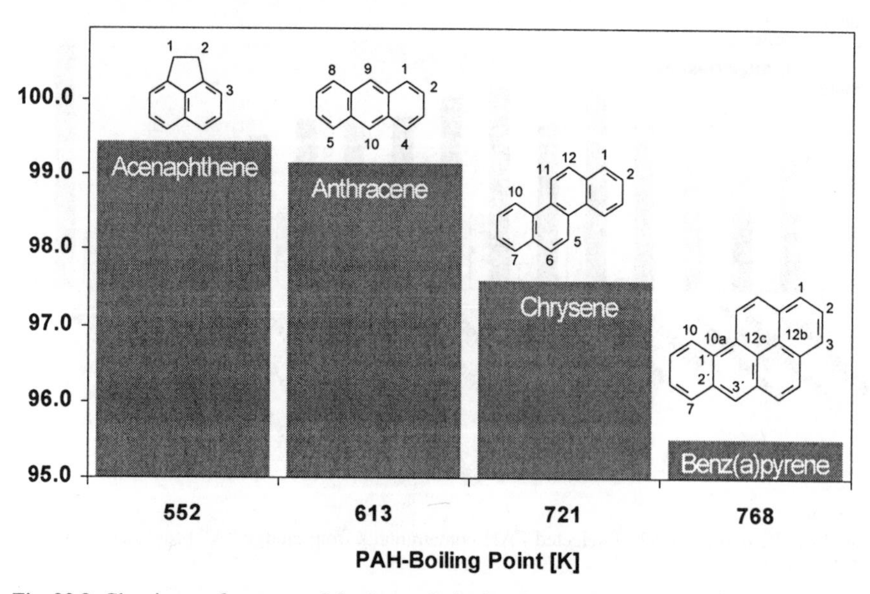

Fig. 33.8. Cleaning performance of the Steam Stripping Process to selected PAH of sludge "A" processed at 609 K; boiling points indicated are for pure substances

To examine the desorption process of a contaminant (e.g. PAH) from its carrier (e.g. soil), the load of the organic substance q_{OS} [mg OS (kg SM)$^{-1}$] of the specific contaminant is used instead of its molar quantity n [mol]. During the Steam Stripping Process the contaminants are transferred into the gas phase, which is attained by steam distillation of non-adsorbed contaminants as well as by desorption of soil-linked contaminants. The different mechanisms can not be sized separately by experiments. Thus for the description of the overall kinetic, a specification analogous to a proper desorption process is chosen in equation 33.4 (Höhne and Niemeyer 2000):

The experiments were carried out with a tube reactor showing a constant length. The increase in the process temperature leads to a reduced steam density, which consequently results in a lowered residence time of the fluid in the tube reactor and thus in a lowered desorption efficiency. Simultaneously an increase in temperature favours desorption kinetics which support an enhanced contaminant transfer into the gaseous phase. Additionally a higher temperature shifts the desorption equilibrium towards a better separation.

For the description of this complex system the model parameters were determined from experimental data obtained at a system pressure of 1 10^5 Pa and different residence times from 0.5 s (T = 609 K) up to 0.7 s (T = 443 K). The reduction in the contaminant-load Δq_{OS}, which is attained during the residence time $\tau = \Delta t$ in the tube reactor, represents residue "A" at different process temperatures as seen in fig. 33.9.

$$\frac{dq_{OS}}{dt} = k \cdot q_{OS,0} \cdot \exp\left(\frac{-E_A}{R \cdot T}\right) \qquad (33.4)$$

$$\ln\left(\frac{\Delta q_{OS}}{\Delta t}\right) = \ln\left(k \cdot q_{OS.0}\right) - \frac{E_A}{R} \cdot \frac{1}{T} = k' - \frac{E_A}{R} \cdot \frac{1}{T} \qquad (33.5)$$

The linear graphs in fig. 33.9 indicate that the model described above produces a satisfactory approximation of the temperature dependency of the process which has the most dominant effect on kinetics. As the molecular size increases the slope of the fitted curve ($-E_A R^{-1}$), and thus the transfer energy, rises. Therefore the decrease in the reduction of the PAH-load with increasing molecular size can be described.

The value 'k', which represents the intercept of the ordinate, shows the natural logarithm of the product of the pre-exponential factor k and the respective initial load of the contaminated soil $q_{OS,0}$ (refer to equation 33.5). As presented in table 33.2, the values of the pre-exponential factor increase with growing molecular size.

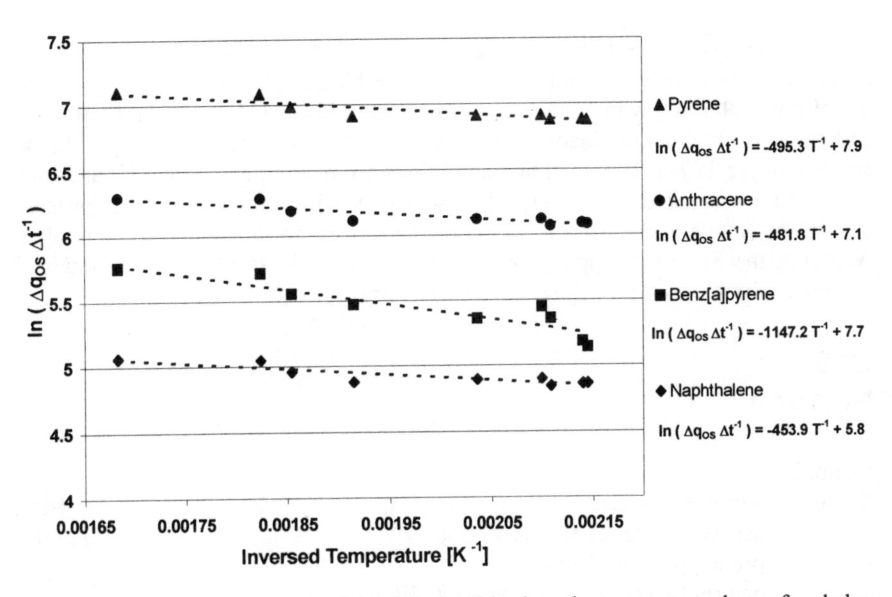

Fig. 33.9. Dependence of the reduction of the PAH-load on the temperature shown for sludge "A"

Table 33.2. Characteristic values of the kinetic approximation of the PAH-reduction in sludge "A" by the Steam Stripping Process

	Naphthalene	Anthracene	Pyrene	Benz[a]pyrene
Activation energy E_A [J mol^{-1}]	3,774	4,006	4,118	9,538
Pre-exponential factor k [s^{-1}]	3.94	4.17	4.25	12.83

The activation energies E_A are low and in the range of physisorption mechanisms. This is in compliance with the partitioning mechanisms already described by Chiou et al. (1979) and Schlebaum (1999). Therefore the uptake of organic chemicals by soil is estimated to be essentially a kind of dissolution in the soil organic matter which is characterised by the occurrence of relatively small sorption energies.

33.5 Conclusion

Own results prove the successful treatment of sludges with soils from a wide range of different origins (sludges from soil washing to sludges from the processing of raw materials) with a large variety of contaminants (MHC, PAH, PCB, Hg), and their contamination loads (e.g. 1,000–225,000 mg MHC (kg SM)$^{-1}$ (Höhne and Niemeyer 1999). The operating conditions are moderate (T < 610 K) and thus costs can be handled effectively. The selective utilisation of different process conditions enable a specific adaptation of the process to the system to be treated. Applying the Stream Stripping Process decontaminated solid matter is obtained for reuse which closes recycling loops in industry.

33.6 Notation

Symbols

d_p	particle diameter	[µm]
E_A	activation energy for desorption	[J mol^{-1}]
k	pre-exponential factor	[s^{-1}]
k'	natural logarithm of the product of k and $q_{OS,0}$	
n	molar quantity	[mol]
p	system pressure	[Pa]
p_i*	saturation vapouur pressure of component i	[Pa]
q_{OS}	load of contaminant (e.g. MHC, PAH)	[mg OS (kg SM)$^{-1}$]
$q_{OS,0}$	load of contaminant in the feed	[mg OS (kg SM)$^{-1}$]
$q_{OS,transferred}$	load of contaminant removed by the treatment	[mg OS (kg SM)$^{-1}$]
R	universal gas constant	[J (K mol)$^{-1}$]

t	time	[s]
τ	residence time	[s]
T	temperature	[K]
$T_{b,i}$	boiling point of pure substance i	[K]
z	reaction order	[-]
η	cleaning performance	[mass %]

Abbreviations

MHC	mineral oil hydrocarbons
OS	organic substance (e.g. MHC, PAH)
PAH	polycyclic aromatic hydrocarbons
SM	dry solid matter
TOC	total organic carbon

References

Anderson WC (1993) Innovative Site Remediation Technology: Thermal Desorption. Vol 6, American Academy of Environmental Engineers, Annapolis, MD, USA, Springer, Berlin, Germany, ISBN 3-540-59066-8

Chiou CT, Peters LJ, Freed VH (1979) A Physical Concept of Soil-Water Equilibria for Nonionic Organic Compounds. Science 206: 831–832

Chiou CT (1989) Theoretical Considerations of the Partition Uptake of Nonionic Organic Compounds by Soil Organic Matter. In: Sawhney BL, Brown K (eds) Reactions and Movement of Organic Chemicals in Soils. Soil Science Society of America, special publication no 22, Atlanta, USA – ISBN 0-89118-788-X, pp 1–30

Eschenbach A (1995) Einfluss von Pleurotus ostreatus, Kompost, Sphinggomonas paucimobilis und der Kontaminationsdauer auf den Verbleib und Abbau ^{14}C-markierter polyzyklischer aromatischer Kohlenwasserstoffe (PAK) in Altlastböden. Thesis, Technische Universität Hamburg-Harburg, Germany, pp 187–190

Dzombak DA, Luthy RG (1984) Estimating Adsorption of Polycyclic Aromatic Hydrocarbons on Soils. Soil Science 137: 292–308

Gmehling J, Onken U, Rarey-Nies JR (1981) Vapour-Liquid Equilibrium Data Collection: Aqueous Systems (Supplement 2, Pt 1b), Dechema, Frankfurt a M, Germany – ISBN 3-921567-91-2

Hassett JJ, Banwart WL (1989) The Sorption of Nonpolar Organics by Soils and Sediments. In: Sawhney BL, Brown K (eds) Reactions and Movement of Organic Chemicals in Soils. Soil Science Society of America, special publication no 22, Atlanta, USA, ISBN 0-89118-788-X, pp 31–44

Höhne J, Niemeyer B (1999) Dampfstripping-Prozess – Eine leistungsstarke Hybridtechnologie. In: Alvermann G, Luther G, Niemeyer B (eds) Kontaminierte Schlämme – Behandlungs- und Nutzungsmöglichkeiten feinkörniger Reststoffe. GKSS-Workshop, Geesthacht, Germany, 24 June 1999, pp 129–133

Höhne J, Niemeyer B (2000) Schadstoffdesorption aus feinkörnigen Böden durch Dampfstrippen. In: Stegmann R (ed) Bodenreinigung – 12 Jahre Forschung und Entwicklung, Technische Universität Hamburg-Harburg, Germany, 29–31 March 2000, pp 133–135

Hudel K (1995) Reinigung organisch kontaminierter Böden und Rückstände durch Wasserdampfextraktion. Thesis, Rheinisch-Westfälische Technische Hochschule Aachen, Germany

Hwang YL, Olson JD, Keller GE (1992) Steam Stripping for Removal of Organic Pollutants from Water. 2. Vapour-Liquid Equilibrium Data. Industrial Engineering Chemical Research 31: 1759–1768

"Landolt-Börnstein", Martienssen W, Hall KR (eds) (1999) Numerical Data and Functional Relationships in Science and Technology. Vol 20, A: Vapour Pressure of Chemicals. Springer, Berlin, Germany, ISBN 3-540-64735-X

Means JC, Wood SG, Hassett JJ, Banwart WL (1980) Sorption of Polynuclear Aromatic Hydrocarbons by Sediments and Soils. Environ Sci Technol 14: 1524–1528

Mersmann A (1980) Thermische Verfahrenstechnik. Springer, Berlin, Germany, pp 26–33

Niemeyer B, Höhne J, Eschenbach A (2000) Steam Stripping – Influence Parameters on the Decontamination of Fine-grained Particles. CHISA 2000, 14th International Congress of Chemical and Process Engineering, Prague, Czech Republic, 27–31 August 2000, CD-ROM of full texts, contribution No 0529

Schlebaum W (1999) Organic Contaminants in Soil – Desorption Kinetics and Microbial Degradation. Thesis, Landbouwuniversiteit Wageningen, The Netherlands, pp 4–7

Seewald H, Jüntgen H (1977) Kinetik der nicht-isothermen Desorption organischer Stoffe von Aktivkohle. Berichte der Bunsengesellschaft 81: 638–645

Sullivan TP (1999) Thermal Desorption: The Basics. Chemical Engineering Progress 10: 49–56

US Environmental Protection Agency (1979) Polynuclear Aromatic Hydrocarbons. EPA Method 610. Federal Register 44, No 233, pp 69514–69517

Wietstock P, Wrage B, Luther G, Niemeyer B, Gronholz C (1997) Process for Separation of Volatile Pollutants from Contaminated Materials, Soil, Wood, Sludge and Sediments. In: Calmano W, Roeters P (eds), International Conference on Contaminated Sediments (ICCS), Vol 1, Rotterdam, The Netherlands, 7–11 September 1997, pp 444–451

Wietstock P, Wilichwoski M, Niemeyer B (1998) Entfernung von PAK aus feinkörnigen Böden durch Niedertemperatur-Desorption. Chemie-Ingenieur-Technik 70: 1167–1168

Wilichowski M (1994) Aufbereitung mineralölkontaminierter Böden durch Bodenwäsche und Flotation. Thesis, Technische Universität Hamburg-Harburg, Germany, pp 17–20

34 Degradation of Chlorinated Arenes by Electroreduction

J. Voss, M. Altrogge, D. Golinske, O. Kranz, D. Nünnecke, D. Petersen, E. Waller
Institute of Organic Chemistry, University of Hamburg, Matin-Luther-King-Platz 6, 20146 Hamburg, Germany

34.1 Introduction

Public discussion on serious environmental problems caused by "the chlorine chemistry" arose with Rachel Carson's book "Silent Spring", in which the disastrous ecological aftermath of massive use of persistent polychlorinated hydrocarbons as insecticides was pointed out. As a consequence, production and application of these compounds was banned. However, methods had to be developed not only to dispose of materials that contain these substances, but also to remedy the numerous contaminated sites.

Besides mechanical, thermal and biological treatment, three types of chemical processes are possible for the degradation:

- Conventional chemical reactions (Hitchman et al. 1995), e.g. with an emulsion of metallic sodium (Knorre et al. 1979; Bilger 1990), catalytic hydrogenation (Barren et al. 1993; Ferrughelli and Horvath 1992; Lingaiah et al. 1999; Marques et al. 1993; 1994), with Raney-Nickel (Liu et al. 1998; Tsukinoki et al. 1995), or with complex hydrides (Epling and Florio 1988; Lassov et al. 1998; Seyed-Mohammad et al. 1992; Simagina et al.1996; Stiles 1994; Yang and Pittman 1998; Yoon et al. 1993).
- Photochemical decomposition (Freeman and Hatlevig 1993; Wada et al. 1998).
- Electrolysis (Bersier et al. 1994; Bunce et al. 1997).

High temperature combustion, which is widely used, at present seems to be the most important method among the chemical processes (VDI 1987). It bears, however, the risk of producing chlorinated dibenzodioxins and dibenzofurans if the temperature is not high enough (Ecklund 1986). On the other hand, microbial degradation of compounds with a high degree of chlorination, e.g. hexachlorobenzene (HCB), TCDD and congeners, is extremely difficult if not impossible (Schmidt 1998). Therefore, alternatives such as electroreduction or electrooxidation deserve attention and require research and development efforts.

In principle, electroreduction of chloroarenes to the parent compounds is a well known reaction (Farwell et al. 1975). It has been studied for decades. But most investigations were directed to a better understanding of the electrode processes and the reaction mechanisms (Savéant 1994), whereas its potential as a means of degradation of xenobiotics has been less well recognised until recently (Walsh and Mills 1994; Bersier et al. 1994; Bunce et al. 1997).

In the following, the results of our experiments on the electroreduction of chlorinated arenes are compiled. They were targeted on either the complete dechlorination of relevant classes of compounds, or at least a reduction of the chlorine content of the substrates to an extent which would make them biodegradable.

34.2 Direct Electrolysis

The most simple method to perform a chemical transformation by use of electricity is *direct electrolysis* of a substrate in a solvent-supporting-electrolyte (SSE) at an electrode surface. We have applied this method for the reductive dehalogenation of chlorinated benzenes, toluenes, biphenyls (PCB), naphthalenes (PCN), dibenzofurans, dibenzodioxins, anisoles, and phenols. Lead or carbon was used as cathode material and tetraethylammonium bromide in methanol as SSE. Most of the experiments were performed batchwise in a glass cell (fig. 34.1). In some cases, a commercially available continuous flow cell (fig. 34.2) was used.

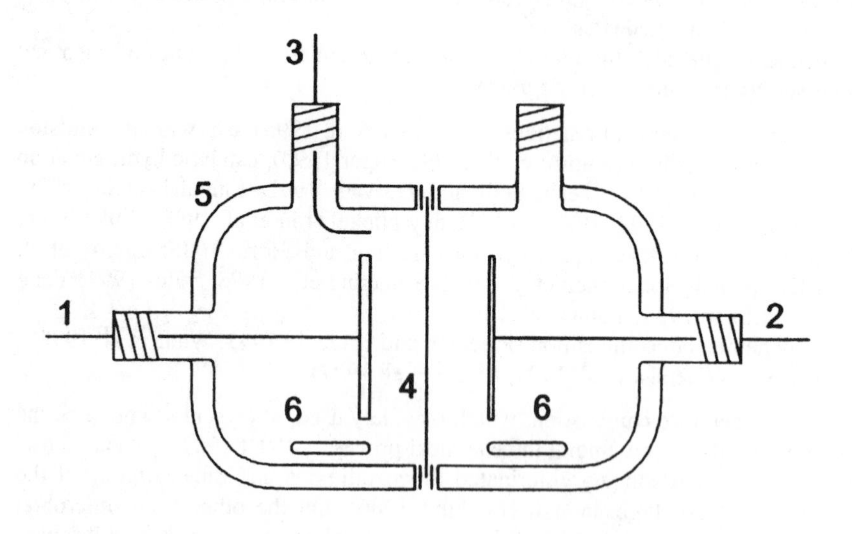

1 Graphite cathode
2 Platinum net anode
3 Silver wire reference electrode
4 Anion exchange membrane
5 Cooling jacket
6 Magnetic stirrer

Fig. 34.1. Batch cell for the electrochemical investigations

34.2.1 Chlorobenzenes

The twelve congener chlorobenzenes including HCB were electrolysed at a lead cathode (Petersen et al. 1990). Without exception, benzene and chlorobenzene were formed as the main products according to eq. (34.1).

$$C_6H_{6-n}Cl_n \xrightarrow[-(n-1)HCl]{2(n-1)e^-,\ 2(n-1)H^+} C_6H_5Cl \xrightarrow[-HCl]{2e^-,2H^+} C_6H_6 \qquad n = 2 - 6 \tag{34.1}$$

Minor amounts of 1,3- and 1,4-dichlorobenzene were also produced by electrolysis of pentachlorobenzene and HCB. A typical batch electrolysis of 4 g of a mixture of 1,4-dichloro-, 1,2,4-trichloro-, 1,3,5-trichloro- and 1,2,3,4-tetrachlorobenzene led to 92 % chlorobenzene and 8 % benzene, with 67 % current efficiency.

34.2.2 Chlorotoluenes

Chlorotoluenes are produced industrially. They are widely used as chemical intermediates and polychlorinated toluenes have also been proposed as substitutes for PCBs (Zanaveskin and Aver'yanov 1998). Therefore, they may occur at contaminated sites or in soils. Since their electrochemical dechlorination on a preparative scale has not yet been described, we have studied this class of compounds (Kranz 2000).

A mixture of the three monochlorotoluenes is cleanly dechlorinated in methanol at a potential of –2.30 V (*vs.* Ag/AgBr) to yield 79 % toluene. The electroreduction of selected dichloro- and trichlorotoluenes is possible at –2.00 V resulting, at that potential, in the formation of monochlorotoluenes (54–84 %) and toluene (4–23 %). A certain regioselectivity is observed, e.g. predominantly 4-chlorotoluene is formed from 2,4-dichlorotoluene. The current efficiencies (23–46 %) are lower than in the chlorobenzene series due to the +I-effect of the methyl substituent, which causes a shift of the reduction potential to more negative values and, consequently, more hydrogen evolution.

34.2.3 Chlorobiphenyls

Polychlorinated biphenyls (PCB) belong to the most notorious xenobiotics. On account of their pronounced chemical and thermal stability, they have found widespread applications, e.g. as transformer oils, until their production and use was prohibited by law in most countries. Nevertheless, large amounts of PCB or oils with PCB contaminants have to be disposed.

We have shown that the chlorine content of PCB can be removed almost completely by electroreduction. In particular, highly chlorinated biphenyls such as the technical products Aroclor® or Clophen® are smoothly dechlorinated.

The monochlorobiphenyls were transformed into biphenyl with 95 % yield and 40 % current efficiency under our standard conditions at –1.96 V (Petersen et al. 1990). 2,2′-Dichloro-, 4,4′-dichloro- and 2,4,5-trichlorobiphenyl yielded 99 % biphenyl with traces of monochlorobiphenyl as byproduct (Voss 1990). Even electrolysis of a suspension of decachlorobiphenyl, which is only sparingly soluble in methanol, led to > 90 % biphenyl (Voss 1993). For our results with Aroclor® and Clophen® (see Chapter 34.5).

34.2.4 Polychloronaphthalenes

Polychlorinated naphthalenes (PCN) known e.g. as Halowax®, Nibrenwachs® or See Kay Wax® (ICI) exhibit chemical and physical properties which resemble those of the PCB. Accordingly, they have been produced and used for similar purposes. Furthermore, small amounts (870 mg kg^{-1}) have been found in Arocolor® and Clophen® , and traces have even been detected in lake sediments, and human fatty tissue and milk (Williams et al. 1993). Their toxicity is significant. Some congeners exhibit toxicity equivalents of the same order of magnitude (TE = 0.002) as many PCB or octachlorodibenzo-1,4-dioxin.

We have electrodehalogenated the monochloro-, four different dichloro-, 1,2,4-trichloro-, 2,3,6,7-tetrachloro- and octachloronaphthalene (Voss and Waller 1996; Waller 1997). The result was rather uniform: 1,4-Dihydronaphthalene, which is also formed by electroreduction of naphthalene itself, was the main product (65–81 %) in all cases. Naphthalene (3–13 %) and tetralene (7–21 %) were found as byproducts. The current efficiencies were 2–16 %. A typical result is demonstrated in eq. (34.2).

(34.2)

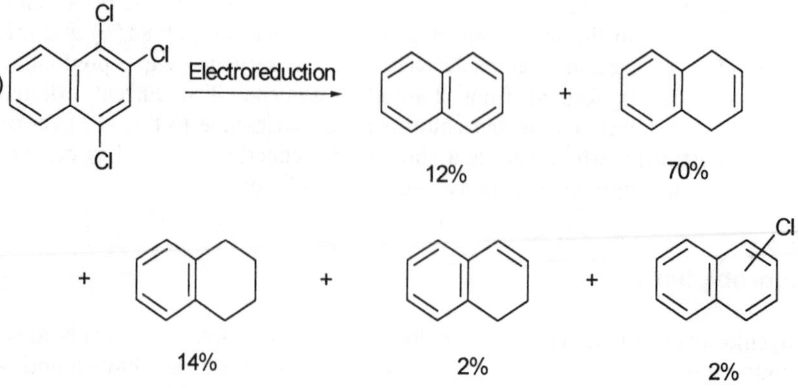

Cl Cl Cl

Electroreduction

12% + 70%

+ 14% + 2% + Cl 2%

In a special experiment 1-chloro- and octachloronaphthalene were electrolysed as a dilute solution (0.1–2 %) in paraffin oil. The result was nearly identical in spite of the decreased availability of the substrate at the cathode.

1-Chloronaphthalene was also electrolysed in the flow cell (fig. 34.2). In this case, a constant current was maintained (galvanostatic method). A turnover of 5–7 mg min^{-1} was achieved with a current density of 700 A m^{-2} and 1,4-dihydronaphthalene and naphthalene were found as products.

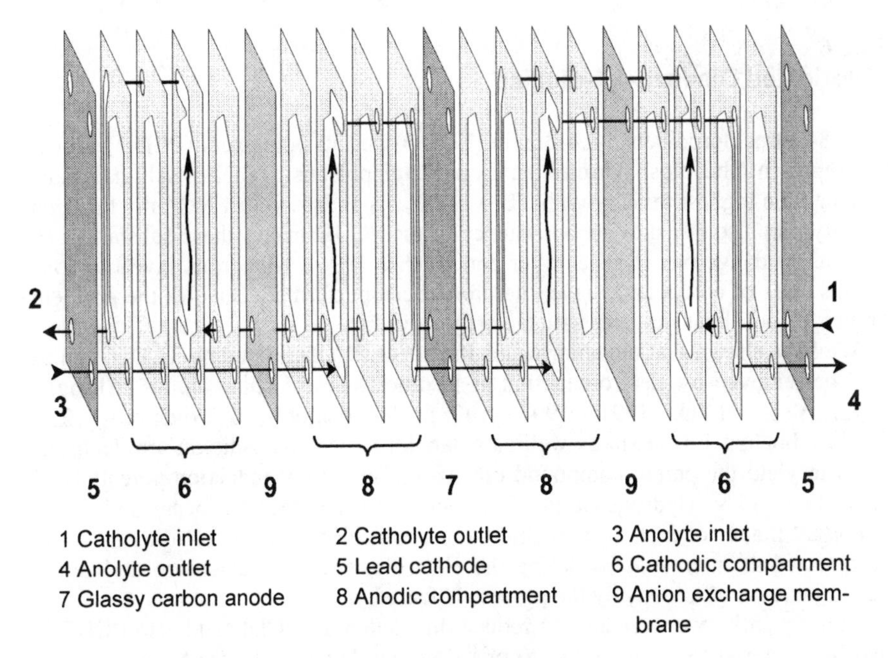

Fig. 34.2. Flow cell for the electrochemical investigations

34.2.5 Chlorinated Dibenzofurans

Due to the +M-effect of oxygen, the annelated benzene rings of dibenzofurans exhibit high electron densities. Nevertheless, electroreduction of polychlorinated dibenzofurans (PCDF) is possible. The four monochloro derivatives are reduced quantitatively at –2.05 to –2.20 V to form 1,4-dihydrodibenzofuran (Waller 1997; Voss et al. 1988, 1999) as the main product, which had been erroneously assigned the structure of 2-biphenylol in our earlier publications (Voss et al. 1990, 1991, 1993; Altrogge 1992), besides dibenzofuran. The same is true for 1,3-dichloro-, 2,4,8-trichloro- and octachlorodibenzofuran. Eq. 34.3 shows the result of a galvanostatic electrolysis in methanolic potassium hydroxide (Kranz 2000). The current efficiency was low (2 %) and the main product was dibenzofuran in this SSE.

(34.3) Cl Cl O —Electroreduction→ O + O

77% 23%

34.2.6
Chlorinated Dibenzo-1,4-dioxins

2,3,7,8-Tetrachlorodibenzo-1,4-dioxin (TCDD) and its congeners (PCDD) probably represent the most harmful environmental poisons. TCDD itself has been assigned the highest toxicity with TE = 1.000 as standard value. Not only the high toxicity, but also the extreme persistence of the PCDD makes these substances so immensely dangerous (Ballschmiter and Bacher 1996). Numerous investigations on this class of compounds have, therefore, been performed. But still the problem of their disposal has not been satisfactorily solved.

We have shown that the chloro substituents of PCDD, including TCDD, can as well be removed by electroreduction as in other types of chloroarenes (Altrogge 1992; Voss et al. 1991, 1993, 1998). Monochloro-, 2,3-dichloro-, 2,7-dichlorodibenzo-1,4-dioxin as well as octachlorodibenzo-1,4-dioxin are dechlorinated to yield the parent compound dibenzo-1,4-dioxin at reduction potentials of –2.30 to –2.35 V. Hydrogenation of the benzene rings does not occur and, as an important matter of fact, no tetrachloro- or pentachlorodibenzo-1,4-dioxins could be detected as intermediates during the electrolysis of the octachloro derivative, i.e. no increase of toxicity by the process is feared.

Consequently, we were able to reduce the content of PCDD and also PCDF in real-life material to a level close to or below the detection limit (Altrogge 1992; Voss et al. 1993) (see Chapter 34.5).

34.2.7
Chloroanisoles and Chlorophenols

Although chloroanisoles have not been produced industrially, they have been detected as environmental contaminants (Führer and Ballschmiter 1998). We have studied their electrodehalogenation (Kranz 2000) with respect to their relationship with chlorophenols and chlorophenylethers such as dichlorophenoxyacetic acid, which are significant xenobiotics.

Electron-uptake of monochloroanisoles is retarded due to the +M-effect of the methoxy substituent. Total dehalogenation is, therefore, not possible. The best result, 79 % conversion to anisole, is achieved with 3-chloroanisole. However, dichloro-, trichloro- and pentachloroanisole are transformed into monochloroanisole (mostly 4-chloroanisole) with good yields and acceptable current efficiencies. In this case carbon fibre instead of lead was suitable as cathode material, which has also been used for the electrodechlorination of dichlorophenoxyacetic acids (Tsyganok et al. 1998).

The situation is less favourable for the electroreduction of chlorophenols because phenolate anions are formed under the basic conditions of the electrolysis. These are extremely reluctant to undergo reduction. Chlorophenols have, on the other hand, been electrooxidised (Polcaro and Palmas 1997).

We have found out that monochlorophenols are dechlorinated to only 2–24 % phenol depending on the position of the substituent (*ortho* > *meta* >> *para*). Dichloro- and trichlorophenols are, however, converted to the monochloro derivatives with yields of up to 80 %. Pentachlorophenol, which has also been studied by other authors (Ross et al. 1997), yields 44 % dichloro- and 43 % monochlorophenol plus 4 % phenol under optimised conditions, i.e. galvanostatic electrolysis with current densities of 255 A m^{-2} at lead cathodes in methanolic potassium hydroxide as SSE (Kranz 2000).

34.3 Indirect Electrolysis

34.3.1 Principle of the Indirect Electrolysis

The first step of an electrochemical reaction consists of the transfer of an electron between the surface of an electrode and a substrate. This step is a heterogeneous process, which is impeded in many cases. The result of this, is the necessity of large overpotentials.

To overcome these disadvantages, one can make use of the indirect electrolysis (Steckhan 1987), which combines the heterogeneous electron transfer with a homogeneous redox reaction. Indirect electrolysis can be regarded as an electrocatalytic process with a mediator as catalyst.

Fig. 34.3 schematically demonstrates the function of a mediator, showing a reduction as an example. First the mediator is reduced (activated) at the cathode. By a homogeneous redox reaction, an electron is then transferred from the reduced mediator to the substrate. The radical anions of the substrate can undergo product forming chemical reactions such as proton abstraction, and displacement of leaving groups etc. The oxidised mediator is regenerated at the cathode.

Compared to a chemical reaction, the indirect electrolysis is a redox reaction without consumption of redox reagents. Instead, the redox equivalents are provided by the electrodes. From this point of view the process is ecologically friendly and saves resources.

The application of indirect electrolyses offers the following advantages:

- Overpotential at the electrode is avoided.
- No passivation or decomposition of the electrode takes place.
- The selectivity is higher, caused by coordination of the substrate to the mediator.
- The cell voltage is lower.
- The use of cheaper and less toxic electrode materials is possible.

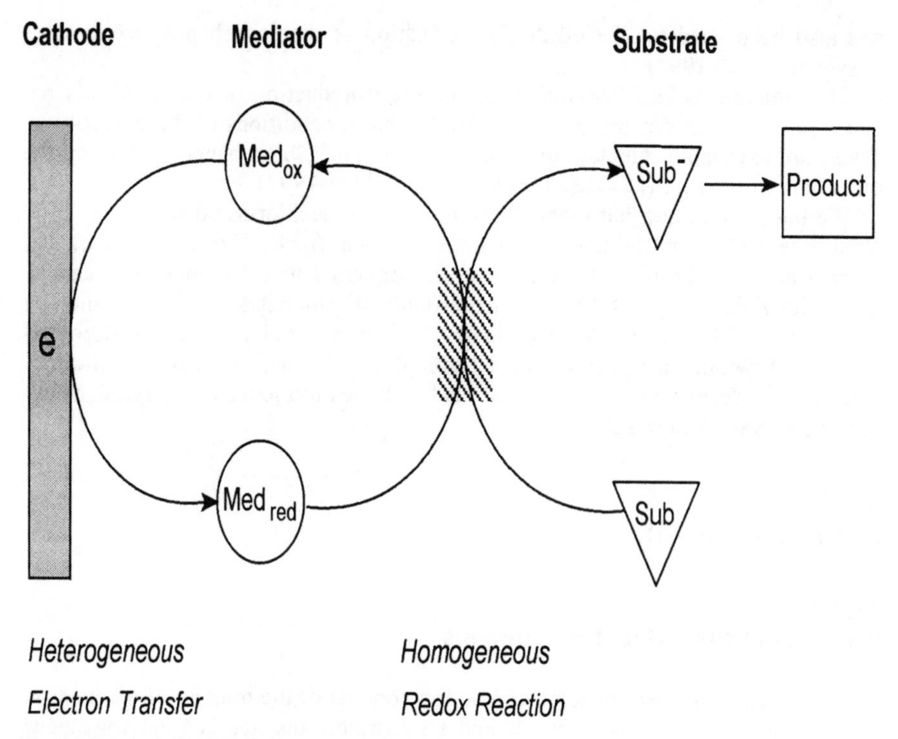

Fig. 34.3. Principle of the indirect electrolysis exemplified by a reduction

34.3.2
Results and Discussion

We chose two mediators for our investigations (Nünnecke 2000), which are well known from the literature: (1,4,8,11-Tetraazacyclotetradecane)nickel(II) chloride [Ni(cyclam)Cl_2, **1**] (Becker et al. 1981; Olivero et al. 1998) and (2,2′-bipyridyl)nickel(II) chloride [Ni(bipy)Cl_2, **2**] (Durandetti et al. 1996). These mediators have been used for carbon-carbon coupling reactions with chlorinated arenes in dimethylformamide.

Since our main target was dehalogenation of the substrates, we performed the indirect electrolysis under protic conditions in methanol. Chlorinated naphthalenes served as model substrates for organic pollutants in a contaminated soil.

As mentioned above, the direct electrolysis of chlorinated naphthalenes leads to the formation of hydrogenated byproducts, which are toxic for microorganisms (Wilkes et al. 1996). With the intention to avoid any hydrogenated species we investigated the reductive dechlorination under the catalytic effect of **1** and **2** in methanol with tetraethylammonium bromide.

The indirect electrodehalgenation of a mixture of 1-chloronaphthalene and 2-chloronaphthalene with both mediators led to unsubstituted naphthalene as the exclusive product (> 99 %). In the case of **1** as mediator, the electrolysis was however not complete. Some starting material remained unreduced, but neither with **1** nor with **2** (see fig. 34.4) as mediator was any formation of hydrogenated naphthalenes observed.

1 **2**

Table 34.1. Half wave potential $E_{½}$ of the mediators **1** and **2**

Mediator		Oxidation stage	$E_{½}$ *vs.* Ag/AgBr [V]	Ref.
$Ni(cyclam)Cl_2$	1	$Ni^{(II)} \rightarrow Ni^{(I)}$	–1.04	Beley et al. 1984
$Ni(bipy)Cl_2$ + bipy[a]	2	$Ni^{(II)} \rightarrow Ni^{(0)}$	–0.92	Amatore and Jutand 1988

[a] If $Ni(bipy)Cl_2$ was used as mediator the free ligand bipyridyl was added in excess. The active complex formed during the electrolysis could, therefore, be $Ni^0(bipy)_2$.

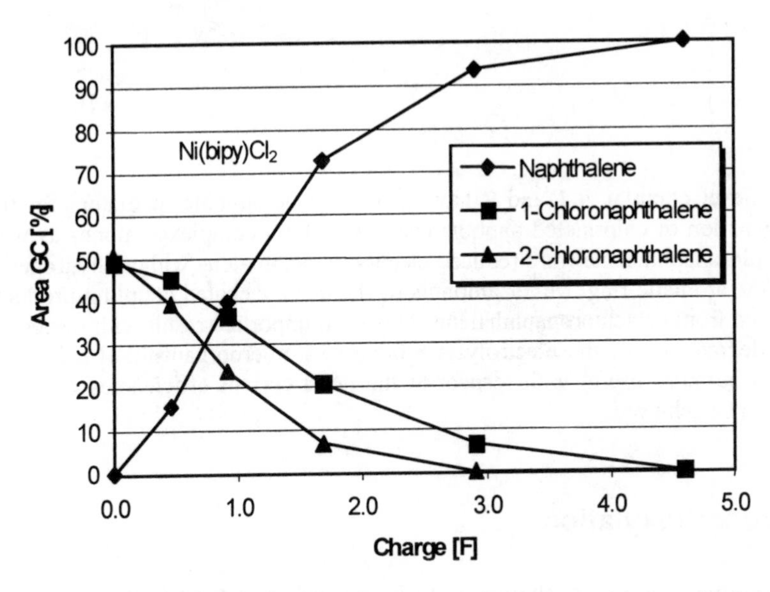

Fig. 34.4. Indirect electroreduction of a mixture of 1- and 2-chloronaphthalene with **2** at –1.40 V *vs.* Ag/AgBr in methanol/0.1 M TEAB

The rates of the electrochemical degradation depend on the mediator used. This is due to the different reaction mechanisms of the two mediators. The active state of Ni(cyclam)Cl_2 is a nickel(I) species, whereas Ni(bipy)Cl_2 reacts via a nickel(0) mechanism. The half wave potentials are given in table 34.1.

We also examined the dechlorination of a highly chlorinated pollutant. The electrolysis of octachloronaphthalene with **2** again led to naphthalene as the main product (cf. Appendix). Minor amounts of the dechlorinated coupling products 2,2′-binaphthyl, 1,2′-binaphthyl and 1,1′-binaphthyl were found as byproducts [eq. (34.4)].

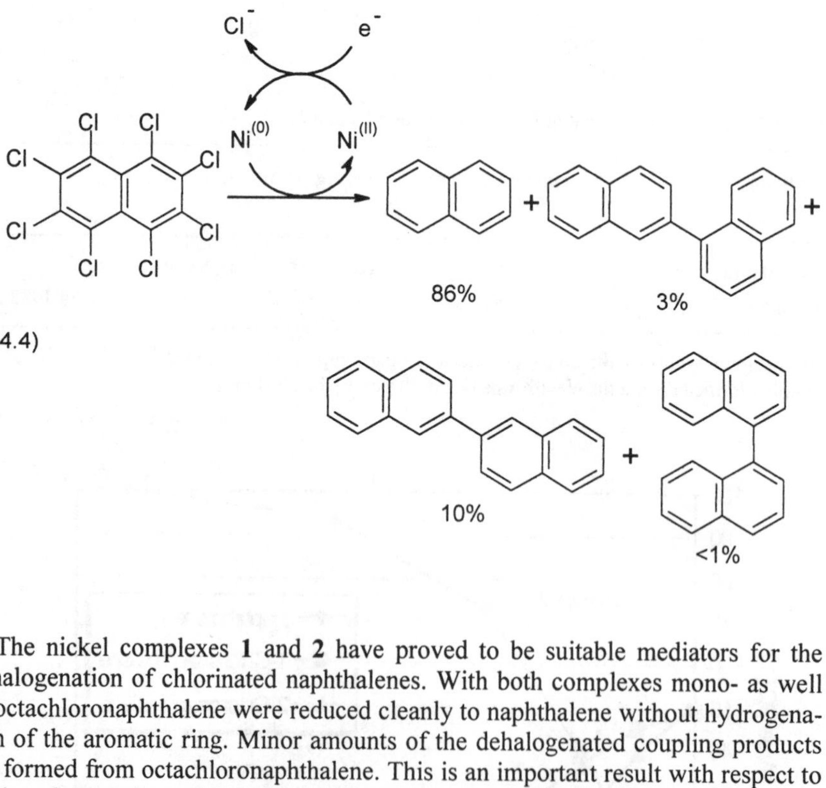

(34.4)

The nickel complexes **1** and **2** have proved to be suitable mediators for the dehalogenation of chlorinated naphthalenes. With both complexes mono- as well as octachloronaphthalene were reduced cleanly to naphthalene without hydrogenation of the aromatic ring. Minor amounts of the dehalogenated coupling products are formed from octachloronaphthalene. This is an important result with respect to further degradation of the electrolysis products by microorganisms. Mediator **2** seems to be more useful with respect to the good current efficiencies of 44 %, which can be achieved.

34.4 Electrocarboxylation

Electrocarboxylation is an alternative to hydrodehalogenation. Carboxylic acids are formed instead of hydrocarbons (Chaussard et al. 1990). Normally, the carboxylic acids are more soluble in aqueous media when compared with aromatic hydrocarbons. Therefore, the bioavailability of these electrolysis products should be increased and consequently their biodegradation be facilitated. Furthermore, the toxicity of the carboxylic acids against microorganisms is, in general, decreased.

The electrolyses are run in the presence of carbon dioxide as electrophile in the catholyte. The choice of electrode material is crucial. In our experiments a combination of a zinc cathode and a magnesium sacrifical anode in an *undivided* batch cell (fig. 34.5), with dimethylformamide as solvent, gave the best results.

The magnesium anode is dissolved, and the magnesium cations formed during the electrolysis ensure the necessary conductivity and the complexation of the carboxylic acids. We have studied several classes of chloroarenes (Golinske et al. 2000) and our results on chlorinated benzenes and dibenzofuranes are exemplified here.

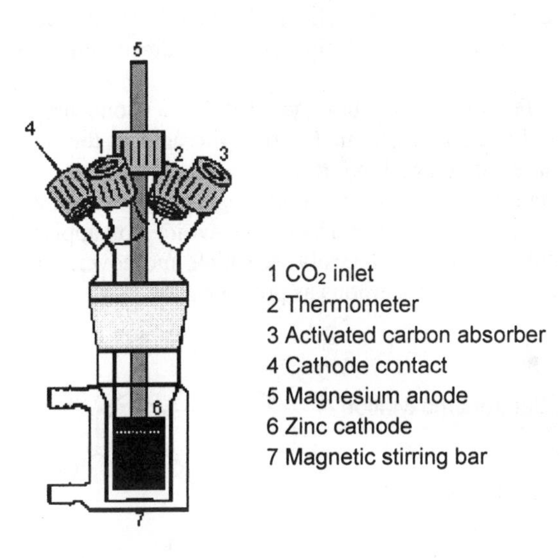

Fig. 34.5. Batch cell for electrocarboxylations

34.4.1 Chlorobenzenes

We chose 1,2,4,5-tetrachloro- and hexachlorobenzene (HCB) as substrates. The product selectivity of the electrolysis of tetrachlorobenzene was low. In spite of the aprotic SSE 35 % hydrodechlorination products, mainly 1,4-dichloro- and 1,2,4-trichlorobenzene were formed, besides 37 % carboxylic acids with 2,5-dichloroterephthalic acid as the major component (17 %). The electrocarboxylation of HCB yielded 65 % pentachlorobenzoic acid, 10 % tetrachloroterephthalic acid and 8 % tetrachloroisophthalic acid but only 1 % pentachlorobenzene. Although only one or two chloro substituents were removed from the substrate, this is a promising result because 92 % of the HCB was substrate, this is a promising result because 92 % of the HCB was transformed into carboxylic acids.

34.4.2
Chlorinated dibenzofurans

Direct electrolysis of chlorinated dibenzofurans in protic solvents produces a relatively large amount of dihydrodibenzofuran besides dibenzofuran (Chapter 34.2.5). Electrocarboxylation of 1-chloro-, 3-chloro- and 4-chlorodibenzofuran yielded, as expected, 57–84 % of the corresponding dibenzofuranmonocarboxylic acids as predominant products. Surprisingly, also dibenzofuran-1,4-dicarboxylic acid was formed in considerable amounts. According to a control experiment, the latter is formed from dibenzofuran itself, which obviously occurs as a reaction intermediate and can be detected as a minor component in the product mixtures. Eq. (34.5) shows a representative example (cf. Appendix). 2-Chlorodibenzofuran represents an exception. Besides 50 % of dicarboxylic acids, only 18 % monocarboxylic acids are produced – without any of the expected dibenzofuran-2-carboxylic acid.

Electrocarboxylation of 1,3-dichlorodibenzofuran yielded 76 % monocarboxylic acids in total with 46 % 3-chlorodibenzofuran-1-carboxylic acid as the main product. Another 13 % of dehalogenated dicarboxylic acids were obtained.

In summary, one can conclude that the undesirable hydrogenation of aromatic rings during the electrodehalogenation of chloroarenes can be avoided by applying carbon dioxide as the electrophile during the electrolysis, which, moreover, leads to carboxylic acids instead of aromatic hydrocarbons as products.

Electrocarboxylation

O Cl → O CO_2H 84%

(34.5)

+ CO_2H O CO_2H 9% + O 4%

34.5
Electroreduction of Real-Life Material

The promising results which we had achieved with the electroreduction of defined model compounds as well as with mixtures and dilute solutions of chloroarenes as substrates (Chapter 34.2–34.4) prompted us to investigate the dehalogenation of real-life material.

34.5.1
Industrially Produced PCB

Electrolysis of up to 10 g Clophen A30® (Bayer AG) with a total content of 42 % chlorine in the batch cell (fig. 34.1) resulted in the formation of 92 % biphenyl. Small amounts (< 6 %) of 4-chlorobiphenyl and hydrogenated compounds, e.g. phenylcyclohexane, were found as byproducts. Aroclor 1254® and Aroclor 1260® (Monsanto) with 54 % and 60 % chlorine content led to a very similar result. Due to the prolonged time of electrolysis, the proportion of hydrocompounds was higher and the current efficiency lower at the expense of hydrogen evolution (Altrogge 1992; Voss et al. 1990; Voss et al. 1993).

34.5.2
Oil Leakage from a Waste Disposal

A suspension of leaked oil with a total content of 12,200 $\mu g\ kg^{-1}$ of tetrachloro-, pentachloro- and hexachlorodibenzofurans and -dibenzo-1,4-dioxins was electrolysed for a prolonged period of time (passage of the 15 fold amount of theoretically required electricity). By this procedure, the concentration of contaminants was lowered to 10 $\mu g\ kg^{-1}$ (Altrogge 1992; Voss et al. 1990, 1993).

34.5.3
Contaminated Soil Extracts

We have studied the electrodechlorination of three different soil extracts. – An extract of "Kieselrot", a red coloured slag used as surface material for sports and playgrounds, was the first example. The extract contained a total contamination of 500 $\mu g\ kg^{-1}$ tetrachloro-, pentachloro-, hexachloro- and heptachlorodibenzofurans and -dibenzo-1,4-dioxins with 230 $\mu g\ kg^{-1}$ pentachlorodibenzofuran as the main component and was dissolved in tetradecane. In a batch electrolysis at –2.30 V (1.5 h with an amperage of 0.185 A), the chloro compounds were completely dehalogenated, i.e. they were undetectable by the extremely sensitive GC/MS/SIR technique. The recoverage proved to be > 95 % in this case.

The second example was a methanolic Soxhlet extract of 30 g from a cellulose containing soil found at a paper mill site. The total contamination was 14,700 $\mu g\ kg^{-1}$ of the same kind of material as in the first example. It was dechlorinated by electroreduction to a level below the detection limit. In particular the TCDD content was below 10 $pg\ kg^{-1}$ after the electrolysis (Altrogge 1992; Voss et al. 1993).

Highly contaminated soil received from Bitterfeld, the centre of the chemical industry in the former GDR, was the third example of real-life material. This soil contained halogenated nitrodiphenyl ethers, 4-chloronitrobenzene and chlorophenols as summarised in table 34.2.

The main contaminant 2,4-dichloro-4′-nitro diphenyl ether (Nitrofen®) has been used as a pesticide in large amounts in agriculture since the early 1960s.

Table 34.2. Main contaminants in soil sample "Bitterfeld" (BVV Chemie GmbH, Bitterfeld)

Compound	Concentration [mg kg^{-1} (soil dry matter)]
2,4-Dichloro-4′-nitro diphenyl ether (Nitrofen®)	8600
4-Chloronitrobenzene	750
2,4,6-Trichlorophenol	2000
2,4-Dichlorophenol	1800

Since the biodegradation of diaryl ether pesticides and chlorinated phenols is limited to low concentrations in the mg kg^{-1}-range (Schmidt 1998), an electrochemical treatment was carried out. The toluene extract of the soil was electrolysed under galvanostatic conditions (1 A/510 A m^{-2}) with a lead cathode in methanolic potassium hydroxide as SSE. However, under electroreductive conditions chlorinated nitroarenes behave differently to the compounds described before. First the nitro group is transformed to an amino group [eq. (34.6)]. Subsequently dechlorination occurs, which is, however, impeded (Baizer and Lund 1983).

Electroreduction

Nitrofen

6%

(34.6)

76%

18%

The electroreduction of Nitrofen® yields, therefore, 4-chloro-4′-amino diphenyl ether as the main product (eq. 6) and 4-chloronitrobenzene is only reduced to 4-chloroaniline without dechlorination. – The two chlorinated phenols present in the soil extract are partially dechlorinated as expected (Chapter 34.2.7).

The results on these real-life materials show that the electrochemical reduction can be directly applied to contaminated soil extracts. Depending on the matrices, the current efficiency was lower than in the experiments with pure substrates. Even in the case of the chlorinated phenols, the method yielded a detoxified electrolysis extract which may be delivered to further biological treatment.

References

Altrogge M (1992) Elektrochemische Enthalogenierung chlorierter Biphenyle, Dibenzofurane und Dibenzo-p-dioxine. PhD thesis, Universität Hamburg

Amatore C, Jutand A (1988) Rates and Mechanism of Biphenyl Synthesis Catalyzed by Electrogenerated Coordinatively Unsaturated Nickel Complexes. Organometallics 7: 2203–2214

Baizer MM, Lund H (1983) Organic Electrochemistry. 2nd ed, Marcel Dekker Inc, New York, Basel

Ballschmiter K, Bacher R (1996) Dioxine, chapt 3. Verlag Chemie, Weinheim

Barren JP, Baghel SS, McCloskey PJ (1993) Reductive Dechlorination of Chlorinated Aromatics. Synth Commun 23: 1601–1609

Becker JY, Kerr JB, Pletcher D, Rosas R (1981) The Electrochemistry of Square Planar Macrocyclic Nickel Complexes and the Reaction of Ni(I) with Alkyl Bromides: Nickel Tetraamine Complexes. J Electroanal Chem 117: 87–99

Beley M, Collin JP, Ruppert R, Sauvage JP (1984) Nickel(II)-Cyclam: An Extremely Selective Electrocatalyst for Reduction of CO_2 in Water. J Chem Soc Chem Commun: 1315–1316

Bersier PM, Carlsson L, Bersier J (1994) Electrochemistry for a Better Environment. Top Curr Chem 170: 113–229

Bilger E (1990) Enthalogenierung mit metallischem Natrium bei hohen Temperaturen ("Degussa-Verfahren"). DECHEMA Jahrestagung, Frankfurt/Main

Bunce NJ, Merica SG, Lipkowski J (1997) Prospects for the Use of Electrochemical Methods for the Destruction of Aromatic Organochlorine Wastes. Chemosphere 35: 2719–2726

Chaussard J, Folest JC, Nedelec JY, Perichon J, Sibille S, Troupel M (1990) Use of Sacrifical Anodes in Electrochemical Functionalization of Organic Halides. Synthesis: 369–381

Durandetti M, Nédélec JY, Périchon J (1996) Nickel-Catalyzed Direct Electrochemical Cross-Coupling between Aryl Halides and Activated Alkyl Halides. J Org Chem 61: 1748–1755

Ecklund G, Petersen JR, Strömberg B (1986) Phenol and HCl at 55 °C yield a Large Variety of Chlorinated Toxic Compounds. Nature 320: 155–156

Epling GA, Florio EM (1988) Borohydride-enhanced Dechlorination of Chlorobenzenes and Toluenes. J Chem Soc Perkin Trans: 703–706

Farwell SO, Beland FA, Geer RD (1975) Reduction Pathways of Organohalogen Compounds Part I, Chlorinated Benzenes, Reduction Pathways of Organohalogen Compounds Part II Polychlorinated Biphenyls. J Electroanalyt Chem 61: 303–314, 315–324

Ferrughelli DF, Horvath IT (1992) Hydrodechlorination of Chloroaromatics: the Use of a Bifunctional Homogeneous Rhodium Catalyst for the Conversion of Chloroaromatics into Saturated Hydrocarbons. J Chem Soc Chem Commun: 806–807

Freeman PK, Hatlevig SA (1993) The Photochemistry of Polyhalocompounds, Dehalogenation by Photoinduced Electron Transfer, New Methods of Toxic Waste Disposal. Top Curr Chem 168: 47–91

Führer U, Ballschmiter K (1998) Bromochloromethoxybenzenes in the Marine Troposphere of the Atlantic Ocean: A Group of Organohalogens with Mixed Biogenic and Anthropogenic Origin. Environ Sci Technol 32: 2208–2214

Golinske D, Voss J, Adiwidjaja G (2000) Electrocarboxylation of Chlorinated Aromatic Compounds. Collect Czech Chem Commun 65: 862–880

Hitchman ML, Spackman RA, Ross NC, Agra C (1995) Disposal Methods for Chlorinated Aromatic Waste. Chem Soc Rev 24: 423–430

Jakobsson E, Eriksson L, Bergmann Å (1992) Synthesis and Crystallography of 1,2,3,4,6,7-Hexachloronaphthalene and 1,2,3,5,6,7-Hexachloronaphthalene. Acta Chem Scand 46: 527–532

Knorre H, Langer M, Pohl G (1979) A Process for Regenerating Spent Lubricating Oils. Degussa AG, Patent DE 2813200C2

Kranz O (2000) Elektroreduktion chlorierter Aromaten und Dipropylether in protischen Lösungsmitteln. PhD thesis, Universität Hamburg

Lassov L, Lee HK, Andy Hor TS (1998) Catalytic Dehalogenation of Highly Chlorinated Benzenes and Aroclors Using $PdCl_2$(dppf) and $NaBH_4$: Efficiency, Selectivity, and Base Support. J Org Chem 63: 3538–3543

Lingaiah N, Uddin MA, Muto A, Sakata Y (1999) Hydrodechlorination of Chlorinated Hydrocarbons over Metal-Carbon Composite Catalysts Prepared by a Modified Carbothermal Reduction Method. Chem Commun: 1657–1658

Liu GB, Tsukinoki T, Kanda T, Mitoma Y, Tashiro M (1998) A New Method for Dechlorination of Chlorobiphenyls Using a Raney Ni-Al Alloy in Dilute Aqueous Alkaline Solution. Tetrahedron Lett 39: 5991–5994

Marques CA, Selva M, Tundo P (1993) Facile Hydrodehalogenation with Hydrogen and Pd/C Catalyst under Multiphase Conditions. J Org Chem 58: 5256–5260

Marques CA, Selva M, Tundo P (1994) Facile Hydrodehalogenation with H_2 und Pd/C Catalyst under Multiphase Conditions: 2. Selectivity and Kinetics. J Org Chem 59: 3830–3837

Nünnecke D (2000) Elektrochemische Enthalogenierung chlorierter Aromaten mittels Nickel(II)-Komplexen als Mediatoren in Methanol. PhD thesis, Universität Hamburg

Nünnecke D, Voss J (1999) Electroreduction of Organic Compounds, 32. Electrodehalogenation of Chloroarenes in Methanol Mediated by Nickel Complexes. Acta Chem Scand 53: 824–829

Olivero S, Rolland JP, Duñach E (1998) Electrochemical Studies of Ni(cyclam)$^{2+}$-Catalyzed Annulation Reactions. Organometallics 17: 3747–3753

Petersen D, Lemmrich M, Altrogge M, Voss J (1990) Elektrochemische Enthalogenierung von chlorierten Benzolen und Biphenylen in Methanol. Z Naturforsch 45b: 1105–1107

Polcaro AM, Palmas S (1997) Electrochemical Oxidation of Chlorophenols. Ind Eng Chem Res 36: 1791–1798

Ross NC, Spackman RA, Hitchman ML, White PC (1997) An Investigation of the Electrochemical Reduction of Pentachlorophenol with Analysis by HPLC. J Appl Electrochem 27: 51–57

Savéant JM (1994) Mechanism and Reactivity in Electron Transfer Induced Aromatic Nucleophilic Substitution, Recent Advances. Tetrahedron 50: 10117–10165

Schmidt S (1998) Biodegradation of Dioxins and Furans. In: Wittch RM (ed) Landes Bioscience Chapt. 8. Austin, Texas, USA

Seyed-Mohammad HT, Tabaei H, Pittman jr CU, Mead KT (1992) Dechlorination of Polychlorinated Biphenyls, 4-Chlorobiphenyl, and Chloro-p-xylene with Alkoxyborohydrides. J Org Chem 57: 6669–6671

Simagina VI, Litvak VV, Stoyanova IV, Yakovlev VA, Mastikhin VM, Afanasenkova IV, Likolobov VA (1996) Low Temperature Hydrodechlorination of Chlorobenzene and 2,3-Dichlorodibenzo-p-dioxin by Hydrides Catalyzed by Transition Metal Compounds. Izv Akad Nauk Ser Khim 6: 1391–1394

Steckhan E (1987) Organic Syntheses with Electrochemically Regenerable Redox Systems. Top Curr Chem 142: 1–70

Stiles M (1994) Nickel Complexes as Soluble Catalysts for Reductive Dehalogenation of Aromatic Halides. J Org Chem 59: 5381–5385

Tsukinoki T, Kakinami T, Iida Y, Ueno M, Ueno Y, Mashimoto T, Tsuzuki H, Tashiro M (1995) Hydrogenation of Halophenols to Cyclohexanols Using Raney Nickel–Aluminium Alloys in Saturated $Ba(OH)_2$ Solution under Mild Conditions. J Chem Soc Chem Commun: 209–210

Tsyganok AI, Yamanaka I, Otsuka K (1998) Electrocatalytic Dehalogenation of Chloroaromatics on Palladium-loaded Carbon Felt Cathode in Aqueous Medium. Chem Lett: 303–304

Uchino M, Asagi K, Yamamoto A, Ikeda S (1975) Preparation and Properties of Aryl(dipyridyl)nickel Halide Complexes. J Organomet Chem 84: 93–103

VDI (1987) Dioxin – Eine technische analytische, ökologische und toxikologische Herausforderung. VDI-Berichte 636

Voss J, Altrogge M, Francke W (1993) Elektrochemische Dehalogenierung chlorierter Aromaten – Von Modellsubstanzen zu praxisrelevanten "Real-Life-Proben". In: Stegmann R (ed) Bodenreinigung, Hamburger Berichte 6. Economica Verlag, Bonn, pp 255–274

Voss J, Altrogge M, Wilkes H, Francke W (1991) Electrochemical Dehalogenation of Chlorinated Dibenzofurans and Dibenzo-p-dioxins in Methanol. Z Naturforsch 46b: 400–402

Voss J, Golinske D, Kranz O, Nünnecke D (1999) Direkte und indirekte Elektroreduktion chlorierter Dibenzofurane und Dibenzodioxine. In: Russow J, Sandstede G, Staab R (eds) Elektrochemische Reaktionstechnik und Synthese. Von den Grundlagen bis zur industriellen Anwendung, Monographie 14. Gesellschaft Deutscher Chemiker, Frankfurt/Main, pp 223–229

Voss J, Petersen D, Lemmrich M, Altrogge M (1990) Chlorabspaltung aus Aromaten – Wertprodukte aus Problemstoffen. Chem Ind (Düsseldorf) 9: 17

Voss J, Waller E (1996) Elektrochemische Enthalogenierung von Mischungen chlorierter Aromaten in Ölmatrices. In: Stegmann R (ed) Hamburger Berichte 10. Economica Verlag, Bonn, pp 65–73

Voss J, Waller E, Kränke P (1998) Electroreduction of 2- and 3-Chlorodibenzofuran in Deuterated Methanol. J Prakt Chem 340: 430–436

Wada Y, Yin H, Kitamura T, Yanagida S (1998) Photoreductive Dechlorination of Chlorinated Benzene Derivatives Catalyzed by ZnS Nanocrystallites. Chem Commun: 2683–2684

Waller E (1997) Elektrochemische Enthalogenierung polychlorierter Naphthaline sowie Elektroreduktion monochlorierter Dibenzofurane. PhD thesis, Universität Hamburg

Walsh F, Mills G (1994) Electrochemical Methods for Pollution Control. Chem Tech Eur 1: 13–18

Wilkes H, Wittich RM, Timmis KN, Fortnagel P, Francke W (1996) Degradation of Chlorinated Dibenzofurans and Dibenzo-*p*-dioxins by *Sphingomonas* sp. Strain RW1. Appl Environ Microbiol 62: 367–371

Williams DT, Kennedy B, LeBel GL (1993) Chlorinated Naphthalenes in Human Adipose Tissue from Ontario Municipalities. Chemosphere 27: 795–806

Yang C, Pittman jr CU (1998) Dechlorination of Pentachlorophenol and 1,2,4-Trichlorobenzene Using $NaBH_4$ and $NaBH_4$/LiCl at 125–135°C in Glyme Solvents. Synth Commun 28: 517–525

Yoon NM, Choi J, Lee HJ (1993) Facile Reduction of Aryl Halides with Borohydride Exchange Resin-Nickel Acetate. Bull Korean Chem Soc 14: 543–545

Zanaveskin LN, Aver'yanov VA (1998) Polychlorobiphenyls: Problems of the Pollution of Environment and Technological Neutralisation Methods. Russ Chem Rev 67: 713–724

Natural Attenuation

35 Natural Attenuation – Non-Destructive Processes

U. Förstner and J. Gerth
Technical University Hamburg-Harburg, Environmental Science and Technology, Eissendorfer Str. 40, 21073 Hamburg, Germany

35.1 Introduction

During the 1990s, natural attenuation grew from a laboratory research phenomenon to a commonly used approach for the cleanup of contaminated groundwater (MacDonald 2000). The concept of natural attenuation/intrinsic remediation, relies on natural, subsurface processes rather than traditional, engineered procedures. Applications of natural attenuation concepts typically involve biological, chemical and geotechnical approaches. Common objectives are the characterisation of the site with regard to the efficiency of the expected retardation/degradation mechanisms, proof of applicability of the natural attenuation concept (i.e. time frame) and elucidation of questions about the persistence of pollution sources.

Major advantages of the concept are – as for most *in situ* procedures – avoidance of secondary wastes and reduction of hazards for people who are exposed, compared to *ex situ* treatments. Problems may arise from the long periods required to reach the remediation goals and from the subsequent surveillance. In the opinion of the U.S. Environmental Protection Agency, intensive education will be needed to receive public acceptance of this concept (EPA 1999).

Regarding the effects on pollutant reduction in the subsurface environment, *destructive processes,* such as biological and abiotic degradation, humification and biological-chemical transformation, can be distinguished from *non-destructive processes*, such as sorption, immobilisation, dilution and volatilisation (Track and Michels 2000). When relying on natural attenuation processes for site remediation, the US EPA prefers processes that degrade or destroy contaminants. Also, the EPA generally expects that natural attenuation will only be appropriate for sites that have a low potential for contaminant migration (EPA 1999). The destructive processes mainly induce a reduction of contaminant mass and discharge; the non-destructive processes – by physical-chemical mechanisms – may lead primarily to a decrease in concentration and mobility of pollutants. Sorption can either be reversible, or in connection with diffusion in micropores, may affect the long-term retardation of pollutants. Due to sorption processes, a decrease in bioavailability can be attained.

The present contribution will mainly focus on the type of non-destructive processes that so far have not received much attention in discussions on the potential applications of the natural attenuation concept. This may partly be due to complex mechanisms leading to the effects of retardation, irreversible sorption, immobilisation, etc. and due also to the fact that it is much more difficult to prove the efficiency of non-destructive processes.

Before we report on the actual state of knowledge about non-destructive processes for attenuating both organic and inorganic contaminants, a short view is presented on the most recent developments in natural attenuation or intrinsic remediation strategies.

35.2 Limitations and Potentials of Natural Attenuation Strategies

According to EPA data of 1999, natural attenuation use in the US Superfund program increased during the 1990s from an application of 6 % to more than 25 % for groundwater contamination sites. However, opinions about whether natural attenuation is an appropriate strategy for managing groundwater contamination are highly polarzed. According to an up-coming report of the US National Research Council (NRC 2000), the rush to use natural attenuation is driven largely by the high costs of constructing and operating engineered groundwater cleanup systems; another factor is the limited effectiveness of these techniques:

> Although considerable effort has been invested in groundwater and soil cleanup, the technologies available for the cleanups are relatively rudimentary (NRC 2000).

Linda Greer, a scientist at the Natural Resources Defence Council, like others, anticipates that at a number of sites, those responsible for the cleanup

> will be back to the drawing board,

if natural attenuation fails to work as predicted. Walter Kovalick of EPA's Technology Innovation Office suggests

> there is a possibility that natural attenuation could suffer bioremediation's early fate: too many promises have been made, and expectations are too high (both citations from MacDonald 2000).

Natural attenuation is potentially disadvantageous because (EPA 1999)

- longer time frames may be required to achieve remediation objectives compared to time frames for active remediation measures at a given site;
- site characterisation is expected to be more complex and costly;
- toxicity and/or mobility of transformation products may exceed that of the parent compound;
- long-term performance monitoring will generally be more extensive and last longer;
- institutional controls may be necessary to ensure long-term protection;
- potential exists for continued contaminant migration, and/or cross-media transfer of contaminants;

Table 35.1. Areas of application for Monitored Natural Attenuation (EPA 1999)

Soil/groundwater	Applicable for both media
Monitoring	Prerequisite: unsuitable sites are eliminated
Cross media transfer	Not permitted
Time frame	Reasonable in the range of other methods
Source control measures	MNA mostly in combination with other (more active) methods

- hydrological and geochemical conditions amenable to natural attenuation may change over time and could result in renewed mobility of previously stabilised contaminants (or naturally occurring metals), adversely impacting remedial effectiveness; and
- more extensive education and outreach efforts may be required in order to gain public acceptance of natural attenuation.

With the increased practical experience and scientific understanding of the processes involved in the so-called "Monitored Natural Attenuation (MNA)", the EPA (1999) has provided some general rules regarding the "Areas of Application", which are summarised in table 35.1 (from Teutsch and Rügner 2000).

In late 1997, the United States National Research Council formed a natural attenuation committee because of concerns expressed by some members of the National Academy of Engineers about the controversies surrounding natural attenuation. A central part of the committee's task was to assess current scientific understanding of the subsurface fate of different contaminant classes in the absence of human intervention (NRC 2000). Based on a review of scientific and technical literature, field reports, and protocols, the committee rated the likelihood that natural attenuation will succeed as a remediation strategy as being "high", "moderate", or "low" for different types of contaminants (organic examples in table 35.2). "High", "moderate" or "low" ratings mean that scientific knowledge and field evidence are sufficient to expect that natural attenuation will protect human health and the environment at more than 75 %, 50 % or 25 % of contaminated sites, respectively. As table 35.2 shows, 3 types of organic contaminants are highly likely to be treated successfully with natural attenuation; 2 are moderately likely to be treated successfully; and 11 have a low likelihood for successful treatment.

Regarding the *processes* proposed for site-specific remediation, the National Research Council (NRC 2000) committee chose a narrower focus than the EPA (1999) policy. Firstly, this was because transformation (including biological and chemical degradation) and immobilisation are more difficult to understand and have a more limited knowledge base than dilution and dispersion processes. Secondly, this was due to the committee's belief that a narrower definition could reduce doubts that community members often have about natural attenuation, in particular, that a number of community environmental advocates would not accept natural attenuation as a remedial strategy because of the inclusion of dilution and dispersion in the formal definition of this phenomenon (MacDonald 2000).

The committee developed a three-part recommended process for the evaluation of the occurrence of natural attenuation at contaminated sites:

- Provision of a conceptual model of the site,
- The search for "footprints" (i.e. changes in water chemistry left by the attenuation reactions),
- Monitoring of the site.

Table 35.2. Current level of understanding and likelihood of success for organic contaminants using natural attenuation (rating see text; from MacDonald 2000)

Chemical class	Dominant attenuation processes	Current level of understanding	Likelihood of success
Hydrocarbons			
BTEX	Biotransformation	High	High
Gasoline, fuel oil	Biotransformation	Moderate	Moderate
Nonvolatile aliphatic compounds	Biotransformation, immobilisation	Moderate	Low
PAHs	Biotransformation, immobilisation	Moderate	Low
Creosote	Biotransformation, immobilisation	Moderate	Low
Oxygenated hydrocarbons			
Low-molecular-weight alcohols, ketones, esters	Biotransformation	High	High
MTBE	Biotransformation	Moderate	Low
Halogenated aliphatics			
Tetrachloroethylene, TCE, carbon tetrachloride	Biotransformation	Moderate	Low
TCA	Biotransformation	Moderate	Low
Methylene chloride	Biotransformation, abiotic transformation	High	High
Vinyl chloride	Biotransformation	Moderate	Low
Dichloroethylene	Biotransformation	Moderate	Low
Halogenated aromatics			
Highly chlorinated			
PCBs, tetrachlorodibenzofuran, pentachlorphenol, multichlorinated benzenes	Biotransformation, immobilisation	Moderate	Low
Less chlorinated			
PCBs, dioxins	Biotransformation	Moderate	Low
Monochlorobenzene	Biotransformation	Moderate	Moderate
Nitroaromatics			
TNT, RDX	Biotransformation, abiotic transformation, immobilisation	Moderate	Low

Due to variations of the types of contaminants, the chemistry of the groundwater, and the geological characteristics of the site, the search for footprints of natural attenuation must consider the unique conditions of the site (the National Research Council study gives such footprints of natural attenuation reactions for a dozen field sites that the committee reviewed as case studies; NRC 2000).

The NRC report made recommendations on the future guidelines for protocols on natural attenuation. A significant weakness of some protocols is the use of "scoring systems" that rate the likelihood that natural attenuation will occur based on site data.

> Scoring systems are generally too simple to represent the complex processes involved and often are used erroneously in judging the suitability of a site for natural attenuation.

Instead, the national-consensus guidelines should apply the type of three-part evaluation process shown above.

35.3 Natural Attenuation of Organic and Inorganic Contaminants

The NRC assessment does not generally give preference to either biotransformation or immobilisation mechanisms. From the necessity of documenting the efficiency of the process at a given site, biological degradation of typical species of organic contaminants will find easier approval by environmental regulators, at least in the near future. Therefore, regarding the dominant processes in natural attenuation, this introductory chapter makes a distinction between organic and inorganic contaminants.

Natural attenuation of organic contaminants is generally demonstrated using a wealth of evidence pointing to reductions in contaminant mass (Brady and Borns 1997). The four most effective components used to convince a regulatory agency are: evidence of contaminant loss in the field, variations in electron donor/acceptor levels, appearance of degradation byproducts, and soil microcosm studies done in the laboratory.

Biological degradation of contaminants is the dominant destructive process in soil and groundwater. Several factors in soil and groundwater influence the velocity of pollutant transformations (table 35.3). Above all, the role of redox processes should be emphasised, which directly or indirectly influences other factors such as pH-value, sorption processes, and biocoenosis of soils and groundwater.

Important milieu parameters for estimating natural attenuation include the concentration of oxygen, nitrate, iron (II), sulfate, methane, and – of less significance – manganese, calcium, bicarbonate and pH. Interpretation of such data can rely on combinations of in-situ microcosms, laboratory batch studies and field observations (Christensen et al. 1994). More advanced methodological concepts include geochemical speciation models, mainly for iron (oxide, sulfide, carbonate, Fe-complexes). Aquifer sediment has a far larger redox buffer capacity than the groundwater and any evaluation of capacities must include sediment analysis; the strength of sediment redox characterisation is that only the sediment accumulates information about past processes and provides a basis for evaluating future

Table 35.3. Factors influencing biological transformation of contaminants (Track and Michels 2000)

Factors	Effects
Redox potential	Concentration and ratios of electron donors/acceptors determine both the pathways and efficiency of degradation
pH-value	Organisms and enzymes exhibit pH-dependent activity optima
Temperature	Influences composition of biocoenosis and velocity of degradation
Solubility, volatility, particle surface, sorption, occlusion	Limit bioavailability of a contaminant. Limitation may negatively influence velocity of degradation. For toxic substances, limitation of bioavailability could even be useful
Water content	Sufficient water content is essential for degradation of contaminants as well as for the transport of educts and products
Auxiliary (co-)substrates	Enable co-metabolic transformation of contaminants, which cannot productively be degraded, or do not induce biological degradation
Nutrients	Essential for growth and reproduction of microorganisms
Biocoenosis	Pollutant-tolerant and degradation-active organisms and consorties determine the rate of material turnover. Development will be decisively driven by potential selection advantage from degradation processes (e.g. energy surplus)
Co-contaminants	Accompanying contaminants can preferentially be degraded, but may influence bioavailability and may inhibit biological degradation
Age	Age of a soil contamination influences bioavailabilty of pollutants. By that time, suitable biocoenoses can develop

capacities. However, in pollution plumes, redox changes over depth may be very dramatic. Changes may occur within a few centimeters, suggesting that mixing of samples over depth may be detrimental to accurate assessment of redox conditions. In addition, local, small-scale, low-permeable hydrogeological heterogeneities may have natural redox conditions not associated with the plume and should not be mistaken as representing redox conditions in the plume (Christensen et al. 2000a).

For metals, the operative process dominating natural attenuation is sorption (Brady and Borns 1997). Sorption may be defined generally to include the processes of adsorption, co-precipitation, precipitation, and diffusion into the matrix. Sorption may either be *reversible* (i.e. definable by a constant K_D for equilibrium that partitions solid and dissolved phases) or slowly reversible. Slowly reversible sorption processes may be considered as effectively "irreversible" if the time scale for re-release of the contaminant from the solid phase (i.e. by desorption and/or dissolution) is long relative to observation or some time scale of interest.

In demonstrating the effectiveness of natural attenuation, the above-mentioned evidence for organics probably cannot be applied to inorganics (Brady and Borns 1997). Here, standard geochemical codes can be applied to calculate whether contaminant levels are limited by the formation of an insoluble phase. However, geochemical modelling to verify uptake by sorption is not advanced enough to be a stand-alone demonstration of metal sorption. Instead, uptake by sorption can be indicated by

- demonstrating that the sorbing phase (e.g. Fe-hydroxides, calcium carbonate) is present in soils and sediments through a solubility calculation or direct observation; and
- showing that an appreciable fraction of the compound is associated with that phase. The latter is most directly done through sequential leaching procedures which dissolve specific minerals, along with any sorbed material.

35.4 Non-Destructive Processes in the Natural Attenuation Concept

In the practice of natural attenuation, non-destructive, "intrinsic" bonding mechanisms and their temporal development so far have found much less recognition compared to destructive processes, such as biological degradation. Yet these so-called "diagenetic" effects, which – apart from chemical processes – involve enhanced (mechanical) consolidation of soil and sediment components by compaction, loss of water, and mineral precipitation in the pore space, may induce an essential reduction of the reactivity of solid matrices. With these attenuation mechanisms – summarised by the term "retention" – a decrease of pollutant mobility by a factor of 100 or more can be achieved, as determined from sorption/desorption characteristics, or by comparing solution/solid distribution coefficients in recent and aged samples.

35.4.1.1 Soil and Sediment Quality Criteria – Evidence for Irreversible Adsorption

Implications for the practical decision process have been demonstrated by Chen et al. (2000) with the example of sediment quality criteria (SQCs). In the United States thus far, regulations such as the EPA's sediment quality criteria, sediment quality advisory levels (SQALs), and chemical-specific fate scores have been directly or indirectly derived from the linear equilibrium model (EPA 1997). The purpose of SQCs is to ensure that the pore water concentration of a certain compound does not exceed that specified in the final chronic water quality criteria (FCVs). In their conclusive study, Chen et al. (2000) investigated the characteristics of irreversible adsorption with five chlorinated benzenes and four natural sediments. Fig. 35.1 illustrates the impact of the proposed irreversible adsorption isotherm on sediment quality criteria; the solid line and dotted line are the predicted isotherms of 1,4-dichlorobenzene in Lake Charles sediment with the irreversible model and the equilibrium model, respectively. It is shown that the SQC of 1,4-dichlorobenzene would be nearly 2 orders of magnitude less when the resistant fraction in the sediment is taken into account.

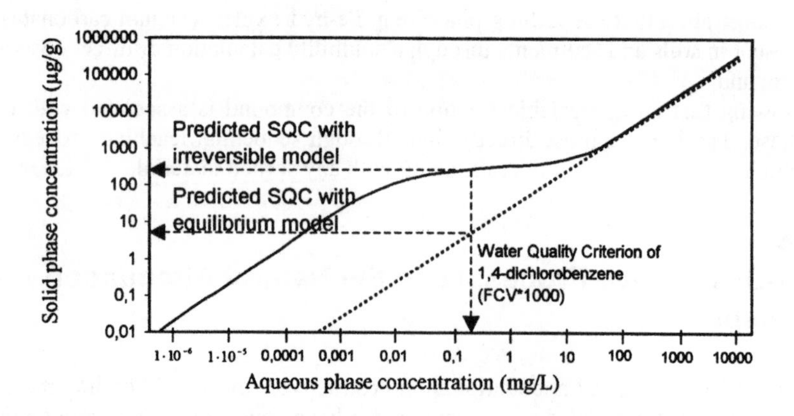

Fig. 35.1. Implications of irreversible adsorption on sediment quality criteria (Chen et al. 2000)

35.4.1.2
Effects on Bioavailability of Inorganic and Organic Contaminants

"Bioavailability" of contaminants is the key term in ecological risk assessment (Kördel et al. 1997), and plays a decisive role in the natural attenuation concept not only with respect to the conditions for optimal degradation (Mahro 2000), but also in relation to intrinsic and temporal enhancement of retention of contaminants in solid matrices. Here, examples are given of typical findings on both inorganic and organic contaminants in sediments and soils:

Geochemical influences on assimilation of sediment-bound metals have recently been evaluated by Griscom et al. (2000) in a series of experiments using the suspension-feeding mussel *Mytilus edulis* and the facultative deposit feeder *Macoma balthica.* Oxidised and reduced radio-labelled sediments were fed to the animals and the assimilation efficiencies (AEs) of ingested metals were determined. For oxic sediment, Cd and Co AEs in *M. edulis* decreased 3–4-fold with increased sediment exposure time to the metals. Smaller but nevertheless significant effects were also noted for Zn and Se but not Ag. Sequential extractions of the oxidised sediments showed a transfer of metals into more resistant sediment components over time, but the rate did not correlate with a decrease in metal assimilation efficiencies. The results imply that metals associated with sulfides and anoxic sediments are bioavailable, that the bioavailability of metals from sediments decreases over exposure time, that organic carbon content generally has a small effect on AEs, and that AEs of sediment-bound metals differ among species (Griscom et al. 2000).

Aging effects on the bioavailability of organic contaminants have been studied comprehensively during recent years by the group of Alexander at Cornell University. Initial findings were that as the residence time of compounds such as phenanthrene and 4-nitrophenol in soil increases, they become increasingly unavailable to microorganisms and resistant to mild extraction (Hatzinger and Alexander 1995). Subsequently, chemical extraction procedures were developed to predict bioavailability of soil-aged organic chemicals (Kelsey et al. 1997; Tang et al. 1999) and the role of nanoporosity and hydrophobicity in sequstration. Bioavailability of typical organic contaminants has also been studied with model solids (Nam and Alexander 1998). The conclusions are that

> correlations are needed to assess the role and contribution of *organic matter, clay content, nanoporosity, surface area,* or other soil properties in governing the rate and extent of decline in *bioavailability,* so that *predictions of diminished exposure* will be possible (Chung and Alexander 1998).

35.4.2
Intrinsic/Temporal Effects on Metal Retention in Sediments and Soils

Metals can be removed from soil solutions and groundwater by:

- sorption to mineral surfaces and/or soil organic matter (SOM);
- formation of insoluble solids;
- uptake by plants and organisms; and occasionally by
- volatilisation (e.g. methylation of mercury) (Brady and Borns 1997).

Metal speciation primarily depends on the ambient biogeochemical conditions of the soil, sediment or groundwater/porewater, i.e. pH, redox state (electron availability), and alkalinity. The presence of chelating (e.g. EDTA, natural organic acids) or solid-forming (e.g. phosphate) ligands is critically important (ionic strength is probably a secondary factor).

35.4.2.1
Interactions of Dissolved Metal Species, Organic Acids, and Calcium

First of all, retention of metals in solid matrices depends on their specific dissolved species. Experiments performed by Förstner et al. (1982) on soil passage of copper and zinc in the presence of humic acids have shown that the more strongly complexed copper is only slightly retained, whereas the ionic or labilely complexed zinc is fixed by sorption or precipitation in the upper section of the sand filter column. Regarding the chemical form of copper in soil solutions, typical dependencies exist on the molecular weight of organic acids (DOC) and their interactions with calcium; the observation that up to 50 % of the DOC in solution was not removed by calcium indicates that a significant portion of copper is potentially highly mobile in soils and can be prone to DOC-facilitated transport (Romkens and Dolfing 1998). From experiments with iron oxide, organic sorbents and mixed systems, Martinez and McBride (1999) have shown that aging for about 200 days increased the copper mobility in the FeOOH-system, while in the long term, the organic system dominated Cu mobility. However, even when metals are coprecipitated with iron (hydr)oxides and aged for long periods, residual solubilities of free Cd, Cu, and Zn can be high enough to reach reported phytotoxicity values or allow excessive uptake by crops (Martínez and McBride 1998).

35.4.2.2
Experience on Temporal Retardation Effects from Sediments and Soils

For inorganic pollutants, mainly heavy metals and arsenic, the effect of natural attenuation predominantly consists of an enhanced retention via processes such as sorption, precipitation, coprecipitation, occlusion, and incorporation in reservoir minerals. From extraction experiments on sediment samples, which had been treated for seven months with metal solutions, Ros Vicent and Duursma (1976) demonstrated that strontium and cadmium remained in exchangeable positions, whereas, for example, zinc, cesium and cobalt were successively occluded in the crystal lattice of minerals. During his investigations on the early diagenetic stages of sediments from the Rhine river, Salomons (1980) found that the proportion of cadmium that was not desorbed with sodium chloride solution (in seawater concentration), increased from 24 % after one-day to 40 % after 60 days of contact between sediment and metal solution. Sediment samples from the river barrage of Vallabreques/Rhône, which had been contaminated by artificial radionuclides from the nearby reprocessing plant, exhibited characteristic differences with respect to the extractability of geogenic and anthropogenic manganese isotopes in the reductive elution step (Förstner and Schoer 1984).

35.4.2.3
Priority Sorbents: Organic Matter, Clay Minerals, Iron Oxides

Experiments were carried out with lead and cadmium on sediment samples from the oxidised surface layer of mudflats in the South San Francisco Bay estuary. The reaction systems were equilibrated for 24 h at the appropriate pH to achieve approximately 90 % metal adsorption, as determined from prior experiments. These experiments indicate a slow release of adsorbed cadmium within a time frame of 96 h, whereas lead was substantially non-labile over the 264-h duration of the experiment (Lion et al. 1982). It was suggested that the proportion of solid organic matter constitutes the main cause for the observed irreversibility of metal sorption. This was experimentally confirmed of selected materials for copper, and – less distinctly – for nickel and cadmium (Förstner 1987). Other findings indicate a temporarily enhanced incorporation of heavy metals in clay minerals (Helios-Rybicka and Förstner 1986). To increase fixation of zinc, nickel, cadmium and arsenic in contact with various soil constituents, long-term diffusion into the crystal lattice of goethite, among other processes, has been suggested by Brümmer, Gerth and Tiller (1988) and Gerth et al. (1993).

35.4.2.4
Implications of Sulfide Precipitation for Sediments and Groundwater

When the redox potential decreases to the level where SO_4^{2-} is reduced to S^{2-}, Fe^{2+} and trace metals react with the S^{2-} to form insoluble precipitates that may become "permanently" buried in the sediments. The chemical basis for the primacy of the sediment sulfide phase for metal binding is assumed to be that, at equilibrium, S^{2-} successfully outcompetes all other common dissolved or particle-associated ligands for metal ions (DiToro et al. 1990). The best strategy for disposing contaminated sediments, therefore, is to isolate them in a permanently reducing environment (Kester et al. 1983).

Sulfate reduction and metal sulfide precipitation also have potential in the remediation of acid mine drainage. This approach has been used by Benner et al. (1999) with the installation of a permeable, reactive barrier into an aquifer contaminated with effluent from mine tailings. The barrier contained a mixture of municipal compost (20 vol.-%), leaf mulch (20 %), wood chips (9 %), gravel (50 %), and limestone (1 %). The limitation that acidic waters can consume generated alkalinity, thereby strongly hampering the ability of sulfate reducing bacteria, has been overcome by promoting sulfate reduction within the aquifer, prior to increased acidification of the effluent upon discharge to the surface. Calculations, based on the long-term performance of column studies, indicate that the reactive barrier has a theoretical treatment lifetime of > 15 years. However, factors that may limit direct transfer of these results from the laboratory into a field setting include preferential flow or preferential mass flux, low temperatures inhibiting bacterial activity, and the possible diffusion of oxygen into the system with time.

It has been suggested that similar concepts could be applied to contaminated aquifers affected by landfill leachates (Köster et al. 2000). In fact, the heavy metals Cd, Cu, and Pb in leachate plumes may be solubility-controlled by sulfides or carbonates; also phosphates and hydroxides may limit metal solubility (Christensen and Nielsen 1987). In addition, reductive dehalogenation of chlorinated hydrocarbons from the landfill can be expected under such conditions (Schulte-Ebbert and Schöttler 1995). However, the presence and significance of sulfate-reducing zones, iron- and manganese-reducing zones, and nitrate-reducing zones may depend highly on local conditions and will be variable over time (Christensen et al. 1994). Therefore, a more comprehensive database on landfill plumes must be developed before the natural attenuation concept becomes a routine approach to remediation of groundwater pollution at landfills (Christensen et al. 2000b).

35.4.3
Organic Pollutants: Hysteresis Effects on Organic Geosorbents

> Understanding the causes of slow sorption/desorption has been hampered by the *heterogeneity of natural particles* as a sorptive and diffusive medium; future research should focus not only on predicting the rates of slow sorption/desorption but also on overcoming the constraints of *slow desorption for remediation purposes* (Pignatello et al. 1996).

Since the middle of the nineties, there has been increasing evidence that the various constituents of the soil matrix vary over a very wide range in respect to their sorption kinetic and intensity, competition for binding sites, and extractability of sorbed pollutants. In the first instance, such *mixed sorption phenomena* of geosorbents, with their hysteresis and aging effects, complicate the interpretation of macroscopic data on the diffusion of hydrophobic organic substances into and from different soil and sediment matrices (Huang and Weber 1997) as well as the quantification of biological availability and transformation of pollutants (Bosma et al. 1997). However, such effects, which are mainly attributed to a steric inhibition of diffusion processes in the variable-size fine pores of organic substances, can be considered as an important attenuation mechanism in addition to abiotic and biotic pollutant degradation.

35.4.3.1
Heterogeneity of Natural Particles as a Sorptive and Diffusive Medium

Soils, sediments, and aquifer solids are heterogeneous at various sample, aggregate, and particle scales. In their "geosorbent" approach, Luthy et al. (1997) have not drawn distinctions between various types of soils and sediments. Rather, such materials are referred to generically as simply a mixture of sorbents. Adherent or entrapped non-aqueous-phase liquids (NAPLs; e.g. solvents, oils and tars) and combustion residue particulate carbon (e.g. chars, soot, and ash) can also function as sorbents. Complex assemblages of these constituents can cause complex mass transfer phenomena, and the term "sequestration" refers to some combination of diffusion limitation, adsorption, and partitioning.

Some geosorbents exhibit typical nonlinear sorption behaviour (Farrell and Reinhard 1994; Huang and Weber 1998). The observed trends, which include increasing apparent hysteresis and decreasing desorption rates and extractabilities of hydrophobic organic contaminants (HOCs) as a function of the sorbate residence time on geosorbents, are incompatible with a simple phase partitioning process. Geosorbents have been characterised as comprising several domains or components that exhibit distinctly different sorption reactivities (Grathwohl 1990; Weber et al. 1992). A qualitative comparison of hypothesised mechanisms and macroscopic observations, which may be useful in the assessment of sorption mechanisms of nonpolar organic compounds with geosorbents, is given in table 35.4 (after Luthy et al. 1997).

From these findings, it has been suggested that the organic matrices of soils and sediments can be divided into two primary categories. These categories manifest mechanistically different sorption behaviours, i.e. an amorphous, gel-like "soft carbon" matrix or domain, and a condensed, glasslike "hard carbon" matrix or domain. Because the relaxation speed of the "glassy" (by analogy to the glassy state of a polymer) structures is both slow and dependent on solute concentration, diffusion of solute molecules into and out of condensed organic matter could be extremely slow. Furthermore, the associated sorption process would most likely be nonlinear, hysteretic, and subject to solute-solute competition. Therefore, absorption into condensed organic matter (2nd column in table 35.4) or diffusion into hydrophobic microporous regions of minerals (5th column) may require extended times (Luthy et al. 1997). In contrast, and by analogy to a rubbery polymer, the soft carbon or amorphous organic matter domain may exhibit partitioning behaviour associated with linear local isotherms, rapid diffusion, no competition for sorption, or sorption reversibility (1st, 3rd and 4th column in table 35.4). Other work (e.g. Xing and Pignatello 1997) has hypothesised that the presence of microvoids of nanometer or smaller scale within natural organic matter (NOM), primarily associated with the more condensed fraction, may also play a significant role in hysteretic desorption behaviour (see review in the paper by LeBoeuf and Weber 2000).

Table 35.4. Comparison of mechanisms and macroscopic observations to assess sorption of nonpolar organic compounds with geosorbents (after Luthy et al. 1997)

Mechanism	Absorption into Amorphous or "Soft" Natural Organic Matter[a]	Absorption into Condensed or "Hard" OM or Combustion R.	Adsorption onto Water-Wet Organic Surfaces (e.g., Soot)	Adsorption onto Exposed Water-Wet Mineral Surfaces	Adsorption into Microvoids or Microporous Minerals[b]
Kinetics	fast (< min) if disaggregated	slow (> days) s/d hysteresis	fast (< minutes)	fast (< minutes)	slow (> days) s/d hysteresis
Isotherm	linear	nonlinear if variable pore size	nonlinear	linear because competition H_2O	nonlinear if variable pore size
Activation energy	low	high	low	low	high
Heat of Sorption	low	moderate to high[1]	low to high[2]	low	moderate to high[3]
Competition	no	yes	yes	no	yes
Sorbate	steric effects not important	steric effects important[4]	steric effects important[5]	steric effects important[5]	steric effects important[6]
Solvent extraction	high	low	high	high	low

[a] or NAPL
[b] (e.g., zeolites) with porous surfaces at water saturation < 100 %
[1] increasing with density of organic matter
[2] depending from hydrophobicity
[3] increasing with decreasing micropore size
[4] for diffusion through matrix
[5] insofar as they allow planar interaction region between sorbate and sorbent
[6] insofar as they influence sorbate ability to diffuse through constricted pores in the sorbent

35.4.3.2 Irreversible Sorption of Organic Contaminants – Experiments and Models

Chemical desorption from soil and sediment is of central importance to most environmental concerns (Kan et al. 1998):

> Desorption affects chemical fate, toxicity, and associated risk to human and aquatic life as well as the efficiency of most remediation technologies.

Evidence of typical effects of hysteresis or irreversible adsorption has been provided by experimental studies on the adsorption and desorption of the organic pollutants naphthalene, phenanthrene and *p*-diclorobenzene to sediment and soil samples by the research group of Kan at Rice University (Kan et al. 1994).

In addition to irreversible adsorption, several other mechanisms have been proposed to explain the observed resistance of desorption. Heterogeneous adsorption with varied adsorption sites is one of the preferred explanations. Here, a fraction of the chemical is assumed to adsorb to sites with high adsorption energy or specificity (e.g. soot), a condensed soil organic phase, or specific adsorption sites on an organic polymer (Weber and Huang 1996). Preferential sorption of planar contaminants, such as chlorobenzenes and PAHs, onto soot-like material has been found in sediments from Lake Ketelmeer in The Netherlands (Jonker and Smedes 2000). In fact, differentiation between the types of organic matter could be an essential issue for the interpretation of slow sorption processes. Investigations by Karapanagioti and Sabatini (2000) on organic matter, conducted with samples taken from different depths and locations in an alluvial aquifer, demonstrated that opaque organic matter fractions dominate the sorption process. The investigations also showed that quantifying this fraction alone could virtually predict the sample K_{OC} value. Complementary mass spectrometric and spectroscopic techniques on sediment samples from Milwaukee Harbor indicated that PAH concentrations on coal- and wood-derived particles were several orders of magnitude higher than on silica particles (Ghosh et al. 2000). The authors suggested that these particles might be removed by density separation from heavier clay, silt, and sand.

Comprehensive investigations by Farrell et al. (1999) on the mechanisms contributing to slow desorption of hydrophobic organic compounds from mineral solids support the hypothesis that micropore blockage by precipitated minerals may limit contaminant desorption rates under water-saturated conditions. Entrapment by obstructed mineral pores is consistent with several previously observed phenomena, including:

- the increase in desorption resistance following soil desiccation (White et al. 1997),
- the effectiveness of low pH aqueous solutions as extraction solvents (Steinberg et al. 1987),
- the increased effectiveness of high-temperature extraction techniques using polar solvents such as methanol (e.g. Kelsey et al. 1997),
- the lack of correlation between desorption rates and compound properties for a particular adsorbent (e.g. Carmichael et al. 1997), and
- the lack of correspondence between soil properties, such as NOM content, cation exchange capacity, or specific surface area with the mass of slow-desorbing contaminants (Pavlostathis and Mathavan 1992).

35.5 Conclusions for Practical Application

> Contaminant entrapment may serve as a natural attenuation mechanism, since the contaminants are functionally removed from partitioning with the bulk groundwater flow (Farrell et al. 1999).

Even if the trapped contaminants are eventually released, they may not pose a threat if their release rates are sufficiently slow for dilution or other mechanisms to keep them at levels below concern.

In a similar way, Brady and Borns (1997) concluded from the presentations of the Sandia Workshop on "Natural Attenuation of Metals and Radionuclides", that slowly-reversible sorption of contaminants from solid phases exposed to uncontaminated groundwater may contribute to *dilution* of the contaminant:

> the extent of dilution will be determined by the rate of contaminant release into solution relative to the velocity of groundwater flow.

This approach does not necessarily contradict the concerns of the US EPA (1999) about simple dilution, or the even more strict position of the NRC assessment (2000) of dilution and dispersion as primary natural attenuation mechanisms. However, a definition of both types of "dilution" in respect to soil and groundwater remediation practice has still to be made.

The basic dilemmas for environmental regulators when considering immobilisation and related processes as practical natural attenuation mechanisms can be demonstrated from examples of heavy metal contaminants, as discussed during the previously-mentioned *Sandia Workshop* (Brady and Borns 1997).

- *Bioavailability.* Unlike the biodegradation of some organic contaminants, which results in the contaminant of concern "going away", typically metals will remain in the subsurface; they are still present, though may be unavailable for biological uptake.
- *Site Specificity.* Because of the inherent site specificity of natural attenuation, it is unlikely that feasibility can be evaluated in the general case. A better approach might be to examine classes of contaminated sites for their suitability for natural attenuation.
- *Monitoring.* Although drastic changes in the composition of natural waters are more the exception than the rule, it will mostly be impossible for site-owners to demonstrate that remobilisation will never occur.
- *Alternative Techniques.* The sequestering of metals out of the aqueous phase often makes their engineered extraction problematic:

> In many cases the technical impracticability of metal extraction is a direct result of the natural attenuation process

Based on the existing experience, of which examples have been presented in Chapter 35.4, methodological approaches should be developed, which could reduce the complex interactions between contaminants and heterogeneous matrices in soils or sediments into a set of easily understandable parameters. Such information should be arranged in a way, that it could be used

1. in a conceptual model of a site and
2. for monitoring the site.

The third component of the site evaluation procedure, as proposed by NRC (2000; see Chapter 35.2), involves the detection of "footprints" of expected natural attenuation processes in the form of changes in water chemistry left by the attenuation reactions. The use of these "footprints" may turn out to be very difficult or impossible in the case of immobilisation, irreversible adsorption etc., since by their nature, these mechanisms take place on a microscopic scale within matrix components. Here, the existence of typical sorbents and sorption reactions for critical contaminants should be demonstrated from a geochemical site survey and from standardised experiments on relevant samples.

At this stage, on the basis of characteristic findings in various research disciplines, a series of example investigations can be listed, which, along with other investigations, can bridge the gap between theoretical process studies and the practical needs for field applications. Such a list, summarising the important research, is presented below *(literature topics in italics).*

1. *Selection of experimental design* for the determination of chemical release rates from soils: Opdyke and Loehr (1999)
2. *Characterisation of organic matrices and sorption/desorption behaviour of hydrophobic organic compounds (HOCs) in soils and sediments*, for example:
   - characterisation of samples by organic petrology methods: Karapanagioti and Sabatini (2000),
   - quantification of the dilute sedimentary soot phase: Gustafsson et al. (1997),
   - microscale location of HOCs by spectroscopic *methods: Ghosh et al. (2000),*
   - characterisation of natural organic matter macromolecules in relation to the sequestration of HOCs, e.g. by differential scanning calorimetry (DSC): LeBoeuf and Weber (2000),
   - identification of specific adsorption and absorption contributions to overall isotherm patterns by the Dual Reactive Domain Model (DRDM): LeBoeuf and Weber (1997),
   - evaluation of "Hysteresis Indices" (HIs) for quantifying differences between the sorption and desorption isotherms for each macromolecular sorbent: Huang and Weber (1997).
3. *Assessment of reduced bioavailability of hydrophobic organic compounds in aged soil samples* using:
   - chemical-extractive methods: Kelsey et al. 1997,
   - unstable and stable isotopes: Eschenbach et al. (1998); Wais et al. (1995), Richnow et al. (1999), and
   - biological methods: e.g. Chung and Alexander (1998*).*
4. *Characterisation of long-term reactivity and bioavailability of heavy metals* in sediments and aquifer solids by:
   - acid producing potential (APP; Kersten and Förstner 1992),
   - relationship between acid volatile sulfide (AVS) and simultaneously extractable metals (SEMs; DiToro et al. 1992),
   - redox buffer capacities: Heron and Christensen (1995),
   - formation of metal hydroxide surface precipitates using molecular-scale techniques: Roberts et al. (1999), Thompson et al. (1999).
5. *Quantification of bioavailable metals in soils by phyto- and zootoxicity tests and microbial biosensors:* Corbisier et al. (1999), Vangronsveld et al. (2000). Combination with extraction procedure: Reid et al. (2000).
6. *Quantification of irreversible sorption phenomena for modification of sediment and soil quality criteria*: Chen et al. (2000).

Other methodological approaches for the assessment of solid-associated pollutants relate to mechanical effects, such as compaction and loss of water in the unsaturated zones of soils, and to hydrodynamic parameters, such as "erosion stability" of sediments. While the former effects typically support reduced chemical mobility of contaminants, resuspension of contaminated sediments by flood events, wave action, navigation traffic and dredging activities may affect large areas ("floodplains") in the lower reaches of the catchment area. Combined approaches for sediment stabilisation include physical "ripening" processes (such as dehydration and shrinkage, increase of permeability, change of consistency from soft to friable or hard (Vermeulen et al. 2000)), a wide spectrum of chemical effects (enhancement of sorption processes), and also biological mechanisms such as phytostabilisation, i.e. precipitation or adsorption near or at plant roots (Joziasse and Gun 2000).

References

Benner SG, Blowes DW, Gould WD, Herbert RB, Ptacek CJ (1999) Geochemistry of a permeable reactive barrier for metals and acid mine drainage. Environ Sci Technol 33: 2793–2799

Bosma TNF, Middeldorf PJM, Schraa G, Zehnder AJB (1997) Mass transfer limitations of biotransformation: Quantifying bioavailability. Environ Sci Technol 31: 248–252

Boyd SA, Sun S (1990) Residual petroleum and polychlorobiphenyl oils as sorptive phases for organic contaminants in soils. Environ Sci Technol 24: 142–144

Brady PV, Borns DJ (1997) Natural Attenuation of Metals and Radionuclides: Report from a Workshop held by Sandia National Laboratories, SAND97-2727, UC-800. Albuquerque, New Mexico and Livermore, California

Brümmer G, Gerth J, Tiller KG (1988) Reaction kinetics of the adsorption and desorption of nickel, zinc, and cadmium by goethite I, Adsorption and diffusion of metals. J Soil Sci 39: 37–52

Carmichael LM, Christman RF, Pfaender FK (1997) Desorption and mineralisation kinetics of phenanthrene and chrysene in contaminated soils. Environ Sci Technol 31: 126–132

Castilla HJ, Werth CJ, McMillan SA (2000) Structural evaluation of slow desorbing sites in model and natural solids using temperature stepped desorption profiles, 2. Column results. Environ Sci Technol 34: 2966–2972

Chefetz B, Deshmukh AP, Hatcher PG, Huthrie EA (2000) Pyrene sorption by natural organic matter. Environ Sci Technol 34: 2925–2930

Chen W, Kan AT, Tomson MB (2000) Irreverible adsorption of chlorinated benzenes to natural sediments: Implications for sediment quality criteria. Environ Sci Technol 34: 385–392

Christensen TH, Bjerg PL, Banwart S, Jakobsen R, Heron G, Albrechtsen H-J (2000a) Characterization of redox conditions in pollution plumes. In: Contaminated Soil 2000, Vol 1. Thomas Telford London, pp 181–188

Christensen TH, Bjerg PL, Kjeldsen P (2000b) Natural attenuation: A feasible approach to remediation of ground water pollution at landfills? Ground Water Monitoring & Remediation 20: 69–77

Christensen TH, Kjeldsen P, Albrechtsen HJ, Heron G, Nielsen PH, Bjerg PL, Holm PE (1994) Attenuation of pollutants in landfill leachate polluted aquifers. Crit Rev Environ Sci Technol 24, 119–202

Christensen TH, Nielsen BG (1987) Retardation of lead in soil. In: Heavy Metals in the Environment, Vol 1. CEP Consultants Edinburgh, New Orleans, pp. 319–323

Chung N, Alexander M (1998) Differences in sequestration and bioavailoability of organic compounds aged in dissimilar soils. Environ Sci Technol 32: 855–860

Corbisier P, van der Lelie D, Borremans B, Provoost A, de Lorenzo V, Brown N, Lloyd J, Hobman J, Csöregi E, Johannsson G, Mattiasson B (1999) Whole cell- and protein-based biosensors for the detection of bioavailable heavy metals in environmental samples. Analytica Chimica Acta 387: 235–244

DiToro DM, Mahony JD, Hansen DJ, Scott KJ, Hicks MB, Mayr SM, Redmond MS (1990) Toxicity of cadmium in sediments: the role of acid volatile sulfide. Environ Toxicol Chem 9: 1487–1502

DiToro DM, Mahony JD, Hansen DJ, Scott KJ, Carlson, AR, Ankley GT (1992) Acid volatile sulfide predicts the acute toxicity of cadmium and nickel in sediments. Environ Sci Technol 26: 96–101

EPA (1997) The incidence and severity of sediment contamination in surface waters of the United States, Vol 1. EPA 823-R-97-006, US Environmental Protection Agency, Washington DC

EPA (1999) Use of Monitored Natural Attenuation at Superfund, RCRA Corrective Action, and Underground Storage Tank Sites. OSWER Directive 9200.4-17P, United States Environmental Protection Agency, Office of Solid Waste and Emergency Response, Washington DC

Eschenbach A, Wienberg R, Mahro B (1998) Fate and stability of nonextractable residues of [^{14}C]PAH in contaminated soils under environmental stress conditions. Environ Sci Technol 32: 2585–2590

Farrell J, Grassian D, Jones M (1999) Investigation of mechanisms contribution to slow desorption of hydrophobic organic compounds from mineral solids. Environ Sci Technol 33: 1237–1243

Farrell J, Reinhard M (1994) Desorption of halogenated organics from model solids, sediments, and soil under unsaturated conditions, 1. Isotherms. Environ Sci Technol 28: 53–62

Förstner U (1987) Changes in metal mobilities in aquatic and terrestrial cycles. In: Patterson JW, Passino R (eds) Metal Speciation, Separation and Recovery. Lewis Publ Chelsea, Mich, pp 3–26

Förstner U, Schoer J (1984) Diagenesis of chemical associations of Cs-137 and other artificial radionuclides in river sediments. Environ Technol Lett 5, 295–306

Förstner U, Schöttler U, Nähle C (1982) Sorption of heavy metals in sand filters in the presence of humic acids. In: Schmidt KH (ed) Artificial Recharge. DVWK-Bulletin 13, pp 95–125

Gerth J, Brümmer GW, Tiller KG (1993) Retention of Ni, Zn and Cd by Si-associated goethite. Z Pflanzenernähr Bodenk 156: 123–129

Ghosh U, Gilette IS, Luthy RG, Zare RN (2000) Microscale location, characterisization, and association of polycyclic aromatic hydrocarbons on harbor sediment particles. Environ Sci Technol 34: 1729–1736

Grathwohl P (1990) Influence of organic matter from soils and sediments from various origins on the sorption of some chlorinated aliphatic hydrocarbons: Implications on K_{OC} correlations. Environ Sci Technol 24: 1687–1693

Griscom SB, Fisher NS, Luoma SN (2000) Geochemical influences on assimilation of sediment-bound metals in clams and mussels. Environ Sci Technol 34, 91–99

Gustafsson O, Haghseta F, Chan C, MacFarlane J, Gschwend PM (1997) Quantification of the dilute sedimentary soot phase: Implications for PAH speciation and bioavailability. Environ Sci Technol 31: 203–209

Hatzinger PB, Alexander M (1995) Effect of aging of chemicals in soil on their biodegradability and extractability. Environ Sci Technol 29: 537–545

Helios-Rybicka E, Förstner U (1986) Effect of oxyhydrate coatings on the binding energy of metals by clay minerals. In: Sly PG (ed) Sediment and Water Interactions. Proc 3rd Symp Geneva, Springer New York, pp 381–385

Heron G, Christensen TH (1995) Impact of sediment-bound iron on redox buffering in a landfill leachate polluted aquifer (Vejen, Denmark). Environ Sci Technol 29: 187–192

Huang W, Weber WJ jr (1997) A distributed reactivity model for sorption by soils and sediments, 10. Relationships between desorption, hysteresis, and the chemical characteristics of organic domains. Environ Sci Technol 31: 2562–2569

Huang W, Weber WJ jr (1998) A distributed reactivity model for sorption by soils and sediments, 11. Slow concentration-dependent sorption rates 32: 3549–3555

Jonker MTO, Smedes F (2000) Preferential sorption of planar contaminants in sediments from Lake Ketelmeer, The Netherlands. Environ Sci Technol 34: 1620–1626

Joziasse J, van der Gun J (2000) In-situ remediation of contaminated sediments: Conceivable and feasible?! In: Contaminated Soil 2000, Vol 1, Thomas Telford, London, pp 516–522

Kan AT, Fu G, Hunter M, Chen W, Ward CH, Tomson MB (1998) Irreversible sorption of neutral hydrocarbons to sediments: Experimental observations and model predictions. Environ Sci Technol 32: 892–902

Kan AT, Fu G, Tomson MB (1994) Adsorption/desorption hysteresis in organic pollutant and soil/sediment interaction. Environ Sci Technol 28: 859–867

Karapanagioti HK, Sabatini DA (2000) Impacts of heterogeneous organic matter on phenanthrene sorption: Different aquifer depths. Environ Sci Technol 34: 2453–2460

Kelsey JW, Kottler BD, Alexander M (1997) Selective chemical extractants to predict bioavailability of soil-aged organic chemicals. Environ Sci Technol 31: 214–217

Kersten M, Förstner U (1991) Geochemical characterization of the potential trace metal mobility in cohesive sediment. Geo-Marine Letts 11: 184–187

Kester DR, Ketchum BH, Duedall IW, Park PK (Eds)(1983) Wastes in the Ocean, Vol 2: Dredged-Material Disposal in the Ocean. Wiley New York

Kördel W, Dassenakis M, Lintelmann J, Padberg S (1997) The importance of natural organic material for environmental processes in waters and soils. Pure Appl Chem 69: 1571–1600

Köster R, Berg H, Bundschuh T (2000) Natural Attenuation anorganischer Schadstoffe – Geochemische Einflußfaktoren der Immobilisierung. In: DECHEMA (ed) "Natural Attenuation – Möglichkeiten und Grenzen naturnaher Sanierungsstrategien", DECHEMA, Frankfurt/M, pp 59–69

LeBoeuf EJ, Weber WJ jr (1997) A distributed reactivity model for sorption by soils and sediments, 8. Sorbent organic domains: Discovery of a humic glass transition and an argument for a polymer-based model. Environ Sci Technol 31: 1697–1702

LeBoeuf EJ, Weber WJ jr (2000) Macromolecular characteristics of natural organic matter, 1. Insights from glass transition and enthalpic relaxation behavior. Environ Sci Technol 34: 3623–3631

Lion LW, Altman RS, Leckie JO (1982) Trace metal adsorption characteristics of estuarine particulate matter: Evaluation of contribution of Fe/Mn oxide and organic surface coatings. Environ Sci Technol 16: 660–666

Luthy RG, Aiken GR, Brusseau ML, Cunningham SD, Gschwend PM, Pignatello JJ, Reinhard M, Traina SJ, Weber WJ jr, Westall JC (1997) Sequestration of hydrophobic organic contaminants by geosorbents. Environ Sci Technol 31: 3341–3347

MacDonald JA (2000) Natural attenuation for groundwater cleanup. Environ Sci Technol 34: 346A–353A

Mahro B (2000) Bioavailability of contaminants. In: Rehm H-J, Reed G (eds) Biotechnology, 2nd completely revised edition, Vol 11b: Environmental Processes II. Wiley-VCH Weinheim, pp 61–88.

Martínez CE, McBride MB (1998) Solubility of Cd^{2+}, Cu^{2+}, Pb^{2+}, and Zn^{2+} in aged coprecipitates with amorphous iron hydroxides. Environ Sci Technol 32: 743–748

Martínez CE, McBride MB (1999) Dissolved and labile concentrations of Cd, Cu, Pb, and Zn in aged ferrihydrite-organic matter systems. Environ Sci Technol 33: 745–750

Nam K, Alexander M (1998) Role of nanoporosity and hydrophobicity in sequestration and bioavailability: Tests with model solids. Environ Sci Technol 32: 71–74

National Research Council (2000) Natural Attenuation for Groundwater Remediation. US National Academy Press, Washington DC (prepublication copy, p 229)

Opdyke DR, Loehr RC (1999) Determination of chemical release rates from soils: Experimental design. Environ Sci Technol 33: 1193–1199

Pavlostathis SG, Mathavan GN (1992) Desorption kinetics of selected volatile organic compounds from field contaminated soils. Environ Sci Technol 26: 532–538

Pignatello JJ, Xing B (1996) Mechanisms of slow sorption of organic chemicals to natural particles. Environ Sci Technol 30: 1–11

Reid BJ, Paton GI, Bundy JG, Jones KC, Semple KT (2000) Determination of soil-associated organic contaminant bioavailability using a novel extraction procedure in conjunction with *lux*-marked microbial biosensors. In: Contaminated Soil 2000, Vol 2. Thomas Telford, London, pp 870–871

Richnow HH, Eschenbach A, Mahro B, Kästner M, Annweiler E, Seifert R, Michaelis W (1999) Formation of nonextractable residues – a stable isotope approach. Environ Sci Technol 33: 3761–3767

Roberts DR, Scheidegger AM, Sparks DL (1999) Kinetics of mixed Ni-Al precipitate formation on a soil clay fraction. Environ Sci Technol 33: 3749–3754

Romkens PFAM, Dolfing J (1998) Effect of Ca on the solubility and molecular size distribution of DOC and Cu binding in soil solution samples. Environ Sci Technol 32: 363–369

Ros Vicent J: cit Duursma EK (1976) Radioactive tracers in estuarine studies. In: Burton JD, Liss PS (eds) Estuarine Chemistry. Academic London, pp 159–183

Salomons W (1980) Adsorption processes and hydrodynamic conditions in estuaries. Environ Technol Lett 1: 356–365

Schulte-Ebbert U, Schöttler U (1995) Systemanalyse des Untersuchungsgebietes "Insel Hengsen". In: Schöttler U, Schulte-Ebbert U (eds) Schadstoffe im Grundwasser, Vol 3: Verhalten von Schadstoffen im Untergrund bei der Infiltration von Oberflächenwasser am Beispiel des Untersuchungsgebietes "Insel Hengsen" im Ruhrtal bei Schwerte. Wiley-VCH Weinheim, pp 475–513

Steinberg, SM, Pignatello JJ, Sawhney BL (1987) Persistence of 1,2-dibromoethane in soils: Entrapment in intraparticle micropores. Environ Sci Technol 21: 1201–1208

Tang J, Robertson BK, Alexander M (1999) Chemical-extraction methods to estimate bioavailability of DDT, DDE, and DDD in soil. Environ Sci Technol 33: 4346–4351

Teutsch G, Rügner H (2000) Entwicklung von Bewertungskriterien natürlicher Schadstoffabbauprozesse in Grundwasserleitern als Grundlage für Sanierungsentscheidungen bei Altstandorten. Zwischenbericht für die Landesanstalt für Umweltschutz Baden-Württemberg, Tübingen

Thompson HA, Parks GA, Brown GE jr (1999) Dynamic interaction of dissolution, surface adsorption and precipitation in an aging cobalt(II)-clay-water system. Geochim Cosmochim Acta 63: 1767–1779

Track T, Michels J (2000) Resümee des 1. Symposiums. In: "Natural Attenuation – Möglichkeiten und Grenzen naturnaher Sanierungsstrategien", DECHEMA, Frankfurt/M, pp 3–15

Vangronsveld J, Spelmans N, Clijsters H, Adriaensens R, Carleer R, Van Poucke D, van der Lelie D, Mergeay M, Corbisier P, Bierkens J, Diels L (2000) Physico-chemical and biological evaluation of the efficacy of *in situ* metal inactivation in contaminated soils. In: Contaminated Soil 2000, Vol 2. Thomas Telford, London, pp 1155–1156

Vermeulen J, van Dijk S, Grotenhuis T, Joziasse J, Rulkens W (2000) Accelerated physical ripening of PAH and oil contaminated sediment to distinguish critical steps in remediation. In: Contaminated Soil 2000, Vol 2. Thomas Telford, London, pp 1184–1185

Wais A, Haider K, Spiteller M, DeGraf AA, Burauel P, Führ F (1995) Using ^{13}C-NMR spectroscopy to evaluate the binding mechanis of bound pesticide residues in soils. J Environ Sci Health, Part B, B30: 1–24

Weber WJ, Huang W (1996) A distributed reactivity model for sorption by soils and sediments, 4. Intraparticle heterogeneity and phase-distribution relationships under nonequilibrium conditions. Environ Sci Technol 30: 881–888

Weber WJ, McGinley PM, Katz LE (1992) A distributed reactivity model for sorption by soils and sediments, 1. Conceptual basis and equilibrium assessment. Environ Sci Technol 26: 1955–1962

White JC, Kelsey JW, Hatzinger PB, Alexander M (1997) Factors affecting sequestration and bioavailability of phenanthrene in soils. Environ Toxicol Chem 16: 2040–2045

Xing B, Pignatello JJ (1997) Dual-mode sorption of low-polarity compounds in glassy poly(vinyl chloride) and soil organic matter. Environ Sci Technol 31: 792–79

36 Natural Attenuation as an Approach to Remediation of Groundwater Pollution at Landfills

T.H. Christensen, P. L. Bjerg, P. Kjeldsen
Department of Environmental Science and Engineering, Groundwater Research Centre, Building 115, Technical University of Denmark, DK-2800 Lyngby, Denmark

36.1 Introduction

In recent years natural attenuation has gained acceptance in several countries, in particular in the USA, as an approach to groundwater remediation for plumes of petroleum hydrocarbons and, in some cases, chlorinated aliphatic solvents. These are the most frequently found pollutants in groundwater, and natural attenuation is used as a remedy in an increasing number of cases.

Leachate pollution plumes from old landfills or landfills with inadequate leachate collection systems are also numerous. Natural attenuation, however, has not yet gained significant focus as a possible remediation technology for leachate plumes. Although this may appear to be a very attractive potential remediation technology for leachate plumes, landfills have special features that affect how natural attenuation can be assessed (Christensen et al. 2000a) and this makes it difficult to use the approaches and protocols used in the natural attenuation of petroleum hydrocarbon and chlorinated solvent plumes.

This paper presents some of the features special to landfills in the context of natural attenuation and draws primarily on experiences gained at the Groundwater Research Centre at the Technical University of Denmark from more than ten years of research on leachate plumes at two old landfills in Denmark.

36.2 What is Natural Attenuation of Groundwater Contaminant Plumes?

The basic principle of natural attenuation as a remediation technique is that natural processes in the subsurface provide sufficient mass reduction of critical pollutants so that no substantial risk exists to downgradient groundwater users or discharge areas. Monitoring of the downgradient groundwater quality is usually required after implementation of natural attenuation as a safeguard to users of the groundwater and ecological habitats in discharge areas. This is because the time frame associated with natural attenuation is usually long and the evidence provided is primarily circumstantial.

The critical issue in applying natural attenuation is to demonstrate quantitatively that the pollutants in the plume are subject to subsurface mass reduction. The published guidelines for natural attenuation of petroleum hydrocarbons (e.g. Wiedemeier et al. 1995; ASTM 1998) and for chlorinated aliphatic compounds (Wiedemeier et al. 1996) suggest several avenues of evidence that can be used. The direct evidence could be decreasing mass of key pollutants observed passing monitoring fences which intercept the plume at different distances down gradient from the source. Indirect evidence could be the removal of electron acceptors (e.g. O_2) from the plume or increasing concentrations of "used" electron acceptors (e.g. Fe^{2+} in the groundwater from reduction of Fe^{3+} associated with the aquifer sediment). The presence of degradation products also provides strong evidence (e.g. Cl^- from anaerobic dehalogenation of chlorinated solvents). In some cases, microcosm degradation studies may be performed with aquifer material from the plume to support the other pieces of evidence.

The scientific database supporting natural attenuation is very substantial for the petroleum hydrocarbons and is also significant for the chlorinated solvents, while the data available is far more scarce for landfill leachate pollutants.

36.3 The Basis for Suggesting Natural Attenuation of Leachate Plumes

Christensen et al. (2000b) presented a review of the attenuation processes for landfill leachate pollutants in groundwater showing that the information available suggests that leachate pollution plumes often are relatively short and that attenuation seems to be substantial. Table 36.1 summaries key characteristics of leachate plumes described in the literature (as compiled by Christensen et al. 2000b).

Most of the existing research reports have focused on the fate of organic chemicals in leachate plumes, and only a few reports have addressed the fate of general organic matter in terms of DOC, general inorganic components and heavy metals in leachate plumes. However, those reports that exist all seem to suggest that all groups of pollutants are subject to natural attenuation (Christensen et al. 2000b). Most of the investigated landfills contained a mixture of wastes including organic municipal waste and were located on sandy aquifers that were naturally aerobic. One exception to this is the North Bay Landfill, Canada (Reinhard et al. 1984) located on a clay till.

The development of plumes with low redox potential seems to be a common feature for leachate contaminated aquifers (Christensen et al. 2000b). Methanogenic conditions may prevail close to the landfill and further out in the plume sulphate-reduction, iron-reduction, manganese-reduction and denitrifying conditions have been observed. In some cases, after mixing with the surrounding groundwater, the outskirts of the plume became aerobic. Strongly reducing conditions have developed over decades of leachate entering into the aquifer. Anaerobic biological activity in the plumes seems ubiquitous (Ludvigsen et al. 1999).

Table 36.1. Characteristics of landfill leachate plumes described in the literature as compiled by Christensen et al. (2000b)

Name	LA[1] [ha]	LP[2]	Type[3]	Length[4] [m]	Width[5] [m]	No[6]	Compounds addressed
Borden, Canada	4.5	1940–1976	Sand	~700	~600	600	Inorganics Redox pairs Organic matter
NorthBay, Canada	28	1962–?	Silt	~400	~100	300	Inorganics Redox pairs Organic matter XOCs
Woolwich, Canada	3.5	1965–?	-	~600	-	500	Chloride Organic matter XOCs
Grindsted, Denmark	10	1936–1972	Sand	~400 More plume?	~400 ?	350	Inorganics Redox pairs Organic matter XOCs
Vejen, Denmark	6	1962–1981	Sand	~400	~100	160	Inorganics Redox pairs Heavy metals Organic matter XOCs
Augsburg, Germany	26	1955–1990?	Sand, Gravel	800?	500?	38	Inorganics Redox pairs Heavy metals XOCs
Unnamed, Germany	15	1954–1970	Coarse Sand	~3000	~500	20	Inorganics Organic matter
Noordwijk, Netherlands	~6	1960–1973	Coarse sand	-	-	22	Inorganics XOCs

[1] Landfill Area
[2] Landfilling period
[3] Type of Aquifer
[4] Length of plume
[5] Width of plume
[6] Number of samples for delineation (approximately)
[7] USA

Table 36.1. (continuation)

Name	LA[1] [ha]	LP[2]	Type[3]	Length[4] [m]	Width[5] [m]	No[6]	Compounds addressed
Army Creek, Delaware[7]	24	1960–1968	Sand	~700	-	20	Inorganics Redox pairs Organic matter XOCs
Babylon, Connecticut[7]	10	1947–1975	Coarse sand	3000	600	120	Inorganics
Bristol (mun.), Vermont[7]	3	1968?–>1992	Sand, gravel	200	150	30	Inorganics Redox pairs XOCs
Bristol (priv.), Vermont[7]	4	1968–1995?	Sand, gravel	100	100	30	Inorganics Redox pairs XOCs
Islip, Connecticut[7]	7	1960–1973	Coarse sand	1500	400	75	Inorganics
KL, Michigan[7]	27	1960–1979	Sand	1600	-	60	Inorganics Redox pairs Organic matter XOCs
Reno County, Kansas[7]	55	1960?–1987	Sand	600	-	-	Inorganics Redox pairs
Shelby County, Tennessee[7]		1968–1988	Sand	<200	-	30	Inorganics Redox pairs Heavy metals XOCs
Unnamed[7]	4	1956–1977	Sandy silt /gravel	+800	200	16	Inorganics Redox pairs Organic matter XOCs

[1] Landfill Area
[2] Landfilling period
[3] Type of Aquifer
[4] Length of plume
[5] Width of plume
[6] Number of samples for delineation (approximately)
[7] USA

The low redox levels have in many cases led to reductive dissolution of iron from the sediment resulting in zones of strongly elevated concentrations of dissolved iron (e.g. Bjerg et al. 1995). While the leachate leaving landfills has usually not been well characterised it is expected that old landfills dating back to the seventies and earlier would contain a mix of almost anything. Yet concentrations of most pollutants have been low in plumes and decreasing with distance from the landfill. Fig. 36.1 shows how the concentrations of several groups of organic chemicals in the Vejen Landfill leachate plume decreased with distance, even after correction for dilution (Lyngkilde and Christensen 1992). Sorption was considered but could not explain the distribution of the compounds leading to the suggestion that degradation of the organic compounds had taken place under strongly reducing conditions (methanogenic, sulphate-reducing and iron-reducing conditions).

Laboratory and field injection studies (Nielsen et al. 1995a, b; Johnston et al. 1996; Rügge et al. 1999a; Bjerg et al. 1999) have also shown that several organic compounds are degradable in leachate contaminated groundwater, albeit often at low rates and in some cases not within the experimental time frame, if degraded at all.

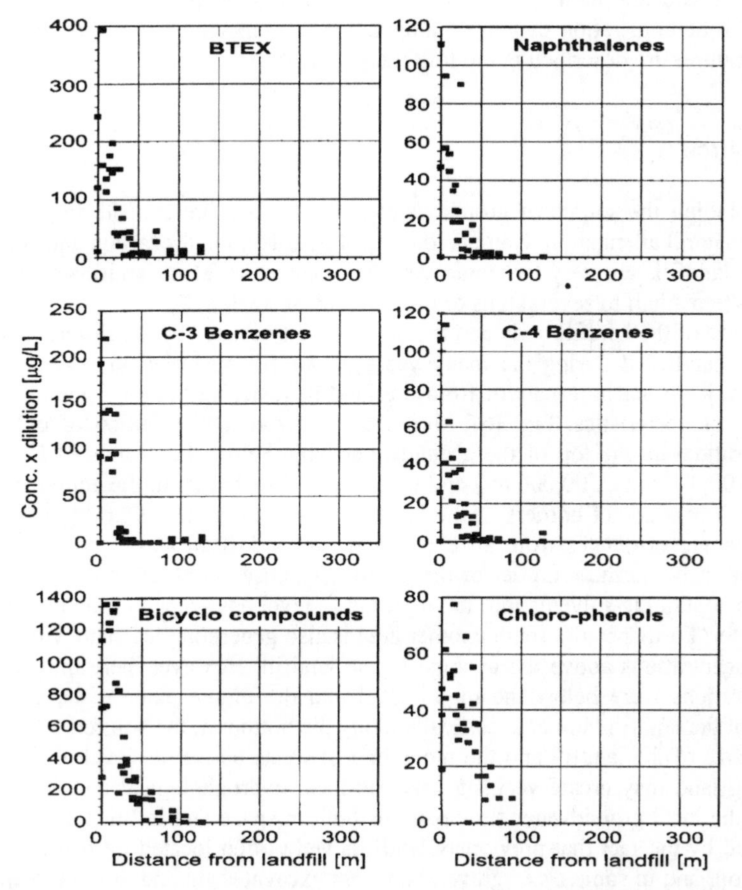

Fig. 36.1. Plots of concentrations of organic chemicals in the Vejen Landfill leachate plume as function of landfill after correction for dilution according to chloride content (Lyngkilde and Christensen 1992a). The first 150 m of the plume, where the organic chemicals degrade, has been characterised as methanogenic, sulphate-reducing and iron-reducing (Lyngkilde and Christensen 1992b).

The published reports suggest that natural attenuation takes place in leachate plumes, but it should be kept in mind that only a few studies have been published in the open literature (in the order of 15) and nearly all plumes studied have been in sandy aquifers that naturally have been aerobic. Thus the process of natural attenuation is more or less unknown in aquifers containing low solid contents of oxidised iron and manganese and with an insignificant supply of oxygen and nitrate.

36.4 Complicating Features of Landfills and Leachate Plumes

Although scientific evidence suggests that natural attenuation processes are significant in leachate plumes, landfills and leachate plumes have complex features that make demonstration of natural attenuation at a specific site a difficult task. These features are described in the following paragraphs.

36.4.1 The Source

Understanding the source of groundwater pollution is a prerequisite for demonstrating natural attenuation. Compared to most petroleum hydrocarbon and solvent sources, landfills are very large and very heterogeneous. Most landfills will contain between a half to several tens of million tons of waste.

The size of the landfill body and the heterogeneity caused by the different types of waste landfilled during the many years of landfill operation may generate a highly variable leaching pattern from the landfill fostering multiple plumes with different characteristics. Fig. 36.2 shows the horizontal distribution of chloride concentrations in the top of the groundwater table below the Grindsted Landfill (about 300,000 tons, 100,000 m^2) and the vertical distribution in the aquifer along the two downgradient borders of the landfill (Kjeldsen et al. 1998a). A very distinct leaching originates from an area where wet industrial waste has been disposed of at the northern border of the landfill (Kjeldsen et al. 1998b). The dense leachate containing solvents into the high mg l^{-1} level has seeped to the bottom of the aquifer. Further to the south another area is also generating leachate with chloride concentrations above the average of the landfill. However, here the solvent concentrations were below the mg l^{-1} level and the plume exits at the western border of the landfill but not specifically along the bottom of the aquifer.

The size of the landfill and the mounding of waste above the level of the surrounding land may create very site specific local hydrogeological conditions affecting the local groundwater flow patterns both in space and in time. This is exacerbated by the fact that previously landfills were often located on derelict land where soil, and in some cases gravel, had been excavated. In addition the seepage of leachate through underlying strata may locally affect the hydraulic conductivity by mineral precipitation, bioclogging or bubbles originating from gaseous degradation products (methane and carbon dioxide). Fig. 36.3 shows the flow directions at the Grindsted Landfill monitored over a 9 month long period about 20 years after closure of the landfill (Kjeldsen et al. 1998a). It is notable that flow will occur in many directions and that the flow directions vary by more than 70 degrees. This suggests that the potential area of leachate contamination may be much wider than expected from the general hydrogeology of the area and that different plumes with different flow directions may develop. This may make long term monitoring for the purpose of demonstrating stationarity a difficult task. The mounding of the groundwater table beneath landfills (as often observed by Christensen et al. 1994) may also lead to a vertical hydraulic gradient, which together with gradients introduced by the higher density of the leachate add a vertical component to the flow of the leachate plume. This may further lead to leachate infiltration into deeper naturally anoxic aquifers potentially having a lower capacity for natural attenuation.

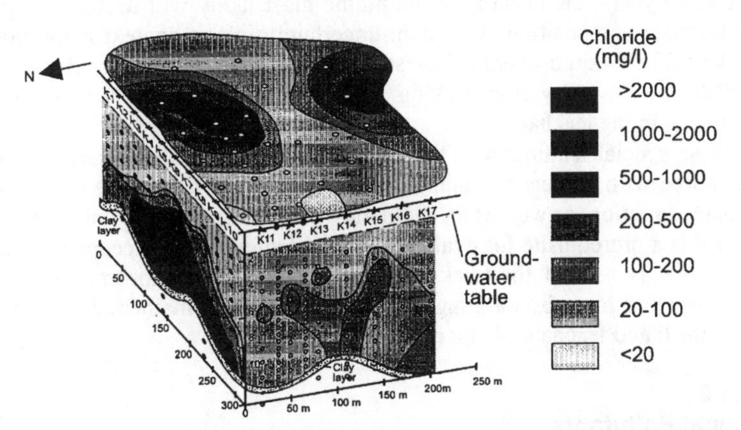

Fig. 36.2. Chloride concentrations at the groundwater table below the Grindsted Landfill and in vertical transects of the aquifer along the downgradient border of the landfill showing variations in leaching patterns (Kjeldsen et al. 1998a).

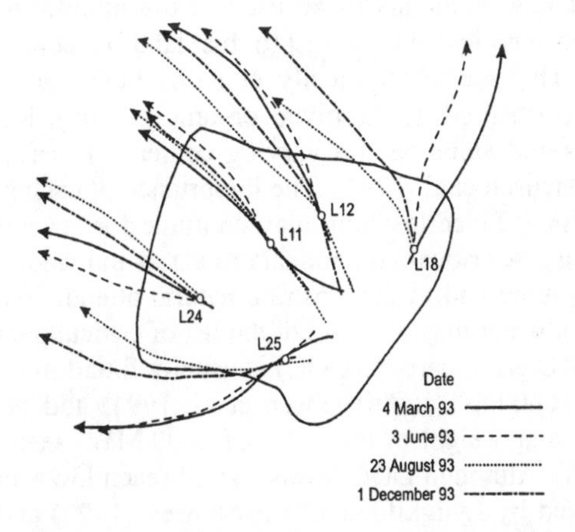

Fig. 36.3. Variations in flow directions at the Grindsted Landfill determined from piezometer readings over a 9 month period (Kjeldsen et al. 1998a).

The size of the landfill and the difficulties in locating new landfills often will prohibit removal of the waste and thus the removal of the source of the leachate. If a site was available to accept waste excavated from an old landfill, the cost of removing the old landfill and the environmental emissions from the operation would also be significant. In most cases the source for the leachate plumes will probably not be removed, and the landfill body and the leachate plume should be considered as a continuum in the context of natural attenuation. This implies a very long time frame, since landfills even after twenty to thirty years typically still contain more than 90 % of the pollution potential within the landfill body (see e.g. Belevi and Baccini 1992). However, the leaching in terms of compound concentrations are considered to peak within a few decades after the landfill operation has terminated and, although the plume from the landfill may persist for more than a hundred years, the hazards of the plume most likely will decrease in parallel to the leachate composition. The main uncertainty in this context is the presence in the landfill of liquid chemical waste in closed containers that after years in the landfill one day may start leaking, contributing unexpectedly to the content of chemicals in the leachate.

These special features with the landfill itself suggest that a careful mapping of the source with respect to main leachate release areas, leachate composition and seepage direction, as well as an evaluation of eventual "time bombs" present in the landfill is a prerequisite for evaluating natural attenuation. Accurate accounting of the source is needed for tracking main plumes, for demonstration of pollutant mass loss and for substantiating conceptual as well as mathematical modelling of the landfill and leachate plume continuum.

36.4.2 Critical Pollutants

Petroleum hydrocarbon and chlorinated solvent plumes contain relatively few or fairly comparable compounds. Data on composition of source liquids, scientific data on degradation pathways and a long history of experience have contributed to identification of key pollutants to be used in the monitoring of plumes and for demonstration of mass loss. In contrast to this, landfills contain a variety of wastes and pollutants. The leachate typically contains high concentrations of general organic matter (measured as dissolved organic carbon), high concentrations of common cations and anions and a varying content of heavy metals and organic chemicals (Christensen et al. 1994). The importance of each group may depend on the waste landfilled. Since leaching may continue for several decades, it is difficult to pinpoint a few critical components to act as indicators of the hazards associated with the plume and to demonstrate natural attenuation. Local legal quality criteria for groundwater may also add to the list of critical parameters.

The dissolved organic carbon (DOC) is usually found in old landfill leachate at a level of 200–1000 mg l^{-1} (Christensen et al. 1994) and fulvic-like compounds (moderate molecular weight of the order of 2000 MW) seem to dominate (Chritensen et al. 1998). Although DOC levels should reach low values in the attenuated plume as observed by Lyngkilde and Christensen (1992) and Bjerg et al. (1995) and, therefore, should be monitored for in the plume. However it is not likely that DOC will be the most critical component.

Common cations and anions will be subject to dilution and in many cases precipitation and ion exchange providing attenuation of the high concentrations found in the leachate. Most of these components have water quality criteria in the tens to 100 mg l^{-1} level and it is unlikely that common cations and ions will be the most critical components in the leachate. However, nitrogen species may be a critical component in the long term. Ammonium concentrations into the 500 mg N l^{-1} range are not unusual for leachate (Christensen et al. 1994) and the nitrogen content of landfilled waste is so high that high ammonium concentrations in the leachate can be expected for several decades. Monitoring of NH_4^+ as well as N_2O, NO_2^- and NO_3^- in the plume of the Grindsted Landfill (fig. 36.4) showed that ammonium was strongly attenuated by ion exchange and oxidation. However the oxidation seemed to take place partly under anaerobic conditions (Bjerg et al. 1995) and currently the mechanism is not understood. The zones of dinitrogen oxide and nitrate following the disappearance of ammonium suggest that oxidation takes place, but the mechanism (manganese-reduction may be involved) must be understood to substantiate the apparent natural attenuation of ammonium.

Heavy metals are usually found in relatively modest concentrations in leachate, likely due to precipitation as sulfides and sorption to solid surfaces in the waste. In addition, the attenuation capacity is substantial in aquifers, so the migration away from the landfill will be very limited as observed at the Vejen landfill (Andersen et al. 1991). This is in spite of the possibility of complex formation with dissolved organic carbon. Christensen et al. (1996) showed that DOC in leachate-polluted groundwater did form complexes with Cd, Ni and Zn, but that the relative mobility of the metals estimated from distribution coefficients was still very modest.

Specific organic chemicals including BTEX, PAH, chlorinated solvents and pesticides, are likely to be some of the critical components due to their low concentration criteria in groundwater. Some of these components will be strongly attenuated by sorption (higher PAHs) and some will degrade in the strongly anaerobic zones downgradient from the landfill. However, degradation rates may be slow and the retention times in the active redox zones may be insufficient to assure complete degradation. For example, chlorinated aliphatic compounds may migrate to higher redox levels before sufficient dehalogenation has occurred. Also intermediates, for example vinylchloride, may appear in the leachate plume and become a critical component. While these issues are also well known for chlorinated solvent plumes subject to natural attenuation, leachate plumes may also contain unknown polar and ionic organic chemicals, e.g. pesticides or metabolites from pesticides. Pesticides have often been observed in landfill leachate (Christensen et al. 1994) and in leachate plumes (Gintautas et al. 1992; Lyngkilde and Christensen 1992) and may be very critical since they usually are considered to be recalcitrant in anaerobic leachate plumes (Rügge et al. 1999b). Some groups of pesticides (e.g. phenoxyacids) may be degradable under aerobic conditions, while other pesticides (e.g. atrazine) are persistent in groundwater in general.

Complete identification of all peaks obtained by chromatographic analyses of samples from the plume may be impossible or at least very demanding. Therefore it may be a useful supplement to apply biotests as a non-specific indicator of toxic organic chemicals in the leachate and in the plume (Baun et al. 2000).

Currently published experiences on leachate plumes are too few to allow for a complete list of key parameters to be included in demonstrating natural attenuation of leachate plumes. However, such a list will most likely include ammonium, benzene, naphthalene, well-known pesticides, TCE and degradation products such as DCE and vinylchloride, and possibly biotest screening for non-volatile polar organic chemicals.

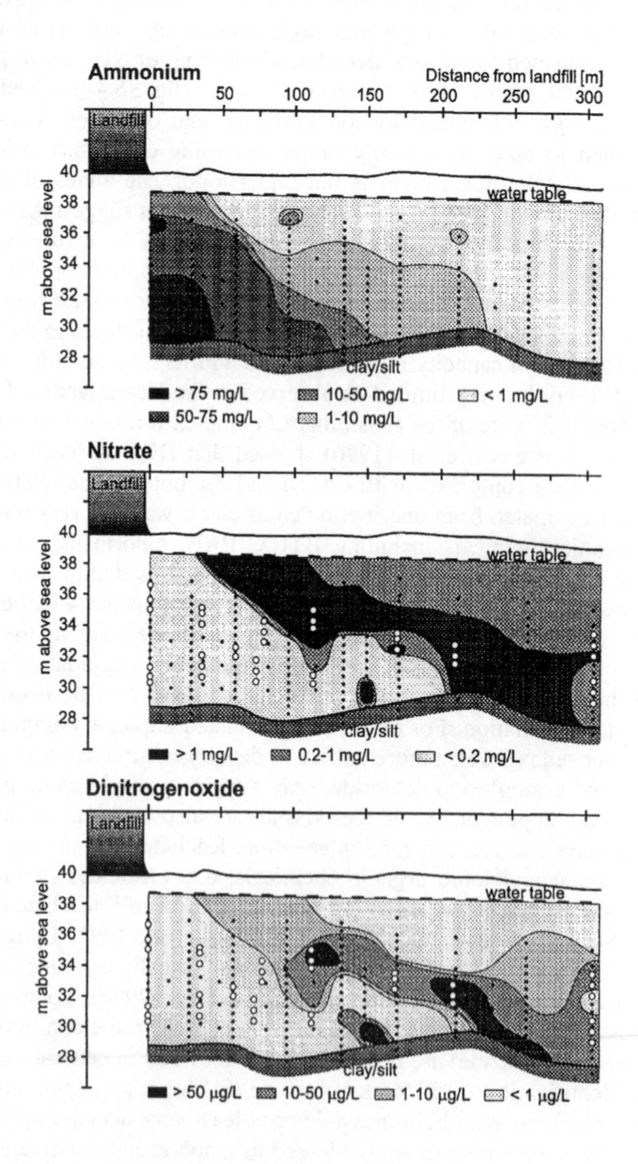

Fig. 36.4. Distribution of ammonium, nitrate and dinitrogenoxide in a vertical, longitudinal transect downgradient of Grindsted Landfill, Denmark. All nitrogen compounds measured as mg N l^{-1} (modified from Bjerg et al. 1995; Ludvigsen et al. 1998).

36.4.3 Mass Reduction

Documenting that reduction of mass of critical components has taken place in the plume traditionally requires that the plume or main plumes have been located. This seems elementary, but it actually takes a dense sampling network to obtain a reasonable picture of the three-dimensional leachate plume even though it may be only a few hundred metres long. This is in particular the case if the local hydrogeology is heterogeneous and strongly affected by the presence of the landfill. If attenuation is substantial, the periphery of the plume must be mapped in terms of indicator parameters such as chloride (although deicing salts used on nearby roads may interfere; Bjerg et al. 1995), specific conductivity or perhaps alkalinity (the leachate and several of the attenuation processes add alkalinity to the plume). Estimates of infiltration through the landfill, groundwater flow velocities and age of the landfill may assist in forming the conceptual model for the plume migration although the local hydraulic dilution may be difficult to estimate. Time series may be required to prove the stationarity of the plume. Attenuation will also take place in a dynamic plume, but the long term available attenuation processes are easier to identify in a stationary plume.

The mass flow of critical components can be determined by integrating over monitoring fences intercepting the plume at different distances downgradient from the landfill. This is a demanding procedure, probably requiring more than a hundred sampling points, since the different critical components will probably be attenuated in different zones of the plume. However, this monitoring approach will show where the mass reduction takes place and the main attenuating processes may be identified from general knowledge of pollutant behaviour in aquifers. This is possible if the redox conditions in the active zone have also been identified and first estimates on hydraulic retention times provided.

Determination of the redox environments is in itself no routine procedure (Christensen et al. 2000c). However, it might not be required to address each theoretical redox environment since they may actually overlap in real leachate plumes. But it may be necessary to identify the strongly reduced environments (methanogenic, sulfate-reducing, iron-reducing), the moderately anaerobic environments (manganese-reducing and nitrate-reducing) and the aerobic environments. Fig. 36.5 shows the redox environments identified in two longitudinal transects 30 m apart in the plume at the Grindsted Landfill (Bjerg et al. 1995). Hydrogen measurements, groundwater composition, sediment analysis and bioassays were useful tools in evaluating the redox environments in the Grindsted Landfill leachate plume (Bjerg et al. 1995; Jakobsen et al. 1998; Heron et al. 1998; Ludvigsen et al. 1998). However, the fairly limited experience regarding redox buffer capacity of the groundwater sediments (see Heron and Christensen (1995) concerning the redox buffering in the Vejen Landfill leachate plume) and the long time frame to be considered (decades to maybe more than a 100 years) call for care in estimating the long term development in the redox conditions.

Indirect documentation of attenuation has been successfully done for other types of pollution plumes. In petroleum hydrocarbon plumes the presence of reduced zones in the plume has been used as evidence of degradation of the hydrocarbons. In this case the hydrocarbons are the main electron donors and, at the same time, the critical components. Such an approach is, however, not feasible for leachate plumes, since the main electron donors (supposedly the general dissolved organic matter from the waste creates the reduced zones of the plume) do not constitute the critical pollutants. In plumes of chlorinated aliphatic compounds the increase in chloride in the plume has been used as evidence of degradation. Although chlorinated solvents also may be a problem in leachate plumes, a release of chloride by reductive dehalogenation will be completely masked by the high chloride concentration in the leachate. Indirect evidence of attenuation of the chlorinated aliphatic compounds may be obtained from the relative composition of water samples (PCE; TCE, DCE; VC). Kromann et al. (1998) showed in landfill leachate with no dilution that the rates as well as the mass ratios would vary, although the degradation sequence for PCE was typical. Perhaps differences in the isotopic composition of chloride originating from inorganic waste and from dechlorination could also be used (Sturchio et al. 1998). Degradation products of BTEXs as well as differences in isotopic composition may also be useful indicators of their anaerobic transformation, but this has not yet been demonstrated for leachate plumes.

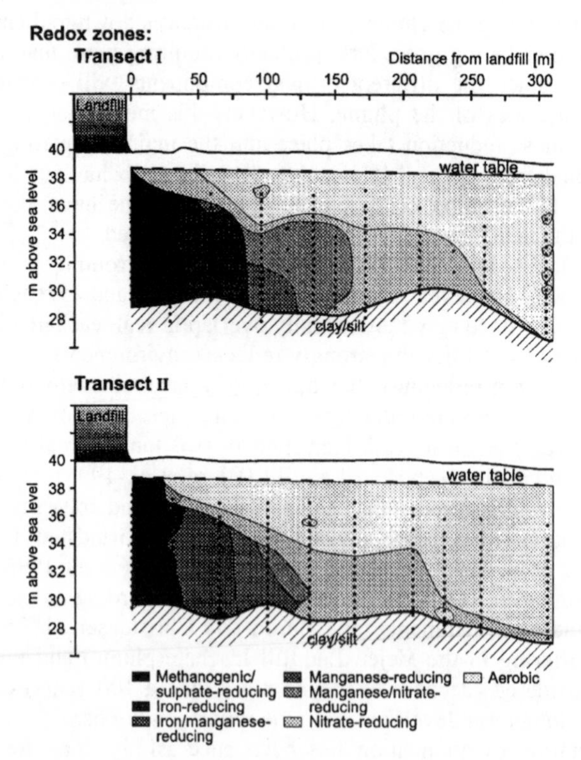

Fig. 36.5. Redox environments determined in two longitudinal transect of the leachate plume at the Grindsted Landfill based on groundwater sample composition (Bjerg et al. 1995)

The potential for mass reduction by degradation can also be demonstrated by performing microcosm studies with aquifer sediments and groundwater from the plume. This may be done by spiking batch reactors or in-situ microcosms (Johnston et al. 1996; Bjerg et al. 1999 and fig. 36.6) with the critical leachate components. The microcosms should represent the redox conditions of the active parts of the plume and should not be amended except with the critical components of interest. For some of the critical components (e.g. benzene and naphthalene) the degradation rates may, however, be so slow under strongly anaerobic conditions (half lives of years) that significant results may take years to develop (Bjerg et al. 1999). Furthermore the actual lag phases even for degradable compounds may be very long, which makes the interpretation of such data difficult and the application questionable in actual cases that are not research projects.

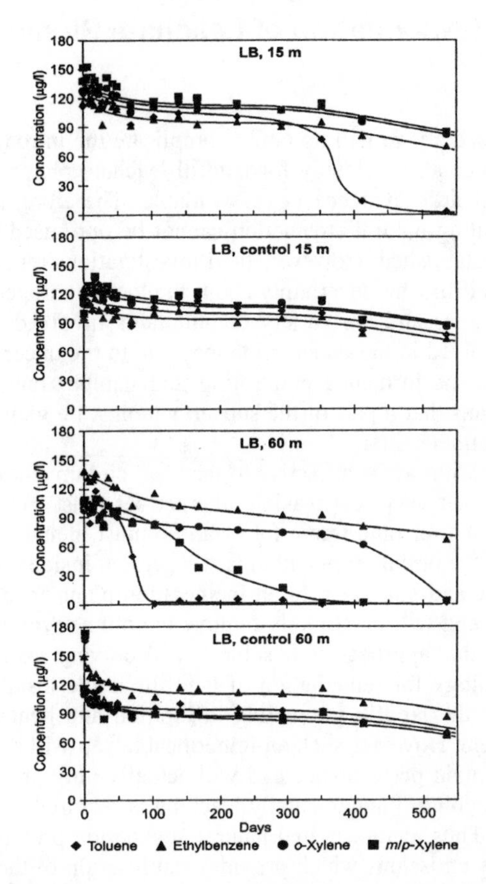

Fig. 36.6. Laboratory microcosm studies on toluene, ethylbenzene and xylene degradation under strongly anaerobic conditions in groundwater and sediment sampled at the Grindsted landfill, Denmark (Bjerg et al. 1999)

As indicated by the above discussion, demonstration of reduction of the mass of critical components in the leachate is no simple and quick task. However, as the database on evidence of natural attenuation at landfills develops, the development of more pragmatic approaches should be possible. For example, for older landfills relative comparison of the composition of water samples from a broad range of sampling points (e.g. methane, sulphate, iron, ammonium, DOC, H_2, BTEXs, naphthalene, biotest, chloride, specific conductivity, alkalinity, pH) could potentially provide enough relative evidence of natural attenuation, although the precise location of the plume and precise estimates of mass removal rates cannot be obtained. Performance monitoring will, in such cases, need to be stringent to provide sufficient safeguard for unknown factors.

36.5 Why is Natural Attenuation of Leachate Plumes Attractive?

The features discussed above will no doubt complicate the introduction of natural attenuation as a remediation strategy for landfill leachate plumes. This relates to the general fact that little evidence is yet available. Therefore, the procedure to follow in demonstrating natural attenuation cannot be optimised before more experience has been established. However, field investigations remain costly due to the large size of landfills, the uncertainty about the local hydrogeology, the possibility of several plumes and the variety of pollutants involved. In addition, the long time frame involved in the remediation may add to the uncertainties, increase the requirements for performance monitoring and hamper public acceptability. This effectively means that a part of the subsurface must be viewed and accepted as a long lasting treatment filter.

On the other hand, it may be attractive to face and circumvent these difficulties in view of the fact that very few feasible alternatives exist. Any alternative expected to involve a shorter time frame for clean-up must include the landfill body itself, since most of the pollution potential is still present inside the body of waste. Removing the waste and hauling it to an incineration plant or to another landfill removes the source and will most likely remove the plume within a few years by natural recovery. If this approach is feasible it will be very expensive. The only other proven technology for remediation of landfills is the installation of an impermeable cap over the landfill preventing infiltration and hence preventing the generation of leachate. However such an impermeable cap will require permanent maintenance to maintain performance and will actually only preserve the landfill and postpone the leaching. Fundamentally leaching is required to obtain stabilisation of the landfill. Thus, demonstrated natural attenuation provides a sustainable technology with few emissions which provides stabilisation of the landfill as well as remediation of the leachate pollution plume.

Natural attenuation seems to be an attractive approach to remediation of many, but not all, old landfills. However, there still are important questions to answer before this becomes a routine approach. First of all, a more comprehensive database on actual landfill plumes must be developed and further research is necessary to develop an easy-to-use protocol.

References

Andersen TV, Holm PE, Christensen TH (1991) Heavy metals in a municipal landfill leachate pollution plume. In: Heavy Metals in the Environment. JG Farmer (ed) vol 2, pp 252–255, International Conference, Edinburgh 16–20 September, CEP Consultants Ltd, Edinburgh, UK

ASTM (1998) Standard guide for remediation of ground water by natural attenuation at petroleum release sites. Designation E 1943–98 American Society for Testing and Materials, West Conshohocken, PA

Baun A, Jensen SD, Bjerg PL, Christensen TH, Nyholm N (2000) Toxicity of organic chemical pollution in groundwater downgradient of a landfill (Grindsted, Denmark). J Env Sci Tech 43: 1647–1652

Belevi H, Baccini, P (1992) Long-term leachate emissions from municipal solid waste landfills. In: Landfilling of Waste: Leachate. Christensen TH, Cossu R, Stegmann R, (eds) pp 431–440, Elsevier Applied Science, London, UK

Bjerg PL, Rügge K, Cortsen J, Nielsen PH, Christensen TH (1999) Degradation of aromatic and chlorinated aliphatic hydrocarbons in the anaerobic part of the Grindsted Landfill leachate plume: In situ microcosm and laboratory batch experiments. Ground Water 37 (1): 113–121

Bjerg PL, Rügge K, Pedersen JK, Christensen TH (1995) Distribution of redox sensitive groundwater quality parameters downgradient of a landfill (Grindsted, Denmark). J Env Sci Tech 29: 1387–1394

Christensen JB, Jensen DL, Grøn C, Filip, Christensen TH (1998) Characterization of the dissolved organic carbon in landfill leachate polluted groundwater. J Water Research 32: 125–135

Christensen JB, Jensen DL, Christensen TH (1996) Effect of dissolved organic carbon on the mobility of cadmium, nickel and zinc in leachate polluted groundwater. J Water Research 30: 3037–3049

Christensen TH, Kjeldsen P, Albrechtsen HJ, Heron G, Nielsen PH, Bjerg PL, Holm PE (1994) Attenuation of landfill leachate pollutants in aquifers. Critical Reviews in Environmental Science and Technology 24: 119–202

Christensen TH, Bjerg PL, Kjeldsen P (2000 a) Natural attenuation: A feasible approach to remediation of ground water pollution at landfills? Ground Water Monitoring and Remediation 20 (Winter): 69–77

Christensen TH, Kjeldsen P, Bjerg PL, Jensen DL, Christensen JB, Baun A, Albrechtsen HJ, Heron G (2000 b) Biogeochemistry of landfill leachate plumes. Applied Geochemistry (in press)

Christensen TH, Bjerg PL, Banwart SA, Jakobsen R, Heron G, Albrechtsen G (2000 c) Characterization of redox conditions in groundwater contaminant plumes. Journal of Contaminant Hydrology 45: 165–241

Gintautas PA, Daniel SR, Macalady DL (1992) Phenoxyalkanoic acid herbicides in municipal andfill leachates. J Env Sci Tech 26: 517–521

Heron G, Christensen TH (1995) Impact of sediment-bound iron on redox buffering in a landfill leachate polluted aquifer (Vejen, Denmark). J Env Sci Tech 29: 187–192

Heron G, Bjerg PL, Gravesen P, Ludvigsen L, Christensen TH (1998) Geology and sediment geochemistry of a landfill leachate contaminated aquifer (Grindsted, Denmark). J Contaminant Hydrology 29: 301–317

Jakobsen R, Albrechtsen HJ, Rasmussen M, Bay H, Bjerg PL, Christensen TH (1998) H_2 concentrations in a landfill leachate plume (Grindsted, Denmark) – extending the use of H_2 measurements through calculations of in situ energetics of terminal electron acceptor processes. J Env Sci Tech 32: 2142–2148

Johnston JJ, Borden RC, Barlaz MA (1996) Anaerobic biodegradation of alkylbenzenes and trichloroethylene in aquifer sediment down gradient of a sanitary landfill. J Contaminant Hydrology 23: 263–283

Kjeldsen P, Bjerg PL, Rügge K, Christensen TH, Pedersen JK (1998 a) Characterization of an old municipal landfill (Grindsted, Denmark) as a groundwater pollution source: Landfill hydrology and leachate migration. J Waste Management and Research 16: 14–22

Kjeldsen P, Grundtvig A, Winther P, Andersen JS (1998 b) Characterization of an old municipal landfill (Grindsted, Denmark) as a groundwater pollution source: Landfill history and leachate compostition. J Waste Management and Research 16: 3–13

Kromann A, Ludvigsen L, Albrechtsen HJ, Christensen TH, Ejlertsson J, Svensson BH (1998) Degradability of chlorinated aliphatic compounds in methanogenic leachates sampled at eight landfills. J Waste Management and Research 16: 54–62

Ludvigsen L, Albrechtsen HJ, Heron G, Bjerg PL, Christensen TH (1998) Anaerobic microbial redox processes in a landfill leachate contaminated aquifer (Grindsted, Denmark). J Contaminant Hydrology 33: 273–291

Ludvigsen L, Albrechtsen HJ, Ringelberg D, Ekelund F, Christensen TH (1999) Microbial distribution and composition in a landfill leachate contaminated aquifer (Grindsted, Denmark). J Microbial Ecology 37: 197–207

Lyngkilde J, Christensen TH (1992 a) Fate of organic contaminants in the redox zones of a landfill leachate pollution plume (Vejen, Denmark). J Contaminant Hydrology 10: 291–307

Lyngkilde J, Christensen TH (1992 b) Redox zones of a landfill leachate pollution plume (Vejen, Denmark). J Contaminant Hydrology 10: 273–289

Nielsen PH, Albrechtsen HJ, Heron G, Christensen TH (1995 a) In situ and laboratory studies on the fate of specific organic compounds in an anaerobic landfill leachate plume. I: Experimental conditions and fate of phenolic compounds. J Contaminant Hydrology 20: 27–50

Nielsen PH, Bjarnadottir H, Winter PL, Christensen TH (1995 b) In situ and laboratory studies on the fate of specific organic compounds in an anaerobic landfill leachate plume. II: Fate of aromatic and chlorinated aliphatic compounds. J Contaminant Hydrology 20: 51–66

Reinhard M, Goodman NL, Barker JF (1984) Occurrence and distribution of organic chemicals in two landfill leachate plumes. J Env Sci Tech 18: 953–961

Rügge K, Bjerg PL, Christensen TH (1995) Distribution of organic compounds from municipal solid waste in the groundwater downgradient of a landfill (Grindsted, Denmark). J Env Sci Tech 29: 1395–1400

Rügge K, Bjerg PL, Pedersen JK, Mosbæk H, Christensen TH (1999 a) An anaerobic field injection experiment in a landfill leachate plume (Grindsted, Denmark). 1. Experimental set-up, tracer movement and fate of aromatic and chlorinated compounds. Water Resources Research 35: 1231–1246

Rügge K, Bjerg PL, Mosbæk H, Christensen TH (1999 b) Fate of MCPP and atrazine in an anaerobic landfill leachate plume (Grindsted, Denmark). Water Research 33: 2455–2458

Sturchio NC, Clausen JL, Heraty LJ, Huang L, Holt BD, Abrajano TA Jr (1998) Chlorine Isotope Investigation of Natural Attentuation of Trichloroethene in an Aerobic Aquifer. J Env Sci Tech 32: 3037–3042

Wiedemeier T, Wilson JT, Kampbell DH, Miller RN, Hansen JE (1995) Technical protocol for implementing intrinsic remediation with long-term monitoring for natural attenuation of fuel contamination dissolved in groundwater. Volume I, Air Force Center for Environmental Excellence, Technology Transfer Division, Brooks Air Force Base, San Antonio, Texas

Wiedemeier T, Swanson MA, Moutoux DE, Gordon EK, Wilson JT, Wilson BH, Kampbell DH, Hansen JE, Haas P, Chapelle FC (1996) Technical protocol for evaluating natural attenuation of chlorinated solvents in groundwater. Air Force Center for Environmental Excellence, Technology Transfer Division, Brooks Air Force Base, San Antonio, Texas

37 Natural Attenuation of Inorganic Pollutants – a Critical View

J. Gerth, F. Dankwarth and U. Förstner
Technical University Hamburg-Harburg, Department of Environmental Science and Technology, Eissendorfer Str. 40, 21071 Hamburg, Germany

37.1 Introduction

Natural Attenuation (NA) has just recently been re-assessed by the US National Research Council (NRC) and is considered as an established remedy with a high likelihood of success (at more than 75 % of contaminated sites) for only a few types of organic contaminants such as BTEX, low-molecular-weight alcohols, ketones and esters, and methylene chloride. For inorganics such as metals, metalloids, oxyanions and radionuclides, the likelihood of success is rated as moderate or low (at more than 50 % or at less than 25 % of contaminated sites, respectively, table 37.1). It is assumed that any given site will have the right conditions for natural attenuation of the particular contaminant. The ratings are based on field evidence and the current understanding of the attenuation processes (MacDonald 2000). It seems that natural attenuation as a remedy for inorganics, in particular, is not given much potential.

The impact of most inorganic pollutants is attenuated by sorption processes, dispersion, dilution, biotransformation, and in part also by volatilisation. Sorption processes are very effective and include adsorption/desorption (reversible binding at the solid/water interface), absorption (diffusion of pollutants into the solid matrix), precipitation and co-precipitation (incorporation into a freshly formed solid), and occlusion (covering of adsorbed pollutants during mineral growth). The most important factors for retention processes are pollutant concentration, the composition of the solid matrix, solution composition (e.g. complexing agents) and E/pH conditions (Brady and Borns 1997).

The NRC's view is that natural attenuation should be selected only when the mechanisms responsible for destroying or immobilising the contaminant are scientifically recognised, documented to be working now at the site, and sustained for as long as the contamination source is present (MacDonald 2000). Natural attenuation processes are always site specific, so every site needs to be evaluated individually. The accepted mechanisms for inorganics, including radionuclides, are immobilisation and/or biotransformation. For tritium, decay is acceptable. In most of the cases listed in table 37.1, the attenuation process is only moderately understood. Although recognised as being important, dilution and dispersion are

Table 37.1. Natural attenuation of inorganic chemicals: processes, level of understanding and likelihood of success (MacDonald 2000)

	Dominant attenuation processes	Current level of understanding	Likelihood of success
Metals:			
Ni, Cu, Zn, Pb	Immobilisation	Moderate	Moderate
Cd	Immobilisation	Moderate	Low
Cr	Biotransf., Immob.	Moderate	Low to moderate
Hg	Biotransf., Immob.	Moderate	Low
Metalloids:			
As, Se	Biotransf., Immob.	Moderate	Low
Oxyanions:			
Nitrate	Biotransformation	High	Moderate
Perchlorate	Biotransformation	Moderate	Low
Radionuclides:			
^{60}Co, ^{137}Cs	Immobilisation	Moderate	Moderate
^{3}H	Decay	High	Moderate
^{90}Sr	Immobilisation	High	Moderate
^{99}Tc	Biotransf., Immob.	Low	Low
$^{238,239,240}Pu$	Immobilisation	Moderate	Low
$^{235,238}U$	Biotransf., Immob.	Moderate	Low

not recommended under the regulatory definition of natural attenuation because of concerns of community advocates (MacDonald 2000).

In this article, we report research data on arsenic in model systems and in soils and sediments of an old tannery site. The applicabilitiy of the natural attenuation concept to inorganic pollutants is discussed in view of the detected binding forms and mobilisation processes.

37.2 Mechanisms and their Significance

Natural attenuation of inorganic pollutants is governed by two processes:

- release from the source, and
- attenuation of the mobilised forms.

Pollutant release from the source is determined by the type of binding. A rough classification would be differentiating between irreversible and reversible binding under natural conditions. The latter is in equilibrium with the dissolved species and is reduced with decreasing solution concentration. It is also affected by other changes in solution composition. Irreversibly bound pollutants are not in equilibrium with dissolved species. However, the characterisation of binding is difficult, because normally a clear distinction between irreversibly and reversibly bound forms is not possible. In most cases, binding strength lies between these two extremes. Characterisation becomes even more complicated when the integrity of the matrix depends on the E/pH conditions, as is the case with iron oxides.

On their way through the vadose zone, or during transport through aquifers, mobile pollutants are diluted by dispersion and can be retarded by coming into contact with reactive surfaces under appropriate geochemical conditions. For inorganic pollutants, which also occur as a natural background in non-polluted systems, dilution eventually leads to "natural" concentration levels. A great advantage of dilution can be seen in the fact that inorganic pollutants are more likely to become immobilised by sorption to high energy binding sites, which are limited in number. This is demonstrated in fig. 37.1, showing the pH dependent sorption of arsenate to hydrous ferric oxide (goethite) at different initial solution concentrations. In the system investigated, arsenate is 100 % bound at concentrations of up to 100 $\mu mol\ l^{-1}$. At higher levels, arsenate divides between sorbed and dissolved forms. The curves for initial solution concentrations of 500 and 1000 $\mu mol\ l^{-1}$ show a similar absolute sorption of about 150 $\mu mol\ g^{-1}$ at around pH 6.

The complete removal from solution at low initial concentrations indicates absorption. In another set of experiments at low initial solution concentration, arsenate was allowed to react with goethite for equilibration times of 16 hours and 7 days. In both cases, total sorption was again 100 %. Equilibration at different pH was followed by a 1-hour extraction using 1 M NaOH. As demonstrated in fig. 37.2, the quantity of arsenate not extracted is clearly greater after 7 days than after 16 hours of equilibration time during the sorption step. The difference is indicative of an arsenate immobilisation, probably by penetration into dead-end pores of the solid.

Similar effects can be observed with metal cations like nickel, zinc or cadmium when they react with silicate clay minerals and, in particular, iron oxides. (Gerth 1985; Gerth et al. 1993). This type of immobilisation indicates early diagenesis which is possible at low pollutant concentrations only. It can be viewed as the key mechanism for the natural attenuation of inorganics in contact with uncontaminated surfaces. But can this mechanism serve as a basis for a remediation concept? An answer to this question can only be obtained from observations in the field and the analysis of contaminated solids.

37.3 Arsenic in Contaminated Soils and Sediments

37.3.1 Contaminated Sites and Binding Forms

Former infiltration fields and sludge deposits of a tannery were investigated for arsenic binding forms and mobility. The contaminated land of 15 ha is located in the periphery of a town and has been a nature reserve for the last 30 years. Arsenic was used as a preservative and, at an early stage in the production history, also as a reducing agent (As_2S_3) in the dehairing and liming step of the tanning process. In the subsequent neutralisation step, arsenic was washed off and discharged with the effluent.

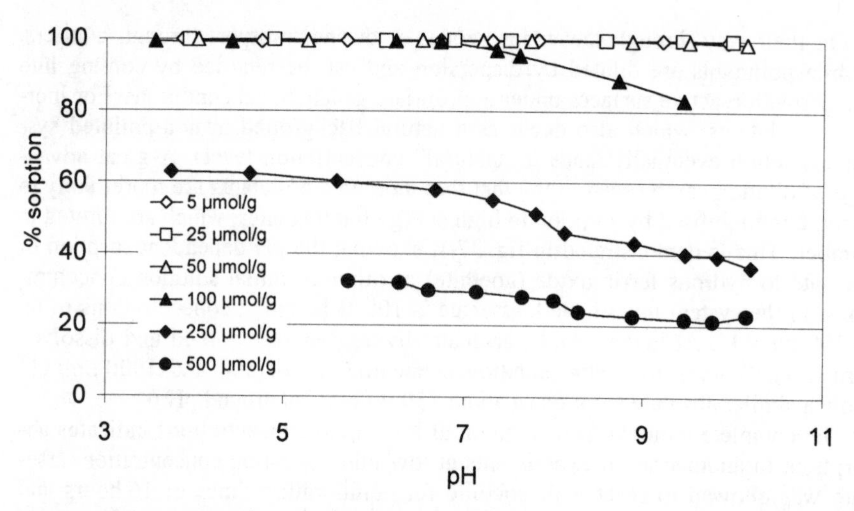

Fig. 37.1. Relative sorption of arsenate to hydrous ferric oxide (goethite: surface area 60 m^2 g^{-1}, 2 g l^{-1}; reaction time 16 hours) as a function of pH and total arsenate concentration (between 5 to 500 µmol g^{-1} equivalent to initial solution concentrations of 10, 50, 100, 200, 500 and 1000 µmol l^{-1}, respectively)

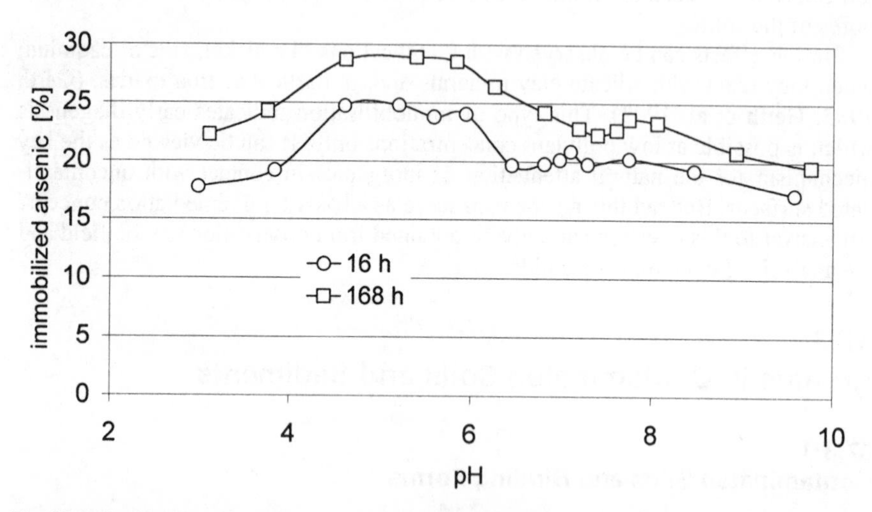

Fig. 37.2. Residual sorption of arsenate (initial concentration 1 µmol l^{-1}) to hydrous ferric oxide (goethite: surface area 60 m^2 g^{-1}, 2 g l^{-1}; reaction time 16 hours and 7 days) after 1-hour extraction with 1 M NaOH versus the pH in the equilibrium solution

Arsenic can be found in different binding forms. Part of the pollutant is bound in a *calcite* rich sludge, which is partly water saturated because it was deposited as a 4 m layer in a 2 ha wetland section of the site. The material is highly contaminated with up to 2000 mg kg^{-1} arsenic and is the major source of groundwater contamination in the area. In a downgradient monitoring well, 120 μg l^{-1} arsenic was found.

Calcitic waste material containing 240 mg kg^{-1} arsenic was also deposited under non-saturated conditions in a former irrigation ditch, and has been there for about 6 decades. The underlying material is a fossil Bs horizon of a former podzol. This horizon contains only background levels of arsenic and has therefore not received any contaminant from the matrix above. In the presence of calcium, arsenic is probably bound as a stable calcium arsenate phase (Sadiq 1997). Bothe and Brown (1999) postulate the formation of arsenate apatite.

Infiltration fields (10 ha) and *settling ponds* (3 ha) were constructed for wastewater treatment on glaciofluvial sand about 4 to 6 m above groundwater. The fields are polder type basins (e.g. 30 by 50 m) arranged at different levels to both sides of a central irrigation ditch in the highest part of the area (elevation: 22 m above sea level). Most of the area is covered with grass and shrubs. The dominant soil type is cambic podzol. An unknown amount of arsenic was discharged with the wastewater which contaminated the soil profile. In the topsoil (Ah horizon), 40–90 mg kg^{-1} arsenic is bound to *organic matter,* consisting of humic material and sedimented fine leather particles. The subsoil (Cs horizon) contains *hydrous ferric oxide* bound arsenic with concentrations between 100 and 500 mg kg^{-1}. The iron oxide phase is poorly crystalline goethite. Both iron and arsenic become mobilised to some extent under mildly reducing conditions after heavy rainfalls, and form co-precipitates upon re-aeration in the lower soil profile.

It would be expected that dilution increases with the distance from the pollutant source, however, at the bank of a brook receiving the contaminated groundwater of the former infiltration fields, a massive precipitate of *ferrihydrite* can be found containing 200 to 400 mg kg^{-1} arsenic. Reduced iron and mobile arsenic form a co-precipitate when they come into contact with air. This demonstrates that, even at distances of several hundred meters from the source, arsenic can accumulate by co-precipitation upon sudden changes in the geochemical conditions.

At a depth of about 8 m in the water saturated zone, the conditions are sulfidic. The subsoil in parts of the settling ponds also contains *sulfides* due to a silt layer which hinders oxygen from penetrating to the lower part of the profile. Arsenic was not abundant in these highly reduced substrates. Apparently, this pollutant does not form sulfidic phases which are stable against long term leaching. Kim et al. (2000) suggest that the carbonation of arsenic sulfide minerals, including orpiment (As_2S_3) and realgar (As_2S_2), is an important process in leaching arsenic into groundwater under anaerobic conditions. It is believed that arseno-carbonate complexes of types $As(CO_3)_2^-$, $As(CO_3)(OH)_2^-$, and $AsCO_3^+$ are formed (Lee and Nriagu 2000). Such processes may also be responsible for the mobilisation of arsenic in the calcitic matrix under reducing conditions.

37.3.2 Assessment of Source Properties

The field data demonstrate that arsenic retention depends on the binding stability in the contaminant source material. There is a strong arsenic release under saturated conditions in particular. In order to assess the role of natural attenuation, different procedures were applied to characterise potentially mobile and immobile fractions in soil samples and sediments of the old tannery site. The results also serve as a basis for producing measures to minimise arsenic leaching.

37.3.2.1
Methods

Soil samples were taken from 4 profiles on the infiltration fields and mixed to give a representative sample of each the Ah and Cs horizon. Samples of the ferrihydrite and the calcitic waste material were taken from a single location. Total contents were determined by aqua regia extraction in a microwave system. For mineral composition, samples were investigated using X-ray diffraction.

Extraction agents (H_2O, and oxalate/ascorbic acid (pH 3.25)) were used in combination with column leaching tests under non-saturated and saturated conditions. Moisture content was varied in column tests to account for intermittent wetting and drying in the field. The major proportion of arsenic is bound by iron oxides present in different modifications. This mineral fraction was characterised for stability against reductive dissolution using a modified version of the method of Postma (1993), with 40 ml of 0.1 M ascorbic acid to extract 2 g batches of sample for different shaking times. The results give an estimate of the arsenic mobilisation potential during temporary water saturation in the vadose zone. Arsenic mobility, as a function of pH, was studied in batch experiments with 4 g of air-dried soil suspended in 40 ml of water, and additions of HNO_3 or NaOH to obtain pH values between 3 and 9 after 18 hours shaking. A separate series of acid/base additions was conducted in the presence of 0.01 M $CaCl_2$. This was to test if calcium can be used to reduce arsenic mobility. After filtering through a 0.45 µm membrane filter, the solutions were analysed by hydride generation graphite furnace AAS.

37.3.2.2
Mobility of Sorbed Arsenic

Calcitic Waste: Effect of Reduction. Field data of the calcitic waste demonstrated that arsenic is bound in an immobile form when the material is deposited under non-saturated conditions. In order to test the stability of the binding against reducing conditions, 1 kg of the surface material, contaminated with 240 mg kg^{-1} arsenic, was mixed with 2 kg of water and placed in a sealed 4 l container. Arsenic in solution increased from zero to 85 µg l^{-1} within 60 days. Mobile iron and organic carbon indicate that reduction processes take place (fig. 37.3). This is comparable with the arsenic concentration of 120 µg l^{-1} in the groundwater sample from the highly contaminated landfill, which itself contained up to 2 g kg^{-1} arsenic. According to Dutré and Vandecasteele (1998), mobile arsenic is in equilibrium with calcium arsenite ($CaHAsO_3$) under these conditions. The experiment demonstrates the conditional character of arsenic immobilisation, which is only guaranteed if this material is kept non-saturated.

Soil Samples: Extraction with Water. Data of the contaminated soil samples Ah, Cs and the ferrihydrite sample from the river bank are given in table 37.2, together with the results of a batch extraction experiment. Water extractable arsenic from samples Ah and Cs clearly exceeds the permissable regulatory value of

10 µg l^{-1}. A column experiment, carried out on sample Cs under non-saturated conditions, yielded leachate concentrations between 35 and 50 µg l^{-1}. In this case, the results of the batch water extraction are similar. Therefore, this simple technique seems appropriate for the assessment of arsenic leachate concentration in this type of matrix. The low content of water extractable arsenic in ferrihydrite refers to the incorporation of arsenic by co-precipitation, whereas in samples Ah and Cs, the higher content of extractable arsenic is indicative of a high proportion of adsorbed forms.

Soil Samples: Simulation of Reducing Conditions. In the vadose zone, wetting can induce mildly reducing conditions, leading to an increase in arsenic mobility. Easy-to-mobilise arsenic was analysed by simulating reducing conditions using 0.1 M ascorbic acid. Fig. 37.4 shows the reductive dissolution of sample Cs and the ferrihydrite sample in a batch extraction experiment.

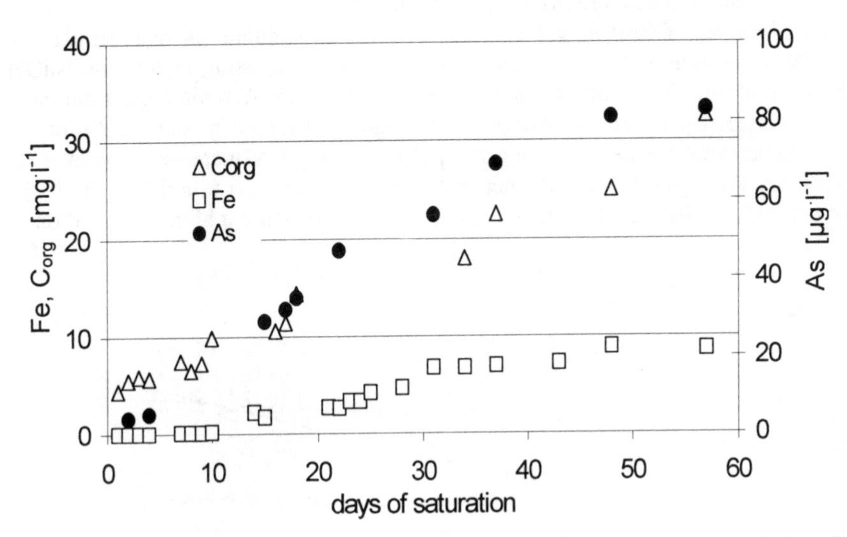

Fig. 37.3. Mobilisation of arsenic, iron and organic carbon from calcitic tannery sludge after water saturation as a function of time

Table 37.2. Selected data of soil substrates (from Dankwarth et al. 2000)
$As/Fe_{asc/ox}$ extractable in hot ascorbic acid/oxalate, *As_t* "total" arsenic in microwave extract with aqua regia, *As_{H2O}* water extractable arsenic according to DIN 38414 (S4 test), *C_t* total carbon, the *pH_{CaCl2}* in all samples was between 4.6 to 5.0

Sample	Depth [cm]	C_t [%]	$Fe_{asc/ox}$	$As_{asc/ox}$ [mg kg^{-1}]	As_t	As_{H2O} [µg l^{-1}]
Ah	0 – 30	5.5	2563	32.5	37.3	26.2
Cs	120 – 150	0.13	12406	93.8	109	35.0
Ferrihydrite	0 – 10	n.d.	60806	214	224	1.4

Soil Samples: Simulation of Reducing Conditions. In the vadose zone, wetting can induce mildly reducing conditions, leading to an increase in arsenic mobility. Easy-to-mobilise arsenic was analysed by simulating reducing conditions using 0.1 M ascorbic acid. Fig. 37.4 shows the reductive dissolution of sample Cs and the ferrihydrite sample in a batch extraction experiment.

Potentially mobile arsenic concentrations can be estimated from the change in high-rate to low-rate dissolution kinetics. About 23 % and 60 % of total arsenic in sample Cs and the ferrihydrite sample, respectively, is mobilised during the fast dissolution step within 7 days. This means that ferrihydrite is essentially labile. In comparison, sample Cs seems relatively stable even though it still releases considerable amounts of arsenic, even during short periods of reduction. In this case, the easy-to-mobilise arsenic is bound to a ferrihydrite fraction, while the immobile arsenic is associated with slowly dissolving goethite. The results confirm the view that the mechanism responsible for arsenic mobilisation is the reductive dissolution of hydrous ferric oxide (McCreadie et al. 2000).

Soil Samples: Effect of pH and Ca. In order to simulate changes in pH, the samples were equilibrated at different pHs for 48 hours, using HNO_3 and NaOH for pH titration. The results are summarised in fig. 37.5. Arsenic concentration is lowest between pH 4 and 6. Almost no change is observed in sample Ah or the ferrihydrite sample when lowering the pH to 3. Mobility increases, however, in sample Cs. At higher pH, a little increase of arsenic release is found for sample Ah and the ferrihydrite sample, whereas sample Cs shows again a high mobilisation.

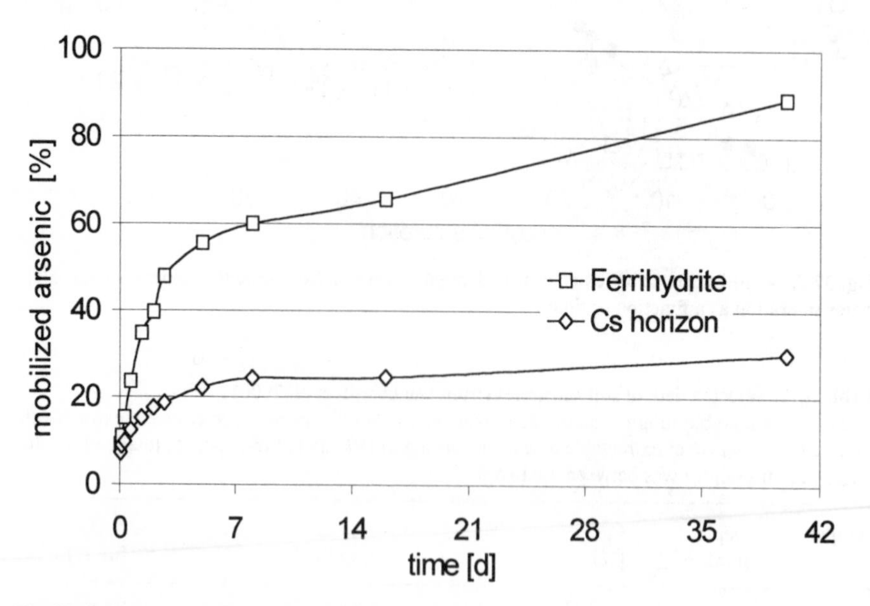

Fig. 37.4. Percent arsenic release during reductive dissolution of ferric hydroxide rich samples using 0.1 M ascorbic acid

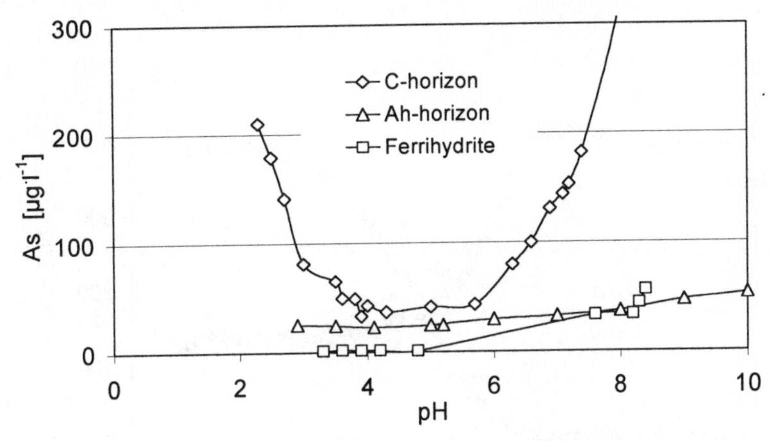

Fig. 37.5. Arsenic concentration after 24 hours of equilibration for different soil samples as a function of pH

Arsenic mobility is reduced by a factor of at least 4 when adding 0.01 M $CaCl_2$, as is demonstrated with sample Cs in fig. 37.6. Similar results were obtained in 0.1 M calcium solution. This effect is also observed with organic rich topsoil material and when using other alkaline earth metal ions. It can probably be related to a complex formation of calcium with arsenate on hydroxyl groups of organic and oxidic surfaces. Precipitation of a defined solid is unlikely because the loss in solution concentration is not complete.

The reduction of arsenic concentration using 0.01 M $CaCl_2$ could also be observed in a column experiment on sample Cs under non-saturated conditions (fig. 37.7).

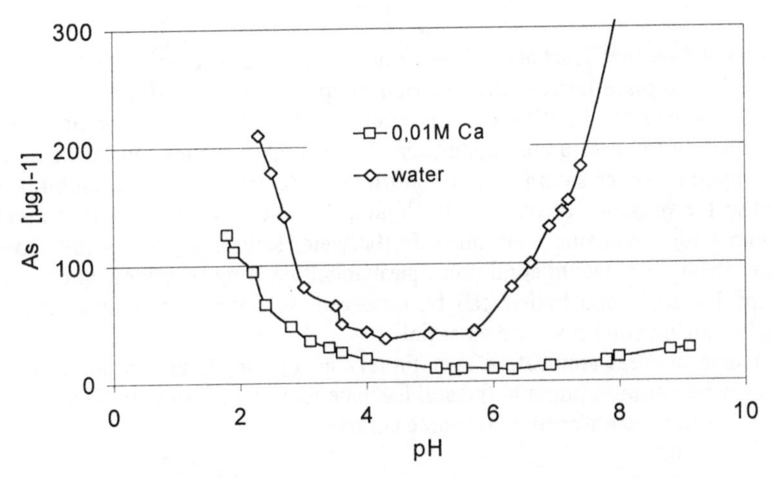

Fig. 37.6. Arsenic concentration in solution after equilibrating sample Cs for 24 hours in water and in 0.01 M $CaCl_2$ at different pHs

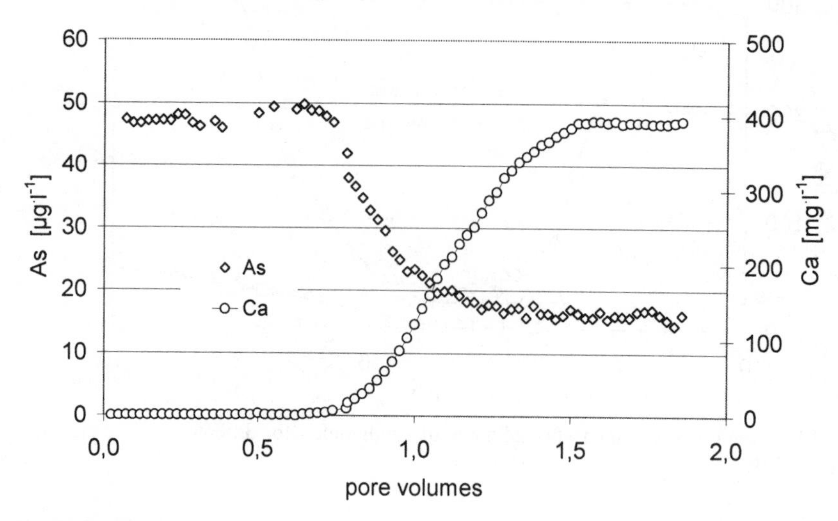

Fig. 37.7. Effect of 0.01 M $CaCl_2$ solution on arsenic concentration in column seepage of iron oxide rich sample Cs

The results suggest that calcium could be applied as part of source control measures. As can be seen in fig. 37.7, arsenic is reduced from 50 to 15 µg l^{-1}, which is close to the regulatory value of 10 µg l^{-1}. Arsenic in contaminated soil can probably be stabilised by liming and by providing the highest possible level of calcium in soil solution.

37.4 Natural Attenuation of Arsenic

It could be shown that leachates of moderately contaminated soil substrates, including highly sorptive ferric hydroxide rich samples with up to 100 mg kg^{-1} arsenic, clearly exceed the limiting concentration of 10 µg l^{-1} by a factor of 2 to 4 under aerated and non-saturated conditions. Much higher values can be expected during temporary water saturation. A significant reduction of arsenic mobility is observed in the presence of dissolved calcium. Arsenic is well retained in a calcitic matrix under oxidising conditions. In the water saturated zone of this material, under strongly reducing conditions, permanent leaching of arsenic at concentrations of 100 µg l^{-1} and higher can be observed. In order to minimise arsenic leaching, oxidising conditions are essential.

For arsenic, natural attenuation is definitely no option. If no action is taken, regulatory concentration limits in the soil leachate and groundwater will be clearly exceeded. There is no alternative to source control.

37.5 Critical Remarks

For the acceptance of natural attenuation, it is a major obstacle that inorganic pollutants cannot be decomposed and remain in the subsurface. The persistence of geochemical conditions favourable to strong sorption is not guaranteed for long periods of time. Reaction conditions are often generated by the waste material itself (e.g. carbonate matrix or sulfidic conditions). In this case, they are not in equilibrium with the surroundings and cannot be termed "natural". Due to a geochemical gradient, the area with optimum reaction conditions may shrink permanently and cause an ongoing release of pollutants. It is difficult to predict such processes for longer time-spans. If retention fails, dilution will be the only mechanism capable of meeting the regulatory target. The total amount of pollutant remains unchanged. Dilution is not the only result of pollutant transport. As demonstrated, inorganic pollutants can also accumulate upon sudden changes in the ambient geochemistry.

The definition of immobilisation is particularly critical to the assessment of natural retention as a remediation strategy. The degree of immobilisation is difficult to determine. Methods are required that consider the properties of the particular element and matrix, like leaching, chemical extraction and digestion. Although the results may be powerful, they are of limited use only for the assessment of future development. Generally, even those fractions that are identified as irreversibly bound have to be considered potentially mobile over long periods of time. Eventually, inorganic pollutants are attenuated to the natural background concentration. A fast mobilisation of "immobile" fractions can occur upon sudden changes in water saturation or by the introduction of reactive compounds (organic matter). These aspects are obstacles to the implementation of natural attenuation for inorganic pollutants (Brady and Borns 1997).

Natural attenuation of inorganic pollutants should, if at all, be implemented only as a target approach based on clearly defined standards. For some inorganic pollutants, regulatory targets can probably be achieved on the basis of "natural" processes under appropriate environmental conditions. These standards are quality criteria for the soild matrix and the aqueous phase. Dilution can only be allowed to the extent that regulatory concentration limits are not exceeded at a certain point of compliance. A site investigation is the basis for the development of a conceptual model of natural attenuation. Modelling is connected with a high degree of uncertainty because long-term reactions cannot be ascertained in advance.

Remediation concepts for inorganic pollutants should not be based on "natural" processes alone. The behaviour of most inorganics in the soil environment and in aquifers is at least moderately understood. Therefore, "active" measures should be taken to preserve and improve the retention potential of the matrix. As could be demonstrated, the retention of arsenic in contaminated soil can be easily improved by slight changes in the geochemical reaction conditions.

References

Bothe JV, Brown PL (1999) Arsenic immobilisation by calcium arsenate formation. Environmental Science and Technology 33: 3806–3811

Brady PV, Borns DJ (1997) Natural attenuation of metals and radionuclides: Report from a workshop held by Sandia National Laboratories. Sandia Report Sand97-2727, UC-800, Albuquerque, New Mexico 87185

Dankwarth F, Gerth J, Förstner U (2000) Minimise arsenic in contaminated soil as a natural attenuation approach. International Conference on Heavy Metals in the Environment, 6–10 August, University of Michigan, Ann Arbor, Michigan, USA

Dutré V, Vandecasteele C (1998) Immobilisation Mechanism of Arsenic in Waste Solified Using Cement and Lime. Environmental Science and Technology 32: 2782–2787

Gerth J (1985) Untersuchungen zur Adsorption von Ni, Zn und Cd durch Bodentonfraktionen unterschiedlichen Stoffbestandes und verschiedene Bodenkomponenten. Dissertation Universität Kiel, Germany, p 267

Gerth J, Brümmer GW, Tiller KG (1993) Retention of Ni, Zn and Cd by Si-associated goethite. Z Pflanzenernähr Bodk 156: 123–129

Kim MJ, Nriagu J, Haack S (2000) Carbonate ions and arsenic dissolution by groundwater. Environmental Science and Technology 34: 3094–3100

Lee JS, Nriagu JO (2000) Arsenic Carbonate Complexes in Groundwater. International Conference on Heavy Metals in the Environment, 6–10 August, University of Michigan, Ann Arbor, Michigan, USA

MacDonald JA (2000) Evaluating natural attenuation for groundwater cleanup. Environmental Science and Technology 34: 346A–353A

McCreadie H, Blowes DW, Ptacek DJ, Jambor JL (2000) Influence of reduction reactions and solid-phase compositions on porewater concentrations of arsenic. Environmental Science and Technology 34: 3159–3166

Postma D (1993) The reactivity of iron in sediments: a kinetic approach. Geochimica et Cosmochimica Acta 57: 5022–5034

Sadiq M (1997) Arsenic chemistry in soils: An overview of thermodynamic predictions and field observations. Water, Air and Soil Pollution 93: 117–136

Appendix

38 Determination of Mineral Oil Hydrocarbons by Means of Thin-Layer Chromatography – Analytical Directions

C. Reimers
Technical University of Hamburg-Harburg, Department of Waste Management, Harburger Schloßstraße 37, 21079 Hamburg, Germany

38.1 Range of Application

The detection limit of thin layer chromatography is at 0.1 µg absolute for squalane and lubricating oil respectively, and at 0.4 µg absolute for diesel fuel. Under the conditions indicated below, this equates to detection levels of 170 mg kg^{-1} squalane and lubricating oil, and 670 mg kg^{-1} diesel fuel. Smaller contents can be determined by spreading greater volumes. At a spread volume of 20 µl 8 mg kg^{-1} lubricating oil and 30 mg kg^{-1} diesel fuel can be detected.

38.2 Principle of the Method

The determination of medium and low volatile mineral oil hydrocarbons (aliphatics and naphthene) in soil materials is based on their separation from other substance classes such as aromatic hydrocarbons, lipids and other compounds and subsequent determination by summation. The extraction of the soil samples is carried out using a mixture of *n*-hexane and acetone (1:1) with ultrasonic support. The total petroleum hydrocarbons are developed on silica gel 60 phases in *n*-hexane and appear after colouring as *one* spot at a hRf-value of about 80. Polycyclic aromatics are recognisable distinctly from the aliphatic compounds. Prerequisite for the in-situ-detection of the aliphatics on silica gel layers is their visibility caused by suitable colouring agents.

As detection reagents suitable are anilinnaphalinsulfonic acid (Gitler 1972), bromothymolblue (Goebgen and Brockmann 1977), sodiumfluoresceinate (Mamlock 1981), berberinechloride (Jork et al. 1990), such as acid violet (Engelhardt and Engel 1997). The evaluation can be carried out visually or densitometrically by means of a thin-layer-scanner.

38.3 Chemical Reagents

- Hexane,
- acetone,
- diesel fuel (1): 200 mg in 10 ml hexane or heptane/acetone, 1:100 to be diluted (0.2 $\mu g\ \mu l^{-1}$),
- lubricating oil (2): 200 mg in 10 ml hexane or heptane/acetone, 1:200 to be diluted (0.1 $\mu g\ \mu l^{-1}$),
- methanol,
- ethanol,
- citric acid mono hydrate,
- anilinonahphalenesulfonic acid: 100 mg 8-anilinonaphthalene-1-sulfonic acid as ammoniumsalt are dissolved in the following mixture and well shaken: a) 20 ml of a 0.2 M NaOH-solution and 20 ml ethanol, b) 57 ml of a solution of 2.1 g citric acid monohydrate and 0.8 g NaOH-flakes in 100 ml water.

38.4 Equipment

- Ultrasonic device,
- spreading device with syringe,
- centrifuge glasses: 50 ml with screw lid and PTFE-coated lid inlays,
- laboratory centrifuge,
- HPTLC-plates: silica gel 60 phases with concentration zone, clean-up in hexane or heptane, subsequently in methanol or *i*-propanole, then 10 min of air-drying,
- TLC-development chambers,
- UV-observation device,
- TLC-device for plate immersion,
- hair-dryer,
- plate heater,
- TLC-scanner.

38.5 Extraction

- 5–10 g homogenised naturally humid soil to be accurately weighed into a 50 ml centrifuge glass with screw lid (Teflon sealing). Drying of the soil is not necessary.
- 10 or 20 ml of a mixture consisting of hexane (or heptane) /acetone 1/1 are to be added. *n*-Hexane, *i*-hexane, cyclohexane can be used but *n*-heptane is less neurotoxic.
- Shake vigorously and extract in an ultrasonic bath (30 min, full performance). Force fields, water level and temperature of the ultrasonic bath are to be care-

fully monitored! They should all be constant as changes in extraction parameters cause irreproducible results.

- Centrifuge for 5 min at 5000 rpm, decant supernatant.
- Use directly or dilute according to concentration.
- Each series of analyses should contain determination of a sample blank with acetone/heptane.

38.6 Thin-Layer Chromatographic Analysis

38.6.1 Separation System

For the analyses high performance thin-layer (HPTLC) plates with a concentration zone are used. This results in better separation performance at smaller spot diameters as well as lower detection sensitivity due to better concentration compared to common thin-layer plates. This is due to smaller grain size and a more dense grain size distribution. Plates with concentration zones can have a clean-up effect when the matrices are complex.

HPTLC silica gel 60 phases are developed in hexane or heptane at chamber saturation. Maintaining chamber saturation is very important otherwise the mineral oil hydrocarbons move up to the solvent front.

38.6.2 Spreading

The extracts and standards are spread out by a device which uses nitrogen at the concentration zone. The distance from the lower plate margin is 10 mm and 5–10 mm at the sides. The distance between the bands and the length of the bands are 3 mm each. The amount of extract to be spread out is variable and is used directly or diluted according to the respective concentration as a standard 20 µl is spread. For manual spreading 0.5 µl fix-volume-capillaries are used and a volume of 1 µl is spread out.

0.2–3 µg absolute are spread of the diesel fuel on the plate, respective 1–15 µl (1), or 0.1–2 µg absolute of the lubricating oil are spread respective 1–20 µl (2). The extracted contamination is spread out analogously (according to the type of contamination).

38.6.3 Standards

Usually a diesel fuel, lubricating oil or a mixtures of both are used as reference substances. For optimum performance the standard should be very similar, or hopefully identical to the contamination. If the contamination is extracted out of the soil and is used as a calibration substance the concentrations to be spread out are selected according to the type of contamination. Higher concentrations are necessary for contaminants of higher volatility (see diesel fuel) and lower concentrations for contaminants of lower volatility (see lubricating oil).

38.6.4 Development

The plates are developed in double-trough chambers or chambers with a V-shaped bottom to reduce the mobile phase. The chamber is conditioned prior to use with solvents for a sufficient time span. During conditioning the chamber is lined with filters that are saturated with the solvent. The plate is developed as long as the solvent front appears to be about 1.5–2 cm under the upper plate margin. The front is marked with a pencil and the plate is air-dried. The development in non polar solvent is executed relatively quickly (about 5–10 min).

38.6.5 Derivatisation

After the development the plates are air-dried for 10 min, then heated for another 10 min at 105 °C on a plate heater and then immersed into anilinonaphthalenesulfonic acid for 1 second.

Heating is necessary for derivatisation in order to be able to guarantee that the solvent is completely eliminated, otherwise the whole plate fluoresces during evaluation. The principle of the derivatisation is the accumulation of the colouring agents onto lipophilic components and thus a change in the fluorescence behaviour of the colouring agents.

38.6.6 Evaluation

The TLC-plate is subjected to a visual observation by means of an UV-observation device. In this manner an overview of the type and the extent of the contamination can be firstly obtained. Furthermore it is possible to recognise irregularities on the TLC-plate that must be considered during measurement. Subsequently the plates are evaluated either densitometrically at a fluorescence excitation of 366 nm/400 nm or visually by estimation of the content.

38.6.7 Calculation

$$C = \frac{100 \cdot Content_{Extract} \cdot 10^{3} \cdot f \cdot (VF)}{Weight_{Soil} \cdot V_{Extraction} \cdot TM} \qquad (38.1)$$

C	TPH-content [mg (kg solid matter)$^{-1}$]
TM	dry mass
VF	dilution factor (if the extract was diluted)
F	sprayed volume [µl]
V	extraction volume [ml]
Weight	weight of soil [g]

References

Engelhardt H, Engel P (1997) J of Planar Chromatogr 10: 336–341

Goebgen HG, Brockmann J (1977) Die Anwendung der Dünnschichtchromatographie in der Abwasseranalytik am Beispiel der Bestimmung schwerflüchtiger Mineralöl-Kohlenwasserstoffe. Vom Wasser 48: 167–178

Jork H, Funk, Fischer, Wimmer (1990) Thin-Layer Chromatography, Reagents and Detection Methods. Vol 1a, VCH-Verlag, Weinheim, Germany, pp 213–215

Lepom P (1997) Bestimmung von Mineralölkontaminationen in Böden und Abfall – Entwicklung eines gaschromatogaphischen Analyseverfahrens. In: Borsdorf H, Flachowsky J (eds) UFZ-Bericht 12/97, pp 41–48

Mamlock M (1981) Berberine Hydrochloride for Detection in Thin-Layer Chromatography. J Chromatogr Sci 19: 53

39 Determination of the PAH-Content in Soils by HPTLC – Analytical Directions

C. Reimers
Technical University of Hamburg-Harburg, Department of Waste Management, Harburger Schloβstraße 37, 21079 Hamburg, Germany

39.1 Range of Application

PAH determination by means of HPTLC is a screening method for the determination of the PAH in soil materials. PAH-contents of more than 1.6 mg kg^{-1} can be determined under the conditions indicated under 39.5 and 39.6. Lower concentrations can be determined by concentration of the eluate and spreading of greater volumes.

39.2 Measuring Principle

A defined entity of the homogenised, wet sample is extracted by means of acetone/*n*-hexane with ultrasonic extraction. The organic phase is separated, washed with water and dried. Subsequently the extract is further cleaned by means of solid-phase-extraction on silica gel which separates the PAHs from the mineral oil hydrocarbons (TPH). The qualitative and quantitative determination of the PAH-content is carried out by means of a thin-layer chromatographic separation on RP-18 phases with fluorescence detection. The peak areas and content of the different PAHs are determined against an external PAH-standard.

39.3 Interference and Retaliatory Actions

Fluorescing soil components can be falsely detected. For this reason a clean-up of the extract by means of solid-phase extraction on activated silica gel is recommended.

During extraction of the soil material it can, under certain circumstances, come to the formation of an emulsion. As a preventative measure it is recommended to pre-treat the extract by centrifuge and to separate the organic phase by means of a Pasteur-pipette.

The extract has to be washed twice with 6 ml of water to eliminate the acetone. During the quantitative analysis of highly contaminated samples superposition of substance response can occur so that a higher analysis result may be obtained. In this case the definition of the peaks is no longer guaranteed. Therefore the extract should be diluted before it is reanalysed .

39.4 Equipment

- Molar sieve for the drying of the solvents,
- TLC-scanner,
- spreading device with syringe,
- centrifuge glasses: 50 ml with screw lid and PTFE-coated lid inlays,
- laboratory centrifuge,
- TLC-plates: RP-18 phases with concentration zone without fluorescence indicator, clean-up in hexane or heptane, subsequently in methanol or *i*-propanole, then 10 min of air-drying,
- development chambers,
- UV-observation device,
- TLC-device for plate immersion,
- SPE-glass columns (8 ml),
- ultrasonic device,
- freezer.

39.5 Extraction and Clean-up

5–10 g of the homogenised wet soil sample is weighed into a 50 ml centrifuge glass and 10 or 20 ml of acetone followed by a similar quantity of *n*-hexane are added and the whole is shaken vigorously. Next the samples are extracted for 30 min. in an ultrasonic bath at the highest setting and then centrifuged for 5 min at 5000 rpm. The organic phase is separated, washed twice with 6 ml of water and then 2–3 spatulas of sodium sulfate are added before being well shaken. The dried extract is then cleaned by means of SPE via activated silica gel (glass columns) (use solvents that have been dried over a molar sieve). Finally the column is equilibrated with *n*-hexane and the PAHs are eluted with *n*-hexane/dichloromethane (1/1; v/v) from the column.

39.6 Spreading of the Extracts/Standards

The spreading-out onto the TLC-plate is carried out by means of a spraying device. 1–9 µl of the TVO-standard and 6–14 µl of the mixed standard are spread out for calibration *(superposed on the same band)*. 1–10 µl of the soil extract is sprayed according to their PAH-content – this can be determined from the colour of the respective extract. If necessary, greater amounts can be sprayed. For highly contaminated samples it is recommended to dilute the extract first.

39.7 TLC-Development

After spreading the TLC-plate is pre-cooled in the freezer. The development is carried out in three sucessive solvents at -20 °C (chamber saturation):

- *n*-hexane,
- *n*-hexane/*tert.* butylmethylether (80/20; v/v),
- acetonitrile/dichloromethane/water (9/1/1; v/v/v).

All solvents are cooled for at least 10 min. in the freezer before application. After development the TLC-Platte is immersed for *1 s* into *n*-hexane/paraffin (4/1; v/v), in order to increase the intensity of fluorescence and to stabilise the PAH on the plate. The plate is dried with cold air (hair-dryer) immediately after immersion and can then be used for further analyses.

The third solvent should not be stored for more than 45 min. at a temperature of -20 °C, otherwise the water will crystallise out of solution.

39.8 Thin-Layer Chromatographic Analysis

The TLC-plate is subjected to a visual observation by means of an UV-observation device. In this way an initial overview over the type and the extent of the contamination can be obtained. After the visual observation of the TLC-plate a densitometrical evaluation by means of a selective fluorescence detection is carried out (table 39.1). The working ranges and detection limits are listed in table 39.2.

Table 39.1. Selected measuring wave lengths for the quantitative determination of PAH in soil materials

Measuring wave length/filter	PAH to be determined
366 / K 400 nm	BAP / IPY / BKF / BBF / BGI and FLA
254 / K 340 nm	BAA and CHR
265 / M 360 nm	PYR / FLA and ANT / PHE
302 / K 400 nm	DAH / IPY
265 / K 340 nm [a]	ACE / FLU

[a] ACE and FLU are only determined qualitatively for the indicated wave length.

Table 39.2. Working ranges and detection limits

Analyte	Working range [ng]	Detection limit [ng absolute]
DAH, IPY	11.2–18.6	2.5
IPY	0.8– 3.6	0.2
BGI	0.8– 3.6	0.2
BAP, BKF	0.8– 3.6	0.2
BBF	0.8– 3.6	0.3
FLA	4.0–18	1
CHR, BAA	12.9–30	2
FLA, PYR	14.3–26.5	5
ANT, PHE	12.6–29.5	5

39.9 Calculation

$$C = \frac{100 \cdot Content_{Extract} \cdot 10^{-6}\, mg \cdot f_1 \cdot f_2 \cdot (VF)}{Weight_{soil} \cdot 10^{-3}\, kg \cdot TM} \qquad (39.1)$$

C	PAH-content [mg kg^{-1} $_{solid\ matter}$]
TM	dry mass
VF	dilution factor (if the eluate was diluted)
100	Factor for the conversion into dry mass
f_1	Factor to take into account the spread-out volume in µl of total eluate
f_2	Factor to take into account the volume that is used during SPE in ml of total extract

39.10 PAH-Abbreviations

Table 39.3. PAH-abbreviations

Substance	Abbreviation	Substance	Abbreviation
Dibenz[ah]anthracene	DAH	Benz[a]anthracene	BaA
Indeno[1,2,3-cd]pyrene	IPY	Pyrene	PYR
Benzo[ghi]perylene	BGI	Fluoroanthene	FLA
Benzo[a]pyrene	BaP	Anthracene	ANT
Benzo[k]fluoroanthene	BkF	Phenanthrene	PHE
Benzo[b]fluoroanthene	BbF	Acenaphthene	ACE
Perylene	PER	Fluorene	FLU
Chrysene	CHR		

39.11 Standard Concentrations

39.11.1 Standard According to German Drinking Water Regulation (PROMOCHEM, PAH 7)

The PAH-content is 2 μg ml^{-1} for IPY, BGI, BkF, BbF and BaP or 10 μg ml^{-1} for fluoroanthene. A 1:5 dilution is used in the analyses so that 1–10 μl of this solution is 0.4–4.0 ng for IPY, BGI, BkF, BbF and BaP or 2–20 ng of fluoroanthene in absolute terms. For the calibration 1–9 μl were used.

39.11.2 Mixed Standard

The mixed standard contains the PAHs and concentrations as listed in table 39.4. For calibration 6–14 μl was used for the spread out volume.

Table 39.4. PAH-concentrations and volume to be spread out

Substance	Concentration [μg 10 ml^{-1}]	Spread out volume [μl]	PAH-content [ng]
DAH	10.7	1–10	1.1–10.7
BaA	10.8	1–10	1.1–10.8
CHR	10.7	1–10	1.1–10.7
PYR	10.4	1–10	1.0–10.4
ANT	10.6	1–10	1.1–10.6
PHE	10.5	1–10	1.1–10.5
ACE	11.4	1–10	1.1–10.3
FLU	11.3	1–10	1.1–10.2

40 Determination of the Total Petroleum Hydrocarbon Content in Soils by Gas-Chromatography

C. Reimers
Technical University of Hamburg-Harburg, Department of Waste Management, Harburger Schloβstraße 37, 21079 Hamburg, Germany

40.1 Introduction

Due partly to careless handling of mineral oils and mineral oil products during transport, storage and disposal, soil is often contaminated and ground water threatened. To be able to assess the extent of the pollution, to identify the contaminants and to evaluate the efficiency of a certain treatment process, extensive chemical analyses are necessary.

Since analysing the individual substances is time-consuming and costly, it is common practice to analyse summation parameters. For this reason, various analytical methods are often applied simultaneously for a rapid and appropriate examination.

When analysing petroleum, it is important to determine the total petroleum hydrocarbon content (TPH). The content of hydrocarbons originating from petroleum is commonly determined as a summation parameter by means of IR-spectroscopy using fluoro-chlorinated hydrocarbons following the DEV H18 procedure (DIN 38409 1981).

It can be seen as a great disadvantage though, that perhalogenated solvents must be used in this procedure. Furthermore, differentiation is hardly possible between different oils, and it is not much better between anthropogenic and biogenic hydrocarbons. Statements about the boiling range of contaminants are not possible either. Such statements are only possible after at least partial separation of the complex hydrocarbon mixture. Because of these disadvantages, great efforts have been made for some time to develop more meaningful and more environmentally friendly methods. The IR-spectroscopic method according to DEV H18 has been replaced by the more meaningful and more environmentally friendly gas-chromatographic determination, according to DEV H53 (hydrocarbon index in water) (1998) and to the ISO/TC/190/SC3/6 (hydrocarbon index in soil) (1998). Further approaches for the determination of the TPH-content in soils include the NEN 5733 (1995) and the CEN/TC929/WG5 (2000) as well as a proposal according to Lepom (1997).

40.2
Method

Gas chromatographic determination is applicable to soils with a content of total petroleum hydrocarbons of more than 100 mg (kg dry mass)$^{-1}$. Substances with a boiling point between 175 and 525 °C (for the most part alkanes between C10 and C40) can be analysed. Petrol cannot be analysed quantitatively using gas chromatographic determination. This method gives information about the qualitative composition of the mineral oil contamination by means of the boiling range and the peak pattern.

In these methods, the total petroleum hydrocarbons are extracted by means of a mixture of acetone/*n*-heptane (2/1), or by another non-polar solvent. The solvent contains the internal standards C9 or C10 and C40. Subsequently, the clean up is carried out via silica gel or magnesium silicate (shaking). A gas-chromatographic determination using an FID, which determines the components lying between the reference substances C9 or C10 and C40 respectively, is carried out.
Halogenated or otherwise-substituted hydrocarbon compounds of a lower polarity can be analysed as well.

40.2.1
Determination of Petroleum Hydrocarbons

In our examinations, the samples were extracted using ultrasonic extraction and the shaking machine. Subsequently, the samples were centrifuged, washed with water, and dried with sodium sulfate. The clean up of the extract was carried out by shaking with silica gel. Internal standards C9 and C40 were used (see fig. 40.1).

The bleeding of the column is performed by injection of *n*-heptane and is used for the correction of the peak area. Following the correction, the total peak area is determined. This is the area between the added internal standards *n*-nonane and *n*-tetracontane that mark the retention time window. The starting point S of the integration is directly at the end of the *n*-nonane/decane peak, and the ending point E is directly at the beginning of the *n*-tetracontane peak (see fig. 40.2.). The calibration was carried out externally against a mixture of diesel fuel and lubricating oil (1/1).

For the gas-chromatographic analysis, it is important to have a discrimination-free feeding system, low bleeding of the column and sufficient thickness of the film to prevent the column from being overloaded when heavily contaminated samples are examined.

40.3
Results

The medium retrieval rates were determined for different soil materials that were artificially contaminated with mineral oil products in different concentrations (see table 40.1). Retrieval rates of up to 75–99 % were obtained.

The retrieval rates of the diesel fuel were between 84 and 99 %. Lubricating oil had retrieval rates ranging between 62 and 94 % for the soils chosen. The retrieval rates depended on the type of soil and contamination.

The examination of real samples has proven that soil samples from suspected contaminated sites contain hydrocarbons with more than 40 C-atoms. It is being considered presently whether the detectable C-range should be extended to for

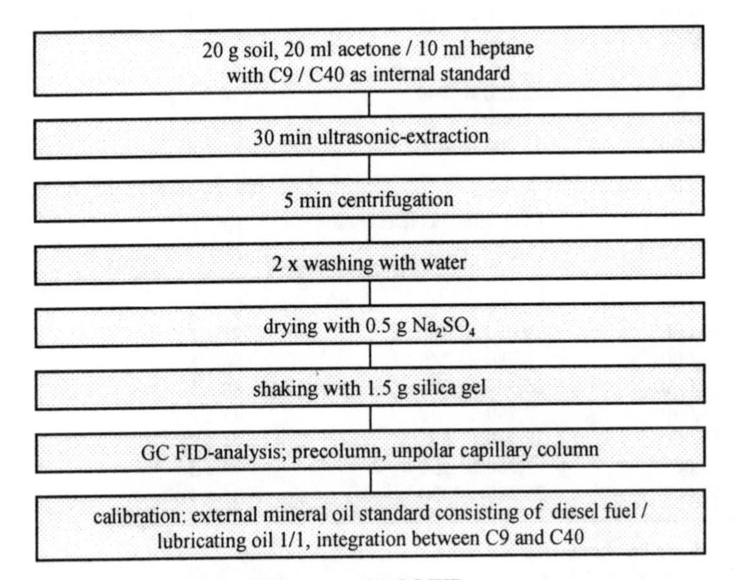

Fig. 40.1. Determination of the TPH-content by GC/FID

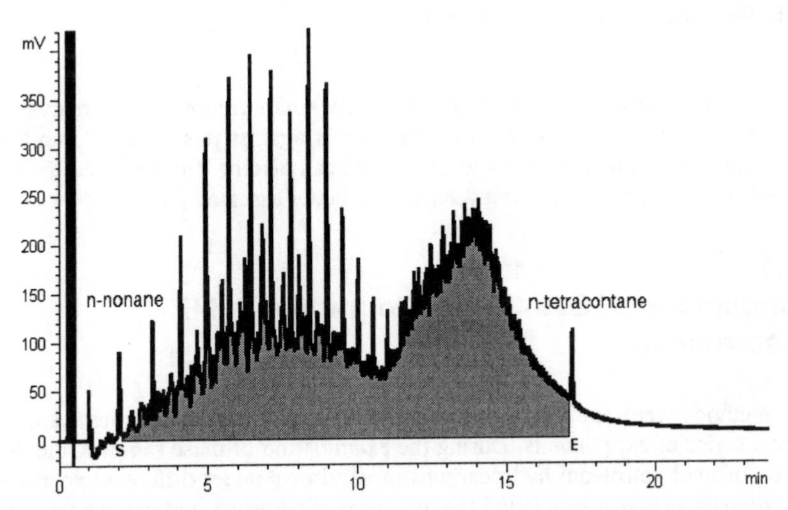

Fig. 40.2. Separation and determination of TPH-content in the range of C9 (*n*-nonane) and C40 (*n*-tetracontane). Detection by FID.

Table 40.1. TPH-content and retrieval rates of contaminated soils artificially contaminated with diesel fuel and lubricating oil, determined by GC/FID

Soil	Lubricating oil [mg kg^{-1}]	Determined concentration [mg kg^{-1}]	RR [%]
1[a]	3,000	2,660 ± 220	89 ± 7
1[a]	30,000	22,630 ± 550	75 ± 2
1[b]	3,000	2,820 ± 10	94 ± 1
1[b]	30,000	27,630 ± 500	92 ± 2
2[a]	3,000	2,030 ± 90	68 ± 3
2[a]	30,000	22,730 ± 940	76 ± 3
2[b]	3,000	1,850 ± 20	62 ± 1
2[b]	30,000	22,700 ± 2470	76 ± 8
Soil	**Diesel fuel [mg kg^{-1}]**	**Determined concentration [mg kg^{-1}]**	**RR [%]**
1[a]	3,000	2,790 ± 20	93 ± 1
1[a]	30,000	29,020 ± 260	97 ± 1
1[b]	3,000	2,580 ± 240	86 ± 8
1[b]	30,000	26,540 ± 380	88 ± 1
2[a]	3,000	2,790 ± 70	93 ± 2
2[a]	30,000	28,220 ± 850	94 ± 3
2[b]	3,000	2,980 ± 90	99 ± 3
2[b]	30,000	25,320 ± 1320	84 ± 4

1. Pseudo-vergleyte Parabraunerde (according German definitions)
[a] Schwach lehmiger Sand, 1.1 wt. % organic carbon.
[b] Stark lehmiger Sand, 0.14 wt. % organic carbon.
2. Gley
[a] Schwach schluffiger Ton, 2.3 wt. % organic carbon.
[b] Schluffig toniger Lehm, 11.0 wt. % organic carbon.

example, 44 C-atoms. However, the relatively small environmental relevance of these high boiling-temperature hydrocarbons is against this idea (Lepom 1996). Hydrocarbons in the mobile range C9 to C25 can also be differentiated from C26 to C40 so that the risk to groundwater can be better assessed (Giese 1999).

40.4 Comparison of Gas-Chromatography and IR-Spectroscopy

The methods were applied to real contamination, where the samples were taken from a series of experiments. During the examination of these samples, the degradation of total petroleum hydrocarbons in windrows under different experimental conditions was investigated, and the influence of additives and oxygen was examined. In five windrows (soil without additives, soil with 10 weight-% compost, soil with 10 weight-% construction waste, soil with ET 20 (a solution with nutrients and total petroleum hydrocarbon degrading microorganisms), and soil without additives as turned windrow), the degradation was observed over a period of time (Koning et al. 1999). The results are shown in table 40.2. The TPH-content of this series of experiments was determined concurrently by IR-spectroscopy, GC and TLC (Reimers et al. 2000). A comparison of IR-spectroscopy and gas chromatography methods is shown in fig. 40.3.

For determination of the TPH-content by IR-spectroscopy (as well as by TLC), the results were based on calibration against the isolated contamination. The different methods correlate clearly very well. Furthermore, it is not only obvious from these examinations but also from other investigations that the results are not directly comparable.

In different co-operative tests (Giese 1996; DEV H53 1998), it was proven that the contents determined according to the DEV H18 procedure (DIN 38409 1981) are generally lower (by up to a factor of about 1.7) than the contents determined by GC (or TLC).

The reasons for these differences are:

- The use of empirically determined extinction coefficients for test calibration according to the DEV H18 procedure (DIN 38409 1981).
- Al_2O_3 (DEV H18) discriminates against long chain aliphatic compounds and aromatic hydrocarbons.
- Heptane and hexane/acetone mixtures respectively are more effective than freon matrices, which are difficult to penetrate because of their polarity.

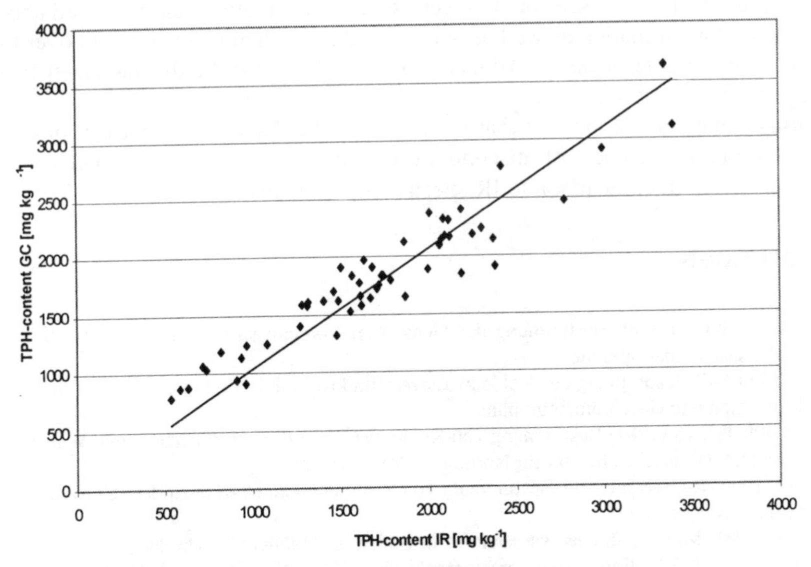

Fig. 40.3. Comparison of the results obtained by IR-spectroscopy and GC

Table 40.2. Determination of the TPH-content in a soil in different windrows and its dependence on time, sampling depth and additives. The soil was contaminated with diesel fuel, with a low content of tar (PAH).

Windrow	Conditions	Date of sampling 03/99	03/99	04/99	05/99	06/99	07/99	09/99
M1 (1–2)	without additive	2000	1890	2140	1640	1960	1620	1910
M2 (0–1)	compost	2140	2120	1790	1460	1770	880	1240
M2 (1–2)	compost	2200	1830	1320	1660	1720	1970	2310
M3 (0–1)	construction waste	2200	2690	2180	2170	2140	1400	1590
M3 (1–2)	construction waste	3270	3110	2480	2770	3130	2120	3640
M4 (0–1)	ET20	1890	1520	1650	1700	2090	1190	2320
M4 (1–2)	ET20	2340	1940	1520	1830	1830	1910	2370
M5 (0–1)	turned windrow	1780	1760	1130	910	1250	1060	1030
M5 (1–2)	turned windrow	1970	1470	1350	1530	1610	2040	1580

(0–1): depth of sampling 0–1 m; *(1–2):* depth of sampling 1–2 m

40.5 Conclusions

Gas chromatography has the advantage that environmentally friendlier solvents can be used than those used for IR-spectroscopy. Information can be gained about the type of contamination as well as about the changes that have occurred over the course of a remedial measure. Biogenic hydrocarbons can be distinguished from total petroleum hydrocarbons.

Disadvantageous is the fact that petrol cannot be determined, but can only be analysed via occurrence of discrete peaks on the solvent tailing before the *n*-nonane peak. In comparison to IR-spectroscopy, the results correlate well.

References

CEN/TC929/WG5 (2000) Bestimmung des Gehalts an Kohlenwasserstoffen von C10 bis C40 mittels Gaschromatographie.

DEV H53 (1998) Bestimmung des Kohlenwasserstoffindex. Teil 4: Verfahren nach Lösungsmittelextraktion und Gaschromatographie.

DIN 38 409 Teil 18 (1981) Bestimmung von Kohlenwasserstoffen. Normenausschuß Wasserwesen im DIN (Deutsches Institut für Normung e V), Germany

Giese G (1996) Ringversuch zur Bestimmung von Summenparametern in Böden. Umweltbehörde Hamburg, Germany

Giese G (1999) Mineralölkohlenwasserstoffe in Böden u. Abfällen: Ersatz der „H18"-Methode durch eine FCKW-freie, gaschromatographische Alternative. Umweltbehörde Hamburg, Germany, unpublished

ISO/TC/190/3/6 (1998) First working draft Soil quality: Determination of mineral oil content by gas chromatography

Koning M, Braukmeier J, Cohrs I, Lüth J, Stegmann R (1999) Untersuchungen zur Optimierung der biologischen Bodenhaltung im Mietenverfahren im Hinblick auf eine gezielte Wiederverwertung des zu behandelnden Bodenmaterials. DBU Projekt 09815

Lepom P (1996) Bestimmung von Mineralölkontaminationen in Böden und Abfall – Stand der Normung bei ISO und CEN. Umweltbundesamt Berlin, Germany

Lepom P (1997) Bestimmung von Mineralölkontaminationen in Böden und Abfall – Entwicklung eines gaschromatogaphischen Analyseverfahrens. In: Borsdorf H, Flachowsky J (eds) UFZ-Bericht 12/97, pp 41–48

NEN 5733 Draft (1995) Soil. Determination of mineral oil content in soil and sediment by gas chromatography

Reimers C, Koning M, Cohrs I, Scholz A (2000) Vergleichende Untersuchungen von Mineralölkohlenwasserstoffen in Bodenproben mit Hilfe der IR-Spektroskopie, Gaschromatographie und Dünnschichtchromatographie, Technical University of Hamburg-Harburg, Department of Waste Management, Germany, unpublished

41 Application of Test Systems for a Balance-Based Examination of Biodegradation of Contaminants in Soil

K. Hupe[1], M. Koning[2], J.-C. Lüth[2], I. Cohrs[2], J. Heerenklage[2], R. Stegmann[2]
[1]Engineering Agency for Waste Management, Prof. Dr.-Ing. Rainer Stegmann and Partner, Nartenstraße 4a, 21079 Hamburg, Germany
[2]Technical University Hamburg-Harburg, Department of Waste Management, Harburger Schloßstraße 37, 21079 Hamburg, Germany

For the set-up of a remediation concept, suitable field simulation pre-tests on a laboratory, as well as on a technical scale, should principally be carried out. The application of test systems and laboratory reactors is also necessary for basic examination and a general understanding of the degradation processes within soils. When conceptualising biological remediation measures the respective laboratory scale pre-tests should be designed in a way, that answers the questions posed concerning the possibility to comply with the respective microbial, physio-chemical and process-geological conditions. The most important, but also most difficult, task is the evaluation of the possibility for remediation of soils. Laboratory results about microbial degradation of relevant contaminants in problem sites have to be transferable into practice and consequently provide data about degradation. This includes degradation rates, final concentrations that can be yielded, formed metabolites and end products respectively. They should be selected between different, complex test systems according to the initial question.

The complexity of the soil system frequently leads to the fact that it is not the actual biodegradability, but the chemical/physical parameters (e.g. transport and sorption processes) which influence the degradation rate in a decisive way. Of particular importance in this context is the bioavailability which is not only strongly dependent on the type of soil matrix, (i.e. its substantial composition and its particle size distribution) but also on the duration of the impact of contamination. An initial decision about the applicability of a biological measure is made using a bioavailability test, which still needs to be standardised. If the result is positive, laboratory tests have to be carried out for a further optimisation of the biodegradation and about the influence of soil physical and chemical parameters (DECHEMA 1992).

The following section contains a short overview of different tests and laboratory reactor systems. These have been developed in past years for the examination of biological contaminant degradation in the course of a treatment of contaminated soils. A detailed description of the systems used by the research centre SFB 188 "Treatment of Contaminated Soils" follows.

Test systems can be principally differentiated into open and closed

Table 41.1. Overview over available open test systems

System	Source (Hupe 1998)
DEWAR-flask for self heating tests	LAGA 1995
Great lysimeter	Bartha et al. 1992
NATEC lysimeter	Figge 1992
Linde test pile (temperature isolated container)	Müller-Marckgraf 1994
Laboratory meso-cosm system	Huesemann 1994
Mini container test system	Eisenbeis et al. 1995
Under floor lysimeter "Rhizolab"	Hwu et al. 1998

Table 41.2. Overview over available closed test systems

System	Source (Hupe 1998)
Isermeyer device	Isermeyer 1952
Voith respirometer (sapromat) with registration of the O_2-consumption	Steinecke 1976
Wösthoff device	Anderson 1982
Heinemeyer device	Heinemeyer et al. 1989
Glass jar with different registration units for CO_2-production	DECHEMA 1992; Hupe et al. 1993; Schaefer et al. 1992
Bühler respirometer BSB digi	Bühler Analysentechnik, Bodelshausen
Micro-Oxymax	Columbus Instruments, Ohio, USA
Arthur respirometer	Arthur Technology, Wisconsin, USA
250 ml-test system for slurry, pile and in-situ examinations	Fu et al. 1996
Column system for modelling the degradation activity and bacteria growth	Harms 1996
Micro system for ^{14}C-examinations (125 ml)	Marinucci and Bartha 1979
NATEC-^{14}C-degradation unit	Keskin 1994
3 litre glass-bioreactor system for ^{14}C-examinations	Kästner et al. 1995
1 litre glass-bioreactor system for ^{14}C-examinations	Eschenbach 1995
4 litre-submersbioreactor with closed gas circuit	Müller-Marckgraf 1994
Miniature ecosystem for ^{14}C-examinations	Hosler et al. 1988
45 litre microcosm reactor system	Govind et al. 1994
300 litre silo-reactor with extern bulk material circulation	Scholz et al. 1996
NATEC mesocosm system for simulation of a terrestrial ecosystem	Figge 1992; Keskin 1994
Anaerobic test system	Rilling and Eckstein 1996
Solid phase stirring reactor	Saner et al. 1997
1 litre bioreactor with a PC-controlled pressure compensation system	Zytner et al. 1997

systems. Some different test systems have been described by the DECHEMA (1992). Open systems are suitable for the examination of chemical/physical processes, such as:

- Change in abiotic factors and environment conditions in the soil material.
- Transport and sorption processes in the soil material.
- Hydrological behaviour in suspensions.

Some open test systems are listed in table 41.1.

Closed systems are used for mass balance-based examination of the degradation of contaminants in soil materials or in a suspension in order to take into account all paths of elimination, attenuation and emission. Closed test systems cannot only be differentiated according to size and their mode of operation – statically or dynamically, solid state or slurry state, aerobic or anaerobic – but also according to their instrumentation. Table 41.2 gives a selection of different available test systems.

With increasing complexity of the test system the possibility to gain a detailed description of the degradation processes in the soil material or in the suspension increases respectively. Jörg et al. (1996) described different test systems for the examination of the biodegradability in a liquid environment under aerobic, anoxic and anaerobic conditions. Heerenklage and Stegmann (1997) introduced test systems at different scales including available on-line-instrumentation that can be used for investigation under aerobic and anaerobic conditions. An exchange of practical experiences on the development and the use of soil bioreactors took place on the workshop "Bioreactors for the Treatment of Contaminated Soil" in Braunschweig in April 1997 (Scholz et al. 1997).

41.1 Test Systems for Balancing Contaminant Degradation – Experiences from the Research Centre SFB 188 "Treatment of Contaminated Soils"

41.1.1 Test Systems for Carrying out Pre-tests for the Optimisation of the Microbial Degradation of Contaminants

The examination of the microbial degradation of organic contaminants in soil materials requires the application of different closed test systems according to the problem and to the investigation aim.

Closed glass jars (volume 1.5 l) are the most simple of the test systems in use. The tests require 50–100 g dry mass of the sample material. The measurement of the microbial activity in the soil material was undertaken according to the Isermeyer-method (Isermeyer 1952) whereby a beaker containing caustic soda is put into each of the jars and the CO_2 absorbed to the caustic soda is titrimetrically determined.

The beakers in the jars are discontinuously exchanged against beakers containing fresh caustic soda and the CO_2 formed during microbial activity is determined subsequently via titration by means of hydrochloric acid. In order to analyse the change in solid matter parameters, single jars are removed and the complete solid matter material is used for analysis. This makes it necessary to set up several parallel glass jars to be able to describe the degradation processes in the respective jars over a test duration of some weeks.

The respirometer (Sapromat D12, Fa. Voith) is a system in which the oxygen consumption is measured. It consists of a thermostated (mostly at 20 °C) water-bath, in which 12 closed test units can be operated in parallel. Each test unit consists of a test flask (volume 0.25 l), an oxygen source and a switch manometer. The single units are connected to each other via tubes and cables. On the head of the test flasks there is a flask containing soda lime pellets. In each of the test flasks is placed 50–100 g of humid (40–60 % of maximum water holding capacity) soil material. The CO_2 formed by the microbial activity in the soil material is absorbed to the soda lime pellets. This causes a negative pressure in the test unit. As soon as a specific low pressure is reached the electric circuit is closed via the switch manometer. This causes the oxygen source to electrolytically release O_2 from a copper sulfate solution. On-line data logging of the oxygen consumption is possible through the connection of the test units to a PC. This system can be applied for the determination of the microbial activity (biomass, basal respiration) in the sample materials or for optimisation of the milieu conditions (e.g. water content, amount of additives). Usually only the respiration activity is measured. In addition the contaminant reduction in the soil materials can be measured by a simple "before/afterwards"-analysis, i.e. the concentration of the contaminants is analysed before and after the experiment.

41.1.2
Test Systems for Balance-Based Investigations of the Biodegradation

While glass jars and the respirometer are simple test systems for pre-tests for the optimization of environment conditions or for the examination of specific factors of influence, balance-based investigations increase the instrumentation expenditure. For balance-based examinations different static (fixed bed reactors: 3 l, 6 l, 90 l) and dynamic (reactors stirred with blades: 6 l and suspension reactors: 4 l, 12 l) bioreactors are used in the research centre SFB 188.

41.1.2.1
Static Fixed Bed Reactors

The static bioreactor systems (fixed bed reactors) are used for the simulation of the conditions in treatment piles or in reactors on a technical scale. They have been continuously improved on the basis of the experimental results. A sieve is mounted in the lower part of the reactor on which the soil material is placed (0.5–1 $kg_{\text{dry mass}}\ l^{-1}_{\text{reactor volume}}$).

Sampling stubs are mounted on the reactors at different height. A controlled gas supplement for the aeration takes place from the bottom to the top or by means of a regulator or by means of a needle valve (manually operated). For a sufficient volumetric gas flow this is usually set to 1 [$l\ kg_{\text{dry mass of soil material}}^{-1}\ h^{-1}$]. To guarantee an approximately constant water content of the soil material the gas is saturated by bubbling through a flask containing destilled water before entering the reactor. The CO_2-concentration in the effluent gas is determined by means of an infrared -photo-spectrometer (IR) and the gas release of volatile contaminants as

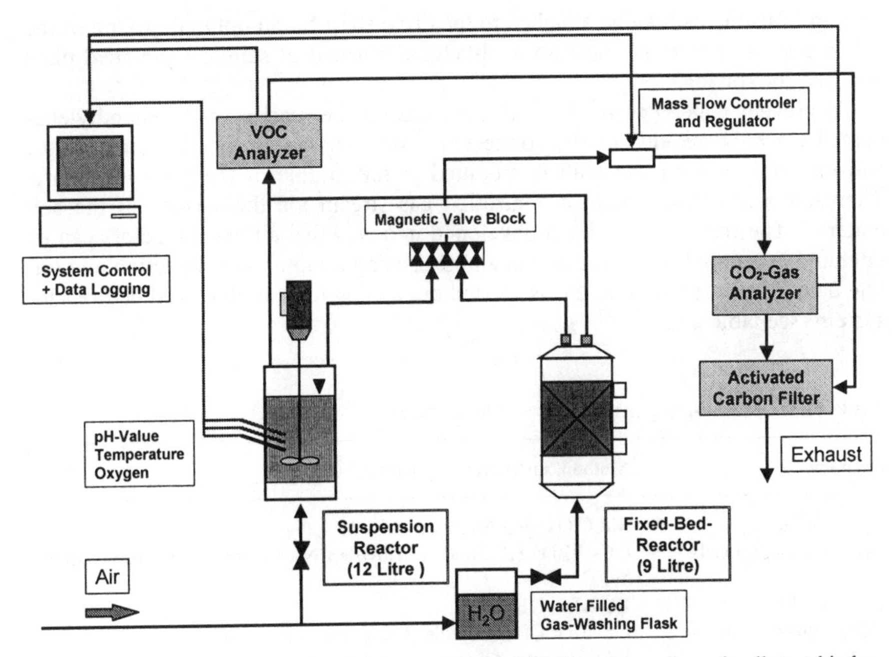

Fig. 41.1. Experimental assembly of the reactor-system for the observation of pollutant biodegradation in the Suspension Reactors and in the Fix-Bed-Reactors

total carbon content (VOC: volatile organic carbons) by means of a flame ionisation detector (FID) are measured on-line. For the on-line measurement of the volumetric flow of the flue gas an inductive volumetric flow meter is used.

The data logging and control of the flue gas of the two parallel operated reactors (up to 16 parallel preparations) is carried out by a PC. Each reactor can be independently switched in series with the measurement instruments by means of PC-controlled magnetic valves. As a control gas samples can be taken out of the reactor by means of a syringe via a septum and subsequently be injected into a gas-chromatograph (GC). Detailed specifications of the applied gas measuring instruments and apparatus can be seen in table 41.3. Fig. 41.1 shows the experimental assembly of the reactor-system.

41.1.2.2
Dynamic Reactor Stirred with Blades

In addition to the static, dynamic (mixed) bioreactors are also available. Investigations concerning the application of bioreactors in dry processes are carried out in reactors stirred with blades (see fig. 41.2). The horizontally operated reactors stirred with blades consist of a cylindrical glass enclosure (volume 5 l). The reactors are half filled (approximately 2.5 l) with soil material – the weight is generally 1.6–1.8 kg wet weight mass of soil material. The drive shaft is gas tight sealed in the front of the reactors and is driven by an infinitely variable motor. Different stirring instruments may be attached to the drive shaft for an optimal mixing of the soil materials. For most applications blades fabricated of stainless punched plate are used for stirring.

To prevent adherence of the soil materials on the reactor wall, polyethylene-coated VA-bars are additionally connected to the transmission shaft to strip off the soil material. A sampling stub is mounted in the middle of the glass enclosure. Thereactors are treated with a controlled gas stream via the surface of the soil material. The reactors have been developed in a way that allows 3 reactors can be operated in parallel by means of a chain gear using a motor. Gas injection and on-line data logging of the flue gas is carried out analogously to the fixed bed reactor system (see table 41.3).

Table 41.3. Gas measuring instruments and apparatuses

Parameter	Method, instrument, apparatus
O_2 – on-line	M&C Oxygen Analyser PMA 10
CO_2 – discontinuously	GC-WLD: HP 5890 II, Hayesep N-column, molar sieve column, WLD
CO_2 – on-line	IR: ADC DB2E
VOC – discontinuously	GC-FID: CP 9000, idle Widepore-column, FID
VOC – on-line	FID: Bernath-Atomic – Total carbon-Analysator 3002
Volumetric flow – on-line	Bronkhorst
Data logging	Regulation program: WorkBench PC 2.0 on a 386-PC under MS-DOS

VOC volatile organic carbons, *GC* Gas-chromatograph, *WLD* thermal conductivity detector, *HP* Hewlett Packard, *IR* Infrared-photospectrometer, *ADC* Analytical Development Company, *FID* Flame ionisation detector, *CP* Chrompack, *PC* personal computer

Fig. 41.2. Photography of the parallelly operated dynamic reactors stirred with blades

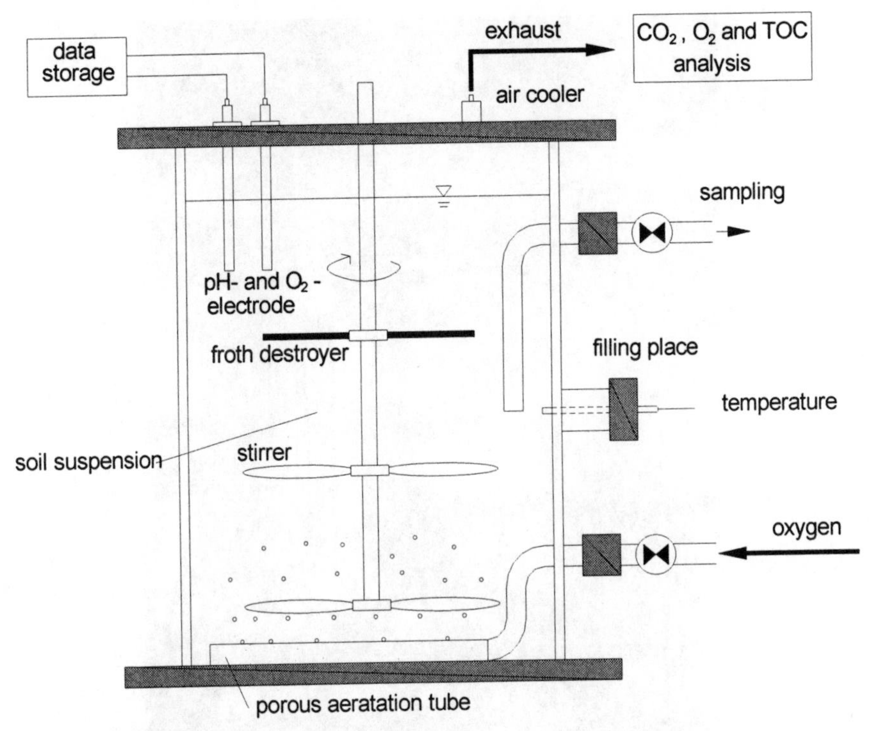

Fig. 41.3. Scheme of a 4.3 litres suspension reactor

41.1.2.3
Dynamic Suspension Reactors

The suspension reactors in use are completely mixed agitating vessels of 4.3 and 12 l of volume (see fig. 41.3 and 41.4). The 4.3 l reactors consist of cylindrical glass flasks, that have stubs for sampling on their sides and are shut off with PVC-plates on the top and bottom. In each of these plates there is a circular groove that contains a grease-lubricated O-ring which tightly encloses the rim of the glass flasks.

Furthermore, the top plates have ducts which house a shovel blade stirrer, electrodes (pH, temperature, oxygen content), a filler neck and a flue gas cooler. The top and bottom plates are compressed together by three rods so that the reactors are gas tight. To prevent an adsorption of organic substances to the bottom plates, PTFE-foils are clamped between the bottom plates and the glass flask. For operation the filled reactors are hinged to a skeleton to which an electromotor and a chain gear is attached. This makes it possible to operate all 6 reactors simultaneously.

The cylindrical 12 litre-suspension reactors are fabricated from high quality steel and are separately driven by means of electromotors. Contrary to the 4.3 litre reactors sampling is undertaken via sampling stubs on the reactor base or in the

Fig. 41.4. Photography of the 4.3 and 12 litres suspension reactors

lower part of the reactor jacket which are flushed by means of air prior to sampling.

The aeration of the suspension reactors is carried out via perforated pipes (4.3 litre reactors) or via spreaders (12 litre reactors) that are mounted just above the reactor base. In order to attain a finely dispersed oxygen feed the perforated aeration apparatus are additionally coated with rubber hoses which have fine holes. By means of the installed stirring apparatus in both reactor systems solid matter contents up to 30 mass-% can be continuously kept in suspension. Propellers, that are mounted to the rods of the stirrer blades just above the surface of the suspension mechanically destroy the foam that may be formed during the biological treatment of contaminated soil materials. This consequently avoids obstruction of the flue gas tubes. With exception of the oxygen measurements the data logging of the flue gas of the suspension reactors is carried out in analogy to the dry systems. Table 41.3 gives an overview over the gas measuring instruments and apparatuses used.

All of the reactor systems presented (static and dynamic) can be heated by means of a heating mantle infinitely variable between 20 °C and 70 °C so that also investigations in the thermophilic range can be carried out.

41.1.3
Balance of the Microbial Degradation of Contaminants

For a complete description of the degradation of the contaminant in the soil material it is not sufficient to describe the decrease in contamination alone. In addition an extensive balance of the degradation of the soil contamination in closed reactor systems is necessary. For balancing the contaminant degradation of undoped contaminants they must be differentiated between investigations on artificial (fresh) contaminants and investigations on real contaminated soil materials from contaminated sites.

For artificial mineral oil contaminants in "dry" treatment systems a balance-model was developed by the DFG research centre (e.g. Lotter 1995). This model has been continuously advanced in the course of the experiments that have been carried out and has been adapted to suspension systems (Hupe et al. 1995; Koning et al. 1997). The balance calculation of this model is based on the simplified assumption, that the higher CO_2, VOC and biomass concentrations in the contaminated sample – in comparison to a non contaminated sample – are caused by the addition of oil. For this reason the respective differences between the contaminated and non contaminated sample are used for balancing the degree of degradation of the contamination. This model describes the continuously changing conditions in the soil after the addition of oil only to a certain degree, but provide a first approximation with good a indication concerning the potential degradation of the mineral oil contamination.

The carbon content of the measured hydrocarbon concentration is estimated via the carbon content of the original oil. For a carbon balance the difference values of the balance-parameters are related to the carbon content of the initial contamination of the sample material. The parameters analysed for the determination and the balance of the contaminant degradation are presented in table 41.4.

Table 41.4. Examination parameters and analysis methods for balance parameters

Parameter	Analysis method
Hydrocarbon contamination	Gravimetrically according to DIN 38 409 H17 IR-spectroscopycally according to DIN 38 409 H18 after ultrasonic extraction GC-FID according to ISO/TC190/SC3/WG6 (first working draft)
Biomass	Analogously to SIR-method Protein determination (Lowry et al. 1951)
CO_2-production	discontinuously: GC-WLD on-line: IR, ADC DB2E
Gas release (VOC)	discontinuously: GC-FID on-line: FID TOC-Analysator Bernath-Atomic

VOC volatile organic carbons, *SIR* Substrate-Induced-Respiration (Andreson and Domsch 1978), *GC* Gas-chromatograph, *WLD* thermal-conductivity detector, *IR* Infrared-photospectrometer, *FID* Flame ionisation detector

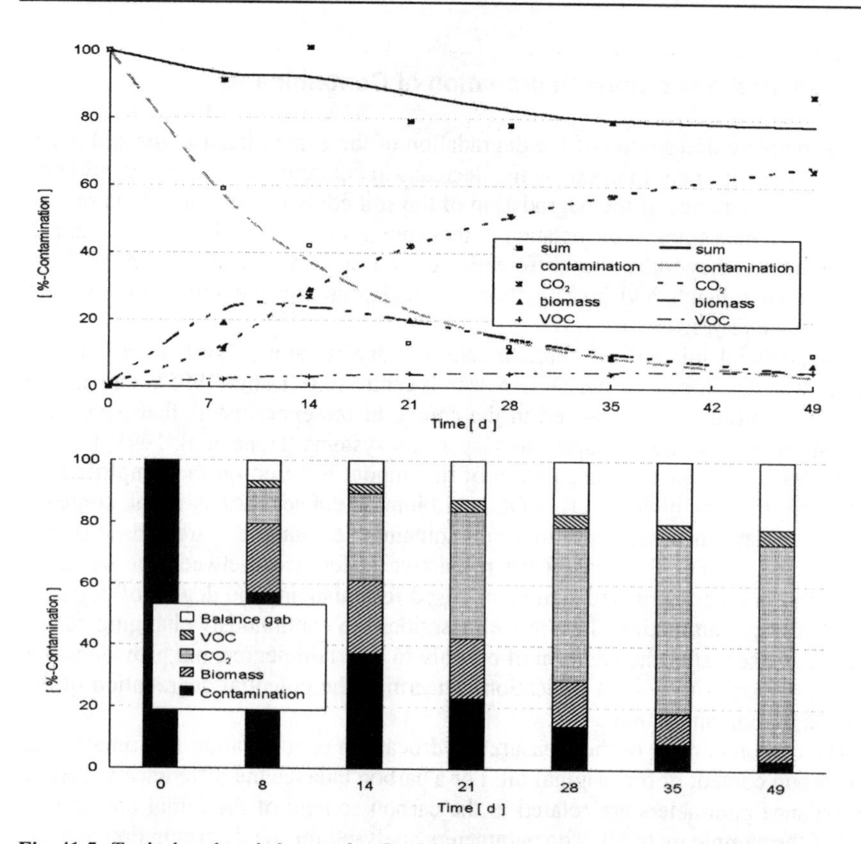

Fig. 41.5. Typical carbon balance of a fixed bed experiment on artificially contaminated soil material (Soil: A_h-horizon, contamination: diesel fuel 1 mass-%, additive: compost 20 mass-%, temperature: 20 °C)

Fig. 41.5 demonstrates a typical carbon balance that has been determined for a fixed bed reactor experiment.

When balancing the contaminant degradation in real contaminated soils there is a great difficulty in determining reproducible reference values, i.e. relating the measured values for CO_2, VOC and biomass on the contaminant degradation only. An approach to solving the problem can be seen in parallel tests using a not contaminated reference soils (Mahro et al. 1993). For this procedure it is necessary to find a soil material, that is as similar as possible to the contaminated soil material with regard to the chemical/physical soil characteristics as well as to the microbial conditions. A potential reference soil could be a non contaminated soil from the boundary area of a contaminated site to be treated. However, because soil contamination alters the environmental conditions in the soil – especially for longer durations of impact (old contaminated sites) – it is unlikely that a respective reference soil can be found without a limited comparability of the contaminated and non contaminated soils.

Alternatively it is possible to use a model soil that has nearly the same soil characteristics with regards to physio-chemical and microbial parameters. For tests of of chemical agents in the region of the European Community (EC) five regionally representative reference soils have been identified in by the German Umweltbundesamt (national environment council) (Kuhnt et al. 1991). The chosen soils cannot be seen as generally applicable reference soils for comparative and balancing estimates in the domain of contaminated sites. However, they were intended for EC-wide standardised application for the evaluation of the soil behaviour of new chemicals. Generally the same problems concerning comparability also exist for these and all other possible reference soils.

The problem of reference soils can be evaded by means of the addition of substances marked with isotopes that are typical for contaminated sites (e.g. PAH) to a real contaminated soil material. It has to be taken into consideration, however, that the transferability of the produced results is limited in comparison to a real aged contamination limit by the differing sorption behaviour as well as the differing bioavailability of freshly doped, marked substances . An artificial ageing of the substances to be doped or a simulation of the ageing process could be beneficial in obtaining accurate information the destination and the long-term behaviour of a contamination at a problem site.

Marked tracer substances using ^{13}C and ^{14}C have been carried out at the research centre SFB 188 using the above introduced fixed bed reactor systems for examining the balance of contaminant degradation in real contaminated soils. By means of ^{14}C-marked PAH (anthracene and hexadecane) Kästner et al. (1995) were able to prove that the errors that occurred during the mass balancing of the degradation of contaminants by Lotter (1995) can be mainly attributed to the formation of bound residues. Further examinations on the formation and long-term stability of bound residues by means of ^{14}C-marked tracer substances have been carried out by Eschenbach et al. (1997) and others.

In comparison to the ^{14}C-method, examinations using ^{13}C-marked tracer substances have the advantage that examination of structure-chemical parameters is possible. In this way additional information can be gained about the molecular structure of bound residues, their types of bonding and their stability in natural environments. A description of the application of ^{13}C-marked tracer substances in the research centre SFB 188 can be found among others in the work of Richnow et al. (1998).

41.2 Perspectives

The test systems and balancing approaches described have proven to be worthwhile in practical application for the evaluation of biological degradation processes in mineral oil contaminated soil materials. These tests are among others used as model systems for optimisation of the microbial treatment in piles (Stegmann et al. 1997). On a technical scale it could be observed that by means of an optimisation of the environmental conditions in the test systems (especially the primary contaminant), degradation can be clearly accelerated based upon the comparative examinations on the contaminant degradation in the test systems and in treatment piles (Koning et al. 1998). This effect is used as a basis for further work to develop a procedure to predict and evaluate as early as possible the biodegradation potential of mineral oil contaminated soils.

Further information regarding this topic can be extracted from Chapter 22 and 26.

References

DECHEMA (1992) Labormethoden zur Beurteilung der biologischen Bodensanierung. 2. Bericht des Interdisziplinären Arbeitskreises "Umweltbiotechnologie-Boden", Ad-hoc-Arbeitsgruppe "Labormethoden zur Beurteilung der biologischen Bodensanierung" (Ltg Klein J, Dott W) DECHEMA, Frankfurt a M, Germany

Eschenbach A, Wienberg R, Mahro B (1997) Bildung und Langzeitstabilität von nicht extrahierbaren PAK-Rückständen im Boden. altlasten spektrum 6/97: 292–297

Heerenklage J, Stegmann R (1997) Test system for the simulation of biological degradation processes. In: Christensen T, Cossu R, Stegmann R (eds) Proceedings Sardinia'97, Sixth International Landfill Symposium. CISA, Cagliari, Italy, pp 281–291

Hupe K (1998) Optimierung der mikrobiellen Reinigung mineralölkontaminierter Böden in statischen und dynamischen Systemen. abfall aktuell, in press

Hupe K, Heerenklage J, Lotter S, Stegmann R (1993) Anwendung von Testsystemen zur Bilanzierung und Optimierung des biologischen Schadstoffabbaus. In: Stegmann R (ed) Bodenreinigung, Hamburger Berichte 6. Economica Verlag, Bonn, Germany, pp 97–119

Hupe K, Lüth J-C, Heerenklage J, Stegmann R (1995) Blade-mixing reactors in the biological treatment of contaminated soils. In: Hinchee RE, Sayles GD, Skeen RS (eds) Biological unit processes for hazardous waste treatment. Battelle Press, Columbus, pp 153–159

Jörg R, Schäfer A, Fink A-B, Boley A, Müller W-R (1996) Weiterentwicklung von Testverfahren zur Untersuchung der biologischen Abbaubarkeit im wäßrigen Milieu unter aeroben, anoxischen und anaeroben Bedingungen. Fachgespräch des BMU "Beitrag der Biotechnologie zu einer nachhaltigen, umweltgerechten Entwicklung", 17. Dezember 1996, Wissenschaftszentrum, Bonn, Germany

Kästner M, Lotter S, Heerenklage J, Breuer-Jammali M, Stegmann R, Mahro B (1995) Fate of ^{14}C-labeled anthracene and hexadecane in compost-manured soil. Applied Microbial Biotechnology 43: 1128–1135

Koning M, Hupe K, Lüth J-C, Knecht S, Timmermann D, Paul M, Stegmann R (1997) Biological soil treatment in solid phase and slurry reactors. In: Alleman BC, Leeson A (eds) In situ and on-site bioremediation. Vol 5, Battelle Press, Columbus, pp 137–143

Koning M, Hupe K, Lüth J-C, Lubitz B, Stegmann R (1998) Transfer of experimental measurands from the laboraty into the remediation practice – comparative investigations into the biological degradation of contaminants in fixed-bed reactors and technical remediation windrows. In: Telford T (ed) Contaminated Soil '98, London, UK, pp 1091–1092

Kuhnt H, Hertling T, Schmotz W, Vetter L (1991) Auswahl von Referenzböden für die Chemikalienprüfung im EG-Bereich. Forschungsbericht, UFOPLAN-Nr: 10602058, Berlin, Germany

Lotter S (1995) Biologische Bodenreinigung – Untersuchungen zur Kohlenstoffumsetzung, Kohlenstoffbilanzierung und Prozeßmodellierung. In: Stegmann R (ed) Hamburger Berichte 9, Economica Verlag, Bonn, Germany

Mahro B, Hupe K, Kästner M (1993) Mass balances in biological soil treatment. Workshop W1, 4th KfK/TNO Conference on Contaminated Soil, 3–7 Mai 1993, Berlin, Germany

Richnow HH, Eschenbach A, Mahro B, Seifert R, Wehrung P, Albrecht P, Michaelis W (1998) The use of 13C-labeled polycyclic aromatic hydrocarbons for the analysis of their transformation in soil. Chemosphere 36 (10): 2211–2224

Scholz J, Müller BG, Schwedes J, Deckwer W-D (1997) Workshop "Bioreaktoren in der Bodensanierung". Chemie Ingenieur Technik 7: 876–878

Stegmann R, Hupe K, Koning M, Lüth J-C (1997) Reinigung kontaminierter Böden in Bioreaktoren. Arbeits- und Ergebnisbericht des Sonderforschungsbereiches 188 der DFG 1995–97, pp 243–288

42 Electroreduction

J. Voss, M. Altrogge, D. Golinske, O. Kranz, D. Nünnecke, D. Petersen, E. Waller
Institute of Organic Chemistry, University of Hamburg, Martin-Luther-King-Platz 6, 20146 Hamburg, Germany

42.1 Indirect Electroreduction of Octachloronaphthalene

Electrolysis was carried out at 40 °C in a batch cell with a volume of 2 ×100 ml. The cathodic and anodic compartment were each filled with 100 ml methanol containing 0.1 M tetraethylammonium bromide as the supporting electrolyte. After the addition of 106 mg (0.26 mmol) of octachloronaphthalene (Jakobsson et al. 1992) and 100 mg (0.35 mmol) of the mediator **2** (Uchino et al. 1975) to the catholyte, nitrogen was bubbled through the solution to avoid any oxidation of the reduced mediator by oxygen. In order to achieve sufficient turn-over rates the potential was adjusted to –1.40 V *vs.* Ag / AgBr. The progress of the electrolysis was monitored by GC. For work-up the catholyte was poured into 100 ml water, acidified with half conc. hydrochloric acid and extracted twice with 150 ml hexane each time. The organic layers were dried with magnesium sulfate. The contents of naphthalene derivatives were determined per GC. The solvent was evaporated *in vacuo* and the products were weighed and analysed by GC. After the passage of 11700 A s at a current of 50-900 mA, a total chemical yield of 91 % (current efficiency 3 %) was obtained, which consisted of 86 % naphthalene, 10 % 2,2′-binaphthyl, 3 % 1,2′-binaphthyl and < 1 % 1,1′-binaphthyl (Nünnecke 2000).

42.2 Electrocarboxylation of 3-Chlorodibenzofuran

The electrolysis was carried out with 1.80 g (8.84 mmol) substrate at about –5 to 0 °C in an undivided cell (see Chapter 34, fig. 42.5) The solvent was dry DMF. A low concentration of tetrabutylammonium bromide was added to adjust the initial conductivity. After bubbling CO_2 for about 30 min to expel the residual air, a low overpressure (0.1 bar) of CO_2 was applied. The electrodes were connected to a power supply and a constant current of $I = 0.25$ A was passed through the stirred mixture. Every hour, a sample was taken to monitor the decrease of the substrate concentration and the increase of the product concentration with respect to the charge consumption. After the end of the electrolysis, the reaction mixture was acidified with 6 M hydrochloric acid under ice-cooling. All products and residual substrate were extracted three times with diethyl ether. The organic layer was extracted with 6 M hydrochloric acid to remove residual DMF and then dried over magnesium sulfate. An aliquot part (50 %) of the organic extract was treated with 2 M aqueous sodium hydroxide to separate carboxylic acids from neutral products. The aqueous layer was again acidified (under pH control) and extracted with diethyl ether, which after evaporation, led to the carboxylic acids. Samples taken during the electrolysis were acidified with 6 M hydrochloric acid and extracted with diethyl ether. A solution of diazomethane in diethyl ether was added dropwise to the organic layer to form the methyl carboxylates. Identification of the products was performed by GC/MS. An aliquot part (25 %) of the organic extract was again derivatized with an etheral solution of diazomethane and supplied with an internal standard for quantitative GC-analysis. A total chemical yield of 85 % was obtained (current efficiency 31 %), which contained 85 % dibenzofuran-3-carboxylic acid, 9 % dibenzofuran-1,4-dicarboxylic acid, 4 % dibenzofuran and 3 % 3-chlorodibenzofuran (Golinske et al. 2000).

References

Golinske D, Voss, J, Adiwidjaja G (2000) Electrocarboxylation of Chlorinated Aromatic Compounds. Collect Czech Chem Commun 65: 862-880

Jakobsson E, Eriksson L, Bergmann Å (1992) Synthesis and Crystallography of 1,2,3,4,6,7-Hexachloronaphthalene and 1,2,3,5,6,7-Hexachloronaphthalene. Acta Chem Scand 46: 527-532

Nünnecke D (2000) Elektrochemische Enthalogenierung chlorierter Aromaten mittels Nickel(II)-Komplexen als Mediatoren in Methanol. PhD thesis Universität Hamburg

Uchino M, Asagi K, Yamamoto A, Ikeda S (1975) Preparation and Properties of Aryl(dipyridyl)nickel Halide Complexes. J. Organomet Chem 84: 93-103

Index